T0134686

Lecture Notes in Computer Science 13378

More information about this series at https://link.springer.com/bookseries/558

Osvaldo Gervasi · Beniamino Murgante ·
Sanjay Misra · Ana Maria A. C. Rocha ·
Chiara Garau (Eds.)

Computational Science and Its Applications – ICCSA 2022 Workshops

Malaga, Spain, July 4–7, 2022
Proceedings, Part II

 Springer

Editors
Osvaldo Gervasi ⓘ
University of Perugia
Perugia, Italy

Beniamino Murgante ⓘ
University of Basilicata
Potenza, Potenza, Italy

Sanjay Misra ⓘ
Østfold University College
Halden, Norway

Ana Maria A. C. Rocha ⓘ
University of Minho
Braga, Portugal

Chiara Garau ⓘ
University of Cagliari
Cagliari, Italy

ISSN 0302-9743 ISSN 1611-3349 (electronic)
Lecture Notes in Computer Science
ISBN 978-3-031-10561-6 ISBN 978-3-031-10562-3 (eBook)
https://doi.org/10.1007/978-3-031-10562-3

This Springer imprint is published by the registered company Springer Nature Switzerland AG
The registered company address is: Gewerbestrasse 11, 6330 Cham, Switzerland

Preface

These six volumes (LNCS 13377–13382) consist of the peer-reviewed papers from the workshops at the 22nd International Conference on Computational Science and Its Applications (ICCSA 2022), which took place during July 4–7, 2022. The peer-reviewed papers of the main conference tracks are published in a separate set consisting of two volumes (LNCS 13375–13376).

This year, we again decided to organize a hybrid conference, with some of the delegates attending in person and others taking part online. Despite the enormous benefits achieved by the intensive vaccination campaigns in many countries, at the crucial moment of organizing the event, there was no certainty about the evolution of COVID-19. Fortunately, more and more researchers were able to attend the event in person, foreshadowing a slow but gradual exit from the pandemic and the limitations that have weighed so heavily on the lives of all citizens over the past three years.

ICCSA 2022 was another successful event in the International Conference on Computational Science and Its Applications (ICCSA) series. Last year, the conference was held as a hybrid event in Cagliari, Italy, and in 2020 it was organized as virtual event, whilst earlier editions took place in Saint Petersburg, Russia (2019), Melbourne, Australia (2018), Trieste, Italy (2017), Beijing, China (2016), Banff, Canada (2015), Guimaraes, Portugal (2014), Ho Chi Minh City, Vietnam (2013), Salvador, Brazil (2012), Santander, Spain (2011), Fukuoka, Japan (2010), Suwon, South Korea (2009), Perugia, Italy (2008), Kuala Lumpur, Malaysia (2007), Glasgow, UK (2006), Singapore (2005), Assisi, Italy (2004), Montreal, Canada (2003), and (as ICCS) Amsterdam, The Netherlands (2002) and San Francisco, USA (2001).

Computational science is the main pillar of most of the present research, and industrial and commercial applications, and plays a unique role in exploiting ICT innovative technologies. The ICCSA conference series provides a venue to researchers and industry practitioners to discuss new ideas, to share complex problems and their solutions, and to shape new trends in computational science.

Apart from the 52 workshops, ICCSA 2022 also included six main tracks on topics ranging from computational science technologies and application in many fields to specific areas of computational sciences, such as software engineering, security, machine learning and artificial intelligence, and blockchain technologies. For the 52 workshops we have accepted 285 papers. For the main conference tracks we accepted 57 papers and 24 short papers out of 279 submissions (an acceptance rate of 29%). We would like to express our appreciation to the Workshops chairs and co-chairs for their hard work and dedication.

The success of the ICCSA conference series in general, and of ICCSA 2022 in particular, vitally depends on the support of many people: authors, presenters, participants, keynote speakers, workshop chairs, session chairs, organizing committee members, student volunteers, Program Committee members, advisory committee members, international liaison chairs, reviewers, and others in various roles. We take this opportunity to wholehartedly thank them all.

We also wish to thank our publisher, Springer, for their acceptance to publish the proceedings, for sponsoring some of the best papers awards, and for their kind assistance and cooperation during the editing process.

We cordially invite you to visit the ICCSA website https://iccsa.org where you can find all the relevant information about this interesting and exciting event.

July 2022   Osvaldo Gervasi
Beniamino Murgante
Sanjay Misra

Welcome Message from Organizers

The ICCSA 2021 conference in the Mediterranean city of Cagliari provided us with inspiration to offer the ICCSA 2022 conference in the Mediterranean city of Málaga, Spain. The additional considerations due to the COVID-19 pandemic, which necessitated a hybrid conference, also stimulated the idea to use the School of Informatics of the University of Málaga. It has an open structure where we could take lunch and coffee outdoors and the lecture halls have open windows on two sides providing optimal conditions for meeting more safely.

The school is connected to the center of the old town via a metro system, for which we offered cards to the participants. This provided the opportunity to stay in lodgings in the old town close to the beach because, at the end of the day, that is the place to be to exchange ideas with your fellow scientists. The social program allowed us to enjoy the history of Malaga from its founding by the Phoenicians...

In order to provoke as much scientific interaction as possible we organized online sessions that could easily be followed by all participants from their own devices. We tried to ensure that participants from Asia could participate in morning sessions and those from the Americas in evening sessions. On-site sessions could be followed and debated on-site and discussed online using a chat system. To realize this, we relied on the developed technological infrastructure based on open source software, with the addition of streaming channels on YouTube. The implementation of the software infrastructure and the technical coordination of the volunteers were carried out by Damiano Perri and Marco Simonetti. Nine student volunteers from the universities of Málaga, Minho, Almeria, and Helsinki provided technical support and ensured smooth interaction during the conference.

A big thank you goes to all of the participants willing to exchange their ideas during their daytime. Participants of ICCSA 2022 came from 58 countries scattered over many time zones of the globe. Very interesting keynote talks were provided by well-known international scientists who provided us with more ideas to reflect upon, and we are grateful for their insights.

Eligius M. T. Hendrix

Organization

ICCSA 2022 was organized by the University of Malaga (Spain), the University of Perugia (Italy), the University of Cagliari (Italy), the University of Basilicata (Italy), Monash University (Australia), Kyushu Sangyo University (Japan), and the University of Minho, (Portugal).

Honorary General Chairs

Norio Shiratori	Chuo University, Japan
Kenneth C. J. Tan	Sardina Systems, UK

General Chairs

Osvaldo Gervasi	University of Perugia, Italy
Eligius Hendrix	University of Malaga, Italy
Bernady O. Apduhan	Kyushu Sangyo University, Japan

Program Committee Chairs

Beniamino Murgante	University of Basilicata, Italy
Inmaculada Garcia Fernandez	University of Malaga, Spain
Ana Maria A. C. Rocha	University of Minho, Portugal
David Taniar	Monash University, Australia

International Advisory Committee

Jemal Abawajy	Deakin University, Australia
Dharma P. Agarwal	University of Cincinnati, USA
Rajkumar Buyya	Melbourne University, Australia
Claudia Bauzer Medeiros	University of Campinas, Brazil
Manfred M. Fisher	Vienna University of Economics and Business, Austria
Marina L. Gavrilova	University of Calgary, Canada
Sumi Helal	University of Florida, USA, and University of Lancaster, UK
Yee Leung	Chinese University of Hong Kong, China

International Liaison Chairs

Ivan Blečić	University of Cagliari, Italy
Giuseppe Borruso	University of Trieste, Italy

Elise De Donker	Western Michigan University, USA
Maria Irene Falcão	University of Minho, Portugal
Robert C. H. Hsu	Chung Hua University, Taiwan
Tai-Hoon Kim	Beijing Jiaotong University, China
Vladimir Korkhov	St Petersburg University, Russia
Sanjay Misra	Østfold University College, Norway
Takashi Naka	Kyushu Sangyo University, Japan
Rafael D. C. Santos	National Institute for Space Research, Brazil
Maribel Yasmina Santos	University of Minho, Portugal
Elena Stankova	St Petersburg University, Russia

Workshop and Session Organizing Chairs

Beniamino Murgante	University of Basilicata, Italy
Chiara Garau	University of Cagliari, Italy
Sanjay Misra	Ostfold University College, Norway

Award Chair

| Wenny Rahayu | La Trobe University, Australia |

Publicity Committee Chairs

Elmer Dadios	De La Salle University, Philippines
Nataliia Kulabukhova	St Petersburg University, Russia
Daisuke Takahashi	Tsukuba University, Japan
Shangwang Wang	Beijing University of Posts and Telecommunications, China

Local Arrangement Chairs

Eligius Hendrix	University of Malaga, Spain
Inmaculada Garcia Fernandez	University of Malaga, Spain
Salvador Merino Cordoba	University of Malaga, Spain
Pablo Guerrero-García	University of Malaga, Spain

Technology Chairs

| Damiano Perri | University of Florence, Italy |
| Marco Simonetti | University of Florence, Italy |

Program Committee

| Vera Afreixo | University of Aveiro, Portugal |
| Filipe Alvelos | University of Minho, Portugal |

Raffaele Garrisi	Polizia di Stato, Italy
Jerome Gensel	LSR-IMAG, France
Maria Giaoutzi	National Technical University of Athens, Greece
Arminda Manuela Andrade Pereira Gonçalves	University of Minho, Portugal
Andrzej M. Goscinski	Deakin University, Australia
Sevin Gümgüm	Izmir University of Economics, Turkey
Alex Hagen-Zanker	University of Cambridge, UK
Shanmugasundaram Hariharan	B.S. Abdur Rahman Crescent Institute of Science and Technology, India
Eligius M. T. Hendrix	University of Malaga, Spain and Wageningen University, The Netherlands
Hisamoto Hiyoshi	Gunma University, Japan
Mustafa Inceoglu	Ege University, Turkey
Peter Jimack	University of Leeds, UK
Qun Jin	Waseda University, Japan
Yeliz Karaca	UMass Chan Medical School, USA
Farid Karimipour	Vienna University of Technology, Austria
Baris Kazar	Oracle Corp., USA
Maulana Adhinugraha Kiki	Telkom University, Indonesia
DongSeong Kim	University of Canterbury, New Zealand
Taihoon Kim	Hannam University, South Korea
Ivana Kolingerova	University of West Bohemia, Czech Republic
Nataliia Kulabukhova	St. Petersburg University, Russia
Vladimir Korkhov	St. Petersburg University, Russia
Rosa Lasaponara	National Research Council, Italy
Maurizio Lazzari	National Research Council, Italy
Cheng Siong Lee	Monash University, Australia
Sangyoun Lee	Yonsei University, South Korea
Jongchan Lee	Kunsan National University, South Korea
Chendong Li	University of Connecticut, USA
Gang Li	Deakin University, Australia
Fang (Cherry) Liu	Ames Laboratory, USA
Xin Liu	University of Calgary, Canada
Andrea Lombardi	University of Perugia, Italy
Savino Longo	University of Bari, Italy
Tinghuai Ma	Nanjing University of Information Science and Technology, China
Ernesto Marcheggiani	Katholieke Universiteit Leuven, Belgium
Antonino Marvuglia	Public Research Centre Henri Tudor, Luxembourg
Nicola Masini	National Research Council, Italy
Ilaria Matteucci	National Research Council, Italy
Nirvana Meratnia	University of Twente, The Netherlands
Fernando Miranda	University of Minho, Portugal
Giuseppe Modica	University of Reggio Calabria, Italy
Josè Luis Montaña	University of Cantabria, Spain

Maria Filipa Mourão	Instituto Politécnico de Viana do Castelo, Portugal
Louiza de Macedo Mourelle	State University of Rio de Janeiro, Brazil
Nadia Nedjah	State University of Rio de Janeiro, Brazil
Laszlo Neumann	University of Girona, Spain
Kok-Leong Ong	Deakin University, Australia
Belen Palop	Universidad de Valladolid, Spain
Marcin Paprzycki	Polish Academy of Sciences, Poland
Eric Pardede	La Trobe University, Australia
Kwangjin Park	Wonkwang University, South Korea
Ana Isabel Pereira	Polytechnic Institute of Bragança, Portugal
Massimiliano Petri	University of Pisa, Italy
Telmo Pinto	University of Coimbra, Portugal
Maurizio Pollino	Italian National Agency for New Technologies, Energy and Sustainable Economic Development, Italy
Alenka Poplin	University of Hamburg, Germany
Vidyasagar Potdar	Curtin University of Technology, Australia
David C. Prosperi	Florida Atlantic University, USA
Wenny Rahayu	La Trobe University, Australia
Jerzy Respondek	Silesian University of Technology, Poland
Humberto Rocha	INESC-Coimbra, Portugal
Jon Rokne	University of Calgary, Canada
Octavio Roncero	CSIC, Spain
Maytham Safar	Kuwait University, Kuwait
Chiara Saracino	A.O. Ospedale Niguarda Ca' Granda, Italy
Marco Paulo Seabra dos Reis	University of Coimbra, Portugal
Jie Shen	University of Michigan, USA
Qi Shi	Liverpool John Moores University, UK
Dale Shires	U.S. Army Research Laboratory, USA
Inês Soares	University of Coimbra, Portugal
Elena Stankova	St Petersburg University, Russia
Takuo Suganuma	Tohoku University, Japan
Eufemia Tarantino	Polytechnic Universiy of Bari, Italy
Sergio Tasso	University of Perugia, Italy
Ana Paula Teixeira	University of Trás-os-Montes and Alto Douro, Portugal
M. Filomena Teodoro	Portuguese Naval Academy and University of Lisbon, Portugal
Parimala Thulasiraman	University of Manitoba, Canada
Carmelo Torre	Polytechnic University of Bari, Italy
Javier Martinez Torres	Centro Universitario de la Defensa Zaragoza, Spain
Giuseppe A. Trunfio	University of Sassari, Italy
Pablo Vanegas	University of Cuenca, Equador
Marco Vizzari	University of Perugia, Italy
Varun Vohra	Merck Inc., USA
Koichi Wada	University of Tsukuba, Japan
Krzysztof Walkowiak	Wroclaw University of Technology, Poland

Zequn Wang	Intelligent Automation Inc, USA
Robert Weibel	University of Zurich, Switzerland
Frank Westad	Norwegian University of Science and Technology, Norway
Roland Wismüller	Universität Siegen, Germany
Mudasser Wyne	National University, USA
Chung-Huang Yang	National Kaohsiung Normal University, Taiwan
Xin-She Yang	National Physical Laboratory, UK
Salim Zabir	France Telecom Japan Co., Japan
Haifeng Zhao	University of California, Davis, USA
Fabiana Zollo	Ca' Foscari University of Venice, Italy
Albert Y. Zomaya	University of Sydney, Australia

Workshop Organizers

International Workshop on Advances in Artificial Intelligence Learning Technologies: Blended Learning, STEM, Computational Thinking and Coding (AAILT 2022)

Alfredo Milani	University of Perugia, Italy
Valentina Franzoni	University of Perugia, Italy
Osvaldo Gervasi	University of Perugia, Italy

International Workshop on Advancements in Applied Machine-Learning and Data Analytics (AAMDA 2022)

Alessandro Costantini	INFN, Italy
Davide Salomoni	INFN, Italy
Doina Cristina Duma	INFN, Italy
Daniele Cesini	INFN, Italy

International Workshop on Advances in Information Systems and Technologies for Emergency Management, Risk Assessment and Mitigation Based on the Resilience (ASTER 2022)

Maurizio Pollino	ENEA, Italy
Marco Vona	University of Basilicata, Italy
Sonia Giovinazzi	ENEA, Italy
Benedetto Manganelli	University of Basilicata, Italy
Beniamino Murgante	University of Basilicata, Italy

International Workshop on Advances in Web Based Learning (AWBL 2022)

Birol Ciloglugil   Ege University, Turkey
Mustafa Inceoglu   Ege University, Turkey

International Workshop on Blockchain and Distributed Ledgers: Technologies and Applications (BDLTA 2022)

Vladimir Korkhov   St Petersburg State University, Russia
Elena Stankova   St Petersburg State University, Russia
Nataliia Kulabukhova   St Petersburg State University, Russia

International Workshop on Bio and Neuro Inspired Computing and Applications (BIONCA 2022)

Nadia Nedjah   State University of Rio De Janeiro, Brazil
Luiza De Macedo Mourelle   State University of Rio De Janeiro, Brazil

International Workshop on Configurational Analysis For Cities (CA CITIES 2022)

Claudia Yamu   Oslo Metropolitan University, Norway
Valerio Cutini   Università di Pisa, Italy
Beniamino Murgante   University of Basilicata, Italy
Chiara Garau   Dicaar, University of Cagliari, Italy

International Workshop on Computational and Applied Mathematics (CAM 2022)

Maria Irene Falcão   University of Minho, Portugal
Fernando Miranda   University of Minho, Portugal

International Workshop on Computational and Applied Statistics (CAS 2022)

Ana Cristina Braga   University of Minho, Portugal

International Workshop on Computational Mathematics, Statistics and Information Management (CMSIM 2022)

Maria Filomena Teodoro   University of Lisbon and Portuguese Naval Academy,
                             Portugal

International Workshop on Computational Optimization and Applications (COA 2022)

Ana Maria A. C. Rocha   University of Minho, Portugal
Humberto Rocha   University of Coimbra, Portugal

International Workshop on Computational Astrochemistry (CompAstro 2022)

Marzio Rosi   University of Perugia, Italy
Nadia Balucani   University of Perugia, Italy
Cecilia Ceccarelli   Université Grenoble Alpes, France
Stefano Falcinelli   University of Perugia, Italy

International Workshop on Computational Methods for Porous Geomaterials (CompPor 2022)

Vadim Lisitsa   Sobolev Institute of Mathematics, Russia
Evgeniy Romenski   Sobolev Institute of Mathematics, Russia

International Workshop on Computational Approaches for Smart, Conscious Cities (CASCC 2022)

Andreas Fricke   University of Potsdam, Germany
Juergen Doellner   University of Potsdam, Germany
Salvador Merino   University of Malaga, Spain
Jürgen Bund   Graphics Vision AI Association, Germany/Portugal
Markus Jobst   Federal Office of Metrology and Surveying, Austria
Francisco Guzman   University of Malaga, Spain

International Workshop on Computational Science and HPC (CSHPC 2022)

Elise De Doncker   Western Michigan University, USA
Fukuko Yuasa   High Energy Accelerator Research Organization
                               (KEK), Japan
Hideo Matsufuru   High Energy Accelerator Research Organization
                               (KEK), Japan

International Workshop on Cities, Technologies and Planning (CTP 2022)

Giuseppe Borruso   University of Trieste, Italy
Malgorzata Hanzl   Lodz University of Technology, Poland
Beniamino Murgante   University of Basilicata, Italy

Anastasia Stratigea	National Technical University of Athens, Grece
Ginevra Balletto	University of Cagliari, Italy
Ljiljana Zivkovic	Republic Geodetic Authority, Serbia

International Workshop on Digital Sustainability and Circular Economy (DiSCE 2022)

Giuseppe Borruso	University of Trieste, Italy
Stefano Epifani	Digital Sustainability Institute, Italy
Ginevra Balletto	University of Cagliari, Italy
Luigi Mundula	University of Cagliari, Italy
Alessandra Milesi	University of Cagliari, Italy
Mara Ladu	University of Cagliari, Italy
Stefano De Nicolai	University of Pavia, Italy
Tu Anh Trinh	University of Economics Ho Chi Minh City, Vietnam

International Workshop on Econometrics and Multidimensional Evaluation in Urban Environment (EMEUE 2022)

Carmelo Maria Torre	Polytechnic University of Bari, Italy
Maria Cerreta	University of Naples Federico II, Italy
Pierluigi Morano	Polytechnic University of Bari, Italy
Giuliano Poli	University of Naples Federico II, Italy
Marco Locurcio	Polytechnic University of Bari, Italy
Francesco Tajani	Sapienza University of Rome, Italy

International Workshop on Ethical AI Applications for a Human-Centered Cyber Society (EthicAI 2022)

| Valentina Franzoni | University of Perugia, Italy |
| Alfredo Milani | University of Perugia, Italy |

International Workshop on Future Computing System Technologies and Applications (FiSTA 2022)

| Bernady Apduhan | Kyushu Sangyo University, Japan |
| Rafael Santos | INPE, Brazil |

International Workshop on Geodesign in Decision Making: Meta Planning and Collaborative Design for Sustainable and Inclusive Development (GDM 2022)

Francesco Scorza	University of Basilicata, Italy
Michele Campagna	University of Cagliari, Italy
Ana Clara Mourão Moura	Federal University of Minas Gerais, Brazil

International Workshop on Geomatics in Agriculture and Forestry: New Advances and Perspectives (GeoForAgr 2022)

Maurizio Pollino	ENEA, Italy
Giuseppe Modica	University of Reggio Calabria, Italy
Marco Vizzari	University of Perugia, Italy

International Workshop on Geographical Analysis, Urban Modeling, Spatial Statistics (Geog-An-Mod 2022)

Giuseppe Borruso	University of Trieste, Italy
Beniamino Murgante	University of Basilicata, Italy
Harmut Asche	Hasso-Plattner-Institut für Digital Engineering gGmbH, Germany

International Workshop on Geomatics for Resource Monitoring and Management (GRMM 2022)

Alessandra Capolupo	Polytechnic of Bari, Italy
Eufemia Tarantino	Polytechnic of Bari, Italy
Enrico Borgogno Mondino	University of Turin, Italy

International Workshop on Information and Knowledge in the Internet of Things (IKIT 2022)

Teresa Guarda	State University of Santa Elena Peninsula, Ecuador
Filipe Portela	University of Minho, Portugal
Maria Fernanda Augusto	Bitrum Research Center, Spain

13th International Symposium on Software Quality (ISSQ 2022)

Sanjay Misra	Østfold University College, Norway

International Workshop on Machine Learning for Space and Earth Observation Data (MALSEOD 2022)

Rafael Santos	INPE, Brazil
Karine Reis Ferreira Gomes	INPE, Brazil

International Workshop on Building Multi-dimensional Models for Assessing Complex Environmental Systems (MES 2022)

Vanessa Assumma	Politecnico di Torino, Italy
Caterina Caprioli	Politecnico di Torino, Italy
Giulia Datola	Politecnico di Torino, Italy

Federico Dell'Anna   Politecnico di Torino, Italy
Marta Dell'Ovo   Politecnico di Milano, Italy

International Workshop on Models and Indicators for Assessing and Measuring the Urban Settlement Development in the View of ZERO Net Land Take by 2050 (MOVEto0 2022)

Lucia Saganeiti   University of L'Aquila, Italy
Lorena Fiorini   University of L'aquila, Italy
Angela Pilogallo   University of Basilicata, Italy
Alessandro Marucci   University of L'Aquila, Italy
Francesco Zullo   University of L'Aquila, Italy

International Workshop on Modelling Post-Covid Cities (MPCC 2022)

Beniamino Murgante   University of Basilicata, Italy
Ginevra Balletto   University of Cagliari, Italy
Giuseppe Borruso   University of Trieste, Italy
Marco Dettori   Università degli Studi di Sassari, Italy
Lucia Saganeiti   University of L'Aquila, Italy

International Workshop on Ecosystem Services: Nature's Contribution to People in Practice. Assessment Frameworks, Models, Mapping, and Implications (NC2P 2022)

Francesco Scorza   University of Basilicata, Italy
Sabrina Lai   University of Cagliari, Italy
Silvia Ronchi   University of Cagliari, Italy
Dani Broitman   Israel Institute of Technology, Israel
Ana Clara Mourão Moura   Federal University of Minas Gerais, Brazil
Corrado Zoppi   University of Cagliari, Italy

International Workshop on New Mobility Choices for Sustainable and Alternative Scenarios (NEWMOB 2022)

Tiziana Campisi   University of Enna Kore, Italy
Socrates Basbas   Aristotle University of Thessaloniki, Greece
Aleksandra Deluka T.   University of Rijeka, Croatia
Alexandros Nikitas   University of Huddersfield, UK
Ioannis Politis   Aristotle University of Thessaloniki, Greece
Georgios Georgiadis   Aristotle University of Thessaloniki, Greece
Irena Ištoka Otković   University of Osijek, Croatia
Sanja Surdonja   University of Rijeka, Croatia

International Workshop on Privacy in the Cloud/Edge/IoT World (PCEIoT 2022)

Michele Mastroianni	University of Campania Luigi Vanvitelli, Italy
Lelio Campanile	University of Campania Luigi Vanvitelli, Italy
Mauro Iacono	University of Campania Luigi Vanvitelli, Italy

International Workshop on Psycho-Social Analysis of Sustainable Mobility in the Pre- and Post-Pandemic Phase (PSYCHE 2022)

Tiziana Campisi	University of Enna Kore, Italy
Socrates Basbas	Aristotle University of Thessaloniki, Greece
Dilum Dissanayake	Newcastle University, UK
Nurten Akgün Tanbay	Bursa Technical University, Turkey
Elena Cocuzza	University of Catania, Italy
Nazam Ali	University of Management and Technology, Pakistan
Vincenza Torrisi	University of Catania, Italy

International Workshop on Processes, Methods and Tools Towards Resilient Cities and Cultural Heritage Prone to SOD and ROD Disasters (RES 2022)

Elena Cantatore	Polytechnic University of Bari, Italy
Alberico Sonnessa	Polytechnic University of Bari, Italy
Dario Esposito	Polytechnic University of Bari, Italy

International Workshop on Scientific Computing Infrastructure (SCI 2022)

Elena Stankova	St Petersburg University, Russia
Vladimir Korkhov	St Petersburg University, Russia

International Workshop on Socio-Economic and Environmental Models for Land Use Management (SEMLUM 2022)

Debora Anelli	Polytechnic University of Bari, Italy
Pierluigi Morano	Polytechnic University of Bari, Italy
Francesco Tajani	Sapienza University of Rome, Italy
Marco Locurcio	Polytechnic University of Bari, Italy
Paola Amoruso	LUM University, Italy

14th International Symposium on Software Engineering Processes and Applications (SEPA 2022)

Sanjay Misra	Østfold University College, Norway

International Workshop on Ports of the Future – Smartness and Sustainability (SmartPorts 2022)

Giuseppe Borruso	University of Trieste, Italy
Gianfranco Fancello	University of Cagliari, Italy
Ginevra Balletto	University of Cagliari, Italy
Patrizia Serra	University of Cagliari, Italy
Maria del Mar Munoz Leonisio	University of Cadiz, Spain
Marco Mazzarino	University of Venice, Italy
Marcello Tadini	Università del Piemonte Orientale, Italy

International Workshop on Smart Tourism (SmartTourism 2022)

Giuseppe Borruso	University of Trieste, Italy
Silvia Battino	University of Sassari, Italy
Ainhoa Amaro Garcia	Universidad de Alcalà and Universidad de Las Palmas, Spain
Maria del Mar Munoz Leonisio	University of Cadiz, Spain
Carlo Donato	University of Sassari, Italy
Francesca Krasna	University of Trieste, Italy
Ginevra Balletto	University of Cagliari, Italy

International Workshop on Sustainability Performance Assessment: Models, Approaches and Applications Toward Interdisciplinary and Integrated Solutions (SPA 2022)

Francesco Scorza	University of Basilicata, Italy
Sabrina Lai	University of Cagliari, Italy
Jolanta Dvarioniene	Kaunas University of Technology, Lithuania
Iole Cerminara	University of Basilicata, Italy
Georgia Pozoukidou	Aristotle University of Thessaloniki, Greece
Valentin Grecu	Lucian Blaga University of Sibiu, Romania
Corrado Zoppi	University of Cagliari, Italy

International Workshop on Specifics of Smart Cities Development in Europe (SPEED 2022)

Chiara Garau	University of Cagliari, Italy
Katarína Vitálišová	Matej Bel University, Slovakia
Paolo Nesi	University of Florence, Italy
Anna Vanova	Matej Bel University, Slovakia
Kamila Borsekova	Matej Bel University, Slovakia
Paola Zamperlin	University of Pisa, Italy

Federico Cugurullo   Trinity College Dublin, Ireland
Gerardo Carpentieri   University of Naples Federico II, Italy

International Workshop on Smart and Sustainable Island Communities (SSIC 2022)

Chiara Garau   University of Cagliari, Italy
Anastasia Stratigea   National Technical University of Athens, Greece
Paola Zamperlin   University of Pisa, Italy
Francesco Scorza   University of Basilicata, Italy

International Workshop on Theoretical and Computational Chemistry and Its Applications (TCCMA 2022)

Noelia Faginas-Lago   University of Perugia, Italy
Andrea Lombardi   University of Perugia, Italy

International Workshop on Transport Infrastructures for Smart Cities (TISC 2022)

Francesca Maltinti   University of Cagliari, Italy
Mauro Coni   University of Cagliari, Italy
Francesco Pinna   University of Cagliari, Italy
Chiara Garau   University of Cagliari, Italy
Nicoletta Rassu   Univesity of Cagliari, Italy
James Rombi   University of Cagliari, Italy
Benedetto Barabino   University of Brescia, Italy

14th International Workshop on Tools and Techniques in Software Development Process (TTSDP 2022)

Sanjay Misra   Østfold University College, Norway

International Workshop on Urban Form Studies (UForm 2022)

Malgorzata Hanzl   Lodz University of Technology, Poland
Beniamino Murgante   University of Basilicata, Italy
Alessandro Camiz   Özyeğin University, Turkey
Tomasz Bradecki   Silesian University of Technology, Poland

International Workshop on Urban Regeneration: Innovative Tools and Evaluation Model (URITEM 2022)

Fabrizio Battisti   University of Florence, Italy
Laura Ricci   Sapienza University of Rome, Italy
Orazio Campo   Sapienza University of Rome, Italy

International Workshop on Urban Space Accessibility and Mobilities (USAM 2022)

Chiara Garau	University of Cagliari, Italy
Matteo Ignaccolo	University of Catania, Italy
Enrica Papa	University of Westminster, UK
Francesco Pinna	University of Cagliari, Italy
Silvia Rossetti	University of Parma, Italy
Wendy Tan	Wageningen University and Research, The Netherlands
Michela Tiboni	University of Brescia, Italy
Vincenza Torrisi	University of Catania, Italy

International Workshop on Virtual Reality and Augmented Reality and Applications (VRA 2022)

Osvaldo Gervasi	University of Perugia, Italy
Damiano Perri	University of Florence, Italy
Marco Simonetti	University of Florence, Italy
Sergio Tasso	University of Perugia, Italy

International Workshop on Advanced and Computational Methods for Earth Science Applications (WACM4ES 2022)

Luca Piroddi	University of Cagliari, Italy
Sebastiano Damico	University of Malta, Malta

International Workshop on Advanced Mathematics and Computing Methods in Complex Computational Systems (WAMCM 2022)

Yeliz Karaca	UMass Chan Medical School, USA
Dumitru Baleanu	Cankaya University, Turkey
Osvaldo Gervasi	University of Perugia, Italy
Yudong Zhang	University of Leicester, UK
Majaz Moonis	UMass Chan Medical School, USA

Additional Reviewers

Akshat Agrawal	Amity University, Haryana, India
Waseem Ahmad	National Institute of Technology Karnataka, India
Vladimir Alarcon	Universidad Diego Portales, Chile
Oylum Alatlı	Ege University, Turkey
Raffaele Albano	University of Basilicata, Italy
Abraham Alfa	FUT Minna, Nigeria
Diego Altafini	Università di Pisa, Italy
Filipe Alvelos	Universidade do Minho, Portugal

Rogerio Calazan	IEAPM, Brazil
Michele Campagna	University of Cagliari, Italy
Lelio Campanile	Università degli Studi della Campania Luigi Vanvitelli, Italy
Tiziana Campisi	University of Enna Kore, Italy
Antonino Canale	University of Enna Kore, Italy
Elena Cantatore	Polytechnic University of Bari, Italy
Patrizia Capizzi	Univerity of Palermo, Italy
Alessandra Capolupo	Polytechnic University of Bari, Italy
Giacomo Caporusso	Politecnico di Bari, Italy
Caterina Caprioli	Politecnico di Torino, Italy
Gerardo Carpentieri	University of Naples Federico II, Italy
Martina Carra	University of Brescia, Italy
Pedro Carrasqueira	INESC Coimbra, Portugal
Barbara Caselli	Università degli Studi di Parma, Italy
Cecilia Castro	University of Minho, Portugal
Giulio Cavana	Politecnico di Torino, Italy
Iole Cerminara	University of Basilicata, Italy
Maria Cerreta	University of Naples Federico II, Italy
Daniele Cesini	INFN, Italy
Jabed Chowdhury	La Trobe University, Australia
Birol Ciloglugil	Ege University, Turkey
Elena Cocuzza	Univesity of Catania, Italy
Emanuele Colica	University of Malta, Malta
Mauro Coni	University of Cagliari, Italy
Elisete Correia	Universidade de Trás-os-Montes e Alto Douro, Portugal
Florbela Correia	Polytechnic Institute of Viana do Castelo, Portugal
Paulo Cortez	University of Minho, Portugal
Lino Costa	Universidade do Minho, Portugal
Alessandro Costantini	INFN, Italy
Marilena Cozzolino	Università del Molise, Italy
Alfredo Cuzzocrea	University of Calabria, Italy
Sebastiano D'amico	University of Malta, Malta
Gianni D'Angelo	University of Salerno, Italy
Tijana Dabovic	University of Belgrade, Serbia
Hiroshi Daisaka	Hitotsubashi University, Japan
Giulia Datola	Politecnico di Torino, Italy
Regina De Almeida	University of Trás-os-Montes and Alto Douro, Portugal
Maria Stella De Biase	Università della Campania Luigi Vanvitelli, Italy
Elise De Doncker	Western Michigan University, USA
Itamir De Morais Barroca Filho	Federal University of Rio Grande do Norte, Brazil
Samuele De Petris	University of Turin, Italy
Alan De Sá	Marinha do Brasil, Brazil
Alexander Degtyarev	St Petersburg University, Russia

Federico Dell'Anna	Politecnico di Torino, Italy
Marta Dell'Ovo	Politecnico di Milano, Italy
Ahu Dereli Dursun	Istanbul Commerce University, Turkey
Giulia Desogus	University of Cagliari, Italy
Piero Di Bonito	Università degli Studi della Campania, Italia
Paolino Di Felice	University of L'Aquila, Italy
Felicia Di Liddo	Polytechnic University of Bari, Italy
Isabel Dimas	University of Coimbra, Portugal
Doina Cristina Duma	INFN, Italy
Aziz Dursun	Virginia Tech University, USA
Jaroslav Dvořak	Klaipėda University, Lithuania
Dario Esposito	Polytechnic University of Bari, Italy
M. Noelia Faginas-Lago	University of Perugia, Italy
Stefano Falcinelli	University of Perugia, Italy
Falcone Giacomo	University of Reggio Calabria, Italy
Maria Irene Falcão	University of Minho, Portugal
Stefano Federico	CNR-ISAC, Italy
Marcin Feltynowski	University of Lodz, Poland
António Fernandes	Instituto Politécnico de Bragança, Portugal
Florbela Fernandes	Instituto Politecnico de Braganca, Portugal
Paula Odete Fernandes	Instituto Politécnico de Bragança, Portugal
Luis Fernandez-Sanz	University of Alcala, Spain
Luís Ferrás	University of Minho, Portugal
Ângela Ferreira	Instituto Politécnico de Bragança, Portugal
Lorena Fiorini	University of L'Aquila, Italy
Hector Florez	Universidad Distrital Francisco Jose de Caldas, Colombia
Stefano Franco	LUISS Guido Carli, Italy
Valentina Franzoni	Perugia University, Italy
Adelaide Freitas	University of Aveiro, Portugal
Andreas Fricke	Hasso Plattner Institute, Germany
Junpei Fujimoto	KEK, Japan
Federica Gaglione	Università del Sannio, Italy
Andrea Gallo	Università degli Studi di Trieste, Italy
Luciano Galone	University of Malta, Malta
Adam Galuszka	Silesian University of Technology, Poland
Chiara Garau	University of Cagliari, Italy
Ernesto Garcia Para	Universidad del País Vasco, Spain
Aniket A. Gaurav	Østfold University College, Norway
Marina Gavrilova	University of Calgary, Canada
Osvaldo Gervasi	University of Perugia, Italy
Andrea Ghirardi	Università di Brescia, Italy
Andrea Gioia	Politecnico di Bari, Italy
Giacomo Giorgi	Università degli Studi di Perugia, Italy
Stanislav Glubokovskikh	Lawrence Berkeley National Laboratory, USA
A. Manuela Gonçalves	University of Minho, Portugal

Leocadio González Casado	University of Almería, Spain
Angela Gorgoglione	Universidad de la República Uruguay, Uruguay
Yusuke Gotoh	Okayama University, Japan
Daniele Granata	Università degli Studi della Campania, Italy
Christian Grévisse	University of Luxembourg, Luxembourg
Silvana Grillo	University of Cagliari, Italy
Teresa Guarda	State University of Santa Elena Peninsula, Ecuador
Carmen Guida	Università degli Studi di Napoli Federico II, Italy
Kemal Güven Gülen	Namık Kemal University, Turkey
Ipek Guler	Leuven Biostatistics and Statistical Bioinformatics Centre, Belgium
Sevin Gumgum	Izmir University of Economics, Turkey
Martina Halásková	VSB Technical University in Ostrava, Czech Republic
Peter Hegedus	University of Szeged, Hungary
Eligius M. T. Hendrix	Universidad de Málaga, Spain
Mauro Iacono	Università degli Studi della Campania, Italy
Oleg Iakushkin	St Petersburg University, Russia
Matteo Ignaccolo	University of Catania, Italy
Mustafa Inceoglu	Ege University, Turkey
Markus Jobst	Federal Office of Metrology and Surveying, Austria
Issaku Kanamori	RIKEN Center for Computational Science, Japan
Yeliz Karaca	UMass Chan Medical School, USA
Aarti Karande	Sardar Patel Institute of Technology, India
András Kicsi	University of Szeged, Hungary
Vladimir Korkhov	St Petersburg University, Russia
Nataliia Kulabukhova	St Petersburg University, Russia
Claudio Ladisa	Politecnico di Bari, Italy
Mara Ladu	University of Cagliari, Italy
Sabrina Lai	University of Cagliari, Italy
Mark Lajko	University of Szeged, Hungary
Giuseppe Francesco Cesare Lama	University of Napoli Federico II, Italy
Vincenzo Laporta	CNR, Italy
Margherita Lasorella	Politecnico di Bari, Italy
Francesca Leccis	Università di Cagliari, Italy
Federica Leone	University of Cagliari, Italy
Chien-sing Lee	Sunway University, Malaysia
Marco Locurcio	Polytechnic University of Bari, Italy
Francesco Loddo	Henge S.r.l., Italy
Andrea Lombardi	Università di Perugia, Italy
Isabel Lopes	Instituto Politécnico de Bragança, Portugal
Fernando Lopez Gayarre	University of Oviedo, Spain
Vanda Lourenço	Universidade Nova de Lisboa, Portugal
Jing Ma	Luleå University of Technology, Sweden
Helmuth Malonek	University of Aveiro, Portugal
Francesca Maltinti	University of Cagliari, Italy

Benedetto Manganelli	Università degli Studi della Basilicata, Italy
Krassimir Markov	Institute of Electric Engineering and Informatics, Bulgaria
Alessandro Marucci	University of L'Aquila, Italy
Alessandra Mascitelli	Italian Civil Protection Department and ISAC-CNR, Italy
Michele Mastroianni	University of Campania Luigi Vanvitelli, Italy
Hideo Matsufuru	High Energy Accelerator Research Organization (KEK), Japan
Chiara Mazzarella	University of Naples Federico II, Italy
Marco Mazzarino	University of Venice, Italy
Paolo Mengoni	University of Florence, Italy
Alfredo Milani	University of Perugia, Italy
Fernando Miranda	Universidade do Minho, Portugal
Augusto Montisci	Università degli Studi di Cagliari, Italy
Ricardo Moura	New University of Lisbon, Portugal
Ana Clara Mourao Moura	Federal University of Minas Gerais, Brazil
Maria Mourao	Polytechnic Institute of Viana do Castelo, Portugal
Eugenio Muccio	University of Naples Federico II, Italy
Beniamino Murgante	University of Basilicata, Italy
Giuseppe Musolino	University of Reggio Calabria, Italy
Stefano Naitza	Università di Cagliari, Italy
Naohito Nakasato	University of Aizu, Japan
Roberto Nardone	University of Reggio Calabria, Italy
Nadia Nedjah	State University of Rio de Janeiro, Brazil
Juraj Nemec	Masaryk University in Brno, Czech Republic
Keigo Nitadori	RIKEN R-CCS, Japan
Roseline Ogundokun	Kaunas University of Technology, Lithuania
Francisco Henrique De Oliveira	Santa Catarina State University, Brazil
Irene Oliveira	Univesidade Trás-os-Montes e Alto Douro, Portugal
Samson Oruma	Østfold University College, Norway
Antonio Pala	University of Cagliari, Italy
Simona Panaro	University of Porstmouth, UK
Dimos Pantazis	University of West Attica, Greece
Giovanni Paragliola	ICAR-CNR, Italy
Eric Pardede	La Trobe University, Australia
Marco Parriani	University of Perugia, Italy
Paola Perchinunno	Uniersity of Bari, Italy
Ana Pereira	Polytechnic Institute of Bragança, Portugal
Damiano Perri	University of Perugia, Italy
Marco Petrelli	Roma Tre University, Italy
Camilla Pezzica	University of Pisa, Italy
Angela Pilogallo	University of Basilicata, Italy
Francesco Pinna	University of Cagliari, Italy
Telmo Pinto	University of Coimbra, Portugal

Fernando Pirani	University of Perugia, Italy
Luca Piroddi	University of Cagliari, Italy
Bojana Pjanović	University of Belgrade, Serbia
Giuliano Poli	University of Naples Federico II, Italy
Maurizio Pollino	ENEA, Italy
Salvatore Praticò	University of Reggio Calabria, Italy
Zbigniew Przygodzki	University of Lodz, Poland
Carlotta Quagliolo	Politecnico di Torino, Italy
Raffaele Garrisi	Polizia Postale e delle Comunicazioni, Italy
Mariapia Raimondo	Università della Campania Luigi Vanvitelli, Italy
Deep Raj	IIIT Naya Raipur, India
Buna Ramos	Universidade Lusíada Norte, Portugal
Nicoletta Rassu	Univesity of Cagliari, Italy
Michela Ravanelli	Sapienza Università di Roma, Italy
Roberta Ravanelli	Sapienza Università di Roma, Italy
Pier Francesco Recchi	University of Naples Federico II, Italy
Stefania Regalbuto	University of Naples Federico II, Italy
Marco Reis	University of Coimbra, Portugal
Maria Reitano	University of Naples Federico II, Italy
Anatoly Resnyansky	Defence Science and Technology Group, Australia
Jerzy Respondek	Silesian University of Technology, Poland
Isabel Ribeiro	Instituto Politécnico Bragança, Portugal
Albert Rimola	Universitat Autònoma de Barcelona, Spain
Corrado Rindone	University of Reggio Calabria, Italy
Ana Maria A. C. Rocha	University of Minho, Portugal
Humberto Rocha	University of Coimbra, Portugal
Maria Clara Rocha	Instituto Politécnico de Coimbra, Portugal
James Rombi	University of Cagliari, Italy
Elisabetta Ronchieri	INFN, Italy
Marzio Rosi	University of Perugia, Italy
Silvia Rossetti	Università degli Studi di Parma, Italy
Marco Rossitti	Politecnico di Milano, Italy
Mária Rostašová	Universtiy of Žilina, Slovakia
Lucia Saganeiti	University of L'Aquila, Italy
Giovanni Salzillo	Università degli Studi della Campania, Italy
Valentina Santarsiero	University of Basilicata, Italy
Luigi Santopietro	University of Basilicata, Italy
Stefania Santoro	Politecnico di Bari, Italy
Rafael Santos	INPE, Brazil
Valentino Santucci	Università per Stranieri di Perugia, Italy
Mirko Saponaro	Polytechnic University of Bari, Italy
Filippo Sarvia	University of Turin, Italy
Andrea Scianna	ICAR-CNR, Italy
Francesco Scorza	University of Basilicata, Italy
Ester Scotto Di Perta	University of Naples Federico II, Italy
Ricardo Severino	University of Minho, Portugal

Jie Shen	University of Michigan, USA
Luneque Silva Junior	Universidade Federal do ABC, Brazil
Carina Silva	Instituto Politécnico de Lisboa, Portugal
Joao Carlos Silva	Polytechnic Institute of Cavado and Ave, Portugal
Ilya Silvestrov	Saudi Aramco, Saudi Arabia
Marco Simonetti	University of Florence, Italy
Maria Joana Soares	University of Minho, Portugal
Michel Soares	Federal University of Sergipe, Brazil
Alberico Sonnessa	Politecnico di Bari, Italy
Lisete Sousa	University of Lisbon, Portugal
Elena Stankova	St Petersburg University, Russia
Jan Stejskal	University of Pardubice, Czech Republic
Silvia Stranieri	University of Naples Federico II, Italy
Anastasia Stratigea	National Technical University of Athens, Greece
Yue Sun	European XFEL GmbH, Germany
Anthony Suppa	Politecnico di Torino, Italy
Kirill Sviatov	Ulyanovsk State Technical University, Russia
David Taniar	Monash University, Australia
Rodrigo Tapia-McClung	Centro de Investigación en Ciencias de Información Geoespacial, Mexico
Eufemia Tarantino	Politecnico di Bari, Italy
Sergio Tasso	University of Perugia, Italy
Vladimir Tcheverda	Institute of Petroleum Geology and Geophysics, SB RAS, Russia
Ana Paula Teixeira	Universidade de Trás-os-Montes e Alto Douro, Portugal
Tengku Adil Tengku Izhar	Universiti Teknologi MARA, Malaysia
Maria Filomena Teodoro	University of Lisbon and Portuguese Naval Academy, Portugal
Yiota Theodora	National Technical University of Athens, Greece
Graça Tomaz	Instituto Politécnico da Guarda, Portugal
Gokchan Tonbul	Atilim University, Turkey
Rosa Claudia Torcasio	CNR-ISAC, Italy
Carmelo Maria Torre	Polytechnic University of Bari, Italy
Vincenza Torrisi	University of Catania, Italy
Vincenzo Totaro	Politecnico di Bari, Italy
Pham Trung	HCMUT, Vietnam
Po-yu Tsai	National Chung Hsing University, Taiwan
Dimitrios Tsoukalas	Centre of Research and Technology Hellas, Greece
Toshihiro Uchibayashi	Kyushu University, Japan
Takahiro Ueda	Seikei University, Japan
Piero Ugliengo	Università degli Studi di Torino, Italy
Gianmarco Vanuzzo	University of Perugia, Italy
Clara Vaz	Instituto Politécnico de Bragança, Portugal
Laura Verde	University of Campania Luigi Vanvitelli, Italy
Katarína Vitálišová	Matej Bel University, Slovakia

Daniel Mark Vitiello	University of Cagliari, Italy
Marco Vizzari	University of Perugia, Italy
Alexander Vodyaho	St. Petersburg State Electrotechnical University "LETI", Russia
Agustinus Borgy Waluyo	Monash University, Australia
Chao Wang	USTC, China
Marcin Wozniak	Silesian University of Technology, Poland
Jitao Yang	Beijing Language and Culture University, China
Fenghui Yao	Tennessee State University, USA
Fukuko Yuasa	KEK, Japan
Paola Zamperlin	University of Pisa, Italy
Michal Žemlička	Charles University, Czech Republic
Nataly Zhukova	ITMO University, Russia
Alcinia Zita Sampaio	University of Lisbon, Portugal
Ljiljana Zivkovic	Republic Geodetic Authority, Serbia
Floriana Zucaro	University of Naples Federico II, Italy
Marco Zucca	Politecnico di Milano, Italy
Camila Zyngier	Ibmec, Belo Horizonte, Brazil

Sponsoring Organizations

ICCSA 2022 would not have been possible without tremendous support of many organizations and institutions, for which all organizers and participants of ICCSA 2022 express their sincere gratitude:

Springer International Publishing AG, Germany (https://www.springer.com)

Computers Open Access Journal (https://www.mdpi.com/journal/computers)

Computation Open Access Journal (https://www.mdpi.com/journal/computation)

University of Malaga, Spain (https://www.uma.es/)

University of Perugia, Italy
(https://www.unipg.it)

University of Basilicata, Italy
(http://www.unibas.it)

Monash University, Australia
(https://www.monash.edu/)

Kyushu Sangyo University, Japan
(https://www.kyusan-u.ac.jp/)

University of Minho, Portugal
(https://www.uminho.pt/)

Universidade do Minho
Escola de Engenharia

Contents – Part II

International Workshop on Computational Optimization and Applications (COA 2022)

External Climate Data Extraction Using the Forward Feature Selection
Method in the Context of Occupational Safety   3
*Felipe G. Silva, Inês Sena, Laires A. Lima, Florbela P. Fernandes,
Maria F. Pacheco, Clara B. Vaz, José Lima, and Ana I. Pereira*

Dynamic Analysis of the Sustainable Performance of Electric Mobility
in European Countries   15
Clara B. Vaz and Ângela P. Ferreira

On Computational Procedures for Optimising an Omni-Channel Inventory
Control Model   29
Joost Goedhart and Eligius M. T. Hendrix

A Bibliometric Review and Analysis of Traffic Lights Optimization   43
*Gabriela R. Witeck, Ana Maria A. C. Rocha, Gonçalo O. Silva,
António Silva, Dalila Durães, and José Machado*

A Genetic Algorithm for Forest Firefighting Optimization   55
*Marina A. Matos, Ana Maria A. C. Rocha, Lino A. Costa,
and Filipe Alvelos*

On Tuning the Particle Swarm Optimization for Solving the Traffic Light
Problem   68
*Gonçalo O. Silva, Ana Maria A. C. Rocha, Gabriela R. Witeck,
António Silva, Dalila Durães, and José Machado*

A Reactive GRASP Algorithm for the Multi-depot Vehicle Routing
Problem   81
*Israel Pereira de Souza, Maria Claudia Silva Boeres,
Renato Elias Nunes de Moraes, and João Vinicius Corrêa Thompson*

How Life Transitions Influence People's Use of the Internet:
A Clustering Approach   97
*Martina Benvenuti, Humberto Rocha, Isabel Dórdio Dimas,
and Elvis Mazzoni*

On Monotonicity Detection in Simplicial Branch and Bound
over a Simplex 113
L. G. Casado, B. G.-Tóth, E. M. T. Hendrix, and F. Messine

Virtual Screening Based on Electrostatic Similarity and Flexible Ligands. . . .   127
Savíns Puertas-Martín, Juana L. Redondo, Antonio J. Banegas-Luna,
Ester M. Garzón, Horacio Pérez-Sánchez, Valerie J. Gillet,
and Pilar M. Ortigosa

Solving a Capacitated Waste Collection Problem Using an Open-Source
Tool   140
A. S. Silva, Filipe Alves, J. L. Diaz de Tuesta, Ana Maria A. C. Rocha,
A. I. Pereira, A. M. T. Silva, Paulo Leitão, and H. T. Gomes

A Systematic Literature Review About Multi-objective Optimization for
Distributed Manufacturing Scheduling in the Industry 4.0   157
Francisco dos Santos, Lino A. Costa, and Leonilde Varela

On Active-Set LP Algorithms Allowing Basis Deficiency   174
Pablo Guerrero-García and Eligius M. T. Hendrix

On the Design of a New Stochastic Meta-Heuristic for Derivative-Free
Optimization.   188
N. C. Cruz, Juana L. Redondo, E. M. Ortigosa, and P. M. Ortigosa

Analyzing the MathE Platform Through Clustering Algorithms.   201
Beatriz Flamia Azevedo, Yahia Amoura, Ana Maria A. C. Rocha,
Florbela P. Fernandes, Maria F. Pacheco, and Ana I. Pereira

A Tabu Search with a Double Neighborhood Strategy.   219
Paula Amaral, Ana Mendes, and J. Miguel Espinosa

**International Workshop on Computational Astrochemistry
(CompAs-tro 2022)**

The $S^+(^4S)+SiH_2(^1A_1)$ Reaction: Toward the Synthesis of Interstellar SiS. . .   233
Luca Mancini, Marco Trinari, Emília Valença Ferreira de Aragão,
Marzio Rosi, and Nadia Balucani

A Theoretical Investigation of the Reactions of $N(^2D)$ and CN
with Acrylonitrile and Implications for the Prebiotic Chemistry of Titan   246
Luca Mancini, Emília Valença Ferreira de Aragão,
and Gianmarco Vanuzzo

Formation Routes of CO from $O(^1D)$+Toluene: A Computational Study   260
Marzio Rosi, Piergiorgio Casavecchia, Nadia Balucani, Pedro Recio,
Adriana Caracciolo, Dimitrios Skouteris, and Carlo Cavallotti

Stereo-Dynamics of Autoionization Reactions Induced by $Ne^*(^3P_{0,2})$
Metastable Atoms with HCl and HBr Molecules: Experimental and
Theoretical Study of the Reactivity Through Selective Collisional
Angular Cones   270
   Marco Parriani, Franco Vecchiocattivi, Fernando Pirani,
   and Stefano Falcinelli

An Ab Initio Computational Study of Binding Energies of Interstellar
Complex Organic Molecules on Crystalline Water Ice Surface Models.   281
   Harjasnoor Kakkar, Berta Martínez-Bachs, and Albert Rimola

International Workshop on Computational Methods for Porous Geo-materials (CompPor 2022)

Optimization of the Training Dataset for Numerical Dispersion Mitigation
Neural Network   295
   Kirill Gadylshin, Vadim Lisitsa, Kseniia Gadylshina,
   and Dmitry Vishnevsky

Numerical Solution of Anisotropic Biot Equations in Quasi-static State   310
   Sergey Solovyev, Mikhail Novikov, and Vadim Lisitsa

Effect of the Interface Roughness on the Elastic Moduli   328
   Tatyana Khachkova, Vadim Lisitsa, and Dmitry Kolyukhin

International Workshop on Computational Science and HPC (CSHPC 2022)

Design and Implementation of an Efficient Priority Queue Data Structure . . .   343
   James Rhodes and Elise de Doncker

Acceleration of Multiple Precision Solver for Ill-Conditioned Algebraic
Equations with Lower Precision Eigensolver.   358
   Tomonori Kouya

Study of Galaxy Collisions and Thermodynamic Evolution of Gas Using
the Exact Integration Scheme   373
   Koki Otaki and Masao Mori

Regularization of Feynman 4-Loop Integrals with Numerical Integration
and Extrapolation   388
   E. de Doncker and F. Yuasa

Acceleration of Matrix Multiplication Based on Triple-Double (TD),
and Triple-Single (TS) Precision Arithmetic   406
   Taiga Utsugiri and Tomonori Kouya

International Workshop on Cities, Technologies and Planning (CTP 2022)

Fragile Territories Around Cities: Analysis on Small Municipalities Within
Functional Urban Areas   427
  Chiara Di Dato and Alessandro Marucci

Assessing Coastal Urban Sprawl and the "Linear City Model"
in the Mediterranean – The Corinthian Bay Example   439
  Apostolos Lagarias, Ioannis Zacharakis, and Anastasia Stratigea

I Wish You Were Here. Designing a Geostorytelling Ecosystem for
Enhancing the Small Heritages' Experience   457
  Letizia Bollini and Chiara Facchini

Smart City and Industry 4.0: New Opportunities for Mobility Innovation   473
  *Ginevra Balletto, Giuseppe Borruso, Mara Ladu, Alessandra Milesi,
  Davide Tagliapietra, and Luca Carboni*

Digital Ecosystem and Landscape Design. The Stadium City of Cagliari,
Sardinia (Italy)   485
  Ginevra Balletto, Giuseppe Borruso, Giulia Tanda, and Roberto Mura

Towards an Augmented Reality Application to Support Civil Defense
in Visualizing the Susceptibility of Flooding Risk in Brazilian Urban
Areas.   494
  *Gustavo Vargas de Andrade, Victor Luis Padilha, Adilson Vahldick,
  and Francisco Henrique de Oliveira*

Framework Proposal of Smart City Development in Developing Country,
A Case Study - Vietnam   507
  *Tu Anh Trinh, Thi Hanh An Le, Le Phuc Tam Do, Nguyen Hoai Pham,
  and Thi Bich Nguyet Phan*

Studying Urban Space from Textual Data: Toward a Methodological
Protocol to Extract Geographic Knowledge from Real Estate Ads   520
  Alicia Blanchi, Giovanni Fusco, Karine Emsellem, and Lucie Cadorel

International Workshop on Digital Sustainability and Circular Economy (DiSCE 2022)

Transforming DIGROW into a Multi-attribute Digital Maturity Model.
Formalization and Implementation of the Proposal   541
  *Paolino Di Felice, Gaetanino Paolone, Daniele Di Valerio,
  Francesco Pilotti, and Matteo Sciamanna*

International Workshop on Econometrics and Multidimensional Evaluation in Urban Environment (EMEUE 2022)

A Methodological Approach for the Assessment of the Non-OSH Costs   561
Maria Rosaria Guarini, Rossana Ranieri, Francesco Tajani,
Pierluigi Morano, and Francesco Sica

A Decision-Making Process for Circular Development of City-Port
Ecosystem: The East Naples Case Study   572
Sabrina Sacco and Maria Cerreta

A Cost-Benefit Analysis for the Industrial Heritage Reuse: The Case of the
Ex-Corradini Factory in Naples (Italy).   585
Marilisa Botte, Maria Cerreta, Pasquale De Toro, Eugenio Muccio,
Francesca Nocca, Giuliano Poli, and Sabrina Sacco

Unraveling the Role Played by Energy Rating Bands in Shaping Property
Prices Using a Multi-criteria Optimization Approach:
The Case Study of Padua's Housing Market.   600
Sergio Copiello and Edda Donati

Explicit and Implicit Weighting Schemes in Multi-criteria Decision Support
Systems: The Case of the National Innovative Housing Quality Program
in Italy.   615
Aurora Ballarini, Sergio Copiello, and Edda Donati

Analysis of the Difference Between Asking Price and Selling Price in the
Housing Market   629
Benedetto Manganelli, Francesco Paolo Del Giudice, and Debora Anelli

Spatial Statistical Model for the Analysis of Poverty in Italy According to
Sustainable Development Goals   641
Paola Perchinunno, Antonella Massari, Samuela L'Abbate,
and Lucia Mongelli

Real Estate Sales and "Customer Satisfaction": Assessing Transparency
of Market Advising.   655
Carmelo Maria Torre, Debora Anelli, Felicia Di Liddo,
and Marco Locurcio

Author Index   669

International Workshop
on Computational Optimization
and Applications (COA 2022)

External Climate Data Extraction Using the Forward Feature Selection Method in the Context of Occupational Safety

Felipe G. Silva$^{(\boxtimes)}$ [ID], Inês Sena [ID], Laires A. Lima [ID], Florbela P. Fernandes [ID],
Maria F. Pacheco [ID], Clara B. Vaz [ID], José Lima [ID], and Ana I. Pereira [ID]

Research Center in Digitalization and Intelligent Robotics (CeDRI), Instituto
Politécnico de Bragança, Campus de Santa Apolónia, 5300-253 Bragança, Portugal
{gimenez,ines.sena,laires.lima,fflor,pacheco,clvaz,
jllima,apereira}@ipb.pt

Abstract. Global climate changes and the increase in average temperatures are some of the major contemporary problems that have not been considered in the context of external factors to increase accident risk. Studies that include climate information as a safety parameter in machine learning models designed to predict the occurrence of accidents are not usual. This study aims to create a dataset with the most relevant climatic elements, to get better predictions. The results will be applied in future studies to correlate with the accident history in a retail sector company to understand its impact on accident risk. The information was collected from the National Oceanic and Atmospheric Administration (NOAA) climate database and computed by a wrapper method to ensure the selection of the most features. The main goal is to retain all the features in the dataset without causing significant negative impacts on the prediction score.

Keywords: Weather conditions · Data extraction · Occupational hazard prediction · Machine learning · Data mining

1 Introduction

Workplace safety is a relevant and multifaceted concern for workers, companies, and policymakers like ministers, legislators, and government officials [5] and it is becoming increasingly important around the world.

The need of ensuring minimum working conditions stems from Greek and Roman civilizations [10]. Ensuring good labour conditions has the potential to increase productivity and consequently improve the operational performance of organizations [3].

Safety is integrated empirically into work performance, as the work is dependent on the production factors. Implementing occupational safety measurements and health policies is essential to ensure a healthy and safe work environment, contributing to reduce the work accidents and occupational diseases, to

O. Gervasi et al. (Eds.): ICCSA 2022 Workshops, LNCS 13378, pp. 3–14, 2022.
https://doi.org/10.1007/978-3-031-10562-3_1

increase the competitiveness of companies, and to improve the sociability, processes, productivity, and quality of products or services, and to adopt innovative practices [3].

Occupational accidents and diseases have multiple impacts on the company operations and their costs, workers, and society in general [6]. These accidents affect significantly the quality of life of workers and the economy of the companies. The loss in production capacity is clear, in terms of days off, compensation, and pensions payable by the companies [6].

Occupational accidents have been the subject of several studies over the years, developing theories that try to explain, prevent and reduce the accidents. As such, work-related accidents have been the subject of several studies, with the development of theories that try to explain, prevent and reduce them. Initially, the actions most used by the various sectors in the reduction of work accidents were focused on the detailed investigation of the event that caused the accident and on the implementation of preventive actions, such as the reduction of excessive workload [5], changing processes and procedures in carrying out tasks, improving and evolving machines and equipment, providing workers with information and training on OSH [6]. However, several companies have started to adopt intelligent technologies that have already been implemented to improve other factors and which can be used to increase safety in the workplace, such as robotics, augmented reality, wearables, and predictive analytics [4].

An example of some areas of implementation is a predictive analysis based on Artificial Intelligence to forecast work hazards, namely, in energy, construction, infrastructures, agro-industries among others.

The studies that use predictive analysis generally apply mathematical models, statistical strategies, and computational techniques to predict the accident. Generally, the input parameters are based on the characterization of the accident, such as the day of the incident, incident event, type of injury, incident type, among others.

Some studies have identified global non-organizational variables that may influence safety and health at work, namely climatic conditions. However, there are few works using the variables and scarce strategies have been developed to explore the association between occupational accidents and climate conditions [2, 12].

Thus, this work aims to collect historical information about weather conditions and detect the most relevant elements to correlate in the future with the historical accident data of a company in the retail sector to understand its impact on the risk of accidents. This study is carried out to validate the importance of using this type of data as an input parameter in a forecasting model under development – in a parallel investigation – to minimize the risk of accidents in a Portuguese company in the retail sector.

This paper is divided into five sections. Section 1 describes the framework of the study carried out. Section 2 presents a brief literature review about the relevance of the climate data in accident prediction and the usual procedure used to extract this kind of data. The adopted procedure to collect and process data is described in Sect. 3. In Sect. 4 the results obtained are analyzed. Finally, the study is concluded and future work is presented in Sect. 5.

2 Literature Review

The weather plays a significant role in everyday life and, from the consumer's point of view, is considered an influential factor in consumer buying behavior and in many aspects of decision-making in the retail sector [19]. The literature review presented in this section encompasses studies on accident prediction considering some climatic factors, along with data extraction techniques and methods for data pre-processing, namely feature selection methods.

2.1 Accident Prediction Models

Despite the relevance of this subject, studies regarding the modeling and prediction of accidents in the workplace considering climate data is scarce. In [12], a study was performed to determine which criteria should be adopted in terms of universal or industry-specific safety climate measures. The authors performed a meta-analytic comparison of their relationships with various of safety-related outcomes, such as safety behavior, risk perceptions, accidents, and injuries, among others. Industry-specific safety climate measures were found to exhibit better predictive power in predicting safety behavior and risk perceptions than universal safety climate measures.

Aleksic et al. [1] use temperature and air pressure as possible external variables to explore their impact on work capacity. A considerable number of studies discuss the dependence of working capacities on temperature. Based on existing research, the authors assume that air pressure can have a significant impact on the occurrence of accidents [1].

The prediction of accidents is based on a good analysis of the root causes, dealing with the consequences, and making the right decisions to prevent the occurrence of the same or similar accident.

Regardless of what is the observed and researched factor for the occurrence of the accident, the objective of the previous mentioned works, including this research, is to increase the level of safety by determining and analysing the factors that tend to reduce the risk of occurrence of the accident.

2.2 Data Extraction

Web data extraction is a complex and time-consuming activity since the web is a vast source of structured or unstructured information in different formats. The tools for collecting data from the web are often based on interacting with the original (website) and extracting data through various methods, such as machine learning, logic, and natural language processing. In the case of an HTML source, the content to be extracted can range from elements on the page to structured hidden data. The extracted data can be post-processed, converted to different formats, and stored for later use [18].

The main challenges involving data capture are concerned with the complexity of data handling, being necessary to provide a high degree of automation to minimise human effort and maintain accurate performance. When extraction techniques are applied to extensive databases such as climate monitoring, capture tools need to be able to process a large volume of information in relatively short periods. On the other hand, as it involves temporal data which vary over time, changes can occur suddenly, so one must take into account the flexibility and structural changes in the data acquisition system [9]. Due to the general public's high consultation of climate data, several sources of weather data are available online and offer servers with accessible application programming interfaces (API). Despite that, studies regarding the influence of weather data on the prediction of accidents are scarce.

2.3   Data Pre-processing

Given the extensive data history and numerous real-time measurements, it is advisable to use computational techniques such as machine learning and data mining for processing high-dimensional data [22].

Feature selection, for example, has become a widely used technique to reduce the size of complex databases. It proposes to choose a subset of relevant features based on criteria of greater relevance, resulting in lower computational cost and better interpretation of the data [22]. The advantages are the improvement of machine learning performance, reduction of the database dimension by optimising the initial dataset for only relevant content, enabling the use of simpler models [20].

Feature selection algorithms can be categorized into supervised, unsupervised, and semi-supervised. Supervised selection evaluates the relevance of features based on labeled data. However, a better choice requires more labeled data and, consequently, time-consuming [7,8,11]. In contrast, unsupervised resource selection is based on a less constrained search problem but it may result in several less accurate and valid subsets [16]. For both cases, it is possible to observe that high-dimensional data with small labeled samples results in a considerable hypothesis space with few restrictions. The combination of the two methods results in semi-supervised feature selection, in which labeled and unlabeled data are combined to estimate the relevance of features [21]. Among the methods used for scanning and capturing climate data, we can highlight the supervised selection algorithms: filter models, wrapper models, embedded models, and hybrid models [7,8,11], which are summarized hereafter.

Filter methods are generally applied in data pre-processing steps. The selection depends on the general characteristics of the training data to select the features, regardless of the estimator, so the relevant variables are chosen based on statistical correlation tests (*e.g.* Pearson's correlation, ANOVA, LDA, and χ^2 test) with the response variable. It is a faster option and applicable to cases where the number of features is extensive [23].

Wrapper methods involve building subsets of features and training models. Based on subsets and inferences from previous models, new features can be added for a better approach. Despite being a good alternative as a forecasting method, it is computationally expensive. Some common examples of wrapper methods are forward selection, backward elimination, and recursive feature elimination [23].

Embedded methods combine aspects of filter and wrapper methods by classifying data and applying learning algorithms to quantify the weights of subsets. Lasso, ridge regression, regularized trees, memetic algorithm, and random multinomial logit are popular examples of embedded methods [23].

Finally, hybrid methods incorporate several types of feature selection into the same process. They are based on combining subsets of independent features, providing a better approximation while integrating external data into the selection process [23].

3 Methodology

Multiple climate factors can be related to accidents, and common examples can be identified every day such as people falling due to a wet floor, or changes in blood pressure of workers on hot days. Thus, climatic factors may be related to some occupational accidents that happen in work environments, it is important to include them in the predictive models to improve their accuracy, and consequently helping to define corrective actions to mitigate accidents.

For this study, it was necessary to get access to a climate dataset by exploring some strategies currently used to collect information and to search for available resources on the internet to produce a historical climate database. As main requirements, climate data resources should have free access and contain information from previous years, so that they can later be linked to the dataset of past accident occurrences.

The National Oceanic and Atmospheric Administration (NOAA) database [17] is used to collect the climate data since it contains historical data from different weather stations around the world. Through the API service available on the NOAA website, it was possible to search for climate information from 2017 until the present.

For greater convenience in obtaining the data, a script was developed using Python programming language to interact with the service provided by NOAA, enabling the automated collection of data recorded from January 1 of 2017 until September 1 of 2021. According to the location of the retailing stores, some specific geographical locations were selected to create the climate datasets, which are Bragança, Gaia, Felgueiras, Matosinhos, Sobreda and Guimarães, as shown in Fig. 1.

Since these climate data are derived from real records, some reading errors can occur due to failures of physical sensors installed in each specific weather station. To eliminate these errors, a pre-processing was applied to the data to prepare them for use.

Fig. 1. Selected locations for climate studies.

In order to pre-process the data, record lines with missing information were removed. This missing data occur due to the weather stations that do not emit complete information on specific days or times, resulting in the total or partial lack of data. Due to the high number of records collected, that deletion did not result in shortage of information.

The date and time data were pre-processed to obtain new features to improve the representativeness of information. According to the work shifts of the retailing stores, four periods per day: dawn, morning, afternoon and night, were used to categorize the time of each record. The day and month were used to identify the seasons of the year in which a record was made. The year, month and day were used to identify the day of the week in which a new record was made.

In learning algorithms, excess of features can add noise that decreases the prediction success rate, an effect known as the curse of dimensionality [15]. Thus, the study of strategies for dimensionality reduction was applied to detect and select the information that proved to be most relevant for the generation of a predictive model. This data reduction also brings benefits in terms of space used for data storage, processing time consumed to generate the model and the time required to perform a prediction.

To ensure the selection of the relevant features, a wrapper method was used by implementing the forward feature selection. This strategy searches the features subgroups that achieve better accuracy, using the accuracy as selection criteria after performing the prediction tests. Figure 2 represents the implementation applied in this study.

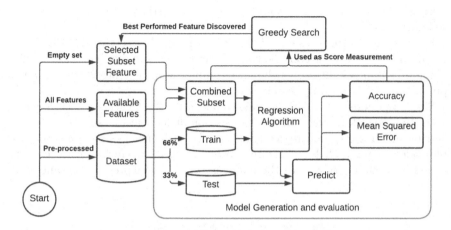

Fig. 2. Forward feature selection method.

The information about the air temperature was used as a target to perform the predictions. The air temperature was chosen because is used in other study [1] and presented a satisfactory accuracy with Linear Regression approach. To execute the tests, the linear regression algorithm provided by SKLearn library [14] was adopted. At each iteration of the algorithm, the behavior of the model was analized when adding a new feature, to detect if the model was getting better or worst.

The wrapper methods find a group of variables that achieve good performance during the prediction, being the results directly affected by the data chosen for training and testing, and also by the learning algorithm used. To reduce this effect, the method was repeated a hundred times for each database to detect the variables that perform better on average, by determining a ranking score that measures the importance of features in each database.

4   Results

Forward Feature Selection initiates with an empty subset and follows two steps: searching for a subset of features to be used; generating a score for the selected features. After these basic steps, the process is repeated until a stop criteria is satisfied [15]. To test all subsets found by the greedy algorithm, the stop criteria was ignored, so the algorithm stops when the available feature set is empty. Using the greedy algorithm as strategy for performing the search and considering the maximum number of input features, $\frac{(N+1)N}{2}$ models were evaluated, where N is the number of input features. In this study, the N is 20, evaluating 210 models during a unique test.

Each test was repeated 100 times for each location. Thus, for all the 6 geographical locations selected, 21000 models were evaluated, totalizing 126000 models, processed in 36 min and 13 s approximately. If the number of features increases, the regression algorithm can increase drastically the time to process the information.

By using 20 features, the total number of possibility sets is 2^{20} [13]. Then, the goal is to discover a good feature subset because finding the best implies testing all the possibilities, which can be impractical. The algorithm evaluates approximately 12% of all possibilities of subsets using the greedy search.

Since, the climate conditions can vary in different locations, each location was tested separately, resulting in the selection of the features subset that most contribute to achieve good regression scores for each specific location. The variables selected for each location are shown in Fig. 3 in which 6 features were defined for each location since the major groups did not present relevant accuracy improvements.

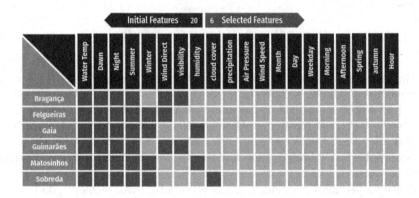

Fig. 3. Selected features for each location

After analyzing the number of features that had improved the accuracy, the variables to be considered were selected. Figure 4 shows the relationship between the number of variables in the best-discovered set and the improvement obtained in the regression accuracy. It is possible to observe that there is a maximum limit of improvement for each location since this score stabilizes from a given size of subset as it is shown in the Fig. 4.

To get the error value, was used 33% of dataset to predict the target. The errors were calculated at the better subsets of each size through the Mean Squared Error (MSE). Figure 5 shows the relationship between the size of the subset and the obtained errors.

Observing the average behavior of the variables detected for each location, of the twenty extracted initially, nine variables will be considered:

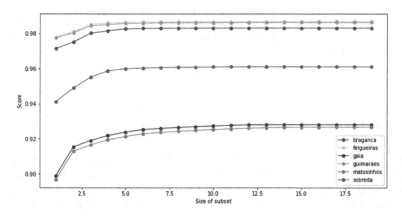

Fig. 4. Relationship between the number of characteristics and the score obtained in the regression

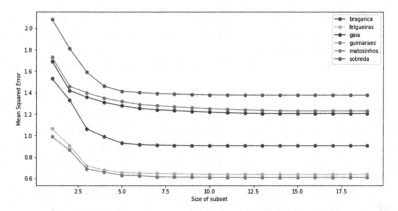

Fig. 5. Relationship between the number of features and the mean square error (MSE) obtained

1. cloud;
2. humidity;
3. visibility;
4. water temperature;
5. wind direction;
6. night;
7. dawn;
8. winter;
9. summer.

As the tests were carried out, the curse of dimensionality effect was detected, which enables to eliminate 14 features from each dataset without causing negative impact on the prediction. This approach enables to obtain an aproximation for the best group of features for learning, because greedy search algorithms can converge to local maximums. The studied locations showed a consensus on four characteristics to be maintained and eleven to be eliminated. The remaining five features were still helpful to improve the assertiveness of the model, standing out in specific places. The four commom characteristics are: water temperature, night, dawn and summer.

5   Conclusions and Future Work

In the scenario used, the Forward Feature Selection approach proved to be advantageous over the statistical methods, such as Pearson's method. When analyzing the obtained results, only by direct correlation with the target feature, it was difficult to define which thresholds were good enough to include or exclude each input feature of the model. The Forward Feature Selection strategy enables to understand what changes were caused by accepting certain features into the subset.

The method also allowed to test predictions, because multiple models are generated and tested at each iteration. In this way, optimizing the model through the selection of features became a simpler and automated task. Each test performed by the algorithm brought feedback on the learning assertiveness that was being obtained.

Despite the existing advantages of the Forward Feature Selection, the use of the algorithm still has a major weakness, which is its computational effort. Although not all possible combinations of features to generate the models are evaluated, its execution time can make it unfeasible, since it creates a considerable quantity of predictive models at each iteration. The selection of the learning algorithm and the number of features directly impact its performance.

By changing the learning algorithm, there is the possibility of increasing the score obtained in the regression and decreasing the mean square error. So far, tests with regression strategies that consume greater computational resources have not been performed, due to the time that would be spent to obtain conclusions.

In the future, we intend to use faster methods that combine filtering and wrapper strategies, such as approaches with embedded feature selection, due to the need to use more demanding learning algorithms and the use of larger datasets and diversified data. Other feature selection techniques will also be applied to analyze the relationship and impact that climate variables have on accident records. In addition to climate information, it is also intended to collect more external information to work place, such as holidays and events, to detect other possible factors that can impact on the occurrence of occupational accidents.

Acknowledgments. This work has been supported by FCT—Fundação para a Ciência e Tecnologia within the Project Scope UIDB/05757/2020 and NORTE-01-0247-FEDER-072598 iSafety: Intelligent system for occupational safety and well-being in the retail sector.

References

1. Aleksic, D., Markovic, M., Vasilijevic, M., Stojic, G., Pavlovic, N., Tanackov, I.: Analysis of impact of meteorological conditions on human factors in estimating the risk of railway accidents. Transport **33**, 57–69 (2018). https://doi.org/10.3846/16484142.2017.1332684
2. Alruqi, W.M., Hallowell, M.R., Techera, U.: Safety climate dimensions and their relationship to construction safety performance: a meta-analytic review. Saf. Sci. **109**, 165–173 (2018). https://doi.org/10.1016/j.ssci.2018.05.019
3. Antão, P., Calderón, M., Puig, M., Michail, A., Wooldridge, C., Darbra, M.R.: Identification of occupational health, safety, security (ohhs) and enviromental performance indicators in port areas. Saf. Sci. **85**, 266–275 (2016)
4. Blanchard, D.: A smarter way to safety (2019). https://www.ehstoday.com/safety-technology/article/21920103/a-smarter-way-to-safety, (last accessed 10 January 2022)
5. Cioni, M., Sabioli, M.: A survey on semi-supervised feature selection methods. Work Employ Soc. **30**, 858–875 (2016)
6. Commission, E.: Communication from the commission to the european parliament, the council, the european economic and social committee and the committee of the regions (2012). https://www.eea.europa.eu/policy-documents/communication-from-the-commission-to-1, (last accessed 20 January 2022)
7. Dy, J.G., Brodley, C.E.: Feature selection for unsupervised learning. J. Mach. Learn. Res. **5**, 845–889 (2004)
8. Ferrara, E., Meo, P.D., Fiumara, G., Baumgartner, R.: Web data extraction, applications and techniques: a survey. Knowl. Based Syst. **70**, 301–323 (2014)
9. Ferreira, A.J., Figueiredo, M.A.: Efficient feature selection filters for high-dimensional data. Pattern Recog. Lett. **33**, 1794–1804 (2012). https://doi.org/10.1016/j.patrec.2012.05.019
10. Freitas, L.: Manual de segurança e saúde do trabalho. Sílabo (2016)
11. Irmak, U., Suel, T.: Interactive wrapper generation with minimal user effort. In: Proceedings of the 15th international conference on World Wide Web - WWW 2006, p. 553. ACM Press (2006). https://doi.org/10.1145/1135777.1135859
12. Jiang, L., Lavaysse, L.M., Probst, T.M.: Safety climate and safety outcomes: a meta-analytic comparison of universal vs. industry-specific safety climate predictive validity. Work Stress **33**, 189–214 (2019)
13. Kohavi, R., John, G.H.: Wrappers for feature subset selection. Artif. Intell. **97**, 273–324 (1997). https://doi.org/10.1016/s0004-3702(97)00043-x, https://www.sciencedirect.com/science/article/pii/S000437029700043X?pes=vor
14. Scikit Learn.: scikit-learn: Machine Learning in python (2022). https://scikit-learn.org/stable/index.html. Accessed 4 Apr 2022
15. Li, J., et al.: Feature selection: a data perspective. ACM Comput. Surv. 50, 1–45 (2017). https://doi.org/10.1145/3136625, https://arxiv.org/abs/1601.07996
16. Liu, H., Motoda, H.: Computational Methods of Feature Selection. Chapman and Hall/CRC. October 2007
17. NOAA: National oceanic and atmospheric administration, National Oceanic and Atmospheric Administration
18. Omondi, A., Lukandu, I.A., Wanyembi, G.W.: A monte carlo-based search strategy for dimensionality reduction in performance tuning parameters. J. AI Data Min. **8**, 471–480 (2020)

19. Rose, N., Dolega, L.: It's the weather: quantifying the impact of weather on retail sales. Appli. Spat. Anal. Policy **15**, 189–214 (2022). https://doi.org/10.1007/s12061-021-09397-0
20. Sheikhpour, R., Sarram, M.A., Gharaghani, S., Chahooki, M.A.Z.: A survey on semi-supervised feature selection methods. Pattern Recogn. **64**, 141–158 (2017)
21. Song, L., Smola, A., Gretton, A., Borgwardt, K.M., Bedo, J.: Supervised feature selection via dependence estimation. In: Proceedings of the 24th International Conference On Machine Learning - ICML 2007, pp. 823–830. ACM Press (2007)
22. Sánchez-Maroño, N., Alonso-Betanzos, A., Tombilla-Sanromán, M.: Filter methods for feature selection - a comparative study. Intell. Data Eng. Auto. Learn. IDEAL **2007**, 178–187 (2007)
23. Weston, J., Elisseff, A., Schoelkopf, B., Tipping, M.: Use of the zero-norm with linear models and kernel methods andré elisseeff bernhard schölkopf mike tipping. J. Mach. Learn. Res. **3**, 1439–1461 (2003)

Dynamic Analysis of the Sustainable Performance of Electric Mobility in European Countries

Clara B. Vaz[1,2][✉] and Ângela P. Ferreira[1,3]

[1] Research Centre in Digitalization and Intelligent Robotics (CeDRI),
Instituto Politécnico de Bragança (IPB), Campus Santa Apolónia,
5300-253 Bragança, Portugal
clvaz@ipb.pt

[2] Centre for Management and Industrial Engineering
(CEGI/INESC TEC), Porto, Portugal

[3] Electromechatronic Systems Research Centre (CISE),
University of Beira Interior, Covilhã, Portugal
apf@ipb.pt

Abstract. As part of the ongoing climate and energy framework, the European Commission raised recently the 2030 greenhouse gas emission reduction target, moving towards a climate-neutral economy. Transportation represents almost a quarter of Europe's greenhouse gas emissions, and it is the remaining sector with increasing emissions, above 1990 levels. Considering also the evolving necessity for the reduction of fossil fuels dependency, Europe's strategy has been designed to support an irreversible shift toward low-emission electric mobility. In this context, the present work assesses the performance of electric mobility in European countries, by using a dynamic analysis in the period 2015–2019, framed in four sustainable dimensions, economy, technology, environment and society. The methodology aggregates several sub-indicators in a composite indicator by using the Data Envelopment Analysis, and evaluates the dynamic change in the sustainable performance through the biennial Malmquist index. Main results indicate that the total productivity change has been improved mainly due to the progression of the frontier that has been observed for all countries from 2018. However, an increasing number of countries have had more difficulties to adopt the best sustainable electric mobility practices, being necessary to design strategies to promote them, mainly in underperforming countries.

Keywords: Electric mobility · Composite indicator · Malmquist index

1 Introduction

Under the European strategy for low-emission mobility framework, Member States have been under guiding principles to increase the efficiency and the sustainability of the transport system, through the deployment of low emission alternative energies and by accelerating the transition towards low and zero emission vehicles [9].

This strategy grips the targeted greenhouse gas (GHG) emissions reduction on the 2030 climate and energy policy framework, as stated in the Regulation (EU) 2018/842 [11]. The transport sector represents almost a quarter of the Union's GHG emissions and it remains as the only sector that has increased steadily the emissions until 2019, when compared with 1990 levels, diverging from the other sectors [15]. Estimates for 2020 indicate a substantial reduction due to decreased activity during COVID-19 pandemic, but European Environmental Agency (EEA) anticipates that transport emissions will deteriorate with the economic recovery and it is projected that domestic transport emissions will stay above 1990 levels till 2029 [10].

Besides, the need for dependency reduction on imported fossil fuels has become crucial due to instability in many fossil fuel-producing countries, which increases the price of energy and reinforces the need to find alternatives.

Under this scenario, the sustainable performance of the electric mobility in European Union is able to drive innovation while enhancing risk management and cost reduction and provides engagement at all levels. By monitoring and analysing the trends towards low-emission mobility, it is possible to exploit deviations and set up guidelines, to provide policy makers with decision tools to design their strategies.

The sustainable model for a given programme, agenda, development or sector, usually requires its evaluation by several perspectives, termed pillars. Early sustainable concepts were evaluated through the Economy growth, Social inclusion and balance of the Environment [20]. Several authors suggested the use of a fourth pillar, the cultural sustainability, which, in a broader scope, can be incorporated in the Social sustainability, focusing on human issues [16]. The economic sustainability is typically described at a country or region level and aims at translating the ability to support a specified level of economic production. With regard to the environmental pillar, it aims to enhance the natural capital and/or the welfare of the population [16].

In the particular sector of the electric mobility, the sustainable framework should also include the dimension of the technological development and innovation. For instance, long-term energy storage is crucial in the penetration of battery electric vehicles (BEV).

This paper aims to evaluate the sustainable performance of the electric mobility in European countries in a dynamic timeline framework, from 2015 until 2019. As stated before, due to reduced activity during COVID-19 pandemic, the time series does not include data from 2020. The methodological approach aggregates sub-indicators in the four sustainable dimensions introduced above: economic, social, environmental and technological into a composite indicator (CI) by using the Data Envelopment Analysis (DEA) technique. Thus, the sustainable performance, CI, is determined through the Benefit of Doubt (BoD) model [3] and to track the change of the sustainable performance of electric mobility, it is used the biennial Malmquist index [18], considering it was originally developed to assess the Malmquist productivity change index [1]. In this context, the proposed approach is innovative because the biennial Malmquist index allows the dynamic analysis of the sustainable performance of electric mobility in European Countries as it can be decomposed into the technical change (frontier shift effect) and efficiency change (catching-up effect) [18]. The frontier shift effect allows to identify the deterioration or progression on the European best practices of sustainable electric mobility while catching-up effect gathers the evolution of each country against the best practices observed in each period.

A previous review performed by the authors [15] identified the main relevant indicators and analysed the methodologies most employed in literature for assessing the deployment of electric mobility. This survey establishes the basis for the selection of the indicators, aiming to convey the affordability, the infrastructure availability, GHG reduction levels and educational level, the latter as a predictor of environmental protection support willingness.

With regard to the electric mobility, in opposition to conventional vehicles using exclusively fossil fuels (diesel and petrol) and/or natural gas (compressed or liquefied) to power an internal combustion engine (ICE), the alternatives can be categorised as electric vehicles (EV) or hybrid electric vehicles (HEV), with different requirements in terms of the support infrastructure.

Electric vehicles can be further subdivided in plug-in electric vehicle (PEV), *i.e.*, electrically-chargeable vehicles, and fuel cell electric vehicles (FCEV).

PEV include full battery electric vehicles (BEV) and plug-in hybrids (PHEV), both requiring a recharging infrastructure to connect them to the electricity grid. BEV are fully powered by one or more electric motors, using electricity stored in an on-board battery and the PHEV have an ICE and a battery-powered electric motor. The battery is recharged by connecting to the grid as well as by the on-board engine. The traction is provided by the electric motor and/or by the ICE, depending on the battery level.

Fuel cell electric vehicles (FCEV) also run on electric motors, but the electricity is generated within the vehicle by a fuel cell, using compressed hydrogen, from dedicated filling stations, and oxygen from the air. The state of art of the production, transport and distribution of hydrogen and the lack of filling stations does explain the negligible share of FCEV in the European passenger car fleet.

Finally, hybrid electric vehicles (HEV) have an internal combustion engine and a battery-powered electric motor. Electricity stored in the on-board battery is generated internally from regenerative braking, cruising and the combustion engine, so they do not need recharging infrastructure. The hybridisation level can range from mild to full, depending on the type of propulsion combination (powered by the combustion engine with the electric motor supporting it, or propulsion shared by both electric motor and combustion engine, respectively).

Cumulative to the different requirements in terms of the infrastructure mentioned above, these technologies have also different impacts on the GHG emissions reduction levels. BEV and FCEV have a tail-pipe CO_2 emissions reduction of 100%, while HEV have a reduction potential ranging from 10% to 40%, depending on the hybridisation level. Finally, PHEV have, on average, 50% to 75% potential for emissions reduction [7]. Regarding the later, it should be mentioned that the emissions reduction potential strongly depends on usage and charging behaviour. Recent studies from the International Council on Clean Transportation indicate that actual PHEV fuel consumption and emissions are higher than the levels in which they are approved, biasing the governmental support at the vehicle purchase and the accounting in the GHG emissions targets [19]. From this point of view, existing EU policies may shift in a near future for the PHEV market. Considering the time span of the data used in this study, the sustainability analysis of the passenger electric mobility is performed considering the electrically-chargeable cars, *i.e.*, BEV and PHEV.

This paper unfolds as follows: the methodology regarding the BoD model, the biennial Malmquist index, and the data are introduced in Sect. 2. Section 3 presents the results and discussion, including the performance assessment and the dynamic analysis, and Sect. 4 concludes this paper.

2   Methodology

DEA is a linear and non-parametric technique to evaluate the relative efficiency of an homogeneous set of Decision Making Units (DMUs) in using multiple inputs to achieve multiple outputs, introduced by Charnes et al. [2]. This enables to identify the "best practices DMUs" in which their linear combination defines the frontier technology. A single summary measure of efficiency is calculated for each DMU by using the frontier technology as a reference. The DMUs which belong to the frontier have an efficiency score equals to 1 while the remaining ones have an efficiency score lower than 1.

2.1   BoD Model

In the case of pure models, in which only use outputs, denominated by sub-indicators, Cherchye et al. [3] introduced the BoD model to compute the composite indicator for each DMU. The BoD model is equivalent to the DEA input oriented model [2], as all sub-indicators are considered as outputs and a single dummy input equal to one is used for all units. The BoD model enables to aggregate several sub-indicators to derive the composite indicator for each DMU, by determining endogenously the weights to aggregate them. Since BoD model does not refer the inputs, it enables to assess its performance rather than its efficiency [4].

Considering a cross-section of m sub-indicators i for each country j ($j = 1, \ldots, s$), being y_{ij}^t the score of that sub-indicator observed for each country j on the period t, and w_i the weight assigned to it. The BoD model (1) assesses the performance for each country under assessment, j_0, to determine its composite indicator, $CI_{j_o}^t$ through the weighted average of the m sub-indicators.

$$CI_{j_0}^t = max \sum_{i=1}^{m} w_i y_{ij_0}^t$$

$$s.t. \quad \sum_{i=1}^{m} w_i y_{ij}^t \leq 1 \qquad \forall j = 1, ..., s \qquad (1)$$

$$w_i \geq 0 \qquad \forall i = 1, ..., m$$

For each country under evaluation j_0 in the period t, the model (1) determines endogenously the optimum weight w_i for each sub-indicator $y_{ij_0}^t$ to maximize its $CI_{j_0}^t$ by comparison with the frontier technology. Thus, the optimum weighting scheme varies with the country under evaluation to maximize its performance.

The model (1) does not avoid to assign zero weights to some sub-indicators, if the unit has a relative poor performance, implying no influence on its evaluation. In the opposite case, if a country has an high relative performance in a given sub-indicator, the model can assign a high weight to this dimension. To avoid these situations, proportional virtual weight restrictions (2) are imposed to the unit under evaluation [22]. Thus, each sub-indicator is required to have a minimum and maximum percentages of contribution on the calculation of the $CI_{j_0}^t$ for the country under evaluation. Since no expert information is available, and setting M the number of sub-indicators, constraint (2) should guarantee that the proportional virtual weight for each sub-indicator should vary between $\frac{1}{M}(1 - k)$ and $\frac{1}{M}(1 + k)$, with $k \in]0, 1[$ for the unit under evaluation. Decision maker should select the k by balancing some flexibility, in which unrestricted model is the most flexible alternative, and consistency which allows that all dimensions are taking into account on the unit under evaluation.

$$\frac{1}{M}(1 - k) \leq \frac{w_i y_{ij_0}^t}{\sum_{i=1}^{M} w_i y_{ij_0}^t} \leq \frac{1}{M}(1 + k) \qquad \forall i = 1, ...M, k \in]0, 1[\qquad (2)$$

2.2 Biennal Malmquist Index

Malmquist productivity change indexes have been the most commonly used approach to estimate total factor productivity change between two data points, corresponding to a DMU in two time periods [18]. The Malmquist productivity index was introduced by Caves et al. [1] and developed further in the context of performance assessments by Färe et al. [13] to accomplish performance comparisons of DMUs over time, by taking the geometric mean of the Malmquist productivity index achieved in the periods t and $t + 1$.

The biennal Malmquist index is introduced by Pastor and Lovell [18] by considering a frontier technology B defined by the convex hull of the frontier technologies concerning the periods t and $t + 1$, given by $B = t, t + 1$.

The Malmquist productivity index is based on radial measures which are defined by Shepard distance functions [21]. The score of $CI_{j_0}^t$ in the model (1) is equal to the DEA radial efficiency score given by the Shephard output distance function [14], being equal to $D^t(1, y_o^t)$. Set y_o^t the sub-indicators vector observed for the DMU$_o$ in period t, the distance function to the period t frontier is given by $D^t(1, y_o^t)$.

Thus, the superscript inside brackets represents the period in each DMU$_o$ is assessed while the superscript outside of brackets denotes the frontier technology used as reference.

The biennial Malmquist index [18] derived for each DMU$_o$ is given by (3).

$$M_o^B = M_o^B(1, y_o^t, y_o^{t+1}) = \frac{D^B(1, y_o^{t+1})}{D^B(1, y_o^t)} = EC_o.TC_o \tag{3}$$

In terms of interpretation, a score of $M_o^B(1, y_o^t, y_o^{t+1}) > 1$ indicates better productivity (global performance) in period $t+1$ than in period t for the DMU$_o$.

The biennial Malmquist index M_o^B enables to measure the total productivity change over time, and can be decomposed into the efficiency change index (EC_o) and the technical change (TC_o) index [18]. The catching-up effect is given by the efficiency change index for each DMU$_o$ according to (4).

$$EC_o = \frac{D^{t+1}(1, y_o^{t+1})}{D^t(1, y_o^t)} \tag{4}$$

The score EC_o compares the efficiency spread between the periods observed for each DMU$_o$. A value of $EC_o > 1$ $(EC_o < 1)$ means that the efficiency spread is smaller in DMU$_o$ observed in period $t + 1$ than the one observed in period t, measuring how much the DMU$_o$ is getting closer from the frontier, *i.e.* catching up effect (farther from the frontier).

The frontier shift is captured by the technical change index for each DMU$_o$, calculated by (5).

$$TC_o = \frac{D^B(1, y_o^{t+1})/D^{t+1}(1, y_o^{t+1})}{D^B(1, y_o^t)/D^t(1, y_o^t)} = \frac{BPG^{B,t+1}(1, y_o^{t+1})}{BPG^{B,t}(1, y_o^t)} \tag{5}$$

The ratio TC_o compares the best practice gap $(BPG^{B,t+1}(1, y_o^{t+1}))$ between the biennial frontier B and the $t + 1$ frontier along the ray defined by $(1, y_o^{t+1})$ with the best practice gap $(BPG^{B,t}(1, y_o^t))$ between the biennial frontier B and the t frontier along the ray defined by $(1, y_o^t)$.

When TC_o is higher (lower) than 1, it indicates that the $t + 1$ frontier is closer to (farther away from) the biennial frontier B along the ray defined by $(1, y_o^{t+1})$ than is the t frontier along the ray defined by $(1, y_o^t)$, *i.e.*, it measures the technological progress (regress) of the frontier, revelling better (worse) productivity.

Globally, the decomposition of M_o^B implies that the sources of better performance can be associated with two factors: less dispersion in the efficiency score of DMU$_o$ between the two periods and/or better productivity associated to the period frontier. The calculation of the biennial Malmquist index and its components is explored in the next section by using a small example.

2.3   Illustrative Example

Suppose that two sub-indicators y_1 and y_2 are achieved for two periods t and $t+1$ concerning four DMUs (A, E, C and D) as shown in the Fig. 1. This example is used to illustrate the calculation of M_o^B, its components EC_o and TC_o, given in Table 1, and their interpretation. The frontier in the period t $(t+1)$ is defined by the DMUs A and C observed in the period t $(t+1)$ while the biennial frontier B is defined by the DMUs A^{t+1} and C^t. Between the two periods, the frontier only has a progression for the DMU A (A^{t+1}), and a regression on the frontier is observed on the other remaining DMUs, which is captured by the technical change index (TC_o).

As shown in the Fig. 1, DMUs A and C are efficient in both periods, so the $EC_o = 1$. The DMU E improves its efficiency between the two periods, as it is getting closer (*i.e.* catching up) from the frontier since the $EC_E > 1$. The opposite occurs for the DMU D as $EC_D < 1$, as it is getting farther from the frontier. For the DMU A in t period, the frontier t has a lower productivity than the frontier B, implying $BPG^{B,t} < 1$. For the DMU A in $t+1$ period, the frontier $t + 1$ has the same productivity than the frontier B, implying $BPG^{B,t+1} = 1$. Thus, $TC_A > 1$ implies a progression in the frontier for the DMU A. The DMU A improves the total productivity $(M_A^B > 1)$ due to the increase of the technical change, keeping the efficiency status between the two periods.

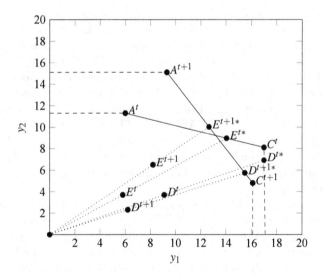

Fig. 1. Illustrative example.

For the DMU E in t period, the frontier t has a lower productivity than the frontier B, implying $BPG^{B,t} < 1$. For the DMU E in $t + 1$ period, the frontier $t + 1$ has a lower productivity than the frontier B, implying $BPG^{B,t+1} < 1$. Thus, $TC_E < 1$ implies a deterioration on the productivity of frontier for the

Table 1. Results of M_o^B, EC_o and TC_o for DMUs A, E, C and D.

DMU	$D^t(1, y_o^t)$	$D^{t+1}(1, y_o^{t+1})$	$D^B(1, y_o^t)$	$D^B(1, y_o^{t+1})$	EC_o	$BPG_o^{B,t}$	$BPG_o^{B,t+1}$	TC_o	M_o^B
A	1.000	1.000	0.748	1.000	1.000	0.748	1.000	1.336	1.336
E	0.413	0.648	0.381	0.592	1.570	0.922	0.914	0.991	1.555
C	1.000	1.000	1.000	0.947	1.000	1.000	0.947	0.947	0.947
D	0.535	0.401	0.535	0.365	0.748	1.000	0.910	0.910	0.681

DMU E. The DMU E improves the total productivity ($M_E^B > 1$) due to the increase of efficiency change.

For the DMUs C and D in t period, the frontier t coincides with the frontier B, implying $BPG^{B,t} = 1$. For these DMUs in $t + 1$ period, the frontier $t + 1$ has a lower productivity than the frontier B, implying $BPG^{B,t+1} < 1$. Thus, $TC_o < 1$ implies a deterioration on the productivity of frontier for the DMU C and D. The DMU C decreases the total productivity ($M_C^B < 1$) due to the decrease of technical change, keeping the efficiency status in both periods. The DMU D decreases the total productivity ($M_D^B < 1$) due to the decrease of the technical change and efficiency change.

2.4   Data

To perform the sustainable performance analysis of the electric mobility in EU-28, several indicators were selected and aggregated in four sustainable dimensions, in accordance with a previous literature survey [15]. Data was collected from literature and online databases, and processed on a yearly basis, for the time period 2015 till 2019. Table 2 presents an overview of the selected indicators organized by the dimensions to sustainable development, the units, and identifies the data sources (Eurostat [12], European Commission (EC) [8] and European Alternative Fuels Observatory (EAFO) [6]). The contextualization and a brief description of the indicators are presented below.

The deployment of the electric mobility is largely correlated with the affordability and the infrastructure availability, under the economic and technological pillars, respectively.

The economic sustainability is supported by two indicators: the energy intensity of the Gross Domestic Product (GDP), given by the ratio between the gross inland energy consumption and the GDP, and the annual average fuel price for petroleum products (euro-super 95 and automotive gas oil). The first one is an anti-isotonic indicator [5], because higher values reflect inefficiencies in the economy and the second one, with regard to the sustainability of the electric mobility, push consumers to invest in solutions that are more economical.

Technological sustainability relates with cheap energy storage systems and long-lasting batteries, able to increase the driving range of BEV, which cannot be translated in a simple indicator. Measurable technological sustainability is given by the number of recharging points normalized by the number of inhabitants and also the market share of electrically-chargeable vehicles, evaluated by the ratio between the newly registered BEV and PHEV and the total newly registered passenger vehicles, per year.

To capture the environmental sustainability, the analysis includes the contribution of GHG emissions from fuel combustion in road transport to the total GHG emissions inventory, consisting in an anti-isotonic attribute. To prevent penalizations of countries with high industrial activity and, consequently, higher GHG emissions, this pillar also includes the industrial production index (IPI), measuring the industry productivity in a given year relative to the base year of 2015. Also, the share of renewable energy in transport with regard to gross final energy consumption is included as renewables impact largely in the environmental sustainability of the sector.

Finally, under the social sustainability dimension, the tertiary education attainment is inserted in the analysis, as a predictor of the willingness of the population to support environmental protection measures. This assumption takes into consideration that people are prone to reduce emissions if they understand the risks from climate change [17].

Table 2. Overview of the selected indicators.

Variable	Description (source)	Source	Units
Economy			
Energy intensity	Energy intensity of GDP in chain linked values (2010)	Eurostat	$(koe/10^3 \; \text{€})$
Fuel price	Fuel price for petroleum products with taxes	EC	$(\text{€}/L)$
Technology			
Recharging points	Number of recharging points per 100 hundred inhabitants	EAFO/Eurostat	n.a.
Market share	Market share of electrically-chargeable vehicles (BEV & PHEV)	EAFO	%
Environment			
GHG emissions	GHG emissions from fuel combustion in road transport	Eurostat	$\frac{\text{tonnes of } CO_2-\text{eq}}{\text{inhabitant}}$
IPI	Industrial production index (base year 2015)	Eurostat	%
Renewable	Share of renewable energy in transport	Eurostat	%
Society			
Education	Tertiary education attainment in the 15-64 age group	Eurostat	%

The descriptive statistics of the data on a yearly basis is presented in Table 3. The overall mean variation between 2015 and 2019 is also included.

Table 3. Descriptive statistics of the variables on a yearly basis.

Sub-indicator	2015		2016		2017		2018		2019		Variation
	Mean	SD	Mean	SD	Mean	SD	Mean	SD	Mean	SD	
Economy											
Energy intensity	173.611	83.817	172.695	82.492	172.025	82.353	167.414	81.090	159.350	73.459	-8%
Fuel price	1.252	0.124	1.154	0.121	1.234	0.127	1.332	0.125	1.325	0.130	6%
Technology											
Recharging points	12.853	20.677	17.455	28.538	23.271	37.080	28.807	45.175	36.601	58.235	185%
Market share	0.009	0.019	0.009	0.013	0.011	0.012	0.018	0.019	0.029	0.034	224%
Environment											
GHG emissions	1.946	1.692	1.975	1.596	1.999	1.581	2.026	1.658	2.042	1.653	5%
IPI	1.000	0.003	1.026	0.028	1.072	0.042	1.096	0.066	1.105	0.079	11%
Renewable	0.065	0.053	0.062	0.047	0.069	0.051	0.078	0.052	0.088	0.053	35%
Society											
Education	0.275	0.069	0.282	0.071	0.289	0.071	0.298	0.072	0.306	0.074	11%

The overall behaviour of the sub-indicators favours the sustainable performance of electric mobility in the EU, except for GHG emissions, still increasing beyond 1990 levels. A substantial increase in the technological sub-indicators, market share of electrically-chargeable vehicles and number of recharging points, represent a solid shift towards low-emission mobility. It is also of relevance the increase of renewable energy in the transport sector, favouring the sustainability of the electric mobility. The decrease in the energy intensity of GDP acts also as a positive trend in pursuing the EU energy targets, through the efficiency increase of the processes.

3   Results and Discussion

This section analyses the sustainable performance of the electric mobility in European countries observed in each year and explores its change over time.

Thus, the sustainable performance of the electric mobility is calculated by the CI^t for each year t abbreviated by $t = 15, 16, 17, 18$ and 19. Further, its dynamic analysis is investigated, by calculating the biennial Malmquist indexes and their components, as described in the methodology.

The CI^t for each year t is calculated from the $M = 8$ sub-indicators listed in Table 2 by using the BoD model (1) and including the constraints given by (2), in which the $k = 0.9$ is adopted to assure the suitable flexibility and consistency. This k score assures that each sub-indicator has a minimum contribution of $\frac{1}{8}(1 - 0.9) = 1.25\%$ and a maximum contribution of $\frac{1}{8}(1 + 0.9) = 23.75\%$ to the CI^t score calculated for each country under evaluation in each period t.

Additionally, all selected sub-indicators must be isotonic, satisfying the assumption of the BoD model that higher sub-indicator scores correspond to better performance. This requires the transformation of GHG emissions and Energy intensity sub-indicators by subtracting the original values from a larger constant number [5] (*i.e.*, a constant equal to 10% higher than the maximum value observed in the sample for all countries and years).

3.1   Performance Analysis

The results concerning the CI^t for each year t, $t = 15, 16, 17, 18$ and 19 and country[1] are presented in Table 4. The best practices frontier in the year t is defined by the best practices countries observed on that year.

It is observed that there are only two benchmarks on sustainable electric mobility which are NET and SWE which keep the efficiency status in all years. Note, that there is no optimum solution for CYP in 2015 with the weighting scheme in use, denoting that it has some low levels of sub-indicators compared to the benchmarks. The direct comparison of CI^t scores for each country enables to understand if the country has came closer (*i.e.* catching up) or further to the frontier. This effect is captured by the efficiency change index given by EC_o that is calculated for each successive years according to (4), which scores are presented in Table 4. We can conclude that, on average, the European countries are getting closer to the frontier of the best practices on sustainable electric mobility until 2018, although they are getting farther from the frontier in 2019. These results do not indicate if the frontier has moved upward or downward which is explored in the dynamic analysis.

3.2   Dynamic Analysis

The biennial Malmquist M_o^B index and its components, the efficiency change, EC_o, and the technical change, TC_o, are determined by using the panel data concerning the selected sub-indicators of 28 European Countries in each pair of successive years (2015/16, 2016/17, 2017/18, 2018/19), as presented in Table 4. Thus, the biennial frontier is defined by the best practices countries in terms of sustainable electric mobility observed on each pair of successive years.

In 2015/16, there are 21 countries that improve the total productivity ($M_o^B > 1$) mainly due to increase of efficiency change ($EC_o > 1$ is observed in 16 countries), and the progression of the frontier ($TC_o > 1$) that is observed in only 11 countries.

In 2016/17, all countries improve their total productivity ($M_o^B > 1$) except CRO. This is mainly due to increase of efficiency change ($EC_o > 1$ is observed on 21 countries), and the progression of the frontier that is observed for 25 countries ($TC_o > 1$). It is observed that the frontier deteriorates ($TC_o < 1$) for CZE, CRO and ITA. Additionaly, DEN, EST, GRE, CRO and UK increase their efficiency spread ($EC_o < 1$).

In 2017/18, all countries improve their total productivity ($M_o^B > 1$) mainly due to increase of technical change ($TC_o > 1$) since the progression of the frontier is observed in all countries. Although, 13 countries decrease their efficiency spread ($EC_o > 1$).

---

[1] The country notation of Eurostat [12] is adopted.

Table 4. Results of CI^t, EC_o, TC_o and M_o^B

Country	CI^t					2015/16			2016/17			2017/18			2018/19		
	CI^{15}	CI^{16}	CI^{17}	CI^{18}	CI^{19}	EC_o	TC_o	M_o^B	EC_o	TC_o	M_o^B	EC_o	TC_o	M_o^B	EC_o	TC_o	M_o^B
BEL	0.80	0.86	0.88	0.89	0.90	1.08	1.00	1.08	1.02	1.04	1.06	1.01	1.03	1.04	1.01	1.03	1.04
BUL	0.13	0.13	0.32	0.31	0.28	0.97	1.22	1.18	2.54	1.14	2.90	0.98	1.15	1.12	0.91	1.26	1.14
CZE	0.46	0.51	0.58	0.58	0.51	1.13	0.95	1.07	1.14	0.95	1.08	1.00	1.24	1.23	0.87	1.24	1.08
DEN	0.95	0.91	0.89	0.94	0.94	0.95	0.95	0.90	0.98	1.04	1.02	1.05	1.10	1.16	1.00	1.03	1.03
GER	0.74	0.84	0.90	0.89	0.86	1.13	0.97	1.09	1.07	1.05	1.12	0.99	1.05	1.04	0.97	1.04	1.00
EST	0.55	0.50	0.46	0.76	0.63	0.91	1.10	1.00	0.92	1.10	1.01	1.66	1.00	1.67	0.83	1.13	0.94
IRE	0.82	0.87	0.91	0.90	0.87	1.06	0.94	1.00	1.04	1.04	1.08	1.00	1.10	1.09	0.96	1.09	1.05
GRE	0.15	0.13	0.12	0.14	0.12	0.84	1.19	1.01	0.98	1.21	1.19	1.10	1.10	1.21	0.90	1.34	1.21
ESP	0.49	0.69	0.81	0.77	0.71	1.40	0.94	1.32	1.17	1.03	1.21	0.95	1.13	1.07	0.92	1.15	1.05
FRA	0.87	0.90	0.91	0.91	0.88	1.03	1.00	1.04	1.01	1.02	1.03	1.00	1.05	1.05	0.97	1.06	1.02
CRO	0.43	0.55	0.49	0.57	0.62	1.28	0.98	1.25	0.88	0.91	0.81	1.18	1.19	1.41	1.08	1.24	1.35
ITA	0.50	0.55	0.62	0.64	0.71	1.10	0.95	1.05	1.13	0.93	1.05	1.04	1.32	1.36	1.10	1.08	1.18
CYP	-	0.63	0.63	0.59	0.51	-	-	-	1.01	1.06	1.07	0.92	1.11	1.02	0.88	1.20	1.05
LAT	0.48	0.55	0.57	0.75	0.62	1.14	0.99	1.13	1.04	1.07	1.11	1.31	1.03	1.35	0.83	1.17	0.97
LIT	0.26	0.30	0.51	0.57	0.49	1.15	1.16	1.33	1.70	1.04	1.76	1.11	1.12	1.24	0.87	1.22	1.06
LUX	0.72	0.82	0.85	0.86	0.84	1.14	0.96	1.08	1.04	1.09	1.13	1.02	1.07	1.09	0.97	1.07	1.04
HUN	0.47	0.47	0.52	0.63	0.59	1.00	1.05	1.05	1.11	1.12	1.24	1.21	1.05	1.27	0.94	1.13	1.06
MAL	0.61	0.49	0.71	0.78	0.75	0.80	0.85	0.68	1.47	1.16	1.71	1.08	1.11	1.20	0.97	1.08	1.05
NET	1.00	1.00	1.00	1.00	1.00	1.00	1.00	1.00	1.00	1.07	1.07	1.00	1.09	1.09	1.00	1.26	1.26
AUS	0.86	0.88	0.90	0.90	0.88	1.03	1.02	1.05	1.02	1.02	1.04	1.00	1.04	1.04	0.98	1.03	1.01
POL	0.25	0.25	0.33	0.36	0.32	0.98	1.04	1.02	1.35	1.08	1.46	1.08	1.14	1.24	0.89	1.31	1.17
POR	0.80	0.83	0.87	0.84	0.80	1.04	0.99	1.03	1.04	1.05	1.09	0.98	1.07	1.04	0.95	1.07	1.01
ROM	0.15	0.14	0.18	0.18	0.29	0.93	1.19	1.11	1.34	1.20	1.60	0.98	1.13	1.11	1.59	1.22	1.94
SLN	0.64	0.71	0.77	0.80	0.73	1.11	0.97	1.08	1.08	1.07	1.16	1.04	1.02	1.06	0.91	1.13	1.02
SLK	0.48	0.49	0.70	0.67	0.54	1.03	0.88	0.90	1.41	1.15	1.63	0.96	1.15	1.11	0.80	1.21	0.97
FIN	0.95	0.89	0.94	0.92	0.96	0.94	1.01	0.96	1.05	1.06	1.11	0.98	1.05	1.03	1.04	1.07	1.11
SWE	1.00	1.00	1.00	1.00	1.00	1.00	1.02	1.02	1.00	1.09	1.09	1.00	1.13	1.13	1.00	1.08	1.08
UK	0.95	0.98	0.98	0.97	0.97	1.03	1.00	1.03	1.00	1.04	1.04	1.00	1.03	1.02	0.99	1.03	1.03
Average	0.61	0.64	0.69	0.72	0.69	1.04	1.01	1.05	1.16	1.07	1.25	1.06	1.10	1.16	0.97	1.14	1.10
$No. > 1$	-	-	-	-	-	16	11	21	21	25	27	13	28	28	6	28	25
$No. < 1$	-	-	-	-	-	9	15	5	5	3	1	13	0	0	20	0	3
$No. = 1$	2	2	2	2	2	2	1	1	2	0	0	2	0	0	2	0	0

In 2018/19, almost all countries improve their total productivity ($M_o^B > 1$) except EST, LAT and SLK. This is mainly due to increase of technical change ($TC_o > 1$) since the progression of the frontier is observed in all countries. Nevertheless, only 6 countries decrease their efficiency spread ($EC_o > 1$) which are BEL, DEN, CRO, ITA, ROM and FIN.

4   Conclusions

This paper presents an innovative approach to assess the dynamics of the sustainable performance of electric mobility in European Countries, from 2015 until 2019.

The sustainable performance is determined through the Benefit of Doubt (BoD) model [3] and it is used the biennial Malmquist index [18] to track the change of the total sustainable performance of electric mobility which can be decomposed into the efficiency change (catching-up effect) and technical change (frontier shift effect).

The catching-up effect gathers the evolution of each country against the best practices observed in each period by comparing the efficiency spread between the periods observed for each country. The frontier shift effect allows to identify the deterioration or progression on the European best practices of sustainable electric mobility by computing the technical change. Thus, it compares the best practice gap between the biennial frontier B and the $t + 1$ frontier along the ray defined by the observed country in $t + 1$ with the best practice gap between the biennial frontier B and the t frontier along the ray defined by the observed country in t. The biennial frontier is defined by the best practices countries in terms of sustainable electric mobility observed on each pair of successive years.

Effectively, the total productivity change has been improved through time mainly due to the progression of the frontier since the best practices of sustainable electric mobility have been improved over time, as progression of the frontier was observed for all countries from 2018. Although the majory of the countries were getting closer to the frontier during 2016 and 2017, the opposite effect occurred in 2018 and 2019. From these results, it is observed that despite an upward of the frontier in the last two years of the analysis, the countries are getting farther from the frontier. This implies that an increasing number of countries have had more difficulties to adopt the best practices of the benchmarks. Our study indicates that is necessary to reinforce the actual policies to promote the electric mobility penetration, mainly in underperforming countries.

Acknowledgment. This work has been supported by FCT—Fundação para a Ciência e Tecnologia within the Project Scope: UIDB/05757/2020.

References

1. Caves, D.W., Christensen, L.R., Diewert, W.E.: The economic theory of index numbers and the measurement of input, output and productivity. Econometrica **50**, 1393–1414 (1982)
2. Charnes, A., Cooper, W.W., Rhodes, E.: Measuring the efficiency of decision making units. Eur. J. Oper. Res. **2**(6), 429–444 (1978)
3. Cherchye, L., Moesen, W., Rogge, N., Van Puyenbroeck, T.: An introduction to 'benefit of the doubt' composite indicators. Soc. Indic. Res. **82**(1), 111–145 (2007)
4. Chung, W.: Using dea model without input and with negative input to develop composite indicators. In: 2017 IEEE International Conference on Industrial Engineering and Engineering Management (IEEM), pp. 2010–2013. IEEE (2017)
5. Dyson, R.G., Allen, R., Camanho, A.S., Podinovski, V.V., Sarrico, C.S., Shale, E.A.: Pitfalls and protocols in DEA. Eur. J. Oper. Res. **132**, 245–259 (2001)
6. EAFO: European Alternative Fuels Observatory (EAFO). https://www.eafo.eu/, (last accessed 20 December 2021)

7. European Automobile Manufacturers' Association (ACEA): Making the transition to zero-emission mobility. progress report (2021). https://www.acea.auto
8. European Commission: Weekly oil bulletin. https://energy.ec.europa.eu/data-and-analysis/weekly-oil-bulletin_en, (last accessed 20 December 2021)
9. European Commission (2016): A european strategy for low-emission mobility. directorate-general for mobility and transport. https://ec.europa.eu/commission/presscorner/detail/en/memo_16_2497, (last accessed 9 April 2022)
10. European Environment Agency (EEA): Indicator assessment: Greenhouse gas emissions from transport in europe. https://www.eea.europa.eu/ims/greenhouse-gas-emissions-from-transport, (last accessed 10 April 2022)
11. European Union law: Regulation (eu) 2018/842 of the european parliament and of the council of 30 May 2018. https://eur-lex.europa.eu/eli/reg/2018/842/oj, (last accessed 9 April 2022)
12. EUROSTAT: European statistical office. https://ec.europa.eu/eurostat/web/main/data/database, (last accessed 28 December 2021)
13. Färe, R., Grosskopf, S., Lindgren, B., Roos, P.: Productivity developments in swedish hospitals: a Malmquist output index approach. In: Charnes, A., Cooper, W.W., Lewin, A., Seiford, L. (eds.) Data envelopment analysis: theory, methodology and applications, pp. 253–272. Kluwer Academic Publishers, Boston (1994)
14. Färe, R., Grosskopf, S., Lovell, C.A.K.: The measurement of efficiency of production. Kluwer Nijhoff Publishing, Boston (1985)
15. Gruetzmacher, S.B., Vaz, C.B., Ferreira, Â.P.: Assessing the deployment of electric mobility: a review. In: Gervasi, O., et al. (eds.) ICCSA 2021. LNCS, vol. 12953, pp. 350–365. Springer, Cham (2021). https://doi.org/10.1007/978-3-030-86976-2_24
16. Holden, E., Linnerud, K., Banister, D.: Sustainable development: our common future revisited. Glob. Environ. Chang. **26**, 130–139 (2014). https://doi.org/10.1016/J.GLOENVCHA.2014.04.006
17. O'Connor, R.E., Bord, R.J., Yarnal, B., Wiefek, N.: Who wants to reduce greenhouse gas emissions? Soc. Sci. Quart. **83**, 1–17 (2002). https://doi.org/10.1111/1540-6237.00067
18. Pastor, J.T., Asmild, M., Lovell, C.A.: The biennial Malmquist productivity change index. Socio Eco. Planning Sci. **45**(1), 10–15 (2011). https://doi.org/10.1016/j.seps.2010.09.001, http://dx.doi.org/10.1016/j.seps.2010.09.001
19. Plötz, P., Moll, C., Bieker, G., Mock, P.: From lab-to-road: real-world fuel consumption and co2emissions of plug-in hybrid electric vehicles. Environ. Res. Lett. **16**, 054078 (2021). https://doi.org/10.1088/1748-9326/ABEF8C
20. Purvis, B., Mao, Y., Robinson, D.: Three pillars of sustainability: in search of conceptual origins. Sustain. Sci. **14**, 681–695 (2019). https://doi.org/10.1017/s0376892900011449, https://doi.org/10.1007/s11625-018-0627-5
21. Shephard, R.W.: Theory of Cost and Production Functions. Princeton University Press, Princeton (1970)
22. Wong, Y.H., Beasley, J.: Restricting weight flexibility in data envelopment analysis. J. Oper. Res. Soc. **41**(9), 829–835 (1990)

On Computational Procedures for Optimising an Omni-Channel Inventory Control Model

Joost Goedhart[1] and Eligius M. T. Hendrix[2(✉)]

[1] Operations Research and Logistics, Wageningen University,
6706 KN Wageningen, The Netherlands
joost.goedhart@wur.nl
[2] Computer Architecture, Universidad de Málaga, 29071 Málaga, Spain
eligius@uma.es

Abstract. Dynamic programming (DP) and specifically Markov Decision Problems (MDP) are often seen in inventory control as a theoretical path towards optimal policies, which are (often) not tractable due to the curse of dimensionality. A careful bounding of decision and state space and use of resources may provide the optimal policy for realistic instances despite the dimensionality of the problem. We will illustrate this process for an omni-channel inventory control model where the first dimension problem is to keep track of the outstanding ordered quantities and the second dimension is to keep track of items sold online that can be returned.

Keywords: Inventory control · Markov decision problems · Stochastic processes · Value iteration · Omni-channel retailing

1 Introduction

Inventory control is a dynamic process. Therefore, dynamic programming has been considered an appropriate technique to derive so-called optimal order policies, see [6]. An illustration of how to implement Value Iteration (VI) is given in [4] for small one and two-dimensional Markov Decision Problem (MDP) cases from perishable inventory control to derive a stationary order policy. One of the early detected challenges is the introduction of lead-time in inventory models, as one should keep track of the pipeline of already ordered quantities [7]. The larger the lead time, the larger the state space which makes the problem intractable. We will illustrate the resulting challenge for a so-called omni-channel retailer model, where a retailer uses multiple channels (online and in-store) to fulfil consumer demand [3]. In such inventory models, inventory is hold to fulfil both in-store demand as well as online orders. A rationing decision is involved that allocates inventory to either online or in-store consumers. As the rationing decision is

This paper has been supported by The Spanish Ministry (RTI2018-095993-B-I00) in part financed by the European Regional Development Fund (ERDF).

O. Gervasi et al. (Eds.): ICCSA 2022 Workshops, LNCS 13378, pp. 29–42, 2022.
https://doi.org/10.1007/978-3-031-10562-3_3

dependent on the inventory level and outstanding orders, the complexity of the action is related to the state space dimension. Additionally, in practice online sales go along with return streams. Modelling this aspect implies an increase in the state space, as we have to keep track of the items sold online which have not been returned yet. The state space increases with the number of days we keep track of unreturned items, as products might be returned the next day or after a longer period.

Our research question is how to implement the corresponding dynamic programming approach and to find the limits of the models that can be solved to optimality within a reasonable computing time and memory usage. We demarcate our question to the situation of an omni-channel retailer confronted with a discrete finite demand for both channels, lost sales and inventory holding cost, ordering cost, fulfilment cost, and a profit margin for both channels.

To investigate the question, we formulate a dynamic model and illustrate its solution procedure via VI. We focus on the computational effort to find the optimal solution and on demarcation of the state space. This paper is organised as follows. We first sketch the concept of deriving the optimal policy by VI in Sect. 2. Section 3 then introduces and investigates several versions of the omni-channel model. Section 4 summarises our findings.

2   The Procedure of VI

To consider an inventory control problem as a Markov Decision Problem (MDP), we should define the state space, decision space, transition probabilities and contributions. For inventory control, this requires to have a good vision on the sequence of events. In our omni-channel case, at the end of the day, after receiving (or not) an outstanding order, the retailer decides on the next order quantity Q and the rationing R. This decision depends on the inventory state I and outstanding order states q symbolised by state vector S; a stationary policy is described by $(Q(S), R(S))$. The rationing R is given as the number of items assigned to the sales in the shop and exposed there, thus R items are stored in-store and $I - R$ items are stored in the backroom from where they can be sold online. The challenge we focus on is that the order quantity is received after a lead time of L periods, requiring to put the outstanding orders in the state space of S, as a vector $q = (q_1, .., q_\ell, \ldots, q_L)$. The inventory dynamics is following the equation

$$I_{new} = (R - d_1)^+ + (I - R - d_2)^+ + q_L, \tag{1}$$

with $x^+ := \max\{x, 0\}$, where d_1 and d_2 represent the realisation of demand in the shop and of demand online respectively. The dynamics of the pipeline inventory is

$$q_\ell = q_{\ell-1}, \ell = 2, \ldots, L \tag{2}$$

and $q_1 = Q$. In more abstract terms, a transformation from one state to the other is given by a function T, transforming the state $S = T(I, q, Q, R, d_1, d_2)$ depending on current state, decisions and realisation of demand. Notice that we implicitly think in time steps of one day instead of in terms of continuous time, which is also reasonable for a retailer situation. Moreover, we think of the state space S as finite and discrete by bounding the inventory by a maximum $\bar{I}$ and also the order quantity by $\bar{Q}$. This means that one can map the state space in an array s, where an element s_j represents a state value for all state variables.

For the optimisation of the inventory control, we have a cost function $C(Q, R, I)$ which depends on the procurement cost and inventory holding cost. Moreover, there is an expected gain $G(I, R)$ of the current state (I, R). This defines the expected profit where the margins for selling online and from the shop differ.

2.1 MDP Background

Under certain circumstances, optimal inventory control can be characterised by the MDP theory, introduced originally by [1]. For the optimal policy (Q, R), there exists a so-called value function $v(S)$ and a scalar π to be interpreted as the expected daily profit such that

$$\forall S \in S, v(S) + \pi = \max_{Q,R} \left(G(I, R) - C(I, Q, R) + E(v(T(S, Q, R, \mathbf{d}_1, \mathbf{d}_2))) \right),$$

where E is the expectation over the stochastic demand $\mathbf{d}_1$ and $\mathbf{d}_2$, which we take as independent events. Working with a discrete finite distribution $p_{1k} = P(\mathbf{d}_1 = k)$ and $p_{2m} = P(\mathbf{d}_2 = m)$, we can see the expected valuation of the future in a discrete way, see [5]. Consider all states in an array $s := (s_0, s_1, \ldots, s_{N-1})$ and define the expected revenue as

$$r(s_i, Q, R) := G(s_i, R) - C(s_i, Q, R),$$

then the Bellman equation comes down to the existence of an array V and constant π such that

$$\forall i, \ V_i + \pi = \max_{Q,R} \left(r(s_i, Q, R) + \sum_{k,m} p_{1k} p_{2m} V_j \right),$$

with V_j interpreted as the value of $v(T(s_i, Q, R, d_{1k}, d_{2m}))$. If we now map the optimal control rule $(Q(S), R(S))$ in an array, then the best thing to do in situation s_i is given by

$$(Q_i, R_i) = \underset{Q,R}{\mathrm{argmax}} \left(r(s_i, Q, R) + \sum_{k,m} p_{1k} p_{2m} V_j \right), \tag{3}$$

with $V_j = v(T(s_i, Q, R, d_{1k}, d_{2m}))$. In theory, the equations imply a fixed point relation around the expected gain π. How to derive the optimal policy now in a computational way? How are we going to be bothered by the curse of dimensionality?

2.2 Value Iteration Procedure

Following the fixed point idea with respect to value π, we will follow a procedure called Value Iteration (VI). From a computational point of view it is sufficient

---

Algorithm 1. Pseudocode of VI for inventory control

1: Set array elements V_i to $r(s_i, 0, 0)$ for $i = 0, \ldots, N - 1$
2: **repeat**
3:     Copy vector V into vector W
4:     **for** $i = 0, \ldots, N - 1$ **do**
5:         **for** Feasible values of Q, R **do**
6:             **for** all demand realisations d_{1k}, d_{2m} **do**
7:                 Get W_j with $s_j = T(s_i, Q, R, d_{1k}, d_{2m})$
8:         $V_j = \max\limits_{Q,R}[r(s_i, Q, R) + \sum_{k,m} p_{1k} p_{2m} W_j]$
9: **until** $\max\limits_{j}(V_j - W_j) - \min\limits_{j}(V_j - W_j) < \varepsilon$

---

to think in copying the last valuation in an array W and to determine the new valuation array V as sketched in Algorithm 1 up to convergence takes place of what is called the span, i.e. the gap between maximum and minimum difference of the two arrays. The convergence is slow when state values that have a low probability of occurrence are included. The challenge is, that this is known in the end, by simulating the optimal policy or alternatively deriving the stationary state probabilities of the Markov Chain. In practice, this leads to ideas of adaptive Dynamic Programming, when we adapt the state space depending on the outcomes of the optimisation.

A large practical computational gain is to compute the expected revenue $r(s_i, Q, R)$ to be calculated beforehand as sketched in Fig. 1, so not to run over all demand realisations during the VI loop. Another important concept is to define, before the iteration process the transition matrix. Instead of doing a loop over possible outcomes of demand as sketched by lines 6 and 7 in Algorithm 1, we construct a matrix $M(Q, R)$, where entrances $M_{ij}(Q, R)$ capture the probability to go from state s_i to state s_j when performing action (Q, R). This means, we can replace part of the evaluation of lines $6 - 8$ in Algorithm 1 by a matrix-vector multiplication multiplying M with array W calling compiled code.

3   Cases

We consider an omni-channel retailer case who sells in both channels for a margin of $m = 45$, where online shipping adds an additional 10 cost per unit. Order cost is $k = 33$ and storing in the shop requires holding cost of $h_1 = 1$ and the backroom $h_2 = 0.5$ per night. The margin and costs define the expected revenue $r(S, Q, R)$. The demand is assumed to be Poisson distributed, but truncated above at the 0.999 quantile in order to have a bounded space on demand. In our base case, we take a mean demand for an item in-store of $\mu_1 = 6$ and online of $\mu_2 = 2$ per day. Based on the probability mass distribution of demand, we can construct a transition matrix which also depends on the rationing decision R following dynamics (1). The order quantity decision Q affects the dynamics of the order pipeline (2), which does not depend on random events.

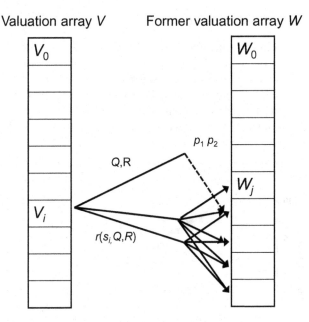

Fig. 1. Data dependence in value iteration

One of the first questions to ask ourselves when implementing a VI is how to bound the state and decision space. In the described model, the maximum inventory $\overline{I}$ does not deviate too much from the maximum order quantity $\overline{Q}$. For the latter, the so-called Economic Order Quantity (EOQ) (see e.g. [6]) is relevant for minimising order cost. In our case, we have to take into account that the rationing to the more expensive channel will be limited up to what is profitable, so the holding costs mainly depend on the back room holding cost h_2. Following general lines, we come to an EOQ $= 32$, which coincides with a replenishment cycle of 4 d. However, demand is stochastic, so a safety stock applies weighting the lost sales of the total demand during the cycle and the expected inventory cost. Therefore, we estimate using the EOQ model and a safety stock estimation that $\overline{I} = \overline{Q} = 45$, is a reasonable upper bound of the state space with respect to the inventory level. Without such an analysis, one depends on a numerical procedure and observe which values will be attained by following the derived optimal policy. We should keep in mind that including state values that have a low probability of occurrence, will not only increase the solution time in a polynomial way, but will also increase the number of iterations necessary for convergence. In our first experiment, we will illustrate this feature. Notice that the rationing decision is always bounded by $R \leq I$.

3.1   First Case, Lead Time $L = 1$

The complexity should be low, if the pipeline of outstanding orders caused by the lead time is low. In our case of a lead time of $L = 1$, one can derive that in fact it is sufficient to capture the state variable S only by inventory I, as $q_l = Q$ can be handled by optimising simultaneously over order quantity Q and rationing R. This means that the ordered quantity does not appear as state variable, but can directly be included in the state transition of inventory level I at the end of the day.

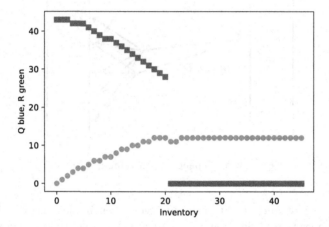

Fig. 2. Optimal order quantity $Q(I)$ and rationing $R(I)$ as function of inventory

The code we developed in python, first calculates the revenue r and a transition matrix $M(Q, R)$ to exploit vector-matrix multiplications and computing the maximum (best decision) using libraries. Following the sketched procedure with a termination criterion of $\varepsilon = 0.1$, the VI converges after 36 iterations and 2 s providing the optimal policies as depicted in Fig. 2. Observing the decisions, we can see that the rationing R is bounded by balancing expected lost sales versus the inventory holding costs and limited to $\overline{R} = 12$; see proof in [3]. The optimal order quantity Q is higher than the EOQ, but prevents putting an order from inventory levels higher than about $I = 20$. For both decisions, one could exploit monotonicity considerations; one can observe that if $Q(I) = 0$, then $Q(I+1) = 0$ and if $R(I) = \overline{R}$, then $R(I+1) = \overline{R}$. We did not make use of such considerations, as we are following matrix multiplication for all potential decisions rather than a do-loop.

We simulated the dynamics of the model for $T = 500,000$ days providing a daily profit of 310. We did an additional experiment varying the size of the state space using $\overline{I} = 50, 100, 150, 200$. In order to inspect the state space, we measured the occurrence of each inventory level as a frequency as depicted in Fig. 3. From the VI follows that the maximum amount to be ordered is $\overline{Q} = 42$.

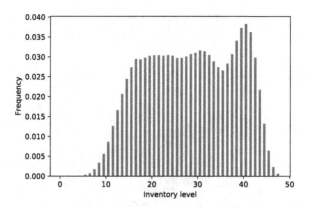

Fig. 3. Frequency of occurrence of inventory level following the optimal decisions

This means that basically high values of the inventory level do not occur, as it does not make sense to order each day a high amount. This is illustrated in Fig. 3. As we measure the inventory level after the order has been delivered, there is no surprise that around $I = 40$, we have a high probability of occurrence. We are more surprised by the equal occurrence of values in the range $I \in [18, 32]$.

Table 1. Computational time varying $\overline{I}$

Upper bound $\overline{I}$	50	100	150	200
Time seconds	3	11	30	61

Varying state limit value $\overline{I} = 50, 100, 150, 200$, one would expect a computational time linear increase in this case. As shown in Table 1, this is not the case. This illustrates, that a strategy could be to start with state boundaries that are too small, observe and possibly simulate the system, and extend the boundaries gradually rather than starting with a space which is too big requiring an enormous computational time. Notice that limits can also be taken lower exploiting extrapolation of the value function. In our example, high values of inventory lead to an optimal policy of not to order, i.e. $Q = 0$. This means that the procedure requires a valuation of high inventory levels, but does not have to evaluate what is the optimal policy for those cases. For the valuation, extrapolation can be used.

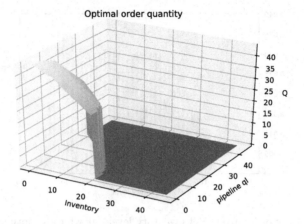

Fig. 4. Optimal order quantity $Q(I, q_2)$ as function of the (pipeline) inventory

3.2 Extending Lead Time $L = 2, 3$

With a longer lead time, we have to take the pipeline inventory into account which has been ordered L days before. For the optimal order quantity, this has quite some impact as sketched in Fig. 4. The rationing decision is less sensitive to the pipeline inventory. Only when inventory levels are low, the pipeline of outstanding orders may influence the ration decision. This is sketched in Fig. 5; only for very low inventory levels the ration is influenced by the order quantity two days ago.

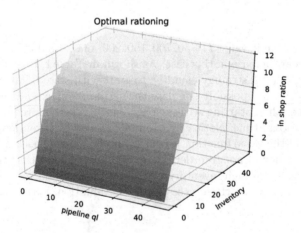

Fig. 5. Optimal ration $R(I, q_2)$ as function of the (pipeline) inventory

To obtain the optimal decision rule, the number of state values has increased to 2, 116 and the VI requires 35 iterations to convergence in 65 s in our implementation. The increase in computational time compared to the case $L = 1$ is less than proportional to the number of state values.

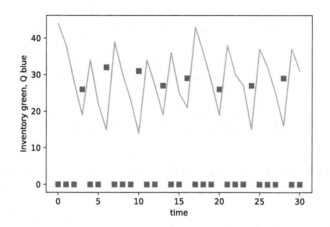

Fig. 6. Inventory development and order quantities, $L = 2$

A simulation first shows us the inventory development and order quantities for the first weeks in Fig. 6. Notice that the pipeline inventory follows the order quantity with a delay. The average daily profit of the optimal rule is 308. This is only slightly less than the situation with a lead time of 1 period (using the same pseudo random demand data).

On one hand, the lead time may provide less predictability of demand and imply higher cost. On the other hand, if we compare the distribution of inventory of a lead time of 2 in Fig. 7 with that of lead time of 1 in Fig. 3, one can observe that the physical inventory is lower in the first case, as part is pipeline inventory for which no inventory cost is incurred.

The state space when extending the lead time is $\overline{Q} = 45$ times bigger leading to a computing time which appears 30 times bigger, which is less than linear, although required memory is linearly increasing. The efficiency gain is due to pre-processing the transition matrix which includes the rationing decision and using library routines to compute the maximum profit rationing value. Notice that the transition of the pipeline inventory is straightforward.

We pushed the implementation to the largest state space we consider by having a lead time $L = 3$ providing 97,336 state values. The computational time increases nearly proportional to the increase in the state space converging in 3330 s seconds after 34 iterations. The outcome of the procedure shows that due to the longer lead-time, the inventory is increased considerably, see Fig. 9. As illustrated in Fig. 8, this also has some impact on the VI procedure. It appears that the orders are going to follow a cyclical behaviour. In that case, also the

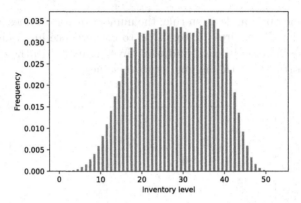

Fig. 7. Frequency inventory level, $L = 2$

convergence should be checked in another way, due to alternating states. This illustrates, that it is of utmost importance to check the results by a simulation.

3.3   Including Online Sales in the State Space

The recent decade, there has been an increased interest in the effect of returning sold online products on the logistics decision, see e.g. [2]. Our question is here, what is the effect on the state space and increasing computational time of including recent sales in the state space? To know the possible returned products, we have to keep track of the number of items sold over the last n days, where n represents the horizon in which customers may return their bought product.

In terms of the profit, the bookkeeping is getting more complicated, but also helps to decide not to sell the product online at all. On a return, we get the margin of the item back, but we loose not only the earlier mentioned shipping costs of 10 units, but also a handling cost of 5 units. This reduces the expected profit for online items drastically. Assume that we have a probability of $p = 40\%$ that an item is returned. The online sales has a profit margin of 35 if the customer keeps the item (60% probability). This should be weighted with the loss of 15 when the product is returned (40% in our example).

For the inventory dynamics, if we consider a fixed probability of returning bought items during the return period, the number of returned items can be modelled using a binomial distribution. The challenge is that this requires to keep track of the number of items bought j periods ago, $(r_1, \ldots, r_j, \ldots, r_n)$. For our exercise, we will consider $n = 1, 2, 3$ and therefore we also have to adapt the probability p_j that a product is returned which has been sold j days ago. We will inspect the computational consequence of adding this state variable to the VI algorithm for $n = 1, 2, 3$. The dynamics of the sold items for $n = 3$ is $r_3 = r_2 - bin(r_2, p_2)$, $r_2 = r_1 - bin(r_1, p_1)$ and $r_1 = \min(I - R, d_2)$, which should be included in the transition matrix computation. For our exercise, we can bound the state space thanks to the truncated demand to $\bar{r}_j = 7$, as that covers

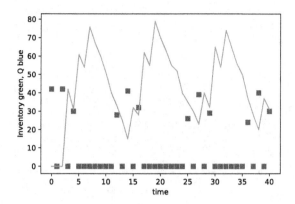

Fig. 8. Inventory development and order quantities, $L = 3$

practically 0.999 of the demand probability for d_2. This implies an increase of the state space with a factor $(\overline{r}_j + 1)^n$.

More cumbersome is now the inventory dynamics. To the dynamics of I_{new} in (1) should now be added the binomial probability distribution of (r_1, p_1), (r_2, p_2) and (r_3, p_3). Of course, this is just one way to model the return of bought products. We implemented the transition matrix, first for a case where $n = 1$ to compare the computational time and the results to that of Case 1 taking all parameters equal. The computational time for our implementation is 8 s compared to the 2 s if no returning of sold products is allowed. However, the state space is 8 times bigger. Convergence requires 36 iterations within 8 s. Again, the time efficiency is reached due to the pre-computation of expected gain and the transition matrix, where not only the probability of demand, but also that of returns is captured to reach state j from state i with decision R. As described, the transition due to the order quantity has a more deterministic character.

The optimal policy now also depends on the number of products sold one day ago. Actually, it anticipates to order less given that there are returns expected of sold products. The behaviour is sketched in Fig. 10. Due to the probability of returns of sold products, the expected profit reduces from 310 to 271 units.

Now extending the state space taking $n = 2$ days as the possibility to return the online bought product, increases the number of state values again with a factor 8. The other parameters are left constant. The algorithm now requires 35 iterations and 102 s. Our next case considered a return time of 3 d, which increases the state space by 8 again. The computational time now increases to 1030 seconds and the profit of the derived policy is basically the same to a two day return period.

The extension of the model allows us now to consider lead time increase and returns from earlier sales. We run a variant for $L = 2$ and $n = 1$ with in total $N = 16,928$ which converged after 35 iterations taking 800 s of computational time. The computation of the transition matrix entails now considering both lead

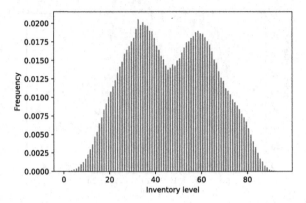

Fig. 9. Frequency inventory level, L=3

time and the probabilities of having returns. The computational time compared to the case of $L = 2$ and $n = 0$ increased more than proportional in the number of state values.

3.4   Summary of the Numerical Results

Table 2. Results of Algorithm 1 on the cases

L	n	N	Tcomp	Profit
1	0	46	2	309
2	0	2,116	65	308
3	0	97,336	3,330	301
1	1	368	8	271
1	2	2,944	102	266
1	3	23,552	1,030	266
2	1	16,928	800	268

The VI algorithm was implemented in python exploiting the NumPy library routines. For an experiment running the cases on a desktop Intel(R) i5-2400 CPU with 8 GB ram, the computational time is reported in Table 2. Remember that all cases use the same cost parameter values and the same (pseudo-) random numbers for validating the expected profit of the resulting policy. The number of state values is represented by N, L is the lead time, n the number of days a customer can return an online bought product, Tcomp is the computational time in seconds and profit the expected daily profit estimated by the simulation.

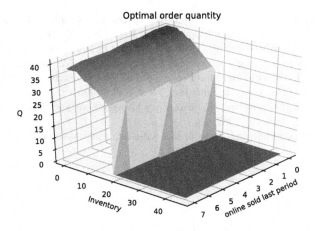

Fig. 10. Optimal order quantity, $L = 1, n = 1$

4   Conclusions

The determination of optimal inventory management policies for higher dimensional state spaces, is often considered not feasible due to an increasing complexity. The computational feasibility of using Value Iteration (VI) mainly depends on a careful bounding of the space using information on the maximum demand and order quantities. In general, the relevant state space to be considered follows from simulating the optimal policy. This means that a strategy to start with smaller space boundaries and gradually increasing them based on the simulation outcomes may help to define the smallest relevant space. In our case, we could analytically estimate the relevant maximum inventory due to revenue considerations on the optimal order quantity following the EOQ model and a safety stock estimation. This has been illustrated in the paper for several cases. The maximum sold items to bound the returns space is mainly determined by the demand of in-shop sales, although it also depends on the rationing decision. Moreover, we found that in python based implementations, one should try to make use of compiled library procedures to replace explicit do-loops.

In our illustration, we have shown that pipeline inventory can be taken into account up to a limited lead-time. An extension towards including past sales to anticipate on returned sold products allows a wider extension due to a strict bounding on the demand levels, leaving out low probability state values which hinder convergence of the VI algorithm.

References

1. Bellman, R.: A Markovian decision process. J. Math. Mech. **6**(5), 679–684 (1957)
2. Bernon, M., Cullen, J., Gorst, J.: Online retail returns management: integration within an omni-channel distribution context. Int. J. Phys. Dis. Logistics Manag. **64**(6), 584–605 (2016)

3. Goedhart, J., Haijema, R., Akkerman, R.: Inventory rationing and replenishment for an omni-channel retailer. Comput. Oper. Res. **140**, 105647 (2022)
4. Hendrix, E.M.T., Kortenhorst, C., Ortega, G.L.: On computational procedures for value iteration in inventory control. IFAC-PapersOnLine **52**(13), 1484–1489 (2019)
5. Puterman, M.L.: Markov Decision Processes: Discrete Stochastic Dynamic Programming, 1st edn. John Wiley & Sons Inc., New York (1994)
6. Silver, E., Pyke, D., Peterson, R.: Inventory Management and Production Planning and Scheduling. Wiley (1998)
7. Zipkin, P.: Old and new methods for lost-sales inventory systems. Oper. Res. **56**(5), 1256–1263 (2008)

A Bibliometric Review and Analysis of Traffic Lights Optimization

Gabriela R. Witeck�ⓘ, Ana Maria A. C. Rocha⁽✉⁾ ⓘ, Gonçalo O. Silvaⓘ,
António Silvaⓘ, Dalila Durãesⓘ, and José Machadoⓘ

Centro ALGORITMI, Department of Production and Systems, University of Minho,
4710-057 Gualtar, Braga, Portugal
gabiwiteck@gmail.com, arocha@dps.uminho.pt,
g.oliveirasilva96@gmail.com, {asilva,
dalila.duraes}@algoritmi.uminho.pt, jmac@di.uminho.pt

Abstract. The significant increase in the number of vehicles in urban areas emerges the challenge of urban mobility. Researchers in this area suggest that most daily delays in urban travel times are caused by intersections, which could be reduced if the traffic lights at these intersections were more efficient. The use of simulation for real intersections can be effective in optimizing the cycle times and improving the traffic light timing to coordinate vehicles passing through intersections. From these themes emerge the research questions: How are the existing approaches (optimization techniques and simulation) to managing traffic lights smartly? What kind of data (offline and online) are used for traffic lights optimization? How beneficial is it to propose an optimization approach to the traffic system? This paper aims to answer these questions, carried out through a bibliometric literature review. In total, 93 articles were analyzed. The main findings revealed that the United States and China are the countries with the most studies published in the last ten years. Moreover, Particle Swarm Optimization is a frequently used approach, and there is a tendency for studies to perform optimization of real cases by real-time data, showing that the praxis of smart cities has resorted to smart traffic lights.

Keywords: Traffic lights · Optimization · Smart cities

1 Introduction

The significant increase in the number of vehicles on the roads emerges the challenge of urban mobility. Researchers in this area suggest that most daily delays in urban travel times are caused by intersections, which could be reduced if the operation of the traffic lights at those intersections were more efficient. Signalized traffic control significantly

This work has been supported by FCT— Fundação para a Ciência e Tecnologia within the R&D Units Project Scope: UIDB/00319/2020 and the project "Integrated and Innovative Solutions for the well-being of people in complex urban centers" within the Project Scope NORTE-01-0145-FEDER-000086.

O. Gervasi et al. (Eds.): ICCSA 2022 Workshops, LNCS 13378, pp. 43–54, 2022.
https://doi.org/10.1007/978-3-031-10562-3_4

reduces vehicle delays at intersections, balances traffic flow, and improves the operational efficiency of an urban street network [1].

In the urban traffic management, the control of the traffic light cycle is fundamental, as its performance directly affects the efficiency of the traffic system, which is controlled by the municipal authorities and its correct installation can improve both traffic flow performance and the safety of all roads users [2]. Due to the characteristics of randomness and complexity of the traffic flow, the traditional light signals working based on a predetermined and predefined clock may not support the actual demand.

Several systems are deployed to control traffic signals, but each system is based on different parameters that can be static (off-line) or dynamic data (real-time). Controlling traffic is not an exhaustive procedure but requires continuous data to be able to continuously function. Many approaches based on different sensors are employed in-vehicle environment perception systems, such as monocular cameras, stereo cameras, and radars. Focusing on the pedestrian and cyclist detection field, vision sensors are preferred, due to the possibility to capture a high-resolution perspective view of the scene with useful color and texture information, compared to active sensors [3]. In this context, optimizing traffic lights may involve implementing the best possible timing settings that control the operation of a traffic signal.

The main objective of traffic signal optimization is to significantly improve the performance of the traffic intersection by minimizing the delay, queue length, the number of stops, and gas emissions. The continuous movement of vehicles generates a huge amount of data that needs to be processed by high-end infrastructure. Recently, optimization approaches have been utilized in traffic control models to increase the performance of traffic signal control systems. Optimizing the transition phase or the shift between the signal timing plans will result in the reduction to its minimum in traveling time, stops, and delays. The positive effects on the environment can lead to a reduction in polluting emissions and fuel consumption. Transition optimization can be an essential way to improve the efficiency of signal systems [4]. Some studies are looking to find out solutions to the problem of congestion, traffic lights, and how to make the process of vehicles on the road safer at a low cost. Thus, emerges a set of research questions regarding the existing approaches to managing traffic lights smartly, the kind of data used, and the benefits of this appliance.

This paper aims to answer three research questions, through a bibliometric literature review. Two important databases (WoS and Scopus) were searched using appropriate keywords to find the most relevant studies on the topic. The outcomes of each research question were discussed.

The paper is organized as follows: Sect. 2 presents the bibliometric review of the literature carried out to support the logic of this study. Section 3 analyzes and discusses the findings of the systematic review research. Finally, Sect. 4 reveals the conclusions and future work.

2   Systematic Literature Review

The research method to carry out a literature review and bibliometric analysis followed four phases [5], which involved the review and evaluation of primary studies relevant to the research. The main is to analyze the state of the art and the current approaches being used in traffic light management using optimization methods, data, and other technologies adopted.

The general process of the proposed study comprises the following steps: (1) Systematic mapping planning; (2) Conduction of the Search; (3) Selection and bibliometric analysis of the primary studies; and (4) Final selection.

2.1 Systematic Mapping Plan

This step consists of two sub-steps: (a) Formulation of the research questions; (b) Selection of the databases and resources.

The research questions (RQ) convey and guide the research directions, as presented below:

RQ1: How are the existing approaches (optimization techniques and simulation) to managing traffic lights smartly?

RQ2: What kind of data (offline and online) was used for traffic lights optimization?

RQ3: How beneficial is it to propose an optimization approach to the traffic system?

The second sub-step is to state the databases and resources. The selection of the databases was based on the content update, availability of the full text of papers, also the quality of the research accuracy. As a result, the database of Scopus and Web of Science (WoS) were eligible. The access to the articles was carried out at Scopus and WoS in the period from 2012 (January) to 2022 (February).

2.2 Conducting the Search

Step 2 aims to define a list of inclusion and exclusion criteria and select appropriate keywords that will be used to find related research work. The list of inclusion and exclusion criteria was established, following the criteria guide to include primary studies that show relevance to the research questions and exclude the studies that do not show relevance to answering the RQ.

The criteria are:

- Exclusion criteria (EC)
    EC1: The study is not written in the English language.
    EC2: The Full-text paper is not available.
    EC3: The research does not cover the topic of smart cities.
    EC4: The document type is not a Conference Paper, Article, Review, Proceedings Paper, or Article in Press.
- Inclusion criteria (IC)
    IC1: The study aims to reduce traffic congestion and delay time.
    IC2: The study is looking for road safety and traffic forecast topics.

Based on the carefully selected keywords (traffic light, optimization, data, smart or intelligent) the database strings must be followed to search the proper databases Web of Science (WOS) and Scopus. Additionally, the search string followed the format: TITLE-ABS-KEY ("traffic light*") AND TITLE-ABS-KEY (optimization) AND TITLE-ABS-KEY (data) AND TITLE-ABS-KEY (smart OR intelligent).

2.3 Selection and Bibliometric Analysis of the Primary Studies

As a result of applying the criteria established in Step 1, the researchers have identified 157 studies. In this step, it was used an open-source scientometric tool called ScientoPy [6]. This tool is a Python script-based tool specialized in temporal scientometric analysis.

Later, the document types, not relevant (EC4), and repeated studies within the databases were eliminated. The papers published in WOS were maintained, and those that were in Scopus were subtracted. In total resulted 93 papers in this phase, see Table 1.

Table 1. Documents uploaded.

Information	Number	Percentage
Total papers uploaded	**157**	
Document type removed – EC4	22	14.0%
Total papers after papers removed	**135**	
Loaded papers from WoS	66	48.9%
Loaded papers from Scopus	69	51.1%
Removed duplicated papers from Scopus	42	60.9%
Total papers after the removal process	**93**	
Papers from WoS	66	71.0%
Papers from Scopus	27	29.0%

Figure 1 shows the pre-process brief graph, that presents the loaded documents for each database and the removed duplicated documents, respectively. Since ScientoPy preprocess script keeps WoS documents over Scopus documents, after the duplication removal filter we see more documents from WoS than the Scopus database [6].

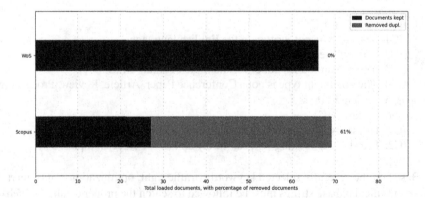

Fig. 1. Loaded and removed documents for each database.

After completing the search phase, the source documents were categorized by several publications considering their type: articles, proceedings papers, conference papers, and reviews (see Fig. 2).

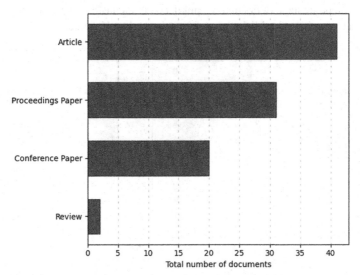

Fig. 2. The number of documents per type.

The author's keywords are shown in Fig. 3 as a word cloud visual representation, where the size of the word or phrase is proportional to the number of documents related to it. The word cloud highlighted the principal topics that are using the optimization related to traffic lights, like intelligent transportation systems and traffic congestion.

Fig. 3. Word cloud for author´s keyword.

All source documents were then linked to authors by their full names, publication year, and country. Figure 4 details the number of publications by Country in recent years. The figure on the left reveals that the United States and China were the countries with the most studies published in the last ten years. On the other hand, the figure on the right side shows the percentage of documents published in the last two years when compared to the average number of published documents in the last 10 years.

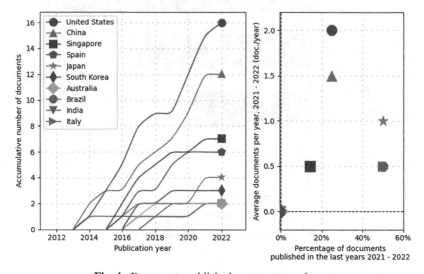

Fig. 4. Documents published per country and year.

After the bibliometric analysis of the results obtained in Step 3, it was necessary to nominate "Basic terms", or keywords, to analyze all the 93 papers found until this step. The terms must be simple and meaningful enough to cover the Research Questions, which can bring back more accurate results in the searched databases. Table 2 shows a list of all nominated keywords.

Table 2. Basic terms for each research question.

Research question	Basic terms
RQ1: How are the existing approaches (optimization techniques and simulation) to managing traffic lights smartly	Optimization, techniques, simulation, methods, tools, simulator, algorithm, system, transport management, traffic monitoring, smart traffic
RQ2: What kind of data (offline and online) was used for traffic lights optimization?	Online, offline, real, virtual
RQ3: How beneficial is it to propose an integrated smart approach that uses online traffic data to detect cyclists, cars, and pedestrians	Results, cyclists, pedestrians, motorized vehicles, road users, benefits, impact, outcome, method, reach, safety, a passerby

2.4 Final Selection

The 93 papers were forwarded to the final selection process. In this step, if the full-text paper was not available (attending the EC2) it will not be considered. This criterion excluded 25 papers.

Then, a total number of 68 paper abstracts were screened. The researchers have reduced the total number of references found by their abstract relevance. The abstract was read and then determined whether the paper is related to the study or not. There-fore, 29 papers were not related to the study. This step follows immediately after abstract screening, which indicates if the abstract is related, and papers received one more screen-ing by reading the full text to determine its relevance. Finally, a total number of 39 papers were reviewed in detail. Table 3 shows the screening and selection process.

Table 3. Papers screened and analyzed

Information	Number
Papers after Step 2	**93**
Papers with full-text not available (EC2)	25
Total papers abstract screened	**68**
Paper not related to the study	29
Total paper full-text screened and analyzed	**39**

3 Results and Discussion

In this section, the total number of 39 papers was reviewed in detail. The papers were studied, and applied the search for "Basic terms", as above mentioned in Table 2. It identified the answers to the research questions.

The RQ1 raised the question regarding the existing approaches to managing traffic lights smartly and it was possible to find many approaches in this matter. It was identified 24 different technologies and algorithms applied to traffic lights optimization, as shown in Table 4.

Among the optimization approaches found in the literature, the ones that have been widely used in traffic light optimization are the Discrete harmony search [7, 8], Genetic Algorithm [9], and Particle Swarm Optimization [10, 11]. The main objective of the optimization approach is to find out an optimal solution for the traffic light problem.

In traffic light optimization, it is also common to apply a simulator, which makes it easy to compare the algorithms on different infrastructures and traffic patterns. The simulations of the results could be found in seven: AIMSUM simulator, Java-Agent-Development-Environment (JADE), MATLAB, Netlogo, NS2, PTV Vissim, and SUMO. The last one (SUMO) was the most used simulator, and it is an open-source traffic sim-ulation package including net import and demand modeling components. The Participle Swarm optimization, a stochastic global optimization method that takes inspiration from the social behaviors of insects and other animals, was the most used.

Table 4. Summary of the technologies and algorithms

RQ1	
Algorithmic game theory (AGT)	Jaya Algorithm
Ant colony optimization (ACO)	Module VisVAP
Artificial bee colony (ABC)	Nearest neighbor (NN)
Breakdown minimization Principle (BMP)	Novel framework dynamic TL control
Cell Transmission Model (CTM)	Particle Swarm optimization
Cellular Genetic Algorithm (CGA)	Proximal Policy Optimization (PPO)
Discrete harmony search (DHS)	Traffic Actuated Optimization Signal Fuzzy control (TAOSF)
Edge for traffic light control (DRLE)	Traffic Actuated Signal Fuzzy Control (TASFC)
Faster R-CNN	VANET
Harmony search (HS)	Wardrop's user Equilibrium (WE)
Intelligent signal light timing model principle	Water cycle algorithm (WCA)
Intelligent traffic signal optimization control platform	YOLO

In [12], a multi-agent-multi-purpose system (MAMP) to solve the traffic light problem was developed, where an ant colony optimization (ACO), by a distributed intelligent traffic system (DITS). The results have shown that for several initial distributions of vehicles, the ACO strategy obtains higher average speeds, smaller average waiting times, and several stopped vehicles than the non-ACO strategy. The application of the ACO combined with the SUMO traffic simulator was also used in [13]. This work confirmed a better performance of the developed algorithm compared to the predefined time controller and other dynamic controllers. Moreover, in the work [14] the Harmony Search (HS) and artificial bee colonies (ABC) to solve the urban traffic light scheduling problem were implemented. The offline data were based on the real traffic in the Jurong area of Singapore. The authors aimed to minimize the total network-wide delay time of all vehicles and the total delay time of all pedestrians at the same time.

The work in [15] mentions that future optimization algorithms for the traffic light scheduling problem should consider the reliability of solutions over multiple traffic scenarios and incorporate simulation strategies that improve reliability. However, the highly dynamic traffic of a city means that no single traffic scenario is a precise representation of the real system, and the fitness of any candidate solution (traffic-light program) will vary when deployed in the city [15].

Managing smart traffic lights can be done by using different approaches and techniques. Furthermore, to perform the optimization and find the optimal values of the traffic light cycle time, some programming languages can be applied, in this study identified two of them: are Phyton and CC++.

To answer the RQ2, the study revealed 18 papers using off-line data, and 21 using real-time data. However, all of them use data from real cases, showing that the praxis of smart cities has resorted to smart traffic lights. The global optimization of traffic-light problems has been applied in smart city case studies when the real-time control of traffic lights is not possible. It requires the optimization and simulation of a traffic scenario that is estimated after collecting data from sensors at the street level.

Although several traffic flows forecasting techniques have been developed and have achieved good forecasting performance, the author´s papers are separate from traffic light control research. In other words, existing studies focus on traffic light optimization or traffic flow prediction, but few studies combine these two techniques. To consider the effect of traffic light control over a period, the expected traffic condition must also be considered.

Table 5. Papers and questions research

Papers	RQ1	RQ2	RQ3
[16]	Particle Swarm optimization, SUMO, VANET	Off-line	The model saves time up to 57% compared to the static plan and in dense traffic conditions with high traffic intensity
[9]	Genetic algorithm, simulator AIMSUM	Off-line	The proposed active control of traffic lights could reduce by half the average delay time needed for vehicles to cross the region under study
[17]	Jaya algorithm, harmony search, and water cycle algorithm	Real-time	Improvement ranges from over 26% and 28% in terms of the minimum and maximum total delay times
[18]	Cell Transmission Model, SUMO	Real-time	The algorithm provides near-optimal solutions with a maximum optimality gap of 5.4%
[19]	Decentralized Reinforcement Learning at the Edge for traffic light control in the IoV (DRLE)	Real-time	DRLE decreases the convergence time by 65.66% compared to PPO and training steps by 79.44% compared to ARS and ES
[20]	Ant Colony Optimization, SUMO, VANET	Real-time	20–25% improvement in delays. Decrease of 14–18% in CO emissions, 15–21% fuel consumption, 7–9% noise rate
[21]	Intelligent traffic signal optimization control platform	Real-time	The overall total delay time was reduced by 23%, and the travel time was reduced by 15%

Eight out of 39 papers presented answers to the RQ3, showing the benefit of using an optimization approach for the traffic light system. Even though the answers did not cover the entire expected content. The respective papers and answers are revealed in Table 5. In general, the results confirm an increase in the overall performance of the system when the application of optimization approaches is introduced. However, most articles do not quantitatively present the results, but only in a qualitative way.

The RQ3 was also focused on looking for papers and solutions by designing a system which is considering both types of road users, motorized vehicles, and cyclists. However, the papers studied had no attention to cyclists in the initial traffic signalization planning. Bicycle traffic and user roads have not been given the same priority as motorized traffic.

4   Conclusions

The number of vehicles has been increasing more and more in urban areas, so the challenge of urban mobility is emerging. The literature in this area suggests that most daily delays in urban travel times are caused by intersections, which could be reduced if traffic lights (in these intersections) were more efficient. Using traffic light simulation to coordinate vehicles passing through intersections can be an effective and efficient strategy when combined with the optimization of traffic light cycle times.

This research aimed to improve the understanding of traffic light management using a bibliometric literature review. Two important databases (WoS and Scopus) were searched using appropriate keywords and started narrowing down to find the most relevant studies based on a clear process of screening the papers. The results of this process culminated in the selection of 39 articles that were within the smart cities area and had the full text available. The outcomes of each research question were clearly discussed.

The bibliometric research revealed that the United States and China were the countries with the most studies published in the last ten years, and the most common document type of publication is articles and proceedings papers.

Later, the selected studies showed 24 existing technologies and algorithms. The results confirm a performance increase in the system with the application of optimization approaches. From these, 18 papers used offline data, and 21 papers used real-time data. But all of them are data from real cases, showing that the praxis of smart cities has resorted to smart traffic lights.

The state-of-the-art in traffic control relies on integrating information technology into traffic systems, referred to as ITS. Such systems usually require investments and work time to upgrade the existing traffic infrastructure by integrating sensors and traffic facilities. Moreover, the papers studied had no attention to cyclists and pedestrians in the initial traffic signalization planning, i.e., bicycle traffic and users' roads have not been given the same priority as motorized traffic.

References

1. Chiou, S.-W.: TRANSYT derivatives for area traffic control optimisation with network equilibrium flows. Transp. Res. Part B Methodol. **37**(3), 263–290 (2003). https://doi.org/10.1016/S0191-2615(02)00013-9

2. Ahmed, F., Hawas, Y.E.: An integrated real-time traffic signal system for transit signal priority, incident detection and congestion management. Transp. Res. Part C Emerg. Technol. **60**, 52–76 (2015). https://doi.org/10.1016/j.trc.2015.08.004

3. Gerónimo, D., López, A.M., Sappa, A.D., Graf, T.: Survey of pedestrian detection for advanced driver assistance systems. IEEE Trans. Pattern Anal. Mach. Intell. **32**(7), 1239–1258 (2010). https://doi.org/10.1109/TPAMI.2009.122

4. Peñabaena-Niebles, R., Cantillo, V., Luis Moura, J.: The positive impacts of designing transition between traffic signal plans considering social cost. Transp. Policy. **87**, 67–76 (2020). https://doi.org/10.1016/j.tranpol.2019.05.020

5. Ramey, J., Rao, P.G.: The systematic literature review as a research genre. In: 2011 IEEE International Professional Communication Conference, pp. 1–7, October 2011. https://doi.org/10.1109/IPCC.2011.6087229

6. Ruiz-Rosero, J., Ramirez-Gonzalez, G., Viveros-Delgado, J.: Software survey: ScientoPy, a scientometric tool for topics trend analysis in scientific publications. Scientometrics **121**(2), 1165–1188 (2019). https://doi.org/10.1007/s11192-019-03213-w

7. Gao, K., Zhang, Y., Sadollah, A., Su, R.: Optimizing urban traffic light scheduling problem using harmony search with ensemble of local search. Appl. Soft Comput. **48**, 359–372 (2016). https://doi.org/10.1016/j.asoc.2016.07.029

8. Simchon, L., Rabinovici, R.: Real-time implementation of green light optimal speed advisory for electric vehicles. Vehicles **2**(1), 35–54 (2020). https://doi.org/10.3390/vehicles2010003

9. Leal, S.S., de Almeida, P.E.M., Chung, E.: Active control for traffic lights in regions and corridors: an approach based on evolutionary computation. Transp. Res. Procedia **25**, 1769–1780 (2017). https://doi.org/10.1016/j.trpro.2017.05.140

10. Olayode, I.O., Tartibu, L.K., Okwu, M.O., Severino, A.: Comparative traffic flow prediction of a heuristic ANN model and a hybrid ANN-PSO model in the traffic flow modelling of vehicles at a four-way signalized road intersection. Sustainability **13**(19), 10704 (2021). https://doi.org/10.3390/su131910704

11. Goel, S., Bush, S.F., Ravindranathan, K.: Self-organization of traffic lights for minimizing vehicle delay. In: 2014 International Conference on Connected Vehicles and Expo (ICCVE), November 2014, pp. 931–936. https://doi.org/10.1109/ICCVE.2014.7297692

12. Jerry, K., Yujun, K., Kwasi, O., Enzhan, Z., Parfait, T.: NetLogo implementation of an ant colony optimisation solution to the traffic problem. IET Intell. Transp. Syst. **9**(9), 862–869 (2015). https://doi.org/10.1049/iet-its.2014.0285

13. Rida, N., Ouadoud, M., Hasbi, A.: Ant colony optimization for real time traffic lights control on a single intersection. Int. J. Interact. Mob. Technol. **14**(02), 196 (2020). https://doi.org/10.3991/ijim.v14i02.10332

14. Gao, K., Zhang, Y., Zhang, Y., Su, R., Suganthan, P.N.: Meta-Heuristics for Bi-Objective Urban Traffic Light Scheduling Problems. IEEE Trans. Intell. Transp. Syst. **20**(7), 2618–2629 (2019). https://doi.org/10.1109/TITS.2018.2868728

15. Ferrer, J., López-Ibáñez, M., Alba, E.: Reliable simulation-optimization of traffic lights in a real-world city. Appl. Soft Comput. **78**, 697–711 (2019). https://doi.org/10.1016/j.asoc.2019.03.016

16. Contreras, M., Gamess, E.: An algorithm based on VANET technology to count vehicles stopped at a traffic light. Int. J. Intell. Transp. Syst. Res. **18**(1), 122–139 (2019). https://doi.org/10.1007/s13177-019-00184-3

17. Gao, K., Zhang, Y., Sadollah, A., Lentzakis, A., Su, R.: Jaya, harmony search and water cycle algorithms for solving large-scale real-life urban traffic light scheduling problem. Swarm Evol. Comput. **37**, 58–72 (2017). https://doi.org/10.1016/j.swevo.2017.05.002

18. Tajalli, M., Mehrabipour, M., Hajbabaie, A.: Network-level coordinated speed optimization and traffic light control for connected and automated vehicles. IEEE Trans. Intell. Transp. Syst. **22**(11), 6748–6759 (2021). https://doi.org/10.1109/TITS.2020.2994468

19. Noaeen, M., et al.: Reinforcement learning in urban network traffic signal control: A systematic literature review. Expert Syst. Appl. **199**, 116830 (2022). https://doi.org/10.1016/j.eswa.2022.116830
20. Balta, M., Ozcelik, I.: Traffic signaling optimization for intelligent and green transportation in smart cities. In: 2018 3rd International Conference on Computer Science and Engineering (UBMK), September 2018, pp. 31–35. https://doi.org/10.1109/UBMK.2018.8566333
21. Wang, Z., Wang, M., Bao, W.: Development and application of dynamic timing optimization platform for big data intelligent traffic signals. In: E3S Web Conference, vol. 136, p. 01008 (2019). https://doi.org/10.1051/e3sconf/201913601008

A Genetic Algorithm for Forest Firefighting Optimization

Marina A. Matos$^{(\boxtimes)}$ ⓘ, Ana Maria A. C. Rocha ⓘ, Lino A. Costa ⓘ,
and Filipe Alvelos ⓘ

ALGORITMI Center, University of Minho, 4710-057 Braga, Portugal
mmatos@algoritmi.uminho.pt, {arocha,lac,falvelos}@dps.uminho.pt

Abstract. In recent years, a large number of fires have ravaged planet Earth. A forest fire is a natural phenomenon that destroys the forest ecosystem in a given area. There are many factors that cause forest fires, for example, weather conditions, the increase of global warming and human action. Currently, there has been a growing focus on determining the ignition sources responsible for forest fires. Optimization has been widely applied in forest firefighting problems, allowing improvements in the effectiveness and speed of firefighters' actions. The better and faster the firefighting team performs, the less damage is done. In this work, a forest firefighting resource scheduling problem is formulated in order to obtain the best ordered sequence of actions to be taken by a single firefighting resource in combating multiple ignitions. The objective is to maximize the unburned area, i.e., to minimize the burned area caused by the ignitions. A problem with 10 fire ignitions located in the district of Braga, in Portugal, was solved using a genetic algorithm. The results obtained demonstrate the usefulness and validity of this approach.

Keywords: Forest fires · Single-objective optimization · Scheduling · Genetic algorithm

1 Introduction

The incidence of forest fires has been high and worrying over the last few years. They represent high risk situations for the health of living beings and forests. Forest fires damage the ecosystem with negative consequences for wildlife and the atmosphere. Australia, between 2019 and early 2020, was devastated by several forest fires caused by record high temperatures, making it the biggest fire recorded in Australian history. About 19 million hectares were burned, killed 33 people and several animals and destroyed 3000 houses [6,11]. In 2018, more than 1.5 million hectares of forest were destroyed in the United States of America,

This work has been supported by FCT Fundação para a Ciência e Tecnologia within the R&D Units Project Scope UIDB/00319/2020 and PCIF/GRF/0141/2019: "O3F - An Optimization Framework to reduce Forest Fire" and the PhD grant reference UI/BD/150936/2021.

where it was considered the second largest area burned since 1984, also caused by the high temperatures that were felt. California faced, in 2017, the deadliest and most destructive fires recorded in its history, with a total of 88 people dead and nearly 9000 lost structures [17,28]. In 2017, Portugal faced several forest fires, which was also a record year (temperatures, dry thunderstorms and deaths). This year was a critical year, with 117 deaths, 254 injuries, major damage to structures (houses and warehouses) and loss of biodiversity in a large forest area [8,18]. There were many other forest fires that caused great damage to our planet, affecting living beings, the planet and the world economy.

To protect and avoid major catastrophes such as forest fires, studies are needed to support firefighting professionals (firefighters, civil protection, etc.). Firefighting is a very important topic in optimization, since improvements in resources and quick actions are crucial requirements for professionals working in this area. Fighting forest fires depends on the number, skills and level of preparedness of firefighting teams [14]. The better and faster the firefighting team performs, the less damage is caused. Some of the challenges also include determining how many and what resources are available to act on a given forest fire and where and when to act [2].

Several works have been presented applying optimization to forest firefighting problems. Resource management in forest fire fighting can be modeled as a scheduling problem, where the machines correspond to resources and the jobs correspond to the fires. Rachaniotis and Pappis [20–22] studied the problem of scheduling a fire fighting resource when there are several fires at the same time, using a specific model for the fire propagation rate. Their objective was to maximize the unburned area, in order to reduce the damage caused. When controlling a forest fire, there must be no delays on the part of the firefighting team, because the shorter the delay, the faster and more effective the fire suppression becomes. A scheduling problem of a single firefighting resource when there are multiple wildfires to be suppressed is considered in [23]. In this work, the objective was to minimize the total delay of arrival at the forest fire site. An illustrative example was solved by a heuristic algorithm and the branch and bound algorithm.

Wu et al. [29] developed an integer linear programming model to determine the optimal routes for forest firefighting. The problem was addressed as an emergency scheduling problem for fires with limited rescue team resources and priority disaster areas. The objective was to minimize the total travel distance of all rescue teams. Araya-Córdova and Vásquez [1] studied the scheduling optimization problem of the disaster emergency unit for the control of forest fires. The objective was to determine a schedule and working times for firefighting, in order to minimize the sum of the total damage and the total cost of waiting for the firefighting teams.

Planning the amount and type of resources needed to fight a given forest fire is also one of the most studied problems. Veiga et al. [26] proposed an integer linear programming model, which addresses the allocation of resources in different time periods during the planning period for extinguishing a fire. A second part of this work involved carrying out a simulation of the problem studied, where they showed that the solution can be obtained quickly up to a planning period of five hours. In fighting forest fires, the longer it takes to fight these fires, the longer they become and the more difficult it is to control them. In [13], a single forest fire suppression processor for simultaneous ignitions was introduced. The aim was to find the optimal sequence of actions in firefighting, minimizing the total fire damage in the burned areas when all fires are suppressed. They proposed a stochastic formulation to solve the problem concluding that the approach was effective and efficient.

This work aims to find an optimal schedule actions to be taken by a single firefighting resource in combating several ignitions. The objective function of the problem is to maximize the unburned area, i.e., to minimize the burned area caused by the ignitions. A problem with 10 forest fires ignitions located in the district of Braga, in Portugal, will be solved using the genetic algorithm.

This article is organized as follows. Section 2 describes the management and suppression of wildfires. The formulation of the scheduling optimization problem and the genetic algorithm are described in Sect. 3. In Sect. 4 the results are presented and discussed. Finally, the conclusions of this study and future work are exposed in Sect. 5.

2   Forest Fires Suppression

The increase in forest fires is currently a worrying factor for society and for planet Earth. Some of them were catastrophic, leading to deaths, pollution, damage to homes and businesses, among other negative aspects. Around 400 million hectares are burned every year worldwide, where only 100 million hectares are forest areas [3,16,25]. A forest fire is based on the combustion of combustible materials existing in forest areas and its main causes are due to human action (most common), climate change and natural agents [9,10]. The spread of a fire depends on the type of fuel, slope of the area, wind strength and direction, weather conditions, ambient humidity and temperature [25].

Forest fires suppression is based on planning what type and amount of suppression resources are needed and where they should be located to best and quickly extinguish the fire [15]. When fighting forest fires, factors, such as wind intensity and ambient temperature, influence the propagation of fire, which makes the suppression problem difficult. Therefore, studying in advance ways to fight fires before they happen is mandatory, in order to obtain safe, effective and fast solutions to reduce the spread of forest fires.

Recently, fire suppression has been the focus of many studies. Cardil et al. [7] developed an analysis of the influence of suppression objectives (fire detection, initial attack and fire control), where fire intensity, fuel type, fire ignition cause, year and homogeneous fire regime areas were also included. This study was designed using data from the Quebec Forest Fire Protection Agency from fires that occurred between 1994 and 2015. They obtained an improvement of 88% in fire suppression and also improved fire detection and control. Overall, this paper has contributed to helping wildfire protection agencies better understand their wildfire suppression systems to better adapt to future changes in the fire regime. A fire management system based on deep reinforcement learning was developed in [19]. This work aimed to simulate a trajectory where agents are trained to select the areas to be treated in order to minimize the spread of fire. An algorithm based on centralized training with decentralized execution was used and the applied approach had a good performance when compared to traditional approaches.

Resource scheduling is one of the main studies that has been addressed in the area of firefighting. A target scheduling model was presented by Zhou and Erdogan [31], where two phases were addressed, the allocation of firefighting resources and the evacuation of residents. This strategy made it possible to optimize resource preparation before the start of the fire and resource allocation decisions during the fire event. The main objectives were to minimize the total cost of operations and property losses and to reduce the number of people at risk of being evacuated. Thus, the type and amount of resources needed are critical to firefighting. An integer linear programming model was proposed by Rodriguez-Veiga et al. [26] in which a problem of resource allocation in different periods of time during the management of a given fire was studied. Zeferino et al. [30] developed a mathematical optimization model to find the optimal location solution for different aircraft that maximizes the coverage of risk areas. Its application was used in a case study in Portugal.

3   Forest Firefighting Optimization

Optimization is widely used in engineering, such as finding the optimal vehicle trajectory, obtaining the shortest route, allocating resources or services among various activities to maximize benefit, controlling wait times on production lines to minimize costs, among others [24]. In this section, the forest firefighting scheduling problem formulation and the Genetic Algorithm (GA) used to solve this problem are described. The firefighting scheduling problem has a combinatorial nature and GA are particularly adequate to explore this kind of search spaces [4].

3.1   Optimization Model

In this paper, a forest firefighting resource scheduling problem is formulated, based on the works of Rachaniotis and Pappis [20–23] who developed and solved several resource scheduling problems in the area of forest fires. Therefore, the resource management in forest firefighting is modeled as a scheduling problem, where the machines correspond to resources and the jobs correspond to the fires. Firefighting resource travel times and fire suppression times correspond to machine setup times and job processing times, respectively. The propagation of fires always depends on the moment when they start to be suppressed by the firefighting resource, that is, the sooner the firefighting starts, the faster and easier it will be to extinguish, minimizing the damage caused.

This paper aims to find an optimal schedule of actions to be taken by a single firefighting resource in combating multiple ignitions that maximizes the unburned area, i.e., minimizes the burned area caused by the ignitions. The forest firefighting scheduling problem to be solved in this paper can be mathematically formulated as follows:

$$\text{Maximize} \sum_{i=1}^{k} V_{[i]} \tag{1}$$

subject to

$$V_{[i]} = V_{[i]0} - a_{[i]} C_{[i]}^2 \geq 0 \qquad \forall i = 1, \ldots, k \tag{2}$$

where the completion time and the processing time for each ignition fire $F_{[i]}$ $(i = 1, \ldots, k)$ are given by

$$C_{[i]} = \sum_{j=1}^{i} P_{[j]}(t) + \sum_{j=1}^{i} T_{[j-1],[j]} \tag{3}$$

$$P_{[i]}(t) = \begin{cases} \alpha_{[i]} t^{\beta_{[i]}} + \gamma_{[i]} t + \delta_{[i]}, & \text{if } t \leq d_{[i]} \\ X_{[i]}, & \text{if } t > d_{[i]} \end{cases} \tag{4}$$

where k is the number of fire ignitions; (1) is the unburned area to be maximized; (2) gives the unburned area of each fire ignition i; (3) is the completion time of each fire ignition; and (4) is the processing time for each fire ignition.

The notation used in the mathematical formulation is described in Table 1.

3.2   Genetic Algorithm

Genetic Algorithms are, probably, the best known and most widely used evolutionary algorithms applied to optimization and became popular with the work of John Holland in 1975 [12]. GA is a global optimization method inspired by the evolutionary theory of the species, namely the natural selection, the survival of the fittest and the inheritance of characteristics from parents to offspring by reproduction. The main components of a GA are the population of chromosomes, the fitness function, the selection, the crossover and the mutation (Fig. 1). GA work with a population of chromosomes that represent potential solutions of the optimization problem. The initial population is generated at random. Different chromosome representations can be used according to the search space of the optimization problem. The fitness function measures the quality of chromosomes and it is related with the objective function. Each generation, the fittest chromosomes in population are more likely to be selected to generate offspring by means of the crossover and the mutation genetic operators. This cycle is repeated until a given stopping criterion is satisfied.

The forest firefighting scheduling problem is a combinatorial problem. GA are particularly adequate to explore this kind of search spaces since different

Table 1. Notation used in the forest firefighting scheduling problem formulation.

Variable	Description
$F_{[i]}$	Fire ignition ($i = 1, 2, \ldots, k$)
$C_{[i]}$	Completion time of $F_{[i]}$ suppression effort
$d_{[i]}$	Containment escape time limit
$V_{[i]}$	Remaining area value (unburned) in fire $F_{[i]}$
$V_{[i]0}$	Value of the area at the time when the overall containment effort begins in fire $F_{[i]}$
$a_{[i]}$	Deterioration rate in each fire $F_{[i]}$
$P_{[i]}(t)$	Processing time elapsed from ignition of fire $F_{[i]}$
t	Time elapsed since ignition of fire $F_{[i]}$
$T_{[i-1],[i]}$	Time required for the processor to travel from suppressed fire $F_{[i]}$ to fire $F_{[i+1]}$
$\alpha_{[i]}$, $\beta_{[i]}$, $\gamma_{[i]}$, $\delta_{[i]}$	Constant parameters depending on the area that $F_{[i]}$ burns (the meteorological conditions, the processor's water, the processor's water capacity and the fire's containment strip width)
$X_{[i]}$	Value that depends on the additional resources that will be dispatched to the specific area

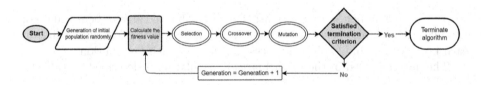

Fig. 1. Genetic algorithm process.

chromosome representations can be used, namely permutation based representations [4]. Therefore, a permutation representation was adopted to represent a
potential solution of this scheduling problem. In this representation, each chromosome is a sequence of integer values that can occur only once. Thus, each
chromosome indicates the order each fire ignition is combated by the firefighting resource. The recombination plays an important role since genetic material
of the chromosomes is combined to ensure that new promising regions of the
search space are explored. However, the genetic operators have to guarantee
that feasible permutations are maintained during the search. Several specialized
genetic operators have been developed that fulfill this requirement such as the
order-based crossover or the inverse mutation [4,27].

In this paper, each chromosome has dimension k (number of fire ignitions) and represents the sequence order of combat of each fire ignition $F_{[i]}$,
$i = 1, 2, \ldots, k$.

4   Experimental Results

In this section, the numerical results for the scheduling of a single firefighting resource to fight ten forest fire ignitions located in the district of Braga in Portugal is presented.

4.1   Implementation Details

The optimization problem and the GA were coded in *Python* language using the *pymoo*: Multi-objective Optimization in Python framework [5]. After some preliminary experiments, the population size was set to 20 and a limit of 200 was imposed in the maximum number of generations. Permutation representation was used to represent solutions of the firefighting problem. Tournament selection, order-based crossover and inverse mutation was used to generate new chromosomes. The default values for the remaining parameters of the GA were used. Since GA is stochastic algorithm, 30 independent runs were performed in order to statistically analyze its performance.

In this forest firefighting scheduling problem, there is a single firefighting resource and 10 fire ignitions located in the district of Braga in Portugal. These fire locations are large forest areas, prone to major fires, chosen using *Google Maps* software. Figure 2 shows the location of Braga fire station (in red) which is where the firefighting resource departs, and the 10 fire ignitions (in light blue).

Fig. 2. Location of the Braga fire station and the 10 fire ignitions.

Table 2 presents the data concerning this problem. In this table, $\alpha_{[i]}$, $\beta_{[i]}$, $\gamma_{[i]}$, $\delta_{[i]}$, $d_{[i]}$ and $X_{[i]}$ are the parameters related to the computation of processing times of each fire $F_{[i]}$. The parameter $a_{[i]}$ is the deterioration rate of fire $F_{[i]}$. The initial area, in hectares, where each coordinate is located and the travel time, in hours, that the firefighting resource takes from the fire station until each ignition are given by $V_{[i]0}$ and $T_{[0],[i]}$, respectively.

Table 2. Data for the forest firefighting scheduling problem in Braga.

$F_{[i]}$	$\alpha_{[i]}$	$\beta_{[i]}$	$\gamma_{[i]}$	$\delta_{[i]}$	$X_{[i]}$	$d_{[i]}$ (h)	$a_{[i]}$	$V_{[i]0}$ (ha)	$T_{[0],[i]}$ (h)
1	0.46	3.20	0.22	0.005	0.20	0.10	0.10	201.25	1.20
2	0.39	3.00	0.24	0.006	1.80	0.80	0.10	791.47	0.98
3	1.51	2.70	0.28	0.008	1.40	0.70	0.10	320.42	0.32
4	1.50	2.50	0.28	0.009	1.00	0.50	0.10	740.19	0.22
5	0.51	3.52	0.24	0.006	1.40	0.70	0.10	8479.81	1.70
6	0.43	3.30	0.26	0.007	0.20	0.10	0.10	354.14	0.25
7	1.66	2.97	0.31	0.009	0.10	0.05	0.10	499.96	0.35
8	1.65	2.75	0.31	0.010	0.60	0.30	0.10	809.45	0.42
9	0.56	3.87	0.27	0.006	0.80	0.40	0.10	3148.38	0.62
10	0.39	2.97	0.24	0.006	1.20	0.60	0.10	779.71	0.48

The travel time matrix of the firefighting resource between each forest fire ignition, in hours, is given by $T_{[i],[j]}$.

$$T_{[i],[j]} = \begin{pmatrix} 0.00 & 0.60 & 1.27 & 1.17 & 1.22 & 1.35 & 1.45 & 1.60 & 1.95 & 1.58 \\ 0.60 & 0.00 & 1.12 & 1.03 & 1.57 & 1.15 & 1.05 & 1.12 & 1.32 & 1.20 \\ 1.27 & 1.12 & 0.00 & 0.18 & 1.77 & 0.40 & 0.53 & 0.73 & 1.07 & 0.60 \\ 1.17 & 1.03 & 0.18 & 0.00 & 1.67 & 0.30 & 0.40 & 0.50 & 0.87 & 0.53 \\ 1.22 & 1.57 & 1.77 & 1.67 & 0.00 & 1.68 & 1.95 & 2.08 & 2.47 & 2.10 \\ 1.35 & 1.15 & 0.40 & 0.30 & 1.68 & 0.00 & 0.27 & 0.50 & 0.85 & 0.43 \\ 1.45 & 1.05 & 0.53 & 0.40 & 1.95 & 0.27 & 0.00 & 0.50 & 0.85 & 0.30 \\ 1.60 & 1.12 & 0.73 & 0.50 & 2.08 & 0.50 & 0.50 & 0.00 & 0.60 & 0.67 \\ 1.95 & 1.32 & 1.07 & 0.87 & 2.47 & 0.85 & 0.85 & 0.60 & 0.00 & 0.93 \\ 1.58 & 1.20 & 0.60 & 0.53 & 2.10 & 0.43 & 0.30 & 0.67 & 0.93 & 0.00 \end{pmatrix}$$

4.2   Results and Discussion

In this section, the results obtained are presented and discussed. Table 3 shows the results of the solutions obtained over the 30 runs. Each table row shows a solution found in a given number of runs (second column). Thus, the first column shows the optimal sequence order to combat the fire ignitions found and the last two columns present the value of the unburned and burned area for that solution, respectively. Among the 30 runs, GA got five different solutions, being one of them found 23 times. The average value of unburned area was

Table 3. Solutions found among the 30 runs.

Solution	No. Runs	Unburned Area (ha)	Burned Area (ha)
$F_5 \to F_9 \to F_1 \to F_{10} \to F_2 \to F_7 \to F_8 \to F_3 \to F_6 \to F_4$	1	16029.383	95.397
$F_1 \to F_9 \to F_5 \to F_{10} \to F_2 \to F_8 \to F_3 \to F_7 \to F_4 \to F_6$	1	16028.506	96.274
$F_5 \to F_9 \to F_1 \to F_{10} \to F_2 \to F_3 \to F_8 \to F_7 \to F_4 \to F_6$	1	16028.414	96.366
$F_5 \to F_9 \to F_1 \to F_{10} \to F_2 \to F_8 \to F_3 \to F_7 \to F_4 \to F_6$	4	16028.080	96.700
$F_5 \to F_9 \to F_1 \to F_{10} \to F_2 \to F_6 \to F_8 \to F_3 \to F_7 \to F_4$	23	16027.333	97.447

16027.576 ha and the burned area was 97.204 ha, over the 30 runs. In each run, 4020 function evaluations were performed by the algorithm.

The best solution found corresponds to the following order of action the firefighting resource: $FS \to F_5 \to F_9 \to F_1 \to F_{10} \to F_2 \to F_7 \to F_8 \to F_3 \to F_6 \to F_4$ where FS denotes the Braga fire station, which is the depart location and F_1 to F_{10} are the fire ignitions. This solution corresponds to a maximum value of unburned area of 16029.383 ha. Figure 3 depicts this solution, where the red circles denote the fire ignitions and the blue lines represent the path that the forest firefighting resource takes to suppress all fires. In the x and y axes the geographic coordinates of each fire ignition are shown.

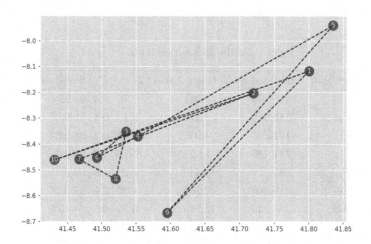

Fig. 3. Optimal sequence.

In Table 4, detailed data for this solution is given. In this table, for each ignition, the unburned and burned areas, in hectares, and the travel and processing times, in hours, are given. The travel time (T) is the time required for the firefighting resource to move from one fire ignition to the next. The processing time

(P) is the time needed to extinguish the fire ignition. The last column of Table 4 shows the percentage of saved area for each ignition.

In this solution, the firefighting resource departs from the Braga fire station (FS) to the first fire ignition in the sequence (F_5). The travel time was 1.7 h, the extinguishing time was 1.4 h and 0.961 ha burned. Then, the resource travels to F_9, taking 2.47 h to arrive, 2.470 ha of area was burned and fire was suppressed after 0.8 h. This was followed by the extinguishing of the F_1 fire, the resource having traveled 1.95 h to its position, it took 1.2 h to suppress it and 3.994 ha burned. Subsequently, the resource took 1.58 h to arrive to F_{10}, took 1.2 h to extinguish it, and 7.921 ha were damaged. Then the fighting resource traveled to the F_2, extinguishing it in 1.8 h and losing 11.449 ha. This was followed by F_7, where the resource needed 1.05 h to get to this one and 0.1 h to extinguish it. The ignition identified as F_8 was the next fire to be suppressed, where 12.210 ha were burned. In the fire F_3, 304.594 ha of forest area were saved out of a total of 320.42 ha. Subsequently, the fire F_6 was extinguished, taking 0.2 h to extinguish and a trip to it of 0.4 h. Finally, the resource moved to the last ignition of the sequence obtained, F_4, having burned 16.589 ha, the travel time from the fire before this one was 0.30 h and it took 1.0 hour to fight it.

At the end of the action to combat forest fires, 95.397 ha of forest were destroyed, in a total of 16124.780 ha, that is, 16029.383 ha of forest area were saved, representing a total of 99.41% of area not destroyed by the 10 ignitions. The total travel time for the firefighting resource was 11.88 h and a total of 8.7 h were required to suppress all fire ignitions. Thus, a total of 20.58 h were required to combat the 10 ignitions, plus 0.22 h to return to Braga fire station.

Table 4. Solution data.

Ignition	Unburned area	Burned area	T	P	Saved area
F_5	8478.849	0.961	1.70	1.4	99.99%
F_9	3145.910	2.470	2.47	0.8	99.92%
F_1	197.256	3.994	1.95	0.2	98.02%
F_{10}	771.789	7.921	1.58	1.2	98.98%
F_2	780.021	11.449	1.20	1.8	98.55%
F_7	489.860	10.100	1.05	0.1	97.98%
F_8	797.240	12.210	0.50	0.6	98.49%
F_3	304.594	15.826	0.73	1.4	95.06%
F_6	340.263	13.877	0.40	0.2	96.08%
F_4	723.601	16.589	0.30	1.0	97.76%
Total	16029.383	95.397	11.88	8.7	99.41%

5   Conclusions and Future Work

The occurrence of forest fires has increased in recent years due to natural factors such as the global warming or other factors caused by society. Forest fires endanger the health of all living beings, bringing negative impacts to planet Earth. Thus, optimizing firefighting actions is a very important topic and depends on the number, skills and level of preparation of firefighting teams.

In this work, a forest firefighting scheduling problem with 10 fire ignitions in the district of Braga in Portugal was studied. The aim was to find the best ordered sequence of actions of one firefighting resource. A genetic algorithm was used to seek the optimal solution of this problem. Thirty independent runs were performed. Five different solutions were obtained in the different runs. These solutions share similarities in terms of the order of suppression of fire ignitions and have close values of unburned area. The best solution found was $FS{\rightarrow}F_5{\rightarrow}F_9{\rightarrow}F_1{\rightarrow}F_{10}{\rightarrow}F_2{\rightarrow}F_7{\rightarrow}F_8{\rightarrow}F_3{\rightarrow}F_6{\rightarrow}F_4$, with an objective function value of 16029.383 ha. These results show the usefulness and validity of this approach. Moreover, this approach can be applied to large instances of the problem providing good solutions.

In the future, we intend to formulate the problem with more than one firefighting resource; to study the effect of other permutation based genetic operators on the performance of the algorithm; to formulate and solve this problem as a multi-objective problem by introducing other objectives such as tardiness; to apply and compare the performance of other optimization algorithms.

References

1. Araya-Córdova, P., Vásquez, Ó.C.: The disaster emergency unit scheduling problem to control wildfires. Int. J. Prod. Econ. **200**, 311–317 (2018)
2. Belval, E.J., Wei, Y., Bevers, M.: Modeling ground firefighting resource activities to manage risk given uncertain weather. Forests **10**(12), 1077 (2019)
3. Bento-Gonçalves, A., Vieira, A.: Wildfires in the wildland-urban interface: key concepts and evaluation methodologies. Sci. Total Environ. **707**, 135592 (2020)
4. Bierwirth, C., Mattfeld, D.C., Kopfer, H.: On permutation representations for scheduling problems. In: Voigt, H.-M., Ebeling, W., Rechenberg, I., Schwefel, H.-P. (eds.) PPSN 1996. LNCS, vol. 1141, pp. 310–318. Springer, Heidelberg (1996). https://doi.org/10.1007/3-540-61723-X_995
5. Blank, J., Deb, K.: pymoo: multi-objective optimization in python. IEEE Access **8**, 89497–89509 (2020)
6. Boer, M.M., de Dios, V.R., Bradstock, R.A.: Unprecedented burn area of Australian mega forest fires. Nat. Clim. Chang. **10**(3), 171–172 (2020)
7. Cardil, A., Lorente, M., Boucher, D., Boucher, J., Gauthier, S.: Factors influencing fire suppression success in the province of quebec (Canada). Can. J. For. Res. **49**(5), 531–542 (2019)
8. Carmo, I.I.V.: O papel dos Instrumentos de Gestão Territorial na prevenção e mitigação dos incêndios florestais: o caso do incêndio de Pedrogão Grande (2017) (in Portuguese). Ph.D. thesis (2021)

9. de Castro, C.F., Serra, G., Parola, J., Reis, J., Lourenço, L., Correia, S.: Combate a incêndios florestais. Escola Nacional de Bombeiros 13 (2003)
10. Costa, A.A.M., et al.: Participação pública e gestão florestal na serra de monte-muro: entre a perceção e a realidade. Sodivir-Edições do Norte Lda, Vila Real (2013)
11. Filkov, A.I., Ngo, T., Matthews, S., Telfer, S., Penman, T.D.: Impact of australia's catastrophic 2019/20 bushfire season on communities and environment. retrospective analysis and current trends. J. Safe. Sci. Resilience 1(1), 44–56 (2020)
12. Holland, J.H.: Adaptation in Natural and Artificial Systems. MIT Press, Cambridge (1975)
13. Kali, A.: Stochastic scheduling of single forest firefighting processor. Can. J. For. Res. 46(3), 370–375 (2016)
14. Khakzad, N.: Optimal firefighting to prevent domino effects. In: Dynamic Risk Assessment and Management of Domino Effects and Cascading Events in the Process Industry, pp. 319–339. Elsevier (2021)
15. Martell, D.L.: A review of operational research studies in forest fire management. Can. J. For. Res. 12(2), 119–140 (1982)
16. Martell, D.L.: Forest fire management. In: Handbook of operations research in natural resources, pp. 489–509. Springer (2007). https://doi.org/10.1007/978-0-387-71815-6_26
17. Nauslar, N.J., Abatzoglou, J.T., Marsh, P.T.: The 2017 north bay and southern california fires: a case study. Fire 1(1), 18 (2018)
18. Nolasco, C.: Terra queimada: portfólio do incêndio de pedrogão grande, castanheira de pera e figueiró dos vinhos. In: Observatório do Risco, pp. 1–23 (2017)
19. Pais, C.: Vulcano: operational fire suppression management using deep reinforcement learning. In: Proceedings of the 19th International Conference on Autonomous Agents and MultiAgent Systems, pp. 1960–1962 (2020)
20. Pappis, C.P., Rachaniotis, N.P.: Scheduling a single fire fighting resource with deteriorating fire suppression times and set-up times. Oper. Res. Int. Journal 10(1), 27–42 (2010)
21. Pappis, C.P., Rachaniotis, N.P.: Scheduling in a multi-processor environment with deteriorating job processing times and decreasing values: the case of forest fires. J. Heuristics 16(4), 617–632 (2010)
22. Rachaniotis, N.P., Pappis, C.P.: Scheduling fire fighting tasks using the concept of deteriorating jobs. Can. J. For. Res. 36(3), 652–658 (2006)
23. Rachaniotis, N.P., Pappis, C.P.: Minimizing the total weighted tardiness in wildfire suppression. Oper. Res. Int. Journal 11(1), 113–120 (2011)
24. Rao, S.S.: Engineering optimization: theory and practice. John Wiley & Sons (2019)
25. Robinne, F.N., Secretariat, F.: Impacts of disasters on forests, in particular forest fires (2020)
26. Rodríguez-Veiga, J., Ginzo-Villamayor, M.J., Casas-Méndez, B.: An integer linear programming model to select and temporally allocate resources for fighting forest fires. Forests 9(10), 583 (2018)
27. Syswerda, G.: Scheduling optimization using genetic algorithms. Handbook of genetic algorithms (1991)
28. Vyklyuk, Y., et al.: Connection of solar activities and forest fires in 2018: Events in the Usa (California), Portugal and Greece. Sustainability 12(24), 10261 (2020)
29. Wu, P., Cheng, J., Feng, C.: Resource-constrained emergency scheduling for forest fires with priority areas: an efficient integer-programming approach. IEEJ Trans. Electr. Electron. Eng. 14(2), 261–270 (2019)

30. Zeferino, J.A.: Optimizing the location of aerial resources to combat wildfires: a case study of Portugal. Nat. Hazards **100**(3), 1195–1213 (2020)
31. Zhou, S., Erdogan, A.: A spatial optimization model for resource allocation for wildfire suppression and resident evacuation. Comput. Ind. Eng. **138**, 106101 (2019)

On Tuning the Particle Swarm Optimization for Solving the Traffic Light Problem

Gonçalo O. Silva$^{(\boxtimes)}$ ⓘ, Ana Maria A. C. Rocha ⓘ, Gabriela R. Witeck ⓘ,
António Silva ⓘ, Dalila Durães ⓘ, and José Machado ⓘ

Centro ALGORITMI, University of Minho, Gualtar, 4710-057 Braga, Portugal
g.oliveirasilva96@gmail.com, gabiwiteck@gmail.com, arocha@dps.uminho.pt,
{asilva,dalila.duraes}@algoritmi.uminho.pt, jmac@di.uminho.pt

Abstract. In everyday routines, there are multiple situations of high traffic congestion, especially in large cities. Traffic light timed regulated intersections are one of the solutions used to improve traffic flow without the need for large-scale and costly infrastructure changes. A specific situation where traffic lights are used is on single-lane roads, often found on roads under maintenance, narrow roads or bridges where it is impossible to have two lanes. In this paper, a simulation-optimization strategy is tested for this scenario. A Particle Swarm Optimization algorithm is used to find the optimal solution to the traffic light timing problem in order to reduce the waiting times for crossing the lane in a simulated vehicle system. To assess vehicle waiting times, a network is implemented using the Simulation of Urban MObility software. The performance of the PSO is analyzed by testing different parameters of the algorithm in solving the optimization problem. The results of the traffic light time optimization show that the proposed methodology is able to obtain a decrease of almost 26% in the average waiting times.

Keywords: Traffic lights problem · Particle swarm optimization · Simulation of urban mobility

1 Introduction

The current increase of world population and the fast economic growth causes an increase in mobility needs, in particular within big cities where a large number of different vehicle types are present [1]. The existing infrastructure was designed with a distant perception of reality and does not support the amount of vehicles that need to transit within these cities. This situation generates traffic jams with

This work has been supported by FCT-Fundação para a Ciência e Tecnologia within the R&D Units Project Scope: UIDB/00319/2020 and the project "Integrated and Innovative Solutions for the well-being of people in complex urban centers" within the Project Scope NORTE-01-0145-FEDER-000086.

O. Gervasi et al. (Eds.): ICCSA 2022 Workshops, LNCS 13378, pp. 68–80, 2022.
https://doi.org/10.1007/978-3-031-10562-3_6

a negative impact on society, mainly on the rise of fuel consumption, together with the increase in greenhouse gas emissions and major delays in transport systems [15]. One of the ways to regulate traffic and improve circulation without the need for infrastructures changes is the optimization of traffic light cycle times. A peculiar situation where traffic lights are deployed is on single-lane roads, often found on road under maintenance, narrow roads or small bridges.

The simulation of Urban MObility software (SUMO) is a microscopic traffic flow simulation platform that includes network and demand modeling components [11]. Since its release as an open source traffic simulation package in 2002, SUMO has been supporting the traffic simulation community with a set of traffic modeling utilities on a variety of optimization problems [1,2,7,14].

In this paper the particle swarm optimization (PSO) algorithm was used to optimize the waiting time in the one-lane-two-ways traffic light problem.

The optimization process involves an interaction between the PSO algorithm and the SUMO simulator, where the traffic light cycle times combined with simulated traffic data are used to evaluate the waiting times. Four experiments are conducted to explore how different parameters affect the behaviour of the PSO algorithm. The results obtained in the experiment with the best performance will be compared to a real scenario in order to assess if there is an improvement in the vehicle waiting times in the one-lane-two-ways traffic light system.

The remaining of this paper is organized as follows. In Sect. 2, a literature review is carried out about optimization techniques and simulators used to solve the traffic light problem. Section 3 presents the traffic light optimization problem and the methodology based in the interaction of PSO and SUMO simulator is described in Sect. 4. In Sect. 5, the implementation details and results are presented and discussed. Finally, in Sect. 6, a final review and insight on future work are made.

2   Related Work

The global optimization of traffic light problems has been applied in case studies of smart cities when the real-time control of traffic lights is not possible or expensive. It requires the optimization and simulation of a traffic scenario that is estimated after collecting data from street level sensors. Among the approaches used, metaheuristics such as particle swarm optimization stand out. PSO is a stochastic global optimization technique that is simple to implement and has proven to be effective in several applications [3,6,9].

In the literature, simulation-optimization models have been used in several traffic light optimization problems. The PSO algorithm together with the VIS-SIM (Verkehr In Städten-SIMulationsmodell) micro simulation software [8] was applied to a real data set of traffic flow in a roundabout with 28 traffic signals. The study achieved a reduction of 55,9% in the average delay time per vehicle and an increase on the number of vehicles transiting the roundabout per unit of time of 9,3%. The SUMO simulator and PSO were used to program the traffic light cycle time in [14]. The results obtained with the simulation-based traffic light cycle optimization showed significant improvements in terms of the number of vehicles completing the simulation and the average travel time required for vehicles to reach their destination. In [1], the genetic algorithm and the particle swarm algorithm were applied in a case study of a road network containing 13 traffic lights in order to produce the minimum total travel time. The SUMO software was used to simulate the road network located in the city center.

Recently, in 2020, [2] introduced the social learning particle swarm optimization (SL-PSO) for the real-time traffic light problem in order to mitigate the falling success rate of the classical PSO in high dimensional optimization problems. Based on real traffic data, an intersection was modeled in SUMO and the SL-PSO showed higher computational efficiency and convergence speed than PSO.

3   The Traffic Light Problem

Along the traffic networks on cities, there are different situations where traffic lights are used. The one-lane-two-ways problem under study in this paper can occur mainly in two situations. On one case, the traffic lights are used at a fixed location and can be found on narrow roads or bridges. In this case, the average number of vehicles for each time of the day can be measured and a time fixed offline optimization can be made to reduce waiting times. On the other, temporary traffic lights are used to regulate traffic on spontaneous events such as road maintenance. In these events, the location of traffic lights changes frequently and this unpredictability requires a more dynamic reading of the demand and dimensions of the network.

This work aims to study the PSO parameters that achieve better and faster results in the first situation described above. Based on a small bridge with only one lane in the surroundings of the city of Braga, some network characteristics were measured, mainly the distance between the two traffic lights and their cycle times. Vehicle demand values where not measured but will be staged and simulated by SUMO.

Thus, in this work, the one-lane-two-ways problem will be optimized by the PSO, to determine the green light times of each of the two traffic lights in order to minimize the average waiting time of the vehicles. The average waiting time (W_{time}), in seconds, of all vehicles that go through the system is given in (1). The waiting time is the amount of time that a vehicle is stopped due to involuntary factors like traffic lights and other vehicles. The decision variable vector y refers to the duration time of the green light phase, in seconds, for each traffic light in the system.

The mathematical formulation of the traffic light optimization problem is given by

$$\text{minimize} \quad W_{time} = \frac{\sum_{i=1}^{N} Wt_i(y)}{N}$$

$$\text{subject to} \quad y \in [20, 120]^2$$

(1)

where W_{time} is the average waiting time value for all the vehicles in the system, $Wt_i(y)$ is the waiting time for each vehicle i considering the green light phase time vector y and N is the total number of vehicles.

4   Methodology

In this section, the methodology is described, presenting the PSO algorithm, the SUMO software as well as the optimization strategy used in this paper.

4.1   Particle Swarm Optimization

Kennedy and Eberhart in 1995 [10], developed the PSO algorithm, a population-based metaheuristic inspired by the movement of birds within flocks, where the population, known as swarm, is composed by particles. The movements of the particles take into consideration their best-known position as well as the global best-known position among the entire swarm. When improved positions are found then they guide the movements of the swarm. The process is repeated until a stopping condition is met.

The candidate solutions of the problem are the positions of each particle, x, that are updated every iteration by its velocity v, as shown in (2).

$$x_{i+1} = x_i + v_{i+1} \qquad (2)$$

The velocity of the particle has three components: the inertia, based on the velocity v_i, the cognitive component, based on the best position that the particle itself found (p), and the social component, based on the best position found by the swarm (g). The cognitive and social components are also multiplied by a random number that is uniformly distributed in the interval $[0, 1]$ ($r_1, r_2 \in U[0, 1]$), respectively. In 1998 Shi and Eberhart [16] introduced a way to influence the inertia, by multiplying its value with the inertia weight, w, balancing the behaviour of the particle between global and local search. Thus, the update of the velocity is computed by

$$v_{i+1} = w * v_i + c_1 * r_1 * (p - x_i) + c_2 * r_2 * (g - x_i) \qquad (3)$$

In order to further explore the effect of the inertia weight, Shi and Eberhart [17] used a decreasing linear inertia weight over the iterations to improve PSO performance given by

$$w_{iter} = \frac{iter_{max} - iter}{iter_{max}} (w_{max} - w_{min}) + w_{min} \qquad (4)$$

where $w_{max} = 0{,}9$ and $w_{min} = 0{,}4$.

Later, a constriction factor was introduced into PSO to reduce the overall velocity values in order to ensure convergence of the algorithm [4,5]. Thus, the new update of the velocity of the particle is given by

$$v_{i+1} = K * [w * v_i + c_1 * r_1 * (p - x_i) + c_2 * r_2 * (g - x_i)] \tag{5}$$

where the constriction factor is calculated as

$$K = \frac{2}{|2 - \varphi - \sqrt{\varphi^2 - 4 * \varphi}|}, \text{ where } \varphi = c_1 + c_2, \varphi > 4 \tag{6}$$

The pseudocode of PSO is presented in Algorithm 1.

Algorithm 1. Pseudocode of PSO

1: Inicialize swarm
2: **while** stopping condition not satisfied **do**
3:      **for** each particle **do**
4:          Evaluate objective function
5:          Update best position
6:      **end for**
7:      Update global best position
8:      **for** each particle **do**
9:          Update velocity (using (3) or (5))
10:         Update position (using (2))
11:     **end for**
12: **end while**

4.2 Simulation of Urban MObility

Eclipse SUMO [12] is an open source, purely microscopic and multi-modal traffic simulator that allows to simulate a variety of traffic management topics. The simulations are deterministic where each vehicle is modelled with its own characteristics and route throughout the network. Like any other simulation software, it allows the assessment of infrastructure and policy changes before implementation.

SUMO makes use of input files in order to assembly the simulation model, in particular, a configuration file (`.sumocfg.xml`), a network file (`net.xml`) and a route file (`.rou.xml`). The configuration file is used to load together the network and route descriptions from the input files, as well as, to detail processing decisions and to select the necessary outputs. The network file describes the traffic infrastructure like road and intersections where the vehicles will run during the simulation. A SUMO network is a graph where the edges are the streets, with the position, shape and speed limit of every lane, and the nodes are the intersections. In case of the intersections having traffic lights, their logic is also described in the file. The route file describes all vehicle characteristics and all the different possible routes. A route is a set of edges and nodes. In order to create the network and route files, SUMO provides a graphical network editor named *netedit*, which allows to create different networks from scratch.

4.3 Simulation-Optimization Strategy

Similarly to [7], the optimization strategy was defined as a two-step routine: the optimization algorithm and the simulation process. In the first step, the PSO algorithm is applied to find the optimal, or near-optimal, traffic light configuration in order to minimize the average waiting time. In the second step, the SUMO software is employed to evaluate the traffic light configurations created by the PSO, returning the waiting time of each vehicle. Figure 1 shows the simulation and optimization process. Every time the PSO finds a new configuration for the traffic lights (i.e., green light phase times), this configuration is sent as input to the SUMO simulator for testing. In a similar way, after every simulation run, the output results (i.e., the waiting time for each vehicle in the system) are sent back for further optimization.

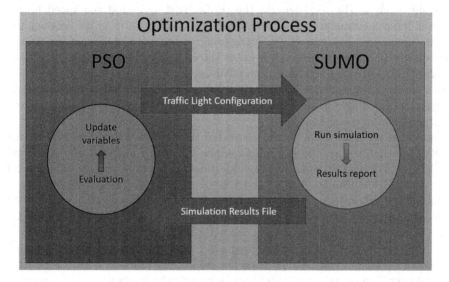

Fig. 1. Interaction between the optimization algorithm and the simulator. (Color figure online)

5 Simulation-Optimization Results

First, the technical details of SUMO configuration and the PSO parameters are analyzed for the proposed simulation and optimization methodology. Then, the obtained results from their application are presented and discussed. A *python* code was developed to program the PSO algorithm, to call the SUMO simulator and also ensure correct communication between the two parts. A PC running Windows 10 operating system equipped with AMD Ryzen 7 4800H CPU @ 2.90 GHz, 16 GB RAM was used.

5.1 SUMO Input Configuration

As previously mentioned, this paper aims to study the one-lane-two-ways road with alternating circulation. Naturally, the first step was to represent the network in the SUMO tool, *netedit*. With the real world scenario in mind, the problem network was built as shown in Fig. 2.

Fig. 2. Network representation. (Color figure online)

During the design process, several characteristics of the road were taken into account, such as the road length, the traffic light logic and the speed limits. Two different maximum speed limits were set, 30 km/h for the road between the traffic lights and 50 km/h for the remaining of the road. As usual, an all-red mandatory phase was set to ensure that the lane was empty before any of the traffic lights turned green to avoid collisions. Based on the time necessary to drive through the distance between the two traffic lights (200 m), plus a safety margin, a time of 30 s was set for this phase. All this network settings were saved in the SUMO network file.

In order to create the SUMO route file the type of vehicle as well as the routes were defined as follows. Only one type of car was outlined, the passenger vehicle, with default attribute values with some minor changes to its acceleration and deceleration abilities, $3\,m/s^2$ and $6\,m/s^2$, respectively. Note that the heavy vehicles are prohibited to cross the bridge due to its dimensions and weight. In this scenario, there were only two different routes, one where the vehicles can only travel from the left to the right, *route*01, and another that is reversed from the first one, *route*02. The arrival frequency was arbitrarily defined, although the model is able to calculate it based on real average demand data. Thus, a car arrives every ten seconds for *route*01 and for *route*02 a frequency of one car every thirteen seconds was defined. Considering the total duration of the simulation of one hour of traffic (3600 s), the arrival time of each vehicle during the simulation was assigned according to a random uniform distribution. Therefore, a total of 628 vehicles belong to the simulation network being 352 to circulate on *route*01 and 276 to travel on *route*02. A simulation stopping criterion was defined for when all vehicles leave the network to make sure all waiting times are considered on the output results.

5.2 PSO Implementation Details

In all experiments, the number of particles was set to 20, as used in an extensive PSO inertia weight study [13], the algorithm stopping condition was set to 50 iterations, to limit execution times, and each experiment was executed 30 times since PSO is a stochastic algorithm. Furthermore, the parameters used were chosen after a small exercise with different values and according to previous studies [5,18].

In order to better understand PSO particles behaviour in this context, four experiments where carried out. This experiments were mainly focused on the effect that different particle velocity parameters can have on PSO convergence. Table 1 shows the parameters used in each experiment, mainly different values of inertia weight (w), cognitive parameter (c_1) and social parameter (c_2). In Experiment 1 and 2, two different fixed values of the inertia weight were used to find out the effect of a relative low and high inertia respectively. After that, in Experiment 3, a linear decreasing inertia weight was used between the two previous values [0,4; 0,9]. In the last experiment, a constriction factor of $K = 0,729$ was used to further assess if lower values of particle velocity help the algorithm to find best solutions.

Table 1. PSO parameters for the different experiments

Experiment	w	c_1	c_2
1	0,4	2	2
2	0,9	2	2
3	0,9 to 0,4	2	2
4	1	2,05	2,05

5.3 Results and Discussion

The results obtained with the four experiments are presented in Table 2. The second to fourth columns show the average values over the 30 runs for the average waiting time, the standard deviation and the average time spent for each run. The last three columns present the values obtained for the best run, in particular the best average waiting time and the green phase times of each traffic light, y_1 and y_2, respectively.

Table 2. Simulation-optimization results.

Experiment	Average values among 30 runs			Best solution		
	Avg W_{time}	Std Dev	Avg Runtime	W_{time}	y_1	y_2
1	39,294	0,471	1242,888	**38,71**	44,782	32,170
2	39,746	0,298	1230,873	39,14	45,278	31,754
3	39,376	0,341	1273,502	38,77	44,427	32,547
4	**39,228**	0,443	**1209,352**	38,73	44,951	31,997

From Table 2, the proposed strategy that obtained better results on average was Experiment 4 (in bold). However, the best solution was obtained in Experiment 1 resulting in a time of $y_1 = 44,782$ and $y_2 = 32,170$ s for the green phase times of each traffic light. All experimental results seem similar in terms of running time, although a slightly shorter runtime can be observed in Experiment 4, perhaps due to the delay of the last vehicles leaving the system being lower, ending the simulations earlier.

To better analyze the behaviour of the particles during the optimization, Fig. 3 depicts the particle positions along the iterations for the best run of each experiment. In the first three experiments, some particles are still far from the best solution found, specially in Experiment 2 where the high inertia caused a slower convergence. The effect of the social and cognitive parameters is the likely cause of some particles escaping the best solution area. In Experiment 4, the particles seem to continuously converge towards one point, the global one.

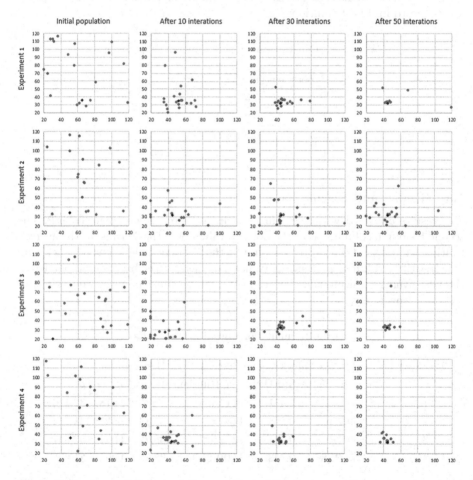

Fig. 3. Comparison of the population position of the best run, of each experiment on different iterations. The red diamond marks the global best found so far. (Color figure online)

In the following, a comparison to a real scenario simulation considering green phase times of 120 s ($y_1 = 120$ and $y_2 = 120$) on each traffic light was conducted. The SUMO simulator was run one time with the same configuration values given in Sect. 5.1 and with these green phase time values. After the simulation, a value of $W_{time} = 52{,}04$ s was obtained for the average waiting time. Thus, when comparing this value with the best solution obtained with the simulation-optimization strategy, $W_{time} = 38{,}71$ s, there is a reduction of 25,6% in the waiting time. Figure 4 presents the evolution of the number of vehicles in the network during those two types of simulations.

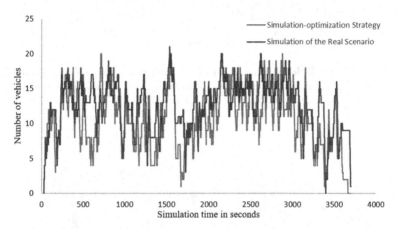

Fig. 4. Number of vehicles in the system along the simulation time of 3600 s, in the simulation-optimization strategy and in the simulation of the real scenario.

The results showed that the number of vehicles in the system was similar in both scenarios, although a reduction in the average waiting time should translate into reduced traffic congestion. However, there were vehicles that could not enter the system at the scheduled time because the queue at traffic lights reached and exceeded the network limits. Figure 5 represents the evolution of the number of vehicles waiting to enter the system for both scenarios, revealing a higher number of vehicles on the real scenario. Therefore, the simulation-optimization strategy effectively reduced traffic congestion.

The comparison between the waiting time along the simulation time of 3600 s in the simulation-optimization strategy and in the simulation of the real scenario can be seen in Fig. 6. There is a clear difference between the range of values of the two figures. In the simulation-optimization strategy, the waiting time is more evenly distributed across all vehicles, reaching a maximum of 107 s, while a large increase to 184 s is registered in the real scenario simulation.

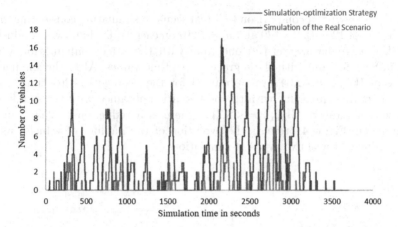

Fig. 5. Number of vehicles waiting in queue to enter the system along the simulation time of 3600 s, in the simulation-optimization strategy and in the simulation of the real scenario.

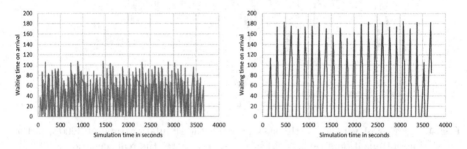

Fig. 6. The waiting time along the simulation time of 3600 s, in the simulation-optimization strategy (on the left) and in the real scenario simulation (on the right).

6   Conclusions and Future Work

Mobility needs, especially in large urban cities, are always on the rise due to the fast economic growth and to the increasing world population. Current stress on existing infrastructure causes traffic jams that are harmful to the health of all citizens with increasing greenhouse gas emissions and noise pollution. The adjustment of the traffic light cycle times to the network requirements is seen as a good solution to this problem.

This paper aimed to optimize the waiting time in the one-lane-two-ways traffic light problem. Thus, a simulation-optimization strategy, based on PSO algorithm and SUMO simulator, was used for solving the traffic light configuration problem. The objective was to minimize the average waiting time of the vehicles in a simulated system. The SUMO simulation software evaluates the traffic light configurations created by the PSO, returning the waiting time for each vehicle.

Different values to the inertia, social and cognitive parameters were used to assess the convergence of the PSO. Four experiments were conducted, and the best solution for the green phase times of each traffic light with values of $y_1 = 44,782$, $y_2 = 32,170$ and $W_{time} = 38,71$ s was obtained in Experiment 1. When comparing these values with a simulation of a real scenario, under the same conditions of the experiments, there was a reduction of the average waiting time by more than 25% when using the simulation-optimization strategy for optimizing the traffic light cycle times.

In the future, the effect of the traffic lights on the surrounding network will be taken into account. A study on the three-way intersection and its possible combinations will be carried out and the utilization of other optimization algorithms will be considered.

References

1. Abushehab, R.K., Abdalhaq, B.K., Sartawi, B.: Genetic vs. particle swarm optimization techniques for traffic light signals timing. In: 2014 6th International Conference on Computer Science and Information Technology (CSIT), pp. 27–35. IEEE (2014)
2. Celtek, S.A., Durdu, A., Alı, M.E.M.: Real-time traffic signal control with swarm optimization methods. Measurement **166**, 108206 (2020)
3. Chouikhi, N., Ammar, B., Rokbani, N., Alimi, A.M.: PSO-based analysis of echo state network parameters for time series forecasting. Appl. Soft Comput. **55**, 211–225 (2017)
4. Clerc, M.: The swarm and the queen: towards a deterministic and adaptive particle swarm optimization. In: Proceedings of the 1999 Congress on Evolutionary Computation, CEC 1999 (Cat. No. 99TH8406), vol. 3, pp. 1951–1957. IEEE (1999)
5. Eberhart, R.C., Shi, Y.: Comparing inertia weights and constriction factors in particle swarm optimization. In: Proceedings of the 2000 Congress on Evolutionary Computation, CEC 2000 (Cat. No. 00TH8512), vol. 1, pp. 84–88. IEEE (2000)
6. Elloumi, W., El Abed, H., Abraham, A., Alimi, A.M.: A comparative study of the improvement of performance using a PSO modified by ACO applied to TSP. Appl. Soft Comput. **25**, 234–241 (2014)
7. Garcia-Nieto, J., Olivera, A.C., Alba, E.: Optimal cycle program of traffic lights with particle swarm optimization. IEEE Trans. Evol. Comput. **17**(6), 823–839 (2013)
8. Gökçe, M.A., Öner, E., Işık, G.: Traffic signal optimization with particle swarm optimization for signalized roundabouts. Simulation **91**(5), 456–466 (2015)
9. Gong, Y., Zhang, J.: Real-time traffic signal control for roundabouts by using a PSO-based fuzzy controller. In: 2012 IEEE Congress on Evolutionary Computation, pp. 1–8. IEEE (2012)
10. Kennedy, J., Eberhart, R.: Particle swarm optimization. In: Proceedings of the International Conference on Neural Networks, ICNN 1995, vol. 4, pp. 1942–1948. IEEE (1995)
11. Krajzewicz, D., Erdmann, J., Behrisch, M., Bieker, L.: Recent development and applications of SUMO-simulation of urban mobility. Int. J. Adv. Syst. Meas. **5**(3&4), 128–138 (2012)

12. Lopez, P.A., et al.: Microscopic traffic simulation using sumo. In: 2018 21st International Conference on Intelligent Transportation Systems (ITSC), pp. 2575–2582. IEEE (2018)
13. Nickabadi, A., Ebadzadeh, M.M., Safabakhsh, R.: A novel particle swarm optimization algorithm with adaptive inertia weight. Appl. Soft Comput. **11**(4), 3658–3670 (2011)
14. Panovski, D., Zaharia, T.: Simulation-based vehicular traffic lights optimization. In: 2016 12th International Conference on Signal-Image Technology & Internet-Based Systems (SITIS), pp. 258–265. IEEE (2016)
15. Peñabaena-Niebles, R., Cantillo, V., Moura, J.L.: The positive impacts of designing transition between traffic signal plans considering social cost. Transp. Policy **87**, 67–76 (2020)
16. Shi, Y., Eberhart, R.: A modified particle swarm optimizer. In: 1998 IEEE International Conference on Evolutionary Computation Proceedings. IEEE World Congress on Computational Intelligence (Cat. No. 98TH8360), pp. 69–73. IEEE (1998)
17. Shi, Y., Eberhart, R.C.: Parameter selection in particle swarm optimization. In: Porto, V.W., Saravanan, N., Waagen, D., Eiben, A.E. (eds.) EP 1998. LNCS, vol. 1447, pp. 591–600. Springer, Heidelberg (1998). https://doi.org/10.1007/BFb0040810
18. Shi, Y., Eberhart, R.C.: Empirical study of particle swarm optimization. In: Proceedings of the 1999 Congress on Evolutionary Computation, CEC 1999 (Cat. No. 99TH8406), vol. 3, pp. 1945–1950. IEEE (1999)

A Reactive GRASP Algorithm
for the Multi-depot Vehicle Routing
Problem

Israel Pereira de Souza[1]([⊠]) (iD), Maria Claudia Silva Boeres[1] (iD),
Renato Elias Nunes de Moraes[1] (iD), and João Vinicius Corrêa Thompson[2] (iD)

[1] Federal University of Espírito Santo, Vitória, ES 29075-910, Brazil
israelpereira55@gmail.com
[2] Federal Center for Technological Education of Rio de Janeiro, Rio de Janeiro,
RJ 20271-204, Brazil

Abstract. The vehicle routing problem (VRP) is a well know hard to
solve problem in literature. In this paper, we describe a reactive greedy
randomized adaptive search procedures algorithm, for short, reactive
GRASP, using a variable neighborhood descent (VND) algorithm as
local search procedure to solve the multi-depot vehicle routing problem (MDVRP). This algorithm, called RGRASP+VND, combines four
distinct local search procedures and a clustering technique. The Cordeau
et al. dataset, a widely well known MDVRP benchmark, is considered
for the experimental tests. RGRASP+VND achieves better results on
most small instances and a lower average solution for all instances on
the experimental tests when compared to the earlier GRASP approaches
in the MDVRP literature.

Keywords: Reactive greedy randomized adaptive search procedures ·
Multi-depot vehicle routing problem · City logistics

1 Introduction

The VRP is a well know logistical model to minimize costs to deliver goods in
time to customers against minimal transportation costs. It has a significant impact
on logistics systems, optimizing traffic and reducing transportation trajectories,
which can directly impact reducing the final cost of the logistical processes.

VRP is an NP-hard [19] problem with several real-world applications. It was
first introduced by Dantzig and Ramser [10] in 1959 with the proposal of a mathematical programming formulation and algorithms to model the delivery of gasoline to service stations. In the following, Clarke and Wright [6] in 1964 improved
this approach, proposing an effective greedy heuristic. Afterward, the problem was
derived into a variety of versions, such as VRP with time windows (VRPTW),
multi-depot VRP (MDVRP), VRP with pickup and delivery (VRPPD), VRP with

Supported by CAPES.

backhauls (VRPB), split delivery VRP (SDVRP), periodic VRP (PVRP), among others. We focus on MDVRP.

MDVRP solving methodologies can be classified into two approaches, which are: the route-first cluster-second, where a single giant tour is built considering all customers and then partitioned into a set of feasible vehicles [4]; and the cluster-first route-second, where the customers are organized into clusters, which are associated to the depots and then, the whole route is built considering the clustered depots [7]. Examples of heuristics that follow the cluster-first route-second approach for the MDVRP are: the multi-phase modified shuffled frog leaping algorithm [20] and a tabu search heuristic with variable cluster grouping [15]. The route-first cluster-second approach is adopted by heuristics as an evolutionary algorithm for the VRP [25], a memetic algorithm for the VRPTW [18], among others. This work concerns the proposal of a heuristic for the MDVRP adopting the cluster-first route-second approach.

Recent literature on MDVRP heuristics involves: a novel two-phase method approach [1]; a biased-randomized variable neighborhood search [26]; a 2-opt guided discrete antlion optimization [3]; a harmony search [22] and an enhanced intelligent water drops algorithm [11]. A GRASP algorithm was used to solve the single truck and trailer routing problem with satellite depots (STTRPSD) [28], of which MDVRP is a special case.

In this work, we describe a Reactive GRASP algorithm for the MDVRP (RGRASP+VND) combining four distinct local search procedures, two of them presented here, and a clustering technique. For purposes of comparison, a literature review was conducted to identify the state-of-the-art algorithms for the MDVRP. For the computational experiments and results analysis, we consider as input, the well-known Cordeau et al. dataset for the MDVRP [8]. State-of-the-art algorithms are identified and compared to RGRASP+VND and the results are discussed in Sect. 5.2.

This paper is structured as follows: Sect. 2 describes the problem with its mathematical formulation. Section 3 presents MDVRP solving approaches, which are the route-first cluster-second and cluster-first route-second, and discusses two clustering techniques. Section 4 describes the RGRASP+VND algorithm for the MDVRP. The analysis of the results obtained from the computational experiments is presented in Sect. 5, which is organized with the algorithm parameters decisions (Sect. 5.1), followed by the comparison of the state-of-the-art algorithms (Sect. 5.2). Section 6 concludes this work.

2   Problem Description

The Multi-depot Vehicle Routing Problem (MDVRP) can be defined using an undirected graph $G = (V, E)$, where $V = \{v_1, v_2, \ldots, v_N, v_{N+1}, v_{N+2} \ldots, v_{N+M}\}$ is the vertex set, $D = \{D_1, D_2, \ldots, D_M\} = \{v_{N+1}, v_{N+2}, \ldots, v_{N+M}\}$ is the depot set and $E = \{(v_i, v_j) : v_i, v_j \in V, i < j\}$ is the edge set. The vertices v_1 to v_N represent the clients or customers and v_{N+1} to v_{N+M}, the depots. Each customer v_i has an associated demand q_i, $i = 1, \ldots, N$. The depots have no associated demand. Moreover, each edge $(v_i, v_j) \in E$ has a Euclidean distance d_{ij}. A fleet of up to K vehicles of capacity Q_k is located in each of the M depots. The MDVRP goal is to minimize a set of vehicle routes such that: i) each vehicle starts and ends at the same depot; ii) each customer is served once by a single vehicle; iii) the total load of a vehicle k, for $k = 1, \ldots, R$, does not exceed its maximum capacity Q_k and iv) the number of vehicles used by a depot does not exceed the maximum number of vehicles K.

Parameters:
N total number of customers.
M total number of depots.
K maximum number of vehicles for a depot.
R number of vehicles used by the M depots.
Q_k maximum capacity of a vehicle $k = 1, \ldots, R$.
T_k maximum distance allowed for a route of vehicle $k = 1, \ldots, R$.
q_i demand of a customer v_i, $i = 1, \cdots, N$.
d_{ij} Euclidean distance between vertices v_i and v_j.

Decision variables:

$$
x_{ijk} = \begin{cases} 1, & \text{if vehicle } k \text{ travels from } v_i \text{ to } v_j. \\ 0, & \text{otherwise.} \end{cases}
$$

The problem is formulated in [17] as:

Minimize:

$$
Cost = \sum_{i=1}^{N+M} \sum_{j=1}^{N+M} \sum_{k=1}^{R} d_{ij} x_{ijk} \tag{1}
$$

$$
\sum_{i=1}^{N+M} \sum_{k=1}^{R} x_{ijk} = 1, \qquad j = 1, 2, \ldots, N \tag{2}
$$

$$
\sum_{j=1}^{N+M} \sum_{k=1}^{R} x_{ijk} = 1, \qquad i = 1, 2, \ldots, N \tag{3}
$$

$$
\sum_{i=1}^{N+M} x_{ihk} - \sum_{j=1}^{N+M} x_{hjk} = 0, \qquad k = 1, 2, \ldots, R, \qquad h = 1, 2, \ldots, N+M \tag{4}
$$

$$
\sum_{i=1}^{N+M} q_i \sum_{j=1}^{N+M} x_{ijk} \leq Q_k, \qquad k = 1, 2, \ldots, R \tag{5}
$$

$$
\sum_{i=1}^{N+M} \sum_{j=1}^{N+M} d_{ij} x_{ijk} \leq T_k, \qquad k = 1, 2, \ldots, R \tag{6}
$$

$$\sum_{i=N+1}^{N+M} \sum_{j=1}^{N} x_{ijk} \leq 1, \qquad k = 1, 2, \ldots, R \tag{7}$$

$$\sum_{j=N+1}^{N+M} \sum_{i=1}^{N} x_{ijk} \leq 1, \qquad k = 1, 2, \ldots, R \tag{8}$$

$$y_i - y_j + (M+N)x_{ijk} \leq N + M - 1 \text{ for } 1 \leq i \neq j \leq N \text{ and } 1 \leq k \leq R \tag{9}$$

The objective function (1) minimizes the total distance travelled by the vehicles. Constraints (2) and (3) guarantee that each customer is served one vehicle. The route continuity is represented by constraint (4). Constraint (5) ensures that the vehicles capacity are respected and the constraint (6) ensures that the vehicle maximum distance is respected. Constraints (7) and (8) verify vehicle availability and subtour elimination is checked by constraints (9).

3   The MDVRP Solution Approach

Heuristics for solving the MDVRP can be classified into cluster-first route-second or route-first cluster-second approaches. In the first approach, the vertices are clustered considering a depot as the center of each cluster. The routes are then determined by properly sequencing the customers in each cluster [7]. In the second approach, a single giant route is built considering all vertices, and then it is split into a set of feasible routes [4].

In this paper, we consider the cluster-first route-second approach. The vertices of the input test problem being solved are clustered into M depots generating M capacitated VRP (CVRP) subproblems.

Several clustering methods can be found in the literature. A classical clustering method is K-means [21]. K-means partitions a set of points into a given number of clusters, assigning them to a center or centroid determined by the nearest distance mean. A group of different clustering methods has been described by Giosa et al. [14] to solve the MDVRP. One of them, the Urgencies clustering method, ranks all customers considering an urgency, and the customer with the most urgency is assigned first. In the same work, Giosa et al. [14] also present a parallel and simplified approach for the Urgencies clustering method. The parallel approach considers all depots on the customer urgency evaluation and the simplified approach only considers two depots for the evaluation. The customer with the highest urgency is assigned to its closest depot. We have chosen to compare the Urgencies clustering method with the parallel approach and the K-means clustering method. The Eq. (10) calculates the urgency u_c of a customer v_c, $c \in 1, 2, \ldots, N$, considering the Urgencies clustering method with parallel approach. The function distance calculates the Euclidean distance between two vertices and the depot D' is the closest depot of the customer v_c. The customer with the highest value of u is assigned to its closest depot.

$$u_c = \left(\sum_{i=1}^{M} \text{distance}(v_c, D_i) \right) - \text{distance}(v_c, D') \tag{10}$$

Clustering examples for the cluster-first route-second approach using the instance pr01, available in the Cordeau et al. dataset [8], generated with K-means and Urgencies clustering methods are plotted in Fig. 1. The clusters are highlighted in different colors. We can see, for this instance, that the clustering solutions are quite similar, differing only by 1 additional customer in the red and green clusters, when considering the K-means solution, against 2 additional customers in the blue cluster, when considering the Urgencies solution. Each cluster represents a CVRP subproblem and the union of all CVRP subproblems composes an MDVRP problem with the cluster-first route-second approach.

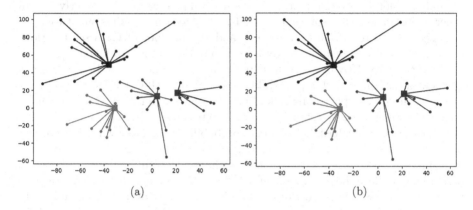

(a)   (b)

Fig. 1. (a) Plotted K-means clustering method on pr01 (b) Plotted Urgencies clustering method on pr01 (points are customers and square are depots). (Color figure online)

In this paper, we compare the K-means clustering method and the Urgencies clustering method with the parallel approach [14] applied to the cluster-first route-second approach, which is discussed in Sect. 5.1. Both clustering methods were implemented considering the depot's maximum capacity, which is the sum of the vehicle capacities in the depot. If a customer that would be assigned to a cluster that has a demand that surpasses the depot's maximum amount of demand, it is assigned instead to the next closest cluster which has a free load to support it.

4   The RGRASP+VND Algorithm

In this section, we describe the Reactive GRASP with variable neighborhood descent (VND) algorithm (RGRASP+VND), implemented to solve the MDVRP.

The Greedy Randomized Adaptive Search Procedures (GRASP) [12] is a multi-start metaheuristic with construction and local search iterations. In each GRASP iteration, a solution is built using a greedy randomized adaptive algorithm, which is then submitted to a local search procedure. The best overall solution is kept as the algorithm result. The GRASP construction procedure iteratively evaluates a set of candidate elements to be integrated into the solution, also checking feasibility. Those with the smallest incremental cost are inserted in an ordered list called the restricted candidate list (RCL), which is limited by the α percentage parameter. For instance, when $\alpha = 0.1$, the list will contain the 10% best candidates.

Thus, the next element selected to integrate the solution being constructed is randomly selected from this list. In this way, one of the best RCL components is chosen to belong to the solution, but not necessarily the top candidate. The RCL is updated and its elements reevaluated until the solution is complete. If feasibility is not achieved, repairing procedures and/or new trials are activated, in order to obtain a feasible solution. The GRASP algorithm requires only two input parameters: the RCL size constraint α and the number of GRASP iterations (*max_iterations*) and can be easily implemented for a parallel scheme [23].

The Reactive GRASP [24] considers not a fixed value for the RCL α parameter, but a random α with a probability selected for each GRASP iteration. The probabilities are computed and self adjusted along the iterations. More formally, an initial value for α is selected from the discrete set $A = \{\alpha_1, \alpha_2, \cdots, \alpha_m\}$, with m predetermined α values. Let p_i be the probability associated with the choice of α_i, for $i = 1, 2, \cdots, m$. The initial values $p_i = 1/m$, for $i = 1, 2, \cdots, m$ correspond to a uniform distribution. The original authors, Prais and Ribeiro, suggest to periodically update the p_i values for $i = 1, 2, \cdots, m$, taking into account information based on the average cost of the solutions obtained over the iterations [24]. Let *block_iterations* be the number of iterations without probabilities update, $F(S^*)$ be the cost of the best solution found so far and A_i the average of the solutions obtained in the iterations using $\alpha = \alpha_i$. After each *block_iterations* of GRASP, the p_i probabilities are updated following the absolute qualification rule, defined as $p_i = \frac{q_i}{\sum_{j=1}^{m} q_j}$, with q_i normalized values given by $q_i = \frac{F(S^*)}{A_i}^{\delta}$.

Beyond the additional parameter *block_iterations*, the Reactive GRASP has also the parameter δ, used to attenuate the p_i updated probabilities, which is commonly adopted as $\delta = 10$ in a diversity of problems in the literature [5, 16, 24].

4.1   A Reactive GRASP for the MDVRP

We have utilized the Reactive GRASP using the absolute qualification rule [24] to update the probabilities. However, this rule fails when not all predetermined values for α_i are selected along with the algorithm execution. In this case, no solutions average (A_i) and their respective q_i would be generated. To solve this problem we generate one solution for each α_i, for $i = 1, 2, \ldots, m$, at the beginning of the algorithm. Considering that the probability for each α_i to be selected starts uniformly, generating one solution for each value will not prejudice the rule. We also constructed a greedy randomized algorithm by adding a Restrict Candidate List (RCL) to the widely known Clarke-Wright savings algorithm [6].

The RGRASP+VND pseudocode is given by Algorithm 1. The algorithm receives as input the maximum number of iterations (*max_iterations*), the number of α values m, the number of iterations which the set probabilities $p_i \in P, i = 1, 2, \ldots, m$, will be updated (*block_iterations*), the δ parameter, which is used to attenuate the p_i updated probabilities, the graph $G = (V, E)$, the number of vertices N, the number of depots M, the maximum number of vehicle of each depot K and the set of vehicle capacities Q. It returns the best solution found for an MDVRP instance problem.

Algorithm 1: Reactive GRASP pseudocode

Input: *max_iterations*, m, *block_iterations*, δ, $G = (V, E)$, N, M, K, Q
Output: *BestSolution*

1 *Clusters* ← ClusterizeInstance(G, N, M, K, Q)
2 A ← ConstuctAlphaSet(m)
3 *BlockSolutions* ← generate one solution for each α_i from A
4 P ← UpdateAlphaProbabilities(*BlockSolutions*)
5 *BlockSolutions* ← ∅
6 **for** *it* ← *1 to max_iterations* **do**
7 | **if** *it % block_iterations == 0* **then**
8 | | P ← UpdateAlphaProbabilities(*BlockSolutions*)
9 | | *BlockSolutions* ← ∅
10 | α ← select a random α_i from A considering the probabilities set P
11 | *Solution* ← GreedyRandomizedSolution(*Clusters*, K, Q, α)
12 | *Solution* ← LocalSearch(*Solution*)
13 | *BestSolution* ← UpdateSolution(*Solution*, *BestSolution*)
14 | *BlockSolutions* ← *BlockSolutions* ∪ *Solution*

The RGRASP+VND algorithm starts clustering the instance problem by the function ClusterizeInstance on line 1. The set of α values A is constructed by function ConstructAlphasSet, that generates m α values uniformly distributed in (0, 0.9], on line 2. Lines 3 and4 fix the Reactive GRASP absolute qualification rule problem, by creating one solution for each $\alpha_i \in A, i = 1, \ldots, m$, on line 3 and updating the set of probabilities P on line 4. This history of solutions (*BlockSolutions*) is emptied on line 5, which will later store all solutions of a block of iterations, used to update the each probability $p_i \in P$ for $\alpha_i \in A$, $i = 1, \ldots, m$. Line 6 controls the amount of solutions which will be generated by the algorithm, defined by *max_iterations* given as input. Line 7 checks if the number of generated solutions is equal the *block_iterations*, by performing the rest of integer division (%) with the current iteration number (*it*) and the *block_iterations* number, if it succeeds, the probabilities $p_i \in P$ of selecting an $\alpha_i \in A$ are updated, for $i = 1, \ldots, m$ by line 8 and the history of solutions (*BlockSolutions*) is emptied on line 9. Line 10 selects a random $\alpha_i \in A$, considering the probability of selection of each α_i, defined by P. Line 11 constructs a *Solution* by the GreedyRandomizedSolution described by the Algorithm 2. The local search procedures described on Sect. 4.2 are applied by line 12 on the solution. Line 13 updates the best solution found. Line 14 stores the solution created on the current block of iterations history.

The GreedyRandomizedSolution pseudocode is described on Algorithm 2. The algorithm receives as input the clustered instance (*Clusters*), the maximum number of vehicles K of each depot, the vehicle capacity set Q and the RCL size constraint α and returns a feasible *Solution*.

Algorithm 2: GreedyRandomizedSolution pseudocode

Input: *Clusters*, K, Q, α
Output: *Solution*
1 *Solution* $\leftarrow \emptyset$
2 Initialize the set of candidate elements
3 Evaluate the incremental costs of the candidate elements
4 **while** *exists at least one candidate element* **do**
5   RCL $\leftarrow$ ConstructRestrictedCandidateList(*Clusters*, K, Q, α)
6     $s \leftarrow$ Select a random element from the RCL
7   *Solution* $\leftarrow$ *Solution* $\cup \{s\}$
8   Update the set of candidate elements
9   Reevaluate the incremental costs
10 **if** *Solution is not feasible* **then**
11   *Solution* $\leftarrow$ Repair(*Solution*)

It starts with an empty solution on line 1. Then the set of candidate elements are initialized on line 2 and their incremental costs are evaluated on line 3. While there is a candidate element, the RCL is constructed on line 5 with its size limited by the parameter α. One random element of RCL is chosen on line 6 and incorporated to the solution on line 7. Then the set of candidate elements are updated on line 8 and their incremental cost are reevaluated on line 9. If the generated solution is infeasible, a new solution is generated by line 11.

4.2   Local Search

The local search used by this paper considers four methods: 2-swap, 2-opt, Drop one point intra-depot, Drop one point next depot. The 2-swap and 2-opt are classical heuristics widely used on optimization problems, the next two algorithms are new local search methods presented in this paper.

The 2-swap and 2-opt [9,13] heuristics were considered in the local search. Their movements can be used by a neighborhood search and perform, respectively, two vertices (customers) swaps and two edges swaps while preserving the tour. The neighborhood of a solution s can be defined as a function that maps a solution to a set of solutions. The neighborhood search algorithm takes as input an initial solution s and computes the best solution s' in the neighborhood of s. Two classical heuristics to perform a neighborhood search are the first improvement and best improvement. The first improvement enumerates systematically and updates the solution when an enhancement is found and the best improvement only updates the solution with the best solution, that is the solution with the highest enhancement of the neighborhood [23]. For this paper we chose the best improvement heuristic for the 2-opt and 2-swap movements.

The 2-opt, 2-swap and Drop one point intra-depots methods are intra-depots, which are performed considering only one cluster. The method Drop one point next depot is a inter-depot method, which consider all clusters of the solution and can adjust it's clusters.

To reference the depot that attends the customer c_i, we will use the notation D_{m,c_i}, which represents the depot D_m that attends the customer c_i.

Drop One Point Intra-depot. For each cluster associated with depot D_i of the solution S, every city c_i which is clustered to depot D_i is removed and reinserted in the best viable position, that is the one that not surpass the vehicle maximum capacity, in a different route of the depot D_i which the highest enhancement is obtained. The Algorithm 3 shows the Drop one point intra-depot pseudocode.

Algorithm 3: Drop one point intra-depot

   Input: S

   Output: S

1 **foreach** *depot $D_i \in S$* **do**

2   **foreach** *route $h_\ell \in D_i$* **do**

3   **foreach** c_i on route h_ℓ **do**

4   **do**

5              $S \leftarrow$ remove and viably reinsert c_i on route h_b, $h_b \neq h_l$ and $h_b \in D_i$, such as its cost is minimum.

6   **if** *the cost of S reduces;*

The Drop one point intra-depot, on line 1, iterates for each depot D_i of S. The line 2 iterates for each route h_l of D_i. The line 3 iterates over all cities c_i on route h_l. The line 5 removes c_i from its current route and reinsert it in a different route that gives the highest enhancement to S without making an infeasible move. This movement is only performed if it improves the solution cost which is checked on line 6.

Drop One Point Next Depot. The Drop one point next depot algorithm seeks to adjust the clusters of the solution, for this reason, it's an inter-depot algorithm. For each city c_i, it's next closest depot D_{next,c_i} is calculated, differently than the one c_i is currently allocated. Then if an improvement is obtained, c_i is removed from S and viably reinserted on the depot D_{next,c_i} where the highest enhancement is obtained. The Algorithm 4 shows the Drop one point intra-depot pseudocode.

Algorithm 4: Drop one point next depot

Input: S, N, M, K, Q
Output: S
1   $max\_demands \leftarrow \sum_{k=1}^{K} Q_k$
2   **for** $m \leftarrow 1$ to M **do**
3     $cluster\_demands[m] \leftarrow$ sum of demands of all customers clustered to D_m
4   **for** $i \leftarrow 1$ to N **do**
5     $D_{current,c_i} \leftarrow$ get the depot c_i is currently associated.
6     $m \leftarrow$ index associated to depot $D_{current,c_i}$.
7   **if** $cluster\_demands[m] + q_i <= max\_demands$ **then**
8       $D_{next,c_i} \leftarrow$ get D_{next,c_i} the closest depot of c_i, where $D_{next,c_i} \neq D_{current,c_i}$.
9   **do**
10         $S \leftarrow$ remove c_i from S and viably reinsert c_i on D_{next,c_i} on the route and position where the highest enhancement is obtained.
11         $cluster\_demands[m] = cluster\_demands[m] - q_i$
12         $m \leftarrow$ index associated to depot D_{next,c_i}.
13         $cluster\_demands[m] = cluster\_demands[m] + q_i$
14   **if** *the cost of S reduces;*

The `Drop one point next depot`, on line 1, calculates the maximum amount of demand that a depot can support. The line 2 iterates over all depots of the instance and the line 3 calculates the amount of demand the cluster has, which is the sum of demands of all customers on the cluster. The line 4 iterates over all customers c_i of the instance. The line 5 get the depot ($D_{current,ci}$) which the customer c_i is associated. The line 6 get the index m of the depot $D_{current,ci}$. The line 7 checks if the depot with index m can support the customer c_i. The line 8 get the closest depot of c_i, different than the depot c_i is already allocated, called $D_{next,ci}$. The line 10 removes c_i and reinsert it on the depot $D_{next,ci}$ on the position that the highest enhancement is obtained, this movement is performed only if the cost of S is improved, which is checked on line 14. The lines 11–13 are used to update the amount of demand the clusters are holding.

5 Computational Experiments and Discussion

This section discusses the experimental results of the described algorithms. We tested the RGRASP+VND algorithm, considering as input, the Cordeau et al. (1997) benchmark dataset [8], available at the Vehicle Routing Problem Repository VRP-REP[1].

Section 5.1 presents the strategies and parameters adopted for the algorithm and Sect. 5.2 discusses the literature review and the RGRASP+VND comparison.

---

[1] *Available at:* http://www.vrp-rep.org/ *(access date: 03/22/2022).*

5.1   Conceiving the RGRASP+VND Algorithm Structure

In order to assess the RGRASP+VND algorithm, results from preliminary tests are registered, considering the combination of two different clustering strategies together with four local search procedures performed individually and combined by the VND algorithm. The goal is to identify which combination of strategies achieved better results.

The procedures implemented for the RGRASP+VND algorithm are: the clustering methods K-means and Urgencies; the Clarke-Wright savings constructive algorithm; the local search strategies 2-swap, 2-opt intra routes, drop one point next depot and drop one point intra-depot, identified in this section respectively as T_1, T_2, T_3 and T_4, and performed each with best improvement; and VND. The four local search strategies are performed in two ways: sequentially, denoted as LS(T_1, T_2, T_3, T_4) or combined by the VND, denoted as VND(T_1, T_2, T_3, T_4), both considering the sequence of execution given by the identification order T_i, $i = 1, \ldots, 4$.

Table 1. Average solution values on eight instances of the Cordeau et al. dataset [8] for MDVRP on the implemented procedures sequences.

	Cluster method	Constructive method	Local search	Average sol.
PS_1	K-means	Clarke-Wright	LS	2245.17
PS_2	K-means	Clarke-Wright	VND	2238.34
PS_3	Urgencies	Clarke-Wright	LS	2093.78
PS_4	Urgencies	Clarke-Wright	VND	**2092.80**

Four different sequences of all procedures were implemented: PS_1, PS_2, PS_3 and PS_4 (see Table 1), each performed over ten instances (p01 to p10) from the Cordeau et al. dataset [8]. Using the K-means clustering method, the Clarke-Wright savings method could not generate a feasible solution for the instances p04 and p07. However, using the Urgencies clustering method the constructive method could generate a feasible solution for all ten instances. To compare the average results, only the eight instances in which both methods could generate feasible solutions are considered. Thus, Table 1 presents the average solutions for eight instances of the Cordeau et al. dataset considering the four procedures sequences.

The algorithm has reached the better average solutions with the Urgencies clustering method against the K-means clustering method for the procedure sequences. The VND achieved better results than the sequential local search. Thus, we have selected the Urgencies clustering method and the VND local search strategy. Ultimately, the parameters used on this algorithm for the experimental tests on Cordeau et al. dataset are: i) Cluster method: Urgencies; ii) $max\_iterations = 10{,}000$; iii) $\delta = 10$; iv) $m = 100$; v) $block\_iterations = 100$.

5.2   Computational Results

In this section, the experiments with the RGRASP+VND are discussed. A literature review was conducted to identify the GRASP state in the MDVRP literature. In this study, the search string: *"grasp" AND ("mdvrp" OR "multi-depot vehicle routing problem")* was used in March of 2022 at the Scopus, IEEE Xplore, and Web of Science databases without any time restriction. Only two distinct papers were obtained and only one (GRASP/VND) [28] has presented results considering instances from the literature.

Table 2. Algorithms solution comparison for the MDVRP instances.

Instance	BKS	GRASP/VND [28]			RGRASP+VND		
		Avg.	*Best*	%Dev	*Avg.*	*Best*	%Dev
p01	576.87*	610.79	592.21	2.66	588.53	**581.10**	**0.73**
p02	473.53*	556.35	529.64	11.85	480.35	**480.35**	1.44
p03	640.65*	694.08	**648.68**	1.25	664.52	660.32	3.07
p04	999.21*	1071.49	1055.26	5.61	1056.64	**1051.10**	5.19
p05	751.26	803.93	**769.37**	**2.41**	794.68	789.89	5.14
p06	876.5*	963.10	924.68	5.50	904.38	**901.67**	**2.87**
p07	881.97*	955.76	**925.80**	4.97	940.96	931.09	5.57
p12	1318.95*	1409.02	**1326.85**	**0.60**	1363.32	1332.71	1.04
p15	2505.42	2809.46	**2553.80**	**1.93**	2699.66	2649.15	5.74
p21	5474.84	6187.00	**5903.63**	**7.83**	6174.64	6135.68	12.07

* proved optimality [2].

The RGRASP+VND and the GRASP/VND both consider the VND local search structure and share the GRASP metaheuristic. However, there are several differences between them. The GRASP/VND uses the route-first cluster-second approach (see Sect. 3), it also adopts the classical GRASP version, where the α parameter calibration is necessary (see Sect. 4) and a different set of local search heuristics for the VND structure were applied. The RGRASP+ VND considers the cluster-first route-second approach and the Reactive GRASP, which is a GRASP adaptation to enhance the α parameter automatically. Moreover, the set of local search heuristics employed for the VND structure is different.

The RGRASP+VND was coded in Python 3.8 and the experiments were carried out on an Intel i5-2310 processor with 4 GB RAM, running under a Linux system Ubuntu 18.04. Table 2 compares the RGRASP+VND and the GRASP/VND algorithms. The Instance column denotes the instance, the BKS column presents the best known solution for the instance, and values followed by * present the optimum solution, proved by R. Baldacci and A. Mingozzi [2]. The following columns consider the algorithm results: the column *Avg.* presents the average solution of 10 runs; the column *Best* presents the best solution in 10 runs and the column *%Dev* presents the algorithm percentage deviation, which is evaluated for each instance as $\frac{Best-BKS}{BKS} \times 100$.

The RGRASP+VND obtained a better solution than GRASP/VND for the instances p01, p02, p04, and p06. On the other hand, GRASP/VND outperforms the RGRASP+VND algorithm for the instances p03, p07, p12, p15, and p21. We noticed that for smaller instances the RGRASP+VND outperforms GRASP/VND, while the opposite occurs for larger instances. The most remarkable result achieved by RGRASP+VND was for the instance p02 which the best solution was only 6.82 units far from the BKS (%Dev = 1.44) against the GRASP/VND, which got the best result with 56.11 units far from the BKS (%Dev = 11.85), with a difference of 49.29 units between the algorithms. The worst result obtained by RGRASP+VND was on the biggest instance p21, in which the best solution value was 660.84 units far from the BKS (%Dev = 12.07) against the GRASP/VND with 428.79 units far from the BKS (%Dev = 7.83), with a difference of 232.05 units between the algorithms.

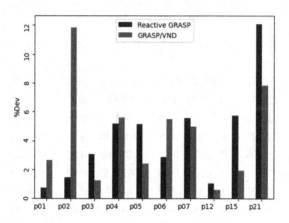

Fig. 2. Comparison of deviation percentage for the RGRASP+VND and the GRASP/VND [28].

Figure 2 presents, for each instance, the RGRASP+VND and GRASP/VND percentage deviations. The highest difference in the methods solutions was for the instance p02, where the RGRASP+VND outperforms the GRASP/VND, with a 10.41% deviation difference. For instances p01 and p04, RGRASP+VND was better than GRASP/VND, but with a small deviation percentage difference (1.93% and 0.42% respectively). For instance p06, the RGRASP+VND was better than GRASP/VND, with a 2.63% deviation difference. GRASP/VND was better than RGRASP+VND for the instances p03, p05, p07, p12, p15, p21, with 1.82%, 2.73%, 0.60%, 0.44%, 3.81%, 4.24% deviation differences respectively.

A literature review was conducted to study the MDVRP state-of-the-art. In this study, the search string: *("mdvrp" OR "multi-depot vehicle routing problem") AND ("metaheuristic" OR "meta-heuristic" OR "exact algorithm" OR "matheuristic")* was used on Scopus database limited to years 2018 to March of 2022. The algorithms found and also those used for comparison were ranked considering their average results on the Cordeau et al. set of instances of 1997, which is the most used MDVRP benchmark. The two methods with the best average results were: A Hybrid Genetic Algorithm [27] and A biased-randomized variable neighborhood search [26]. Both methods outperform the reported GRASP results for the Cordeau et al. instances package.

Table 3 compares the RGRASP+VND and the state-of-the-art algorithms: HGSADC [27], BR-VNS [26]. The HGSADC achieves the best results, where it obtained the BKS on all instances of the experimental tests, excluding the p03 and p04, and their average deviation was lower than 0.20% on all instances. The BR-VNS had the second best results and their average deviation was lower than 2%. The RGRASP+VND was the worse algorithm in this comparison and its average deviation has reached 12.07% on the biggest instance (p21).

Table 3. RGRASP+VND comparison with state-of-art methods for the MDVRP instances.

Instance	BKS	HGSADC [27]		BR-VNS [26]		RGRASP+VND	
		Avg.	%Dev	*Avg.*	%Dev	*Avg.*	%Dev
p01	576.87*	**576.87**	0.00	**576.87**	0.00	588.53	2.02
p02	473.53*	**473.53**	0.00	473.87	0.07	480.35	1.44
p03	640.65*	**641.19**	0.08	**641.19**	0.08	664.52	3.73
p04	999.21*	**1001.04**	0.18	1003.49	0.43	1056.64	5.75
p05	751.26	**750.03**	0.00	751.94	0.25	794.68	5.95
p06	876.5*	**876.5**	0.00	**876.5**	0.00	904.38	3.18
p07	881.97*	**881.97**	0.00	885.74	0.43	940.96	6.69
p12	1318.95*	**1318.95**	0.00	**1318.95**	0.00	1363.32	3.36
p15	2505.42	**2505.42**	0.00	2511.43	0.24	2699.66	7.75
p21	5474.84	**5474.84**	0.00	5576.25	1.85	6174.64	12.07

* proved optimality [2].

GRASP metaheuristic is not known as performing better than state-of-the-art MDVRP algorithms. However, the RGRASP+VND algorithm brings contributions to the GRASP literature for the MDVRP which currently is minimal. It shows improvement over the GRASP/VND for most small instances of the experimental tests, where the Cordeau et al. dataset is considered. We also highlight that the RGRASP+VND presents more stable results, where it achieves better average results against the GRASP/VND for every instance on the experimental tests (see Table 2).

6   Conclusions

This work introduced the Reactive GRASP metaheuristic with VND local search strategy for the MDVRP, combining four distinct local search procedures, two of them presented in this paper, and a clustering technique. Through the literature review, we observed that the GRASP metaheuristic is not known as performing better than the highest performing algorithms in the literature. However, the RGRASP+VND algorithm brings contributions to the GRASP literature of the MDVRP. It achieves better results on most small instances of the Cordeau et al. dataset [8] and also achieves better average solutions for all instances on the experimental tests against the GRASP/VND [28] (see Table 2).

In future work, the application of other constructive algorithms can be investigated for the RGRASP+VND algorithm. Furthermore, the RGRASP+VND parametrization can be improved as well with more local search procedures. The application of RGRASP+VND could also be investigated on other VRP versions, for instance, those with time window constraints.

References

1. Baghbadorani, R.R., Ghanavati, A., Zajkani, M., Haeri, M.: A novel two-phase approach to solve multi-depot vehicle routing problem. In: 2021 25th International Conference on System Theory, Control and Computing (ICSTCC), pp. 390–394. IEEE (2021)
2. Baldacci, R., Mingozzi, A.: A unified exact method for solving different classes of vehicle routing problems. Math. Program. **120**(2), 347–380 (2009)
3. Barma, P.S., Dutta, J., Mukherjee, A.: A 2-opt guided discrete antlion optimization algorithm for multi-depot vehicle routing problem. Decis. Mak. Appl. Manage. Eng. **2**(2), 112–125 (2019)
4. Beasley, J.E.: Route first-cluster second methods for vehicle routing. Omega **11**(4), 403–408 (1983)
5. Boudia, M., Louly, M.A.O., Prins, C.: A reactive grasp and path relinking for a combined production-distribution problem. Comput. Oper. Res. **34**(11), 3402–3419 (2007)
6. Clarke, G., Wright, J.W.: Scheduling of vehicles from a central depot to a number of delivery points. Oper. Res. **12**(4), 568–581 (1964)
7. Comert, S.E., Yazgan, H.R., Kır, S., Yener, F.: A cluster first-route second approach for a capacitated vehicle routing problem: a case study. Int. J. Procure. Manage. **11**(4), 399–419 (2018)
8. Cordeau, J.F., Gendreau, M., Laporte, G.: A tabu search heuristic for periodic and multi-depot vehicle routing problems. Netw. Int. J. **30**(2), 105–119 (1997)
9. Croes, G.A.: A method for solving traveling-salesman problems. Oper. Res. **6**(6), 791–812 (1958)
10. Dantzig, G.B., Ramser, J.H.: The truck dispatching problem. Manage. Sci. **6**(1), 80–91 (1959)
11. Ezugwu, A.E., Akutsah, F., Olusanya, M.O., Adewumi, A.O.: Enhanced intelligent water drops algorithm for multi-depot vehicle routing problem. PLOS ONE **13**(3), e0193751 (2018)

12. Feo, T.A., Resende, M.G.: Greedy randomized adaptive search procedures. J. Global Optim. **6**(2), 109–133 (1995)
13. Flood, M.M.: The traveling-salesman problem. Oper. Res. **4**(1), 61–75 (1956)
14. Giosa, I., Tansini, I., Viera, I.: New assignment algorithms for the multi-depot vehicle routing problem. J. Oper. Res. Soc. **53**(9), 977–984 (2002)
15. He, Y., Miao, W., Xie, R., Shi, Y.: A tabu search algorithm with variable cluster grouping for multi-depot vehicle routing problem. In: Proceedings of the 2014 IEEE 18th International Conference on Computer Supported Cooperative Work in Design (CSCWD), pp. 12–17. IEEE (2014)
16. Iori, M., Locatelli, M., Moreira, M.C.O., Silveira, T.: Reactive GRASP-based algorithm for pallet building problem with visibility and contiguity constraints. In: Lalla-Ruiz, E., Mes, M., Voß, S. (eds.) ICCL 2020. LNCS, vol. 12433, pp. 651–665. Springer, Cham (2020). https://doi.org/10.1007/978-3-030-59747-4_42
17. Kulkarni, R., Bhave, P.R.: Integer programming formulations of vehicle routing problems. Eur. J. Oper. Res. **20**(1), 58–67 (1985)
18. Labadi, N., Prins, C., Reghioui, M.: A memetic algorithm for the vehicle routing problem with time windows. RAIRO Oper. Res. **42**(3), 415–431 (2008)
19. Lenstra, J.K., Rinnooy Kan, A.H.G.: Complexity of vehicle routing and scheduling problems. Networks **11**(2), 221–227 (1981)
20. Luo, J., Chen, M.-R.: Multi-phase modified shuffled frog leaping algorithm with extremal optimization for the MDVRP and the MDVRPTW. Comput. Ind. Eng. **72**, 84–97 (2014)
21. MacQueen, J., et al.: Some methods for classification and analysis of multivariate observations. In: Proceedings of the 5th Berkeley Symposium on Mathematical Statistics and Probability, Oakland, CA, USA, vol. 1, pp. 281–297 (1967)
22. Misni, F., Lee, L.: Harmony search for multi-depot vehicle routing problem. Malays. J. Math. Sci. **13**(3), 311–328 (2019)
23. Moscato, P., Cotta, C.: An accelerated introduction to memetic algorithms. In: Gendreau, M., Potvin, J.-Y. (eds.) Handbook of Metaheuristics. ISORMS, vol. 272, pp. 275–309. Springer, Cham (2019). https://doi.org/10.1007/978-3-319-91086-4_9
24. Prais, M., Ribeiro, C.C.: Reactive GRASP: an application to a matrix decomposition problem in TDMA traffic assignment. INFORMS J. Comput. **12**(3), 164–176 (2000)
25. Prins, C.: A simple and effective evolutionary algorithm for the vehicle routing problem. Comput. Oper. Res. **31**(12), 1985–2002 (2004)
26. Reyes-Rubiano, L., Calvet, L., Juan, A.A., Faulin, J., Bové, L.: A biased-randomized variable neighborhood search for sustainable multi-depot vehicle routing problems. J. Heuristics **26**(3), 401–422 (2018)
27. Vidal, T., Crainic, T.G., Gendreau, M., Lahrichi, N., Rei, W.: A hybrid genetic algorithm for multidepot and periodic vehicle routing problems. Oper. Res. **60**(3), 611–624 (2012)
28. Villegas, J.G., Prins, C., Prodhon, C., Medaglia, A.L., Velasco, N.: GRASP/VND and multi-start evolutionary local search for the single truck and trailer routing problem with satellite depots. Eng. Appl. Artif. Intell. **23**(5), 780–794 (2010)

How Life Transitions Influence People's Use of the Internet: A Clustering Approach

Martina Benvenuti[1] , Humberto Rocha[2(✉)] , Isabel Dórdio Dimas[2] ,
and Elvis Mazzoni[1]

[1] Department of Psychology, Alma Mater Studiorum-University of Bologna, Cesena,
Italy
{martina.benvenuti2,elvis.mazzoni}@unibo.it
[2] CeBER, Faculty of Economics, University of Coimbra, Av Dias da Silva 165,
3004-512 Coimbra, Portugal
{hrocha,idimas}@fe.uc.pt

Abstract. This research aimed, firstly, to define a conceptual model
that considers potential resources/challenges (Physical, Cognitive, Emotional, Social, Material, Environmental, Digital) and describes how those
influence the Internet use and modify human behavior during life transitions (e.g., changing school, finding a job). Secondly, starting on that
model, user profiles were outlined. Instead of grouping study participants into pre-defined groups, clustering techniques were used to group
users with similar profiles. The main advantage of this methodological approach is that the participant groups, i.e., different user profiles,
emerged intrinsically from the data. A cross-sectional study was proposed
based on the compilation of an Online questionnaire. The sample consists of 1.524 participants. Three clusters emerged with different mean
ages: young adult users (mean age = 33.83), youngest users (25.79), and
oldest users (36.80). Differences were identified between all dimensions
measured, particularly between youngest users and oldest users.

Keywords: Life transition · Internet use · Clustering

1 Introduction

Integration between being online and offline is an important part of the psychology of human beings [15,44]. While, in some cases, it is possible to find a
balance between these two aspects, in others integration might be problematic
[27]. Accordingly, the theoretical perspective adopted for this research assumes
that the Internet can become either a problematic or a functional tool depending on how it is used and the reasons behind that use [6,34,35]. Indeed, this
research is underpinned by a theoretical framework that simultaneously considers both the positive and negative outcomes generated by the use of Internet.
Ekbia and Nardi explained that Internet technologies could enable situations

of inverse instrumentality, a process involving the objectification of users [16], whereby their behavior is regulated in a predictable manner, drawing them in or pushing them away from their activities. On the other hand, studies by Leont'ev [28] and later Kaptelinin and Nardi [24] proposed the construct of functional organ to describe how a tool (e.g., the Internet) allows people to achieve better and more powerful results which would not be attainable individually without that tool. Therefore, the goals of this research are firstly to define an integrative and flexible conceptual model that describes how life transitions (e.g., changing school, finding a job, moving to another city, etc.) linked to specific life periods influence people's use of the Internet (both in problematic and functional ways). Secondly, the model is adopted as a starting point for outlining user profiles that describe different ways of using the Internet and its applications. Furthermore, differences between these profiles are explored in terms of life transitions and challenges linked to these transitions. The benefit of this research is a proposed developmental model that connects user profiles to specific life periods characterized by transitions that generate challenges. Thus, the focus is not on the life stages themselves, as is the case in previous research about the use of the Internet by specific age groups (adolescents, emerging adults, adults, etc.), but rather on the transitions that people face, and on whether and how Internet use might facilitate these transitions.

Internet Use and Life Transitions: A Theoretical Perspective

Human development is characterized by transition and transformation processes in which chronological age is a dominant force (age always increases over time). In their *Lifespan Model of Developmental Challenge*, Hendry and Kloep [21] address the transitions and transformations arising from potential resources and challenges [26].

Potential resources characterize each individual from the beginning of their life. However, the distribution of those resources amongst individuals is uneven. Furthermore, there are some resources that seem "personal" to the individual (such as money) and others that are more socially defined (such as access to the education system). Thus, different micro- and macro-systems, with their different climates, laws, health systems, etc., generate different opportunities [9], such as access to education employment. From this perspective, resources and challenges are interdependent: [...] *challenges are defined by resources, and vice versa. Only by knowing an individual's resources can we decide whether a particular task is a challenge, and only by knowing a particular task can we decide if an individual has the resources to deal with it.* [...] [26]. Starting from these perspectives, it is also essential to consider the potential *digital* resources/challenges within the system (Fig. 1), since in recent years the use of the Internet has grown exponentially in every context of human life affecting relevant behaviors, interactions, and communications [2]. This is particularly true for the young, for whom technological literacy is crucial for interacting, finding information, work, recreation, and for a wide variety of other activities [5,43,47]. Figure 1 is a revised and extended version of Kloep, Hendry and Saunders diagram [26], that adds the transition processes to the interdependence of potential challenges and potential resources.

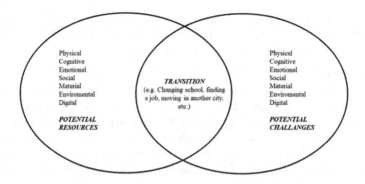

Fig. 1. Potential (digital) challenges and resources diagram, with transition as an intersection process.

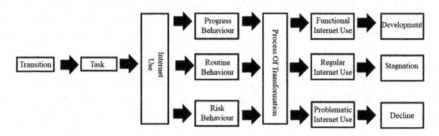

Fig. 2. Transition and transformation of internet use model.

The interdependence between potential resources and potential challenges inevitably affects the way in which the transition process takes place, develops and progresses. The Internet plays an important role in these evolutionary dynamics. Thus, it is pointless to speak of functional and/or problematic Internet use if we do not consider an evolutionary point of view regarding human transitions (Fig. 2).

Emerging Adults [3,4] namely people between 18 to 29 years of age, face a critical period of life in terms of human transitions [19], like shifting location for study or work purposes. These challenges involve many types of tasks, such as finding accommodation. To carry out these tasks quickly and easily, digital environments have become fundamental and irreplaceable. Nevertheless, in line with Hendry and Kloep's model [21,22], facing challenges not always leads to development. Thus, the use that young people make of the Internet could lead to three different kinds of behaviors: (1) Progress; (2) Routine or (3) Risk (see Fig. 2). So, according to these different kinds of behavior, Internet use could become: (1) functional, and thereby leading a person to overcome the transition and consequently bring about a development; (2) regular, leading a person to stagnation; (3) problematic, leading a person to a decline.

Summarizing, this is a person-centered model that offers an integrative and flexible perspective that includes potential digital resources/challenges and Internet usage, both in problematic and functional ways.

The Research Study

Previous research has already established that use of the Internet can be very helpful (functional) for facing transitions [17,36] but can also prove useless or even harmful (dysfunctional), decreasing the potential to face transitions [34]. Thus, the present study sets out to analyze the two poles of this continuum of functional-dysfunctional Internet use by using two specific concept/dimensions: Functional Internet Use (FIU) for the functional side and Problematic Internet Use (PIU) for the dysfunctional one.

Furthermore, previous studies have already found that functional or dysfunctional Internet use is determined by many factors, such as Self-Esteem, Self-Control, Online Social Support, Offline Social Support, Mindfulness, Cognitive Absorption, Life Satisfaction and Job Satisfaction [7,34]. Moreover, following the above assumption and the *Transition and Transformation of Internet Use Model* (Fig. 2), this study has sought to answer the following Research Questions: (RQ1) *Can transitions play a role in defining how and why people use digital technologies in different ways?* And since certain periods of life are associated with specific transitions and professional needs, (RQ2) *can the period of life determine different profiles based on digital technology use?*

Internet Use Habits: Devices, Social Networking Sites and Web Applications

Considering the environments people use to connect is crucial in order to gain a better understanding of the dynamics underlying Internet use during life transitions. In this regard, Social Networking Sites (SNSs) play an important role. During the transition from school to university or from school/university to work, the online contacts in a person's SNS networks could help them bridge gaps between their background knowledge and skills and those required in the new context in which that person will operate [18,36]. Particularly, the use of SNSs and Web Applications could often help people to cope with transitions because they represent a fundamental sort of pre-socialization with the new social context (e.g., that in which people move to attend university or join a new organization) by helping to create contacts with those who are already part of that environment. In this regard, it was expected that: (H1). Individuals' life periods that are specifically characterized by transitions and the attendant challenges posed influence how much they use the Internet, the devices they use to navigate it, and the applications they use.

Problematic Internet Use

This study adopts the concept of Problematic Internet Use (PIU) as used by Caplan [11,12], i.e. the individuals' predisposition to develop maladaptive Internet-related cognitive traits: (1) preference for online social interactions (POSI), (2) deficient mood regulation, (3) deficient self-regulation (compulsive use subscale and cognitive preoccupation subscale), (4) and negative outcomes. Factors considered in this study related to PIU are:

1. Self-Esteem: individuals with lower self-esteem are more prone to develop PIU symptoms [13];
2. Self-Control: people with low self-control might be led to PIU behavior [12];
3. Online Social Support and Offline Social Support: people with low offline social support tend to look for contacts online and consequently spend excessive time online [48];
4. Mindfulness: defines the presence or absence of attention related to what happens in the present [46]. It is possible to assume that this factor, combined with self-regulation, affects PIU because one of the most common experiences during time spent online is unawareness of time passing [30];
5. Cognitive Absorption: a high level of cognitive absorption when using the Internet allows better use of it (Heightened Enjoyment, Control and Curiosity) but, at the same time, it can lead to compulsive phenomena (Temporal Dissociation and Focused Immersion) [1].

Internet use during life transitions (e.g., from adolescence to emerging adulthood) could be dysfunctional because it is driven by exploration, challenges and changes [21]. In these periods, people are in the phase of selecting life prospects [4,21] and they tend to have different experiences, even in online environments, that may occur in a dysfunctional way, e.g. individuals characterized by low extroversion and low self-esteem are perceived as less popular both online and offline (hypothesis of social compensation) [49]. Thus, if individuals do not receive appropriate social support in daily life, they tend to create a parallel life to activate contacts and build relationships online in order to compensate for this shortage, and this could lead to a dysfunctional use of the Internet [49].

In this regard, considering the above assumptions, it was expected that: (H2). Dysfunctional use of the Internet could change during the individual's lifetime in response to challenges linked to their life transitions.

Functional Internet Use
Recent research has shown how the use of SNSs could lead to higher levels of well-being, possibly leading people to Functional Internet Use (FIU). Valkenburg, Peter and Schouten have highlighted how the frequency of use of SNSs indirectly affected Self-Esteem and psychological well-being in a sample of adolescents [45]. This frequency of use is affected by the frequency of positive feedback (e.g. "Likes" on Facebook or "Re-tweets" on Twitter) the sample received on their SNSs profiles. Moreover, in another study, analyzing the relationship between social capital (i.e. the potential benefits of creating and maintaining interpersonal relationships), Self-Esteem and the use of SNSs in American college students (and also in Italian students - see [36], it turned out that those with low Self-Esteem are more driven to use Facebook to maintain social capital than those with higher Self-Esteem [41]. In this regard during the transition from school to university or from school/university to work, for example, Social-Support and the use of SNSs play an important role. During these transitions, people

using SNSs face a gap between the knowledge and skills they bring to the new organization and those required to work/study there. Thus, considering these assumptions, functional use of the Internet (FIU), namely use that facilitates completion of a challenge/task during a life transition, involves a number of factors: Self-Esteem, Online Social-Support, Number of Online Contacts (total sum of the contacts, including acquaintances and friends, that a person has on his/her social network/s profile/s), Life Satisfaction, and Job Satisfaction. It is expected that: (H3). Functional use of the Internet could change during life, in accordance with challenges linked to life transitions.

2   Materials and Methods

2.1   Data Collection

To verify the above hypotheses, a cross-sectional study was proposed based on the compilation of an anonymous online questionnaire [40], following approval from the local university bioethics committee. Respondents were recruited through announcements made on the main SNSs in Italy; efforts were made to achieve a gender-balanced sample population.

2.2   Sample Description

The sample consists of 1,524 participants, 1,050 female (68.9%) and 474 males (31.1%), with a mean age of 31.3 years (SD = 11.8).

2.3   Measures

The questionnaire items are grouped according to the three main areas analyzed in this research study, namely: (1) Measures of Problematic Internet Use, (2) Measures of Functional Internet Use, (3) Measures of Internet Use Habits.

2.4   Measures of Problematic Internet Use

Problematic Internet Use (PIU)
In order to measure PIU, the Italian version of Caplan's [12] Generalized Problematic Internet Use Scale 2 (GPIU2) was used. This consists of 15 items, measured on an 8-point Likert scale (1 = "definitely disagree" and 8 = "definitely agree"), in response to the instruction: "Indicate your degree of agreement-disagreement with the following statements". Cronbach's alpha for this scale was .92.

Self-Esteem
For this measure, the validated Italian version [38] of the Rosenberg Self-Esteem Scale was used [39]. It comprises 10 items on a 4-point Likert scale (1 = "strongly agree" to 4 = "strongly disagree"). Cronbach's alpha for this scale was .81.

Self-Control
Thirteen items from the Brief Self-Control Scale edited by Tangney, Baumeister and Boone [42], and later re-validated by Maloney, Grawitch, and Barber [33], were used. The selected items are those most predictive for Self-Control and are based on a 5-point Likert scale (1 = "not at all", 5 = "very much"). Cronbach's alpha for this scale was .83.

Online and Offline Social-Support
The Offline Social-Support Scale [48], which in turn was adapted from a previous study by Leung and Lee [29], was used. Cronbach's alpha for this scale was .93. Online Social-Support was also evaluated using the Online Social-Support Scale [48], indicating support from the online environment. For the 11 items covered in both scales, participants rated their agreement to the general statement *"How often is each of the following kinds of support available to you if you need it?"* according to a Likert scale from 1 = "never" to 5 = "all the time". Cronbach's alpha for this scale was .98.

Mindfulness
For this aspect, the Mindfulness Attention Awareness Scale (MAAS) was chosen [10,30]. This consists of a mono-dimensional score, with 15 items measured on a 6-point Likert scale from 1 = "almost always" to 6 = "almost never". Cronbach's alpha for this scale was .78.

Cognitive Absorption
The Cognitive Absorption Scale [1] was used as a measure of Internet engagement. The authors define the construct as " [...] *a state of deep involvement with software* [...]" [1]. The scale comprises 20 items measured on a 7-point Likert scale from 1 = "strongly disagree" to 7 = "strongly agree". Cronbach's alpha for this scale was .82.

2.5   Measures of Functional Internet Use

Functional Internet Use (FIU)
A brief scale for measuring FIU was specifically created for the purposes of this study. The scale is composed of 4 items: (1) *"being connected increases my ability to reach certain goals"*, (2) *"being connected improves my productivity"*, (3) *"being connected is useful for carrying out my activities"*, (4) *"being connected improves my performance"*. FIU was measured on a 7-point Likert scale, from 1 = "strongly disagree" to 7 = "strongly agree". Cronbach's alpha for this scale was .90.

Online Social-Support
As described earlier, Online Social-Support was evaluated using the Online Social-Support Scale [48].

Number of Online Contacts
The number of online contacts refers to: (1) contacts (i.e., all those individual subjects have on your online profiles); (2) acquaintances (i.e., those the subject

does not interact with regularly, whether online or in everyday life); and (3) friends (i.e., those with whom the subject habitually interacts, beyond simple online or offline contact in everyday life). Different thresholds were set for each of the three categories: 20,000 for online contacts, 10,000 for acquaintances, and 2,000 for friends. For the combined Online Contacts measure (total of all three categories), a threshold of 20,000 was set.

Life Satisfaction
This study adopted the Satisfaction with Life Scale [14], consisting of 5 items (5-point Likert scale from 1 = "strongly disagree" to 5 = "strongly agree"). All five are framed in a positive way (e.g., "*In general my life is close to my ideal*"). Cronbach's alpha for this scale was .87.

Job Satisfaction
Job satisfaction was measured using the Brayfield and Rothe job satisfaction scale [8], revalidated by Judge, Locke, Durham and Kluger [23]. This has five items ranked on a 10-point Likert scale ranging from 1 = "strongly disagree", to 10 = "strongly agree". As one of the aims of the study was to collect data from students as well, the scale's items included both job and academic satisfaction (e.g., "*I feel quite happy with my job/my studies*"). Cronbach's alpha for this scale was .81.

2.6   Measures of Internet Use Habits

Use of Devices
Questions referring to the use of devices (computers, tablets, smartphones, and consoles) during the day were also included (e.g., "*Indicate how many hours a day you use a tablet*").

Use of Social Networking Sites and Web Applications
Participants had to rank the five Internet tools (Facebook, Instagram, YouTube, WhatsApp, and Email) that they use the most during the day.

Time Connection and Interaction
The participants were asked to indicate how many hours are willing to devote to online activities during working and free time.

2.7   Data Analysis – Clustering

The hypotheses in this study were tested by considering groups of users with similar profiles. Instead of grouping study participants into pre-defined groups, clustering techniques were used to group users with similar profiles. The main advantage of this methodological approach is that the participant groups, i.e., different user profiles, emerged intrinsically from the data.

Indeed, in this study, clustering was approached in such a way that users with similar features/behaviors were grouped together into a cluster that differed from users in other clusters [20]. Two of the most commonly used clustering techniques are the K-means [31] and partitioning around medoids [25]. K-means clustering finds, iteratively, k centroids that define k clusters by assigning each individual to the cluster with the nearest centroid. The coordinates of each centroid correspond to the mean of the coordinates (features) of the users in the cluster, which prevents it from being used when categorical variables exist. Furthermore, K-means clustering is known to be sensitive to outliers [37]. Thus, K-medoids clustering is the most frequently adopted alternative method when the mean or median is not clearly defined or robustness to outlier data is required. K-medoids is similar to K-means but has medoids, i.e., representative individuals, instead of centroids. K-medoids minimizes the sum of medoids to cluster member distances. As medoids are actual data points in the data set, the K-medoids algorithm can be used in situations where the mean of the data is not present within the data set. Hence, K-medoids is useful for clustering categorical data where a mean is impossible to define or interpret.

Dissimilarity between individuals is measured as the distance between them. Assuming that we have data from n subjects with p variables (attributes, measures) that can be organized in matrix format as

$$\begin{bmatrix} x_{11} & x_{12} & x_{13} & \cdots & x_{1p} \\ x_{21} & x_{22} & x_{23} & \cdots & x_{2p} \\ \vdots & \vdots & \vdots & \ddots & \vdots \\ x_{n1} & x_{n2} & x_{n3} & \cdots & x_{np} \end{bmatrix},$$

a dissimilarity matrix is typically computed

$$\begin{bmatrix} 0 & & & & \\ d(2,1) & 0 & & & \\ d(3,1) & d(3,2) & 0 & & \\ \vdots & \vdots & & \ddots & \\ d(n,1) & d(n,2) & \cdots & d(n,n-1) & 0 \end{bmatrix},$$

where $d(i,j)$ is the dissimilarity between subjects i and j. Note that 0 imply that the subjects are equal and thus close to 0 means subjects are similar. Obviously this is a symmetric matrix with zeros in the diagonal ($d(i,i) = 0$). After computing this dissimilarity matrix, the goal is to group similar subjects.

Depending on the type of data available, different dissimilarity measures can be used. The most commonly used dissimilarity measure is the common distance (Euclidean) when variables are continuous. A measure commonly used when other than continuous variables are present is the Mahalanobis distance [32]. This is based on the correlations between variables with which different patterns can be identified and analyzed. It differs from Euclidean distance in that it takes into account the correlations of the data set and is invariant to scale, i.e., it does not depend on the scale of the measurements. If $x_i = (x_{i1}, x_{i2}, \ldots, x_{ip})$ and

$x_j = (x_{j1}, x_{j2}, \ldots, x_{jp})$ are two data points, the Mahalanobis distance is defined as

$$d(x_i, x_j) = \sqrt{(x_i - x_j)^T S^{-1}(x_i - x_j)} \tag{1}$$

where S is the covariance matrix. Note that if S is the identity matrix, the Mahalanobis distance reduces to the Euclidean distance:

$$d(x_i, x_j) = \sqrt{\sum_{k=1}^{p}(x_{ik} - x_{jk})^2}. \tag{2}$$

Because the data in this study also contained binary and categorical variables, an implementation of Partitioning around medoids (PAM) algorithm in MATLAB was used to identify emergent clusters in our data. Thus, the Mahalanobis distance was used as a dissimilarity measure. In order to find the optimal number of clusters, the K-medoids clustering algorithm was run for k = 2, 3, 4,.... The number of clusters chosen corresponds to the number of clusters for which the aggregate distance, from the medoids to the remaining elements that compose each cluster, no longer significantly reduces.

3   Results

Based on previous research, to test H1, the following variables were included in the construction of clusters: age, time connection, number of online contacts, use of devices (computers, tablets, smartphones, and consoles), use of networking tools (Facebook, Instagram, YouTube, WhatsApp, Email). For our data set, the optimal number of clusters was K = 3 (Table 1).

As Table 1 shows, Cluster 2 is composed of the youngest users who had, on average, the greatest number of contacts. This is also the profile that spent most time online. Users included in this profile were frequent users of YouTube, WhatsApp, and Facebook. They mostly accessed the Internet via computers and smartphones. Cluster 3 comprises the oldest users in the sample, who had, on average, the lowest number of contacts. Compared to the other profiles, this is the profile that spent least time online, but had more time available for interacting during work/study. Users included in this profile mostly accessed the Internet via computers, smartphones and tablets; they were infrequent users of Facebook and frequent users of WhatsApp and Email. Finally, Cluster 1 includes the young adult users. They spent less time online than the other profiles and were less willing to interact online both during work time and free time. These users accessed the Internet mostly via computers and smartphones and were frequent users of Facebook, WhatsApp and Email.

Table 1. Clusters description (variables presented were included in the cluster construction).

	Cluster 1	Cluster 2	Cluster 3
N	522	623	379
Quantitative variables	*M (SD)*	*M (SD)*	*M (SD)*
Age	33.83 (11.85)	25.79 (9.25)	36.80 (11.84)
Hours online	7.39 (5.90)	8.85 (6.35)	6.80 (5.98)
Hours interacting free time	6.63 (6.12)	8.35 (6.39)	8.67 (7.80)
Hours interacting work time	5.03 (5.50)	5.24 (6.02)	7.38 (7.06)
Number of Online contacts	849.89 (1501.29)	1080.37 (1885.78)	767.83 (1843.44)
Categorical variables	*n (%)*	*n (%)*	*n (%)*
Use of devices			
Computer	487 (93.3)	600 (96.3)	359 (94.7)
Tablet	161 (30.8)	158 (25.4)	243 (64.1)
Smartphone	479 (91.8)	561 (90)	335 (88.4)
Console	29 (5.6)	65 (10.4)	31 (8.2)
Use of SNS			
Facebook	512 (98.1)	580 (93.1)	87 (23)
Instagram	123 (23.6)	132 (21.2)	67 (17.7)
YouTube	4 (0.8)	623 (100)	105 (27.7)
WhatsApp	438 (83.9)	503 (80.7)	301 (79.4)
Email	397 (76.1)	391 (62.8)	265 (69.9)

Table 2. Differences between variables resulting from clusters.

Variables	Cluster 1		Cluster 2		Cluster 3		F	η^2
	M	SD	M	SD	M	SD		
PIU	2.28^a	1.12	2.82^b	1.35	2.03^c	1.08	79.14***	.10
Self-esteem	22.18^b	3.02	21.77^a	3.38	22.57^b	2.93	10.88***	.02
Self-control	45.98^a	7.26	42.52^b	7.74	46.56^a	8.04	45.72***	.06
Online social-support	2.85^a	1.07	3.04^b	1.02	2.73^a	1.12	34.45***	.05
Offline social-support	3.90	.04	3.88	.03	3.94	.04	1.24	.00
Mindfulness	4.21^a	.70	4.07^b	.68	4.31^a	.75	14.21***	.02
Cognitive absorption	3.80^a	.76	4.01^b	.72	3.73^a	.81	30.10***	.04
Life satisfaction	4.58^a	1.17	4.25^b	1.25	4.73^a	1.23	13.66***	.02
Job satisfaction	6.81	1.91	6.66	1.86	6.86	1.83	13.75***	.02
FIU	14.45^a	6.61	14.52^a	6.04	16.04^b	6.67	7.55***	.01

Note: Means with different letters are significantly different at the level of $\alpha <$ .05 according to the post-hoc test of Tuckey HSD. *** p < .001.

Testing of H2 and H3 was performed through ANOVAs (Table 2). F-tests indicated significant overall differences by cluster type at $p < .001$ on all variables considered, with the exception of offline social support. Post-hoc analyses indicated that Cluster 2 presented significant higher values on PIU, Online Social-Support and cognitive absorption, and significant lower values on Self-Esteem (while the difference with Cluster 1 was not significant), Self-Control, mindfulness, life satisfaction and functional Internet use (not significant different from Cluster 1). Moreover, post-hoc analyses also indicated that Cluster 1 is significantly different from Cluster 3 in two of the variables considered: PIU, for which it presents higher values, and FIU, where it presents lower values.

4   Discussion and Conclusions

This research set out to clarify when and how the use of the Internet and its applications/SNSs could become problematic or functional during a specific period of life characterized by different challenges. Since certain periods of life are associated with specific transitions and professional needs (see [4, 21, 22]), this may determine different profiles of digital technology use and, at the same time, the potential resources that characterize people's digital life may result in their functional or problematic use of technologies in dealing with those transitions. Thus, this research aimed to answer the questions: (RQ1) *Can transitions play a role in defining how and why people use digital technologies in different ways?* And, (RQ2) *can the period of life determine different profiles based on digital technology use?*

In order to achieve these objectives, three hypotheses were tested using cluster analysis for H1 and ANOVA for H2 and H3.

H1 (life periods of individuals, characterized by the transitions and challenges they face, influence how much they use ICT) was confirmed. As described in Table 1, Cluster 1 (mean age = 33.83) includes young adult users. They spent less time online than the other profiles and were less willing to interact online during both work/study time and free time; they access the Internet mostly via computers and smartphones and were frequent users of Facebook, WhatsApp and Email. Indeed, many subjects in Cluster 1 already had a job and mainly use social networks and chats to maintain contact with their peer group. Cluster 2 (mean age = 25.79) is the profile that had, on average, the greatest number of online contacts and was also the profile that spent most time online, via computer and smartphone. Indeed, this age group, which coincides with the younger emerging adults [3,4], is recognized as experiencing a period of life characterized by great challenges and changes. From this viewpoint, the Internet plays a significant role as a means of constructing a bridge between school and university or school and the working environment [7, 36]. Moreover, users included in this profile were frequent users of YouTube, WhatsApp, and Facebook. In this regard, SNSs are also fundamental as a means of pre-socialization during transitions towards the new environment and connecting with those who are already a part of that system. Finally, Cluster 3 (36.80) is constituted by the oldest users in the sample, who had, on average, the lowest number of contacts compared to the other clusters. This profile spent least time online, but is most willing to spend time interacting during work time. This suggests that the adult cluster uses the Internet principally for work and less during free time. Indeed, these users mostly accessed the Internet via computers, smartphones and tablets, and they were frequent users of WhatsApp and Email.

H2 (dysfunctional use of the Internet could change during life based on the challenges linked to life transitions) was partially confirmed: the results show that Cluster 2 (the youngest subjects) has higher values on PIU, Online Social Support and cognitive absorption (although the difference with Cluster 1 was not significant), Self-Control, mindfulness, life satisfaction and functional Internet use (not significantly different from Cluster 1). These results tend to confirm the social compensation hypothesis: since Emerging Adults are constructing their professional identity, they are characterized by instability, exploration, and change. These dynamics could affect self-esteem, since this profile can encounter many failures, posing the need for further social support and for sharing information and knowledge. This behavior poses a potential risk if individuals perceive online contexts as the simplest way to compensate for these shortcomings.

Finally, H3 (functional use of the Web could change during life based on the challenges people are facing) is confirmed, as the highest significant scores in FIU are those of Cluster 3. Indeed, this group is the one having already achieved many life goals, is the most stable and, following the model described in Figs. 1 and 2, has the most potential resources at their disposal.

Acknowledgments. Thanks to Jeffrey Earp for language revision of the original manuscript. This study has been funded by national funds, through FCT, Portuguese Science Foundation, under project UIDB/05037/2020.

References

1. Agarwal, R., Karahanna, E.: Time flies when you're having fun: cognitive absorption and beliefs about information technology usage. MIS Q. **24**(4), 665–694 (2000)
2. Anderson, E.L., Steen, E., Stavropoulos, V.: Internet use and problematic internet use: a systematic review of longitudinal research trends in adolescence and emergent adulthood. Int. J. Adolesc. Youth **22**(4), 430–454 (2017)
3. Arnett, J.J.: Emerging Adulthood: The Winding Road Through the Late Teens and Twenties. Oxford University Press, New York (2004)
4. Arnett, J.J.: Human Development: a Cultural Approach. Pearson Education, Boston (2012)
5. Aslanidou, S., Menexes, G.: Youth and the Internet: Uses and practices in the home. Comput. Educ. **51**(3), 1375–1391 (2008)
6. Bagozzi, R.P., Dholakia, U.M., Pearo, L.R.K.: Antecedents and consequences of online social interactions. Media Psychol. **9**(1), 77–114 (2007)
7. Benvenuti, M., Mazzoni, E., Piobbico, G.: Being online in emerging adulthood: between problematic or functional use of the internet. In: Wright, M.F. (ed.) Identity, Sexuality, and Relationships among Emerging Adults in the Digital Age. IGI Global, Hershey, PA (2019)

8. Brayfield, A.H., Rothe, H.F.: An index of job satisfaction. J. Appl. Psychol. **35**(5), 307 (1951)
9. Bronfenbrenner, U.: The Ecology of Human Development. Harvard University Press, Harvard (1979)
10. Brown, K.W., Ryan, R.M.: The benefits of being present: mindfulness and its role in psychological well-being. J. Pers. Soc. Psychol. **84**(4), 822 (2003)
11. Caplan, S.E.: Relations among loneliness, social anxiety, and problematic internet use. Cyberpsychol. Behav. **10**, 234–241 (2007)
12. Caplan, S.E.: Theory and measurement of generalized problematic Internet use: a two-step approach. Comput. Hum. Behav. **26**(5), 1089–1097 (2010)
13. Davis, R.A., Flett, G.L., Besser, A.: Do people use the Internet to cope with stress. In: 110th Annual Convention of the American Psychological Association, Chicago (2002)
14. Diener, E.D., Emmons, R.A., Larsen, R.J., Griffin, S.: The satisfaction with life scale. J. Pers. Assess. **49**(1), 71–75 (1985)
15. Durkin, K., Baldes, M.: Young people and the media. Br. J. Dev. Psychol. (Spec. Issue) **27**, 1–12 (2009)
16. Ekbia, H., Nardi, B.A.: Inverse instrumentality: how technologies objectify patients and players. In: Leonardi, P., Nardi, B., Kallinikos, J. (eds.) Materiality and organizing: social interactions in a technological world. Oxford University Press, Oxford (2012)
17. Ellison, N., Steinfield, C., Lampe, C.: The benefits of Facebook "Friends": social capital and college students' use of online social network sites. J. Comput. Mediat. Commun. **12**, 1143–1168 (2007)
18. Frozzi, G., Mazzoni, E.: On the importance of social network sites in the transitions which characterize "emerging adulthood". ICST Trans. E-Educ. E-Learn. **11**, 1–11 (2011)
19. Goldscheider, F.: Recent changes in U.S. young adult living arrangements in comparative perspectives. J. Family Issues **18**, 708–724 (1997)
20. Han, J., Kamber, M., Tung, A.K.: Spatial clustering methods in data mining: a survey. In: Miller, H.J., Han, J. (eds.) Geographic Data Mining and Knowledge Discovery. CRC Press, London (2001)
21. Hendry, L.B., Kloep, M.: Lifespan Development: Resources, Challenges and Risks. Thomson Learning, London (2002)
22. Hendry, L., Kloep, M.: Adolescence and Adulthood: Transitions and Transformations. Macmillan International Higher Education, New York (2012)
23. Judge, T.A., Locke, E.A., Durham, C.C., Kluger, A.N.: Dispositional effects on job and life satisfaction: the role of core evaluations. J. Appl. Psychol. **83**(1), 17 (1998)
24. Kaptelinin, V., Nardi, B.A.: Acting with Technology: Activity Theory and Interaction Design. The MIT Press, Cambridge, MA (2006)
25. Kaufman, L., Rousseeuw, P.J.: Finding Groups in Data: An Introduction to Cluster Analysis, vol. 344. Wiley (2009)
26. Kloep, M., Hendry, L., Saunders, D.: A new perspective on human development. Proc. Conf. Int. J. Arts Sci. **1**(6), 332–343 (2009)
27. LaRose, R., Lin, C.A., Eastin, M.S.: Unregulated Internet usage: addiction, habit, or deficient self-regulation? Media Psychol. **5**(3), 225–253 (2003)
28. Leont'ev, A.N.: The problem of activity in psychology. J. Russ. East Eur. Psychol. **13**(2), 4–33 (1974)

29. Leung, L., Lee, P.S.: Multiple determinants of life quality: the roles of Internet activities, use of new media, social support, and leisure activities. Telematics Inform. **22**(3), 161–180 (2005)
30. MacKillop, J., Anderson, E.J.: Further psychometric validation of the mindful attention awareness scale (MAAS). J. Psychopathol. Behav. Assess. **29**(4), 289–293 (2007)
31. MacQueen, J.: Some methods for classification and analysis of multivariate observations. In: Proceedings of the 5th Berkeley Symposium on Mathematical Statistics and Probability, vol. 1, no. 14, pp. 281–297 (1967)
32. Mahalanobis, P.C.: On the generalized distance in statistics. In: Proceedings of the National Institute of Science of India, Calcutta (1936)
33. Maloney, P.W., Grawitch, M.J., Barber, L.K.: The multi-factor structure of the brief self-control scale: discriminant validity of restraint and impulsivity. J. Res. Pers. **46**(1), 111–115 (2012)
34. Mazzoni, E., Cannata, D., Baiocco, L.: Focused, not lost: the mediating role of temporal dissociation and focused immersion on problematic Internet use. Behav. Inf. Technol. **36**(1), 11–20 (2017)
35. Mazzoni, E., Baiocco, L., Cannata, D., Dimas, I.: Is Internet the cherry on top or a crutch? Offline social support as moderator of the outcomes of online social support on problematic internet use. Comput. Hum. Behav. **56**, 369–374 (2016)
36. Mazzoni, E., Iannone, M.: From high school to university: impact of social networking sites on social capital in the transitions of emerging adults. Br. J. Edu. Technol. **45**(2), 303–315 (2014)
37. Park, H.S., Jun, C.H.: A simple and fast algorithm for K-medoids clustering. Exp. Syst. Appl. **36**(2), 3336–3341 (2009)
38. Prezza, M., Trombaccia, F.R., Armento, L.: La scala dell'autostima di Rosenberg: Traduzione e validazione Italiana. Giunti Organizzazioni Speciali (1997)
39. Rosenberg, M.: Rosenberg Self-Esteem scale (RSE). Acceptance and Commitment Therapy. Measures Package, 61. Society and the Adolescent Self-image (1965)
40. Shaughnessy, J., Zechmeister, E., Zeichmeister, J.: Research Methods in Psychology, 10th edn. McGraw-Hill, New York (2014)
41. Steinfield, C., Ellison, N.B., Lampe, C.: Social capital, self-esteem, and use of online social network sites: a longitudinal analysis. J. Appl. Dev. Psychol. **29**, 435–445 (2008)
42. Tangney, J.P., Baumeister, R.F., Boone, A.L.: High self-control predicts good adjustment, less pathology, better grades, and interpersonal success. J. Pers. **72**(2), 271–324 (2004)
43. Thorsteinsson, E.B., Davey, L.: Adolescents' compulsive Internet use and depression: a longitudinal study. Open J. Depression **3**(1), 13 (2014)
44. Turkle, S.: Alone Together: Why We Expect More from Technology and Less from Each Other. Basic Books (2017)
45. Valkenburg, P.M., Peter, J., Schouten, A.P.: Friend networking sites and their relationship to adolescents' well-being and social self-esteem. CyberPsychol. Behav. **9**(5), 584–590 (2006)
46. Walach, Harald, Buchheld, Nina, Buttenmüller, Valentin, Kleinknecht, Norman, Schmidt, Stefan: Measuring mindfulness—the Freiburg Mindfulness Inventory (FMI). Pers. Ind. Differ. **40**(8), 1543–1555 (2006)
47. Wallace, P.: Internet addiction disorder and youth: there are growing concerns about compulsive online activity and that this could impede students' performance and social lives. EMBO Rep. **15**(1), 12–16 (2014)

48. Wang, E.S.T., Wang, M.C.H.: Social support and social interaction ties on internet addiction: integrating online and offline contexts. Cyberpsychol. Behav. Soc. Netw. **16**(11), 843–849 (2013)
49. Zywica, J., Danowski, J.: The faces of Facebookers: investigating social enhancement and social compensation hypotheses; predicting Facebook^{TM} and offline popularity from sociability and self-esteem, and mapping the meanings of popularity with semantic networks. J. Comput. Mediat. Commun. **14**(1), 1–34 (2008)

On Monotonicity Detection in Simplicial Branch and Bound over a Simplex

L. G. Casado[1]($\boxtimes$)(iD), B. G.-Tóth[2](iD), E. M. T. Hendrix[3](iD), and F. Messine[4](iD)

[1] Informatics, CeiA3, Almería University, Almería, Spain
leo@ual.es
[2] Informatics, University of Szeged, Szeged, Hungary
boglarka@inf.szte.hu
[3] Computer Architecture, Universidad de Málaga, Málaga, Spain
eligius@uma.es
[4] LAPLACE-ENSEEIHT, University of Toulouse, Toulouse, France
frederic.messine@laplace.univ-tlse.fr

Abstract. The concept of exploiting proven monotonicity for dimension reduction and elimination of partition sets is well known in the field of Interval Arithmetic Branch and Bound (B&B). Part of the concepts can be applied in simplicial B&B over a box. The focus of our research is here on minimizing a function over a lower simplicial dimension feasible set, like in blending and portfolio optimization problems. How can monotonicity be detected and be exploited in a B&B context? We found that feasible directions can be used to derive bounds on the directional derivative. Specifically, Linear Programming can be used to detect the sharpest bounds.

Keywords: Global optimization · Simplex · Branch and bound

1 Introduction

Monotonicity considerations to remove subsets and to reduce dimension has a long tradition in Interval Arithmetic based branch and bound, see [4,8]. The basic property is to be able to remove interior boxes where the function is monotone over the box and to reduce the dimension with respect to monotone components when a box facet is at the boundary of the search space.

Ideas of monotonicity were not investigated in the simplicial branch and bound overview book [10]. More recently, [2,3,5,6] extended monotonicity considerations towards simplicial partition sets. One of the main observations is that if the function to be minimized, f, is monotonically increasing in a direction from a facet of a simplicial partition set S towards its opposite vertex, then S can be reduced to F and consequently, we have a simplicial dimension reduction.

This paper has been supported by The Spanish Ministry (RTI2018-095993-B-I00) in part financed by the European Regional Development Fund (ERDF).

O. Gervasi et al. (Eds.): ICCSA 2022 Workshops, LNCS 13378, pp. 113–126, 2022.
https://doi.org/10.1007/978-3-031-10562-3_9

As was shown in [2], if F is not border, i.e. not included in one of the faces of the feasible set with the same dimension, then S can be removed from consideration.

The latter question is mainly convenient if the feasible set is a box, as considered in traditional simplicial B&B. However, if the feasible set is a simplex with a dimension lower than that of the function to be minimized, the determination of monotonicity and of a facet being border is more challenging. The focus of this paper is mainly on these research questions. How to demonstrate monotonicity and how to capture that a facet is border.

To investigate these question, Sect. 2 introduces mathematical properties of monotonicity over simplicial sets. Section 3 then discusses some special cases together with LP and MIP models to find monotone directions. Section 4 describes how to keep track of border facets, while Sect. 5 summarises our findings.

2   Mathematical Notation and Properties

2.1   Notation

We consider the minimization of a continuously differentiable function $f : \mathbb{R}^n \to \mathbb{R}$, over a feasible set Δ, which is an $p-$simplex, i.e. $\Delta := \mathrm{conv}(\mathcal{W})$ is defined by a set of $p + 1$ affine independent vectors that serve as vertices

$$\mathcal{W} := \{v_0, \dots, v_p\} \subset \mathbb{R}^n, p < n. \tag{1}$$

The idea is to find or enclose all global minimum points of

$$\min f(x), x \in \Delta. \tag{2}$$

The consideration of $n-$dimensional functions over $m < n$ lower dimensional simplicial feasible area appears for instance in blending problems [1]. Our context is that of a branch and bound algorithm to enclose all minimum points of f on Δ. In contrast to the algorithms described in [10], the used partition sets are $m-$simplices S, where $m \leq p$. This means $S := \mathrm{conv}(\mathcal{V})$ with $\mathcal{V}$ a set of $m + 1 = |\mathcal{V}|$ vertices. The branch and bound algorithm works with a set Λ of partition sets, which as a whole include all global minimum points.

Although we usually limit our context to the use of longest edge bisection, where the longest edge (v, w) of a partition set S is bisected using mid-point $x := \frac{v+w}{2}$, we pose the monotonicity question in a larger context where any partition method may be used, as described in [7]. The set of evaluated points that serve as vertices of the partition sets is denoted by X. Specifically, we focus on dimension reduction due to monotonicity considerations, where a set $\mathcal{V}$ of vertices of $m-$simplex S is reduced to $\mathcal{V} \setminus \{v\}$ and S is replaced by one (or more) of its facets $F := \mathrm{conv}(\mathcal{V} \setminus \{v\})$ for some $v \in \mathcal{V}$. Notice that F is an $(m-1)$-simplex. It may be clear that for $m = 0$, the $0-$simplex $S = \mathrm{conv}(\{v\})$ is an individual point and does not have faces. Its dimension cannot be reduced.

The centroid of $m-$simplex $S = \text{conv}(\{v_0, v_1, \ldots, v_m\})$ is given by $c := \frac{1}{m+1} \sum_{j=0}^{m} v_j$ and the relative interior is defined by

$$\text{rint}(S) = \{x = \sum_j \lambda_j v_j, \lambda_j > 0, j = 0, \ldots, m, \sum_{j=0}^{m} \lambda_j = 1\}. \tag{3}$$

The relative boundary of a simplex S is defined by removing the relative interior from it. Given a simplicial partition set S, we are interested in whether its (simplicial) facets F are border with respect to the feasible set Δ. In general, we can define a simplex to be border with respect to a simplicial feasible set.

Definition 1. *Given p-simplex feasible area Δ. An $m-$simplex S with $m < p$ is called* border *with respect to Δ if there exists an m-simplex face φ of Δ, such that $S \subseteq \varphi$.*

One of the main questions is how to determine whether a facet of a simplex is border in a numerically efficient way. Border facets and the concept of monotonicity are used to reject a simplex or to reduce its simplicial dimension.

Relevant information is an enclosure G of the gradient $\nabla f(x) \subseteq G := [\underline{G}, \overline{G}], \forall x \in S$. Interval vector G can be calculated by Interval Automatic Differentiation over the interval hull of a simplex S, see [9,11]. Now consider directional vector d as the difference between two points in S, then the corresponding directional derivative $d^T \nabla f(x)$ is also included in the inner product

$$d^T G = \left[\underline{d^T G}, \overline{d^T G}\right] = \left[\sum_{i=1}^{n} \min\{d_i \underline{G}_i, d_i \overline{G}_i\}, \sum_{i=1}^{n} \max\{d_i \underline{G}_i, d_i \overline{G}_i\}\right]. \tag{4}$$

2.2   Mathematical Properties on Monotonicity

The monotonicity is based on directional derivative bounds of (4). Notice that condition $0 \notin G$ is necessary to have monotonicity, but not sufficient. The question is which direction d to consider. The most general result for an $m-$simplex is the following.

Proposition 1. *Let $S \subseteq \Delta$ be an $m-$simplex with gradient enclosure G. If $\exists\, x, y \in S$, such that direction $d = x - y$ has corresponding directional derivative bounds (4) with $0 \notin [\underline{d^T G}, \overline{d^T G}]$ then $\text{rint}(S)$ does not contain a global minimum point of (2).*

Proof. Consider $z \in \text{rint}(S)$. As z is in the relative interior, there exists a feasible direction d in which lower function values can be found, i.e. $\exists \varepsilon \in \mathbb{R}$ small enough, such that $z + \varepsilon d \in S$ and $f(z + \varepsilon d) < f(z)$. So z cannot be a minimum point of f. □

The elaboration for an algorithm depends on the choice of the direction d and the way to compute it.

Corollary 1. *Let $S \subseteq \Delta$ be an m−simplex as partition set in a branch and bound algorithm with corresponding gradient enclosure G. If the conditions of Proposition 1 apply and S has no border facets, then S can be rejected.*

The argument is that the relative boundary of S may contain a global minimum point, but the same point is enclosed in the relative boundary of another partition set.

Given that the minimum is not in $\mathrm{rint}(S)$, we have to decide which of the facets to focus on. In [2], we made use of the following property in the design of a specific algorithm.

Proposition 2. *Given m−simplex $S = \mathrm{conv}(\mathcal{V})$ with centroid c and a facet F generated by removing vertex v from $\mathcal{V}$. Consider direction $d = v − c$. If $\underline{d^T G} > 0$, then the facet F contains all minimum points in S, i.e. $\mathrm{argmin}_{x \in S} f(x) \subseteq F$.*

Practically, this means that S can be replaced by F if $\underline{d^T G} > 0$. However, a similar reasoning applies as in Corollary 1; if F is a non-border facet, then simplex S can be removed from further consideration in a branch and bound context. The idea is again that faces of F may contain the minimum. However, because we are dealing with a partition, the same points are also included in other simplicial partition sets.

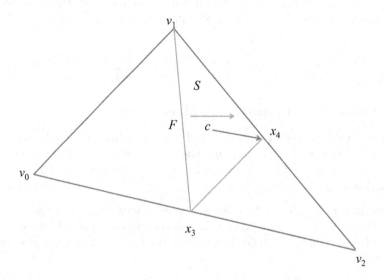

Fig. 1. Three partition sets generated by bisection, using bisection points x_3 and x_4. We focus on monotone directions in S.

Example 1. For the illustration of the concept, consider the simplices in Fig. 1. It shows three partition sets generated by bisection, using bisection points x_3

and x_4. Consider $S = \text{conv}(\mathcal{V})$ with $\mathcal{V} = \{v_1, x_3, x_4\}$, where we assume that the orange direction provides a monotonously increasing direction. According to Proposition 1, the interior of S does not contain a global minimum point. Let the blue direction provide a direction $d = v - c$ for which the lower bound of (4) is positive for facet $F = \text{conv}(\mathcal{V} \setminus \{v\})$ with $v = x_4$. According to Proposition 2, facet $F = \text{conv}(\{v_1, x_3\})$ contains all minimum points on S. Now, the border considerations show that we even can remove S, as there is another partition set at its left, that encloses all minimum points.

There are two questions we address in this paper.

- Is there a way to show that a direction d in which f is monotonic on S exists?
- The direction $d = v - c$ for a facet F may not be monotonically increasing, but can there be another direction from facet F to vertex v?

3   Cases of Directional Derivatives

To prove that there exists a monotone direction in an m−simplex, at least we should have $0 \notin G$. This is a necessary, but not sufficient condition for an m−simplex, $m < n$. To prove that such a direction exists, according to Proposition 1, we need to find a direction $d = x - y$, with $x, y \in S$ corresponding to a positive lower bound of the directional derivative

$$0 < \underline{d^T G} = \sum_{i=1}^{n} \min\{d_i \underline{G}_i, d_i \overline{G}_i\}. \tag{5}$$

Finding such a direction can be done by searching for the steepest monotone direction $\max_d \underline{d^T G}$. Consider the terms $z_i = \min\{d_i \underline{G}_i, d_i \overline{G}_i\}$. This means we can write $\underline{d^T G} = \sum_{i=1}^{n} z_i$. If we fix one of the point $x \in S$ in $d = x - y$, then the term $z_i(y) = \min\{(x_i - y_i)\underline{G}_i, (x_i - y_i)\overline{G}_i\}$ is a concave function in y, as it is the minimum of two affine functions. Therefore, the lower bound on the directional derivative $\underline{g}(y) := \overline{(x - y)^T G}$ is a concave function being the sum of concave terms. Similarly, it can be shown that the upper bound $\overline{g}$ on the directional derivative is a convex function. We will illustrate this with an example and then show how an LP problem can be formulated to find a maximum of $\underline{g}(y)$.

Example 2. Consider a simplex $S = \text{conv}(\mathcal{V})$ with $\mathcal{V} = \{v_0, v_1, v_2\}$ in $\mathbb{R}^6$ with $v_0 = 0, v_1 = (1, -2, 3, -4, 5, 6)^T$ and $v_2 = (0, 3, -2, 5, -4, -5)^T$. In $d = x - y$, we take as fixed point $x = v_0$ and vary y over the edge between v_1 and v_2 as suggested in Proposition 2, so $y = \lambda v_1 + (1 - \lambda)v_2, 0 \le \lambda \le 1$. Figure 2 sketches the oncave piece-wise linear shape of $\underline{g}$ as function of λ and the convex shape of $\overline{g}$.

Looking for the existence of a positive value of $\underline{d^T G}$ for some direction d, we can fix x to the centroid in the directional vector, i.e. $d = c - y$. The maximization

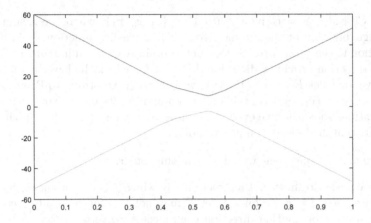

Fig. 2. Shape of concave $\underline{g}(y) = \underline{g}(\lambda v_1 + (1-\lambda)v_2$ as function of λ in orange and convex $\overline{g}$ in purple (Color figure online).

of concave piece-wise linear function $\underline{g}(y)$ over $y \in S$ can be formulated as the following LP problem.

$$\max \sum_{i=1}^{n} z_i$$

$$\text{s.t.} \quad z_i \leq (c_i - y_i)\underline{G}_i, \quad i = 1, \ldots, n$$

$$z_i \leq (c_i - y_i)\overline{G}_i, \quad i = 1, \ldots, n$$

$$\sum_{j=1}^{m} \lambda_j = 1 \tag{6}$$

$$y = \sum_{j=0}^{m} \lambda_j v_j$$

$$\lambda_j \geq 0, j = 0, \ldots, m$$

If there is no monotone direction, then $y = c$ and $\sum_{i=1}^{m} z_i = 0$.

Example 3. For the illustration of the concept, consider the 2-simplex S defined by three vertices $\mathcal{V} = \{(4, 0, 1)^T, (0, 0, 0)^T, (3, 2, 1)^T\}$ in 3-dimensional space, projected in 2D for Fig. 3. Its centroid is given by $\frac{1}{3}(7, 2, 2)^T$. Now let the bounds of the gradient be given by $\underline{G} = (-3, 1, 0)^T$ and $\overline{G} = (1, 2, 1)^T$. For none of the directions $d = v_j - c$, we have that lower bound $\underline{d}^T\underline{G}$ is positive. Running the LP (6) provides us with a positive directional derivative bound of $\frac{2}{3}$ for the point $y = (\frac{7}{3}, 0, \frac{7}{12})^T$. The monotone direction $c - y$ is drawn by an orange arrow in Fig. 3. This means that f is monotone on S. This illustrates that checking a finite number of directions over the simplex is not necessarily sufficient to prove that f is monotone. The LP (6) provides a numerical proof of f being monotone over S or not. Notice again that $y = c$ is a feasible solution of the LP yielding an objective function value of zero as soon as monotonicity cannot be proven.

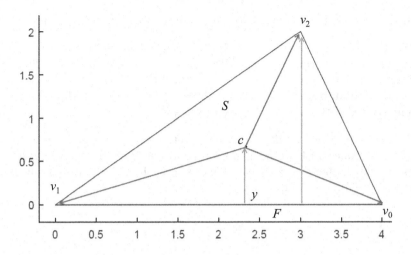

Fig. 3. 2D image of the 2-simplex S. The point y provides a positive lower bound on the directional derivative in direction $d = c - y$.

Following the line of reasoning of the example, according to Proposition 1, we can conclude that simplex S can be replaced by its facets. Now at least one of the facets is of interest, although that cannot be concluded by only considering the directions $d = v - c$ for this specific example. The question is now whether there is a monotone direction from a vertex v to a point y on the facet $F = \mathrm{conv}(\mathcal{V} \setminus \{v\})$ for which $d = v - y$ has a positive lower bound on the directional derivative $\underline{d}^T G > 0$. If such direction can be found, according to a similar reasoning as in Proposition 2, the minimum of the simplex is in F.

To answer the question whether such a direction would exist, we can solve the following LP for a specific facet F maximizing the lower bound of a directional derivative. Consider now the ordered vertex set $\mathcal{V} := \{v_0, v_1, \ldots, v_m\}$ and the vertex set of F given as $\{v_1, \ldots, v_m\} := \mathcal{V} \setminus \{v_0\}$. Focusing on the direction $v_0 - y$ with $y = \sum_{j=1}^m \lambda_j v_j$, we can demonstrate that there is a (maximum) positive directional derivative, if it exists, by solving the LP

$$\max \sum_{i=1}^n z_i$$

$$\text{s.t.} \quad z_i \leq \underline{G}_i(v_{0i} - y_i), \ i = 1, \ldots, n$$

$$z_i \leq \overline{G}_i(v_{0i} - y_i), \ i = 1, \ldots, n$$

$$\sum_{j=1}^m \lambda_j = 1 \tag{7}$$

$$y = \sum_{j=1}^m \lambda_j v_j$$

$$\lambda_j \geq 0, j = 1, \ldots, m.$$

If the result is positive, we have proven all minima of f over S are on F.

Example 4. We can now show how LP (7) is working, following the illustration in Fig. 3 for facet $F = \text{conv}\{v_0, v_1\}$. Although the lower bound on the directional derivative of $d = v_2 - c$ is not positive, the LP will provide a solution $y = (3, 0, \frac{3}{4})^T$ with an objective function value of 2. The corresponding direction is also illustrated with an arrow between $(3, 0, \frac{3}{4})^T$ and v_2 in Fig. 3.

In a procedure for searching for such a facet, in the worst case we need to solve LP (7) $m+1$ times. Instead, we might solve only one Mixed Integer Programming problem (MIP) where a binary variable δ_j selects the facet corresponding to the most positive directional derivative.

$$\max \sum_{i=1}^{n} z_i$$

$$\text{s.t.} \quad z_i \leq \underline{G}_i d_i, \quad i = 1, \ldots, n$$

$$z_i \leq \overline{G}_i d_i, \quad i = 1, \ldots, n$$

$$\sum_{j=0}^{m} \lambda_j = 1$$

$$d = \sum_{j=0}^{m} \delta_j v_j - \sum_{j=0}^{m} \lambda_j v_j \tag{8}$$

$$\sum_{j=0}^{m} \delta_j = 1.$$

$$\lambda_j \geq 0, j = 0, \ldots, m$$

$$\delta_j \in \{0, 1\}, j = 0, \ldots, m$$

MIP (8) can be used to replace LP problem (7) if such a monotonously increasing direction towards one of the vertices exists. If this is not the case, the solution is 0 with the direction $d = 0$. However, there still may exist an increasing direction according to LP (6).

Example 5. For our example in Fig. 3 with $S = \text{conv}\{v_0, v_1, v_2\}$ the MIP (8) finds indeed the positive objective function value of $\sum_{i=1}^{n} z_i$, stating that the facet corresponding to $\delta_2 = 1$ provides the maximum derivative lower bound.

The two steps, looking whether a monotone direction exists and identifying which facet contains all the minima (if any), can also be done in one step. Thus, instead of solving LP (6) and (7) $m+1$ times (or MIP (8)), we can solve directly one MIP as follows. Let $d = x - y$, where $x, y \in S$, so $x = \sum_{j=0}^{m} \lambda_j v_j$ and $y = \sum_{j=0}^{m} \mu_j v_j$. Consider direction $d = x - y = \sum_{j=0}^{m} (\lambda_j - \mu_j) v_j$.

$$\max \sum_{j=0}^{m} \delta_j$$

$$\text{s.t.} \quad z_i \leq \underline{G}_i d_i, \; i = 1, \ldots, n$$

$$z_i \leq \overline{G}_i d_i, \; i = 1, \ldots, n$$

$$\sum_{j=0}^{m} \lambda_j = 1$$

$$\sum_{j=0}^{m} \mu_j = 1$$

$$d = \sum_{j=0}^{m} (\lambda_j - \mu_j) v_j \quad (9)$$

$$\sum_{i=1}^{n} z_i \geq \varepsilon$$

$$\mu_j + 1 \geq 2\delta_j \quad \forall j = 1, \ldots, m$$

$$\sum_{j=0}^{m} \delta_j \leq 1.$$

$$\mu_j, \lambda_j \in [0,1], j = 1, \ldots, m$$

$$\delta_j \in \{0,1\}, j = 1, \ldots, m$$

A solution with $\mu_k = 1$ and $\mu_j = 0, j \neq k$ represents a direction d pointing to vertex v_k. A solution with $\lambda_k = 1$ and $\lambda_j = 0, j \neq k$ represents a direction pointing from vertex v_k. We connect μ_j with binary variables δ_j such that $\delta_j = 1$ implies $\mu_j = 1$. The inequality $\sum_{i=1}^{n} z_i \geq \varepsilon$ with $\varepsilon > 0$ assures d is a monotone direction. Moreover, it forces $\mu \neq \lambda$, because otherwise $z = 0$ as well. If no monotone direction exists, (9) has no feasible solution.

The objective is to maximize the sum of δ_j, meaning that we aim at finding a monotone direction with $\mu_k = 1$ corresponding to $\delta_k = 1$. In this case, we know that facet $F = \text{conv}(\mathcal{V} \setminus \{v_k\})$ contains all minima according to Proposition 2. If the objective is zero, there is no facet containing all the minima, i.e. there is no $\delta_k = 1$, but there is a monotone direction d.

Example 6. Following our example in Fig. 3 with $S = \text{conv}\{v_0, v_1, v_2\}$ the MIP (9) finds the positive objective function value 1 for $\sum_{i=1}^{n} z_i$, stating that the facet corresponding to $\delta_2 = 1$ provides a positive directional derivative for $d = v_2 - v_0$. Notice, that this direction is not the maximum directional derivative, but still positive, which is the main question.

Interestingly, solving the LP-s and MIP-s in Matlab, the necessary time for this example was counter-intuitive: LP (6) 0.521, LP (7) 0.033, while MIP (9) 0.048 s.

We investigated whether the counter-intuitive result of a smaller solution time for the MIP than for the LP is a general trend. Therefore, we compared the solution time and effectiveness of formulations (6), (7), (8) and (9). We took 447 simplices from a branch and bound process over the functions Hartman 3, 4 and 6 that, as the name suggests, have dimension 3, 4 and 6. We have used the routines `linprog` and `intlinprog` in Matlab setting IntegerTolerance to $1e-6$ and ConstraintTolerance to $1e-8$. The result is given in Table 1.

Table 1. Time and effectiveness of the LP and MIP formulations

	LP (6)	LP (7)	MIP (8)	MIP (9)
Hartman 3				
Monotone dir exists	81.6%	–	–	77.6%
No monotone dir	18.4%	–	–	18.4%
Mon.neg.dir.from_F exists	–	73.5%	73.5%	69.4%
No Mon.dir.from_F exists	–	24.5%	24.5%	18.4%
Best result not found		2.0%	2.0%	4.1%
Time	0.018	0.029	0.009	0.012
Hartman 4				
Monotone dir exists	92.3%	–	–	80.2%
No monotone dir	7.7%	–	–	7.7%
Mon.neg.dir.from_F exists	–	67.0%	60.4%	57.1%
No Mon.dir.from_F exists	–	33.0%	33.0%	23.1%
Best result not found	–	–	6.6%	12.1%
Time	0.014	0.030	0.008	0.016
Hartman 6				
Monotone dir exists	98.0%	–	–	87.3%
No monotone dir	2.0%	–	–	2.0%
Mon.neg.dir.from_F exists	–	83.1%	73.0%	72.6%
No Mon.dir.from_F exists	–	16.9%	16.9%	14.7%
Best result not found	–	–	10.1%	10.7%
Time	0.015	0.038	0.009	0.011

In each line of the table we give for the LP and MIP formulations the percentage of the effectiveness measured as proven monotonicity. For instance, for problem Hartman 3, LP (6) proved in 81.6% that there is a monotone directional derivative.

We can prove there is a monotone decreasing direction from any facet by solving LP (7) for all vertices, or by solving any of the MIP-s once. Comparing the formulations, LP (7) is the strongest, while MIP (9) is the weakest due to the ε in its formulations, which is hard to set together with the tolerances. The percentage development shows that, as the found monotone directions are less and less solving LP (7), MIP (8) and MIP (9), and the percentage where the best results are not found goes up in the same order.

Surprisingly, the average computing time is the smallest for MIP (8), followed by MIP (9) or LP (6), and the slowest is LP (7). The latter is no surprise as in that case we added up the time needed to solve LP (7) for all facets, or until it found a monotone decreasing direction from a facet.

4   Keeping Track of Border Facets

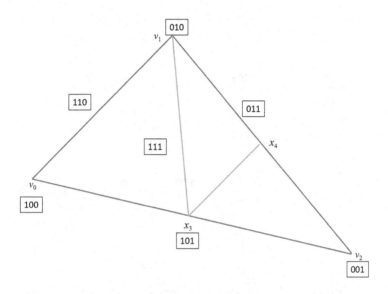

Fig. 4. Faces of the feasible set with bisection points x_3 and x_4.

For a box constrained feasible set, finding the border status of a partition set is relatively easy, as it is determined by lower and upper bounds on the components and the correspondence with the simplicial partition sets. To determine the border status of a given facet in a simplicial feasible set, we use a labelling system to find out which minimum dimensional face of Δ the F is included in. This is done by assigning to each face φ of feasible set $\Delta = \text{conv}(\mathcal{W})$ a label $\mathcal{B}(\varphi)$, starting with the vertex faces labeled $\mathcal{B}(v_j) = 0...010...0$ where the only 1 is the jth bit, for $j = 0, \ldots, n$. Each face φ, which is a convex combination of vertices $\mathcal{V} \subseteq \mathcal{W}$, the corresponding bit-string $\mathcal{B}(\varphi)$ has a value 1 for each vertex $v \in \mathcal{V}$ in the same position as in $\mathcal{B}(v)$. For instance, in Fig. 4, the edge (v_1, v_2) has label 011, and the simplex Δ has label 111. In fact, for an m-simplex face φ of the feasible set, its label is given by the bitwise OR operation (**BitOr**) of the label of all its vertices. The complete face graph is given in Fig. 5 for a 4-vertex simplex.

In a bisection refinement, the label can easily be determined. After bisecting the original set of vertices $\mathcal{W}$, we store the bisection points in set X. For instance in Fig. 4, we have $X = \{v_0, v_1, v_2, x_3, x_4\}$. We label all generated points $x \in X$, which serve as vertices for the partition sets. The label of point x is the same as the label of the minimum dimensional face φ of the feasible set x is in. For instance, the label of x_3 is 101, the same as the label of face $\mathrm{conv}(v_0, v_2)$. During bisection, a new vertex $x = \frac{v+w}{2}$ gets label $\mathcal{B}(x) = \mathbf{BitOr}(\mathcal{B}(v), \mathcal{B}(w))$.

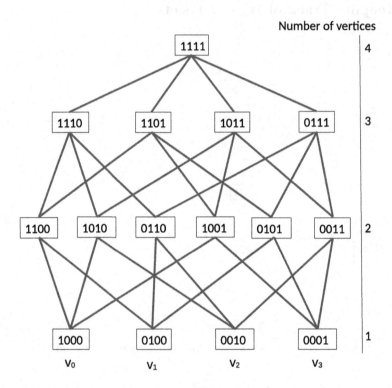

Fig. 5. Face graph for a 3-simplex. Binary boxed labels for each face.

Given an $m-$simplex $S = \mathrm{conv}(\mathcal{V})$, the question is what is the label of the (smallest dimensional) face φ it is included in. This is determined by the label $\mathcal{B}(\varphi) = \mathbf{BitOr}(\mathcal{B}(\mathcal{V}))$, to be interpreted as a bitor on all its vertex labels. The number of ones of a bitstring $\mathcal{B}(\varphi)$ is denoted by $|\mathcal{B}(\varphi)|$, giving the number of vertices of φ. According to Definition 1, $m-$simplex S is border if there exists an $m-$simplex face φ of the feasible set including S ($m < p$).

Proposition 3. *Given $m-$simplex $S = \mathrm{conv}(\mathcal{V})$ with $m < p$, if $|\mathbf{BitOr}(\mathcal{B}(\mathcal{V}))| = m + 1$, then simplex S is border.*

Proof. Consider the face φ, which is the minimal dimensional face containing S, i.e. label $\mathcal{B}(\varphi) = \mathbf{BitOr}(\mathcal{B}(\mathcal{V}))$. As $|\mathbf{BitOr}(\mathcal{B}(\mathcal{V}))| = m + 1$, we have $|\mathcal{B}(\varphi)| = m + 1$, thus φ is an $m-$simplex. Therefore, S is enclosed by an $m-$simplex face of the feasible set and thus is border.

For example in Fig. 4, the edge (x_3, v_2) is border, as $|\mathbf{BitOr}(\mathcal{B}(\{x_3, v_2\}))| =$ $|\mathbf{BitOr}(101, 001)| = |101| = 2$ corresponding to face $\mathrm{conv}(\{v_0, v_2\})$. In contrast, edge (x_3, x_4) is not border, because $|\mathbf{BitOr}(\mathcal{B}(\{x_3, x_4\}))| = |111| = 3 \neq 2$. In fact, the minimum dimensional face it is included in, is Δ itself.

5 Conclusions

The interest in monotonicity in simplicial branch and bound is relatively recent. Given bounds on the gradient, the essential idea is that we have to check bounds on the directional derivative for a feasible direction related to the simplicial dimension of a partition set. In this paper, we show that the determination of monotonicity of a function over a simplicial partition set can be done by solving an LP problem. Moreover, it is possible for a facet to find the highest lower bound based on a specific LP. The outcome determines, whether it is possible to reduce the dimension of the simplicial partition set or to decide to remove it from further consideration. Several steps can be combined by solving a specific MIP problem.

For the decision on the removal of a simplex, it is relevant whether a facet is border with respect to the feasible set. For a box constrained feasible set this is relatively easy. This paper shows that by consistently labeling points that serve as vertices, it is possible to determine a facet is border or not.

In our future work, we implement the LP type of tests in a simplicial branch and bound framework to investigate whether the number of generated simplices is decreasing compared to an algorithm, where one direction is tested per facet.

References

1. Casado, L.G., Hendrix, E.M.T., García, I.: Infeasibility spheres for finding robust solutions of blending problems with quadratic constraints. J. Global Optim. **39**(4), 577–593 (2007). https://doi.org/10.1007/s10898-007-9157-x
2. G.-Tóth, B., Casado, L.G., Hendrix, E.M.T., Messine, F.: On new methods to construct lower bounds in simplicial branch and bound based on interval arithmetic. J. Global Optim. **80**(4), 779–804 (2021). https://doi.org/10.1007/s10898-021-01053-8
3. G.-Tóth, B., Hendrix, E.M.T., Casado, L.G.: On monotonicity and search strategies in face based copositivity detection algorithms. Cent. Eur. J. Oper. Res. **30**, 1071–1092 (2021). https://doi.org/10.1007/s10100-021-00737-6
4. Taft, E., Hansen, E., Nashed, Z., Walster, W. G.: Gobal Optimization Using Interval Analysis. 2nd edn.,p. 728. CRC Press, Boca Raton (2003). https://doi.org/10.1201/9780203026922
5. Hendrix, E.M.T., Tóth, B., Messine, F., Casado, L.G.: On derivative based bounding for simplicial branch and bound. RAIRO **55**(3), 2023–2034 (2021). https://doi.org/10.1051/ro/2021081
6. Hendrix, E., Salmerón, J., Casado, L.: On function monotonicity in simplicial branch and bound. In: LeGO 2018, Leiden, The Netherlands, p. 4 (September 2018). https://doi.org/10.1063/1.5089974

7. Horst, R.: On generalized bisection of n-simplices. Math. Computat. **66**(218), 691–699 (1997). https://doi.org/10.1090/s0025-5718-97-00809-0

8. Kearfott, R.B.: An interval branch and bound algorithm for bound constrained optimization problems. J. Global Optim. **2**(3), 259–280 (1992). https://doi.org/10.1007/BF00171829

9. Moore, R.E., Kearfott, R.B., Cloud, M.J.: Introduction to Interval Analysis. Society for Industrial and Applied Mathematics, USA (2009). https://doi.org/10.1137/1.9780898717716

10. Paulavičius, R., Žilinskas, J.: Simplicial Global Optimization. Springer, New York (2014). https://doi.org/10.1007/978-1-4614-9093-7

11. Rall, L.B. (ed.): Examples of software for automatic differentiation and generation of Taylor coefficients. In: Automatic Differentiation: Techniques and Applications. LNCS, vol. 120, pp. 54–90. Springer, Heidelberg (1981). https://doi.org/10.1007/3-540-10861-0_5

Virtual Screening Based on Electrostatic Similarity and Flexible Ligands

Savíns Puertas-Martín[1,3](✉)[iD], Juana L. Redondo[1][iD],
Antonio J. Banegas-Luna[2][iD], Ester M. Garzón[1][iD], Horacio Pérez-Sánchez[2][iD],
Valerie J. Gillet[3][iD], and Pilar M. Ortigosa[1][iD]

[1] Supercomputing - Algorithms Research Group (SAL), University of Almería,
Agrifood Campus of International Excellence, ceiA3, 04120 Almería, Spain
{savinspm,jlredondo,gmartin,ortigosa}@ual.es
[2] Structural Bioinformatics and High Performance Computing Research Group
(BIO-HPC), Universidad Católica de Murcia (UCAM), 30107 Murcia, Spain
ajbanegas@alu.ucam.edu, hperez@ucam.edu
[3] Information School, University of Sheffield, Sheffield S1 4DP, UK
v.gillet@sheffield.ac.uk

Abstract. Virtual Screening (VS) is a technique aimed at reducing
the time and budget required when working on drug discovery campaigns. The idea consists of applying computational procedures to prefilter databases to a subset of potential compounds, to be characterized
experimentally in later phases.

The problem lies in the fact that the current VS methods make sim-
plifications, meaning they are not exhaustive. One particular common
simplification is to consider the molecules as rigid. Such an assumption
greatly reduces the computational complexity of the optimization problem to be solved, but it may result in poor or inefficient predictions. In
this work, we have extended the features of Optipharm, a recently developed piece of software, by applying a methodology that considers the
flexibility of the molecules. The new OptiPharm has several strengths
over its previous version. More precisely, (i) it includes a prefilter based
on molecule descriptors, (ii) simulates molecule flexibility by computing
different poses for each rotatable bond, (iii) reduces the search space
dimension, and (iv) introduces circular limits for the angular variables
to enhance searchability. As the results show, these improvements help
OptiPharm to achieve better predictions.

Keywords: Ligand based virtual screening · Molecule's flexibility ·
Optimization

1 Introduction

Virtual Screening (VS) methods can be divided into structure-based (SBVS) and
ligand-based (LBVS) methods [4,12,14] This work focuses on similarity LBVS
methods [2,16]. In these techniques, the starting point is a source drug whose

© The Author(s), under exclusive license to Springer Nature Switzerland AG 2022
O. Gervasi et al. (Eds.): ICCSA 2022 Workshops, LNCS 13378, pp. 127–139, 2022.
https://doi.org/10.1007/978-3-031-10562-3_10

shape, electrostatic potential, or other descriptor is known. This source ligand or crystal will be the target, and the virtual screening methods try to find the more similar molecules in an extensive database or chemolibrary. When calculating the electrostatic similarity between the target and a compound in the database, the more used methodology in the literature consists of optimizing in terms of shape by using Rapid Overlay of Chemical Structures (ROCS) [11], selecting a number N of compounds with the highest shape similarity values, and then evaluate them in terms of electrostatic similarity. Based on the assumption that a more realistic description of the compound bioactivity during the optimization procedure may help to obtain better predictions, a new version of OptiPharm was implemented in [10], which involves the direct optimization of the electrostatic similarity. As the results showed, the new methodology provided better predictions in electrostatic potential than the classical ones. In this work, we go a step forward and propose new improvements in OptiPharm aimed to reach even better predictions. To do so, we firstly include the flexibility of the molecules in the optimization procedure.

Protein flexibility is necessary for metabolism, transport, and function biological effects. Except for simple molecules such as O_2, both ligands and receptors are flexible molecules, which means that there is not a single three-dimensional representation of these molecules, but many. The conformational richness increases exponentially with the molecule's size, i.e., the more atoms (and therefore bonds, angles, and torsions) it possesses, the more degrees of freedom there are. These degrees of freedom are not additive but multiplicative, giving rise to many possible conformational states (see Fig. 1).

(a) A molecule of the target DB00331 that has some rotable bonds.   (b) A set of conformation generated from the rigid DB00331 molecule.

Fig. 1. A rigid molecule (a) can generate different conformations (b). An example for the DB00331 target from the DrugBank database is shown here. The Target structure has been painted green for both figures. (Color figure online)

For this reason, most of the studies are based on ligands where flexibility is considered to assume the protein to be almost rigid or with partial flexibility, so that they only rotate a maximum of the possible rotatable bonds [1,5,7]. In some cases, the solution is to perform the Virtual Screening considering the molecule as rigid, and then apply a process where the flexibility is studied for the number of rotatable bonds allowed by the algorithm [5]. This process sometimes consists of varying in a discrete way each of the angles of the rotatable bonds to find the best solution. Following any of these methods, the computational time is reduced, but many solutions keep unexplored. What we propose in this work is a previous analysis of the molecules. First, many descriptors are calculated, and the most representatives are selected to compute the difference between the target and the molecules in the database. Then best molecules are filtered and selected as flexible molecules and are applied a conformational generation process. This methodology allows exploring all the conformation of each compound widely but saves time by discarding uninteresting molecules.

Apart from considering the flexibility of the molecules, the new OptiPharm incorporates mechanisms of interest that enhance the search and helps to reduce the computational cost. As the results will show, all these improvements help provide better predictions in the molecules.

The rest of the paper is organized as follows. Section 2 describes the scoring function considered in this study and resumes the main ideas of OptiPharm, focusing on the new procedures and strategies developed. Section 3 summarizes the computational and scientific context taken into consideration for the experiments. Finally, Sects. 4 and 5 show the main results and conclusions inferred.

2   Methods

2.1   Electrostatic Similarity Scoring Function

The electrostatic similarities are obtained by numerical solution of the Poisson equation [3], viz:

$$\nabla\{\epsilon(r)\nabla\phi(r)\} = -\rho_{mol}(r) \tag{1}$$

where $\phi(r)$ is the electrostatic potential, $\epsilon(r)$ is the dielectric constant, and $\rho_{mol}(r)$ is the molecular charge distribution. Electrostatic similarity between two compounds is compared by determining E_{AB}:

$$E_{AB} = \int \phi^A(r)\phi^B(r)\Theta^A(r)\Theta^B(r)\mathbf{dr} \approx h^3 \sum_{ijk} \phi^A_{ijk}\phi^B_{ijk}\Theta^A_{ijk}\Theta^B_{ijk} \tag{2}$$

where Θ is a masking function to ensure potentials interior to the compound are not considered part of the comparison. The integral appearing in (2) is a volume integral, computed using a grid-spacing parameter, h.

Notice that the accuracy obtained from (2) depends on the number of atoms in the two compared molecules. To measure the similarity between compounds, regardless of the number of atoms that they are composed of and the descriptor used, the Tanimoto Similarity [6] value is computed as follows:

$$TcE = \frac{E_{AB}}{E_{AA} + E_{BB} - E_{AB}} \tag{3}$$

where E_{AB} is the A molecule overlaid onto B molecule. E_{AA} and E_{BB} is the overlap of the molecules A and B, respectively.

2.2 OptiPharm Algorithm

OptiPharm is a recent software designed explicitly for LBVS problems. It implements a global evolutionary optimizer capable of calculating the similarity between two compounds, a target and a query. To do so, it uses different methods in the optimization process to gradually adjust the position of the query while the target fixes its position. The interested reader is referred to [9,10] for an in-depth description of the original algorithm. In this work, we present a new version of OptiPharm. In the following, we briefly describe the new contributions.

To explore the solution space, OptiPharm works with a user-defined population of size M, which applies reproduction, selection, and improvement methods to each member of the population. A member or solution of this population represents the rotation and translation of the query molecule. Originally ten parameters were used to represent this modification, which means to work in a 10-dimensional search space. This paper presents a new version of OptiPharm, where the search space is reduced to 6 dimensions. The main change consists of replacing the use of quaternions with a semi-sphere parametrization, which simplifies the definition of the rotation axis. Consequently, searchability is enhanced due to the reduction of the search space dimension. Nevertheless, not only that, this new system avoids the repetition of the same rotation axis already explored.

This new mechanism provides improved freedom of exploration. In addition to reducing input parameters, the new version incorporates some problem knowledge, such as a mechanism to keep the angular variables between 0 and 2π in a continuous circular. So, if during the search an angle α takes a value greater than 2π, it is updated to the $\alpha - 2\pi$ value. In the previous version of OptiPharm, this value was updated to the maximum value of 2π.

2.3 Methodology

Procedure for Rigid Molecules

The process is trivial when working with rigid molecules and will be referred to throughout this paper under the name *Rigid*. As explained in the previous section, OptiPharm allows getting the best overlapping between two molecules, the target, and the query, to maximize the electrostatic similarity score. Consequently, when this procedure is repeated for each molecule in the database, their similarity score can be known. After that, the last step sorted the molecules by their similarity value. This procedure returns the most similar ones of interest since they can be successful potential drugs because they are the most similar to the target. Figure 2 shows this process to obtain a ranked list of compounds.

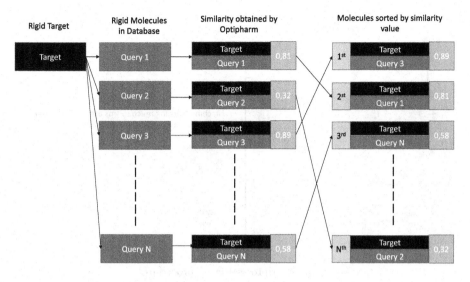

Fig. 2. Procedure to rank rigid molecules.

Procedure for Molecular Conformations

Working with flexible molecules with some rotational bonds implies modifying the methodology to obtain the similarity between a target molecule and a query molecule in the database. Multiple alternative conformers of this molecule are constructed by modifying the rotatable bonds with various rotation angles. This procedure simulates the flexibility in the rotatable bonds of a given molecule. However, the number of molecules in the database grows dramatically, and consequently, so does time.

A solution to this problem is to first discard those not promising compounds in the database and then generate conformations to the remaining molecules. This process, which we have called *Flexible* throughout the document, is explained in Fig. 3. In this figure, first, the descriptors are calculated for each molecule in the database. In this work, more than 4,000 different descriptors are obtained. However, many are not relevant or are repetitive, so different machine learning metrics are applied to discard them. In particular, variance and correlation are applied, and those descriptors whose values are 0 in most cases have been removed. This filter reduces the number of descriptors to a more limited number.

The obtained descriptors are then used to filter the molecules in the database. As can be seen in Fig. 4, the Euclidean distance between the query molecule and each of the targets is calculated. Once they are available, the compounds are ordered from smallest to largest by the distance value, and the best M compounds are selected based on an empirical cutoff value. Finally, several conformations are generated for the selected query compounds and the target.

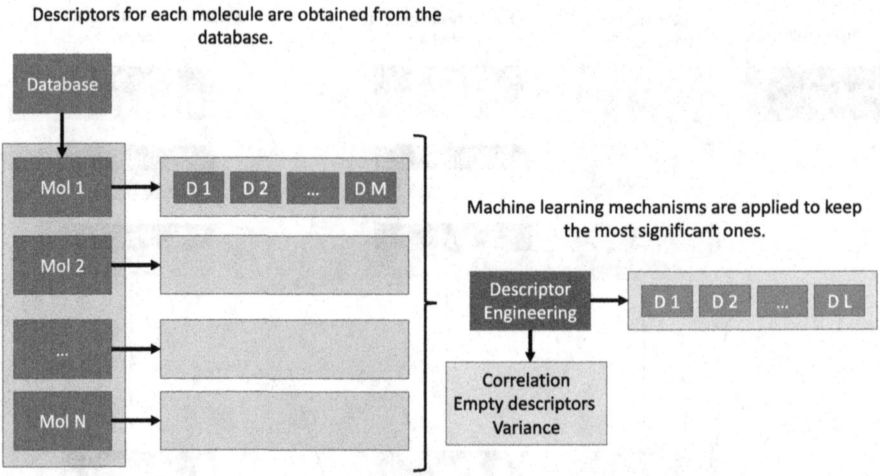

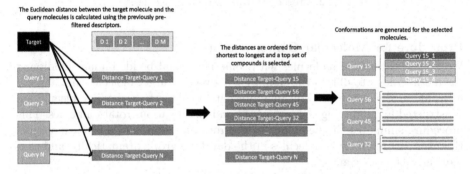

Fig. 3. Descriptor selection procedure.

Fig. 4. Molecule filtering process based on Euclidean distance and subsequent generation of conformations of selected molecules.

After generating multiple conformations for target and query, an optimization procedure is run using OptiPharm. Figure 5 shows such a procedure using an example where only three conformations have been generated for both the target and the query.

As shown in Fig. 5, an extensive comparison is performed, which involves running $nt \times nq$ times OptiPharm algorithm instead of just one for rigid molecules. In this exhaustive comparison, nt represents the number of conformations of the target molecule, and nq is the number of conformations of the query molecule. Once the maximum similarity of each of the comparisons has been calculated, the algorithm searches for the highest value and provides it as the final similarity result between the two flexible molecules.

Once the flexible target has been compared with all molecules in the database, they are ordered based on their conformation computed similarity value.

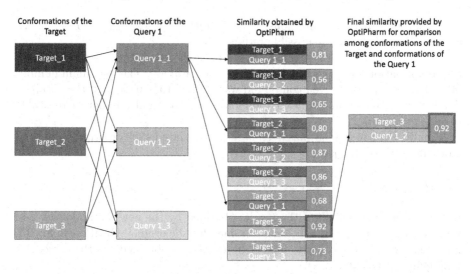

Fig. 5. Procedure for obtaining maximum similarity when working with conformations of molecules.

Consequently, new query compounds with a high similarity value can be identified while they are not detected when working with rigid molecules.

3 Materials

Hardware. The experiments of this work have been carried out using a cluster of 8x Bull Sequana X440-A5: 2 AMD EPYC Rome 7642 (48 cores) and 512 GB of RAM memory and 240 GB SSD.

FDA Database. The database used in this work was obtained from Drugbank, v.5.0.1 [15]. Specifically, a subset of 1,751 molecules validated by the Food and Drug Administration (FDA) has been used. The FDA is a federal agency of the U.S. Department of Health and Human Services responsible for protecting and promoting public health by controlling, among other things, prescription and over-the-counter pharmaceutical drugs (medicines). The original database has been downloaded from https://go.drugbank.com.

Software. The new version of OptiPharm, described in Sect. 2.2 is the optimization algorithm used to find the maximum similarity between two compounds. It has been configured to consider the hydrogen atoms of each molecule. In addition, all the heavy atom radii have been set to 1.7Å. Furthermore, all compound pairs are centered and aligned. Consequently, the molecule centroids have been located at the coordinates center of the search space. Finally, each molecule has been aligned so that its longest axis has been oriented at X-axis and the shortest along the Z-axis. The input parameter set used in OptiPharm have been:

$N = 200,000$ function evaluations, $M = 5$ starting poses, $t_{max} = 5$ iterations, and $R = 1$ as the smallest possible radius.

Additionally, software OMEGA [8] has been the generator selected to obtain the conformations of targets and queries in the database. The maximum number of conformations for a given compound was limited to 500, though the obtained number was smaller in many compounds due to a small number of rotatable bonds.

Finally, Dragon (6.0.38) has been used to calculated 4885 descriptors for each molecule in the database.

4   Results

In this section, we will show the results obtained by the new methodology, and we will compare them with the ones obtained for the original OptiPharm in [10]. For illustration, we will only depict the outcomes obtained for the molecules DB00381 and DB00876.

Table 1. Top-10 most similar compounds in electrostatic to the target DB00381.

Rigid		Flexible				
Query	Tc_E	Target conformation	Query conformation	Tc_E	Rk_R	Tc_E^R
DB00630	0.377	32	DB09237_43	0.762	1406	0.118
DB00409	0.377	32	DB01214_481	0.727	1384	0.121
DB00751	0.374	32	DB00622_175	0.718	999	0.207
DB00933	0.374	32	DB00557_212	0.704	1186	0.178
DB00998	0.370	32	DB00383_159	0.700	398	0.264
DB00334	0.367	32	DB01244_44	0.689	1549	0.105
DB00891	0.359	32	DB00979_23	0.683	517	0.254
DB00611	0.358	32	DB00571_483	0.679	240	0.280
DB00540	0.358	32	DB01359_440	0.666	1268	0.153
DB00647	0.357	32	DB00748_58	0.665	260	0.278

Following our methodology, we first calculate the 4885 descriptors for all molecules in the database. Subsequently, this group is reduced to 757 by applying the statistical metrics. With these numbers, it is computed the Euclidean distance between each molecule and the target. Later, the molecules are shorted according to that distance in ascending order. Only those molecules ranked within 10% of the shortest distance were selected. For the 175 molecules that remain, several conformations are generated. In particular, the sub-database obtained for each target consists of 47,983 and 38,833 conformations for the targets DB00381 and DB00876, respectively. In addition, target DB00381 resulted in 383 conformations, and DB00876 in 154.

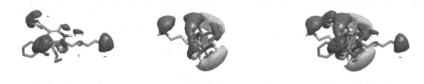

(a) Rigid Target DB00381   (b) Rigid Query DB00630   (c) Rigid overlapping

Fig. 6. Maximum similarity solution for Target DB00381 when working with rigid molecules. Figures (a) and (b) represent the target and query compounds and their electrostatic fields. Figure (c) represents the optimal overlapping between the two compounds where Tc_E is maximum.

(a) Flex. Target DB00381   (b) Flex. Query DB09237   (c) Flexible overlapping

Fig. 7. Maximum similarity solution for Target DB00381 when working with flexible molecules. Figures (a) and (b) represent the more similar conformations of target and query compounds and their electrostatic fields. Figure (c) represents the optimal overlapping between the two compounds where Tc_E is maximum.

Tables 1 and 2 compare the main results obtained for the methodologies *Rigid* and *Flexible* for Targets DB00381 and DB00876. In particular, they show the 10 queries with the greatest similarity provided for both versions. More precisely, for *Rigid* methodology, we provide its name, *Query*, and the corresponding similarity value Tc_E. For *Flexible*, we indicate the pair target-query that has obtained the best match, i.e. we identify those two molecules by also indicating their corresponding conformation number. This information is depicted in columns *Target conformation* and *Query conformation*. Finally, we show at column Tc_E the scoring function value obtained for each match. For the sake of comparison, we also indicate in column Rk_R the position that the *Query conformation* occupies in the list obtained by *Rigid*, and its corresponding scoring value in column Tc_E^R.

The results show an improvement in the quality of the solutions. As can be seen in Table 1, the most similar compound found following the *Flexible* methodology (0.762) improves twice the value of the *Rigid* (0.377). Moreover, the most similar compounds in the rankings are different, i.e., DB00630 for the *Rigid* and DB09237 for the *Flexible*. Additionally, this result can be seen

Table 2. Top-10 most similar compounds in electrostatic to the target DB00876.

Rigid		Flexible				
Query	Tc_E	Target Conformation	Query conformation	Tc_E	Rk_R	Tc_E^R
DB00774	0.532	69	DB00338_126	0.861	1164	0.225
DB00880	0.530	35	DB08897_40	0.861	1577	0.159
DB01153	0.527	18	DB00736_54	0.860	888	0.261
DB00690	0.526	69	DB06766_6	0.835	1403	0.197
DB00897	0.522	120	DB00966_424	0.832	772	0.276
DB00819	0.513	69	DB01129_440	0.829	713	0.285
DB01101	0.512	35	DB04843_72	0.813	1378	0.201
DB00425	0.507	1	DB04880_2	0.800	502	0.321
DB01002	0.505	103	DB00642_47	0.775	704	0.286
DB00809	0.503	35	DB06274_105	0.522	618	0.301

(a) Rigid Target DB00876   (b) Rigid Query DB00774   (c) Rigid overlapping

Fig. 8. Maximum similarity solution for Target DB00876 working with rigid molecules. Figures (a) and (b) represent the target and query compounds and their electrostatic fields. Figure (c) represents the optimal overlapping between the two compounds where Tc_E is maximum.

graphically in Figs. 6 and 7 where the molecules, their optimal overlapping, and electrostatic fields are represented using VIDA [13].

If the last two columns of *Flexible* are analyzed, it is clear that there is no good overlapping when rigid molecules are used. As seen in the Rk_R column, most of the top molecules in *Flexible* are below the position 1000th in the *Rigid* list.

Table 2 shows results along the same line as Table 1. In this study case, the top solution improves from 0.532 to 0.861 finding different compound as well. Figures 8 and 9 show graphically the most similar compounds for both methods.

The improvement obtained in the previous results is due to the solutions, including flexibility. However, as we previously stated, this increases the computational time considerably. However, the time has been ostensibly reduced thanks to the *Flexible* methodology employed in this work. The average time for an optimization process is 29 s for these target molecules in the

 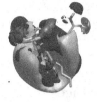

(a) Flex. Target DB00876   (b) Flex. Query DB00338   (c) Flexible overlapping

Fig. 9. Maximum similarity solution for Target DB00876 working with flexible molecules. Figures (a) and (b) represent the more similar conformations of target and query compounds and their electrostatic fields. Figure (c) represents the optimal overlapping between the two compounds where Tc_E is maximum.

current hardware. If we focus on the target DB00381 with a database of $38,833$ conformations from the initial 175 molecules, each molecule generated on average 221 conformations, although the maximum could be 500 for each one). If this value is extrapolated to the whole database, $(1751 * 221 =)386,971$ conformations could be generated. So, the time saved with the filter applied by discarding unpromising compounds is 116 days per target, and whether 500 conformations are generated, 264 computation days would be saved.

5 Conclusions and Future Work

In this work, we have improved the software OptiPharm by considering molecule flexibility. Apart from that, the new version includes several mechanisms to reduce the computational effort. In particular, we have reduced the number of optimization parameters and the range of freedom in some of them. Consequently, the search space decreases, and the number of function evaluations needed to find the optimal similarity drops. Besides, we have analyzed and applied descriptors to filter the initial database. We have used statistical metrics such as variance, correlation, or Euclidean distance.

The results have shown that the new OptiPharm can obtain solutions with higher scoring values than the original one, meaning that new query compounds with a high similarity value can be identified. These compounds are not detected when working with rigid molecules. In addition, the descriptor filter allows to drastically reduce the run time, saving for the study at hand up to 264 days.

Future work proposes implementing a conformation generation algorithm as an internal procedure of OptiPharm and including new scoring functions.

Acknowledgement. This work was supported by the Spanish Ministry of Economy and Competitiveness through the CTQ2017-87974-R, RTI2018-095993-B-I00 and EQC2019-006418-P grants; by the Junta de Andalucía through the grant Proyectos de excelencia (P18-RT-1193), by the Programa Regional de Fomento de la Investigación (Plan de Actuación 2018, Región de Murcia, Spain) through the "Ayudas a la realización de proyectos para el desarrollo de investigación científica y técnica por grupos competitivos (20988/PI/18)" grant; by the University of Almeria throught the "Ayudas a proyectos de investigación I+D+I en el marco del Programa Operativo FEDER 2014-20" grant (UAL18-TIC-A020-B). Savíns Puertas Martín is a fellow of the "Margarita Salas" grant (RR_A_2021_21), financed by the European Union (NextGenerationEU).

References

1. Axenopoulos, A., Rafailidis, D., Papadopoulos, G., Houstis, E.N., Daras, P.: Similarity search of flexible 3D molecules combining local and global shape descriptors. IEEE/ACM Trans. Comput. Biol. Bioinf. **13**(5), 954–970 (2016). https://doi.org/10.1109/TCBB.2015.2498553
2. Bahi, M., Batouche, M.: Deep learning for ligand-based virtual screening in drug discovery. In: 2018 3rd International Conference on Pattern Analysis and Intelligent Systems (PAIS), pp. 1–5 (2018). https://doi.org/10.1109/PAIS.2018.8598488
3. Böttcher, C., Belle, O.V., Belle, B.: Theory of Electric Polarization. Elsevier, Amsterdam (1974). https://doi.org/10.1016/B978-0-444-41019-1.50006-7
4. Fatumo, S., Adebiyi, M., Adebiyi, E.: In silico models for drug resistance. In: Kortagere, S. (eds.) In Silico Models for Drug Discovery. Methods in Molecular Biology, vol. 993. Humana Press, Totowa (2013). https://doi.org/10.1007/978-1-62703-342-8_4
5. Hu, J., Liu, Z., Yu, D.J., Zhang, Y.: LS-align: an atom-level, flexible ligand structural alignment algorithm for high-throughput virtual screening. In: Bioinformatics, vol. 34, pp. 2209–2218. Oxford University Press (2018). https://doi.org/10.1093/bioinformatics/bty081
6. Jaccard, P.: Distribution de la flore alpine dans le bassin des dranses et dans quelques régions voisines. Bulletin de la Société Vaudoise des Sciences Naturelles **37**, 241–272 (1901)
7. Kalászi, A., Szisz, D., Imre, G., Polgár, T.: Screen3D: a novel fully flexible high-throughput shape-similarity search method. J. Chem. Inf. Model. **54**(4), 1036–1049 (2014). https://doi.org/10.1021/ci400620f
8. OMEGA 4.1.0.2: OpenEye Scientific Software: Santa Fe, NM, USA (2019). http://www.eyesopen.com
9. Puertas-Martín, S., Redondo, J.L., Ortigosa, P.M., Pérez-Sánchez, H.: OptiPharm: an evolutionary algorithm to compare shape similarity. Sci. Rep. **9**(1), 1398 (2019). https://doi.org/10.1038/s41598-018-37908-6
10. Puertas-Martín, S., Redondo, J.L., Pérez-Sánchez, H., Ortigosa, P.M.: Optimizing electrostatic similarity for virtual screening: a new methodology. In: Informatica, pp. 1–19 (2020). https://doi.org/10.15388/20-INFOR424
11. ROCS: OpenEye Scientific Software: Santa Fe, NM. http://www.eyesopen.com
12. Tanrikulu, Y., Krüger, B., Proschak, E.: The holistic integration of virtual screening in drug discovery. Drug Discov. Today **18**(7–8), 358–364 (2013). https://doi.org/10.1016/j.drudis.2013.01.007
13. VIDA 4.4.0.4: OpenEye Scientific Software: Santa Fe, NM. http://www.eyesopen.com
14. Vázquez, J., López, M., Gibert, E., Herrero, E., Luque, F.J.: Merging ligand-based and structure-based methods in drug discovery: an overview of combined virtual screening approaches. Molecules **25**(20), 4723 (2020). https://doi.org/10.3390/molecules25204723

15. Wishart, D.S., et al.: DrugBank 5.0: a major update to the DrugBank database for 2018. Nucl. Acids Res. **46**(D1), D1074–D1082 (2018). https://doi.org/10.1093/nar/gkx1037

16. Yang, Y., et al.: Ligand-based approach for predicting drug targets and for virtual screening against COVID-19. Brief. Bioinform. **22**(2), 1053–1064 (2021). https://doi.org/10.1093/bib/bbaa422

Solving a Capacitated Waste Collection Problem Using an Open-Source Tool

A. S. Silva[1,2,3] , Filipe Alves[1,4(✉)] , J. L. Diaz de Tuesta[2] ,
Ana Maria A. C. Rocha[4] , A. I. Pereira[1] , A. M. T. Silva[3] ,
Paulo Leitão[1] , and H. T. Gomes[2]

[1] Research Centre in Digitalization and Intelligent Robotics (CeDRI),
Instituto Politécnico de Bragança, 5300-253 Bragança, Portugal
{adriano.santossilva,filipealves,apereira,pleitao}@ipb.pt
[2] Centro de Investigação de Montanha (CIMO), Instituto Politécnico de Bragança,
5300-253 Bragança, Portugal
{jl.diazdetuesta,htgomes}@ipb.pt
[3] Laboratory of Separation and Reaction Engineering - Laboratory of Catalysis
and Materials (LSRE-LCM), Faculty of Engineering, University of Porto,
Rua Dr. Roberto Frias, 4200-465 Porto, Portugal
adrian@fe.up.pt
[4] ALGORITMI Center, University of Minho, 4710-057 Braga, Portugal
arocha@dps.uminho.pt

Abstract. Increasing complexity in municipal solid waste streams
worldwide is pressing Solid Waste Management Systems (SWMS), which
need solutions to manage the waste properly. Waste collection and transport is the first task, traditionally carried out by countries/municipalities
responsible for waste management. In this approach, drivers are responsible for decision-making regarding collection routes, leading to inefficient
resource expenses. In this sense, strategies to optimize waste collection
routes are receiving increasing interest from authorities, companies and
the scientific community. Works in this strand usually focus on waste
collection route optimization in big cities, but small towns could also
benefit from technological development to improve their SWMS. Waste
collection is related to combinatorial optimization that can be modeled
as the capacitated vehicle routing problem. In this paper, a Capacitated
Waste Collection Problem will be considered to evaluate the performance
of metaheuristic approaches in waste collection optimization in the city
of Bragança, Portugal. The algorithms used are available on Google OR-
tools, an open-source tool with modules for solving routing problems.
The Guided Local Search obtained the best results in optimizing waste
collection planning. Furthermore, a comparison with real waste collection
data showed that the results obtained with the application of OR-Tools
are promising to save resources in waste collection.

This work has been supported by FCT - Fundação para a Ciência e Tecnologia
within the R&D Units Project Scope: UIDB/05757/2020, UIDB/00690/2020, UIDB/50
020/2020, and UIDB/00319/2020. Adriano Silva was supported by FCT-MIT Portugal PhD grant SFRH/BD/151346/2021, and Filipe Alves was supported by FCT PhD
grant SFRH/BD/143745/2019.

Keywords: Waste collection · Vehicle routing problem · Optimization · Google OR-Tools

1 Introduction

Waste generation is growing exponentially, and solid waste management systems are now under pressure to manage the high content of waste. Advances in the industry combined with population growth in cities are increasing the complexity of municipal solid waste (MSW) streams worldwide. Due to the increased complexity shown by the waste, management costs are becoming a problem of public concern. Investments in this sector are often left aside in low and middle-income countries, which can lead to environmental and health problems of inappropriate MSW management [20]. In this sense, solid waste management systems (SWMS) need solutions to manage this high increase in both complexity and amount of waste [11]. An alternative could be to reduce the costs associated with collection and transport, to have more funds available for the assembly of facilities with more sophisticated technology to deal with the waste. Collection and transport represent the first instance in SWMS and constitute 60–80% of total costs [14].

The fleet of vehicles responsible for waste collection and transport over a specific area can also include different types of vehicles, each one assigned to particular kinds of waste [1]. Further challenges are associated with waste transport fleets, including route planning, cost-management, labor allocation, traffic congestion (for big cities), periodic maintenance of the fleet, and task assignments. Traditionally, solid waste collection is carried out without prior analysis related to the demand for planning collection routes. In this strategy, drivers are responsible for route planning, and there is a predetermination of the number of trips per week that trucks will make [7]. This approach has several limitations since waste generation, fundamentally stochastic, is often considered a constant entity. The inefficient collection of solid waste gives rise to unnecessary expenditures, and can also have a negative impact on air quality around waste dumpsters, which can disturb the population (obnoxious effect) [13,15].

In this regard, cities can benefit from innovative solutions achievable by using Information and Communication Technologies (ICT). The use of these technologies can upgrade the city's infrastructure to the smart city level, saving resources that could be used to improve other aspects of SWMS. Practical examples would be the optimization of collection routes using search algorithms, and also the use of Internet of Things (IoT) technology. With such a tool, it is possible to apply vehicle-guided techniques to improve the efficiency of operations. IoT can be defined, in general terms, as an extension of traditional computer networks, in which devices and gadgets capable of sensing, computing, and communicating are used. Several types of smart objects can be used in a collaborative approach to build a system for smart cities, improve decision-making with real-time information, and apply algorithms to solve mathematical problems instantly [2,3,21].

To optimize the collection routes using a merged system composed of optimization algorithms and IoT, which showed the best results in the literature, route optimization algorithms need to be applied and validated. In this work, three metaheuristics available in Google OR-tools will be assessed to optimize the waste collection in 20 paper waste dumpsters in the city of Bragança, Portugal. The level of waste during the test days was determined based on a stochastic approach, that takes into consideration the region where the dumpster is inserted.

The rest of the paper is organized as follows: Sect. 2 brings the related literature and real data regarding municipal solid waste generation; Sect. 3 presents the methodology employed; Sect. 4 summarizes the results; and finally, Sect. 5 shows the main findings of the present study and future work.

2   Related Literature

In this section, the most relevant and updated literature associated with the waste collection topic is presented. Documents were obtained by searching in Web of Science (WoS) and Scopus databases for keywords "waste collection" and "optimization" in the fields "Abstracts, keywords, and Titles". 995 and 1333 documents were found in this first search in WoS and Scopus databases, respectively. After removing duplicates using an open-source python-based scientometric analysis, the ScientoPy tool [22], the WoS entries were reduced to 631 and Scopus to 780, which represents 63.4% and 58.5% of the initial databases. Figure 1 presents the accumulative number of documents associated with the most relevant keywords throughout the years.

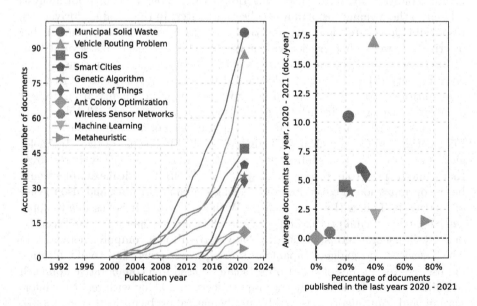

Fig. 1. The 10 most relevant keywords in the merged database. (Color figure online)

From Fig. 1, it is possible to verify the sharp growth of keywords (mainly from the year 2012), such as solid waste, vehicle routing problem, smart cities and specific algorithms, which reveals the priority and applicability of this research.

2.1 Solid Waste Generation

According to the latest available data, global waste generation reached 2.01 billion tons in 2016, which corresponds to 0.74 kg of daily waste generated per capita. This value can vary between 0.11 kg to 4.54 kg, depending on the income level of the country [10]. Countries in the Pacific, East Asia, Europe, and Central Asia regions account for 43% of the world's waste generation. Sub-Saharan, Middle East, and North Africa regions are the regions with the lowest waste generation, accounting for 15% of the world's waste. The overall scenario shows that waste generation is strictly connected to economic development.

Municipal solid waste represents an average of 7–10% of the total waste generation in EU countries, being one of the most complex to manage [10]. On the other hand, in Portugal, this type of waste has a high share of the total waste generated. The way that a particular European country deals with this stream is an indicator of the overall quality of waste management systems in that country. In fact, Portugal is the EU Country with the highest share of municipal solid waste (32.8%), followed by Latvia (32.6%) and Croatia (23.3%). The waste generated in cities presents a complex composition, has a very high public visibility, and has a direct impact on human health. Municipal solid waste data shows that each European citizen was responsible for an average of 505 kg of municipal waste generation in 2020. Eurostat data show an increase of 8.2% compared to the first record in 1995 for the total waste generated in Europe. In this scenario, Portugal is one of the countries with the highest increase in waste generation, with 45.7% more urban waste generated compared to the first record [10].

The company responsible for waste collection in the Northeast region of Portugal is Resíduos do Nordeste (see www.residuosdonordeste.pt), covering 13 municipalities and around 134000 inhabitants. In this context, Bragança is the city with the largest number of inhabitants, therefore with the highest amount of waste generated (27.4% of the total waste collected in 2020). In 2020 the company collected a total amount of 56973.74 tons of waste, from which 4719 came from the selective collection of waste. Despite the low amount of waste collected in the selective collection, data records show an increasing number of recycling points installed by the company. Each point is composed of 3 dumpsters, for the deposition of cards and paper (blue), glass (green), and metal and plastic (yellow). The number of recycling points increased from 616 to 939 from 2016 to 2020, and the amount of waste collected increased from 3039 to 4719 t. Figure 2 shows the evolution of the most significant types of waste collected by the company in a selective collection from 2014 to 2020. The trend of waste collected for different types of waste shows that paper waste is the most generated, with the highest increase in collection among the other types.

2.2   Waste Collection Approaches

From the database containing 1411 records, those with the keyword "vehicle routing problem" (see Fig. 1) or similar were filtered for further analysis of literature, which resulted in a small database composed of 88 documents. From these 88, 35 papers were found with the potential to contribute to a brief literature review on the topic of vehicle routing problems in a municipal solid waste collection from 2017 to 2021. A fundamental step for solving the vehicle routing problem is the mathematical formulation. Each study considers specific objective functions and constraints in an attempt to mimic real-case scenarios in specific regions. The basis for most of the problems reported in the literature is the vehicle routing problem (VRP), a variant of the classic traveling salesman problem (TSP). The difference between TSP and VRP is the number of vehicles scheduled to perform the task, and in this regard, VRP necessarily uses more than one vehicle to collect items.

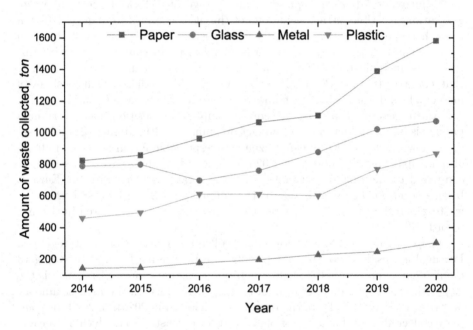

Fig. 2. Evolution of waste collected by Resíduos do Nordeste per type.

Other important issues related to the solid waste management system are presented in the literature. Chao et al. [12] for example addressed the optimization of collection routes in urban areas, considering the recycling operation in the sorting facility. A multi-objective approach was used to maximize the total profits of the system and balance the workload of recycling operations at the sorting facility over time. Zhang et al. [25] studied the optimization of waste transport routes with the objective of minimizing total distances traveled and maximizing the satisfaction of residents. Resident satisfaction was used in their work as the penalty cost against a time window constraint, as the activity frequency of residents is highly correlated with the time of the day.

For waste collection problems, the authors generally consider capacity constraints on trucks and time windows for collection. These considerations are widely known in VRP literature, with the respective variant of the problems namely Capacitated Vehicle Routing Problem (CVRP), Vehicle Routing Problem With Time Windows (VRPTW), and CVRPTW (both constraints considered). However, VRP can be used in several situations of scheduled collection, which gives rise to several innovative formulations of the problem that can be seen in [17]. Among these innovative formulations, one has shown particular interest in this work: Waste Collection Problem (WCP).

This class of problem was first introduced by Beltrami and Bodin in 1974, using a heuristic algorithm to address the collection of waste in New York City. WCP is an extension of VRP modeled to be closer to the problem faced by companies/municipalities regarding route planning for the collection of solid wastes. In the WCP, a set of homogeneous or heterogeneous fleets is designed to collect waste from multiple dumpsters, with a single depot. The problem is then to collect the highest amount of waste in the shortest distance. Therefore, three main agents can be identified in this formulation: the fleet, the central depot, and the dumpsters. At an early stage, to properly address the problem faced by one company, it is necessary to acquire information about how the system works in a real scenario so that the formulation is as complete as possible.

Similar to the classic VRP, WCP has different variations based on the type of waste to be collected. For instance, practices adopted for residential waste collection are different from those used in commercial waste. Some constraints used for VRP problems can be considered in the context of waste collection, which generates problems such as WCP with time windows or time-dependent WCP. An extensive and interesting literature review on solid waste collection optimization objective functions, and constraints can be found in [8].

2.3 Optimization Algorithms

Due to the large problem sizes and numerous constraints in waste collection problems, the optimization of collection routes of real-life waste management systems requires the use of non-exact methods to generate a near-optimal solution for the routes. Undoubtedly, the most popular approach to solving waste collection optimization is through metaheuristics. These methods are able to provide a sufficiently good solution to an optimization problem using limited computational capacity or incomplete information. The mechanism behind this algorithm is based on mimicking nature, once they are inspired by biological evolution or physical sciences. Metaheuristic algorithms are classified into trajectory-based (for example, simulated annealing and tabu search) and population-based approaches (for example, ant colony optimization and genetic algorithms).

Ant Colony Optimization (ACO) is based on the behavior of the ants following the path of other ants due to the deposition of pheromones and is recognized to be advantageous in terms of convergence speed when compared to other algorithms. Mancera-Galván et al. [16] employed two ACO algorithms to study waste collection routes, achieving improvements in route collection considering objective length minimization and routes re-design.

Genetic Algorithms (GA) is a widely known metaheuristic inspired by the evolutionary theory of the species, namely the natural selection, underlying concepts of survival of the fittest and the inheritance of characteristics from parents to offspring by reproduction. Three operators are often used for this purpose, namely crossover, reproduction, and mutation. GA is simple to be implemented, and for this reason, several works have explored their use to solve VRP in a solid waste collection context [4,9].

Simulated annealing (SA) is a widely used optimization technique inspired by metallurgy annealing. Babaee et al. [5] addressed the vehicle routing problem through SA with the goal of minimizing the total operational cost considering time windows. Tabu Search (TS) is a metaheuristic that uses memory structures (tabu list) to store ineligible candidate solutions to generate other candidates, prohibiting exploration of a solution in the tabu list. A waste collection synchronization mechanism was developed by Shao et al. [23] using TS to achieve higher profits. Guided Local Search (GLS) is a metaheuristic search method that operates by associating of cost and penalty of each solution, using properties of both TS and SA algorithms. In the work done by Barbucha et al. [6], GLS was combined with an asynchronous team concept to build an agent-based GLS algorithm to solve the CVRP.

3   Methodology

The main goal of this work is to evaluate the performance of different optimization algorithms available in the open-source tool (Google OR-tools) to find optimal routes for paper waste collection in the city of Bragança, located in the Northeast region of Portugal (inland city). The study was carried out considering a period of Da days and k dumpsters (real locations provided by the company). The daily level inside each dumpster was determined using a demographic factor, calculated by analyzing the individual dumpsters nearby. The locations of dumpsters j that needed to be collected on the day i along with waste levels were used as input to the system, and the output was the total distances traveled and the load carried using different search strategies to find the best route. Figure 3 shows a representation of the system, its input and, consequently, its output.

3.1   Capacitated Waste Collection Problem

The type of waste collected in selective collection with higher generation is paper, which is why paper waste dumpsters were chosen to be the focus of this study. The problem addressed here can be summarized as a capacitated waste collection vehicle routing problem (CWCP). This is a variant of VRP where vehicles with limited carrying capacity need to pick up or deliver items to multiple locations. The items have a quantity, such as weight or volume, and the vehicles have a

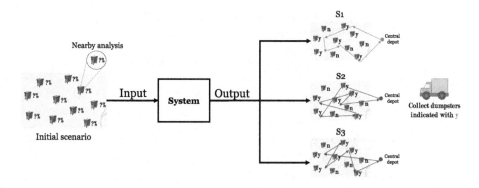

Fig. 3. Representation of the waste collection system.

maximum capacity that they can carry. The problem is to pick up or deliver the items for the lowest cost, without ever exceeding the capacity of the vehicles.

These routes can be described using a graph where the dumpsters are the vertices and the roads are the arcs. A cost is associated with each arc, which can be the distance (cost associated with this approach), the travel time, or for example, the amount of fuel used. The objective is to minimize the total distance traveled for the waste collection route, which reduces the costs associated with fuel consumption and also carbon emissions. The CWCP approach, can be represented as a weighted directed graph $G = (V, A)$ where $V = \{0, 1, 2, \ldots, n\}$ represents the set of vertices and $A = \{(l, j) : l \neq j \text{ for } l, j = 0, \ldots, n\}$ represents the set of arcs. In this formulation, the vertex 0 represents the depot, and the others represent all the waste collection locations, where N represents the index set of V. For each arc (l, j) a non-negative value c_{lj} is associated, which corresponds to the distance between the vertex l and j, in terms of cost. Associated with each location l a demand $d_l \geq 0$ is defined ($d_0 = 0$) such that $d_l \leq Q$, assuming that the set of trucks, K, have the same capacity Q. Thus, the CWCP solution must take into account the following assumptions:

- Trucks have limited capacity Q.
- The number of trucks is supposed to be limited (K) and homogeneous.
- Every location is visited exactly once by exactly one truck.
- Trucks only collect one type of waste (*i.e.* paper).
- Each route begins and ends at the central depot.
- Dumpsters must be served once each 2 days.
- Waste must be transported to the central depot when the full capacity of the truck is reached.
- The routes of trucks and their sequence order are calculated during the construction of the solution, depending on the algorithm used.

To solve the CWCP described above, the mathematical formulation of the problem will be presented based on a general formulation found in the literature [24]. In order to find the sequence order of visits to dumpster locations, the decision variables are defined as follows:

$$x_{ljk} = \begin{cases} 1, \text{ if truck } k \text{ visits location } j \text{ after location } l \\ 0, \text{ otherwise.} \end{cases}$$

$$y_{lk} = \begin{cases} 1, \text{ if truck } k \text{ visits location } l \\ 0, \text{ otherwise.} \end{cases}$$

The mathematical formulation of the CWCP is given by:

$$\text{Minimize} \sum_{l \in N} \sum_{j \in N} c_{lj} \sum_{k \in K} x_{ljk} \tag{1}$$

subject to:

$$\sum_{k \in K} y_{lk} = 1, \quad \forall l \in N \setminus \{0\} \tag{2}$$

$$\sum_{k \in K} y_{0k} \leq K \tag{3}$$

$$\sum_{j \in N} x_{ljk} = \sum_{j \in N} x_{jlk} = y_{lk}, \quad \forall l \in N, \forall k \in K \tag{4}$$

$$\sum_{l \in N} d_l y_{lk} \leq Q, \quad \forall k \in K \tag{5}$$

$$u_{jk} \geq u_{lk} + Q(x_{ljk} - 1) + d_j, \quad \forall l, j \in N \setminus \{0\}, l \neq j, \tag{6}$$
$$\text{such that } d_l + d_j \leq Q, \forall k \in K$$

$$d_l \leq u_{lk} \leq Q, \quad \forall l \in N \setminus \{0\}, \forall k \in K \tag{7}$$

$$x_{ljk} \in \{0, 1\}, \quad \forall l, j \in N, \forall k \in K \tag{8}$$

$$y_{lk} \in \{0, 1\}, \quad \forall l \in N, \forall k \in K \tag{9}$$

Thus, the objective function presented in (1) allows minimizing the total distance while respecting the constraints. The constraints (2) ensure that every location is visited once and is left by the same truck, while the set of constraints (3) guarantees that every truck leaves the depot only once. In turn, constraints (4) guarantee continuity of the route, i.e., the number of trucks arriving at every location and entering the depot is equal to the number of trucks leaving. In the constraints (5) capacity constraints are stated, making sure that the sum of the demands of the locations visited in a route are less than or equal to the capacity of the truck performing the service. The subtour elimination restrictions are expressed by constraints (6), ensure that the solution contains no cycles disconnected from the depot. Additional decision variables, u_{lk}, are used in the subtour elimination constraints and represent the truck load k after visiting the location l (7). Finally, (8) and (9) specify the definition domain of the decision variables x_{ljk} and y_{lk}.

3.2   Waste Level Throughout Days

Due to the lack of data, it is still not possible to create models to accurately predict the waste levels over the days in the city of Bragança. However, the initial level within the chosen dumpsters can be calculated using a uniform probability distribution as the initial approach. Over the period studied, changes in waste level were considered dynamic and determined based on the analysis of the proximity of each dumpster using the software Google Earth to find the populated area. The parameter filling velocity (fv) was calculated with this image analysis. This parameter is the key to determining the waste level during the days as it constitutes the percentage in the volume of daily waste oscillation each day. Equation (10) illustrates the expressions used to determine the filling velocity (fv_j) of dumpster j, $j = 1, ..., dumpsters$, the daily waste level (L), and the amount of waste (Lc) in m^3.

$$fv_j = \frac{FA_j}{TA}\sigma_M \wedge L_{i,j} = L_{i-1,j} + fv_j \wedge Lc_{i,j} = L_{i,j}\frac{TV}{100} \tag{10}$$

where FA_j represents the filled area with buildings around dumpster j, TA is the total area around each dumpster, and σ_M is the parameter that introduces stochastic behavior in waste oscillation. The $L_{i,j}$ and $Lc_{i,j}$ represent the waste level and amount in day i for dumpster j, respectively, and TV represents the total volume of the dumpster. In Fig. 4, all collection points are illustrated on the map of the city of Bragança (left side, marked in yellow) along with an example of how FA_j was determined for one location (right side).

Fig. 4. Location of all dumpsters (left side) and example of a filled area with buildings around a dumpster (right side).

3.3 Open Source Solver - Google OR-Tools

There are many difficulties in developing algorithms to solve classic VRP problems, whether in the online software or closed source, which present specific requirements and barriers for learning and/or application in real data/scenarios.

Recently, an open-source tool has emerged that presents several libraries and different solvers for different variants of problems and/or types of VRP, the Google OR-Tools [19]. This tool emerged in 2019, developed by researchers and programmers from the Google company, who present in that cloud system, several learning libraries, statistical data, vehicle routing problems, programming environment, flexibility, functionality, and easy to use in different programming languages (such as python). Furthermore, this tool has some pre-defined algorithms, such as (meta-)heuristics and several strategies for defining objectives, variables, constraints, and/or parameters. In addition, the integration of skills with external systems or online services is allowed (Google Maps API, distance matrix API, and others), as well as visualization systems.

Google OR-Tools is a fast and portable package, extremely practical to solve complex combinatorial optimization problems, allowing tests and/or applications in real-world problems. Google OR-Tools can solve many types of VRP, including problems with pickups and deliveries, and/or multiple capacity dimensions, initial loads, skills, scheduling problems, and so on. Finally, there is extensive documentation available online about OR-Tools, full of examples and libraries, which reveals its promising capacity in the field of operational research systems.

3.4   Route Optimization

The main goal of this work is to study the performance of three optimization algorithms, more specifically three metaheuristics, available in open-source Google OR-Tools to optimize paper waste collection routes in the city of Bragança. The proposed mechanism for the collection was to collect all dumpsters once every 2 days, to put the algorithms under stress. The metaheuristics to be assessed are Guided Local Search, Tabu Search, and Simulated Annealing. Algorithm 1 illustrates the complete procedure adopted to perform the study.

Algorithm 1. Dynamic waste collection

1: $L_{0,j} \leftarrow rand(0, CM)$
2: **for** $i = 1$ to $days$ **do**
3:      **for** $j = 1$ to $dumpsters$ **do**
4:          $fv_j = \dfrac{FA_j}{TA} * rand(C_{min}, C_{max})$
5:          $L_{i,j} = L_{i-1,j} + fv_j$
6:          **if** $mod(i, 2) \neq 0$ **then**
7:              $Lc_{i,j} = L_{i,j} \dfrac{TV}{100}$
8:              $L_{i,j} = 0$
9:              $col_{i,j} = j$
10:         **end if**
11:     **end for**
12: **end for**
13: **for** i in $days$ **do**
14:     $DM_i \leftarrow distance\_matrix(col_{i,j})$
15:     $TD_i, load_i \leftarrow GLS, TS, SA(DM_i, Lc_{i,j})$
16: **end for**

To invoke the metaheuristics GLS, TS and SA, the within Algorithm 1, two input values are required: the distance matrix and the load inside the dumpsters. The parameter $col_{i,j}$ carries the information of which dumpsters should be collected over the days and is used by the *distance_matrix()* function in order to return DM_i, which will carry the distance matrix information daily. The load inside each dumpster on collection days is stored in $Lc_{i,j}$, which is calculated by multiplying the waste level $col_{i,j}$ by the total volume TV. The maximum level that each dumpster can have on day 0 is given by CM. Daily changes can range from C_{min} to C_{max}, which represents the lower and upper threshold for waste generation, respectively.

4   Numerical Results

The results obtained for optimizing the routes using Guided Local Search, Tabu Search, and Simulated Annealing available in the Google OR-Tools, considering default parameter values, will be presented and discussed. The collection decision-making was chosen to put the algorithms under stress, so that on collection days, the algorithm has to find the best route considering all points of collection. In this regard, trucks need to collect all waste once each 2 days.

A period of 30 days $(days = 30)$ was considered. The fleet of trucks is composed of 3 trucks $(K = 3)$ with a maximum capacity of $Q = 16\,\mathrm{m}^3$, and each dumpster $(dumpsters = 20)$ have a maximum capacity of $TV = 2.5\,\mathrm{m}^3$. Both data used for capacities are real, and the dumpsters/truck ratio in this system is close to the real operations of the company responsible for the collection. The maximum level for each dumpster CM was defined as 80%, C_{min} is 20% and C_{max} was chosen to be 80%. The total area of a circle with radius 150m defines the value of TA $(TA = 70685.83\,\mathrm{m}^2)$.

In order to get the real distances for each collection point (distance matrix) DM_i uses the Google Maps API module, serving as input to the information in $col_{i,j}$, that contains the latitude and longitude of the dumpster location j in day i that needs collection.

4.1   Results and Discussion

The proposed system for waste collection is considering the collection of all dumpsters once every 2 days, so the daily amount of waste collection will be the same. The relevance of levels in the algorithm is part of decision-making since the metaheuristics take into consideration the levels to find the optimal route. However, the solution in terms of the total distance to the optimal route obtained by each metaheuristic is different, which will be the focus of this discussion. The total distance traveled and the total execution time in each approach are represented in Fig. 5.

Traveled distances demonstrate that GLS is the algorithm that returns the shortest path for the collection of waste in the system considered here. For instance, the result in GLS was 1.77% and 3.42% lower than the results obtained in TS and SA, respectively. GLS and SA present similar results in terms of execution time, with a little advantage for GLS. This parameter is highly important

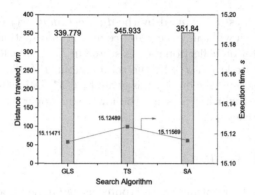

Fig. 5. Total distances traveled and computational times in the algorithms.

for future work since the real system will be composed of a higher number of locations, increasing the computational time to process and find the best route.

The tool available in Google OR-tools enables the user to visualize the path that each vehicle is assigned to follow, along with the load collected at distinct points. With this in mind, a graphical representation could show what are the practical differences between each route. Thus, Fig. 6 provides the illustration of the routes found using the three metaheuristics on the first collection day (day 2).

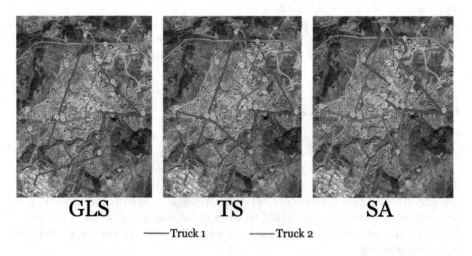

Fig. 6. Graphical representation of collection routes for different algorithms.

The representation shows that each algorithm has its own routes to collect the waste on the first day of collection. Due to the amount of waste to be collected on this day, only 2 trucks were assigned for collection in all algorithms.

4.2   Comparison with Real Scenario

The results obtained for route optimization demonstrate few differences using distinct algorithms. However, it is important to remember that all algorithms are working with route optimization, which should be enough to overcome real-case scenarios of traditional waste collection. To prove that all algorithms are delivering the optimized route, a comparison with real data from the company can be performed.

Real data provided by the company include the average distance traveled by the trucks, the amount of fuel they spent, and the annual amount of waste paper collected in 2020. The real system is composed of a larger number of recycling points and trucks, and the same fleet collects paper and plastic on different days. For this reason, a correlation parameter named collection cost (CC) was necessary for comparison purposes. The CC of the real system was calculated based on the average cost in one month and the amount of waste collected in m^3. The amount of waste collected was provided in mass, so the conversion was performed based on data on paper waste density [18]. The results obtained for the real scenario are shown along with CC obtained for optimized routes in Fig. 7.

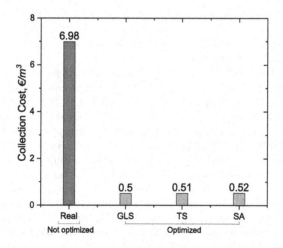

Fig. 7. Collection costs in real and optimized scenarios.

The difference between the estimated CC with real data and the optimized results is astonishing, however, is worth mentioning that this result was calculated based on estimates of the average collection of waste and density value that may differ from the real one. Despite this detail, the results serve to demonstrate how powerful route optimization can be to save resources in an SWMS. Furthermore, the average gas price in 2020 was considered for this analysis.

5   Conclusions and Future Work

This research shows the extreme usefulness of a recent and popular open-source optimization tool, Google OR-Tools, for solving a real capacitated waste collection vehicle routing problem, CWCP, in the city of Bragança.

In the developed approach three metaheuristics were used, GLS, SA, and TS, to optimize the waste collection of 20 paper waste dumpsters once every 2 days. Additionally, the tool allowed the evaluation and graphical representation of each route solution. The experimental results showed the quality of solutions achieved by the three approaches. In terms of performance, it is possible to assess that GLS was the best optimization algorithm to find the shortest path in the waste collection system. Additionally, a cost comparison with a real scenario and the optimized scenario showed an effective cost-saving, proving the potential application of OR-Tools for route optimization in SWMS.

After the optimization procedure is properly implemented and validated in the lower-scale system used in this study, the next step is to jump to a study in a real large-scale waste collection system in the city of Bragança.

References

1. Akbarpour, N., Salehi-Amiri, A., Hajiaghaei-Keshteli, M., Oliva, D.: An innovative waste management system in a smart city under stochastic optimization using vehicle routing problem. Soft. Comput. **25**(8), 6707–6727 (2021). https://doi.org/10.1007/s00500-021-05669-6
2. Aleyadeh, S., Taha, A.E.M.: An IoT-based architecture for waste management. In: 2018 IEEE International Conference on Communications Workshops (ICC Workshops), pp. 1–4 (2018). https://doi.org/10.1109/ICCW.2018.8403750
3. Ashwin, M., Alqahtani, A.S., Mubarakali, A.: IoT based intelligent route selection of wastage segregation for smart cities using solar energy. Sustain. Energy Technol. Assess. **46**, 101281 (2021). https://doi.org/10.1016/j.seta.2021.101281
4. Assaf, R., Saleh, Y.: Vehicle-routing optimization for municipal solid waste collection using genetic algorithm: the case of Southern Nablus city. Civil Environ. Eng. Rep. **26**(3), 43–57 (2017). https://doi.org/10.1515/ceer-2017-0034
5. Babaee Tirkolaee, E., Abbasian, P., Soltani, M., Ghaffarian, S.A.: Developing an applied algorithm for multi-trip vehicle routing problem with time windows in urban waste collection: a case study. Waste Manage. Res. **37**(1_suppl), 4–13 (2019). https://doi.org/10.1177/0734242X18807001
6. Barbucha, D.: An agent-based guided local search for the capacited vehicle routing problem. In: O'Shea, J., Nguyen, N.T., Crockett, K., Howlett, R.J., Jain, L.C. (eds.) KES-AMSTA 2011. LNCS (LNAI), vol. 6682, pp. 476–485. Springer, Heidelberg (2011). https://doi.org/10.1007/978-3-642-22000-5_49
7. Delgado-Antequera, L., Caballero, R., Sánchez-Oro, J., Colmenar, J.M., Martí, R.: Iterated greedy with variable neighborhood search for a multiobjective waste collection problem. Expert Syst. Appl. **145**, 113101 (2020). https://doi.org/10.1016/j.eswa.2019.113101

8. Hannan, M., et al.: Solid waste collection optimization objectives, constraints, modeling approaches, and their challenges toward achieving sustainable development goals. J. Clean. Prod. **277**, 123557 (2020). https://doi.org/10.1016/j.jclepro.2020.123557
9. Karakatič, S.: Optimizing nonlinear charging times of electric vehicle routing with genetic algorithm. Expert Syst. Appl. **164**, 114039 (2021). https://doi.org/10.1016/j.eswa.2020.114039
10. Kaza, S., Yao, L., Bhada-Tata, P., Van Woerden, F.: What a Waste 2.0: A Global Snapshot of Solid Waste Management to 2050. World Bank Publications (2018). http://hdl.handle.net/10986/30317
11. Kinobe, J.R., Bosona, T., Gebresenbet, G., Niwagaba, C., Vinnerås, B.: Optimization of waste collection and disposal in Kampala city. Habitat Int. **49**, 126–137 (2015). https://doi.org/10.1016/j.habitatint.2015.05.025
12. Lei, C., Jiang, Z., Ouyang, Y.: A discrete-continuous hybrid approach to periodic routing of waste collection vehicles with recycling operations. IEEE Trans. Intell. Transp. Syst. **21**(12), 5236–5245 (2020). https://doi.org/10.1109/TITS.2019.2951571
13. Liang, Y.C., Minanda, V., Gunawan, A.: Waste collection routing problem: a minireview of recent heuristic approaches and applications. Waste Manage. Res. (2021). https://doi.org/10.1177/0734242X211003975
14. Lu, X., Pu, X., Han, X.: Sustainable smart waste classification and collection system: a bi-objective modeling and optimization approach. J. Clean. Prod. **276**, 124183 (2020). https://doi.org/10.1016/j.jclepro.2020.124183
15. Ma, Y., Zhang, W., Feng, C., Lev, B., Li, Z.: A bi-level multi-objective location-routing model for municipal waste management with obnoxious effects. Waste Manage. **135**, 109–121 (2021). https://doi.org/10.1016/j.wasman.2021.08.034
16. Mancera-Galván, E.A., Garro, B.A., Rodríguez-Vázquez, K.: Optimization of solid waste collection: two ACO approaches. In: Proceedings of the Genetic and Evolutionary Computation Conference Companion, pp. 43–44 (2017). https://doi.org/10.1145/3067695.3082043
17. Mańdziuk, J.: New shades of the vehicle routing problem: emerging problem formulations and computational intelligence solution methods. IEEE Trans. Emerg. Topics Comput. Intell. **3**(3), 230–244 (2018). https://doi.org/10.1109/TETCI.2018.2886585
18. Palanivel, T.M., Sulaiman, H.: Generation and composition of municipal solid waste (MSW) in Muscat, Sultanate of Oman. APCBEE Proc. **10**, 96–102 (2014). https://doi.org/10.1016/j.apcbee.2014.10.024
19. Perron, L., Furnon, V.: Or-tools. https://developers.google.com/optimization/
20. Rızvanoğlu, O., Kaya, S., Ulukavak, M., Yeşilnacar, M.İ: Optimization of municipal solid waste collection and transportation routes, through linear programming and geographic information system: a case study from Şanlıurfa, Turkey. Environ. Monit. Assess. **192**(1), 1–12 (2019). https://doi.org/10.1007/s10661-019-7975-1
21. Rosa-Gallardo, D.J., Ortiz, G., Boubeta-Puig, J., García-de-Prado, A.: Sustainable WAsTe collection (SWAT): one step towards smart and spotless cities. In: Braubach, L., et al. (eds.) ICSOC 2017. LNCS, vol. 10797, pp. 228–239. Springer, Cham (2018). https://doi.org/10.1007/978-3-319-91764-1_18
22. Ruiz-Rosero, J., Ramirez-Gonzalez, G., Viveros-Delgado, J.: Software survey: ScientoPy, a scientometric tool for topics trend analysis in scientific publications. Scientometrics **121**(2), 1165–1188 (2019). https://doi.org/10.1007/s11192-019-03213-w

23. Shao, S., Xu, S.X., Huang, G.Q.: Variable neighborhood search and Tabu search for auction-based waste collection synchronization. Transp. Res. Part B Methodol. **133**, 1–20 (2020). https://doi.org/10.1016/j.trb.2019.12.004

24. Toth, P., Vigo, D.: The vehicle routing problem. In: SIAM (2002). https://doi.org/10.1137/1.9780898718515

25. Zhang, S., Zhang, J., Zhao, Z., Xin, C.: Robust optimization of municipal solid waste collection and transportation with uncertain waste output: a case study. J. Syst. Sci. Syst. Eng. **31**, 204–225 (2021)

A Systematic Literature Review About Multi-objective Optimization for Distributed Manufacturing Scheduling in the Industry 4.0

Francisco dos Santos[1,2]($\boxtimes$) ⓘ, Lino A. Costa[1] ⓘ, and Leonilde Varela[1] ⓘ

[1] Centro ALGORITMI, Universidade do Minho, 4710-057 Braga, Portugal
francisco_dos_santos@outlook.pt, {lac,leonilde}@dps.uminho.pt
[2] IP-UNIKIVI, Uíge, Angola

Abstract. Multi-objective optimization problems are frequent in many engineering problems, namely in distributed manufacturing scheduling. In the current Industry 4.0 this kind of problems are becoming even more complex, due to the increase in data sets arising from the industry, thus requiring appropriate methods to solve them, in real time. In this paper, the results of a Systematic Literature Review are presented to reveal the state of the art in this scientific domain and identify the main research gaps in the current digitalization era. The results obtained allow to realize the importance of the multi-objective optimization approaches. Typically, when addressing large scale real problems, the existence of many objectives usually benefits with the establishment of some level of trade-off between objectives. In this paper, a summarized description and analysis is presented, related to several main issues arising currently in companies requiring the application of multi-objective optimization based distributed scheduling, for enabling them to fulfill requisites imposed by the Industry 4.0. In this context, issues related to energy consumption, among other customer-oriented objectives are focused to enable properly support decision-making through the analysis of a set of 33 main publications.

Keywords: Multi-objective optimization · Distributed manufacturing scheduling · Industry 4.0 · Decision-making support

1 Introduction

In this paper, a summarized description and analysis of the main results obtained through the application of a Systematic Literature Review (SLR), a widely used and well-known methodology currently used [1], is carried out. SLR is an important research methodology, consisting not only of a simple text review on a given subject, but rather, on an approach that also permits answering several research questions, properly formulated. Thus, it is a "methodology that allows not only to locate contents or existing studies, through the selection and evaluation of different contributions, but also to analyze and synthesize data, duly organized, by themes, in order to report them properly and to

O. Gervasi et al. (Eds.): ICCSA 2022 Workshops, LNCS 13378, pp. 157–173, 2022.
https://doi.org/10.1007/978-3-031-10562-3_12

highlight relevant contents". Thus, this methodology allows the extraction of important conclusions, in a clear and well-structured way, with the purpose of showing some new evidence, about something, or in some way, that has not yet been explored [2]. In this sense, a SLR on multi-objective optimization for solving distributed manufacturing scheduling problems in the context of the Industry 4.0 or I4.0, for short, is focused in this study. Scientific publications, related to various issues, that apply multi-objective optimization to real problems in industry, namely, regarding its transition to the I4, are analyzed. In this context, energy consumption in industry, customer criteria satisfaction, among other objectives and issues that arise in the context of multi-objective optimization in solving industrial problems are explored to properly support the decision-making process. A total of 33 publications were deeply studied.

Optimization refers to finding one or more solutions that correspond to extreme values or compromises of one or more objectives. Multi-objective optimization (MOO) involves a problem solving in which an optimal optimization of multiple conflicting objectives is intended to be achieved. Objectives are in conflict if there are trade-offs or compromises between these objectives, that is, if there is no single feasible solution that is optimal for all objectives [3]. These problems with two or three conflicting objectives are usually called multi-objective optimization problems (MOOP). These are different from single-objective optimization problems, which are simpler in terms of complexity and demand from the decision maker [4]. Problems with more than three objectives conflicting with each other, known as many-objective problems, are very difficult to solve but are common in almost all areas of knowledge and usually [5].

In the I4.0, many difficulties have been encountered, namely related to the multi-objective nature of many scheduling problems that involve conflicting objectives. Frequently, decision-making based on multi-criteria or multi-objective approaches has been adopted to address these problems at a more real scale, therefore, complex, because they are critical problems to be solved, requiring a proper decision support process, based on an analysis of the various existing alternatives [6]. In most cases, solving multi-criteria decision-making problems is focused on finding solutions for a given set of alternatives [7]. However, every Pareto optimal solution is an equally acceptable solution to a multi-objective optimization problem. Therefore, in the process of choosing a single preferred solution among all alternatives can be a difficult task due to a high cognitive effort required [8].

On the other hand, the manufacturing scheduling (MS) problem is an optimization problem in which several tasks are assigned to machines, organized in a certain way, in a production system, at given instants, while trying to optimize objectives, such as makespan, cost or load balancing, among others [9, 10]. In the context of I4.0, scheduling or MS can be aided or achieved through an AI-based approach, namely for solving distributed manufacturing scheduling (DMS) problems. Such an approach, combined with other emerging production technologies that enhance mass customization, through the current cyber-physical systems (CPS), underlying the I4.0, allow an innovative approach in production [11]. Through the use of a DMS approach, it is expected to achieve several advantages, namely the reduction of production time, while aiming to meet product delivery times, among other performance measures to be optimized.

With the advent of the I4.0, there has been an increasing need to handle large volumes of varied data and in many of these cases these are problems with many conflicting objectives, a fact that further increases the complexity of the problems, making it even more difficult to be solved, particularly in terms of optimization. Thus, difficult decision-making processes in real production environments emerge in this context. The transformation of data into information or knowledge has been fundamental in several areas of engineering to support decision-making, including in more complex contexts, given the existence of many alternative scenarios, with a large number of solutions [5].

In this paper, it is intended to present the results of a SLR, about multi-objective optimization as a tool to support decision-making in solving MOOP, occurring in the context of DMS, currently, in the I4.0 era. Therefore, the aim is to identify gaps and opportunities for research in this scientific domain.

This paper is structured as follows. In Sect. 2, the SLR methodology is briefly described. Section 3 presents the results and a deeper analysis of a main set of 33 publications found. Finally, in Sect. 4, the main conclusions are addressed, along with some planed future work.

2   Methodology

In this research, the SLR methodology is used. SLR is a systematic method and, which should be comprehensive and reproducible, in order to better identify, assess and synthesize existing works, with a greater or lesser degree of complexity, proceeding to their registration and that have been produced by a researcher, a professional or simply a scholar [1, 2, 12]. SLR, in addition to allowing to contribute to the development of theory, also allows to facilitate the interpretation of existing contributions, through complements and extensions of research areas.

SLR uses a pre-planned research strategy, and this type of review is intended to answer specific research questions about existing studies [2]. In this work the following research questions were formulated:

- RQ1: Is MOO being applied to solve DMS problems in the current I4.0 era?
- RQ2: What are the main issues that arise when studding MOO-DMS in the I4.0?
- RQ3: What else can be further explored for solving MOO-DMS problems?

To answer these research questions, in this work, the SLR methodology by [13] was adopted. This methodology is illustrated in Fig. 1 and involves the following three main steps: (i) Input, (ii) Processing and (iii) Output.

Further to this adaptation, it was also applied the proposals of [1] and [2].

In the input phase (1), the objectives of the search are defined, the search terms are identified, and the keywords for the subject under study are selected, as well as the platforms where the different materials to be studied can be searched. After the selection of the research platforms, it is necessary that the input is adequate and with quality, thus promoting quality output (3).

The processing phase (2) is where the whole process is done in order to obtain a desired result. Such processing includes five steps, from planning to the last step, which

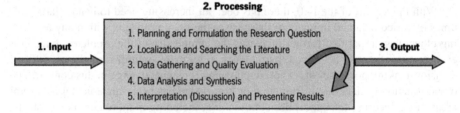

Fig. 1. Stages of the literature review (adapted from [13]).

serve as the basis for presenting the results of the research, and these steps can be, iteratively, restarted if necessary [13]. Typically, this phase encompasses all or most of the following steps [2, 13]: (1) Planning and Formulating the Research Question; (2) Locating and Searching the Literature; (3) Data Collection and Quality Assessment; (4) Data Analysis and Synthesis; (5) Interpretation (Discussion) and Result Presentation.

The first step of this phase is related to the planning and formulation of research questions in order to guide the SLR. The second stage is associated with locating and searching the literature. In this paper, the Scopus[1] and Web of Science[2] (WOS) platforms were used, and to complement the b-on[3] Portuguese library platform. The b-on provides access to the full text of papers, in several indexed scientific journals, within the Portuguese academic community.

The research was carried out in the time span from 2010 to 2022. Table 1 presents the search platforms used, the search terms, as well as the results obtained. The following keywords or search terms were applied according to the specificity of each search platform: "multiobjective optimization OR multi-objective optimization", "distributed scheduling", "industry 4.0" and "manufacturing". The results are presented in two parts: without filtering and refining the search. In total, without filters and using only the search terms, 615 articles were obtained. By further refining the search, by applying the other search options, it was possible to reduce the search to a total of 74 articles. For the b-on platform, in the first part, the search term is applied without any search option and without any filter applied, presenting 230 articles as the search result. In the second part, search options are also applied, thus refining, and restricting the focus of the search and resulting in only 46 articles remaining for analysis. For the Scopus platform, 282 articles were identified using the search terms alone. After applying the search options, 12 articles were obtained. Finally, for the WOS platform, 103 articles were found in the first part, and with the search options applied and considering only the articles with open access, the output was reduced to 16 articles.

The third stage, data collection and quality assessment, is related directly with the collection of literature in the different platforms. It is in this stage that the inclusion and exclusion criteria were defined. The exclusion criteria were chosen based on the research objectives and defined keywords. Some search terms were applied only to the

---

[1] https://www.scopus.com.

[2] https://www.webofscience.com.

[3] https://www.b-on.pt.

Table 1. Search terms.

Platform	Search terms	Total without search options	Search options		Last execution by	
			Limiters	Expanders	Screen and search	Results with search options
B-On	AB (multiobjective optimization OR (multi-objective optimization)) AND TX distributed scheduling AND TX industry 4.0 AND TX manufacturing	230	Full Text; Peer Reviewed; Publication Date: 2010.01–2022.02; Research areas: Science, Applied Sciences, Engineering, Computer Science, Mathematics, Technology, Information Technology	Search also in the full text of the articles; Apply equivalent subjects	Advanced search	46
Scopus	(TITLE-ABS-KEY (multiobjective AND optimization) AND ALL (distributed scheduling) AND TITLE-ABS-KEY (industry 4.0) AND ALL (manufacturing))	282	Publication Date: 2010.01–2022.02; Research areas: Engineering, Computer Science, Business, Management and Accounting, Decision Sciences,	N/A	Advanced search	12
Web of Science	Multiobjective optimization OR multi-objective optimization (Title) and distributed scheduling (Topic)	103	Publication Date: 2010.01–2022.02; Research areas: Computer Science Information Systems, Engineering Electrical Electronic, Telecommunications, Computer Science Software Engineering, Computer Science Artificial Intelligence, Mathematics Interdisciplinary Applications, Computer Science Theory Methods, Engineering Multidisciplinary, Multidisciplinary Sciences, Operations Research Management Science; Open Access Articles		Advanced search	16

title and others to the abstract or the entire text of the article. Articles that were not peer-reviewed were excluded. Finally, as an inclusion factor only articles written in English were considered, with full text available, limitation of search time between 2010 and 2022, resulting in a reduction in the number of articles to 46, 12 and 16 for the b-on, Scopus and WoS platforms, respectively, as mentioned before and presented in Table 1.

The data analysis and synthesis are the fourth stage. This phase was performed by categorizing by title, year of publication, publisher as well as reading the abstracts of the selected articles, as summarized in the Fig. 2. Reading the abstracts allowed to understand the subject addressed in the article. Those articles related to the industry were included and those articles outside the industry context were excluded. Finally, in the fifth stage (Interpretation (Discussion) and Result Presentation), a summary of each selected article was done. Overall, 615 documents were identified on the three search platforms, only 33 are within the scope of this study, as shown in Fig. 2.

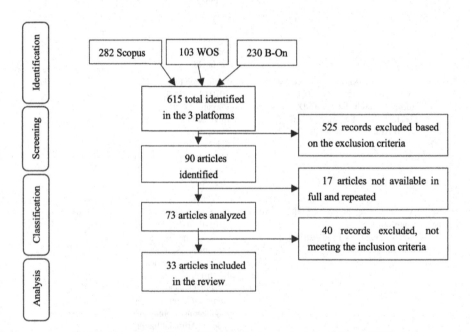

Fig. 2. Summary of the article selection process (adapted from [14]).

In this sense, Table 2 presents the summary of the processing phase, where it is presented what was applied, the inclusion and exclusion criteria, among others.

The output phase is the last of the three phases. It is the phase where the results of the research process are presented. In a succinct way, a discussion about the subject under study is described. In this phase, the content of materials of great interest to the research in focus is presented, eliminating materials of little interest to the study.

Table 2. Summary of the research phases.

Phases of the review	Description
(1) Planning and Formulating the Research Question	- What are the most studied issues when it comes to multi-objective optimization in distributed manufacturing scheduling in the I4.0 era? - Is multi-objective optimization in the I4.0 era applied to distributed manufacturing scheduling problems? - What contributions already exist to aid decision making applied to multi-objective optimization in distributed manufacturing scheduling problems? - What else can be explored that is yet to be explored in multi-objective optimization in distributed manufacturing scheduling problems?
(2) Localization and Literature Search	Scopus, Web of Science, B-On, search time interval: 2010–01/3–2022 considered keywords: multi-objective optimization, distributed scheduling, industry 4.0, decision-making
(3) Data Collection and Quality Assessment	Exclusion: Not peer reviewed, not industry related, article not complete, article not available Inclusion: any MOO subject in industry, academic articles, only written in English, research time limitation between 2010 and 2022 Total articles: 33
(4) Data Analysis and Synthesis	Excel was used to organize and categorize the subjects of the articles according to the keywords, the subject of the article, the year of publication, and also the publisher
(5) Interpretation (Discussion) and Result Presentation	Interpretation on the use of multi-objective optimization in distributed manufacturing scheduling problems in the Industry 4.0, based on the survey of the literature done in order to answer the formulated research questions

3 Results and Discussion

As already mentioned in the previous sections, from the results obtained, 33 articles were considered that met the conditions set to answer the research questions. The characteristics of these selected articles were analyzed. All selected articles were organized in an Excel file to facilitate quantitative analysis, information such as year of publication, publishers, keywords, as well as the number of citations.

An important aspect to consider is the year of publication. In this sense, Fig. 3a presents the number of publications per year, within the scope of the subject under research. In this figure, it is represented the publications in the time interval between January 2010 and March 2022. Through this figure, it is possible to observe the variation of publications per year, showing that in the first four years there is no article among the 33 presented, and having been reached in 2020 the highest number of publications, with a total of 10 publications, and in the last year there is not a large number of publications because it refers to only three months. Although, it can be observed that the research about the subject under study is growing. On the other hand, it can be seen that the 33 articles are from different publishers, being the publisher with the largest number of articles the Taylor & Francis Ltd and Elsevier, both with 7 articles, followed by 6 articles from Springer and the remaining 5 from different publishers, as illustrated in Fig. 3b.

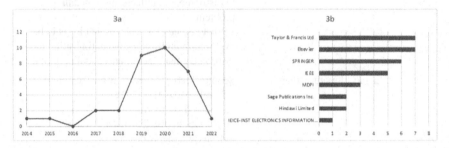

Fig. 3. (a) Number of articles per year. (b) Publishers of the articles.

In the context of multi-objective optimization, several topics are addressed, but when referring to distributed manufacturing scheduling problems it comes out that this subject has not been fully addressed, but still on a small scale, which decreases even more when referring to problems occurring in the I4.0. Since this topic is quite recent, the goal of this work is pertinent to identify different multi-objective optimization problems applied to manufacturing scheduling problems in the I4.0 context. Based on the review carried out, it can be concluded that currently problems that associate these three topics are still little explored, because as it can be seen, in different collected materials only six articles address two topics (Multi-objective Optimization and Scheduling) while only two address all three topics (Multi-objective Optimization, Scheduling and Industry 4.0).

The results are summarized in Table 3, where the 33 articles are classified. In this sense, the crossing of the topics using the following keywords are presented: Multiobjective Optimization, Scheduling and Industry 4.0, identified in the title, keywords or abstract of the articles.

3.1 Multi-objective Algorithms in Industry

Based on the reading of the selected articles, it can be realized that the I4.0 is accepted in the manufacturing industry as it guides an innovative and promising production paradigm. A new feature of I4.0-based manufacturing systems is the application of advanced, intelligent machines that have communication, self-optimization, and self-training capabilities [17]. In [17] a production flow program considering multiple objectives is investigated. An integer mixed programming model was formulated by the

Table 3. Keyword crossing.

Criteria	Keywords	Reference
Multi-objective optimization	Multi-objective, multicriteria	[15–34]
Scheduling	Programing	[15, 17, 18, 21, 25, 27, 29, 32, 35–40]
Industry 4.0	I4.0, industry, manufacturing	[17, 18, 23, 28, 39, 41–44]
Multi-objective optimization & scheduling	—	[15, 17, 18, 21, 25, 27, 29]
Multi-objective optimization & Industry 4.0	—	[17, 18, 23, 28]
Scheduling & Industry 4.0	—	[17, 18, 39]
Multi-objective optimization & scheduling & Industry 4.0	—	[17, 18]

authors and solved by a fireworks algorithm developed and designed with specific strategies, such as spark burst procedure. Simulation experiments were conducted by the authors on a set of test problems, and their experimental results demonstrated that their model and the proposed algorithm achieved satisfactory performance when compared with some state-of-the-art multi-objective optimization algorithms [17].

In view of the growth in data volume and with many dimensions motivated by the emergence of the I4.0, the need for using artificial intelligence (AI) technologies for this complex data analysis is also growing [28]. To address these aforementioned problems, in [28] it is proposed a multi-objective evolutionary algorithm based on decomposition with adaptive exploration. Through dynamic exploration and joint exploration, their proposed algorithm improved the global search capability as well as the diversity of the solutions found by the algorithm.

In [26] several multi-objective meta-heuristic optimization algorithms to generate Pareto optimal solutions for micro-turning and micro-milling applications are used. A comparative study was conducted by the authors to evaluate the performance of the non-dominated sorting genetic algorithm II (NSGA-II), multi-objective lion ant optimization (MOALO) and multi-objective dragonfly optimization (MODA).

A new efficient 2-stage NSGA-II algorithm was proposed in [16] and the results compared with the original NSGA-II algorithm. The authors report that the performance of the 2-stage NSGA-II algorithm was significantly superior to that of the original NSGA-II algorithm.

The optimization of task scheduling in a distributed heterogeneous computing environment is a difficult non-linear multi-objective problem. It plays an important role in decreasing the service response time and cost and also in increasing the quality of service [35]. In [35], the authors consider four conflicting objectives with each other, namely for minimizing task transfer time, task execution course, energy consumption and task queue length. They developed a comprehensive multi-objective optimization model for task scheduling, allowing to reduce the costs from both customer's and supplier's point of view. Their model was assessed by applying two evolutionary multi-objective algorithms: Multi-Objective Particle Swarm Optimization (MOPSO) and Multi-Objective Genetic Algorithm (MOGA). The simulation results showed that the proposed model provided optimal trade-off solutions among the four conflicting objectives, thus significantly reducing the response time of the improvisational work (makespan).

3.2 Energy Consumption in Industry

In general, the operation of industrial systems requires the consumption of energy for the operation of various machines. Energy efficient production scheduling is an area that has attracted attention from researchers due to the massive energy consumption of the manufacturing process [27]. In this regard, in [27], the authors presented a study of an energy-efficient production job scheduling problem with sequence-dependent setup time with the goal of minimizing the production time, total delay and total energy consumption, simultaneously. In [20], a multi-objective optimization method for wind and thermal power systems is proposed. The wind speed in the wind farm is considered as the decision variable and the minimization of fuel costs and emission costs of the thermal power unit as objectives.

Nowadays, with the emergence of the I4.0, it is evident the increase in energy consumption necessary to be compatible with the new technologies used by the fourth industrial revolution. In recent years, attention to energy conservation has been increasing for manufacturing activities and especially for energy-intensive industries [40]. A multi-objective optimization algorithm to improve productivity and reduce costs and energy consumption of autonomous industrial processes was presented in [23], in order to achieve sustainable growth. The analyses were done in an assembly-line production with robotic cells and subsequent material handling systems thus using autonomous guided vehicles (AGVs) as an internal conveyor. In [44] it is also presented a solution approach for multi-objective optimization in vehicle-based automated storage systems. These systems over traditional systems have great advantages since they have a flexible travel pattern of autonomous vehicles, thus allowing the number of vehicles in the system to be varied based on the demand for alternatives [44].

Optimizing the production schedule to be more energy efficient while meeting production goals is a difficult task [24]. Therefore, how to schedule and distribute production tasks to meet production goals while making the best use of the ever fluctuating energy market prices and availability of locally installed energy sources? [24]. Accordingly, a generic algorithm was presented in [24] for optimizing production schedules in terms of energy consumption, peak consumption reduction and makespan, which considers that tasks can be performed in different ways, having different characteristics.

Another approach to energy consumption in industry was presented in [22]. These authors presented a multi-objective stochastic optimization algorithm to analyze energy consumption, construction period, and demonstrated its benefits in the construction process, considering random events.

Multi-robotic services are commonly used to increase the efficiency of I4.0 application, including emergency management in smart factories. The workflow of robotic services consists of data-intensive tasks that are sensitive to computation-intensive delays. When multiple robots in a cloud instance work collaboratively, the optimal resource allocation for the robotic workflow tasks becomes a challenging problem. Because the tasks are interdependent, inconveniences in data exchange between local robots and the remote cloud degrades the quality of services [41]. In their study, these authors addressed the simultaneous optimization of the makespan, energy consumption, and cost while allocating resources for robot workflow tasks. On the other hand, in [30] some problems are described considering the design and line balance of mixed-model disassembly with multi-robotic workstations under uncertainty. The tasks of different models are executed simultaneously by the robots that have different disassembly capabilities. The robots have unidentified task times and energy consumption. The processing times of the tasks are assumed to be interval numbers when considering the uncertainties in the disassembly process. A mixed-integer mathematical programming model to minimize the cycle time, workstation energy cycle and total energy consumption is proposed.

3.3 Planning on Production Lines

Manufacturing process scheduling is a complex process and in the context of the I4.0 it becomes even more complex. Meta-heuristics help to solve these complex problems, despite requiring greater computational power. The use of high-performance architectures present in cloud computing and edge computing, can support the development of better meta-heuristics allowing to deal with complex problems of I4.0 [39]. In this sense, a literature review is presented in [39], providing an overview of meta-heuristics for activity scheduling. Logistics and supply chain systems (SCS) are important parts in industrial systems [43]. In this paper, the author collects and analyses papers that explore several problems in logistics and SCS. Different important approaches that can be found in logistics and SCS are presented, considering game theories, meta-heuristics, multi-objective optimization, robust control, and fuzzy sets.

An approach for cloud-based distributed optimization for factory process scheduling and planning was presented in [42]. The authors proposed three approaches for dynamically managing subpopulation creation and elimination developed by a multi-objective algorithm. In [19] it is also addressed the Distributed Hybrid Flow Shop Scheduling Problem with Multiprocessor Tasks, by considering two objectives simultaneously (the makespan and the total energy consumption). The problem was divided into three subproblems: task assignment between factories, task sequence in each factory, and machine assignment for each task. They then formulated a mixed interlinear programming model and proposed a Novel Multi-Objective Evolutionary Algorithm based on Decomposition (NMOEA/D). In [32] two problems that include schedule assignment (in cases of identical part production) and part sequencing (in cases of different part production) are modeled. The first model was used to minimize cycle time and operating cost, and the second to minimize both the mean and standard deviation of total production cost as well as cycle times. A simulation-based mathematical modeling and optimization method was presented to schedule, assign similar parts, and plot the optimal sequence of different parts in [33]. These authors proposed an optimization method based on data analysis for enriched distributed process planning while considering machine tool selection, cutting tool selection, and machining condition determination.

Due to stochastic uncertainty, the problems of scheduling semiconductor manufacturing systems with the goal of optimizing classical performance indices tend to be increasingly complicated [36]. In [31], the problem of scheduling a semiconductor manufacturing system with uncertain processing times was addressed.

The proposal involved a three-phase multi-objective optimization approach, which enabled to collaboratively optimize performance indices and their robustness measures. Furthermore, in [37] an integrated scheduling problem in a distributed manufacturing system is investigated. A multi-objective stochastic model considering two scheduling criteria as well as operation process uncertainties. A multi-objective brain storm optimization algorithm was developed, and the proposed model demonstrated promising experimental results when compared with two state-of-the-art multi-objective algorithms [37].

In recent years, due to the vast applications in production systems, hybrid flow scheduling problems have gained increasing attention. Most previous studies assume that the job processing times are deterministic and constant [29]. In [29] a stochastic hybrid flow planning problem for minimizing markespan and total delay is described.

With the emergence of cloud computing technologies, services with the same functionalities and different non-functionalities occur in the cloud manufacturing system. Therefore, optimizing the composition of manufacturing services become increasingly important in meeting customer demands. This issue involves multi-objective optimization [31]. A new manufacturing service composition model based on service quality, as well as considerations of crowdsourcing and service correlation, was proposed in [31]. Due to the e-commerce development and growth, warehouse logistics has also been facing emerging challenges in ways that include more order batches and shorter order processing cycles. Therefore, intelligent warehouse systems using autonomous robots for automated storage and intelligent order scheduling are becoming dominant [25]. Accordingly, in [25] a multi-robot cooperative scheduling system in intelligent warehouse is proposed. The cooperative multi-robot scheduling system of intelligent storage aims to drive many robots in an intelligent warehouse to perform distributed tasks in an optimal manner (e.g., in time saving and energy conservation).

There is currently great attention on Cloud Manufacturing (CMfg) as a new service-oriented manufacturing paradigm. To integrate activities and services through a CMfg, two important issues to facilitate the success of CMfg are Service Load Balancing and Transport Optimization [38]. In this regard, a new queuing network for parallel scheduling of multiple processes and customer orders to be supplied is developed in [38].

Service oriented systems based on customer order as well as metrics related to cost and time are important for the success of a cloud manufacturing system [18]. Another important aspect is establishing equity among factories with cloud systems to obtain benefits according to capacity. On the other hand, is also relevant to establish equity among customers who order products based on their preferences. In this regard, in [18] a multi-objective mathematical model capable of considering both customers and service factories and the equity between them is developed. This model is able to simultaneously assist the decision maker regarding rejecting or accepting customer orders, assigning jobs to machines, determining a price for each job as well as scheduling work on the machines. Meanwhile, in [15] a Multi-objective Genetic Algorithm Optimization (GMOA) is proposed to solve flexible ordering problems for a medium-scale screw factory. New smart manufacturing technology facilitates the interconnection of process planning data as well as purchasing scheduling and optimizing the integration of different manufacturing processes [21]. A multi-objective optimization integrated process planning and scheduling with uncertain processing time and date is addressed in [21].

Sustainability has been widely studied in the industrial area, so end-of-life product recovery is an active way to achieve sustainable manufacturing, while extending producer responsibility to closed-loop product service. This is still a challenge to provide flexible and intelligent recovery plans for industrial equipment at different product service periods [34]. However, in [34] the intelligent recovery decision making problem is addressed. The authors proposed a system framework for the implementation of intelligent end-of-life product recovery management based on product state monitoring. Multi-objective optimization models are formulated to identify the age-dependent recovery roadmap that best matches the product condition and meets business objectives. Recovery profits and energy performance are optimized in the models to achieve more environmentally friendly recovery.

4   Conclusions

In this paper, a SLR concerning multi-objective optimization approaches for distributed manufacturing scheduling problems in the I4.0 era was presented. Based on the results obtained, a study considering a set of 33 articles collected from different platforms (Web of Science, Scopus and b-on) was conducted. A careful analysis of these results was performed, six articles emerged as the most relevant for answering the research questions raised in this work.

Moreover, it can be concluded that research about this subject has been growing in recent years, as could be seen through a graph showing the number of publications per year. It was also possible to identify several issues regarding multi-objective optimization in distributed manufacturing scheduling problems in the I4.0 era. There are concerns on reducing energy consumption, along with logistics and supply chain systems issues, that are modeled as multi-objective problems and solved through multi-objective optimization algorithms, namely for production flow scheduling. As response to RQ1, this study shows that there are some recent applications of multi-objective approaches to solve DMS problems. However, this is an area that is starting to grow and, since the large amount of data available in this I4.0 era, it is expected to see in near future new relevant developments.

The recent features of scheduling job shop problems (JSP) in the I4.0 are highlighted as distributed intelligent scheduling in contrast to traditional centralized scheduling. In terms of RQ2, one of the future trends of the JSP scheduling in the I4.0 should be in the development and use of different kind of approaches, namely based on AI, such as intelligent agents. In situations where the traditional centralized workshop scheduling is transformed into the distributed intelligent scheduling, it can significantly reduce the computational workload and the system will become more flexible and agile [11]. In this sense, it can also be said that the I4.0 is already being accepted in the manufacturing industry, as it guides an innovative and promising production paradigm to be further explored. A new and fundamental feature of I4.0-based manufacturing systems is the application of advanced, intelligent machines that have communication, self-optimization, and self-training capabilities [17].

Also, in the context of the I4.0 manufacturing scheduling can be further supported or better achieved through the use of different kind of approaches, namely on AI, and for solving distributed manufacturing scheduling problems. Such approaches, conciliated with other emerging production technologies that enhance mass customization, through the current cyber-physical systems underlying the I4.0, and thus allowing to reach an innovative approach in production [11]. Through the use of a distributed MS approaches, several advantages are expected to be achieved, namely the reduction of production time, while aiming to meet product delivery deadlines, among other performance measures to be optimized. Several algorithms have been applied and proposed to solve multi-objective optimization problems, in the manufacturing scheduling domain, namely in distributed manufacturing scheduling. Based on this review, it was also found that multi-objective optimization is being used in the I4.0 to optimize energy consumption associated to the operation of new technologies. The increasing advances in technology, associated with intelligent manufacturing systems in the I4.0 has been reflected by growing energy requirements. This can be an important feature in near future as well as environmental issues.

There are multiple issues that can be further explored concerning RQ3. In terms of future work, the use of multi-objective optimization approaches and methods to solve distributed manufacturing scheduling problems, in the context of the I4.0 should be further explored. These kind of scheduling problems should continue to be oriented to the current requirements imposed by the I4.0, as they do remain underexplored. Therefore, it is thought to be a rather vast area, and therefore, of great interest to continue to be explored, through the application of multi-objective optimization approaches, by considering diverse kind of internal and external performance measures of companies.

Acknowledgements. This work was supported by national funds through the FCT-Fundação para a Ciência e Tecnologia through the R&D Units Project Scopes: UIDB/00319/2020, and EXPL/EME-SIS/1224/2021.

References

1. Okoli, C.: A guide to conducting a standalone systematic literature review. Commun. Assoc. Inf. Syst. **37**, 879–910 (2015)
2. Thomé, A.M.T., Scavarda, L.F., Scavarda, A.J.: Conducting systematic literature review in operations management. Prod. Plan. Control **27**, 408–420 (2016)
3. Deb, K., Saxena, D.: Searching for Pareto-optimal solutions through dimensionality reduction for certain large-dimensional multi-objective optimization problems. In: 2006 IEEE Congress on Evolutionary Computation (CEC 2006), pp. 3353–3360 (2006)
4. Patil, M.V., Kulkarni, A.J.: Pareto dominance based multiobjective cohort Intelligence algorithm. Inf. Sci. (Ny) **538**, 69–118 (2020)

5. Santos, F., Costa, L.: Multivariate analysis to assist decision-making in many-objective engineering optimization problems. In: Gervasi, O., et al. (eds.) ICCSA 2020. LNCS, vol. 12251, pp. 274–288. Springer, Cham (2020). https://doi.org/10.1007/978-3-030-58808-3_21

6. Yang, M., Nazir, S., Xu, Q., Ali, S. Uddin, M.I.: Deep learning algorithms and multicriteria decision-making used in big data: a systematic literature review. Complexity **2020**, 18 (2020). Article ID 2836064. https://doi.org/10.1155/2020/2836064

7. Rymaszewski, S., Wątróbski, J., Karczmarczyk, A.: Identification of reference multi criteria domain model - production line optimization case study. Procedia Comput. Sci. **176**, 3794–3801 (2020)

8. Rocha, L.C.S., de Paiva, A.P., Rotela Junior, P., Balestrassi, P.P., da Silva Campos, P.H.: Robust multiple criteria decision making applied to optimization of AISI H13 hardened steel turning with PCBN wiper tool. Int. J. Adv. Manuf. Technol. **89**, 2251–2268 (2017)

9. Varela, M.L.R., Silva, S.D.C.: An ontology for a model of manufacturing scheduling problems to be solved on the web. In: Azevedo, A. (ed.) Innovation in Manufacturing Networks. BASYS 2008. IFIP – The International Federation for Information Processing, vol. 266, pp. 197–204. Springer, Boston (2008). https://doi.org/10.1007/978-0-387-09492-2_21

10. Varela, M.L.R., Trojanowska, J., Carmo-Silva, S., Costa, N.M.L., Machado, J.: Comparative simulation study of production scheduling in the hybrid and the parallel flow. Manag. Prod. Eng. Rev. **8**, 69–80 (2017)

11. Zhang, J., Ding, G., Zou, Y., Qin, S., Fu, J.: Review of job shop scheduling research and its new perspectives under Industry 4.0. J. Intell. Manuf. **30**(4), 1809–1830 (2017). https://doi.org/10.1007/s10845-017-1350-2

12. Donato, H., Donato, M.: Stages for undertaking a systematic review. Acta Med. Port. **32**, 227–235 (2019)

13. Levy, Y., Ellis, T.J.: A systems approach to conduct an effective literature review in support of information systems research. Inf. Sci. J. **9**, 181–212 (2006)

14. Bittencourt, V.L., Alves, A.C. Leão, C.P.: Industry 4.0 triggered by Lean thinking: insights from a systematic literature review. Int. J. Prod. Res. **59**, 1496–1510 (2021)

15. Liu, T.K., Chen, Y.P., Chou, J.H.: Developing a multiobjective optimization scheduling system for a screw manufacturer: a refined genetic algorithm approach. IEEE Access **2**, 356–364 (2014)

16. Eftekharian, S.E., Shojafar, M., Shamshirband, S.: 2-Phase NSGA II: an optimized reward and risk measurements algorithm in portfolio optimization. Algorithms **10**, 1–15 (2017)

17. Fu, Y., Ding, J., Wang, H., Wang, J.: Two-objective stochastic flow-shop scheduling with deteriorating and learning effect in Industry 4.0-based manufacturing system. Appl. Soft Comput. J. **68**, 847–855 (2018)

18. Vahedi-Nouri, B., Tavakkoli-Moghaddam, R., Rohaninejad, M.: A multi-objective scheduling model for a cloud manufacturing system with pricing, equity, and order rejection. IFAC-PapersOnLine **52**, 2177–2182 (2019)

19. AbdelAziz, A.M., Soliman, T.H.A., Ghany, K.K.A., Sewisy, A.A.E.-M.: A Pareto-based hybrid whale optimization algorithm with Tabu search for multi-objective optimization. Algorithms **12**, 261 (2019)

20. Guo, X., Gong, R., Bao, H., Lu, Z.: A multiobjective optimization dispatch method of wind-thermal power system. IEICE Trans. Inf. Syst. **E103D**, 2549–2558 (2020)

21. Wen, X., Li, X., Gao, L., Wang, K., Li, H.: Modified honey bees mating optimization algorithm for multi-objective uncertain integrated process planning and scheduling problem. Int. J. Adv. Robot. Syst. **17**, 1–17 (2020)

22. He, W., Li, W., Xu, S.: A Lyapunov drift-plus-penalty-based multi-objective optimization of energy consumption, construction period and benefit. KSCE J. Civ. Eng. **24**(10), 2876–2889 (2020). https://doi.org/10.1007/s12205-020-2072-0

23. Rubio, F., Llopis-Albert, C., Valero, F.: Multi-objective optimization of costs and energy efficiency associated with autonomous industrial processes for sustainable growth. Technol. Forecast. Soc. Change **173**, 121115 (2021)
24. Küster, T., Rayling, P., Wiersig, R., Pozo Pardo, F. D.: Multi-objective optimization of energy-efficient production schedules using genetic algorithms. Optim. Eng. (2021)
25. Yang, S., et al.: A novel maximin-based multi-objective evolutionary algorithm using one-by-one update scheme for multi-robot scheduling optimization. IEEE Access **9**, 121316–121328 (2021)
26. Joshi, M., Ghadai, R.K., Madhu, S., Kalita, K., Gao, X.Z.: Comparison of NSGA-II, MOALO and MODA for multi-objective optimization of micro-machining processes. Materials (Basel). **14**, 1–16 (2021)
27. He, L., Chiong, R., Li, W., Dhakal, S., Cao, Y., Zhang, Y.: Multiobjective optimization of energy-efficient job-shop scheduling with dynamic reference point-based fuzzy relative entropy. IEEE Trans. Ind. Inform. **18**, 600–610 (2022)
28. Qian, W., et al.: An improved MOEA/D algorithm for complex data analysis. Wirel. Commun. Mob. Comput. **2021**, 20 (2021). Article ID 6393638. https://doi.org/10.1155/2021/6393638
29. Fu, Y., Zhou, M., Guo, X., Qi, L.: Scheduling dual-objective stochastic hybrid flow shop with deteriorating jobs via bi-population evolutionary algorithm. IEEE Trans. Syst. Man Cybern. Syst. **50**, 5037–5048 (2020)
30. Fang, Y., Ming, H., Li, M., Liu, Q., Pham, D.T.: Multi-objective evolutionary simulated annealing optimisation for mixed-model multi-robotic disassembly line balancing with interval processing time. Int. J. Prod. Res. **58**, 846–862 (2020)
31. Zhang, W., Yang, Y., Zhang, S., Yu, D., Li, Y.: Correlation-aware manufacturing service composition model using an extended flower pollination algorithm. Int. J. Prod. Res. **56**, 4676–4691 (2018)
32. Vaisi, B., Farughi, H., Raissi, S.: Schedule-allocate and robust sequencing in three-machine robotic cell under breakdowns. Math. Probl. Eng. **2020**, 24 (2020). Article ID 4597827. https://doi.org/10.1155/2020/4597827
33. Ji, W., Yin, S., Wang, L.: A big data analytics based machining optimisation approach. J. Intell. Manuf. **30**(3), 1483–1495 (2018). https://doi.org/10.1007/s10845-018-1440-9
34. Meng, K., Qian, X., Lou, P., Zhang, J.: Smart recovery decision-making of used industrial equipment for sustainable manufacturing: belt lifter case study. J. Intell. Manuf. **31**(1), 183–197 (2018). https://doi.org/10.1007/s10845-018-1439-2
35. Ramezani, F., Lu, J., Taheri, J., Hussain, F.K.: Evolutionary algorithm-based multi-objective task scheduling optimization model in cloud environments. World Wide Web **18**(6), 1737–1757 (2015). https://doi.org/10.1007/s11280-015-0335-3
36. Liu, J., Qiao, F., Kong, W.: Scenario-based multi-objective robust scheduling for a semiconductor production line. Int. J. Prod. Res. **57**, 6807–6826 (2019)
37. Fu, Y., Wang, H., Huang, M.: Integrated scheduling for a distributed manufacturing system: a stochastic multi-objective model. Enterp. Inf. Syst. **13**, 557–573 (2019)
38. Ghomi, E.J., Rahmani, A.M., Qader, N.N.: Service load balancing, task scheduling and transportation optimisation in cloud manufacturing by applying queuing system. Enterp. Inf. Syst. **13**, 865–894 (2019)
39. Coelho, P., Silva, C.: Parallel metaheuristics for shop scheduling: enabling Industry 4.0. Procedia Comput. Sci. **180**, 778–786 (2021)
40. Fu, Y., Tian, G., Fathollahi-Fard, A.M., Ahmadi, A., Zhang, C.: Stochastic multi-objective modelling and optimization of an energy-conscious distributed permutation flow shop scheduling problem with the total tardiness constraint. J. Clean. Prod. **226**, 515–525 (2019)
41. Afrin, M., Jin, J., Rahman, A., Tian, Y.C., Kulkarni, A.: Multi-objective resource allocation for Edge Cloud based robotic workflow in smart factory. Future Gener. Comput. Syst. **97**, 119–130 (2019)

42. Dziurzanski, P., Zhao, S., Przewozniczek, M., Komarnicki, M., Indrusiak, L.S.: Scalable distributed evolutionary algorithm orchestration using Docker containers. J. Comput. Sci. **40**, 101069 (2020)
43. Choi, T.M.: Guest editorial to the special issue on logistics and supply chain systems engineering. IEEE Trans. Syst. Man, Cybern. Syst. **50**, 4852–4855 (2020)
44. Yetkin, E.B.: A multi-objective optimisation study for the design of an AVS/RS warehouse. Int. J. Prod. Res. **59**, 1107–1126 (2021)

On Active-Set LP Algorithms Allowing Basis Deficiency

Pablo Guerrero-García[1]([⊠])[ID] and Eligius M. T. Hendrix[2][ID]

[1] Applied Mathematics, Universidad de Málaga, 29071 Málaga, Spain
pablito@ctima.uma.es
[2] Computer Architecture, Universidad de Málaga, 29071 Málaga, Spain
eligius@uma.es

Abstract. An interesting phenomenon in linear programming (LP) is how to deal with solutions in which the number of nonzero variables is less than the number of rows of the matrix in standard form. An interesting approach is that of basis-deficiency-allowing (BDA) simplex variations, which work with a subset of independent columns of the coefficient matrix in standard form, where the basis is not necessarily represented by a square matrix. By considering a different view on the usual dual-primal non-symmetric interaction, our aim is to show a relation between BDA and the non-simplex active-set methods. The whole is illustrated by several numerical examples. The ideas may aid the understanding of nowadays BDA approaches to sparse implementation and dealing with the Phase I.

Keywords: Linear programming · Phase I · Basis-deficiency-allowing simplex variations · Non-simplex active-set method · Farkas lemma · Sparse matrix

1 Introduction

It is intriguing that since the introduction of the simplex method for linear programming (LP) in 1947, there has been interest in viewing algorithms from a nonlinear optimisation point of view. Most well known and developed are the ideas of interior point methods since the first barrier methods of Dikin in 1967. A generalization of the well-known dual simplex method can be found in [9,35] relaxing operations towards non-simplex steps. One can find more recent considerations in this line in [25,31] with the incorporation of steepest-edge pivoting rules, see also [11,24]. In [9] Gill and Murray presented a non-simplex active-set algorithm for solving a problem (P) starting with a feasible point x_0. Descriptions and analyses can also be found in [6,10,24]. Our interest is to investigate the possibility to develop a non-simplex active-set (NSA) method for a dual linear program in standard form (D). A challenge in practical cases is to have a high degeneracy, which for a simplex-like method provides basis-deficiency, i.e. the dual variable set has less than n variables. Therefore, we aim

This paper has been supported by The Spanish Ministry (RTI2018-095993-B-I00) in part financed by the European Regional Development Fund (ERDF).

for the development of a method which is also called a basis-deficiency-allowing (BDA) simplex variation, see [25, 31, 32].

Our first viewpoint is that of the method presented by Gill and Murray which uses a non-simplex active-set algorithm for solving an LP problem starting with a feasible point x_0. Our question is whether changing the primal vision by the dual vision may help us to be able to handle rank-deficiency instances.

To investigate the question, we introduce the notation in Sect. 2 and the main algorithm in Sect. 3. We sketch the convergence of the algorithm in Sect. 4 and discuss several ways to deal with Phase I in Sect. 5. Section 6 discusses implementation issues and numerical experiments. Section 7 summarises our findings.

2 Notation of the Primal-Dual Vision

Consider the usual non-symmetric primal-dual relation in linear programming. Notice that we deviate from the usual notation here. This means, we exchanged b and c, x and y, n and m, and (P) and (D) in the notation of e.g. [24]. Now consider

$$(P) \quad \min \ \ell(x) := c^T x \, , \ x \in \mathbb{R}^n \qquad (D) \quad \max \ \mathcal{L}(y) := b^T y \, , \ y \in \mathbb{R}^m \tag{1}$$
$$\text{s.t.} \ A^T x \geq b \qquad\qquad\qquad\quad \text{s.t.} \ Ay = c \quad\ \ , \ y \geq 0$$

where $A \in \mathbb{R}^{n \times m}$ with $m \geq n$ and $\text{rank}(A) = n$. Let $\mathcal{F}$ and $\mathcal{G}$ denote the feasible region of (P) and (D), respectively.

The idea is the following. We start with a dual feasible point y_0. From there, we separate the index set $[1 : m] := \{1, 2, \ldots, m\}$ into an ordered basic set $\mathcal{B}_k$ with $m_k := |\mathcal{B}_k|$ elements, with $m_k \leq n$ and its complement representing the columns in $\mathcal{N}_k := [1 : m] \setminus \mathcal{B}_k$, which can have more than $m - n$ zero elements. Let $A_k \in \mathbb{R}^{n \times m_k}$ and $N_k \in \mathbb{R}^{n \times (m - m_k)}$ be sub-matrices of A formed by the columns corresponding to $\mathcal{B}_k$ and $\mathcal{N}_k$, respectively. Notice that $\text{rank}(A_k) = m_k$. We will use the notation $\mathcal{R}(A)$ to represent the range of the columns in A and $\text{null}(A)$ denotes its null space. Column j of A is denoted by a_j. Similarly, a_i^T is row i of matrix A^T. Moreover, $b_{\mathcal{B}_k}$ represents an m_k vector with the elements of $\mathcal{B}_k$ of b and $b_{\mathcal{N}_k}$ has the non-basic elements. The index k will be left out if it is clear from the context.

3 Algorithm

The algorithm works like the simplex method in determining in each iteration a non-basic variable that enters the basis and a basic variable(s) leaving the basis. The big difference is that it can handle basis-deficiency. This means that $m_k \leq n$ and m_k may reduce and grow. A pseudo-code is given in Algorithm 1.

- Notice in line 4, if we have a complementary pair of primal-dual feasible points, then x_k is optimal for (P).
- Line 15 determines step size and basis leaving variable(s), unless in line 12 we found that we have an unbounded dual objective function.

Algorithm 1. Pseudocode active-set algorithm

1: Set $k \leftarrow 0$, y_0 feasible point of (D) with corresponding $\mathcal{B}_0$ and A_0
2: $x_k \leftarrow$ solves $A_k^T x = b_{\mathcal{B}_k}$ with residuals r_k following from $r_{\mathcal{N}_k} = N_k^T x_k - b_{\mathcal{N}_k}$
3: **if** $r_k \geq 0$ **then**
4:     **return** x_k is optimal
5: select $p \in \mathcal{N}_k$ with $r_{kp} < 0$
6: **if** $a_p \notin \mathcal{R}(A_k)$ **then**
7:     $\mathcal{B}_{k+1} \leftarrow \mathcal{B}_k \cup \{p\}$, $\mathcal{N}_{k+1} \leftarrow [1:m] \setminus \mathcal{B}_{k+1}$, set A_{k+1}
8:     $y_{k+1} \leftarrow$ follows from $A_{k+1}y_{\mathcal{B}_{k+1}} = c$, $k \leftarrow k + 1$, goto step 2
9: **else**
10:     solve $A_k\delta = a_p$
11:     **if** $\delta \leq 0$ **then**
12:         **return** Unbounded $\mathcal{L}$
13:     **else**
14:         Direction d_k is initially unit vector e_p and elements of $\mathcal{B}_k$ follow from $-\delta$
15:         Step-size $\tau \leftarrow \min_{q \in \mathcal{B}_k}\{\frac{y_{kq}}{-d_{kq}}|d_{kq} < 0\}$, $Q \leftarrow \operatorname{argmin}_{q \in \mathcal{B}_k}\{\frac{y_{kq}}{-d_{kq}}|d_{kq} < 0\}$
16:         $y_{k+1} \leftarrow y_k + \tau d_k$, $\mathcal{B}_{k+1} \leftarrow \mathcal{B}_k \setminus Q \cup \{p\}$, $\mathcal{N}_{k+1} \leftarrow \mathcal{N}_k \cup Q$
17:         Set A_{k+1}, $k \leftarrow k + 1$, goto step 2

The algorithm does not require a primal feasible starting point. In that sense, it is an exterior method as sketched by the bottom-left part of Fig. 1. It has been studied by several researchers.

In earlier studies we have shown [40] that the algorithm is equivalent to the primal BDA simplex variation given by Pan in [25]. However, we put emphasis on a description which is easy to understand due to a geometrical interpretation and its independence of implementation details. The algorithm also fits in the primal-feasibility search loop of the sagitta method described by Santos-Palomo [37], which is related to a loop used in dual active-set methods for quadratic programming. Moreover, Li [20] extended the algorithm to deal with upper and lower bounds on y.

Practically, one obtains the primal-feasibility Phase I of Dax 1978 [5] idea for Rosen 1960 [35] when one replaces the min-ratio test by a most-obtuse-angle row rule to determine the leaving variable and removing an objective function consideration. In this context, we focused earlier [16] on the classical example of Powell [33] which illustrates the cycling behaviour of a most-obtuse-angle simplex pivoting rule described in [21,31].

The algorithm ends with an optimal solution for (P) and (D). However, the solution does not necessarily correspond to a square basis. Notice that the algorithm maintains dual feasibility and complementary slackness.

Example 1. Consider the following instance of an LP problem (adapted from [30]).

$$A = \begin{pmatrix} -1 & 1 & 1 & -1 & 1 & 0 & 1 & -2 \\ 2 & -1 & 1 & 1 & 0 & 1 & -1 & 1 \\ 1 & -2 & 0 & -1 & 0 & 0 & 0 & -3 \\ -1 & 1 & 0 & 1 & 0 & 0 & 0 & 1 \\ 2 & -1 & 0 & 1 & 0 & 0 & 0 & -1 \end{pmatrix}, c = \begin{pmatrix} 2 \\ 1 \\ 0 \\ 0 \\ 0 \end{pmatrix}, b^T = (2, -3, 4, -1, 0, 0, -2, 3)$$

We take as starting point $y_0 = (0, 0, 0, 0, 2, 1, 0, 0)^T$ with $\mathcal{L}_0 = 0$ and $\mathcal{B}_0 = \{5, 6\}$. Using

$$A_0 = \begin{pmatrix} 1 & 0 \\ 0 & 1 \\ 0 & 0 \\ 0 & 0 \\ 0 & 0 \end{pmatrix}, \text{ we have } x_0 = 0 \text{ and } r_0 = (-2, 3, -4, 1, 0, 0, 2, -3)^T. \text{ Selecting}$$

y_3 to enter the deficient basis, the step size in step 15 will be $\tau = 1$, having $Q = \{6\}$ leaving the basis. This means that $\mathcal{B}_1 = \{3, 5\}$ corresponding to

$$A_1 = \begin{pmatrix} 1 & 1 \\ 1 & 0 \\ 0 & 0 \\ 0 & 0 \\ 0 & 0 \end{pmatrix}, y_1 = (0, 0, 1, 0, 1, 0, 0, 0)^T, \mathcal{L}_1 = 4. \text{ Solving } A_1^T x =$$

$\begin{pmatrix} 4 \\ 0 \end{pmatrix}$, provides primal solution $x_1 = (0, 4, 0, 0, 0)^T$ with residuals $r_1 = (6, -1, 0, 5, 0, 4, -2, 1)^T$. This means the solution is not optimal. We select y_7 for entering the basis. We have as direction of improvement $d_1 = (0, 0, 1, 0, -2, 0, 1, 0)^T$ providing a step size of $\tau = \frac{1}{2}$, $Q = \{5\}$ leaves the basis and $y_2 = y_1 + \frac{1}{2}d_1 = (0, 0, \frac{3}{2}, 0, 0, 0, \frac{1}{2}, 0)^T$. Having

$$A_2 = \begin{pmatrix} 1 & 1 \\ 1 & -1 \\ 0 & 0 \\ 0 & 0 \\ 0 & 0 \end{pmatrix}, \text{ with } \mathcal{B}_2 = \{3, 7\} \text{ as deficient basis and objective function}$$

value $\mathcal{L}_2 = 5$. Now $x_2 = (1, 3, 0, 0, 0)^T$ is the new primal point with residuals $r_2 = (3, 1, 0, 3, 1, 3, 0, -2)^T$. This means we are interested to have $p = 8$ in the basis. However, from matrix A it is easy to see that $a_8 \notin \mathcal{R}(A_2)$. This means that we extend the basis to capture also a_8. The new matrix for $\mathcal{B}_3 = \{3, 7, 8\}$ becomes

$$A_3 = \begin{pmatrix} 1 & 1 & -2 \\ 1 & -1 & 1 \\ 0 & 0 & -3 \\ 0 & 0 & 1 \\ 0 & 0 & -1 \end{pmatrix}. \text{ It is interesting that one obtains a degenerate solution}$$

in the extension case, as variable y_8 cannot take a positive value by adapting the other positive values with a direction δ, given that $a_8 \notin \mathcal{R}(A_2)$. So in this example, we have that solution $y_3 = y_2$. As the solution is degenerate, the primal has alternative solutions of $A_3^T x = b_{\mathcal{B}_3}$, as there are two rows less. Some possible solutions x_3 to this under-determined system of linear equations are the so-called basic least-squares solution $(1, 3, -\frac{2}{3}, 0, 0)^T$ and the so-called minimum-norm least-squares solution $(1, 3, -\frac{6}{11}, \frac{2}{11}, -\frac{2}{11})^T$, where the corresponding residual vectors $r_3 = A^T x_3 - b$ are $(\frac{7}{3}, \frac{7}{3}, 0, \frac{11}{3}, 1, 3, 0, 0)^T$ and $(\frac{21}{11}, \frac{27}{11}, 0, \frac{39}{11}, 1, 3, 0, 0)^T$. This means, both the dual point and the residuals of the primal point are completely nonnegative, indicating that y_3 and x_3 are optimal solutions of the LP problem with $\ell_3 = \mathcal{L}_3 = 5$.

The numerical example illustrates how one can work with a deficient basis. The algorithm not only allows extending the basis, but it can also shrink, if $|Q| > 1$.

Example 2. To illustrate a potential reduction of $|B_k|$, consider the same values for A and b of Example 1 and we consider now $c^T = (1, 1, 0, 0, 0)$. We take as starting point $y_0 = (0, 0, 0, 0, 1, 1, 0, 0)^T$ with the same $\mathcal{L}_0$, $\mathcal{B}_0$, A_0, x_0 and r_0. After selecting y_3 to enter the deficient basis, this example leads to tie in the min-ratio test since $d_0 = (0, 0, 1, 0, -1, -1, 0, 0)^T$. Hence $\tau = 1$, but there a multiple deletion occurs with $Q = \{5, 6\}$. This means that $\mathcal{B}_1 = \{3\}$ corresponding to

$$A_1 = \begin{pmatrix} 1 \\ 1 \\ 0 \\ 0 \\ 0 \end{pmatrix}, y_1 = (0, 0, 1, 0, 0, 0, 0, 0)^T, \mathcal{L}_1 = 4.$$ Solving $A_1^T x = 4$ provides

a set of primal points, which can be described by $x_1 = (a, b, c, d, e)^T$ where $a + b = 4$. For example, using $c = d = e = 0$ we obtain:

 – $a = 4$ and $b = 0$ with residuals $r_1 = (-6, 7, 0, -3, 4, 0, 6, -11)^T$
 – $a = 0$ and $b = 4$ with residuals $r_1 = (6, -1, 0, 5, 0, 4, -2, 1)^T$
 – $a = 2$ and $b = 2$ with residuals $r_1 = (0, 3, 0, 1, 2, 2, 2, -5)^T$

This means that none of them is optimal in spite of having $\ell_1 = 4 = \mathcal{L}_1$, but note that the number of candidates to enter the deficient basis differs in the next iteration. For the illustration we choose the third possibility corresponding to $a = b = 2$, which is the minimum 2-norm solution of $A_1^T x = 4$ (with $\|x_1\|_2 = \sqrt{8} < 4$). In that case, only y_8 is a basis entering candidate variable, $a_8 \notin \mathcal{R}(A_1)$. Consequently, $\mathcal{B}_2 = \{3, 8\}$ corresponds to

$$A_2 = \begin{pmatrix} 1 & -2 \\ 1 & 1 \\ 0 & -3 \\ 0 & 1 \\ 0 & -1 \end{pmatrix}, y_2 = y_1, \mathcal{L}_2 = \mathcal{L}_1.$$ Solving $A_2^T x = \begin{pmatrix} 4 \\ 3 \end{pmatrix}$ provides a set of

primal points, for example:

 – $x_2 = (0, 4, 1/3, 0, 0)^T$, $r_2 = (19/3, -5/3, 0, 14/3, 0, 4, -2, 0)^T$;
 – $x_2 = (47, 77, -30, 10, -10)/31^T$, $r_2 = (-15, 143, 0, 91, 47, 77, 32, 0)/31^T$.

This means that none of them is optimal in spite of having $\ell_2 = 4 = \mathcal{L}_2$. However, there are (again) multiple candidates to enter the deficient basis in the next iteration. Consider the second possibility i.e., the minimum 2-norm solution of $A_2^T x = (4, 3)^T$. Then only y_1 can be selected to enter the basis, $a_1 \notin \mathcal{R}(A_2)$ and then $\mathcal{B}_3 = \{1, 3, 8\}$ corresponds to

$$
A_3 = \begin{pmatrix} -1 & 1 & -2 \\ 2 & 1 & 1 \\ 1 & 0 & -3 \\ -1 & 0 & 1 \\ 2 & 0 & -1 \end{pmatrix}, y_3 = y_2, \mathcal{L}_3 = \mathcal{L}_2. \text{ The solution of } A_3^T x = \begin{pmatrix} 2 \\ 4 \\ 3 \end{pmatrix} \text{ provides}
$$

a set of primal points among which:

- $x_3 = (19, 29, -15, 0, 0)/12^T$, $r_3 = (0, 56, 0, 37, 19, 29, 14, 0)/12^T$;
- $x_3 = (154, 274, -100, 30, -25)/107^T$, $r_3 = (0, 456, 0, 332, 154, 274, 94, 0)/107^T$.

This means that both are optimal with $\ell_3 = 4 = \mathcal{L}_3$. Notice that we have finished with a deficient basis and that there was no improvement in the objective functions for two iterations. This illustrates how the selection of the technique to solve the underdetermined system of equations giving the current primal point affects the overall path.

4 Proof of Convergence

Without going into a formal description, it can be shown that the algorithm converges under some circumstances. We will detail some observations.

One of the observations is that degeneracy of the dual when a basic variable is exchanged for a non-basic one does not occur. This is more or less included in the idea that A_k is not full rank leaving out all zero valued variables in y, i.e. $y_j > 0$ for $j \in \mathcal{B}_k$. In this way, convergence of $\mathcal{L}_k := b^T y_k$ is forced. Notice that degeneracy will occur when extending the basis with a new variable, as the old solution will be taken.

During the iterations, we have $y_{k+1} = y_k + \tau d_k$ with $\tau > 0$. In order to have $\mathcal{L}_{k+1} > \mathcal{L}_k$, so $b^T y_{k+1} > b^T y_k$ it must hold that $b^T d_k > 0$. To prove this, we can consider that $A d_k = 0$. As $A_k \delta = a_p$, by the construction of the search direction vector defined in line 14, we have that $A d_k = -A_k \delta + a_p = 0$. Now consider the residuals where for the basic variables we have $r_{kj} = 0, j \in \mathcal{B}_k$ and specifically $r_p < 0$. From $b = A^T x_k - r_k$, it follows that

$$
b^T d_k = (A^T x_k - r_k)^T d_k = x_k^T A d_k - r_k^T d_k = 0 - r_p > 0 \tag{2}
$$

We are not certain of an improvement, when adding a_p in case $a_p \notin \mathcal{R}(A_k)$. In that case y_k with $y_{kp} = 0$ is still a feasible solution, such that $\mathcal{L}_{k+1} \geq \mathcal{L}_k$, but no guarantee is given on improvement. This step is of course limited up to when $m_k = n$ is reached where $a_p \notin \mathcal{R}(A_k)$ cannot occur anymore. Due to the non-degeneracy, we are certain of an improvement when $m_k = n$.

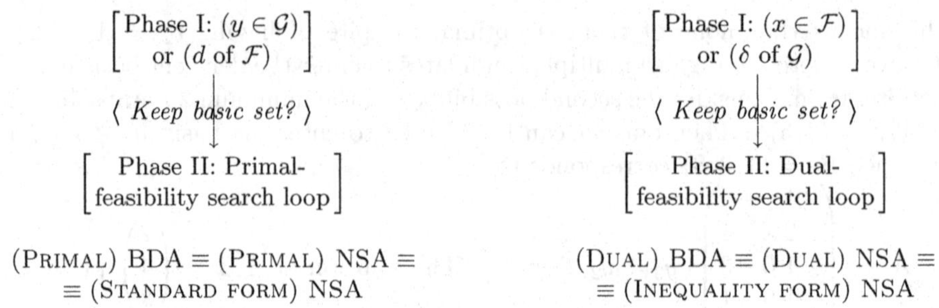

$$\begin{bmatrix} \text{Phase I: } (y \in \mathcal{G}) \\ \text{ or } (d \text{ of } \mathcal{F}) \end{bmatrix}$$

$$\begin{bmatrix} \text{Phase I: } (x \in \mathcal{F}) \\ \text{ or } (\delta \text{ of } \mathcal{G}) \end{bmatrix}$$

⟨ *Keep basic set?* ⟩

⟨ *Keep basic set?* ⟩

$$\begin{bmatrix} \text{Phase II: Primal-} \\ \text{feasibility search loop} \end{bmatrix}$$

$$\begin{bmatrix} \text{Phase II: Dual-} \\ \text{feasibility search loop} \end{bmatrix}$$

(PRIMAL) BDA ≡ (PRIMAL) NSA ≡
≡ (STANDARD FORM) NSA

(DUAL) BDA ≡ (DUAL) NSA ≡
≡ (INEQUALITY FORM) NSA

Fig. 1. Primal (left) and dual (right) two-phase approaches

Note that an anti-cycling rule like Bland's rule is required to deal with degenerate steps. Nevertheless, to prove finite termination (i.e., that no cycling occurs) we only need to have the dual non-degeneracy on iterations in which variables leave the basis.

5   Relation with Farkas Lemma and Phase I Approaches

The next question is to find a starting solution y_0 for the algorithm. For this, we can think of the Phase I approach suggested by Gass based on the single artificial variable Phase I, see [10,32,40]. Given an infeasible point $\hat{y} \geq 0$ such that $A\hat{y} \neq c$, a Phase I can be constructed using a single artificial variable z. Determine the corresponding values of the slack variables $\hat{s} = A\hat{y} - c$ and solve

$$\min z \quad \text{s.t. } Ay - \hat{s}z = c \ , \ y \geq 0 \ , \ z \geq 0 \tag{3}$$

Notice that the solution $(\hat{y}^T, 1)^T$ is feasible for this problem, where we have that $m_0 < n$. This means, one can solve this Phase I using the primal NSA method, e.g. with an infeasible solution $\hat{y} = 0$ and consequently start with the reduced basis $\mathcal{B}_0 = \{m + 1\}$ using a suitable min-ratio tie-breaker. Using $\hat{y} = 0$ leads to a primal problem of Phase I

$$\min c^T d \quad \text{s.t. } \begin{pmatrix} A^T \\ c^T \end{pmatrix} d \geq -e_{m+1} \tag{4}$$

providing a primal null-space descent direction d, see [37,41] and the top leftmost part of Fig. 1.

Let us consider an alternative of what we have described so far, i.e. applying the same algorithmic scheme on two different problems, namely first on (3) and then on (P). We now consider two different algorithmic schemes on the same problem. First we use dual NSA directly on (P), applying instead of the min-ratio test a most-obtuse-angle rule, i.e., replacing $\min_{p \in \mathcal{N}} \frac{r_p}{-w_p}$ by $\min_{p \in \mathcal{N}} w_p$, where $w_p := a_p^T d < 0$. Then we apply the primal NSA to (P). In this "dual-then-primal" approach we start from an arbitrary dual infeasible point y_0, and do not take the primal information of x into account. The nontrivial mathematical equivalence of the two proposals when having to deal with degenerate steps has been proven in [41]. A suitable normalization of w_p should be applied.

A completely different alternative to Phase I is to consider the Nonnegative Least Squares (NNLS) approach described in [4,40]. One can apply the NNLS algorithm [19] or Dax algorithm [7] to the positive semi-definite quadratic problem

$$\min_{y \geq 0} \frac{1}{2} \|Ay - c\|_2^2 \tag{5}$$

which implies solving a sequence of unconstrained least squares problems of the form $\min \|A_k z - c\|_2$ (NNLS) or of the form $\min \|A_k w - s_k\|_2$ (Dax). Such an approach, does not require artificial variables and obtains a feasible direction of the feasible region $\mathcal{F}$ (i.e., a so-called direction of $\mathcal{F}$) or a dual basic feasible solution with $m_k \leq n$. It appears that the computation can be described in terms of a primal null-space descent direction [38]. Instead of (5), one can focus on its Wolfe dual, which is a least distance strictly convex quadratic problem

$$\min \frac{1}{2} \|u\|_2^2 \quad \text{s.t.} \quad A^T u \geq v, \tag{6}$$

where $v := A^T c$. This also implies solving a sequence of least squares problems. To solve (6), one can use any of the primal, dual and primal-dual active-set methods introduced in [36]. Note that a primal direction $d := u - c$ is easy to generate.

We can also adopt the dual point of view (see top rightmost part of Fig. 1), namely using the dual NSA of [9,10] and the Gass Phase I with a single artificial variable like in (3)

$$\min z \quad \text{s.t.} \quad A^T x - \hat{r} z \geq b, \; z \geq 0 \tag{7}$$

where $\hat{r} = A^T \hat{x} - b$. Consider again $\hat{x} = 0$ in $(\hat{x}^T, 1)^T$ representing a primal degenerate vertex and $\mathcal{B}_0 = \emptyset$ and a suitable min-ratio tie-breaker. Note that proving finite convergence also requires an anti-cycling rule and that in this case the dual problem corresponding to the Phase I is

$$\max(b^T, 0)^T \delta \quad \text{s.t.} \quad \begin{pmatrix} A & 0 \\ b^T & 1 \end{pmatrix} \delta = e_{n+1}, \; \delta \geq 0. \tag{8}$$

Notice that again the same algorithmic scheme can be applied to two different problems, namely to (7) and to (P).

An alternative is to use two different algorithmic schemes on the same problem. A first step can be to use primal NSA directly on (P). In this concept, we use another selection rule. Let $v_q := e_i^T (A_k^T A_k)^{-1} A_k^T a_p > 0$, where q corresponds to index i of $\mathcal{B}$. Now we replace $\min_{q \in \mathcal{B}} \frac{y_q}{v_q}$ by $\max_{q \in \mathcal{B}} v_q$. After that, one can apply the dual NSA on (P). This "primal-then-dual" approach can start from an arbitrary primal infeasible point x_0 and does not require any dual information on y. This implies a subtle variation of the Rosen-Dax method described in Sect. 3, since it may take the objective function into account in both phases. Steepest-edge rules seem to have been recently incorporated by Pan et al. [31] into the Phase II of this alternative.

As pointed out in [40], also for the primal a least-squares-based Phase I can be used, solving

$$\min_{x \in \mathbb{R}^n} \frac{1}{2} \|x\|_2^2 \quad \text{s.t.} \ A^T x \geq b, \tag{9}$$

with a quadratic algorithm when $b \notin \mathcal{R}(A^T)$. Alternatively, we can use Cline's method, see [19] implying to use NNLS to

$$\min_{\delta \geq 0} \frac{1}{2} \left\| \begin{pmatrix} b^T \\ A \end{pmatrix} \delta - e_1 \right\|_2^2. \tag{10}$$

This will generate a feasible direction δ of the dual $\mathcal{G}$ or a feasible primal point $x \in \mathcal{F}$ (by Farkas' dual lemma). Note that when $b \in \mathcal{R}(A^T)$ we can directly solve $A^T x = b$ to generate a feasible point $x \in \mathcal{F}$, but then $Ay = c$ would imply $b^T y = x^T A y = x^T c$.

The presented Phase I approaches are mainly sparse-friendly, as we show in Sect. 6. Actually, they facilitate alternative proofs of Farkas lemma which are as elementary as the proof given by Dax in [7]. Note that the NNLS-based case does not require an anti-cycling rule in exact arithmetic. An application of this idea can be found in [13].

6   QR Factorization and Numerical Results

The efficiency of implementing the pseudo-code depends a lot on how to deal with the way to solve the set of linear equations. Mainly step 6 depends a lot on working with a QR factorization or not. We will first discuss this phenomenon in Sect. 6.1.

6.1   QR Factorization Aspects

The implementation of the algorithmic schemes given above for dense problems [26, 39] can be realized using the QR factorization of A_k (adding and deleting columns). However, sparse problems are far more relevant. Therefore, we describe two sparse orthogonal approaches.

The first sparse orthogonal approach is to use a sparse QR factorization of A_k^T adding and deleting rows. The dense counterpart was studied by Powell [34]. Since A_k is formed by a subset of the columns of a fixed matrix A, we can adapt [42] Saunders technique for square matrices [43] to matrices with more rows than columns. A way to do so is to use the static data structure of George and Heath (e.g., see [17] and the references therein) but allowing row downdating on it. This facilitates taking advantage of the intermediate results obtained when dealing with the problem $\min \|A^T x - b\|_2$ by processing A^T row by row in relation to (D). The column order of A_k does not affect the density of a Cholesky factor R_k of $A_k A_k^T$. Computational experience is reported in [15].

A disadvantage of the least-squares-based Phase I approaches compared to single artificial variable approaches is that the necessity to obtain least squares solutions does not allow working with the QR factorization of A_k^T alone. In [6], Dax recommends to use an iterative relaxation technique. However, this does not facilitate making use of the relation between A_k and A_{k+1}. Another sparse orthogonal approach is to adapt the methodology of Björck [2] and Oreborn [23]. They apply a sparse NNLS algorithm via CSNE with the Cholesky factor R_k of $A_k^T A_k$ using a "short-and-fat" matrix A. Their idea is to apply an active set algorithm for the sparse least squares problem

$$\min \frac{1}{2} y^T C y + d^T y \quad \text{s.t.} \quad \underline{y} \le y \le \overline{y} \tag{11}$$

with a positive definite matrix C. In the problem under investigation $\overline{y} = \infty$ and $\underline{y} = 0$ and $d = -A^T c$, with $C = A^T A$ only positive semi-definite. Hence, to maintain a sparse QR factorization of A_k, we adapted [38] using a similar technique as in [3] and prevent creating C. Notice that the column ordering of A determines the order in A_k.

Table 1. NETLIB test problem instances; nnz: number of nonzeros

#	Name	Optimum value	m	n	nnz(A)	nnz(b)	nnz(c)	nnz(LP)
1	AFIRO	.46475314286E+3	51	27	102	5	7	114
2	SC50B	.70000000000E+2	78	50	148	1	5	154
3	SC50A	.64575077059E+2	78	50	160	1	10	171
4	SC105	.52202061212E+2	163	105	340	1	20	361
8	STOCFOR1	.41131976219E+5	165	117	501	27	8	536
6	ADLITTLE	−.22549496316E+6	138	56	424	82	37	543
9	BLEND	.30812149846E+2	114	74	522	30	8	560
7	SCAGR7	.23313898243E+7	185	129	465	133	53	651
10	SC205	.52202061212E+2	317	205	665	1	38	704
12	SHARE2B	.41573224074E+3	162	96	777	36	24	837
14	LOTFI	.25264706062E+2	366	153	1136	8	49	1193
15	SHARE1B	.76589318579E+5	253	117	1179	31	103	1313
17	SCORPION	−.18781248227E+4	466	388	1534	282	76	1892
22	BRANDY	−.15185098965E+4	303	220	2202	2	54	2258
19	SCAGR25	.14753433061E+8	671	471	1725	475	179	2379
20	SCTAP1	−.14122500000E+4	660	300	1872	360	154	2386
16	DUISRAEL	−.89664482186E+6	316	142	2411	171	89	2671
23	ISRAEL	.89664482186E+6	316	174	2443	89	171	2703
29	BANDM	.15862801845E+3	472	305	2494	165	118	2777
31	SCFXM1	−.18416759028E+5	600	330	2732	23	116	2871
30	E226	.18751929066E+2	472	223	2768	189	99	3056
26	SCSD1	−.86666666743E+1	760	77	2388	760	1	3149
28	AGG	.35991767287E+8	615	488	2862	131	432	3425

6.2   Numerical Experiments

We run computational experiments with slight modifications of two of the Phase I approaches comparing to TOMLAB LPSOLVE V3.0 [18] sparse implementation of the usual primal simplex method, see [12]. The chosen sparse technique depends on the used primal null-space descent direction. The details of the computational experience, both for sparse adaptations of classical test problems and for the first 31 smallest NETLIB problems in Table 1[8], can be found in [12] and [14] respectively. The results reveal an advantage in terms of number of iterations, quality of solutions and execution time when suitable pivot strategies are used and with no special anti-cycling tools. Specifically, for the NETLIB instance, 6812 (in 51 min) and 6950 (in 52 min) iterations versus 9586 (in 58 min) with TOM-LAB, as shown in Table 2. In the sparse adaptations of classical test problems, the difference is even more pronounced. A similar computational experience is reported in [21].

Table 2. A comparison of best options for each algorithm

#	TOMLAB V3.0		SAGDISP V1.1		SAGFIAB V1.1	
	Iters	CtSecs	Iters	CtSecs	Iters	CtSecs
1	23	197	24	49	24	55
2	54	225	63	214	63	220
3	59	242	60	170	60	176
4	122	1109	131	775	131	780
6	151	626	144	1021	144	1038
7	196	2219	175	1648	218	2527
8	154	1334	141	1274	142	1307
9	150	741	132	1186	132	1186
10	258	6267	297	4163	297	4212
12	169	1307	196	1653	165	1329
14	345	4932	288	12073	288	11985
15	407	3323	248	2626	231	2351
16	579	6508	311	8145	304	7833
17	414	29967	375	7761	445	8689
19	952	86738	670	28688	696	29720
20	549	23695	384	9255	397	10172
22	588	12402	322	17768	326	18784
23	585	8272	431	13226	425	12781
26	583	4932	129	29017	132	29089
28	523	57743	554	10573	567	10397
29	1191	43754	568	82872	617	92550
30	864	18164	669	46258	608	35262
31	670	31594	500	24030	538	31324
	9586	58'	6812	51'	6950	52'

Sparse orthogonal approaches have in general the future that they can be parallelised and they fit well within a mixed interior-point simplex methodology. This is due to the fact that, in spite of using active-set methods, we work on top of the static structure of the Cholesky factor R of AA^T (our first approach, as in interior-point algorithms using the normal equations approach) or else on that of A^TA (our second approach, as in the interior-point method described in [1]), with $R_k \subset R$. Nonetheless, an implementation based on a dynamic revised LU factorization [27] has been shown by Pan to outperform the commercial primal simplex implementation included in MINOS [22, 28, 29]. However, in this case we face the same difficulties that arise when trying to parallelise the simplex method and we cannot deal with minimum-norm solutions of least-squares sub-problems easily.

7 Conclusions

In 1973 Gill and Murray presented a non-simplex active-set (NSA) method for the primal linear programming problem in inequality form $\min\{c^Tx : A^Tx \geq b\}$. We studied an NSA method for its dual linear program in standard form $\max\{b^Ty : Ay = c, y \geq 0\}$. This leads to an exterior method that facilitates working with a subset of independent columns of A which is not necessarily a square basis. We illustrate the relation with so-called basis-deficiency-allowing (BDA) simplex variations. Moreover, we provide a proof of finite termination of the method under the assumption of dual non-degeneracy.

A focus has been on two BDA features of interest related to the Phase I and its sparse implementation. The latter is carried out by solving either a sequence of sparse compatible systems or a sequence of sparse least squares problems. These "sparseness-friendly" Phase I approaches are related with Farkas' lemma and its proof. This implies an alternative proof to that by Dax in [7] which does not rely on anti-cycling pivoting rules. The computational behaviour of the tested Phase I approaches is highly encouraging.

References

1. Birge, J.R., Freund, R.M., Vanderbei, R.J.: Prior reduced fill-in solving equations in interior point algorithms. Oper. Res. Lett. **11**(4), 195–198 (1992)
2. Björck, Å.: A direct method for sparse least squares problems with lower and upper bounds. Numer. Math. **54**(1), 19–32 (1988)
3. Coleman, T.F., Hulbert, L.A.: A direct active set algorithm for large sparse quadratic programs with simple bounds. Math. Program. **45**(3), 373–406 (1989)
4. Dantzig, G.B., Leichner, S.A., Davis, J.W.: A strictly improving linear programming Phase I algorithm. Ann. Oper. Res. **46/47**(2), 409–430 (1993)
5. Dax, A.: The gradient projection method for quadratic programming. Technical report, Institute of Mathematics, Hebrew University, Jerusalem, Israel (1978)
6. Dax, A.: Linear programming via least squares. Linear Algebra Appl. **111**, 313–324 (1988)
7. Dax, A.: An elementary proof of Farkas' Lemma. SIAM Rev. **39**(3), 503–507 (1997)

8. Gay, D.M.: Electronic mail distribution of linear programming test problems. Committee Algorithms (COAL) Newslett. **13**, 10–12 (1985)

9. Gill, P.E., Murray, W.: A numerically stable form of the simplex algorithm. Linear Algebra Appl. **7**, 99–138 (1973)

10. Gill, P.E., Murray, W., Wright, M.H.: Numerical Linear Algebra and Optimization, vol. 1. Addison-Wesley, Redwood City (1991)

11. Gould, N.I.M.: The generalized steepest-edge for linear programming, Part I: theory. Technical report CORR 83-2, Department of Combinatorics & Optimization, University of Waterloo, Waterloo, Ontario, Canada, January 1983

12. Guerrero-García, P.: Range-space methods for sparse linear programs (in Spanish). Ph.D. thesis, Department of Applied Mathematics, University of Málaga, Spain, Málaga, Spain, July 2002

13. Guerrero-García, P., Santos-Palomo, A.: Gyula Farkas would also feel proud. Technical report MA-02-01, Department of Applied Mathematics, University of Málaga, Spain, Málaga, Spain, November 2002

14. Guerrero-García, P., Santos-Palomo, A.: A comparison of three sparse linear program solvers. Technical report MA-03-04, Department of Applied Mathematics, University of Málaga, Spain, Málaga, Spain, October 2003

15. Guerrero-García, P., Santos-Palomo, A.: Solving a sequence of sparse compatible systems. Technical report MA-02-03, Department of Applied Mathematics, University of Málaga, Spain, Málaga, Spain, October 2003

16. Guerrero-García, P., Santos-Palomo, A.: Phase-I cycling under the most-obtuse-angle pivot rule. Eur. J. Oper. Res. **167**(1), 20–27 (2005)

17. Heath, M.T.: Numerical methods for large sparse linear least squares problems. SIAM J. Sci. Stat. Comput. **5**(3), 497–513 (1984)

18. Holmström, K.: The TOMLAB optimization environment v3.0 user's guide. Technical report, Mälardalen University, Västerås, Sweden, April 2001

19. Lawson, C.L., Hanson, R.J.: Solving Least Squares Problems. Prentice-Hall, Englewood Cliffs (1974)

20. Li, W.: Bound constraints simplex method with deficient basis. Pure Appl. Math. (Xi'an) **20**(2), 171–176 (2004)

21. Li, W., Guerrero-García, P., Santos-Palomo, A.: A basis-deficiency-allowing primal Phase-I algorithm using the most-obtuse-angle column rule. Comput. Math. Appl. **51**(6/7), 903–914 (2006)

22. Murtagh, B.A., Saunders, M.A.: Minos 5.5 user's guide. Technical report SOL 83-20R, Systems Optimization Laboratory, Stanford University, Stanford, CA, July 1998

23. Oreborn, U.: A Direct Method for Sparse Nonnegative Least Squares Problems. Licentiat thesis, Department of Mathematics, Linköping University, Linköping, Sweden (1986)

24. Osborne, M.R.: Finite Algorithms in Optimization and Data Analysis. Wiley, Chichester (1985)

25. Pan, P.Q.: A basis-deficiency-allowing variation of the simplex method for linear programming. Comput. Math. Appl. **36**(3), 33–53 (1998)

26. Pan, P.Q.: A dual projective pivot algorithm for linear programming. Comput. Optim. Appl. **29**(3), 333–346 (2004)

27. Pan, P.Q.: Revised basis-deficiency-allowing simplex algorithm using the LU-factorization for linear programming. Presented at The Sixth International Conference on Optimization: Techniques and Applications, Ballarat, Australia, December 2004

28. Pan, P.Q.: A revised dual projective pivot algorithm for linear programming. SIAM J. Optim. **16**(1), 49–68 (2005)
29. Pan, P.Q.: A primal deficient-basis simplex algorithm for linear programming. Appl. Math. Comput. **196**(2), 898–912 (2008)
30. Pan, P.Q.: Linear Programming Computation, 1st edn. Springer, Cham (2014). https://doi.org/10.1007/978-3-642-40754-3
31. Pan, P.Q., Li, W., Wang, Y.: A Phase-I algorithm using the most-obtuse-angle rule for the basis-deficiency-allowing dual simplex method. OR Trans. (Chinese) **8**(2), 88–96 (2004)
32. Pan, P.Q., Pan, Y.P.: A Phase-1 approach for the generalized simplex algorithm. Comput. Math. Appl. **42**(10/11), 1455–1464 (2001)
33. Powell, M.J.D.: An example of cycling in a feasible point algorithm. Math. Program. **20**(1), 353–357 (1981)
34. Powell, M.J.D.: An upper triangular matrix method for quadratic programming. In: Mangasarian, O., Meyer, R., Robinson, S. (eds.) Nonlinear Programming 4, pp. 1–24. Academic Press, New York (1981)
35. Rosen, J.B.: The gradient projection method for nonlinear programming. Part I: linear constraints. J. Soc. Industr. Appl. Math. **8**(1), 181–217 (1960)
36. Santos-Palomo, A.: New range-space active-set methods for strictly convex quadratic programming. In: Olivares-Rieumont, P. (ed.) Proceedings of the III Conference on Operations Research, vol. 2, p. 27. Universidad de La Habana, Cuba, March 1997
37. Santos-Palomo, A.: The sagitta method for solving linear programs. Eur. J. Oper. Res. **157**(3), 527–539 (2004)
38. Santos-Palomo, A., Guerrero-García, P.: Solving a sequence of sparse least squares problems. Technical report MA-03-03, Department of Applied Mathematics, University of Málaga, Spain, Málaga, Spain, September 2003
39. Santos-Palomo, A., Guerrero-García, P.: Computational NETLIB experience with a dense projected gradient sagitta method. Technical report MA-05-05, Department of Applied Mathematics, University of Málaga, Spain, Málaga, Spain, September 2005
40. Santos-Palomo, A., Guerrero-García, P.: A non-simplex active-set method for linear programs in standard form. Stud. Inform. Control **14**(2), 79–84 (2005)
41. Santos-Palomo, A., Guerrero-García, P.: Sagitta method with guaranteed convergence. Technical report MA-05-02, Department of Applied Mathematics, University of Málaga, Spain, Málaga, Spain, February 2005
42. Santos-Palomo, A., Guerrero-García, P.: Updating and downdating an upper trapezoidal sparse orthogonal factorization. IMA J. Numer. Anal. **26**(1), 1–10 (2006)
43. Saunders, M.A.: Large-scale linear programming using the Cholesky factorization. Technical report CS-TR-72-252, Computer Science Department, Stanford University, Stanford, CA, January 1972

On the Design of a New Stochastic Meta-Heuristic for Derivative-Free Optimization

N. C. Cruz[1][(✉)] [ID], Juana L. Redondo[2] [ID], E. M. Ortigosa[1] [ID],
and P. M. Ortigosa[2] [ID]

[1] Department of Computer Architecture and Technology, University of Granada,
Granada, Spain
{ncalavocruz,ortigosa}@ugr.es
[2] Department of Informatics, University of Almería,
ceiA3 Excellence Agri-food Campus, Almería, Spain
{jlredondo,ortigosa}@ual.es

Abstract. Optimization problems are frequent in several fields, such as the different branches of Engineering. In some cases, the objective function exposes mathematically exploitable properties to find exact solutions. However, when it is not the case, heuristics are appreciated. This situation occurs when the objective function involves numerical simulations and sophisticated models of reality. Then, population-based meta-heuristics, such as genetic algorithms, are widely used because of being independent of the objective function. Unfortunately, they have multiple parameters and generally require numerous function evaluations to find competitive solutions stably. An attractive alternative is DIRECT, which handles the objective function as a black box like the previous meta-heuristics but is almost parameter-free and deterministic. Unfortunately, its rectangle division behavior is rigid, and it may require many function evaluations for degenerate cases. This work presents an optimizer that combines the lack of parameters and stochasticity for high exploration capabilities. This method, called Tangram, defines a self-adapted set of division rules for the search space yet relies on a stochastic hill-climber to perform local searches. This optimizer is expected to be effective for low-dimensional problems (less than 20 variables) and few function evaluations. According to the results achieved, Tangram outperforms Teaching-Learning-Based Optimization (TLBO), a widespread population-based method, and a plain multi-start configuration of the stochastic hill-climber used.

Keywords: Black-box optimization · Direct search · Stochastic meta-heuristic

1 Introduction

Optimization problems are ubiquitous. They usually arise in fields such as Architecture, Engineering, and Applied Sciences in general [3, 7, 19]. Broadly speaking,

O. Gervasi et al. (Eds.): ICCSA 2022 Workshops, LNCS 13378, pp. 188–200, 2022.
https://doi.org/10.1007/978-3-031-10562-3_14

this sort of problem requires finding the extremes (maxima or minima, depending on the goal) of a function. The latter, called objective function in this context, involves different variables and models some aspect of interest. For instance, it can represent the cost of manufacturing a product depending on the providers selected and the quantities bought. In this situation, the points sought would be the minima, i.e., the values of the variables resulting in the minimum function value (cost). Alternatively, if the function modeled the strength of the resulting product, the points of interest would presumably be the maxima of the corresponding function. One of the applications in which optimization stands out is model tuning, where the parameters become variables, and the objective function is the comparison between the achieved and desired output [4,12]. It allows automating processes that used to rely on experts and might be biased by them.

Depending on the objective function, constraints, and variables (e.g., continuous or discrete), there exist different types of optimization problems and methods to address them. For example, linear objective functions and constraints with real bounded variables generally result in problems relatively easy to solve [2,5]. However, this is not always the case, especially when the functions involved do not exhibit a closed analytical form or do not have exploitable mathematical properties (such as linearity and convexity). Fortunately, there exist methods with fewer problem requirements that can even treat it as a black box. They usually rely on intuitive ideas to obtain acceptable results [10,15]. These strategies are known as heuristics when they are problem-specific approaches and meta-heuristics of general-purpose. Similarly, the use of randomness serves to enhance the exploration capabilities at the expense of uncertainty [4].

The formulation of the optimization problem that attracts the attention of this work is as follows:

$$\begin{aligned} \underset{x}{\text{minimize}} \quad & f(x) \\ \text{subject to} \quad & L_i \leq x_i \leq U_i, \ i = 1, \dots, N. \end{aligned} \tag{1}$$

where f is a N-dimensional objective function, i.e., $f : \mathbb{R}^N \rightarrow \mathbb{R}$. The term x refers to any input in $\mathbb{R}^N$ belonging to the region $[L_1, U_1] \times \dots \times [L_N, U_N]$, which is known as the search space. As can be seen, the problem only consists of the objective function and the bounds of each variable. There is no extra information about the mathematical properties of f (e.g., convexity, linearity and smoothness). It can only be evaluated in the N-dimensional domain defined as the search space. Thus, this problem can be classified as a black-box optimization one with box constraints [3,7].

There exist numerous population-based meta-heuristics that can be applied to the problem defined above [1,15]. Traditional genetic algorithms, Differential Evolution [4,13], and the Universal Evolutionary Global Optimizer (UEGO) [6,11] are good examples. However, they have multiple parameters, so they require fine parameter tuning. Teaching-Learning-Based Optimization (TLBO) [4,14] avoids this problem as a population-based method that only needs the population size and the number of iterations. Regardless, this sort of method will generally need numerous function evaluations to converge due to its haphazard exploration strategy. Another option especially conceived for black-box optimization is DIRECT [7]. In contrast to the previous ones, it is a deterministic method. Furthermore, it only expects the number of function evaluations allowed and a tolerance factor that usually has little effect on its performance [8]. However, its rectangle division mechanism is rigid, and it may require an excessive number of function evaluations for degenerate cases.

This work presents an optimizer that balances the virtual lack of parameters, the use of randomness, and extensive exploration capabilities. To achieve this, it takes inspiration from previous optimization algorithms. The resulting method, called Tangram, defines an exploration procedure that moves from the center of the search space towards its corners. It keeps track of the best result achieved so far but uses that reference to divide the search space into smaller regions according to a deterministic division scheme. The new zones are ultimately explored by a local optimizer known as SASS [9], which is stochastic yet has a robust default configuration. Tangram aims to be effective for low-dimensional problems (less than 20 variables) and to be compatible with low budgets of function evaluations.

The rest of the paper is structured as follows: Sect. 2 describes the proposed method in detail. Section 3 explains the experimentation carried out. Finally, Sect. 4 shows the conclusions and states the future work.

2   Method Description

For a given optimization problem in the form of Eq. 1, the algorithm Tangram only expects as input the number of allowed function evaluations. Like DIRECT, the optimizer sees the search space as an N-dimensional unit hypercube, so it is first scaled accordingly from $\mathbb{R}^N$ to $[0,1]^N$. This strategy simplifies the implementation of the method and avoids issues concerning variables of significantly different scales at local search [16]. This idea is depicted in Fig. 1 for a hypothetical problem in a sub-domain in $\mathbb{R}^3$, $[10,20] \times [0,7] \times [1,3]$. The space of solutions remains unaltered.

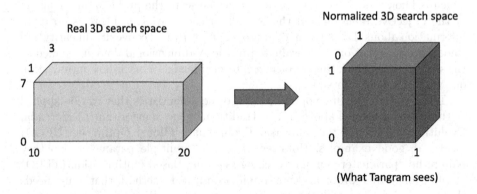

Fig. 1. Normalization of a hypothetical 3D search space.

Again like DIRECT, Tangram starts by evaluating the center of the search space, i.e., $(0.5, \ldots, 0.5) \in \mathbb{R}^N$, which becomes the current result. This process defines its initialization stage, which is shown in Fig. 2a. After that, it compares the budget of function evaluations to the dimensionality of the search space and enters into one of two modes, namely, the standard or the incisive one. Although both keep the same fundamentals, there is a subtle yet significant difference between them. The standard mode is covered first, as it was the only one at preliminary design stages. The local solver, which remains the same, is also explained in detail. Later, the decision criterion and the incisive mode are described in terms of the potential flaws of the standard one.

2.1   Standard Execution Mode

Algorithm 1 outlines Tangram (omitting the selection between modes). As can be seen, the initial steps described above (without the mode selection) are in lines 1 and 2.

Algorithm 1. Tangram optimizer (Standard Mode only)

    Input: Function: $f : \mathbb{R}^N \rightarrow \mathbb{R}$; Int: *evals*

1 Point corners$[2^N]$ = get_Corners_Of_Hypercube(N);

2 Point *result* = $(0.5, \ldots, 0.5) \in \mathbb{R}^N$;

3 **while** *evals* > 0 **do**

4   *result*, = Global_Phase(*result*);   // Change if improved only!

5   Point midpoints$[2^N]$;

6   **for** *corner* $\in$ *corners* **do**

7 | midpoints[*corner*] = $(result + corner)\,/2$;

8   **end**

9   midpoints = sort(midpoints, order=ascending f);

10   **for** *point* $\in$ *midpoints* **do**

11 | midpoints[*point*] = Local_Phase(*point*, radius=$|corner - point|$);

12   **end**

13   *result* = best_Of(*result* $\cup$ *midpoints*);

14 **end**

15 **return** *result*;

The optimization loop defining the standard mode is between lines 3 and 14. It lasts while there are function evaluations remaining and consists of the following stages:

Global Phase (line 4): The local search component, SASS, is launched from the current result to try to improve it. This is the local solver generally used with the aforementioned memetic algorithm, UEGO, and has been chosen because of its effectiveness for different problems. The optimizer is configured so that the maximum step size is equal to the diameter of the search space, i.e., $\sqrt{N}$.

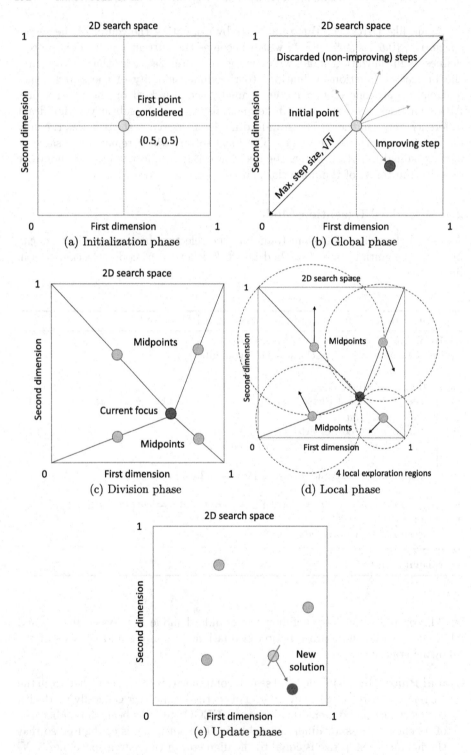

Fig. 2. Main concepts of Tangram.

This strategy allows reaching any solution and also comes from the way in which UEGO handles its initial or first-level species. Figure 2b depicts this process starting from the initial point, i.e., the center of the search space, for a hypothetical 2D problem. Notice how SASS attempts several movements, but the current result only changes when a better point is found.

Division (lines 5 to 9): Tangram computes and evaluates the midpoint between each corner of the search space and the current result. This part emphasizes the exploration of the search space. It allows the method to escape from local optima and identify new promising regions. It is vaguely inspired in how DIRECT keeps a representing point of every region of the search space. However, the division does not keep the regions strictly isolated. Figure 2c represents this stage graphically extending the previous context. It also serves to explain the name chosen for the proposed optimizer. Namely, if the resulting regions were colored in different colors, they would look like a Tangram, i.e., the widespread Chinese dissection puzzle [18].

Local Phase (lines 10 to 12): Tangram launches its local optimizer, SASS, from each of the previous midpoints to improve them. Those having a better value for the objective function are chosen first. By proceeding this way, in case the number of function evaluations allowed is low, the method ensures having explored the most promising regions at least. This aspect is relevant because the optimizer has been conceived for situations in which calling the objective function is computationally demanding. At this stage, in contrast to the global phase, SASS is configured not to take steps bigger than the distance between each starting midpoint and the corner used to define it (line 11). The further the midpoint is from the current global result, the wider initial region it has. Thus, exploration is automatically enhanced while the whole search space always remains virtually covered. Figure 2d summarizes these ideas graphically. Finally, notice that since the local solver updates its current point every time that it finds a better one, the definition of regions is dynamic like in UEGO.

Update (line 13): If any of the points found after the division and local searches outperforms the current solution, that point replaces it. Then, the algorithm returns to the global phase and starts to repeat the previous process if there are function evaluations available. This stage is show in Fig. 2e.

After consuming all the function evaluations allowed, Tangram returns the best solution achieved so far (line 15). However, despite being omitted from Algorithm 1, notice that the method registers every evaluation of the objective function. Hence, this situation can be detected at any stage. If that happens, Tangram assigns an infinite value to any new point and tries to finish as soon as possible.

2.2 Local Search Component (SASS)

Regarding the local search component, as introduced, it is SASS, which was initially proposed in [17]. This name is an acronym for Single-Agent Stochastic

Search. SASS does not have any special requirement for the objective function apart from being fully defined in the search space. For this reason, it is especially suitable for black-box optimization. It was one of the main reasons why it was chosen for UEGO, and the same criterion has been followed for Tangram.

SASS is a stochastic hill-climber of adaptive step size. It starts at any given point, which is treated as its current local solution, and randomly decides a direction to move. The jump size cannot be greater than a given threshold. As explained, it will be $\sqrt{N}$ for the global phase and the distance between the midpoint and the corner used to compute it for any local one. However, the jump size is further scaled depending on the number of improving (accepted) and non-improving (discarded) movements.

In practical terms, every movement consists in generating a new candidate solution, x', according to Eq. (2), where x is the current solution and ξ is a normally-distributed random vector (perturbation). The standard deviation, σ, is globally defined between 1e−5 and 1, starting at the upper bound. Every component has a specific mean or bias factor that is initially set to 0. They form the bias vector, $b = (b_1, \ldots, b_N)$. If the movement amplitude (the module of the perturbation), is greater than allowed, ξ is rescaled by the maximum step size.

$$x' = x + \xi \tag{2}$$

SASS then computes the objective function at x'. If it is better than the current solution, x' replaces it, the iteration is considered successful, the bias vector is updated as $b = 0.2b + 0.4\xi$, and a new iteration starts. Otherwise, the opposite direction is explored by generating and evaluating a new candidate solution, $x'' = x - \xi$. If it outperforms the current one, x'' replaces it, and the iteration is also considered successful. In this situation, the bias vector is updated as $b = b - 0.4\xi$. However, if neither x' nor x'' are better than the current solution, the iteration is tagged as failed, and the bias vector is set to $b = 0.5b$. Figure 3 shows the key aspects of an iteration of SASS in Tangram, which forces any perturbation vector to be within a delimited (yet moving) region.

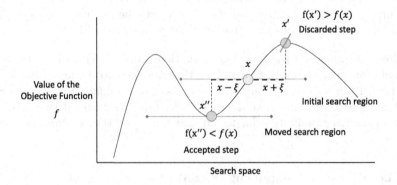

Fig. 3. Depiction of an iteration of SASS.

The global standard deviation, σ, is doubled after five consecutive successful movements (expansion) or halved after three consecutive discarded ones (contraction), which represents the adaptive nature of SASS. This is the recommended configuration, and it is known to perform well. Besides, within Tangram, SASS will terminate after 32 iterations. This arbitrary threshold is assumed enough to converge to the nearby optima according to previous knowledge on the method when used with UEGO. Nevertheless, varying this local budget would just result in a second parameter to tune.

2.3   Incisive Execution Mode

According to the previous explanation, the first function evaluation is always for the center of the search space. Then, the loop of the standard execution mode starts. Every global phase takes 32 evaluations through SASS. After that, the division requires 2^N function evaluations, one for each new midpoint. Let us think of a problem with $N = 20$, which is the highest dimensionality for a problem expected to be addressed with Tangram. It would be approximately 1×10^6 evaluations, which is the recommended budget for a robust execution of UEGO. Later, the local phase would end the current iteration after consuming $32*2^N$ evaluations, i.e., 2^{N+5}.

In this context, if the function evaluation budget is low (less than $33 + 2^N$), Tangram will not even be able to reach the local stage. This means that the local search will have been launched only once. Thus, the candidate solutions competing with the current result will be a subset of midpoints not sharpened by SASS. Accordingly, the probability of having identified a competitive point in the search space is low. The incisive mode of Tangram deals with this problem.

The incisive mode is selected over the standard one when the number of function evaluations allowed is less than $33 + 2^N$. This mode maintains the structure of the standard one but has a relevant modification in the division and local phases, which are merged. Namely, instead of generating and evaluating all the midpoints before moving to the local exploration phase, Tangram launches SASS from every new midpoint iteratively. By proceeding this way, the proposed method avoids the lack of local optimization that results in working with non-sharpened points at the expense of not prioritizing the most promising regions.

3   Experimentation and Results

The goals of Tangram are to be effective for low-dimensionality problems, without tuning requirements, and with a low budget of function evaluations. These aims are well aligned with the experimentation described in [3]. Thus, the 20 continuous box-constrained problems proposed in that work, which feature between 1 and 10 variables, have been addressed with Tangram. As the authors say, they are challenging for black-box methods not exploiting any analytical information. The limit of function evaluations has been computed as in the referred paper, i.e., $30(N+1)$, which is low and makes the test more challenging. It is also compatible with a context where the cost function is computationally demanding (e.g., simulation-based model tuning).

To have an adequate reference, Tangram has been compared to a pure random search (PRS), a plain multi-start configuration of SASS (MSASS), and the population-based optimizer TLBO. The pure random search is expected to be a baseline reference, since any optimizer should be more effective than simply generating and evaluating points. The multi-start SASS is a especially descriptive comparison: Since Tangram can be seen as a rule-based multi-start component linked to SASS, it should serve as a better guide for this local solver rather than simply generating random starts. Finally, TLBO has been included in the comparison due to its fame of being simple to tune and effective [4]. It is also relevant to highlight that both TLBO and MSASS achieved very good results in the model tuning application described in [4], which supports their selection.

All of the optimizers considered have been configured with the same function evaluation limit. This task is more difficult for TLBO, as it depends on a population size and a number of cycles. For this reason, TLBO takes that limit at least but slightly exceeds it in some cases. The development environment used is MATLAB 2020a in Mac OSX (MacBook Pro, Intel i5 2.9 GHz, 8 GB of RAM). Each method has been executed 200 times for each case to handle stochasticity.

Table 1 contains the results of experimentation. More specifically, the first column lists the 20 problems addressed including their name with their dimensionality and optimum value below. The other columns show the results obtained with each optimizer in the form of the average value of the objective function ± the standard deviation below. The cells in bold highlight the best result of the row. As can be seen in the results, the average differs from the theoretical optimum in most cases. Namely, only problems 3, 4, and 5 have a method that virtually offers the best possible value. This situation confirms the challenging nature of the testbeds for this kind of method and the evaluation budget allowed.

In this context, Tangram stands out as the best performing solver, as it offers the best average result for all the instances. In general, the number of function evaluations allows it to run in the standard mode. However, for problems 11, 14, and 15, the incisive mode is selected according to the defined criterion (e.g., for problem 8, the budget is 270, while the threshold is 289, so the incisive mode is activated). Its role is decisive, as the results of the standard mode for instances 11, 14, and 15 would have been $8003.54550 \pm 2889.46980$, 332.05440 ± 61.81830, and 246.38150 ± 63.87150, respectively. Hence, the predominant position of the proposed optimizer would be lost in those degenerate cases. There is another revealing fact to highlight: MSASS performs worse that Tangram even though they share the same local solver. Thus, the design of Tangram seems effective.

Regarding the other methods, as expected, PRS is the worst option due to its complete lack of search orientation. MSASS and TLBO approximately share the second position, with 12 relative victories for the former and 8 for the latter. Regardless, the differences are small in general, and both methods exhibit particularly bad averages in some problems. See for example problems 7 for MSASS and 12 for TLBO, where they are almost equivalent to PRS.

Table 1. Results of the compared optimizers for the 20 test problems.

	Problem	PRS	MSASS	Tangram	TLBO
1	branin (2D: 0.3979)	0.96322 ±0.56865	0.88190 ±0.56618	**0.59423 ±0.28007**	0.87950 ±0.62720
2	camel (2D: −1.0316)	−0.81055 ±0.17976	−0.90562 ±0.16800	**−0.94933 ±0.14442**	−0.90070 ±0.16960
3	ex4_1_1 (1D: −7.4873)	−6.69670 ±1.22100	−7.11600 ±1.01150	**−7.42420 ±0.17009**	−7.28160 ±0.66860
4	ex4_1_2 (1D: −663.4994)	−656.42850 ±13.03620	−661.33590 ±6.22790	**−662.86990 ±1.51650**	−660.71710 ±10.67700
5	ex8_1_1 (2D: −2.0218)	−1.86310 ±0.10054	−2.02000 ±0.00483	**−2.02140 ±0.00115**	−1.99390 ±0.08030
6	ex8_1_4 (2D: 0)	1.01380 ±0.77884	0.81421 ±0.89664	**0.45123 ±0.59453**	0.68060 ±0.79930
7	goldsteinprice (2D: 3)	24.15690 ±20.62910	21.46320 ±25.74360	**12.00310 ±14.97650**	16.31050 ±17.93680
8	hartman3 (3D: −3.8626)	−3.63850 ±0.14490	−3.71340 ±0.16565	**−3.78000 ±0.09150**	−3.67310 ±0.18480
9	hartman6 (6D: −3.3224)	−2.30300 ±0.36590	−2.72140 ±0.30703	**−2.83540 ±0.24749**	−2.59240 ±0.30490
10	least (3D: 14085.1398)	89426.35870 ±48162.68080	85657.53800 ±90868.74070	**54380.77930 ±23588.04900**	66667.12100 ±46139.03410
11	perm0_8 (8D: 1000)	7594.46260 ±2814.81180	4386.96390 ±2552.2268	**4193.35220 ±2640.35760**	4781.10300 ±2837.30440
12	perm_6 (6D: 1000)	2410159.72000 ±1880505.3559	1348865.22460 ±1705484.8604	**1040601.32000 ±1771209.76700**	2010600.02910 ±2541900.01131
13	rbrock (2D: 0)	17.69760 ±27.8399	13.49500 ±60.25080	**4.22410 ±5.3311**	7.91002 ±13.0823
14	schoen_10_1 (10D: −1000)	311.78160 ±54.87700	194.78420 ±79.80000	**180.87800 ±92.37480**	223.21380 ±106.12960
15	schoen_10_2 (10D: −1000)	292.26700 ±43.38010	117.05820 ±96.74010	**87.80030 ±104.40370**	179.37180 ±110.95960
16	schoen_6_1 (6D: −1000)	162.39550 ±115.7606	−45.82550 ±207.03060	**−300.34410 ±228.3106**	108.37580 ±124.07530
17	schoen_6_2 (6D: −1000)	126.52030 ±83.2584	−90.22060 ±186.51570	**−325.30290 ±192.2352**	10.06960 ±156.32220
18	shekel10 (4D: −10.5363)	−1.25000 ±0.51079	−1.86710 ±1.03470	**−3.20280 ±1.34020**	−1.87330 ±1.09240
19	shekel5 (4D: −10.1532)	−0.88923 ±0.47061	−1.60540 ±0.99205	**−3.50500 ±1.39050**	−1.45500 ±0.77001
20	shekel7 (4D: −10.4028)	−1.03920 ±0.41355	−1.65020 ±0.90632	**−3.26410 ±1.45330**	1.73670 ±0.78700

4　Conclusions and Future Work

In this work, a new algorithm for black-box optimization has been described. The method is virtually parameter-free, as it only expects the number of allowed function evaluations, and aims to be effective for low-dimensionality problems (up to 20 variables). It is expected to be valid for model tuning through numerical optimization. Its name is Tangram, and it combines several ideas from the literature. More specifically, it works in a unitary hypercube and tries to apply a deterministic strategy to define regions, like DIRECT. Besides, its behavior can be self-adapted to adapt the search to best-effort explorations in which the algorithm detects that it will not be able to execute all its steps appropriately. The proposal also uses a local optimizer from the literature, SASS, which is the one usually used within the memetic algorithm UEGO. From this population-based method, Tangram replicates the scheme of making SASS focus on different regions by limiting the steps yet keeping one covering the whole search space. Nevertheless, Tangram avoids the sophisticated management of points as a population in UEGO.

Tangram has been used to address 20 benchmark problems from the literature in a context of few function evaluations allowed. Its performance has been compared to a pure random search, a random multi-start configuration of SASS, and TLBO, a widespread population-based method. Tangram outperforms all of them. Among the appreciated aspects, one of the most descriptive ones is the fact that Tangram performs better than SASS in the multi-start configuration, which supports the design of the proposal. Another interesting idea to highlight from the experimentation is how the adaptive nature of Tangram is effectively able to anticipate the potential flaws of its standard mode and opts for the incisive one instead.

For future work, we will study how to automatically detect the local convergence of SASS to save function evaluations that can be reassigned to other parts of the method. We also intend to increase the number of benchmark problems and the optimizers compared, possibly starting with DIRECT and UEGO.

Acknowledgements. This work has been supported by the Spanish Ministry of Science through the projects RTI2018-095993-B-I00, financed by MCIN/AEI/ 10.13039/501100011033/ and FEDER ("A way to make Europe"), as well as INTSENSO (MICINN-FEDER-PID2019-109991GB-I00), by Junta de Andalucía through the projects UAL18-TIC-A020-B, CEREBIO (P18-FR-2378) and P18-RT-1193 by the European Regional Development Fund (ERDF). N.C. Cruz is supported by the Ministry of Economic Transformation, Industry, Knowledge and Universities from the Andalusian government.

References

1. Boussaïd, I., Lepagnot, J., Siarry, P.: A survey on optimization metaheuristics. Inf. Sci. **237**, 82–117 (2013)
2. Boyd, S., Boyd, S.P., Vandenberghe, L.: Convex optimization. Cambridge University Press (2004)

3. Costa, A., Nannicini, G.: RBFOpt: an open-source library for black-box optimization with costly function evaluations. Math. Program. Comput. **10**(4), 597–629 (2018). https://doi.org/10.1007/s12532-018-0144-7

4. Cruz, N.C., Marín, M., Redondo, J.L., Ortigosa, E.M., Ortigosa, P.M.: A comparative study of stochastic optimizers for fitting neuron models. application to the cerebellar granule cell. Informatica **32**, 477–498 (2021)

5. Griva, I., Nash, S.G., Sofer, A.: Linear and nonlinear optimization, vol. 108. Siam (2009)

6. Jelasity, M., Ortigosa, P.M., García, I.: Uego, an abstract clustering technique for multimodal global optimization. J. Heuristics **7**(3), 215–233 (2001)

7. Jones, D.R., Martins, J.R.R.A.: The DIRECT algorithm: 25 years later. J. Global Optim. **79**(3), 521–566 (2021)

8. Jones, D.R., Perttunen, C.D., Stuckman, B.E.: Lipschitzian optimization without the lipschitz constant. J. Optim. Theory Appl. **79**(1), 157–181 (1993)

9. Lančinskas, A., Ortigosa, P.M., Žilinskas, J.: Multi-objective single agent stochastic search in non-dominated sorting genetic algorithm. Nonlinear Anal. Model. Control **18**(3), 293–313 (2013)

10. Lindfield, G., Penny, J.: Introduction to nature-inspired optimization. Academic Press (2017)

11. Marín, M., Cruz, N.C., Ortigosa, E.M., Sáez-Lara, M.J., Garrido, J.A., Carrillo, R.R.: On the use of a multimodal optimizer for fitting neuron models. application to the cerebellar granule cell. Frontiers Neuroinformatics **15**, 663797 (2021)

12. Monterreal, R., Cruz, N.C., Redondo, J.L., Fernández-Reche, J., Enrique, R., Ortigosa, P.M.: On the optical characterization of heliostats through computational optimization. In: Proceedings of SolarPACES 2020, pp. 1–8 (2020)

13. Price, K., Storn, R.M., Lampinen, J.A.: Differential evolution: a practical approach to global optimization. Springer Science & Business Media (2006)

14. Rao, R.V., Savsani, V.J., Vakharia, D.P.: Teaching-learning-based optimization: an optimization method for continuous non-linear large scale problems. Inf. Sci. **183**(1), 1–15 (2012)

15. Salhi, S.: Heuristic Search. Springer, Cham (2017). https://doi.org/10.1007/978-3-319-49355-8

16. Snyman, J.A., Wilke, D.N.: Practical Mathematical Optimization. SOIA, vol. 133. Springer, Cham (2018). https://doi.org/10.1007/978-3-319-77586-9

17. Solis, F.J., Wets, R.J.B.: Minimization by random search techniques. Math. Oper. Res. **6**(1), 19–30 (1981)

18. Wang, F.T., Hsiung, C.C.: A theorem on the Tangram. Am. Math. Mon. **49**(9), 596–599 (1942)

19. Zou, F., Wang, L., Hei, X., Chen, D.: Teaching-learning-based optimization with learning experience of other learners and its application. Appl. Soft Comput. **37**, 725–736 (2015)

Analyzing the MathE Platform Through Clustering Algorithms

Beatriz Flamia Azevedo[1,2](✉) ⓘ, Yahia Amoura[1] ⓘ, Ana Maria A. C. Rocha[2] ⓘ,
Florbela P. Fernandes[1] ⓘ, Maria F. Pacheco[1,3] ⓘ, and Ana I. Pereira[1,2] ⓘ

[1] Research Centre in Digitalization and Intelligent Robotics (CeDRI),
Instituto Politécnico de Bragança, 5300-253 Bragança, Portugal
{beatrizflamia,yahia,fflor,pacheco,apereira}@ipb.pt
[2] ALGORITMI Center, University of Minho, Campus de Azurém,
4800-058 Guimarães, Portugal
arocha@dps.uminho.pt
[3] Center for Research and Development in Mathematics and Applications CIDMA,
University of Aveiro, Aveiro, Portugal

Abstract. University lecturers have been encouraged to adopt innovative methodologies and teaching tools in order to implement an interactive and appealing educational environment. The MathE platform was created with the main goal of providing students and teachers with a new perspective on mathematical teaching and learning in a dynamic and appealing way, relying on digital interactive technologies that enable customized study. The MathE platform has been online since 2019, having since been used by many students and professors around the world. However, the necessity for some improvements on the platform has been identified, in order to make it more interactive and able to meet the needs of students in a customized way. Based on previous studies, it is known that one of the urgent needs is the reorganization of the available resources into more than two levels (basic and advanced), as it currently is. Thus, this paper investigates, through the application of two clustering methodologies, the optimal number of levels of difficulty to reorganize the resources in the MathE platform. Hierarchical Clustering and three Bio-inspired Automatic Clustering Algorithms were applied to the database, which is composed of questions answered by the students on the platform. The results of both methodologies point out six as the optimal number of levels of difficulty to group the resources offered by the platform.

Keywords: E-learning · Data analysis · Educational technology · Machine learning · Clustering

This work has been supported by FCT Fundação para a Ciência e Tecnologia within the R&D Units Project Scope UIDB/00319/2020, UIDB/05757/2020 and Erasmus Plus KA2 within the project 2021-1-PT01-KA220-HED-000023288. Beatriz Flamia Azevedo is supported by FCT Grant Reference SFRH/BD/07427/2021.

1   Introduction

The development of Information and Communication Technologies (ICT) is reflected in the dynamics of the educational fields. This advancement has facilitated and made everyday tasks more accessible, effective, and faster to perform [20]. The ICT are directly involved in the development of teaching and learning processes by supporting innovative pedagogical actions and providing new learning spaces. In this way, it is possible to transform the classical classroom into a virtual one by eliminating the existing space-time barriers [7]. Among the ICT-based teaching methods is e-learning; e-learning platforms such as MathE are intensively involved in higher education teaching and learning practices through the support of online classes. They offer many advantages in terms of communication, students interaction, group development, and greater access to knowledge [6], as well as providing students access to a wide spectrum of information in a multitude of ways. The results of the use of an e-learning platform are reflected in the improvement of students' skills, self-motivation, engagement, and attitude towards educational content. This teaching approach is currently proliferating since the COVID-19 pandemic situation; its independence regarding location, time, effort, and cost makes it the most suitable option for student learning and assessment [16]. This type of pedagogy has particularities that distinguish it from other teaching modalities. Some researchers recognize it as a progression of distance education [5,25]. According to others, it represents a novelty that differs significantly from face-to-face teaching [18].

Promoting an e-learning method requires different types of resources, in particular digital and technological resources. Among the available digital tools are videos, teaching platforms, video conferences, podcasts, social networks, as well as many other resources [26]. The technological resources represent hardware tools including the desktop computer, tablet, smartphone, among others [17].

E-learning offers a range of particularities such as stimulating the development of dialogue and group work [4], strengthening interprofessional relationships among learners [10], promoting collaboration between the participants themselves, allowing the achievement of joint goals in the development of different tasks [13], making synchronous and asynchronous communication easier [22], and allowing learning from anywhere where there is an available internet connection [24]. E-learning also encourages the acquisition of digital competences by the students [11], allowing the adjustment to their personal rhythm [14], increasing their interest and motivation towards learning, as they are able to adapt to their specific learning style [1], giving everyone an unlimited number of learning resources [21]; E-learning also facilitates the monitoring of student activity by the teacher [2].

A solid education in Mathematics, in particular, has great importance both in the areas of exact sciences as well as human and biological sciences. However, Mathematics is frequently the subject of disorientation and complaints from students at all levels of their educational journey. One way to change the students' pragmatic view of mathematics is through interactive learning platforms. Under this scenario, the MathE platform emerges as a digital, innovative, dynamic, and intelligent tool for teaching and learning mathematics. The MathE platform will be better described in Sect. 2.

This paper aims to analyze some of the data collected by the MathE platform over the 3 years the platform has been online. In particular, with this research it is expected to reach conclusions about the best way to reorganize the resources available on the platform into different levels of difficulty. For this, unsupervised learning techniques, namely clustering, will be used for data analysis.

The rest of the paper is divided as follows. In Sect. 2, the concepts of the MathE collaborative learning platform are described. Section 3 explains the methodology applied in this work, which are Hierarchical and Partitioning Clustering techniques. The dataset used is described in Sect. 4 and the obtained results and their discussion are presented in Sec. 5. Finally, Sect. 6 concludes the work and sets forward-looking guidelines for the future of the platform.

2   The MathE Platform

MathE is a collaborative e-learning platform that aims to provide users with greater mathematical skills in higher education by creating a virtual space for learning and exchange. Like other e-learning platforms in mathematics, this platform represents a remarkable transition from the classical Learning Management Systems (LMS) to an interactive Intelligent Tutoring System (ITS). MathE is distinguished by its dynamic teaching environment involving both teachers and students or also external contributors and learners in the field of Mathematics. Furthermore, MathE is a non-commercial tool, being completely free and available 24 h a day, for all individuals interested in improving their knowledge and understanding of Mathematics.

MathE relies on an essential set of resources presented in the form of lessons, exercises, quizzes, videos, and other materials. These resources encourage students to study and practice mathematics outside the classic classroom rhythm, without the need for a teacher to be present.

Currently, there are 99 teachers and 1161 students from different nationalities enrolled on the platform: Portuguese, Brazilian, Turkish, Tunisian, Greek, German, Kazakh, Italian, Russian, Lithuanian, Irish, Spanish, Dutch and Romanian. In its current stage, the platform is organized into three main sections: **Student's Assessment** (composed of multiple-choice questions divided into topics, with two difficulty levels (basic and advanced), which were previously defined by a professor member of the platform); **MathE Library** (composed of valuable and diversified materials related to the topics and subtopics covered by the platform, such as videos, lessons, exercises, training tests and other formats); and **Community of Practice** (provides a virtual place where teachers and students have the opportunity to interact in order to fulfill their common goals, thus consolidating a strong network community). More details about each section are described in [2,3], and can also be found in the Platform Website (mathe.pixel-online.org).

MathE includes fifteen topics in Mathematics, among the ones that are in the classic core of graduate courses: Analytic Geometry, Complex Numbers, Set Theory, Differential Equations, Differentiation (including 3 subtopics: Derivatives, Partial Differentiation, Implicit Differentiation and Chain Rule), Fundamental Mathematics (2 subtopics: Elementary Geometry and Expressions and Equations), Graph Theory, Integration (3 subtopics: Integration Techniques, Double Integration and Definite Integrals), Linear Algebra (5 subtopics: Matrices and Determinants, Eigenvalues and Eigenvectors, Linear Systems, Vector Spaces Linear Transformations, Others), Optimization (2 subtopics: Linear Optimization and Nonlinear Optimization), Probability, Real Functions of a Single Variable - RFSV (2 subtopics: Limits and Continuity and Domain, Image and Graphics), Real Functions of Several Variables - RFSV (1 subtopic: Limits, Continuity, Domain and Image) and Statistics, as presented in Fig. 1. However, it is essential to mention that the platform's content is constantly being updated, and other topics and subtopics may be created whenever necessary.

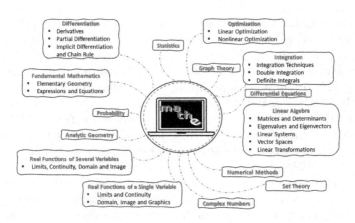

Fig. 1. Topics and subtopics currently available on the MathE Platform.

This paper is focused on the Student's Assessment section of MathE Platform. In this section of the platform, the students can train and test their skills in the *Self Need Assessment* (SNA) and *Final Assessment* (SFA) subsections, respectively. The Self Need Assessment section aims to provide the students with some training assessment to test if a particular topic that he/she enrolled in is already known and understood: suppose that the students or the teachers believe that their understanding needs to be deepened. In this case, the student can choose to answer to another training assessment to measure his/her level of confidence to perform a final assessment. Each training assessment will be randomly generated from the assessments database composed of questions/answers. In this way, the same student will be able to answer different training assessments on the same topic. When answering a training assessment, the students will have immediate access to the obtained mark: the test will randomly select seven questions from a given set, and after the student submits the test, the mark will appear automatically, allowing self-assessment. On the other hand, the purpose of the Final Assessment section is to evaluate the student performance after practicing with training assessment questions (and all the related resources available in MathE platform). In the Final Assessment section, the teacher can select the questions and the assessment will be available for the students at a chosen moment, defined by the teacher. In this case, the student will submit the test and receive feedback on the following day; the teacher will have access to the results at the end of the test, one day before the students [3].

As already mentioned, the questions available on the MathE platform are divided into two levels of difficulty (basic and advanced). The classification into basic or advanced is done by a professor registered on the platform. However, previous studies [3] concluded that two levels are insufficient for separating the available content. Therefore, to further meet the needs of users of the MathE platform in a more suitable and personalized way, it is undergoing profound changes that will make it even more interactive and customized. For this, a digital intelligence system is being developed and, in the near future, the questions will be addressed to students in a personalized way and not randomly, as is currently the case. One of the first needs that were identified is to reorganize the available resources. Thus, this work intends to investigate an optimal number of levels of difficulty to reorganize the questions available in MathE. For this, the data from the questions belonging to SNA, answered in the last 3 years in which the platform was online, were analyzed by hierarchical clustering and automatic clustering techniques to define the number of levels of difficulty that are defined by the algorithms as the optimal number of clusters.

3   Methodology

Clustering is one of the most widely used methods for unsupervised learning. It is used in datasets where there is no defined association between input and output. Thus, clustering algorithms consist of performing the task of grouping a set of elements with similarities in the same group and those with dissimilarities in other groups [27]. The methodology used in this work refers to two types of clustering: Hierarchical Clustering and Partitioning Clustering.

3.1   Hierarchical Clustering

Hierarchical clustering is an unsupervised technique for performing exploratory data analysis. This technique consists of building a binary merge tree, starting from the data elements stored in the leaves and proceeding by merging the closest "sub-sets" two by two until reaching the root of the tree, which contains all the elements of a dataset, denoted as X [23]. The graphical representation of this binary merge tree is called dendrogram. Basically, a dendrogram consists of many U-shaped lines that connect data points in a hierarchical tree. The height of each U represents the distance between the two data points being connected. To draw a dendrogram, we can draw an internal node $s(X')$ composed by the subset $X' \subseteq X$ at height $h(X') = |X'|$. Thereafter, the edges between this node $s(X')$ and its two sibling nodes $s(X_1)$ and $s(X_2)$ with $X' = X_1 \cup X_2$ (and $X_1 \cap X_2 \neq \varnothing$) are drawn. Considering this, each defined subset of X can be interpreted as a cluster. Figure 2a illustrates a generic representation of a dendrogram, whereas Fig. 2b is the equivalent Venn diagram. Note that the set $X \equiv \{a, b, c, 1, 2, 3\}$ is the root note, and the subset $\{a, b, c\}, \{1, 2, 3\}, \{a, b\}, \{a, 2\}$ are the internal nodes. At the end, are the leaves, in this case, represented by the subsets $\{a\}, \{b\}, \{c\}, \{1\}, \{2\}, \{3\}$.

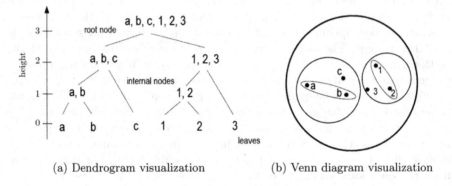

(a) Dendrogram visualization   (b) Venn diagram visualization

Fig. 2. Dendrogram and Venn diagram representation.

In this approach, the Matlab® `dendrogram` function [19] was used to generate the Hierarchical cluster and, consequently, the dendrogram representation. In this particular function, two important informations must be considered:

- If there are 30 or fewer data points in the original dataset, then each leaf in the dendrogram corresponds to one data point.
- If there are more than 30 data points, then the dendrogram collapses lower branches so that there are 30 leaf nodes. As a result, some leaves in the plot correspond to more than one data point.

3.2   Partitioning Clustering

Partitioning clustering decomposes a dataset into a set of disjoint clusters. Considering a dataset of X_m points, a partitioning method constructs K $(X_m \geq K)$ partitions of the data, with each partition representing a cluster C. That is, it classifies the data into K groups by satisfying the following requirements: (1) each group contains at least one point, and (2) each point belongs to exactly one group [12].

In real-world data clustering analysis problems, identifying the number of clusters and, consequently, the appropriate partitioning of the dataset is quite a difficult task. An unappropriated selection of the number of clusters results in poor performance since, in traditional clustering algorithms, the results often depend on the initial starting points [9]. In this context, automatic data clustering techniques that combine clustering and optimization techniques have helped to overcome these challenges and have also offered several improvements in the clustering methods. The automatic clustering process consists of solving an optimization problem, aiming to minimize the similarity within a cluster and maximize the dissimilarity between the clusters.

In this work, the Davies-Bouldin index (DB) [8] will be used as a clustering similarity and dissimilarity measure that will define the number of cluster centroids, which is the number of groups into which the dataset will be divided. DB index is based on a ratio of intra-cluster and inter-cluster distances. It is used to validate cluster quality and also to determine the optimal number of clusters. Consider that cluster C has members $X_1, X_2, ..., X_m$. The goal is to define a general cluster separation measure, S_i and M_{ij}, which allows computing the average similarity of each cluster to its most similar cluster. The lower the average similarity, the better the clusters are separated and the better the clustering results. To better explain how to get the Davies-Bouldin index, four steps are considered [8].

In the first step, it is necessary to evaluate the average distance between each observation within the cluster and its centroid, that is the dispersion parameter S_i, also known as intra-cluster distance, given by Eq. (1),

$$S_i = \left\{ \frac{1}{T_i} \sum_{j=1}^{T_i} |X_j - A_i|^q \right\}^{\frac{1}{q}} \tag{1}$$

where, for a particular cluster i, T_i is the number of vectors (observations), A_i is its centroid and X_j is the jth (observation) vector.

The second step aims to evaluate the distance between the centroids A_i and A_j, given by Eq. (2), also known as inter-cluster distance. In this case, a_{ki} is the kth component of the n-dimensional vector a_i, which is the centroid of cluster i, and N is the total number of clusters. It is worth mentioning that M_{ij} is the Minkowski metric of the centroids which characterize clusters i and j and $p = 2$ means the Euclidean distance.

$$M_{ij} = \left\{ \sum_{k=1}^{N} |a_{ki} - a_{kj}|^p \right\}^{\frac{1}{p}} = ||A_i - A_j||_p \tag{2}$$

In the third step, the similarity between clusters, R_{ij}, is computed as the sum of two intra-cluster dispersions divided by the separation measure, given by Eq. (3), that is the within-to-between cluster distance ratio for the ith and jth clusters.

$$R_{ij} = \frac{S_i + S_j}{M_{ij}} \text{ for } i, j = 1, ..., N \tag{3}$$

Finally, the last step calculates the DB index, that is, the average of the similarity measure of each cluster with the cluster most similar to it (see Eq. (4)). R_i is the maximum of R_{ij} $i \neq j$, so, the maximum value of R_{ij} represents the worst-case within-to-between cluster ratio for cluster i. Thus, the optimal clustering solution has the smallest Davies-Bouldin index value.

$$DB = \frac{1}{N} \sum_{i=1}^{N} R_i. \tag{4}$$

Considering the definition of the DB index, a minimization problem can be defined, whose objective function is the DB index value. Thus, metaheuristics can be used in order to solve this problem as an evolutionary bio-inspired algorithm.

Therefore, in order to compare the results obtained through different approaches, three bio-inspired evolutionary algorithms are used in this work: Genetic Algorithm (GA), Particle Swarm Optimization (PSO) and Differential Evolution (DE). Thus, the main difference between the so-called automatic algorithms that will be used in this paper is the optimization process to define the DB index, since each one of them employs a different bio-inspired optimization algorithm: GA, PSO or DE. More information about these algorithms can be found at [15, 28, 29].

4   Dataset

In this paper, all the questions answered on the section SNA of the MathE platform were analyzed to identify patterns based on the type of student answers to each question, whether correct or incorrect. Thus, the data collected considers information of 6942 answers distributed among 766 questions of 15 topics. These answers were provided by 285 students of different nationalities, over 3 years, in which the platform is online. It is important to highlight that the questions and the topics are constantly being added to the platform, so naturally some topics have more questions answered than others. Table 1 describes the dataset.

The *Topic* column describes all the MathE topics available on the platform. Next, the *Question Available* column describes the number of questions available on the platform for each topic. Thereafter the *Question Answered* column gives the number of different questions answered in each topic. The last three columns refer to the type of answers provided by the students: the first column shows the number of correct answers per topic, followed by the column of the incorrect ones, and the last column corresponds to the sum of the correct and incorrect answers, which is equal to the total number of answers per topic.

To investigate the optimal number of levels of difficulty into which the questions will be divided, the probability of correct answers for each of the 766 questions was evaluated. That is, for each question the number of correct answers divided by the total number that this question was answered. This information was then considered as the input variable of the clustering algorithm, in a first scenario.

Table 1. MathE dataset

Topic	Questions available	Questions answered	Correct answers	Incorrect answers	Total answers
Linear algebra	211	199	1741	1955	**3696**
Fund. math	91	84	365	396	**761**
Graph theory	49	34	29	19	**48**
Differentiation	144	96	193	397	**590**
Integration	127	54	67	94	**161**
Analytic geometry	40	40	156	183	**339**
Complex numbers	41	37	231	277	**508**
Dif. equation	41	30	56	40	**96**
Statistic	41	26	155	175	**330**
R. F. single variable	52	25	28	46	**74**
Probability	46	34	32	54	**86**
Optimization	96	25	11	26	**37**
R. F. several variable	58	15	5	13	**18**
Set theory	40	26	26	16	**42**
Numerical methods	42	41	73	83	**156**
Total	**1119**	**766**	**3168**	**3774**	**6942**

However, it is known that the information contained in the probability variable of a question that was, for example, answered 20 times, is different from a question that was answered only 2 times.

Thus, to achieve greater reliability in the results, it was decided to divide the complete dataset into smaller sets, according to the number of questions answered, resulting in 4 datasets, as described below:

- **dataset** 1: questions answered 1 time, at least.
- **dataset** 2: questions answered 5 times, at least.
- **dataset** 3: questions answered 10 times, at least.
- **dataset** 4: questions answered 15 times, at least.

Note that datasets 2, 3, and 4 are subsets of dataset 1. The dimension of each dataset, according to the topics, is presented in Table 2.

5   Results and Discussion

This section presents a general analysis of the data described in Sect. 4. Thereafter, the results from Hierarchical and Partitioning Clustering Algorithms are presented and discussed.

Table 2. Number of different questions per dataset

Topic	Dataset 1	Dataset 2	Dataset 3	Dataset 4
Linear algebra	199	154	125	70
Fund. math	84	63	29	8
Graph theory	34	0	0	0
Differentiation	96	61	20	0
Integration	54	12	1	0
Analytic geometry	40	38	14	0
Complex numbers	37	33	30	4
Dif. equation	30	6	0	0
Statistic	26	25	22	1
R.F. single variable	25	8	0	0
Probability	34	3	0	0
Optimization	25	0	0	0
R.F. several variable	15	0	0	0
Set theory	26	0	0	0
Num. methods	41	14	1	0
Total	**766**	**417**	**242**	**83**

5.1   General Analysis of the Data

From Table 1, more specifically in the *Total Answers* column, it is possible to observe that the Linear Algebra topic is the most used topic in the platform. Approximately 53% of the total questions answered correspond to this topic. However, this fact is not surprising, considering that Linear Algebra is a subject present in almost all curricula of higher education courses that include mathematics. After Linear Algebra, the most requested topic is Fundamentals of Mathematics, which corresponds to 10% of the total answered questions. Fundamentals of Mathematics includes questions about the essential background for higher education and, in turn, has also substantial demand on the platform. The other topics have a lower rate of use, however, from the data, it is possible to conclude that all topics are consistently being exploited.

Figure 3 compares the performance of the students by topics through the percentage of correct answers and incorrect ones, constituted by the data from columns *Correct Answers, Incorrect Answers* and *Total Answers* of Table 1. Although these values are complementary, presenting them in confrontation allows a better evaluation of the results.

Thus, from Fig. 3, in practically all topics, represented by 1 to 15, at least 30% of the questions were answered correctly. The topics Set Theory (14), Graph Theory (3), and Differential Equation (8) had the highest percentage of correct answers, 62%, 60%, and 58% respectively. It is important to highlight in these three topics the highest rate of 50%, which means that in this topic the rate

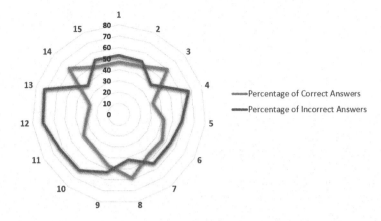

Fig. 3. Percentage of correct and incorrect answered questions, by topics.

of correct answers is higher than the rate of incorrect ones. On the other hand, Optimization (12) and Real Function of Several Variable (13) had the lowest rate of correct answers, it is 30% and 28%, respectively. Moreover, it is important to mention that these two topics had the fewest questions answered, as can be seen in the *Total Answers* column of Table 1.

In order to determine the optimal number of partitions of the dataset, two different clustering methodologies were performed: Hierarchical Clustering and Partitioning Clustering. The results are presented bellow, and they were obtained using an Intel(R) i5(R) CPU @1.60 GHz with 8 GB of RAM using Matlab 2019a ® software [19].

5.2   Hierarchical Clustering Results

The data from all datasets was evaluated by Hierarchical Clustering techniques, as presented in the Sect. 3. Figure 4 presents the results in the form of Dendrogram for the datasets 1–4. Figure 4a shows the Dendrogram generated by all questions answered (dataset 1); the results of the dataset that includes only the questions that were answered at least 5 times (dataset 2) are depicted in Fig. 4b; Fig. 4c refers to questions answered at least 10 times (dataset 3); and the results of questions answered at least 20 times (dataset 4) are presented in Fig. 4d.

When comparing the four dendrograms, it is noticed that the distances between the clusters become smaller and smaller as the datasets become more restrictive. That is, dataset 1 is less homogeneous (similar) than dataset 2 and so on. This makes perfect sense, given the divisors applied to generate each of the datasets used.

In each dendrogram it can be seen precisely the possible divisions of the clusters, which depend on the chosen similarity measure (distance) between the groups. In this work, this division also represents how many levels of difficulty the platform questions will be divided into. In this way, when choosing horizontal cut

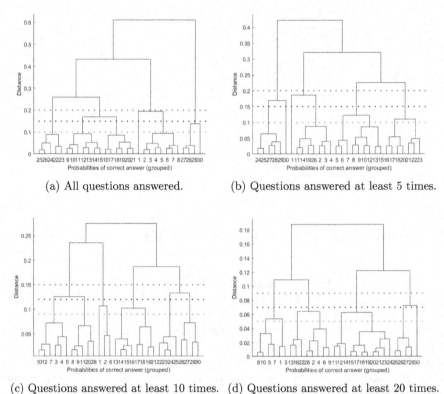

(a) All questions answered.   (b) Questions answered at least 5 times.

(c) Questions answered at least 10 times.   (d) Questions answered at least 20 times.

Fig. 4. Dendrogram results.

lines with small distances, there will be more levels of difficulty and, consequently, very similar questions in terms of difficulty within each cluster. In contrast, larger distances lead to more heterogeneity concerning the difficulty level in each generated cluster. The prominent question is, "What is the best value for the cut line?". From the previous work [3], it is known that two difficulty levels, meaning 2 clusters, are not enough, so we are interested in delimiting a horizontal line that includes 3 clusters (levels) or more.

If 7 or 8 clusters are used (green dashed line), the clusters will be composed of few questions, but of high similarity. As we are dealing with people (students and teachers), and the requirements, preferences, needs, and characteristics would make a lot of difference, it is interesting to have a little heterogeneity between the groups. Therefore, it is not interesting to restrict the dissimilarities of the elements of each cluster so much.

Another possibility is 3–4 levels (red dashed line). However, knowing that 2 is very little, 3 or 4 may not make much difference, and the problems presented in [3] may remain. We need a middle ground. Thus, 6 is a good split possibility. With 6 clusters (black dashed line) it is expected to be able to maintain a balance between the similarities and dissimilarities of the questions at each level.

However, the analyzes obtained so far have a high content of partiality of the authors. For this reason, it was decided to analyze the same datasets by another clustering technique, in this case Partitioning Clustering, through Bio-inspired automatic clustering techniques, whose results are presented below.

5.3   Partitioning Clustering Results

The four datasets mentioned were also evaluated by the Partitioning Clustering. In this case, three Bio-inspired clustering techniques were used to define the optimal number of clusters automatically. Consequently, the main difference in the definition of the number of clusters is in the algorithm used to minimize the DB index, that is GA, PSO and DE, as presented in Sect. 3. Besides, since these algorithms are stochastic, the results may vary from one iteration to another, requiring more than one execution of the algorithm. Moreover, it is interesting to compare the results of the different bio-inspired algorithms.

For all bio-inspired algorithms, the common parameters used were: maximum number of clusters equal to 10; initial population equal to 100, maximum number of iterations equal to 250, which was also the stopping criterion considered. For the GA, a rate of 0.8 was considered for selection and crossover, and 0.3 for a mutation. On the other hand, for PSO, the chosen rates were: global learning coefficient equal to 2, personal learning coefficient equal to 1.5, inertia weight equal to 1 and inertia weight damping equal to 0.99. Finally, for DE, the rates are equal to 0.2 for crossover and the scaling bound factor varies between $[0.2, 0.8]$. Each algorithm was performed 30 times for each dataset, and the smaller DB index obtained was defined as the optimal solution.

Table 3 presents the results of each algorithm, in terms of DB index and the optimal number of clusters (No. Clusters), for each dataset.

Table 3. Clustering bio-inspired algorithm results

Algorithm	Results	Dataset 1	Dataset 2	Dataset 3	Dataset 4
GA	DB index	0.4547	0.4638	0.4981	0.4624
	No. clusters	3	5	6	6
PSO	DB index	0.4546	0.4757	0.6131	≈0
	No. clusters	6	6	6	2
DE	DB index	0.4525	0.4753	0.4686	0.4468
	No. clusters	6	6	9	7

As can be seen, the smallest DB index was obtained by the DE approach on the dataset 1 resulting in an index of 0.4525, with 6 clusters. Moreover, the number 6 appears at least once for each dataset considered. And, for a general analysis, 7 out of the 12 tests performed (each algorithm on each dataset) indicated 6 as the value of the clusters, which is in line with the results and conclusions obtained by the hierarchical clusters. Considering this, the DE approach on the dataset 1 was chosen as the optimal solution. The detailed results of this approach are presented in Table 4, in terms of centroid coordinator, probability intervals (Prob. Inter.), intra-cluster distance (Intra C. Dist.) and inter-cluster distance. Note that, each cluster is defined as C_k, where $k \in [1, 6]$.

Table 4. Detailed results of the optimal DE solution

Results		C1	C2	C3	C4	C5	C6
Centroids		0.0000	0.8338	1.0000	0.4876	0.6664	0.2787
Prob. Inter.			[0, 0.13]	[0.14, 0.37]	[0.38, 0.57]	[0.58, 0.75]	[0.76, 0.94]	[0.95, 1.00]
Intra C. Dist.		0.026	0.0358	0.0000	0.0511	0.0485	0.0634
Inter cluster distance	C1	0					
	C2	0.8338	0				
	C3	1.0000	0.1662	0			
	C4	0.4876	0.3463	0.5124	0		
	C5	0.6664	0.1675	0.3336	0.1788	0	
	C6	0.2787	0.5551	0.7213	0.2033	0.3876	0

Note that the centroids have only one coordinate as they are on a straight line, since the clusters were defined from a single variable. The values in *Prob. Inter.* define the probabilistic intervals that delimit each of the clusters and, consequently, the interval of each level of difficulty, based on the probability of a student hitting a question.

Finally, Fig. 5 illustrates the optimal solution, given by the dataset 1 with the DE algorithm. In this case, each level of difficulty is presented by a different color. And the axis x represents the probability of a question being answered correctly while the axis y represents each topic available on MathE platform. So, the level 1 are the easiest questions, while the level 6 are the most difficult questions, based on the probability of correct answers.

So far, only questions per topic have been analyzed. However, in some topics, there are also subtopics, so the need arose to verify that the results of the topics can be verified for the subtopics. For this, the topic of Linear Algebra, which has 6 subtopics and has the largest number of questions answered, was chosen for further analysis. Thus, the same metrics were applied to the definition of datasets, and the parameters of the algorithms were reproduced exclusively for the questions that compose the subtopics that correspond to the Linear Algebra topic.

Table 5 presents the results for the 4 datasets, where the first dataset is composed of questions answered at least 1 time, in the second are the questions answered at least 5 time, followed by the datasets composed of questions answered 10 times and 20 times, respectively. Again, a trend towards 6 clusters was observed. As can be seen, for the 12 tests performed, in 5 of them, the number 6 was pointed out as the optimal solution. In this case, the optimal solution,

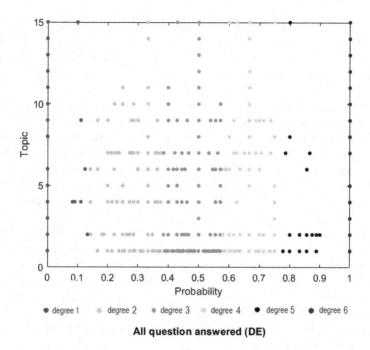

All question answered (DE)

Fig. 5. Clustering optimal solution.

represented by the smallest DB index, was obtained by the PSO algorithm in dataset 1, being 0.4237 the DB index value.

Table 5. Linear algebra clustering results

Algorithm	Results	Dataset 1	Dataset 2	Dataset 3	Dataset 4
GA	DB index	0.4282	0.4363	0.4244	0.4322
	No. clusters	6	5	5	7
PSO	DB index	0.4237	0.4652	0.4614	0.4479
	No. clusters	6	6	6	2
DE	DB index	0.4246	0.4731	0.4409	0.4258
	No. clusters	7	6	8	7

6 Conclusions and Future Work

The MathE platform is an online educational system that aims to help students who struggle to learn college mathematics as well as students who want to deepen their knowledge of a multitude of mathematical topics, at their own pace. The platform has the aim of offering a dynamic and engaging tool to teach and learn mathematics, relying on interactive digital technologies that enable customized study. This work extracted information from the data collected by the MathE platform in order to trace paths for the creation of an intelligent and customized management system for the platform. It is expected that in the near future the platform will be able to make use of intelligent mechanisms, based on optimization algorithms and machine learning, to make autonomous decisions, tailored according to the needs of each user. One of the decisions to be made refers to the distribution of questions according to the students' background and demands. Thus, the information collected through this research will serve as a guide to make the choice of optimal strategies to improve the performance of the platform. Hence, the information from 285 students who used the *Students Assessment Section* on the MathE platform between April 2019 and February 2022 was considered.

Currently, the resources available in the MathE platform are organized into two levels of difficulty, basic and advanced; any user of the platform with teacher profile can define the level of each question. However, in order to improve the resources that are available in the platform, making it autonomous for making certain decisions, some adjustments are necessary. This work aimed to investigate the optimal number of levels of difficulty in which the resources in the MathE platform should be reorganized. For this, the information from the questions answered over the time that the platform is online was analyzed, through the probabilities of correct answers for each question. This information was analyzed through two clustering techniques, namely hierarchical clustering and partitioning clustering.

According to the presented results, it can be concluded that both methodologies reached a consensus that 6 levels of difficulty is the optimal solution for the reorganization of the platform, both for topics and subtopics. With 6 levels of difficulty, it will be possible to work better with the implementation of algorithms that will make the MathE platform autonomous and intelligent. In addition, a more accurate division of the content tends to motivate student users even more, since in this way, they will be able to better follow the advance or retreat of their knowledge when moving through the different levels of difficulty.

Thus, considering the described results, the reorganization of the available questions remains as future work. Besides, it is also expected that future work will be developed in order to distribute the questions in an intelligent and personalized way, respecting the needs of each user.

References

1. Ashwin, T.S., Guddeti, R.M.R.: Impact of inquiry interventions on students in e-learning and classroom environments using affective computing framework. User Model. User-Adap. Inter. **30**(5), 759–801 (2020). https://doi.org/10.1007/s11257-019-09254-3
2. Azevedo, B.F., Amoura, Y., Kantayeva, G., Pacheco, M.F., Pereira, A.I., Fernandes, F.P.: Collaborative Learning Platform Using Learning Optimized Algorithms, vol. 1488. Springer, Cham (2021). https://doi.org/10.1007/978-3-030-91885-9-52

3. Azevedo, B.F., Pereira, A.I., Fernandes, F.P., Pacheco, M.F.: Mathematics learning and assessment using MathE platform: a case study. Educ. Inf. Technol., 1–23 (2021). https://doi.org/10.1007/s10639-021-10669-y
4. Bakhouyi, A., Dehbi, R., Banane, M., Talea, M.: A semantic web solution for enhancing the interoperability of e-learning systems by using next generation of SCORM specifications. In: International Conference on Advanced Intelligent Systems for Sustainable Development, pp. 56–67. Springer, Cham (2019). https://doi.org/10.3991/ijet.v14i11.10342
5. Beinicke, A., Bipp, T.: Evaluating training outcomes in corporate e-learning and classroom training. Vocat. Learn. 11(3), 501–528 (2018). https://doi.org/10.1007/s12186-018-9201-7
6. Benta, D., Bologa, G., Dzitac, I.: E-learning platforms in higher education. Case study. Procedia Comput. Sci. 31, 1170–1176 (2014). https://doi.org/10.1016/j.procs.2014.05.373
7. Cabero Almenara, J., Barroso Osuna, J.M.: Los escenarios tecnológicos en realidad aumentada (ra): posibilidades educativas en estudios universitarios. Aula Abierta - Revistas Eletronicas de la Unviersidad de Oviedo (2018)
8. Davies, D.L., Bouldin, D.W.: A cluster separation measure. IEEE Trans. Pattern Anal. Mach. Intell. PAMI 1(2), 224–227 (1979). https://doi.org/10.1109/TPAMI.1979.4766909
9. Ezugwu, A., Shukla, A., Agbaje, M., Oyelade, O., José-García, A., Agushaka, J.: Automatic clustering algorithms: a systematic review and bibliometric analysis of relevant literature. Neural Comput. Appl. (2020). https://doi.org/10.1007/s00521-020-05395-4, https://hal.archives-ouvertes.fr/hal-03217646
10. Gunasinghe, A., Abd Hamid, J., Khatibi, A., Azam, S.F.: The adequacy of UTAUT-3 in interpreting academician's adoption to e-learning in higher education environments. Interact. Technol. Smart Educ. (2019). https://doi.org/10.1108/ITSE-05-2019-0020
11. Herodotou, C., Rienties, B., Hlosta, M., Boroowa, A., Mangafa, C., Zdrahal, Z.: The scalable implementation of predictive learning analytics at a distance learning university: insights from a longitudinal case study. Internet High. Educ. 45, 100725 (2020). https://doi.org/10.1016/j.iheduc.2020.100725
12. Jin, X., Han, J.: Partitional Clustering. Springer, Boston (2010). https://doi.org/10.1007/978-0-387-30164-8-631
13. Kalpokaite, N., Radivojevic, I.: Teaching qualitative data analysis software online: a comparison of face-to-face and e-learning ATLAS.ti courses. Int. J. Res. Meth. Educ. 43(3), 296–310 (2020). https://doi.org/10.1080/1743727X.2019.1687666
14. Kayser, I., Merz, T.: Lone wolves in distance learning?: an empirical analysis of the tendency to communicate within student groups. Int. J. Mob. Blended Learn. (IJMBL) 12(1), 82–94 (2020). https://doi.org/10.4018/IJMBL.2020010106
15. Kennedy, J., Eberhart, R.: Particle swarm optimization. In: Proceedings of ICNN'95 - International Conference on Neural Networks, vol. 4, pp. 1942–1948 (1995). https://doi.org/10.1109/ICNN.1995.488968
16. Khlifi, Y.: An advanced authentication scheme for e-evaluation using students behaviors over e-learning platform. Int. J. Emerg. Technol. Learn. 15(4) (2020). https://doi.org/10.3991/ijet.v15i04.11571
17. Laskaris, D., Heretakis, E., Kalogiannakis, M., Ampartzaki, M.: Critical reflections on introducing e-learning within a blended education context. Int. J. Technol. Enhanc. Learn. 11(4), 413–440 (2019). https://doi.org/10.1504/IJTEL.2019.102550

18. Luo, N., Zhang, Y., Zhang, M.: Retaining learners by establishing harmonious relationships in e-learning environment. Interact. Learn. Environ. **27**(1), 118–131 (2019). https://doi.org/10.1080/10494820.2018.1506811
19. MATLAB: The MathWorks Inc. https://www.mathworks.com/products/matlab.html (2019a)
20. Moreno-Guerrero, A.J., Aznar-Díaz, I., Cáceres-Reche, P., Alonso-García, S.: E-learning in the teaching of mathematics: an educational experience in adult high school. Mathematics **8**(5), 840 (2020). https://doi.org/10.3390/math8050840
21. Moubayed, A., Injadat, M., Shami, A., Lutfiyya, H.: Student engagement level in an e-learning environment: clustering using k-means. Am. J. Distance Educ. **34**(2), 137–156 (2020). https://doi.org/10.1080/08923647.2020.1696140
22. Mousavi, A., Mohammadi, A., Mojtahedzadeh, R., Shirazi, M., Rashidi, H.: E-learning educational atmosphere measure (EEAM): a new instrument for assessing e-students' perception of educational environment. Res. Learn. Technol. **28** (2020). https://doi.org/10.25304/rlt.v28.2308
23. Nielsen, F.: Hierarchical Clustering, pp. 195–211. Springer, Cham (2016). https://doi.org/10.1007/978-3-319-21903-5-8
24. Rakic, S., Tasic, N., Marjanovic, U., Softic, S., Lüftenegger, E., Turcin, I.: Student performance on an e-learning platform: mixed method approach. Int. J. Emerg. Technol. Learn. **15**(2) (2020). https://doi.org/10.3991/ijet.v15i02.11646
25. Sathiyamoorthi, V.: An intelligent system for predicting a user access to a web based e-learning system using web mining. Int. J. Inf. Technol. Web Eng. (IJITWE) **15**(1), 75–94 (2020). https://doi.org/10.4018/IJITWE.2020010106
26. Shakah, G., Al-Oqaily, A., Alqudah, F.: Motivation path between the difficulties and attitudes of using the e-learning systems in the Jordanian universities: Aajloun university as a case study. Int. J. Emerg. Technol. Learn. (iJET) **14**(19), 26–48 (2019). https://doi.org/10.3991/ijet.v14i19.10551
27. Shalev-Shwartz, S., Ben-David, S.: Understanding Machine Learning: From Theory to Algorithms. Cambridge University Press, Cambridge (2014)
28. Sivanandam, S.N., Deepa, S.N.: Introduction to Genetic Algorithms, 1st edn. Springer, Cham (2008). https://doi.org/10.1007/978-3-540-73190-0
29. Storn, R., Price, K.: Differential evolution-a simple and efficient heuristic for global optimization over continuous spaces. J. Global Optim. **11**(4), 341–359 (1997). https://doi.org/10.1023/A:1008202821328

A Tabu Search with a Double Neighborhood Strategy

Paula Amaral[1,2]([envelope]) [iD], Ana Mendes[1] [iD], and J. Miguel Espinosa[3] [iD]

[1] Nova SST–FCT Nova, Campus de Caparica, 2829-516 Caparica, Portugal
paca@fct.unl.pt
[2] NovaMaths CMA Nova, Campus de Caparica, 2829-516 Caparica, Portugal
[3] IT & Digital Department, GALP, Lisbon, Portugal

Abstract. Tabu Search (TS) is a well known and very successful method heuristic approach for hard optimization problems, linear or nonlinear. It is known to produce very good solutions, optimal or close to optimal for some hard combinatorial optimization problems. The drawback is that we have in general no optimality certificate, but this is the price to be paid for problems where the exact methods are too costly in terms of time and computational memory. One of the drawbacks of TS is that the strategies must be studied and refined for every instance. Many proposals exist to enhance the efficiency of this method heuristic. TS is particularly fragile in cases where there are many local optima of the problem. TS may be slow in the process of escaping a region of attraction of a local optima. If the strategy for evaluating the solutions in the neighborhood takes time to move away from the local optimum then it may compromise the search efficiency. In this paper we propose a double neighborhood strategy with opposite optimization directions (minimization and maximization). While one search for the best solution in the neighborhood the second search for the worse, and two parallel process develop switching from the minimization to maximization and vice-versa, when in consecutive iterations there is no improvement in solution. With this proposal, it is intended that the research can escape the attraction zone of a local optimum, more quickly allowing the research space to be better explored. We present an application to a Knapsack problem.

Keywords: Tabu Search · Neighborhood search · Double neighborhood

1 Introduction

Tabu Search [8–10, 18] is a metaheuristic that has successfully been applied to find good feasible solutions for hard optimization problems [12, 17]. There are many successful applications of TS, for instance in timetabling scheduling [3], in Neural Networks [22], Vehicle Routing [5, 7, 19], as a general solver for the constraint satisfaction problems [14]. In general it can be described as a neighborhood search method incorporating techniques for escaping local optima and avoid cycling [11]. A first level Tabu Search (TS) comprises the following concepts in each iteration:

This work is funded by national funds through the FCT - Fundação para a Ciência e a Tecnologia, I.P., under the scope of the project UIDB/00297/2020 (Center for Mathematics and Applications).

- Current starting solution - Start search point.
- Search Neighborhood - Points that will be inspected from the current solution.
- Move - A basic operation in the definition of the neighborhood.
- Evaluation - A procedure to evaluate the points in the neighborhood.
- Tabu list - The tabu moves that are not allowed in the current iteration
- Aspiration Criteria - May revoke a tabu status.

A general, very basic, iteration of TS will consist in finding a set of points in the neighborhood of the current point. Evaluate these points and chose the one that has the best evaluation, as long as the move associated to this point is not tabu. If it is tabu we can apply an aspiration criteria or not. Next we add the move, or solution, or a related attribute that generated the best evaluated point to the tabu list. After we proceed to the next iteration from the current point. There are many interesting additional refinements that can greatly increase the performance of TS.

In this paper a new approach for the TS is proposed. We use two parallel searches, with two distinct neighborhood N_1 and N_2, and opposite strategies in every iteration. At every iteration if in N_1 it is applied the minimization direction in N_2 we apply a maximization direction. The search strategy switches if a local optima is recognized. In this way, when a local minimum is found, we escape from that local optima by following the path along the maximization direction. When a local maximum is encountered, the direction of search switches to minimization to explore new local optima.

2   Motivation

What led to the study of this new strategy was the fact that many times the tabu search takes time to move away from the local optimum. With this proposal, it is intended that the search can escape the attraction zone of a local optimum, more quickly allowing the search space to be more explored.

To illustrate this idea more clearly, in Fig. 1 is represented a function whose objective is to determine the global minimum. Considering two neighborhoods, the first uses the classical strategy. In this case it will be selected the solution with the minimum value of the neighborhood and in the second it will alternate between the selection of the maximum and of the minimum value.

In the example, for the first neighborhood the tabu search starts at the initial solution, x_0 and considering the assigned interval as a neighborhood, the best point in the neighborhood is x_1. In the second iteration the best point in the neighborhood of x_1, $N(x_1)$ is the point x_2. This spot is a local minimizer. With this classic strategy, many iterations will be needed to escape this local optimum because, the minimum is chose, so the left point in interval representing the neighborhood is picked. That's why we chose to use two neighborhoods. In addition to the usual strategy of choosing the best solution in If we want to escape this region of attraction of x_2 we must climb the hill as fast as possible. A second strategy is implemented, which pursues a sequence of worst solutions until a local maximum is found. In addition to the usual strategy of choosing the best solution in Fig. 1, if we switch from the minimum in the neighborhood to the maximum, the point picked in every iteration will be the right-hand-side of the interval. After a few iterations a local maximum is attained. Once a local maximum is reached, the strategy is reversed (minimum) and the best solutions in the neighborhood are chosen. When a local optimum is obtained, in this case a local minimum, the strategy is reversed again, changing the search direction again to the maximum of the objective function. Thus, with two neighborhoods, the second neighborhood, starting from the initial solution x_0, selects the point x_3 as the next center of the next neighborhood. The search continues and in the second iteration it moves to the solution x_4, as a local maximum has been reached, the second neighborhood strategy is inverted and starts to select the best solution in the neighborhood. In the next iteration as you are minimizing the $N(x_4)$ is the point x_5 and so on until reaching the global optimum, x^*. In this example it can be seen that with the use of this strategy the search can converge more quickly to an optimal point. In addition the neighborhood space will be better explored. In the next section we make a brief description of a classic Tabu Search, before presenting the Double Neighborhood Tabu Search approach.

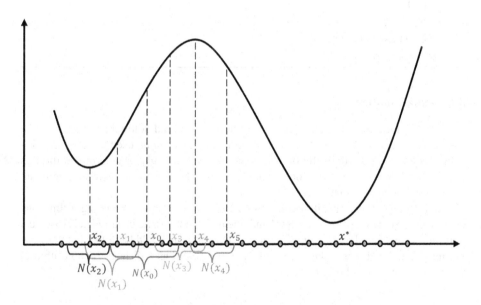

Fig. 1. Example of double neighborhood

3   Tabu Search

The Tabu Search is a metaheuristic that explores the space of solutions, moving at each iteration to the best solution in the neighborhood, not accepting movements that could lead to the solutions already visited. In the Algorithm 1, the main steps of the Tabu Search [17] are presented.

Algorithm 1. Tabu Search

Data: S_0 Initial solution;
Stoping criteria
Result: S^* Best solution found
$T \leftarrow []$;
$S \leftarrow S_0$;
$S^* \leftarrow S$;
repeat

> Generate $N(S)$ neighborhood solutions of S;
> Find best solution $\hat{S}$ in $N(S)$ that is not tabu ;
> **if** $\hat{S}$ *better than* S^* **then**
> | $S^* \leftarrow \hat{S}$
> **end**
> $S \leftarrow \hat{S}$;
> Updates the tabu list T;

until *Stop Criteria*;

3.1   Initial Solution

To start the Tabu Search, it is necessary to define an initial solution, from which the first search neighborhood is defined. Different strategies can be used to determine this solution, for example, the solution can be generated randomly or determined through a heuristic. In the second case, one of the most used heuristics for their simplicity are those based on *greedy* methods.

Generating a random initial solution becomes faster, however the algorithm may take longer to converge. On the other hand, the use of a heuristic usually leads to a local optimum with a smaller number of iterations. However, it should be noted that this does not mean that in all cases the use of better initial solutions will lead the algorithm to converge to better local optima [21].

3.2   Search Space

The search space is the space that contains all feasible solutions to the problem. In Tabu Search, for nonconvex problems, the idea is to explore the solutions space as comprehensively and as completely as possible.

3.3 Structure of the Neighborhood

In each iteration, a current solution is considered, denoted by S, and the set of neighborhood solutions is defined by $N(S)$. Formally $N(S)$ is a subset of the search space that contains all solutions obtained by a transformation of S according to some rule. For each problem there is more than one possible neighborhood structure that can be applied.

There are also several strategies to select the solution in the neighborhood, some of them being [15,21]:

– Select the best solution in the neighborhood, that is, all solutions in the neighborhood are examined and the solution with the best value for the objective function is chosen. For large neighborhoods, this type of exploration turns out to be quite time-consuming, which implies a great computational effort.
– Select the first admissible solution, this strategy consists of choosing the first solution in the neighborhood that satisfies all the constraints of the problem and is not tabu, so only a partial exploration of the neighborhood is performed.
– Another strategy that can be applied is to randomly select a solution from the neighborhood.

The partial evaluation of the neighborhood may imply a deterioration in the quality of the solution, since not all solutions are analyzed in each iteration. However, at a global level this limitation may end up not having a negative influence on the quality of the solution produced, because a good final solution can still be achieved by exploring regions different than those obtained if the full neighborhood was inspected [20].

The construction of efficient tabu search algorithms implies a good balance between the computational burden of each iteration (more complex neighborhoods with a greater number of solutions to explore) and the number of iterations (in an attempt to better explore all solution spaces). More complex neighborhoods may enrich the research, since more solutions are considered at each iteration, but at the same time there is a risk of the computational effort becoming excessive.

Simple Neighborhood Structure. Suppose that a solution is indexed by a vector with natural numbers. A simple neighborhood is created by simple movements, where at each iteration a single variable switches its position. For instance, in a Vehicle Routing Problem (VPR) a vector represents a sequence of costumers visited in one route. Switching the position of a component of the vector, in this case means to change the order by which the costumers in that route is visited, or even changed to another route. In the example of the Fig. 2, customer 5 was moved, which will be placed in the position before customer 2. This action can be presented as a move, that can be recorder in different ways, for instance by (5), the number of the costumer that changed the position, by the pair (5,6) meaning that costumer 5 moved from position 6, or (5, 6, 2) meaning that costumer 5 moved from position 5 to 2.

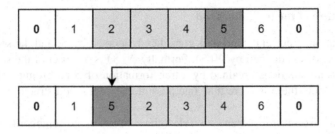

Fig. 2. Example simple neighborhood structure

3.4 Tabu List

To avoid revisiting a previous solution and in order to escape from a local optima, a list representing the moves that are tabu, meaning moves that could not be performed, is defined throughout the algorithm.

For instance, in the previous example, if the tabu {5} is defined it means that in a number of sequential iterations, the position of costumer 5 is not allowed to change. If it is defined as tabu the pair (5, 6) means that costumer 5 cannot be moved back to position 6. If (5, 6, 2) is defined as tabu, it implies that costumer 5 cannot move from position 2 to 6. More complex tabu list elements are more interesting but may lead to a more time consuming method.

3.5 Tabu Tenure

The tabu tenure defines the number of iterations during which a certain tabu movement is kept in the tabu list. The value of the tabu tenure may be fixed or may be dynamically defined. One of the drawback of Tabu Search is related to the need of tuning the tabu tenure, whose choice affects the performance of the algorithm. In [16] the authors presented a method to automatically manage the tabu tenure in the TS using a Decision Expert System based on a Fuzzy Inference Rule Based System (FIRBS). Two concepts "frequency" and "inactivity" related to the number of times a move was attempt and the last time it was called were used in the construction of the FIRBS.

3.6 Aspiration Criteria

The aspiration criteria allows to revoke a tabu status under some predefined condition [13].

3.7 Stopping Criteria

In tabu search there are several ways to end the search process, such as:

- After a fixed number of iterations;
- After a certain number of iterations with no improvement in the value of the best solution;

– When the objective function reaches a predefined value. This criterion prevents the execution of the algorithm when a solution is considered sufficiently good.

Several additional features can be incorporated in a TS, such as Aspiration Criteria, Intensification, Diversification, Multistart. In this section we just mentioned the basic features of TS.

4   Double Neighborhood Tabu Search (DNTS)

As already mentioned as an alternative to having just one type of search, with one strategy, we propose considering two neighborhoods, with to independently processes and with opposite strategies (minimization and maximization).

Algorithm 2. Change of Neighborhood Strategy

The search starts with the neighborhood maximizing the value of the objective function;
if *search is maximizing* **then**
| **if** *a local maximum is reached* **then**
| | The neighborhood strategy switches to minimization;
| **end**
end
if *search is minimizing* **then**
| **if** *a local minimum is reached* **then**
| | The neighborhood strategy switches to maximization;
| **end**
end

4.1   The Double Neighborhood Algorithm

In the classic approach, the common search, consists in selecting the solution of the neighborhood that presents the better value for the objective function. The DNTS, on the other hand, considers simultaneously two neighborhoods (Neighborhood 1 and Neighborhood 2) and varies between the selection of the best solution from the neighborhood and the worst one, that is, it alternates between a direction of minimization and maximization in order to escape as quickly as possible from a region of the search space corresponding to a local optimum.

At the beginning of the Tabu Search, we start by choosing the worst value of the objective function in neighborhood 2. When a local optimum is reached, that is, when there are only better values in the neighborhood than the current solution, the exchange is made and the search begins to choose the values in the neighborhood. The search changes again when after t iterations there was no improvement in the value obtained by the neighborhood. In the Algorithm 2 there is a description of how the exchanges in neighborhood search strategy are carried out, for a minimization problem.

The addition of a new neighborhood, implies the definition of new parameters, as for instance the *tabu tenure*, and the number of iterations during which the strategy (minimization or maximization) is maintained. As the neighborhoods are independent, then *tabu tenure* will also be, i.e. the neighborhoods can be defined with different *tabu tenure* values. The way in which the parameter values are determined may vary according to the problem under study, and are defined at the beginning of the algorithm. We should run the program for different values of *tabu tenure*, to fine-tune the parameters to determine the best values of *tabu tenure* for the problem.

4.2 The Knapsack Problem

For this new approach for the Tabu Search, results obtained for a class of optimization problem are now presented. The tests were performed for a classic optimization problem, the Knapsack Problem, which has a simple to formulation and has many interesting practical applications [4], such as optimizing investments, cargo loading [6], capital budgeting [2].

This is a well-known optimization problem and consists of filling a backpack with objects with a certain weight and value. The purpose of the problem is to fill the backpack with objects with the highest possible value, but the weight of the objects cannot exceed the maximum weight supported by the backpack. The mathematical formulation of the problem is the following:

N = Number of objects;
P = Capacity of the Knapsack;
p_i = weight of object i, for $i = 1, \ldots N$
v_i = value of of object i, for $i = 1, \ldots N$
$$x_i = \begin{cases} 1 & \text{if object is placed in the knapsack} \\ 0 & \text{otherwise} \end{cases} \quad \text{for } i = 1, \ldots N$$

$$\max \sum_{i=1}^{N} v_i x_i$$

s.t .

$$\sum_{i=1}^{N} p_i x_i \leq P$$

$$x_i \in \{0, 1\} \text{ for } i = 1, \ldots N$$

Algorithm 3. Tabu Search with Double Neighborhood for a minimization problem

---

Data: S_0 Initial solution;
Stopping criteria
Result: S^*, fo^* Best solution and best value found
$T_1 \leftarrow []$;
$T_2 \leftarrow []$;
$S_1 \leftarrow S_0$;
$S_2 \leftarrow S_0$;
$Mode_1 = \min$;
$Mode_2 = \max$;
// If $Mode_k = \min$ then search for the solution in the
   neighborhhod with minimum objective function value
// If $Mode_k = \max$ then search for the solution in the
   neighborhhod with maximum objective function value
$fo^* = v(S_0)$;   // $v(S)$ is the objective function value of S
$S^* \leftarrow S_0$;   // S^* is the incumbent solution
;

repeat
    Generate $N_1(S_1)$ neighborhood solutions of S_1;
    Find solution corresponding to $Mode_1$, $\hat{S}_1$ in $N(S_1)$ that is not tabu ;
    Generate $N_2(S_2)$ neighborhood solutions of S_2;
    Find solution corresponding to $Mode_2$, $\hat{S}_2$ in $N(S_2)$ that is not tabu ;
    if $v(\hat{S}_1) < fo^*$ **then**
        $S^* \leftarrow \hat{S}_1$;
        $fo^* = v(\hat{S}_1)$;   // Update the incumbent solution
    end
    if $v(\hat{S}_2) < fo^*$ **then**
        $S^* \leftarrow \hat{S}_2$;
        $fo^* = v(\hat{S}_2)$;   // Update the incumbent solution
    end
    if $Mode_2 = \max \wedge v(\hat{S}_2) < v(S_2)$ **then**
        $Mode_2 = \min$;   // S_2 may be a local maximizer
    end
    if $Mode_2 = \min \wedge v(\hat{S}_2) > v(S_2)$ **then**
        $Mode_2 = \max$;   // S_2 may be a local minimizer
    end
    if $Mode_1 = \max \wedge v(\hat{S}_1) < v(S_1)$ **then**
        $Mode_1 = \min$
    end
    if $Mode_1 = \min \wedge v(\hat{S}_1) > v(S_1)$ **then**
        $Mode_1 = \max$
    end
    $S_1 \leftarrow \hat{S}_1$;
    $S_2 \leftarrow \hat{S}_2$;
    Updates the tabu list T_1 and T_2;
until *Stopping Criteria*;

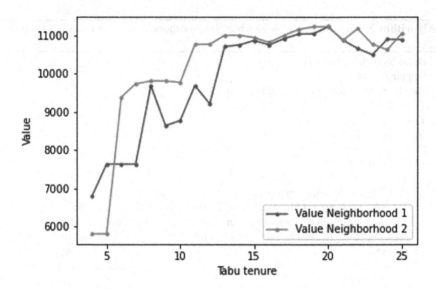

Fig. 3. Variation of the result obtained according to the *tabu tenure*

4.3   Computational Experience

The data sets used to test the backpack problem were taken from [1], where small and large sets were used. The solution was encoded by an N dimensional array of $0\backslash1$. A value in component i represented variable x_i. The neighborhood used for the tests was simple, in this problem the items are or are not included in the backpack, so the neighborhood consisted of changing the state of an item, from 0 to 1 or vice-versa. To fine-tune the *tabu tenure* several values were tested for the *tabu tenure* in each of the data sets and it was chosen the best. It should be noted that the second neighborhood has one more parameter that needs to be adjusted, which is the number of iterations that neighborhood 2 must remains with the same strategy (min or max) before switching it.

During the tests to determine the best value of *tabu tenure*, it was found that the value of this parameter is generally lower in neighborhood 2 than in neighborhood 1. In the Fig. 3 it can be seen that when the value of *tabu tenure* is lower the results of neighborhood 2 are better than those of neighborhood 1. *tabu tenure* has this behavior because with the structure of neighborhood 2 it is less likely that the search will enter a loop, so it is not necessary to limit as many elements as in neighborhood 1, as this is more likely to create loops.

The results obtained for each of the neighborhoods as the iteration in which the best solution was obtained are found in Table 1 and the graphic representation presented above corresponds to the data set with 200 elements in the table. Through its analysis we can conclude that both neighborhoods obtained the optimal value for the example presented, however neighborhood 1 needed a greater number of iterations to obtain the optimal solution. Analyzing the results obtained, it can be concluded that the use of the double neighborhood can present better results with a smaller number of iterations for large data sets. On smaller datasets using a common tabu search seems to be sufficient.

Table 1. Results obtained for the double neighborhood in the knapsack problem

Number of elements	Optimal value	Best value Neigh. 1	Best iteration Neigh. 1	Best value Neigh. 2	Best Iteration Neigh. 2
10	295	295	19	295	31
20	1024	1024	0	1024	0
4	35	35	4	35	6
4	23	23	0	23	0
15	481.07	481.07	0	481.07	0
10	52	52	11	52	22
7	107	107	0	107	0
23	9767	9767	32	9767	26
5	130	130	0	130	0
20	1025	1025	0	1025	0
100	9147	9147	127	9147	110
200	11238	11238	3725	11238	700
500	28857	25806	4968	24500	1998
1000	54503	42499	1219	43701	331
2000	110625	67743	295	67000	209

5   Conclusions

In this paper a new approach to the tabu search was proposed and tested, which consists of the use of two independent neighborhoods with the switching of the search strategy in the inspection of the solutions in the neighborhood. In the instances tested, for a Knapsack problem, it was shown that it was a good strategy and that it can be applied to different types of problems. However, although the results obtained for this approach are better than those obtained by the common tabu search, this strategy requires more computational effort. However, we believe in the merits of this new proposal specially in problems with many local optima.

Acknowledgement. We would like to thank the reviewers for their valuable comments and suggestions that helped to improve our manuscript.

References

1. Instances of 0/1 knapsack problem. http://artemisa.unicauca.edu.co/~johnyortega/instances_01_KP/
2. Abdul-Rahman, S., Yung, V.C.: Comparison of optimization and tabu search approach for solving capital budgeting based on knapsack problem (2019)
3. Amaral, P., Pais, T.C.: Compromise ratio with weighting functions in a tabu search multi-criteria approach to examination timetabling. Comput. Oper. Res. **72**, 160–174 (2016)

4. Assi, M., Haraty, R.A.: A survey of the knapsack problem. In: 2018 International Arab Conference on Information Technology (ACIT), pp. 1–6. IEEE (2018)
5. Cheeneebash, J., Nadal, C.: Using tabu search heuristics in solving the vehicle routing problem with time windows: application to a Mauritian firm. Univ. Maurit. Res. J. **16**, 448–471 (2010)
6. Cho, M.: The knapsack problem and its applications to the cargo loading problem. Anal. Appl. Math **48**, 48–63 (2019)
7. Cordeau, J.F., Laporte, G.: Tabu search heuristics for the vehicle routing problem, pp. 145–163. Springer, Boston (2005). https://doi.org/10.1007/0-387-23667-8_6
8. Gendreau, M.: An introduction to tabu search. In: Handbook of Metaheuristics, pp. 37–54. Springer, Cham (2003). https://doi.org/10.1007/0-306-48056-5_2
9. Glover, F.: Future paths for integer programming and links to artificial intelligence. Comput. Oper. Res. **13**(5), 533–549 (1986)
10. Glover, F., Laguna, M.: Tabu Search. Kluwer Academic Publisher, Dordrecht (1997)
11. Hertz, A., Taillard, É., Werra, D.: A tutorial on tabu search (1992)
12. Hertz, A., de Werra, D.: The tabu search metaheuristic: how we used it. Ann. Math. Artif. Intell. **1**, 111–121 (1990). https://doi.org/10.1007/BF01531073
13. Hvattum, L.M.: On the value of aspiration criteria in tabu search. Int. J. Appl. Metaheuristic Comput. (IJAMC) **7**(4), 39–49 (2016)
14. Nonobe, K., Ibaraki, T.: A tabu search approach to the constraint satisfaction problem as a general problem solver. Eur. J. Oper. Res. **106**(2), 599–623 (1998). https://doi.org/10.1016/S0377-2217(97)00294-4
15. Osman, I.: Metastrategy simulated annealing and tabu search algorithms for the vehicle routing problem. Ann. Oper. Res. **41**, 421–451 (1993)
16. Pais, T.C., Amaral, P.: Managing the tabu list length using a fuzzy inference system: an application to examination timetabling. Ann. Oper. Res. **194**, 341–363 (2012). https://doi.org/10.1007/s10479-011-0867-6
17. Pirim, H., Bayraktar, E., Eksioglu, B.: Tabu search: a comparative study. In: Jaziri, W. (ed.) Tabu Search, chap. 1. IntechOpen, Rijeka (2008). https://doi.org/10.5772/5637
18. Prajapati, V.K., Jain, M., Chouhan, L.: Tabu search algorithm (TSA): a comprehensive survey. In: 2020 3rd International Conference on Emerging Technologies in Computer Engineering: Machine Learning and Internet of Things (ICETCE), pp. 1–8. IEEE (2020)
19. Taillard, É., Badeau, P., Gendreau, M., Guertin, F., Potvin, J.: A tabu search heuristic for the vehicle routing problem with soft time windows. Transp. Sci. **31**, 170–186 (1997)
20. Taillard, E.: Tabu search. In: Siarry, P. (ed.) Metaheuristics, pp. 51–76. Springer, Cham (2016)
21. Talbi, E.G.: Metaheuristics: From Design to Implementation. Wiley, Hoboken (2009)
22. Werra, D., Hertz, A.: Tabu search: a tutorial and an application to neural networks. OR Spektrum **11**, 131–141 (1989)

International Workshop
on Computational Astrochemistry
(CompAs-tro 2022)

The $S^+(^4S)+SiH_2(^1A_1)$ Reaction: Toward the Synthesis of Interstellar SiS

Luca Mancini[1]([✉])(ID), Marco Trinari[1], Emília Valença Ferreira de Aragão[1,2](ID), Marzio Rosi[3](ID), and Nadia Balucani[1](ID)

[1] Dipartimento di Chimica, Biologia e Biotecnologie,
Università degli Studi di Perugia, 06123 Perugia, Italy
{luca.mancini2,emilia.dearagao}@studenti.unipg.it,
nadia.balucani@unipg.it
[2] Master-tec srl, Via Sicilia 41, 06128 Perugia, Italy
emilia.dearagao@master-tec.it
[3] Dipartimento di Ingegneria Civile ed Ambientale, Università degli Studi di Perugia,
06125 Perugia, Italy
marzio.rosi@unipg.it

Abstract. We have performed a theoretical investigation of the $S^+(^4S)$ + $SiH_2(^1A_1)$ reaction, a possible formation route of the $HSiS^+$ and $SiSH^+$ cations that are alleged to be precursors of interstellar silicon sulfide, SiS. Electronic structure calculations allowed us to characterize the relevant features of the potential energy surface of the system and identify the reaction pathways. The reaction has two exothermic channels leading to the isomeric species $^3HSiS^+$ and $^3SiSH^+$ formed in conjunction with H atoms. The reaction is not characterized by an entrance barrier and, therefore, it is expected to be fast also under the very low temperature conditions of insterstellar clouds. The two ions are formed in their first electronically excited state because of the spin multiplicity of the overall potential energy surface. In addition, following the suggestion that neutral species are formed by proton transfer of protonated cations to ammonia, we have derived the potential energy surface for the reactions $^3HSiS^+/^3SiSH^++NH_3(^1A_1)$.

Keywords: Ab initio calculations · Astrochemistry · Silicon sulfide

1 Introduction

Silicon, the seventh most abundant element in the universe, is mostly trapped in the dust grains present in the interstellar medium (ISM) in the form of silicates. Nevertheless, an increasing number of silicon-bearing molecules, have been detected in different regions of the ISM, with a molecular size going up to 8 atoms [1]. The presence of silicon in the gas phase is mainly related to violent events, such as shocks, in which the high energetic material ejected from young stars impact a quiescent zone and allows the sputtering of the refractory core

O. Gervasi et al. (Eds.): ICCSA 2022 Workshops, LNCS 13378, pp. 233–245, 2022.
https://doi.org/10.1007/978-3-031-10562-3_17

of interstellar dust grains [2]. As soon as silicon is released in the gas phase, it can immediately react and it is mostly converted into SiO [3,4]. Therefore, the diatomic molecule SiO is the most abundant silicon-bearing species and it is considered a useful target to probe shock regions [2]. Formation routes of the SiO molecule are well characterized [5–7], while chemical processes leading to the formation of other Si-bearing molecules are still uncertain and poorly known.

Another interesting Si-bearing diatomic molecule is silicon sulfide, SiS, detected for the first time in 1975 by Morris et al. [8] towards the molecular envelope of the carbon-rich AGB star IRC+ 10216. From that first observation, silicon sulfide (SiS) has been detected sporadically, mostly in circumstellar envelopes around evolved stars [9,10] and towards high mass star-forming regions with shocks (such as Sgr B2 and Orion KL [4,11–14]). Recently, a census of silicon molecules has been performed by Podio et al. for L1157-B1, that is, a shocked region driven by a jet originated from a Sun-like protostar [2]. Together with the detection of SiO (and its isotopologues ^{29}SiO and ^{30}SiO), the authors reported the first detection of the SiS molecule in a low-mass star forming region. Interestingly, a surprisingly high abundance of SiS has been inferred in a well-localized region around the protostar, with a SiO/SiS abundance ratios decreasing from typical values of 200 to values as low as 25. Since the abundance of other sulfur-bearing species (e.g. SO) remains constant across the region, the reason for such diversity cannot be ascribed to a local overabundance of sulphur. The analysis of the spatial distribution shows a strong gradient across the shock (being SiS abundant at the head of the cavity and not detected at the shock impact region), suggesting a different chemical origin for SiO and SiS. The reason for such a peculiar diversity is unknown and represents a challenge to our comprehension of interstellar silicon chemistry.

Given the difficulties associated to the experimental investigation of the postulated reactions, theoretical quantum chemistry calculation appears to be pivotal to elucidate the chemistry of silicon-bearing species. Recently, some interesting analysis on the potential role of neutral gas-phase chemistry in the formation and destruction routes of SiS has been proposed. For instance, some of the present authors performed a theoretical characterization of the potential energy surface (PES) of the reactions SiH+S, SiH+S_2 and Si+HS [15,16], together with a kinetic analysis based on the capture theory and Rice-Ramsperger-Kassel-Marcus calculations [17]. In addition, Zanchet et al. [18] examined the reaction of atomic silicon with SO and SO_2 and used the derived rate coefficients to simulate SiS formation in the outflows of star-forming regions like Sgr B2, Orion KL, and L1157-B1. In particular, the PES of the SiOS system has been analyzed through ab initio calculations. Kinetics calculations on the derived PES lead to the conclusion that the two reactions can play a key role for the formation of SiS, while the reaction of silicon sulfide with atomic oxygen represents an important destruction pathway of SiS. Very recently, after the work by Rosi et al. [16], the reaction of the silicon atom with the SH radical has been re-investigated by Mota et al. [19]. Accurate multi-reference calculations were employed to derive the potential energy surface for the system and quasiclassical trajectory calculations were performed to derive the reaction rate coefficient. The inclusion of the data in a gas-grain astrochemical model of the L1157-B1 shock region, suggested a very important role for the SiS formation.

One last example revealing the pivotal role of neutral gas phase chemistry for the formation of interstellar silicon sulfide is represented by the work of R. Kaiser and coworkers, reporting a synergistic theoretical and experimental analysis of different reaction involving silicon species [20,21]. The combined crossed molecular beam and electronic structure calculations analysis provided new data on the formation of silicon sulfide. The synergy with astrochemical modeling allowed a better understanding of the role of neutral gas-phase chemistry toward star forming regions.

In spite of the recent advances summarized above, the two main databases for astrochemical models (KIDA and UMIST) [22,23] still report the electron-ion recombination of the HSiS$^+$ ion as the only significant process leading to the formation of SiS:

$$HSiS^+ + e^- \rightarrow H + SiS \tag{1}$$

However, in those databases most of the ion-molecule reactions considered as possible formation routes of the HSiS$^+$ cation are inefficient according to laboratory experiments [24,25].

As a part of our systematic investigation on interstellar SiS formation, we have now focused on possible ion-molecules reactions. In particular, here we report on the theoretical characterization of the PES for the reaction between the S$^+$ cation (in its ground state, ^{4}S) and the SiH$_2$(^{1}A$_1$) radical. Sulphur cations are easily formed because the sulphur ioniziation energy (10.36 eV) is lower than that of atomic or molecular hydrogen. The SiH$_2$ radicals can be formed by photodissociation or other high-energy processes involving SiH$_4$. This last species can be synthesized by successive hydrogenation processes of Si atoms on the surface of the interstellar dust grains [4,26] and subsequently released in the gas phase during shocks.

In addition to electron-ion recombination, other processes can be invoked for the formation of neutral molecules starting from a protonated charged species, that is, proton transfer to neutral molecules with a large proton affinity like ammonia. As proposed by Taquet et al. [27] proton transfer to ammonia can be considered a good alternative to electron-ion recombination. Therefore, we have also derived the PES for

$$HSiS^+/SiSH^+ + NH_3 \rightarrow SiS + NH_4^+ \tag{2}$$

With the present contribution we aim to add other new pieces to the complex puzzle of the chemistry of interstellar silicon.

2   Computational Details

The SiH$_2$(^{1}A$_1$)+S$^+$(^{4}S) and HSiS$^+$/SiSH$^+$+NH$_3$(^{1}A$_1$) reactions were analyzed adopting a computational strategy previously used with success in several cases [15,28–30]. A first investigation of the potential energy surface was performed through Density Functional Theory (DFT) calculations, using the Becke3-parameter exchange and Lee-Yang-Parr correlation (B3LYP) [31,32] hybrid functional, in conjunction with the correlation consistent valence polarized set aug-cc-pV(T+d)Z [33–35]. Harmonic vibrational frequencies were calculated at the same B3LYP/aug-cc-pV(T+d)Z level of theory, using the Hessian matrix (second derivatives) of the energy, in order to determine the nature of each stationary point, i.e. minimum if all the frequencies are real and saddle point if there is one, and only one, imaginary frequency. Intrinsic Reaction Coordinates (IRC) calculations [36,37] were performed to assign each identified stationary point to the corresponding reactants and products. Subsequently, the energy of each stationary point was computed with the more accurate coupled cluster theory including single and double excitations as well as a perturbative estimate of connected triples CCSD(T) [38–40], with the same basis set aug-cc-pV(T+d)Z. The zero-point energy, computed using the scaled harmonic vibrational frequencies obtained at the B3LYP/aug-ccpV(T+d)Z level of theory, was added to correct both energies [B3LYP and CCSD(T)] to 0 K. All calculations were carried out using GAUSSIAN 09 [41] and the vibrational analysis was performed using MOLDEN [42,43].

3 Results

3.1 The $SiH_2(^1A_1)$ + $S^+(^4S)$ Reaction

The quartet potential energy surface (PES) for the system $SiH_2(^1A_1)+S^+(^4S)$ shows three different minima (MIN1, MIN2, MIN3), linked by two transition states (TS1, linking MIN1 and MIN2; TS2 connecting MIN2 and MIN3). The reaction starts with the barrierless interaction of the S^+ cation with the SiH_2 radical, leading to the formation of the intermediate MIN1, located 280.6 kJ mol^{-1} below the reactant energy asymptote. MIN1 features a new Si-S bond with a bond distance of 2.351 Å. The so formed intermediate can directly dissociate, through a barrierless H-elimination process, leading to the formation of the $^3HSiS^+$ cation and an H atom. The global exothermicity of the process is -122.3 kJ mol^{-1}. Alternatively, MIN1 can isomerize to the MIN2 intermediate by overcoming a barrier of 93.1 kJ mol^{-1}. The related transition state, TS1, clearly shows the migration of a hydrogen atom from Si to S. Two possible combinations of products can be formed starting from the aforementioned MIN2. A barrier of 133.6 kJ mol^{-1} must be overcome in order to form the previously described H+$^3HSiS^+$ products. A transition state, TS3, was identified, showing an increase in the S-H bond distance up to 2.416Å. A different H elimination process can take place starting from MIN2, leading to the barrierless formation of the $^3SiSH^+$ cation and an H atom, located 90.1 kJ mol^{-1} below the reactant energy asymptote. The last isomerization process identified in the PES is related to the migration of a hydrogen atom from the Si atom to the S side of the intermediate MIN2, leading to the formation of MIN3, located 204.3 kJ mol^{-1} below the reactant energy asymptote, in which both H atoms are linked to the S atom of the adduct. A barrier of 99.3 kJ mol^{-1} must be overcome during the migration

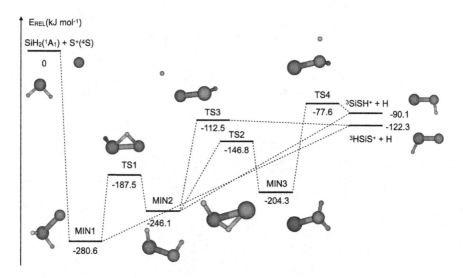

Fig. 1. Schematic representation of the Potential Energy Surface (PES) for the reaction SiH$_2$(^{1}A$_1$)+S$^+$(^{4}S) with the energies evaluated at the CCSD(T)/aug-cc-pV(T+d)Z level of theory.

process. The so formed MIN3 can directly dissociate to H+^{3}SiSH$^+$, overcoming a barrier of 126.7 kJ mol^{-1}. The related transition state, TS4, clearly shows the breaking of a S-H bond, with a bond distance of 2.322Å. A schematic representation of the PES derived for the SiH$_2$(^{1}A$_1$) + S$^+$(^{4}S) reaction is reported in Fig. 1, while in Figs. 2, 3, 4 and 5 the geometry (bond distances in Å and angles in degree) of the stationary points are illustrated, together with those of reactants and products (at the B3LYP/aug-ccpV(T+d)Z level of theory). A list of enthalpy changes and barrier heights (in kJ mol^{-1}, corrected at 0 K) computed at the CCSD(T)/aug-cc-pV(T+d)Z and B3LYP/aug-cc-pV(T+d)Z levels of theory is reported in Table 1.

It is important to stress that, given the multiplicity of the investigated PES, the ^{3}HSiS$^+$/^{3}SiSH$^+$ products are not formed in their ground singlet states, but in the first electronically excited triplet states.

3.2 The ^{3}HSiS$^+$/^{3}SiSH$^+$+NH$_3$(^{1}A$_1$) Reaction

The barrierless ^{3}HSiS$^+$ + NH$_3$(^{1}A$_1$) reaction shows a global exothermicity of -50.4 kJ mol^{-1} at the CCSD(T)/aug-cc-pV(T+d)Z level of theory. The interaction between the HSiS$^+$ cation and ammonia leads to the formation of the MIN4 intermediate, more stable than the reactants by 121.3 kJ mol^{-1}. Once formed, the intermediate can dissociate into silicon monosulfide (^{3}SiS) and ammonium ion. A similar mechanism was derived for the reaction ^{3}SiSH$^+$ + NH$_3$(^{1}A$_1$). In this case the global exothermicity is 82.3 kJ mol^{-1}, while the intermediate MIN5 shows a relative energy of -153.7 kJ mol^{-1} with respect to the reac-

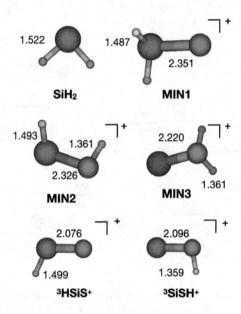

Fig. 2. Bond distances (reported in Angstroms) for the reactant, products and minima identified for the $SiH_2(^1A_1)$ + $S^+(^4S)$ reaction at the B3LYP/aug-cc-pV(T+d)Z level of theory.

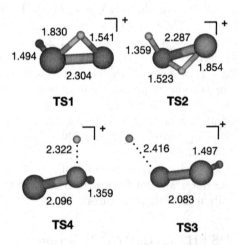

Fig. 3. Bond distances (reported in Angstroms) for the main transition states of the reaction $SiH_2(^1A_1)$ + $S^+(^4S)$, evaluated at the B3LYP/aug-cc-pV(T+d)Z level of theory.

tant energy asymptote. Both the reactions take place without the presence of entrance and/or exit barriers. The schematic PES of the two processes are reported in Fig. 6, while the optimized geometries of the most relevant stationary points are reported in Fig. 7. A list of enthalpy changes and barrier heights (in

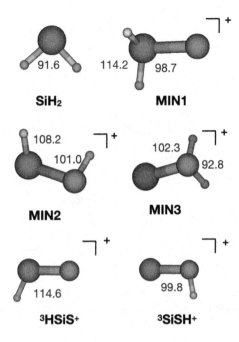

Fig. 4. Main angles (reported in degree) for the reactant, products and minima identified for the SiH$_2$(^{1}A$_1$) + S$^+$(^{4}S) reaction at the B3LYP/aug-cc-pV(T+d)Z level of theory.

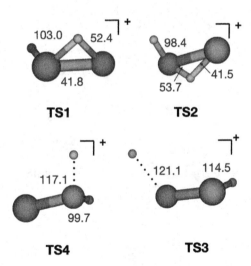

Fig. 5. Main angles (reported in degree) for the main transition states of the reaction SiH$_2$(^{1}A$_1$) + S$^+$(^{4}S), evaluated at the B3LYP/aug-cc-pV(T+d)Z level of theory.

kJ mol^{-1}, corrected at 0 K) computed at the CCSD(T)/aug-cc-pV(T+d)Z and B3LYP/aug-cc-pV(T+d)Z levels of theory is reported in Table 1.

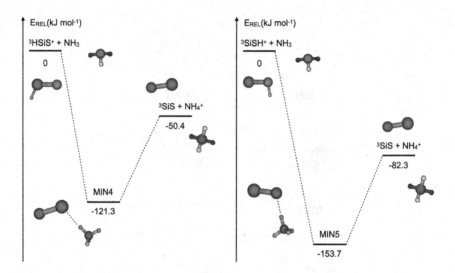

Fig. 6. Schematic representation of the Potential Energy Surface (PES) for the proton transfer processes ^{3}HSiS$^+$/^{3}SiSH$^+$+NH$_3$(^{1}A$_1$) with the energies evaluated at the CCSD(T)/aug-cc-pV(T+d)Z level of theory.

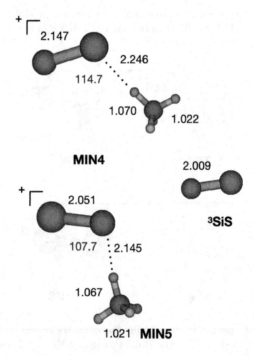

Fig. 7. Bond distances (in Angstroms) and angles (in degrees) for the main stationary points identified for the ^{3}HSiS$^+$/^{3}SiSH$^+$+NH$_3$(^{1}A$_1$) reactions at the B3LYP/aug-cc-pV(T+d)Z level of theory.

Table 1. Enthalpy changes and barrier heights (kJ mol^{-1}, 0 K) computed at the CCSD(T)/aug-cc-pV(T+d)Z and B3LYP/aug-cc-pV(T+d)Z levels of theory for the system SiH$_2$(^{1}A$_1$) + S$^+$(^{4}S).

Reaction path	CCSD(T)		B3LYP	
	ΔH_0^0	Barrier heights	ΔH_0^0	Barrier heights
SiH$_2$ + S$^+$ → MIN1	−280.6	–	−320.4	–
MIN1 → MIN2	34.5	93.1	32.2	83.9
MIN2 → MIN3	41.8	99.3	62.9	109.1
MIN1 → H + ^{3}HSiS$^+$	158.3	–	150.0	–
MIN2 → H + ^{3}HSiS$^+$	123.8	133.6	117.8	120.7
MIN2 → H + ^{3}SiSH$^+$	156.0	–	157.8	–
MIN3 → H + ^{3}SiSH$^+$	114.2	126.7	94.9	101.2
^{3}HSiS$^+$ + NH$_3$ → MIN4	−121.3	–	−116.4	–
MIN4 → ^{3}SiS + NH$_4^+$	70.9	–	80.9	–
^{3}SiSH$^+$ + NH$_3$ → MIN5	−153.7	–	−147.1	–
MIN5 → ^{3}SiS + NH$_4^+$	71.0	–	71.7	–

4   Discussion and Conclusions

In this work, we have reported the results of a theoretical investigation of the reaction between the S$^+$ cation and the SiH$_2$(^{1}A$_1$) radical. The reaction has two exothermic channels leading to the isomeric species ^{3}HSiS$^+$ and ^{3}SiSH$^+$ formed in conjunction with H atoms. The reaction is not characterized by an entrance barrier and, therefore, it is expected to be fast also under the very low temperature conditions of insterstellar clouds. The two ions are formed in their first electronically excited state because of the spin multiplicity of the overall PES. A subsequent proton transfer process with ammonia can well account for the formation of neutral triplet silicon sulfide, SiS, without invoking electron-ion recombination.

As already noted above, the PES derived for the reaction between the SiH$_2$(^{1}A$_1$) radical and the S$^+$ cation (in its ground state ^{4}S) can only lead to the formation of the two aforementioned radicals in the triplet state, ^{3}HSiS$^+$/^{3}SiSH$^+$. The successive proton transfer carried out by an ammonia molecule (whose ground electronic state is ^{1}A$_1$) leads to a triplet potential energy surface. As a consequence, according to the 'correlation rules', the title reaction can only form silicon sulfide in a triplet spin state, ^{3}SiS. However, in systems of this kind, intersystem crossing from the triplet to the singlet PES is certainly possible, being also facilitated by the presence of atoms belonging to elements of the third period of the Periodic Table. In particular, once the first intermediate (MIN1) in the ^{3}PES for the S$^+$+ SiH$_2$ reaction is formed, an intersystem crossing mechanism can allow reaching the singlet PES. Starting from this point both the ^{1}HSiS$^+$/^{1}SiSH$^+$ cations can be formed and, consequently, the neutral

[1]SiS molecule could be produced. To account for such an effect, it is necessary to derive the singlet PES (these calculations are currently under way) and to characterize the seam of crossing. Future work from our group will assess the role of intersystem crossing for this system, as already done for a series of reactions involving atomic oxygen [44–53].

As already mentioned in the 'Introduction' section, in the last years different theoretical investigation have been performed to access the viability of gas phase reactions for the formation of interstellar SiS, including a systematic study undertaken by some of the present authors to characterize the possible formation routes of SiS in low-mass star forming regions via neutral-neutral gas-phase processes [15–17]. With this contribution we plan to add useful information to the complex network of gas phase reactions involving Si-bearing molecules. The sputtering of dust grains in shocked regions leads to the release of silicon in the gas phase. Therefore, if we demonstrate that SiS is totally formed by neutral-neutral and/or ion-neutral gas-phase reactions, its presence and distribution in interstellar objects could become a kind of signpost for that type of chemistry being dominant in the above mentioned regions.

Acknowledgements. This project has received funding from the Italian MUR (PRIN 2020, "Astrochemistry beyond the second period elements", Prot. 2020AFB3FX) and from the European Union's Horizon 2020 research and innovation programme under the Marie Skłodowska-Curie grant agreement No 811312 for the project 'Astro-Chemical Origins' (ACO). The authors thank the Herla Project - Università degli Studi di Perugia (http://www.hpc.unipg.it/hosting/vherla/vherla.html) for allocated computing time. The authors thank the Dipartimento di Ingegneria Civile ed Ambientale of the University of Perugia for allocated computing time within the project "Dipartimenti di Eccellenza 2018-2022".

References

1. The Cologne Database for Molecular spectroscopy. https://cdms.astro.uni-koeln. de/classic/molecules
2. Podio, L., et al.: Silicon-bearing molecules in the shock L1157-B1: first detection of SiS around a Sun-like protostar. Mon. Not. R. Astron. Soc. Lett. **470**(1), L16–L20 (2017)
3. Herbst, E., Millar, T., Wlodek, S., Bohme, D.: The chemistry of silicon in dense interstellar clouds. Astron. Astrophys. **222**, 205–210 (1989)
4. MacKay, D.: The chemistry of silicon in hot molecular cores. Mon. Not. R. Astron. Soc. **274**(3), 694–700 (1995)
5. Schilke, P., Walmsley, C., Pineau des Forets, G., Flower, D.: SiO production in interstellar shocks. Astron. Astrophys. **321**, 293–304 (1997)
6. Le Picard, S.D., Canosa, A., Rebrion-Rowe, C., Rowe, B.: The Si (^3P_J)+ O_2 reaction: a fast source of SiO at very low temperature; CRESU measurements and interstellar consequences. Astron. Astrophys. **372**(3), 1064–1070 (2001). https://doi.org/10.1051/0004-6361:20010542
7. Gusdorf, A., Des Forêts, G.P., Cabrit, S., Flower, D.: SiO line emission from interstellar jets and outflows: silicon-containing mantles and non-stationary shock waves. Astron. Astrophys. **490**(2), 695–706 (2008)

8. Morris, M., Gilmore, W., Palmer, P., Turner, B., Zuckerman, B.: Detection of interstellar SiS and a study of the IRC+ 10216 molecular envelope. Astrophys. J. **199**, L47–L51 (1975)

9. Cernicharo, J., Guélin, M., Kahane, C.: A λ2 mm molecular line survey of the C-star envelope IRC+ 10216. Astron. Astrophys. Suppl. Ser. **142**(2), 181–215 (2000)

10. Prieto, L.V., et al.: Si-bearing molecules toward IRC+ 10216: ALMA unveils the molecular envelope of CWLeo. Astrophys. J. Lett. **805**(2), L13 (2015)

11. Dickinson, D., Kuiper, E.R.: Interstellar silicon sulfide. Astrophys. J. **247**, 112–115 (1981)

12. Ziurys, L.M.: SiS in Orion-KL-evidence for 'outflow' chemistry. Astrophys. J. **324**, 544–552 (1988)

13. Ziurys, L.: SiS in outflow regions-more high-temperature silicon chemistry. Astrophys. J. **379**, 260–266 (1991)

14. Tercero, B., Vincent, L., Cernicharo, J., Viti, S., Marcelino, N.: A line-confusion limited millimeter survey of Orion KL-II. Silicon-bearing species. Astron. Astrophys. **528**, A26 (2011)

15. Rosi, M., et al.: Possible scenarios for SiS formation in the interstellar medium: electronic structure calculations of the potential energy surfaces for the reactions of the SiH radical with atomic sulphur and S2. Chem. Phys. Lett. **695**, 87–93 (2018)

16. Rosi, M., et al.: Electronic structure and kinetics calculations for the Si+SH reaction, a possible route of SiS formation in star-forming regions. In: Misra, S., et al. (eds.) ICCSA 2019. LNCS, vol. 11621, pp. 306–315. Springer, Cham (2019). https://doi.org/10.1007/978-3-030-24302-9_22

17. Skouteris, D., et al.: A theoretical investigation of the reaction H+SiS2 and implications for the chemistry of silicon in the interstellar medium. In: Gervasi, O., et al. (eds.) ICCSA 2018. LNCS, vol. 10961, pp. 719–729. Springer, Cham (2018). https://doi.org/10.1007/978-3-319-95165-2_50

18. Zanchet, A., Roncero, O., Agúndez, M., Cernicharo, J.: Formation and destruction of SiS in space. Astrophys. J. **862**(1), 38 (2018)

19. Mota, V., Varandas, A., Mendoza, E., Wakelam, V., Galvão, B.: SiS formation in the interstellar medium through Si+SH gas-phase reactions. Astrophys. J. **920**(1), 37 (2021)

20. Doddipatla, S., He, C., Goettl, S.J., Kaiser, R.I., Galvão, B.R., Millar, T.J.: Nonadiabatic reaction dynamics to silicon monosulfide (SiS): a key molecular building block to sulfur-rich interstellar grains. Sci. Adv. **7**(26), eabg7003 (2021)

21. Goettl, S.J., et al.: A crossed molecular beams and computational study of the formation of the astronomically elusive thiosilaformyl radical (HSiS, X^2A'). J. Phys. Chem. Lett. **12**(25), 5979–5986 (2021)

22. Wakelam, V., et al.: The 2014 KIDA network for interstellar chemistry. Astrophys. J. Suppl. Ser. **217**(2), 20 (2015)

23. McElroy, D., Walsh, C., Markwick, A., Cordiner, M., Smith, K., Millar, T.: The UMIST database for astrochemistry 2012. Astron. Astrophys. **550**, A36 (2013)

24. Wlodek, S., Bohme, D.K.: Gas-phase oxidation and sulphidation of $Si^+(^2P)$, SiO^+ and SiS^+. J. Chem. Soc. Faraday Trans. 2 Mol. Chem. Phys. **85**(10), 1643–1654 (1989)

25. Wlodek, S., Fox, A., Bohme, D.: Gas-phase reactions of Si^+ and $SiOH^+$ with molecules containing hydroxyl groups: possible ion-molecule reaction pathways toward silicon monoxide, silanoic acid, and trihydroxy-, trimethoxy-, and triethoxysilane. J. Am. Chem. Soc. **109**(22), 6663–6667 (1987)

26. Ceccarelli, C., Viti, S., Balucani, N., Taquet, V.: The evolution of grain mantles and silicate dust growth at high redshift. Mon. Not. R. Astron. Soc. **476**(1), 1371–1383 (2018)

27. Taquet, V., Wirström, E.S., Charnley, S.B.: Formation and recondensation of complex organic molecules during protostellar luminosity outbursts. Astrophys. J. **821**(1), 46 (2016)

28. Skouteris, D., et al.: Interstellar dimethyl ether gas-phase formation: a quantum chemistry and kinetics study. Mon. Not. R. Astron. Soc. **482**(3), 3567–3575 (2019)

29. Mancini, L., et al.: The reaction $N(^2D)+CH_3CCH$ (methylacetylene): a combined crossed molecular beams and theoretical investigation and implications for the atmosphere of titan. J. Phys. Chem. A **125**(40), 8846–8859 (2021)

30. Balucani, N., Skouteris, D., Ceccarelli, C., Codella, C., Falcinelli, S., Rosi, M.: A theoretical investigation of the reaction between the amidogen, NH, and the ethyl, C_2H_5, radicals: a possible gas-phase formation route of interstellar and planetary ethanimine. Mol. Astrophys. **13**, 30–37 (2018)

31. Becke, A.D.: A new mixing of Hartree-Fock and local density-functional theories. J. Chem. Phys. **98**(2), 1372–1377 (1993)

32. Stephens, P.J., Devlin, F.J., Chabalowski, C.F., Frisch, M.J.: Ab initio calculation of vibrational absorption and circular dichroism spectra using density functional force fields. J. Phys. Chem. **98**(45), 11623–11627 (1994)

33. Dunning, T.H., Jr.: Gaussian basis sets for use in correlated molecular calculations. I. The atoms boron through neon and hydrogen. J. Chem. Phys. **90**(2), 1007–1023 (1989)

34. Woon, D.E., Dunning, T.H., Jr.: Gaussian basis sets for use in correlated molecular calculations. III. The atoms aluminum through argon. J. Chem. Phys. **98**(2), 1358–1371 (1993)

35. Kendall, R.A., Dunning, T.H., Jr., Harrison, R.J.: Electron affinities of the first-row atoms revisited. Systematic basis sets and wave functions. J. Chem. Phys. **96**(9), 6796–6806 (1992)

36. Gonzalez, C., Schlegel, H.B.: An improved algorithm for reaction path following. J. Chem. Phys. **90**(4), 2154–2161 (1989). https://doi.org/10.1063/1.456010

37. Gonzalez, C., Schlegel, H.B.: Reaction path following in mass-weighted internal coordinates. J. Phys. Chem. **94**(14), 5523–5527 (1990). https://doi.org/10.1021/j100377a021

38. Bartlett, R.J.: Many-body perturbation theory and coupled cluster theory for electron correlation in molecules. Annu. Rev. Phys. Chem. **32**(1), 359–401 (1981)

39. Raghavachari, K., Trucks, G.W., Pople, J.A., Head-Gordon, M.: A fifth-order perturbation comparison of electron correlation theories. Chem. Phys. Lett. **157**(6), 479–483 (1989)

40. Olsen, J., Jo/rgensen, P., Koch, H., Balkova, A., Bartlett, R.J.: Full configuration-interaction and state of the art correlation calculations on water in a valence double-zeta basis with polarization functions. J. Chem. Phys. **104**(20), 8007–8015 (1996)

41. Frisch, M., et al.: Gaussian 09, rev. A. 02. Gaussian. Inc., Wallingford, CT (2009)

42. Schaftenaar, G., Noordik, J.H.: Molden: a pre-and post-processing program for molecular and electronic structures. J. Comput. Aided Mol. Des. **14**(2), 123–134 (2000). https://doi.org/10.1023/A:1008193805436

43. Schaftenaar, G., Vlieg, E., Vriend, G.: Molden 2.0: quantum chemistry meets proteins. J. Comput.-Aided Mol. Des. **31**(9), 789–800 (2017)

44. Leonori, F., et al.: Experimental and theoretical studies on the dynamics of the O(^{3}P)+ propene reaction: primary products, branching ratios, and role of inter-system crossing. J. Phys. Chem. C **119**(26), 14632–14652 (2015)

45. Gimondi, I., Cavallotti, C., Vanuzzo, G., Balucani, N., Casavecchia, P.: Reaction dynamics of O(^{3}P)+ propyne: II. Primary products, branching ratios, and role of intersystem crossing from ab initio coupled triplet/singlet potential energy surfaces and statistical calculations. J. Phys. Chem. A **120**(27), 4619–4633 (2016)

46. Vanuzzo, G., et al.: Reaction dynamics of O(^{3}P)+ propyne: I. Primary products, branching ratios, and role of intersystem crossing from crossed molecular beam experiments. J. Phys. Chem. A **120**(27), 4603–4618 (2016)

47. Caracciolo, A., et al.: Combined experimental and theoretical studies of the O(^{3}P)+ 1-butene reaction dynamics: primary products, branching fractions, and role of intersystem crossing. J. Phys. Chem. A **123**(46), 9934–9956 (2019)

48. Cavallotti, C., et al.: Theoretical study of the extent of intersystem crossing in the O(^{3}P)+ C$_6$H$_6$ reaction with experimental validation. J. Phys. Chem. Lett. **11**(22), 9621–9628 (2020)

49. Vanuzzo, G., et al.: Crossed-beam and theoretical studies of the O(^{3}P, ^{1}D)+ ben-zene reactions: primary products, branching fractions, and role of intersystem cross-ing. J. Phys. Chem. A **125**(38), 8434–8453 (2021)

50. Fu, B., et al.: Experimental and theoretical studies of the O(^{3}P)+ C$_2$H$_4$ reaction dynamics: collision energy dependence of branching ratios and extent of intersystem crossing. J. Chem. Phys. **137**(22), 22A532 (2012)

51. Balucani, N., Leonori, F., Casavecchia, P., Fu, B., Bowman, J.M.: Crossed molecu-lar beams and quasiclassical trajectory surface hopping studies of the multichannel nonadiabatic O(^{3}P)+ ethylene reaction at high collision energy. J. Phys. Chem. A **119**(50), 12498–12511 (2015)

52. Leonori, F., Occhiogrosso, A., Balucani, N., Bucci, A., Petrucci, R., Casavecchia, P.: Crossed molecular beam dynamics studies of the O(^{3}P)+ allene reaction: primary products, branching ratios, and dominant role of intersystem crossing. J. Phys. Chem. Lett. **3**(1), 75–80 (2012)

53. Casavecchia, P., Leonori, F., Balucani, N.: Reaction dynamics of oxygen atoms with unsaturated hydrocarbons from crossed molecular beam studies: primary products, branching ratios and role of intersystem crossing. Int. Rev. Phys. Chem. **34**(2), 161–204 (2015)

A Theoretical Investigation
of the Reactions of N(^{2}D) and CN
with Acrylonitrile and Implications
for the Prebiotic Chemistry of Titan

Luca Mancini[1(✉)] , Emília Valença Ferreira de Aragão[1,2] ,
and Gianmarco Vanuzzo[1]

[1] Dipartimento di Chimica, Biologia e Biotecnologie, Università degli Studi di
Perugia, 06123 Perugia, Italy
luca.mancini2@studenti.unipg.it
[2] Master-tec srl, Via Sicilia 41, 06128 Perugia, Italy

Abstract. The reactions between acrylonitrile and two different reactive species, namely N(^{2}D) and the CN radical were investigated by performing accurate electronic structure calculations with the aim to unveil the most important aspects of the Potential Energy Surfaces. For each reaction, several product channels involving the elimination of H atoms were identified, allowing the formation of different radical species, depending on the initial site of attack. Both reactions appears to be exothermic and without an entrance barrier, suggesting their possible efficient role in the nitrogen-rich chemistry of the atmosphere of Titan.

Keywords: Ab initio calculations · Astrochemistry · Planetary
atmosphere · Titan

1 Introduction

Chemical reactions involving molecular species containing nitrogen appears to be pivotal in several extraterrestrial environments, including the atmosphere of Titan [1–3]. After the appearance of life, the geomorphology and the characteristic of our planet have undergone massive modifications. As a consequence, in order to have a clear picture of the first stages of our prebiotic chemistry, the study of planets and moons, similar to the primitive Earth, can be of great help. In particular, the main focus of astrobiological studies is Titan, the largest moon of Saturn, as well as one of the few moons of the Solar System, which possess a thick atmosphere. Interesting data about Titan and its unique atmosphere became available with different exploratory missions, including the two Voyager missions [4–7] and, much later, the fascinating Cassini-Huygens mission [8]. According to the different analysis, the main component of Titan's atmosphere appears to be molecular nitrogen (N_2) [6], together with other species, present in

O. Gervasi et al. (Eds.): ICCSA 2022 Workshops, LNCS 13378, pp. 246–259, 2022.
https://doi.org/10.1007/978-3-031-10562-3_18

lower amount, including methane, the second most abundant molecule, and several hydrocarbons (e.g. ethane, ethylene, acetylene, etc.) [5]. The best strategy for the understanding of the chemistry of Titan's atmosphere is characterized by a multidisciplinary approach, which culminates in accurate photochemical models [9–11] that include physical and chemical parameters, derived through theoretical and/or experimental investigations, allowing to obtain the rate constants for the most relevant elementary reactions [12]. Considering the high stability of the N$_2$ molecule, one of the initial steps for the chemistry of N-bearing molecules is considered to be the impact of molecular nitrogen with high energy free electrons, leading to the formation of the "active nitrogen", including nitrogen atoms in their ground and excited states N(^{4}S) and N(^{2}D), respectively. The first electronically excited state of atomic nitrogen, N(^{2}D), is metastable with a relatively long radiative lifetime [13]. It is 230.0 kJ/mol higher than the ground state and appears to be very reactive [14,15]. The second excited metastable state of atomic nitrogen, N(^{2}P), can be present simultaneously with N(^{2}D). Nevertheless, its reactivity appears to be lower and it can be easily converted into N(^{2}D) [13]. Several theoretical and experimental studies have been performed to study the reactivity of N(^{2}D) with different hydrocarbons, with possible applications in the atmosphere of Titan (namely, N(^{2}D)+ C$_2$H$_2$; C$_2$H$_4$; C$_2$H$_6$; C$_6$H$_6$; CH$_3$CCH; HC$_3$N; C$_7$H$_8$) [16–26]. Another abundant reactive species in the atmosphere of Titan appears to be the cyano (CN) radical. According to the last refinements of photochemical models, the main formation pathway for the CN radical is the photodissociation of HCN, which is very abundant [27]. The aforementioned process appears to be particularly efficient between 800 km and 1000 km of altitude [28]. In the same regions, namely at altitudes above 600 km [13],the interaction between EUV photons and N$_2$ leads to the formation of N(^{2}D) with high efficiency. A particularly important N-containing molecule detected on Titan's atmosphere appears to be acrylonitrile (CH$_2$CHCN). In the last years this molecule received a great deal of attention because it was found to be the best candidate for the formation of membrane-like structures in apolar solvents [29], such as the methane-rich lakes, peculiars on Titan's surface. The presence of acrylonitrile was initially inferred by the detection of the CH$_2$CHCNH$^+$ cation using the INMS (Ion and Neutral Mass Spectrometer) [30–33] onboard the Cassini orbiter, while the definitive proof of the presence of neutral CH$_2$CHCN came from the analysis of the data coming from the ALMA facilities (Atacama Large Millimeter/Submillimeter Array) [34]. A follow-up study using higher sensitivity data from the ALMA archive, presented the very first spatially resolved map of the distribution of acrylonitrile in Titan's atmosphere [35]. The main formation mechanism for acrylonitrile is the reaction between the cyano radical and the ethylene molecule (C$_2$H$_4$). Different theoretical and experimental investigations on this reaction [36–43] revealed a rate coefficient of the order of 10^{-10} cm^3 molec^{-1} s^{-1} and demonstrated that the main reaction pathway is related to the elimination of hydrogen atoms, with the subsequent formation of CH_2CHCN. Not much is known about the possible destruction mechanisms of acrylonitrile. The main loss mechanisms are indeed a proton transfer process from $HCNH^+$ (or alternatively $C_2H_5^+$), followed by electron recombination of HC_3NH^+ [44,45]. According to the previously reported photochemical models, the reaction of $N(^2D)$ and CN radical with the CH_2CHCN molecule could represent a feasible destruction mechanism on Titan's atmosphere. Unfortunately the only data available in the models for the title reactions derive from extrapolation processes [12,27,28,46] or comes from analogies with similar systems [28], without considering the presence of different isomers. In the present contribution we report a theoretical characterization of the main reaction pathways for the $N(^2D)+CH_2CHCN$ and $CN+CH_2CHCN$ reactions, with particular attention on the different mechanism for the initial bimolecular attack.

2   Computational Methods

The overall doublet Potential Energy Surface (PES) for the two $N(^2D)+$ CH_2CHCN and $CN+CH_2CHCN$ reactions was investigated adopting a computational strategy previously used with success in several cases [16,17,23,47–49]. A first exploration of the Potential Energy Surface was performed using of Density Functional Theory (DFT) calculations, with the B3LYP [50,51] functional, together with with the correlation consistent valence polarized set aug-cc-pVTZ [52–54]. The same level of theory was used for the analysis of harmonic vibrational frequencies, in order to determine the nature of each stationary point, i.e. minimum if all the frequencies are real and saddle point if there is one, and only one, imaginary frequency. Intrinsic Reaction Coordinates (IRC) calculations [55,56] were performed to assign each identified stationary point to the corresponding reactants and products. The calculation of the harmonic vibrational frequencies allowed us to derive the zero-point energy, to correct the energy to $0\,K$. In order to obtain a better estimation of the energy of $N(^2D)$, we added 230 kJ/mol (related to the experimental separation [57] $N(^4S)$-$N(^2D)$) to the calculated energy of $N(^4S)$. All calculations were carried out using GAUSSIAN 09 [58], while the graphical analysis of the calculated stationary points was performed using MOLDEN [59,60].

3   Results

3.1   The $CN+CH_2CHCN$ Reaction

The first step of the $CN+CH_2CHCN$ reaction is the addition of CN to the terminal carbon of the CH_2CHCN molecule, leading, in a barrierless process, to the formation of the MIN1 intermediate. The global exothermicity of the process

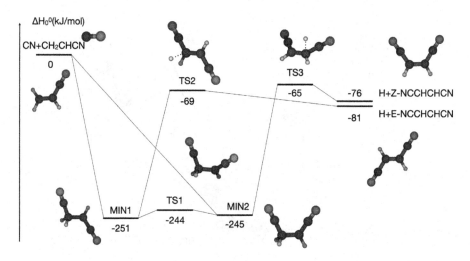

Fig. 1. Two-dimensional representation of the Potential Energy Surface for the reaction CN+CH$_2$CHCN with the energies evaluated using the B3LYP functional and the aug-cc-pVTZ basis set.

is 251 kJ/mol. Once formed, MIN1 can directly undergo a H-elimination process, forming the E-NCCHCHCN radical. The global exothermicity of the process is 81 kJ/mol, while a barrier of 182 kJ/mol is present in the product channel. The frequency analysis of the transition state, TS2, shows the breaking of a C-H σ bond, with a bond distance of 2.092 Å. Alternatively, the aforementioned MIN1 can isomerize to MIN2, through a rotation around the central C-C bond. A small barrier of 7 kJ/mol is represented by the TS1 transition state. Once formed, MIN2 can lead to the formation of atomic hydrogen, together with the Z-NCCHCHCN radical, located 76 kJ/mol under the energy of the CN+CH$_2$CHCN reagents. A transition state, TS3, located 65 kJ/mol below the energy of the reactants, was identified for the H-elimination process. A schematic diagram, representing the PES derived for the CN+CH$_2$CHCN reaction is reported in Fig. 1, while in Fig. 2 are reported the geometry (bond distances in Å) of the most important stationary points, including the CH$_2$CHCN reactant, evaluated through the previous described DFT calculations. A list of enthalpy changes and barrier heights (in kJ/mol, corrected at 0 K) computed at the same level of theory is reported in Table 1.

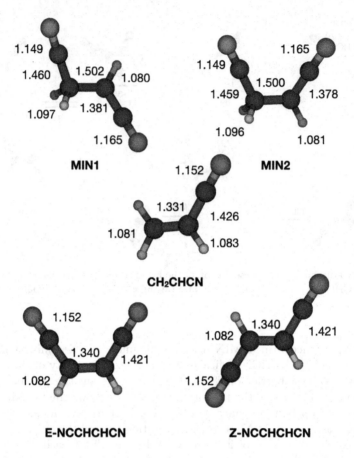

Fig. 2. Bond distances (reported in Angstroms) for some of the most important stationary points identified in the PES for the CN+CH$_2$CHCN reaction.

3.2   The N(^{2}D)+CH$_2$CHCN Reaction

The potential energy surface for the N(^{2}D)+CH$_2$CHCN reaction shows four different minima (MIN1, MIN2, MIN3, and MIN4) linked by three possible transition states (TS4 connecting MIN1 and MIN4; TS1 connecting MIN1 and MIN3, and TS5, connecting MIN3 and MIN2). The first step of the reaction is represented by the barrierless attack of N(^{2}D) to the C$_1$=C$_2$ double bond of the acrylonitrile molecule, leading to the formation of the cyclic intermediate MIN1, located 444 kJ/mol below the reactant energy asymptote. Once formed, MIN1 can directly undergo a H-elimination process. In details, the elimination of a hydrogen atom from the terminal (C$_1$) carbon atom leads to the formation of the cyclic c-CH(N)CHCN co-fragment, located 255 kJ/mol below the reactant energy asymptote through a barrier of 199 kJ/mol. Alternatively, the cyclic c-CH$_2$(N)CCN cofragment can be formed through the fission of the C$_2$-H σ bond, overcoming a barrier of 215 kJ/mol. The global exothermicity of the process is

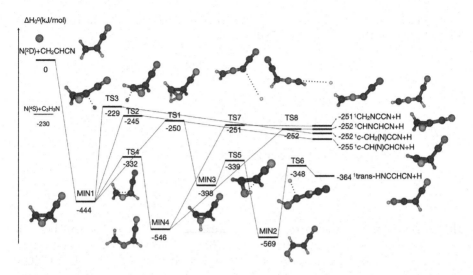

Fig. 3. Two-dimensional representation of the Potential Energy Surface for the reaction N(^{2}D)+CH$_2$CHCNN with the energies evaluated using the B3LYP functional and the aug-cc-pVTZ basis set.

252 kJ/mol. A different fate for the MIN1 intermediate is represented by isomerization processes. A migration of an H atom leads to the formation of the MIN3 intermediate, through the TS1 transition state. Once formed, the MIN3 intermediate, located 398 kJ/mol below the reactant energy asymptote, can directly isomerize leading to the formation of the MIN2 intermediate, in which the nitrogen atom is inserted in the C$_1$-H bond of the acrylonitrile molecule. The global exothermicity of the process is 171 kJ/mol, while a barrier of 59 kJ/mol must be overcome. The related transition state, TS5, located 339 kJ/mol below the energy of the reactants, clearly shows the breaking of the C$_2$-N bond. The elimination of a hydrogen atom from the new formed MIN2 intermediate leads to the formation of the linear co-fragment trans-HNCCHCN, with a relative energy of −364 kJ/mol. A different isomerization process identified starting from the aforementioned MIN1 intermediate is characterized by a ring opening process. The breaking of the C$_1$-C$_2$ σ bond of MIN1, indeed, leads to the formation of the linear intermediate MIN4, showing a relative energy of −546 kJ/mol. A barrier of 112 kJ/mol must be overcome during the process, while the related transition state, TS4, clearly shows the breaking of the C$_1$-C$_2$ σ bond, with a bond distance of 1.905 Å. Finally two possible H-elimination processes can take place starting from MIN4, leading to the formation of two different combination of products, H+CHNCHCN and H+CH$_2$NCCN, located 252 kJ/mol and 251 kJ/mol below the reactant energy asymptote, respectively, depending on whether the H-elimination process takes place from the terminal C$_1$ carbon or from the internal C$_2$ carbon. Two different transition states were located in the PES for the H-elimination processes, representing a barrier almost equal to

Table 1. List of Enthalpy variations and computed energy barriers (reported in kJ/mol, 0 K) for the $CN+CH_2CHCN$ reaction.

Reaction path	B3LYP/aug-cc-pVTZ	
	ΔH_0^0	Barrier heights
$CN+CH_2CHCN \rightarrow MIN1$	−251	–
$CN+CH_2CHCN \rightarrow MIN2$	−245	–
$MIN1 \rightarrow MIN2$	6	7
$MIN1 \rightarrow H + E\text{-}NCCHCHCN$	170	182
$MIN1 \rightarrow H + Z\text{-}NCCHCHCN$	169	180

the energy of the products, related to a weak C-H interaction as can be seen from the bond distances of 4.599 Å and 4.731 Å respectively. A schematic representation of the Potential Energy Surface derived for the $N(^2D)+CH_2CHCN$ reaction is reported in Fig. 3, while in Figs. 4, 5 and 6 are reported the geometry (bond distances in Å) of the most important stationary points evaluated at the B3LYP/aug-cc-pVTZ level of theory. A list of enthalpy changes and barrier heights (in kJ/mol, corrected at 0 K) computed at the B3LYP/aug-cc-pVTZ level of theory is reported in Table 2.

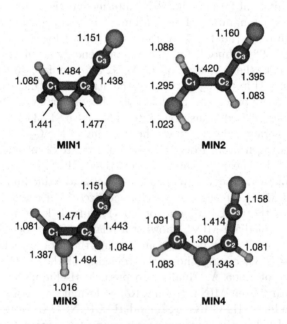

Fig. 4. Bond distances (reported in Angstroms) for the minimum points identified in the PES for the $N(^2D)+CH_2CHCN$ reaction.

Table 2. List of Enthalpy variations and computed energy barriers (reported in kJ/mol, 0 K) for the $N(^2D)+CH_2CHCN$ reaction.

Reaction path	B3LYP/aug-cc-pVTZ	
	ΔH_0^0	Barrier heights
$N(^2D)+CH_2CHCN \rightarrow$ MIN1	-444	$-$
MIN1 $\rightarrow$ MIN4	-102	112
MIN1 $\rightarrow$ MIN3	46	194
MIN1 $\rightarrow$ H $+ {}^1c\text{-}CH_2(N)CCN$	192	215
MIN1 $\rightarrow$ H $+ {}^1c\text{-}CH(N)CHCN$	189	199
MIN4 $\rightarrow$ H $+ {}^1CH_2NCCN$	295	295
MIN4 $\rightarrow$ H $+ {}^1CHNCHCN$	294	294
MIN3 $\rightarrow$ MIN2	-171	59
MIN2 $\rightarrow$ H $+ {}^1trans\text{-}HNCCHCN$	205	221

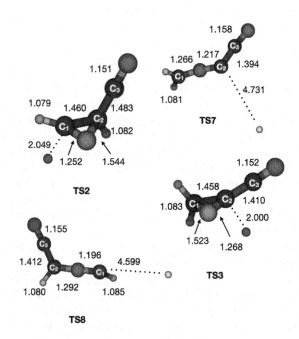

Fig. 5. Bond distances (reported in Angstroms) for transition states identified in the PES for the $N(^2D)+CH_2CHCN$ reaction.

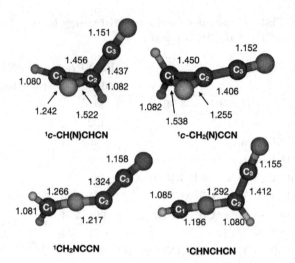

Fig. 6. Bond distances (reported in Angstroms) for the most important products identified in the PES for the $N(^2D)+CH_2CHCN$ reaction.

4   Discussion

The two reactions $N(^2D)+CH_2CHCN$ and $CN+CH_2CHCN$ show some differences, especially considering the initial bimolecular approach. The attack of $N(^2D)$ to the double bond of the acrylonitrile molecule is driven by the high electrophilicity of the nitrogen, attracted by the electron density of the double bond. As a result the $N(^2D)+CH_2CHCN$ reaction, leads to the formation of a cyclic intermediate, characterized by the presence of two C-N bond, involving the two ethylenic carbon atoms, in a cyclic structure. The formation of two main cyclic species is possible, thanks to a H-elimination process directly after the initial attack, depending on wether the hydrogen atom is eliminated from the terminal or internal C atom. Alternatively the C(N)C ring opening process was identified as the responsible of the formation of a linear intermediate, MIN4, which can directly undergo a H-elimination process. A comparison with the reaction of $N(^2D)$ with ethylene (C_2H_4), previously investigated with the same theoretical approach [21], shows several similarities. Nevertheless, in the case of acrylonitrile, the presence of a CN group removes the symmetry of ethylene, leading to an increased number of possible reactive channels and products. Differently, when the co-reactant is the cyano radical, the first step of the reaction is the attack of the CN radical to the terminal carbon of acrylonitrile, leading to the formation of a new C-C σ bond, as shown also from the C-C bond distance of 1.502 Å. Also in this case, several similarities can be noticed with the $CN+C_2H_4$ reaction [37], with the main differences driven by the loss of the symmetry in the acrylonitrile molecule. In general, a complete analysis of the global potential energy surface, including additional channels for the two reactions is needed and will be the focus of future works, including a more accurate analysis of the energies at the CCSD(T) level of theory [61,62]. The data will be validated considering a comparison with the experimental results, obtained through the such as crossed molecular beams (CMB) apparatus, with the aim to unveil the global reaction mechanism.

5 Conclusions

The present contribution is related to the theoretical investigation of the potential energy surface for the two N(^{2}D)+CH$_2$CHCN and CN+CH$_2$CHCN reactions, that are important for the chemistry of planetary atmospheres, in particular in the atmosphere of Titan. Both the N(^{2}D) and the CN addition processes to the CH$_2$CHCN molecule have been found to be barrierless.

The low temperatures of the atmosphere of Titan represent a harsh chemical environments, where only reactions characterized by the absence of an entrance barrier, together with the presence of stationary points below the reactant energy asymptote, can take place. Following this formalism, the two N(^{2}D)+CH$_2$CHCN and CN+CH$_2$CHCN reactions, appears to be feasible and important pathways to add information about the chemistry of nitrogen bearing molecules on Titan's atmosphere. A complete understanding of the overall PES will allow a kinetic analysis in order to derive the rate constants and branching ratios of the two processes. Subsequently, the derived kinetic parameters for the title reactions will be used to improve the chemical models of the atmosphere of Titan.

Acknowledgements. This work was supported by the Italian Space Agency (ASI, DC-VUM-2017-034, Grant n°2019-3 U.O Life in Space). This project has received funding from the European Union's Horizon 2020 research and innovation programme under the Marie Skłodowska Curie grant agreement No 811312 for the project "Astro-Chemical Origins" (ACO). The authors thank the Herla Project - Università degli Studi di Perugia (http://www.hpc.unipg.it/hosting/vherla/vherla.html) for allocated computing time. The authors thank the Dipartimento di Ingegneria Civile ed Ambientale of the University of Perugia for allocated computing time within the project "Dipartimenti di Eccellenza 2018-2022".

References

1. Balucani, N.: Nitrogen fixation by photochemistry in the atmosphere of titan and implications for prebiotic chemistry. In: The Early Evolution of the Atmospheres of Terrestrial Planets, pp. 155–164. Springer, Cham (2013). https://doi.org/10.1007/978-1-4614-5191-4_12

2. Balucani, N.: Elementary reactions of N atoms with hydrocarbons: first steps towards the formation of prebiotic N-containing molecules in planetary atmospheres. Chem. Soc. Rev. **41**(16), 5473–5483 (2012)

3. Balucani, N.: Elementary reactions and their role in gas-phase prebiotic chemistry. Int. J. Mol. Sci. **10**(5), 2304–2335 (2009)

4. Broadfoot, A., et al.: Extreme ultraviolet observations from Voyager 1 encounter with Saturn. Science **212**(4491), 206–211 (1981)

5. Hanel, R., et al.: Infrared observations of the saturnian system from Voyager 1. Science **212**(4491), 192–200 (1981)
6. Lindal, G.F., Wood, G., Hotz, H., Sweetnam, D., Eshleman, V., Tyler, G.: The atmosphere of Titan: an analysis of the Voyager 1 radio occultation measurements. Icarus **53**(2), 348–363 (1983)
7. Coustenis, A., Bezard, B., Gautier, D.: Titan's atmosphere from Voyager infrared observations: I. The gas composition of titan's equatorial region. Icarus **80**(1), 54–76 (1989)
8. Brown, R.H., Lebreton, J.P., Waite, J.H.: Titan from Cassini-Huygens (2009). https://doi.org/10.1007/978-1-4020-9215-2
9. Yung, Y.L., Allen, M., Pinto, J.P.: Photochemistry of the atmosphere of Titan: comparison between model and observations. Astrophys. J. Suppl. Ser. **55**(3), 465–506 (1984)
10. Lavvas, P., Coustenis, A., Vardavas, I.: Coupling photochemistry with haze formation in Titan's atmosphere, part I: model description. Planet. Space Sci. **56**(1), 27–66 (2008)
11. Lavvas, P., Coustenis, A., Vardavas, I.: Coupling photochemistry with haze formation in Titan's atmosphere, part II: results and validation with Cassini/Huygens data. Planet. Space Sci. **56**(1), 67–99 (2008)
12. Hébrard, E., Dobrijevic, M., Bénilan, Y., Raulin, F.: Photochemical kinetics uncertainties in modeling titan's atmosphere: a review. J. Photochem. Photobiol. C **7**(4), 211–230 (2006)
13. Dutuit, O., et al.: Critical review of N, N^+, N_2^+, N^{++}, and N_2^{++} main production processes and reactions of relevance to Titan's atmosphere. Astrophys. J. Suppl. Ser. **204**(2), 20 (2013)
14. Herron, J.T., et al.: Evaluated chemical kinetics data for reactions of N (^{2}D), N_2(^{2}P), and N_2 in the gas phase (1999)
15. Schofield, K.: Critically evaluated rate constants for gaseous reactions of several electronically excited species. J. Phys. Chem. Ref. Data **8**(3), 723–798 (1979)
16. Mancini, L., et al.: The reaction N(^{2}D)+ CH_3CCH (methylacetylene): a combined crossed molecular beams and theoretical investigation and implications for the atmosphere of Titan. J. Phys. Chem. A **125**(40), 8846–8859 (2021)
17. Liang, P., et al.: Combined crossed molecular beams and computational study on the N(^{2}D)+ HCCCN ($X^1\Sigma+$) reaction and implications for extra-terrestrial environments. Mol. Phys. **120**(1–2), e1948126 (2022)
18. Balucani, N., et al.: Combined crossed molecular beam and theoretical studies of the N (^{2}D)+ CH_4 reaction and implications for atmospheric models of Titan. J. Phys. Chem. A **113**(42), 11138–11152 (2009)
19. Balucani, N., et al.: Cyanomethylene formation from the reaction of excited nitrogen atoms with acetylene: a crossed beam and ab initio study. J. Am. Chem. Soc. **122**(18), 4443–4450 (2000)
20. Balucani, N., et al.: Formation of nitriles and imines in the atmosphere of Titan: combined crossed-beam and theoretical studies on the reaction dynamics of excited nitrogen atoms N (^{2}D) with ethane. Faraday Discuss. **147**, 189–216 (2010)
21. Balucani, N., et al.: Combined crossed beam and theoretical studies of the N (2D)+ C_2H_4 reaction and implications for atmospheric models of Titan. J. Phys. Chem. A **116**(43), 10467–10479 (2012)
22. Balucani, N., Cartechini, L., Alagia, M., Casavecchia, P., Volpi, G.G.: Observation of nitrogen-bearing organic molecules from reactions of nitrogen atoms with hydrocarbons: a crossed beam study of N (2D)+ ethylene. J. Phys. Chem. A **104**(24), 5655–5659 (2000)

23. Mancini, L., de Aragão, E.V.F., Rosi, M., Skouteris, D., Balucani, N.: A theoretical investigation of the reactions of N(^{2}D) with small alkynes and implications for the prebiotic chemistry of Titan. In: Gervasi, O., et al. (eds.) ICCSA 2020. LNCS, vol. 12251, pp. 717–729. Springer, Cham (2020). https://doi.org/10.1007/978-3-030-58808-3_52

24. Rosi, M., et al.: A computational study on the attack of nitrogen and oxygen atoms to toluene. In: Gervasi, O., et al. (eds.) ICCSA 2021. LNCS, vol. 12953, pp. 620–631. Springer, Cham (2021). https://doi.org/10.1007/978-3-030-86976-2_42

25. Rosi, M., et al.: A computational study on the insertion of N(^{2}D) into a C—H or C—C bond: the reactions of N(^{2}D) with benzene and toluene and their implications on the chemistry of Titan. In: Gervasi, O., et al. (eds.) ICCSA 2020. LNCS, vol. 12251, pp. 744–755. Springer, Cham (2020). https://doi.org/10.1007/978-3-030-58808-3_54

26. Balucani, N., et al.: A computational study of the reaction N(^{2}D) + C$_6$H$_6$ leading to pyridine and phenylnitrene. In: Misra, S., et al. (eds.) ICCSA 2019. LNCS, vol. 11621, pp. 316–324. Springer, Cham (2019). https://doi.org/10.1007/978-3-030-24302-9_23

27. Loison, J., et al.: The neutral photochemistry of nitriles, amines and imines in the atmosphere of titan. Icarus **247**, 218–247 (2015)

28. Vuitton, V., Yelle, R., Klippenstein, S., Hörst, S., Lavvas, P.: Simulating the density of organic species in the atmosphere of titan with a coupled ion-neutral photochemical model. Icarus **324**, 120–197 (2019)

29. Stevenson, J., Lunine, J., Clancy, P.: Membrane alternatives in worlds without oxygen: creation of an azotosome. Sci. Adv. **1**(1), e1400067 (2015)

30. Cui, J., et al.: Analysis of titan's neutral upper atmosphere from Cassini Ion neutral mass spectrometer measurements. Icarus **200**(2), 581–615 (2009)

31. Vuitton, V., Yelle, R., McEwan, M.: Ion chemistry and n-containing molecules in titan's upper atmosphere. Icarus **191**(2), 722–742 (2007)

32. Magee, B.A., Waite, J.H., Mandt, K.E., Westlake, J., Bell, J., Gell, D.A.: INMS-derived composition of Titan's upper atmosphere: analysis methods and model comparison. Planet. Space Sci. **57**(14–15), 1895–1916 (2009)

33. Müller-Wodarg, I., Griffith, C.A., Lellouch, E., Cravens, T.E.: Titan: Interior, Surface, Atmosphere, and Space Environment, vol. 14. Cambridge University Press, Cambridge (2014)

34. Palmer, M.Y., et al.: Alma detection and astrobiological potential of vinyl cyanide on titan. Sci. Adv. **3**(7), e1700022 (2017)

35. Lai, J.Y., et al.: Mapping vinyl cyanide and other nitriles in Titan's atmosphere using ALMA. Astron. J. **154**(5), 206 (2017)

36. Sims, I.R., et al.: Rate constants for the reactions of CN with hydrocarbons at low and ultra-low temperatures. Chem. Phys. Lett. **211**(4–5), 461–468 (1993)

37. Balucani, N., et al.: Crossed beam reaction of cyano radicals with hydrocarbon molecules. III. Chemical dynamics of vinylcyanide (C$_2$H$_3$CN; X^{1}A') formation from reaction of CN (X^2 Σ+) with ethylene, C$_2$H$_4$ (X ^{1}A$_g$). J. Chem. Phys. **113**(19), 8643–8655 (2000)

38. Balucani, N., et al.: Formation of nitriles in the interstellar medium via reactions of cyano radicals, CN (X^2 Σ+), with unsaturated hydrocarbons. Astrophys. J. **545**(2), 892 (2000)

39. Balucani, N., Asvany, O., Osamura, Y., Huang, L., Lee, Y., Kaiser, R.: Laboratory investigation on the formation of unsaturated nitriles in Titan's atmosphere. Planet. Space Sci. **48**(5), 447–462 (2000)

40. Leonori, F., Petrucci, R., Wang, X., Casavecchia, P., Balucani, N.: A crossed beam study of the reaction CN+ C_2H_4 at a high collision energy: the opening of a new reaction channel. Chem. Phys. Lett. **553**, 1–5 (2012)

41. Balucani, N., et al.: A combined crossed molecular beams and theoretical study of the reaction CN+ C_2H_4. Chem. Phys. **449**, 34–42 (2015)

42. Vereecken, L., De Groof, P., Peeters, J.: Temperature and pressure dependent product distribution of the addition of CN radicals to C_2H_4. Phys. Chem. Chem. Phys. **5**(22), 5070–5076 (2003)

43. Gannon, K.L., Glowacki, D.R., Blitz, M.A., Hughes, K.J., Pilling, M.J., Seakins, P.W.: H atom yields from the reactions of CN radicals with C_2H_2, C_2H_4, C_3H_6, trans-2-C_4H_8, and iso-C_4H_8. J. Phys. Chem. A **111**(29), 6679–6692 (2007)

44. Geppert, W., et al.: Dissociative recombination of nitrile ions: $DCCCN^+$ and $DCCCND^+$. Astrophys. J. **613**(2), 1302 (2004)

45. Vigren, E., et al.: Dissociative recombination of nitrile ions with implications for Titan's upper atmosphere. Planet. Space Sci. **60**(1), 102–106 (2012)

46. Butterfield, M.T., Yu, T., Lin, M.C.: Kinetics of CN reactions with allene, butadiene, propylene and acrylonitrile. Chem. Phys. **169**(1), 129–134 (1993)

47. Rosi, M., et al.: Possible scenarios for SiS formation in the interstellar medium: electronic structure calculations of the potential energy surfaces for the reactions of the SiH radical with atomic sulphur and S_2. Chem. Phys. Lett. **695**, 87–93 (2018)

48. Skouteris, D., et al.: Interstellar dimethyl ether gas-phase formation: a quantum chemistry and kinetics study. Mon. Not. R. Astron. Soc. **482**(3), 3567–3575 (2019)

49. Recio, P., et al.: A crossed molecular beam investigation of the $N(^2D)+$ pyridine reaction and implications for prebiotic chemistry. Chem. Phys. Lett. **779**, 138852 (2021)

50. Becke, A.D.: A new mixing of Hartree-Fock and local density-functional theories. J. Chem. Phys. **98**(2), 1372–1377 (1993)

51. Stephens, P.J., Devlin, F.J., Chabalowski, C.F., Frisch, M.J.: Ab initio calculation of vibrational absorption and circular dichroism spectra using density functional force fields. J. Phys. Chem. **98**(45), 11623–11627 (1994)

52. Dunning Jr., T.H.: Gaussian basis sets for use in correlated molecular calculations. I. The atoms boron through neon and hydrogen. J. Chem. Phys. **90**(2), 1007–1023 (1989)

53. Woon, D.E., Dunning Jr., T.H.: Gaussian basis sets for use in correlated molecular calculations. III. The atoms aluminum through argon. J. Chem. Phys. **98**(2), 1358–1371 (1993)

54. Kendall, R.A., Dunning Jr., T.H., Harrison, R.J.: Electron affinities of the first-row atoms revisited. Systematic basis sets and wave functions. J. Chem. Phys. **96**(9), 6796–6806 (1992)

55. Gonzalez, C., Schlegel, H.B.: An improved algorithm for reaction path following. J. Chem. Phys. **90**(4), 2154–2161 (1989). https://doi.org/10.1063/1.456010

56. Gonzalez, C., Schlegel, H.B.: Reaction path following in mass-weighted internal coordinates. J. Phys. Chem. **94**(14), 5523–5527 (1990). https://doi.org/10.1021/j100377a021

57. Moore, C.E.: Atomic Energy Levels. US Department of Commerce, National Bureau of Standards (1949)

58. Frisch, M., et al.: Gaussian 09, rev. A. 02. Gaussian. Inc., Wallingford, CT (2009)

59. Schaftenaar, G., Noordik, J.H.: Molden: a pre-and post-processing program for molecular and electronic structures. J. Comput. Aided Mol. Des. **14**(2), 123–134 (2000)

60. Schaftenaar, G., Vlieg, E., Vriend, G.: Molden 2.0: quantum chemistry meets proteins. J. Comput. Aided Mol. Des. **31**(9), 789–800 (2017). https://doi.org/10.1007/s10822-017-0042-5
61. Vanuzzo, G., et al.: The reaction N(^{2}D)+CH$_2$CHCN (vinylcyanide): a combined crossed-beams and theoretical study and implications for the atmosphere of Titan. J. Phys. Chem. A (submitted)
62. Marchione, D., et al.: Unsaturated dinitriles formation routes in extraterrestrial environments: a combined experimental and theoretical investigation of the reaction between cyano radicals and cyanoethene (C$_2$H$_3$CN). J. Phys. Chem. A. **126**(22), 3569–3582 (2022). https://doi.org/10.1021/acs.jpca.2c01802

Formation Routes of CO from O(^{1}D)+Toluene: A Computational Study

Marzio Rosi[1(✉)], Piergiorgio Casavecchia[2], Nadia Balucani[2], Pedro Recio[2], Adriana Caracciolo[3], Dimitrios Skouteris[4], and Carlo Cavallotti[5]

[1] Department of Civil and Environmental Engineering, University of Perugia, 06125 Perugia, Italy
marzio.rosi@unipg.it

[2] Department of Chemistry, Biology and Biotechnologies, University of di Perugia, 06123 Perugia, Italy
{piergiorgio.casavecchia,nadia.balucani}@unipg.it

[3] Ann and H.J. Smead, Department of Aerospace Engineering Sciences, University of Colorado, Boulder, CO 80303, USA
adriana.caracciolo@colorado.edu

[4] Master-Tec, 06128 Perugia, Italy

[5] Department of Chemistry, Materials and Chemical Engineering, Politecnico di Milano, 20131 Milan, Italy
carlo.cavallotti@polimi.it

Abstract. The interaction between oxygen atoms in their first electronically excited state ^{1}D with toluene has been characterized by electronic structure calculations. We focused our attention, in particular, on the different pathways leading to the formation of CO. Six different reaction channels have been investigated. Our results suggest that, while for accurate energies high level calculations, as CCSD(T), are necessary, in particular when strong correlation effects are present, for semi-quantitative results DFT methods are adequate and provide information useful when larger systems than toluene as polycyclic aromatic hydrocarbons are under investigation.

Keywords: Computational chemistry · Ab initio calculations · Density functional theory methods · Coupled clusted calculations · Combustion · Fuels

1 Introduction

The bimolecular reaction between atomic oxygen in its ground state ^{3}P and first excited state ^{1}D and toluene is extremely important in combustion chemistry since toluene is the major aromatic found in gasoline. To understand the kinetics of toluene oxidation is necessary to know the primary products and the branching ratios of the bimolecular reactions O(^{3}P, ^{1}D)+toluene as a function of temperature and pressure. Despite extensive kinetic studies [1–4], little is known about the mechanism and branching fractions of this reaction. We have therefore undertaken a synergistic experimental and theoretical investigation of the reactions O(^{3}P, ^{1}D)+toluene with the same approach recently used to identify all primary products and determine the branching fractions for the reactions of O (^{3}P, ^{1}D) with a variety of unsaturated aliphatic and aromatic hydrocarbons [5–10]. This approach consists in experiments performed using the crossed molecular beam technique with soft electron ionization mass spectrometric detection and time-of-flight analysis. The experimental results are combined with electronic structure calculations of the triplet and singlet potential energy surfaces and statistical RRKM/Master Equation kinetic computations. In this way, the product distribution as a function of collision energy, as well as temperature and pressure, with intersystem crossing taken into account, is obtained and provides important information for improved kinetic modeling of toluene combustion.

We have recently investigated both experimentally and theoretically [9, 10] the reaction between O in the ground ^{3}P state and the first excited ^{1}D state and benzene. The main products are atomic hydrogen and carbon monoxide; hydrogen can be formed both in the triplet and in the singlet surfaces while CO can be formed only in the singlet surface starting form O(^{1}D) or via intersystem crossing. Preliminary experimental information on the reaction O(^{1}D, ^{3}P)+toluene suggest that the main products are H, CH$_3$ and CO. For analogy with the reaction O(^{1}D, ^{3}P)+benzene we expect that H and CH$_3$ could be formed both in the singlet and triplet surfaces, while CO should be formed only in the singlet surface.

In this contribution we present the study of the singlet surface O(^{1}D)+toluene focusing our attention to the formation of CO. The interaction with the triplet O(^{3}P) surface via intersystem crossing will be the subject of future investigations.

2 Theoretical Methods

The evaluation of the potential energy surface of O(^{1}D)+toluene has been performed at Density Functional Theory (DFT) level of calculation optimizing the stationary points (minima and transition states) using the ωB97X-D [11, 12] functional, and the 6–311+G(d,p) basis set [13, 14]. The same level of theory has been used to evaluate the vibrational frequencies in order to establish the nature of the stationary point, i.e. minimum if all the frequencies are real, transition state if there is one, and only one, imaginary frequency. Moreover, In order to assign the saddle points to the relative reactants and products we performed intrinsic reaction coordinate (IRC) calculations [15, 16] The energy of all the stationary points was computed at a more accurate level of approximation performing CCSD(T) calculations [17–19] using the correlation consistent aug-cc-pVTZ basis set [20] and the DFT optimized geometries. This computational scheme has been previously successfully applied [21–35]. Both the DFT and the CCSD(T) energies were corrected to 0 K by adding the zero point energy correction computed using the scaled harmonic vibrational frequencies evaluated at DFT level. The energy of O(^{1}D) was estimated by adding the experimental [36] separation O(^{3}P) – O(^{1}D) of 45.3 kcal mol^{-1} to the energy of O(^{3}P) at all levels of calculation. All calculations were done using Gaussian 09 [37] while the analysis of the vibrational frequencies was performed using Molekel [38, 39].

3 Results and Discussion

3.1 O(^{1}D)+Toluene

Figure 1 reports the optimized geometries at ωB97X-D/6–311+G(d,p) level of the iso-mers deriving from the interaction of oxygen in its excited ^{1}D state with toluene. We have three different isomers depending on the carbon atoms interacting with oxygen, which are almost degenerate: species **2** is only 1.4 kcal/mol less stable than isomer **1**, while species **3** is 1.8 kcal/mol above species **1** at ωB97X-D/6–311+G(d,p) level of calculation. At CCSD(T)/aug-cc-pVTZ//ωB97X-D/6–311+G(d,p) level of theory, the relative energies are almost the same: species **2** is 1.8 kcal/mol less stable than **1** while **3** is 1.9 kcal/mol less stable than **1**. O(^{1}D) is able to form two σ bonds with two adjacent carbon atoms, because it has two electrons with opposite spin in different p orbitals. These electrons interact with the two electrons with opposite spin involved in the π bond of the two carbon atoms. The energy required for breaking the π bond is more than compensated by the energy associated to the formation of the two σ bonds between oxygen and carbon, as it is shown in Table 1 where the energetics of the reactions under investigation is reported. We can notice that the formation of species **1**, **2**, or **3** is exother-mic by about 100 kcal/mol both at DFT and CCSD(T) level of calculation. In particular, the formation of **1** is exothermic by 100.4 kcal/mol at DFT level (100.7 at CCSD(T) level); the formation of **2** by 99.0 kcal/mol at DFT level (98.9 at CCSD(T) level) and that of **3** by 98.6 kcal/mol at DFT level (98.8 at CCSD(T) level). The breaking of the π bond is shown by the C—C bond distance reported in Fig. 1 which corresponds to that of a single bond for the carbon atoms interacting with oxygen, being 1.51 Å in **1** and 1.50 Å in **2** and **3**. From Fig. 1 we can notice that the oxygen atom bonds symmetrically to two adjacent carbon atoms, being the bond distances (1.43 Å in **1** and 1.42 Å in **2** and **3**) equal for both C—O bonds and correspond to a single σ bond. The first step towards the formation of CO is always an hydrogen shift between two adjacent carbon atoms. Starting from species **1** or species **2** the shift of an hydrogen from C2 to C1 leads to species **4** reported in Fig. 2. In species **4** the oxygen shows a double bond with C2 being the C—O distance only 1.22 Å. From species **2** we can have also the shift of an hydrogen from C3 to C4 with the formation of species **16**, while species **3** can show the shift of an hydrogen from C3 to C2 or from C4 to C5 giving rise to species **15** or **17**, respectively. Also in species **15**, **16** and **17** the oxygen atom shows a double bond with C3, C3 and C4, respectively.

In Fig. 3 we have reported the potential energy surface (PES) relative to the reac-tions leading to the formation of CO starting from the species reported in Fig. 1. We have two different panels, panel **a** reports the pathways originating from species **1** (and in part species **2**), while panel **b** reports the reactions starting from species **2** and **3**. The energy changes and barrier heights relative to these reactions are shown in Table 1 both at ωB97X-D/6–311+G(d,p) level of calculations and at the more accurate CCSD(T)/aug-cc-pVTZ//ωB97X-D/6–311+G(d,p) level of theory. In Fig. 3 for clarity only the CCSD(T)/aug-cc-pVTZ//ωB97X-D/6–311+G(d,p) energy values are reported.

Table 1. ΔH_0^0 and barrier heights (kcal/mol, 0 K) computed at DFT (CCSD(T)) level for the pathways leading to CO formation starting from O(^{1}D)+toluene. All the energies include zero point energy correction. For the numbering of the isomers see Fig. 3.

	ΔH_0^0	Barrier height
O(^{1}D)+C$_6$H$_5$CH$_3$ → 1	−100.4 (−100.7)	
O(^{1}D)+C$_6$H$_5$CH$_3$ → 2	−99.0 (−98.9)	
O(^{1}D)+C$_6$H$_5$CH$_3$ → 3	−98.6 (−98.8)	
1 → 4	24.2 (−22.8)	46.0 (47.8)
2 → 4	−25.6 (−24.6)	38.3 (40.8)
4 → 6	25.3 (26.4)	35.1 (35.1)
6 → 9	6.8 (5.8)	55.7 (53.6)
9 → 13+CO	−17.8 (−23.4)	29.9 (27.1)
4 → 5	−4.0 (−3.4)	36.5 (37.4)
5 → 8	37.5 (36.0)	42.4 (46.9)
8 → 12	−1.1 (−2.4)	9.4 (6.3)
12 → 14+CO	−19.2 (−24.5)	37.7 (33.1)
4 → 7	33.9 (33.0)	37.1 (38.1)
7 → 10	−5.0 (−2.6)	5.5 (6.1)
10 → 11+CO	−19.2 (−24.5)	29.6 (26.6)
2 → 16	−30.1 (−28.2)	44.8 (45.4)
16 → 20	43.3 (40.7)	45.9 (45.3)
20 → 22	−8.2 (−7.6)	3.4 (4.2)
22 → 14+CO	−18.2 (−23.8)	30.2 (28.7)
3 → 15	−30.3 (−28.0)	44.5 (45.4)
15 → 19	46.0 (43.6)	46.1 (45.1)
19 → 9	−10.8 (−8.1)	2.7 (2.7)
9 → 13+CO	−17.8 (−23.4)	29.9 (27.1)
3 → 17	−28.9 (−26.9)	40.4 (42.9)
17 → 18	40.9 (39.1)	44.3 (44.3)
18 → 21	−10.2 (−7.3)	3.3 (4.0)
21 → 11+CO	−18.1 (−23.7)	31.9 (29.9)

It is worth noting from Table 1 that the DFT and CCSD(T) energies are in reasonable agreement, the difference being about 2 – 3 kcal/mol. Only the dissociation reactions which show, as expected, more relevant correlation effects present a slightly larger energy difference (about 5 kcal/mol). This point suggests that for the study of larger systems it should be possible to use DFT calculations in order to provide at least semi-quantitative results. This approach could be used therefore for the study of the interaction of oxygen atoms with larger polycyclic aromatic hydrocarbons (PAH's) where accurate CCSD(T) calculations are unfeasible due to the size of the systems.

From Fig. 3 and Table 1 we can see that we have six different pathways leading to the formation of CO. Let us consider in details these reactions. We will refer only to the more accurate CCSD(T) energies for simplicity. From Fig. 3, panel **a**, we see that species **1** and **2** isomerize to species **4** through the shift of an hydrogen atom from the C in ortho to the C in ipso, overcoming an energy barrier of 47.8 and 40.8 kcal/mol, respectively. Species **4** is more stable than **1** by 22.2 and **2** by 24.6 kcal/mol, respectively. Species **4** can follow three different pathways in order to evolve towards the CO formation, which are shown in Fig. 3 panel **a** with different colors. Let us start with the blue path, which implies the following reactions: **4 → 6 → 9 → 13**+CO. In this pathway we can notice the opening of the 6 membered ring (**6**) followed by the formation of a 5 membered ring (**9**). This last step implies a relatively high energy barrier (53.6 kcal/mol). In the red pathway we have the following reactions: **4 → 5 → 8 → 12 → 14**+CO. This pathway implies an additional isomerization with respect to the blue one and shows an energy barrier of 46.9 kcal/mol for the isomerization **5 → 8**. The green pathway implies the isomerizations: **4 → 7 → 10 → 11**+CO. This seems to be the preferred reaction path since it implies relatively low barrier heights. CO can be formed also starting from species **2** and **3**, as it is shown in panel **b** of Fig. 3. Let us consider these reactions. We have three different pathways, all of them starting with the shift of an hydrogen atom from one carbon to another adjacent carbon atom. The blue pathway implies the reactions: **2 → 16 → 20 → 22 → 14**+CO. The highest energy barrier in this reaction scheme is the first one which is equal to 45.4 kcal/mol. The other two reaction paths start from isomer **3**. In the red path we have the reactions: **3 → 15 → 19 → 9 → 13**+CO; this last reaction is in common with the blue path of panel **a**. Also in this case the highest energy barrier is the one relative to the first isomerization (45.4 kcal/mol). In the green pathway we have the reactions: **3 → 17 → 18 → 21 → 11**+CO. In this case the highest barrier height is the second one relative to the isomerization **17 → 18** (44.3 kcal/mol).

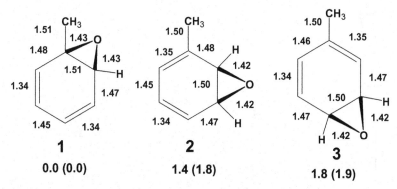

Fig. 1. Optimized geometries of the species obtained for the attack of O(^{1}D) to toluene. Geometry optimization has been performed at ωB97X-D/6–311+G(d,p) level. Bond lengths in Å. Relative energies (kcal/mol) with respect to species **1** have been evaluated at ωB97X-D/6–311+G(d,p) and, in parentheses, at CCSD(T)/aug-cc-pVTZ//ωB97X-D/6–311+G(d,p) level of theory.

In order to establish which is the preferred reaction pathway it is necessary to compute the branching fractions. For this purpose, RRKM calculations are under way. However, in the experimental conditions the presence of O(^{1}D) is only a minor part with respect to O(^{3}P). For this reason the main reaction path leading to the formation of CO should start form the triplet surface and through intersystem crossing should arrive to the singlet surface and to the final products as it is shown in our recent study of O+benzene [9, 10]. These calculations will be the subject of future work.

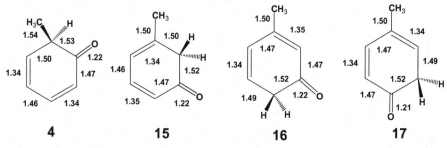

Fig. 2. Optimized geometries of the species involved in the first step of the reactive channels. Geometry optimization has been performed at ωB97X-D/6–311+G(d,p) level. Bond lengths in Å.

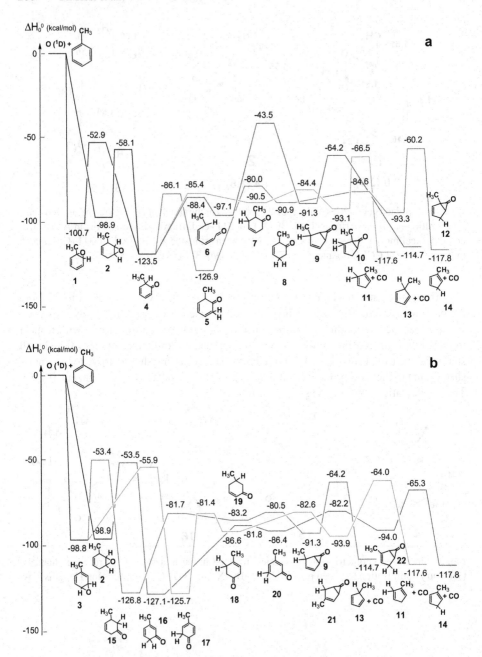

Fig. 3. Potential energy surface for the O(^{1}D)+toluene reaction. Only the pathways leading to the formation of CO are reported. Relative energies (kcal/mol) respect to O(^{1}D)+toluene, computed at CCSD(T)/aug-cc-pVTZ//ωB97X-D/6–311+G(d,p) level with inclusion of ZPE. In panel **a** pathways originating from isomer **1** are reported, while in panel **b** pathways originating from isomers **2** and **3** are shown.

4 Conclusions

The study at *ab initio* level of the interaction of O(^{1}D) with toluene, focusing the attention on the formation routes of CO, shows six possible different reaction channels. Although the potential energy surface provides information on the preferred exit channel through the barrier heights and the number of subsequent isomerization steps, the evaluation of branching fractions through RRKM calculations is necessary for a quantitative comparison. This will be the subject of future investigations The comparison between DFT and CCSD(T) calculations suggests that useful semi-quantitative results can be obtained also at DFT level of approximation. This is an important result in order to extend our study to complex PAHs.

Acknowledgments. The authors acknowledge the MUR (Italian Ministry of University and Research) for "PRIN 2017" funds, project MAGIC DUST, Grant Number 2017PJ5XXX. MR thanks the DICA, University of Perugia, for providing computing resources within the project "Dipartimenti di Eccellenza 2018–2022".

References

1. Brezinsky, K.: The high-temperature oxidation of aromatic hydrocarbons. Prog. Energy Combust. **12**, 1–24 (1986)
2. Atkinson, R., Pitts, J.N.: Rate constants for the reaction of O(^{3}P) atoms with Benzene and Toluene over the temperature range 299–440 K. Chem. Phys. Lett. **63**, 485–489 (1979)
3. Nicovich, J.M., Gump, C.A., Ravishankara, A.R.: Rates of reactions of O(^{3}P) with Benzene and Toluene. J. Phys. Chem. **86**, 1684–1690 (1982)
4. Tappe, M., Schliephake, V., Wagner, H.: Gg. reactions of Benzene, Toluene and Ethylbenzene with Atomic Oxygen O(^{3}P) in the Gas Phase Z. Phys. Chem. **162**, 129–145 (1989)
5. Cavallotti, C., et al.: Relevance of the channel leading to Formaldehyde + Triplet Ethylidene in the O(^{3}P)+Propene reaction under combustion conditions. J. Phys. Chem. Lett. **5**, 4213–4218 (2014)
6. Leonori, F., et al.: Experimental and theoretical studies on the dynamics of the O(^{3}P) + Propene reaction: primary products, branching ratios, and role of intersystem crossing. J. Phys. Chem. C **119**, 14632–14652 (2015)
7. Gimondi, I., Cavallotti, C., Vanuzzo, G., Balucani, N., Casavecchia, P.: Reaction dynamics of O(^{3}P)+Propyne: II. Primary products, branching ratios, and role of intersystem crossing from Ab initio coupled triplet/singlet potential energy surfaces and statistical calculations. J. Phys. Chem. A **120**, 4619–4633 (2016)
8. Caracciolo, A., et al.: Combined experimental and theoretical studies of the O(^{3}P) + 1-Butene reaction dynamics: primary products, branching ratios and role of intersystem crossing. J. Phys. Chem. A **123**, 9934–9956 (2019)
9. Cavallotti, C., et al.: A theoretical study of the extent of intersystem crossing in the O(^{3}P) + C$_6$H$_6$ reaction with experimental validation. J. Phys. Chem. Lett. **11**, 9621–9628 (2020)
10. Vanuzzo, G., et al.: Crossed-beam and theoretical studies of the O(^{3}P, ^{1}D) + Benzene reactions: primary products, branching fractions, and role of intersystem crossing. J. Phys. Chem. A **125**, 8434–8453 (2021)
11. Chai, J.-D., Head-Gordon, M.: Long-range corrected hybrid density functionals with damped atom-atom dispersion corrections. Phys. Chem. Chem. Phys. **10**, 6615–6620 (2008)

12. Chai, J.-D., Head-Gordon, M.: Systematic optimization of long-range corrected hybrid density functionals. J. Chem. Phys. **128**, 084106 (2008)
13. Krishnan, R., Binkley, J.S., Seeger, R., Pople, J.A.: Self-consistent molecular orbital methods. XX. A basis set for correlated wave functions. J. Chem. Phys. **72**, 650–654 (1980)
14. Frisch, M.J., Pople, J.A., Binkley, J.S.: Self-consistent molecular orbital methods 28. Supplementary functions for Gaussian basis sets. J. Chem. Phys. **80**, 3265–3269 (1984)
15. Gonzalez, C., Schlegel, H.B.: An improved algorithm for reaction path following. J. Chem. Phys. **90**, 2154–2161 (1989)
16. Gonzalez, C., Schlegel, H.B.: Reaction path following in mass-weighted internal coordinates. J. Phys. Chem. **94**, 5523–5527 (1990)
17. Bartlett, R.J.: Many-body perturbation theory and coupled cluster theory for electron correlation in molecules. Annu. Rev. Phys. Chem. **32**, 359–401 (1981)
18. Raghavachari, K., Trucks, G.W., Pople, J.A., Head-Gordon, M.: Quadratic configuration interaction. A general technique for determining electron correlation energies. Chem. Phys Lett. **157**, 479–483 (1989)
19. Olsen, J., Jorgensen, P., Koch, H., Balkova, A., Bartlett, R.J.: Full configuration–interaction and state of the art correlation calculations on water in a valence double-zeta basis with polarization functions. J. Chem. Phys. **104**, 8007–8015 (1996)
20. Dunning Jr., T.H.: Gaussian basis sets for use in correlated molecular calculations. I. The atoms boron through neon and hydrogen. J. Chem. Phys. **90**, 1007–1023 (1989)
21. de Petris, G., Cacace, F., Cipollini, R., Cartoni, A., Rosi, M., Troiani, A.: Experimental detection of theoretically predicted N_2CO. Angew. Chem. **117**, 466–469 (2005)
22. De Petris, G., Rosi, M., Troiani, A.: SSOH and HSSO radicals: An experimental and theoretical study of $[S_2OH]^{0/+/-}$ species. J. Phys. Chem. A **111**, 6526–6533 (2007)
23. Bartolomei, M., et al.: The intermolecular potential in NO-N_2 and (NO-N_2)$^+$ systems: Implications for the neutralization of ionic molecular aggregates. PCCP **10**, 5993–6001 (2008)
24. Leonori, F., et al.: Observation of organosulfur products (thiovinoxy, thioketene and thioformyl) in crossed-beam experiments and low temperature rate coefficients for the reaction $S(^1D) + C_2H_4$. PCCP **11**, 4701–4706 (2009)
25. Leonori, F., et al.: Crossed-beam and theoretical studies of the $S(^1D) + C_2H_2$ reaction. J. Phys. Chem. A **113**, 4330–4339 (2009)
26. De Petris, G., Cartoni, A., Rosi, M., Barone, V., Puzzarini, C., Troiani, A.: The proton affinity and gas-phase basicity of sulfur dioxide. ChemPhysChem **12**, 112–115 (2011)
27. Berteloite, C., et al.: Low temperature kinetics, crossed beam dynamics and theoretical studies of the reaction $S(^1D) + CH_4$ and low temperature kinetics of $S(^1D) + C_2H_2$. Phys. Chem. Chem. Phys. **13**, 8485–8501 (2011)
28. Rosi, M., Falcinelli, S., Balucani, N., Casavecchia, P., Leonori, F., Skouteris, D.: Theoretical study of reactions relevant for atmospheric models of Titan: Interaction of excited nirogen atoms with small hydrocarbons. LNCS **7333**, 331–344 (2012)
29. Skouteris, D., Balucani, N., Faginas-Lago, N., Falcinelli, S., Rosi, M.: Dimerization of methanimine and its charged species in the atmosphere of Titan and interstellar/cometary ice analogs. Astron. Astrophys. **584**, A76 (2015)
30. Falcinelli, S., Rosi, M., Cavalli, S., Pirani, F., Vecchiocattivi, F.: Stereoselectivity in autoionization reactions of hydrogenated molecules by metastable gas atoms: the role of electronic couplings. Chem. Eur. J. **22**, 12518–12526 (2016)
31. Troiani, A., Rosi, M., Garzoli, S., Salvitti, C., de Petris, G.: Vanadium Hydroxide Cluster Ions in the Gas Phase: Bond-Forming Reactions of Doubly-Charged Negative Ions by SO_2-Promoted V-O Activation. Chem. Eur. J. **23**, 11752–11756 (2017)
32. Rosi, M., et al.: An experimental and theoretical investigation of 1-butanol pyrolysis. Front. Chem. **7**, 326 (2019). https://doi.org/10.3389/fchem.2019.00326

33. Sleiman, C., El Dib, G., Rosi, M., Skouteris, D., Balucani, N., Canosa, A.: Low Temperature Kinetics and Theoretical Study of the Reaction CN + CH3NH2: A Potential Source of Cyanamide and Methyl Cyanamide in the Interstellar Medium. PCCP **20**, 5478–5489 (2018). https://doi.org/10.1039/c7cp05746f

34. Rosi, M., et al.: Possible scenarios for SiS formation in the interstellar medium: electronic structure calculations of the potential energy surdaces for the reactions of the SiH radical with atomic sulphur and S$_2$. Chem. Phys. Lett. **695**, 87–93 (2018). https://doi.org/10.1016/j.cplett.2018.01.053

35. Balucani, N., Skouteris, D., Ceccarelli, C., Codella, C., Falcinelli, S., Rosi, M.: A theoretical investigation of the reaction between the Amidogen, NH, and the ethyl, C$_2$H$_5$, radicals: a possible gas-phase formation route of interstellar and planetary Ethanimine. Molec. Astrophys. **13**, 30–37 (2018). https://doi.org/10.1016/j.molap.2018.10.001

36. Moore, C.E.: Atomic Energy Levels, Natl. Bur. Stand. (U.S.) Circ. N. 467, U.S., GPO, Washington, DC (1949)

37. Gaussian 09, et al. Gaussian, Inc., Wallingford CT (2009)

38. Flükiger, P., Lüthi, H.P., Portmann, S., Weber, J., MOLEKEL 4.3: Swiss Center for Scientific Computing, Manno (Switzerland), (2000–2002)

39. Portmann, S., Lüthi, H. P., MOLEKEL: An Interactive Molecular Graphics Tool Chimia 54, 766–769 (2000)

Stereo-Dynamics of Autoionization Reactions Induced by Ne*(^{3}P$_{0,2}$) Metastable Atoms with HCl and HBr Molecules: Experimental and Theoretical Study of the Reactivity Through Selective Collisional Angular Cones

Marco Parriani[1], Franco Vecchiocattivi[1], Fernando Pirani[1,2], and Stefano Falcinelli[1(✉)] [ID]

[1] Department of Civil and Environmental Engineering, University of Perugia, Via G. Duranti 93, 06125 Perugia, Italy
franco@vecchio.it, {fernando.pirani,stefano.falcinelli}@unipg.it
[2] Department of Chemistry, Biology and Biotechnologies, University of Perugia, Via Elce di Sotto 8, 06100 Perugia, Italy

Abstract. In this paper are presented mass spectrometric determinations as a function of the collision energy in the 0.03–0.50 eV range as recorded in a crossed molecular beam experiments involving autoionization reactions between Ne*(^{3}P$_{2,0}$) metastable atoms and HCl and HBr molecules. The total and partial ionization cross sections for both investigated systems are presented and discussed in a comparative way. The comparison of the recorded data allows to point out similarities and differences on the collisional stereodynamics of Ne*(^{3}P$_{0,2}$)-HCl and Ne*(^{3}P$_{0,2}$)-HBr systems. In particular, an accurate characterization of the interaction potentials, which is mandatory for a comprehensive description of Ne*-HX (X = Cl and Br) reactive collisions, has been outlined. Such a theoretical analysis suggests that the formation of the proton transfer, NeH$^+$, ions as well as of other possible product ions (i.e. HX$^+$ and NeHX$^+$, parent and associate ions, respectively) comes from reactivity that is selectively open along angular cones showing different orientation and acceptance. In particular, the performed analysis highlights that the proton transfer rearrangement reaction, which is open in both Ne*-HX autoionizing collision, is much more efficient for Ne*+HCl respect to Ne*+HBr autoionization. The present investigation points out that such an efficiency variation is related to the following crucial points: (i) the different charge distribution on HX$^+$ ionic products, (ii) the balance between two distinct microscopic mechanisms that are operative in such processes (a pure physical-photoionization-indirect mechanism and a chemical-oxidation-direct mechanism), which are reactions of interest in combustion chemistry, plasma physics and chemistry, as well as in astrochemistry and for the chemistry of planetary ionospheres.

Keywords: Autoionization · Stereo-dynamics · Proton transfer · Charge transfer · Transition state · Mass spectrometry · Astrochemistry

© The Author(s), under exclusive license to Springer Nature Switzerland AG 2022
O. Gervasi et al. (Eds.): ICCSA 2022 Workshops, LNCS 13378, pp. 270–280, 2022.
https://doi.org/10.1007/978-3-031-10562-3_20

1 Introduction

In general, autoionization reactions, also named Penning ionization or chemi-ionization processes [1–4], are relevant in different fields of scientific research, as combustion chemistry, plasma physics and chemistry, as well as in astrochemistry and in ionosphere of Planets where they have an effect on the transmission of radio and satellite signals [5, 6]. More recently, from a fundamental point of view these processes have attracted the attention of the scientific community interested in studying the characteristics of quantum resonances present in reactive and non-reactive collisions. Such an important research topic has been addressed and developed using the molecular beams technique performed under ultra-cold conditions through very advanced and frontier experiments able to simulate what happens from a chemical point of view in interstellar environments [7].

Chemi-ionizations are well-known elementary gas-phase reactions which imply neutral reagents able to generate more stable ionic species. They are very common in nature and take place in numerous environments of chemical interest. In particular, these reactions are relevant in combustion chemistry, since are considered as primary steps in flames [8, 9].

When chemi-ionizations involve reagent species in an internally excited metastable state as ionizing agent of atomic or molecular targets, we commonly refer to collisional autoionization or Penning ionization processes. The reader interested in a general overview of these processes can refer to various review articles already published [2, 3].

When we consider autoionization reactions involving electronically excited metastable species (this is the case of noble gas atoms, Ng, excited in their first electronic level), our research group demonstrated that a general semiempirical model can be applied for their description. In fact, very recently, we developed an original and innovative theoretical approach which allow a simple and general rationalization and representation of chemi-ionizations as prototype elementary oxidation processes [10–12]. In our model, we can schematize this kind of processes by the Eq. (1) below:

$$
\begin{aligned}
\mathrm{Ng}^* + \mathrm{X} &\to (\mathrm{Ng}\cdots\mathrm{X})^* \to (\mathrm{Ng} \xleftarrow{e^-} \mathrm{X})^* \to \\
&(\mathrm{Ng}\cdots\mathrm{X})^+ + e^- \to \text{ion products}
\end{aligned}
\tag{1}
$$

where, Ng^* is a Noble gas open-shell atom, in an electronically excited metastable state, which collides with the X target being an atomic or molecular species. In this case, the collision leads to the generation of a collision complex $(\mathrm{Ng}\cdots\mathrm{X})^*$ in an excited energetic state which relaxes its excess of energy via an electron emission [2, 3], as shown in Eq. (1). $(\mathrm{Ng}\cdots\mathrm{X})^*$ is the precursor state of the process with an electron excited in a high energetic Rydberg state. For such barrierless reactions it represents the transition state which, under thermal collisions, gives rise to the formation of ionic products: X^+ (parent ion), NgX^+ (associated ionic aggregates), as well as dissociated or rearranged ions. This occurs by an oxidation process which is schematized in Eq. (1) by the left-going arrow representing an electron exchange between the collisional target X and the core of Ng^*.

The autoionization processes involving $\mathrm{Ng}^* + \mathrm{HCl}$ and HBr molecules have attracted great interest from various researcher groups during last forty years, since they are important for the development of excimer lasers. In fact, the chemi-ionization reactions generated by collision with $\mathrm{Ne}^*(^3\mathrm{P}_{0,2})$ metastable atoms are important processes occurring in the formation of the XeCl^* excimer molecule which is produced in Xe + HCl gaseous mixtures discharge containing an excess of Ne. Due to this considerable scientific interest, a number of valuable articles have been published on this topic since the 1980s, either from a theoretical [13–16] and an experimental point of view [17–24].

This paper reports on mass spectrometric determinations as a function of the collision energy in the range of 0.03–0.50 eV for $Ne^*(^3P_{0,2})$-HCl and $Ne^*(^3P_{0,2})$-HBr autoionization processes. The experimental study was performed using a crossed molecular beams apparatus allowing the measure of the total and partial cross sections for all the open ionization channels in the investigated energy range. In order to highlights similarities and differences in the collisional stereodynamics of the two systems, new data recorded for Ne^*-HBr chemi-ionization are analyzed and discussed in comparison with the previous ones obtained for Ne^*-HCl system [23, 24]. With this procedure, we were able to rationalize the behavior of the two Ne^*-HX (X = Cl, Br) systems in term of the main physical properties of the interaction potentials involved in the reactive collisions. This was possible taking into account of the results recorded in other laboratories and obtained in Penning ionization electron spectroscopy (PIES) studies previously performed either by Hotop and coworkers [25, 26] and by Ohno group [27].

Previous mass determinations on Ne^*- HCl chemi-ionization performed in our laboratory showed the formation of the following ionization channels at an averaged collision energy of 45 meV:

$$Ne^* - HCl \rightarrow Ne + HCl^+ + e^- \qquad (\approx 87\%) \qquad (2)$$

$$\rightarrow NeH^+ + Cl + e^- \qquad (\approx 10.5\%) \qquad (3)$$

$$\rightarrow NeHCl^+ + e^- \qquad (\approx 2.5\%) \qquad (4)$$

Very interesting was the formation of the protonated neon ion, NeH^+ coming out from a chemical proton transfer reaction between the colliding reagents. Furthermore, experiments aimed to investigate the isotopic effect have been performed studying by mass spectrometry the chemi-ionization of DCl induced by Ne^* atoms. Such an experiment pointed out a statistical character of the proton transfer process that was confirmed by theoretical calculations performed by SCF, CI, and CASSCF methods [23]. In this paper we extend such an analysis to the Ne^*-HBr chemi-ionization in order to achieve a complete understanding of such autoionization reactions. In particular, we are able to emphasize that the reaction channels, leading to the formation of parent HCl^+ or HBr^+ ions in ground and excited electronic states, of associated ($NeHCl^+$ and $NeHBr^+$) ions and of NeH^+ proton transfer species, are opened along most suitable angular cones having specific orientation and acceptance.

2 Theoretical and Computational Approach

The theoretical and computational approach employed for atom-molecules autoionization processes is founded on the so called Optical Potential Model (OPM) introduced first in 1940 by Hans Bethe [28]. In this model is adopted a complex potential W in order to fully describe the intermolecular potential microscopic autoionization dynamics according to the following equation:

$$W = V - \frac{i}{2}\Gamma \tag{5}$$

V is the real part of the complex potential able to simulate the Ng^* - X collisional inter-action during the approach of the reacting partners, while Γ is an imaginary function of such a OPM that can be used for a proper quantification of the autoionization probability through which the intermediate collisional complex $(Ng\text{---}X)^*$ (i.e., the transition state of the autoionization reaction) evolves toward the formation of the final ion products [29–32].

In our laboratory we have developed a new semiclassical treatment which was recently applied to various atom-atom autoionization processes [33, 34]. In these studies, for the first time the real and the imaginary part of OPM were treated as two consid-ered interdependent analytical functions, following the general treatment whose details can be found by the reader in refs. [11, 12] where are also highlighted differences and peculiarities respect to previous computational models commonly used to treat from a theoretical point of view autoionization reactions [35, 36]. In this model, we are able to clarify that autoionization reactions are prototype of simple oxidation processes and can involve two different microscopic reaction mechanisms: (i) at low energy collision, high internuclear distances are probed by the reacting partners and a typical photo-ionization (physical) mechanism is the responsible of the production of final ions (*indirect mechanism*); (ii) for collision energy in the thermal or hyper-thermal regime a real oxidation (chemical) mechanism become prominent through an electron transfer between Ng^* and X (*direct mechanism*) as is showed in Eq. (1) [12]. Furthermore, our semiclassical treat-ment allowed to obtain a clear physical description of the reactive collision highlighting the direct relation between the angular momentum couplings by Hund's cases and the selectivity of the electronic rearrangements which trigger the two *direct* and *indirect* microscopic mechanisms mentioned above [33].

In this paper, we apply our computational treatment in the analysis of the total and partial ionization cross sections obtained in experiments performed using a beam of $Ne^*(^3P_{0,2})$ as an ionizing agent of HCl and HBr molecules. Our theoretical description is useful not only to understand the behavior of either total ionization cross sections and branching ratios of ionic products, but also to evaluate the more probable orientation of the Ng^*-HX (X = Cl and Br) reagents in their preferential reactive angular cones, establishing if and when the involved intermolecular electric field gradients (owing to anisotropic intermolecular forces) are effective. For this latter purpose we used the same theoretical approach successfully applied in the case of autoionization reactions involving other hydrogenated molecules as H_2O, NH_3 and H_2S [37, 38].

3 Cross Section Measurements

The measurements of total and partial ionization cross sections as well as of branching ratios of various accessible ionization channels following the autoionization between $Ne^*(^3P_{0,2})$ metastable atoms and the molecular HX targets have been performed using mass spectrometry technique coupled with a molecular beam apparatus. This prototype high vacuum machine, is composed by three differentially pumped vacuum chambers able to study the microscopic stereodynamics of autoionization processes in a single collision condition, since operating at about 10^{-7} mbar. Such an apparatus consists of a primary chamber which is the one where the rare gas metastable atoms can be generated using two kinds of sources: (i) an electron bombardment device working at about 150 eV; (ii) a microwave discharge using a brass (water-cooled) resonant cavity working at 2450 MHz and 70–200 kW as typical operating power range.

In the second vacuum chamber, the primary beam of metastable atoms so produced is analyzed in its velocity distribution using the time of flight (TOF) technique by means of a TOF disk: this technique is used to record mass spectra and cross sections as a function of the collision energy that are the physical observables presented in this paper.

The scattering chamber is the third one, hosting the crossed molecular beams volume (which crosses at right angle) and the two diagnostic devices allowing the detection of product ions and the kinetic energy of emitted electrons from the transition state of the studied reactions. For the first measurements, a quadrupole mass filter is used and is located below the scattering volume. For the latter kind of measurements, an electron energy analyzer, placed above the crossed molecular beams allows the so-called Penning Ionization Electron Spectroscopy (PIES) measurements, obtaining a real spectroscopy of the transition state for such processes. The experimental apparatus is schematized in Fig. 1 and is fully described elsewhere [39–41].

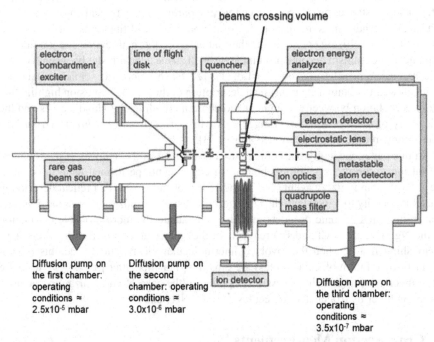

Fig. 1. The scheme of the prototype apparatus operating in the laboratory of Perugia. It consists of two molecular beams crossing at right angle in the third vacuum chamber (see the right side of the figure). Left side: the first vacuum chamber hosting the source of metastable rare gas atoms. The second vacuum chamber is shown in the central part of the diagram: here the velocity selection of the metastable atomic beam is performed by TOF technique. Detectors of the metastable rare gas atom beam and of ion products are channel electron multipliers (CEM) are located in the scattering chamber (on the right) together the quadrupole mass filter and the electron energy analyzed which are the used diagnostic techniques.

4 Results and Analysis

Mass spectrometric determinations on $Ne^*(^3P_{0,2})$-Br autoionizing collisions recorded in the collision energy range of 30–500 meV showed the possible formation of HBr^+ Penning ions, $NeHBr^+$ associative ionization and NeH^+ ions produced by a proton transfer reaction between the collisional partners. The relative abundances for each open ionization channels, as determined at an averaged collision energy of 45 meV, are reported below:

$$Ne^* - HBr \; \rightarrow \; Ne + HBr^+ + e^- \qquad (\approx 96.6\%) \qquad (6)$$

$$\rightarrow NeH^+ + Cl + e^- \qquad (< 1.0\%) \qquad (7)$$

$$\rightarrow NeHBr^+ + e^- \qquad (\approx 2.7\%) \qquad (8)$$

From the mass spectra recorded as a function of the collision energy, we have determined the total ionization cross sections and the branching ratios for the possible ionization open channels. Such data are collected for the two Ne^*-HX (X = Cl and Br) systems in the same experimental conditions and are shown in Fig. 2 in a comparative way.

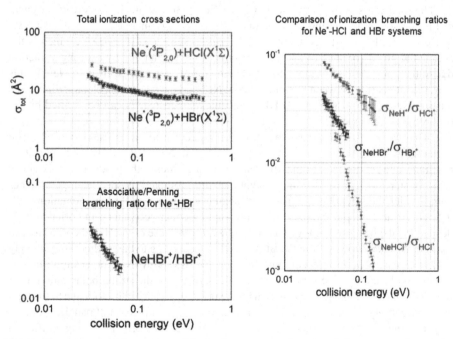

Fig. 2. The comparison between total ionization cross sections (see left-upper panel) and the branching ratios (see left-lower and right panel) recorded for the accessible ionization channels in Ne^* + HX (X = Cl and Br) systems, measured under the same experimental conditions as a function of the collision energy.

From Fig. 2, it can be appreciated that the comparison between the cross sections for the two analyzed systems allows to clearly highlight that for HCl compared to HBr there is a more efficient ionization process in the whole range of the probed collision energies: the ionization cross section is 20 $Å^2$ at a collision energy of 100 meV for Ne^*-HCl, while respect in the case of Ne^*-HBr its value is of about 10 $Å$ ^{2} at the same energy. In addition, the cross sections for HBr show a sharper decrease as the collision energy increases.

Looking at the mass spectrometric data of Eqs. (2)–(4) and (6)–(8), it should be point out that in the case of Ne^*-HCl there is a higher efficiency to react following a proton transfer reaction producing NeH^+ final ions compared to the case of Ne^*-HBr autoionizing collisions. In fact, in Ne^*-HBr autoionization the production of NeH^+ ion is very small as indicated by the relative abundance that is less than 1% at 45 meV of collision energy (see Eq. (7)). Due to this limitation, it was not possible to obtain branching ratios data as a function of the collision energy for the ratio NeH^+/HBr^+ as done in the case of Ne^*-HCl, whose data are reported in the right panel of Fig. 2.

Concerning the data of branching ratios related to associative $(NeHX^+)$/Penning (NeX^+) ionization channels, we can note that the relative probability of the associative ionization is quite similar for the two investigated systems when we are probing the lower portion of the investigated thermal energy range, while for collision energies higher than 45 meV the formation of $NeHCl^+$ ion shows a faster decrease as the collision energy increases respect to $NeHBr^+$.

The total cross sections for Ne^*-HCl and Ne^*-HBr systems in the left-upper panel of Fig. 2 show a decreasing trend as the collision energy increases. This is clear evidence that for both systems the reaction transition state, i.e., $(Ne\cdots HX)^*$, which is the collisional precursor state of entrance channels, is characterized by a strong attraction due to the polarization of outer and floppy cloud of $3s$ external electron of Ne^*. In this way, the interaction between the ionic core of Ne^+ and the permanent electric dipole and quadrupole moments of HCl and HBr molecules is boosted, as can be verified by the data reported in Table 1.

In has to be noted that the symmetry of the molecular orbitals, related to the electron exchange and the electron removing, highlights that in Ne^*-HCl and HBr autoionization reactions, open channels forming different electronic states of the product ions should be strongly stereo-selective. This means that they are promoted by specific angular cones within which the reaction takes place with spatial acceptance and space orientation well-defined. As an example, a representation of angular cones acceptance and orientation is depicted in the left panel of Fig. 3 as determined for the case of the H-end collinear approach of Ne^* atom to HX (X = Cl, Br) molecules, according to the methodology already used in the case of atom-molecule autoionization collisions involving other simple hydrogenated molecules [37, 38]. In the present Ne^*-HX cases, the strong polarization of $3s$ floppy atomic cloud of Ne^* is promoted by the positively charged H atom due to the high electronegativity of the two Cl and Br halogen atoms (see Fig. 3 – left panel). In such a way, this interaction is characterized by a mixing of atomic orbitals having a different symmetry and can stimulate an *indirect* reaction mechanism, which is a pure physical mechanism giving rise to a typical photoionization process. This is realized by the break of the validity of the optical selection rules (which allows the metastability of the Ne^*

Table 1. Ground neutral HCl/HBr and the ionic species HCl^+/HBr^+ in their ground and first excited electronic states: μ (electric dipole moment), Q (electric quadrupole), I (ionization potential), E_A (electron affinity), D_0 (dissociation energy), R_e (equilibrium distance), asymptotic correlation with atomic states and energy levels.

	HCl ($X^1\Sigma^+$)	HBr ($X^1\Sigma^+$)	HCl+ ($X2\Pi_{3/2}$)	HBr+ ($X2\Pi_{3/2}$)	HCl+ ($A2\Sigma_{1/2}$)	HBr+ ($A2\Sigma_{1/2}$)
μ/D	1.103	0.827				
Q/DÅ	3.8	4.3				
I/eV	12.751	11.672				
E_A	< 0	< 0				
D_0/eV	4.434	3.758	4.651	3.900	1.760	1.775
R_e/Å	1.275	1.414	1.315	1.448	1.514	1.684
Asymptotic correlation	H($^2S_{1/2}$) + Cl($^2P_{3/2}$)	H($^2S_{1/2}$) + Br($^2P_{3/2}$)	H($^2S_{1/2}$) + Cl$^+$($^3P_2$)	H($^2S_{1/2}$) + Br$^+$($^3P_2$)	H$^+$ + Cl($^2P_{3/2}$)	H($^2S_{1/2}$) + Br$^+$($^1D_2$)
Energy level/eV	0 (reference)	0 (reference)	12.751	11.672	16.272	15.296

atoms) and generates a virtual photon exchange between reagents accompanied by an electron ejection [11, 12].

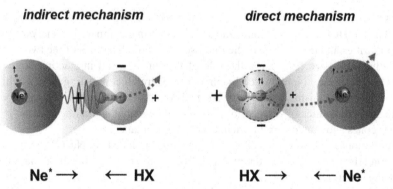

Fig. 3. The schematic view of atomic and molecular orbitals of Ne* and HX (X = Cl and Br) reagents in a collinear configuration of the autoionizing transition state (Ne⋯HX)* evolving to the HX$^+$ ion formation of in their first excited ($A^2\Sigma$) electronic state. The electronic rearrangements involved are indicated together with the shape of acceptance angular cone for the approach of Ne* on: (i) the H side of HX molecule (indirect-physical-photoionization mechanism – left panel); (ii) the X side of HX molecule (direct-chemical-oxidation mechanism – right panel).

On the other hand, those collisions where the Ne* and HX partners are approaching each other along the collinear direction, with Ne* towards the X atom, are characterize by a different microscopic mechanism which is an oxidation (chemical) mechanism also called *direct mechanism*. In this case, during the collision the HX molecule behaves highlighting its role as a reducing agent. Consequently, the behavior of neon is dominated by its ionic core (acting as fluorine atom, at short internuclear distances and higher collision energies) so inducing the oxidation of HX via an electron transfer according to the exchange mechanism [12] schematized in the right panel of Fig. 3.

In this way, one electron is ejected from a bonding orbital of HX and this leads to the formation of HX^+ in the first excited ($A^2\Sigma$) electronic state (see Fig. 3), strongly reducing its binding energy value respect to that of the neutral HX state. As a consequence, this favors the dissociation of the HX^+ Penning ion giving rise to the production of NeH^+ by a proton transfer reaction. Analogous situation has been as recently demonstrated in various autoionizing collisions studied at thermal energy, as for example in the case of Ne^*+N_2 reactions [42].

Moreover, possible autoionization processes can be also induced by those collisions between the Ne^* and HX molecules in which the reacting partners are approaching each other in a perpendicular direction. In this case, Penning Ionization Electron Spectroscopy studies, performed in high resolution conditions, have demonstrated that the collisional autoionization involves the formation of the ($X^2\Pi$) ionic ground state of HX^+ which is related to a precursor state $[Ne\cdots HX^+(X^2\Pi)]$ having a perpendicular geometry [25–27]. The same PIES spectra, recorded by Hotop group [25, 26] and by Ohno and coworkers [27] clearly show that the formation of the precursor state $[Ne\cdots HX^+(A^2\Sigma)]$ in its first excited electronic state preferentially assumes a collinear configuration and will be produced by the two mechanisms shown in Fig. 3.

5   Conclusions

In this article total and partial ionization cross sections are reported and discussed for $Ne^*(^3P_{0,2})$ + HCl and HBr autoionization processes in a comparative analysis pointing out similarities and differences in the microscopic stereodynamics of the two reactions. In the discussion we considered the role of the fundamental interaction components characterizing the autoionizing collision and we were able to distinguish two different microscopic mechanisms (called as *indirect* ad *direct* mechanisms) which are dependent on the stereodynamics of opened ionization channels confined along specific reactive angular cones showing different orientation and acceptance.

Furthermore, the lower efficient autoionization induced by $Ne^*(^3P_{0,2})$ metastable atoms on HBr molecules, respect to that of HCl, is the result of the combination of the following different factors:

a) Polarization effects on Ne^* which trigger the *indirect* reaction mechanism (the physical mechanism depicted in the left panel of Fig. 3); they are attenuated since HBr molecule is less polar with respect to HCl (see the values of dipole moments reported in Table 1);

b) The overlap integral between orbitals of the colliding partners exchanging the electron (Ne^* atoms and HX molecules) tends to reduce since the molecular orbitals of HBr are more diffuse respect to those of HCl;

c) The crucial role of the so called "σ hole" [43–50], localized along HX bond length on the external X part of HX molecules, tends to increase its effect in the case of X = Br; this causes a reduction on the reactivity along the corresponding angular cone of Fig. 3 (right panel) related to the *direct mechanism* (the chemical one) due to the change of the HX molecule polar character.

Acknowledgments. This work was supported and financed with the "Fondo Ricerca di Base, 2018, dell'Università degli Studi di Perugia" (Project Titled: Indagini teoriche e sperimentali sulla reattività di sistemi di interesse astrochimico). Support from Italian MIUR and University of Perugia (Italy) is acknowledged within the program "Dipartimenti di Eccellenza 2018–2022".

References

1. Hotop, H., Illenberger, E., Morgner, H., Niehaus, A.: Chem. Phys. Lett. **10**(5), 493–497 (1971)
2. Siska, P.E.: Rev. Mod. Phys. **65**, 337 (1993)
3. Brunetti, B., Vecchiocattivi, F.: Autoionization dynamics of collisional complexes. In: Ng, C.Y., Baer, T., Powis, I. (eds.) Current Topic on Ion Chemistry and Physics, pp. 359–445. Wiley, New York (1993)
4. Falcinelli, S., Pirani, F., Candori, P., Brunetti, B.G., Farrar, J.M., Vecchiocattivi, F.: Front. Chem. **7**, 445 (2019)
5. Falcinelli, S., Pirani, F., Vecchiocattivi, F.: Atmosphere **6**(3), 299–317 (2015)
6. Alagia, M., Balucani, N., Candori, P., Falcinelli, S., Richter, et al.: Rendiconti Lincei Scienze Fisiche e Naturali **24**, 53–65 (2013)
7. Dulieu, O., Osterwalder, A.: Cold Chemistry, pp. 1–670. Molecular Scattering and Reactivity Near Absolute Zero, Royal Society of Chemistry, Cambridge (2018)
8. Calcote, H.F.: Electrical properties of flames. Symp. Combust. Flame Explos. Phenom. **3**(1), 245–253 (1948)
9. Sugden, T.M.: Excited species in flames. Annu. Rev. Phys. Chem. **13**(1), 369–390 (1962)
10. Falcinelli, S., Vecchiocattivi, F., Pirani, F.: Phys. Rev. Lett. **121**, 163403 (2018)
11. Falcinelli, S., Vecchiocattivi, F., Pirani, F.: Commun. Chem **3**, 64 (2020)
12. Falcinelli, S., Farrar, J.M., Vecchiocattivi, F., Pirani, F.: Acc. Chem. Res. **53**, 2248 (2020)
13. Bettendorff, M., Peyerimhoff, S.D., Buenker, R.J.: Chem. Phys. **66**, 261 (1982)
14. Werner, H.-J., Rosmus, P., Schätzl, W., Meyer, W.: J. Chem. Phys. **80**, 831 (1984)
15. Chapman, D.A., Balasubramanian, K., Sin, S.H.: Phys. Rev. A **38**, 6098 (1988)
16. Candori, P., Falcinelli, S., Pirani, F., Tarantelli, F., Vecchiocattivi, F.: Chem. Phys. Lett. **436**, 322–326 (2007)
17. Snyder, H.L., Smith, B.T., Martin, R.M.: Chem. Phys. Lett. **94**, 90 (1983)
18. de Vries, M.S., Tyndall, G.W., Martin, R.M.: J. Chem. Phys. **80**, 1366 (1984)
19. Tsuji, M., Maier, J.P., Obase, H., Nishimura, Y.: Chem. Phys. **110**, 17–26 (1986)
20. Obase, H., Tsuji, M., Nishimura, Y.: J. Chem. Phys. **87**, 2695–2699 (1987)
21. Simon, W., Yencha, A.J., Ruf, M.-W., Hotop, H.: Z. Phys. D **8**, 71 (1988)
22. Tokue, I., Tanaka, H., Yamasaki, K.: J. Phys. Chem. A **106**, 6068-6074 (2002)
23. Aguilar Navarro, A., Brunetti, B., Falcinelli, S., Gonzalez, M., Vecchiocattivi, F.: J. Chem. Phys. **96**(1), 433–439 (1992)
24. Brunetti, B., Cambi, R., Falcinelli, S., Farrar, J.M., Vecchiocattivi, F.: J. Phys. Chem. **97**(46), 11877-11882 (1993)
25. Yencha, A.J., Ganz, J., Ruf, M.-W., Hotop, H.: Z. Phys. D - Atoms Mol. Clust. **14**, 57–76 (1989)
26. Yencha, A.J., Ruf, M.-W., Hotop, H.: Z. Phys. D - Atoms Mol. Clust. **21**, 113–130 (1991)
27. Imura, K., Kishimoto, N., Ohno, K.: J. Phys. Chem. A **106**, 3759–3765 (2002)
28. Bethe, H.A.: Phys. Rev. **57**, 1125–1144 (1940)
29. Miller, W.H., Morgner, H.: J. Chem. Phys. **67**, 4923–4930 (1977)
30. Gregor, R.W., Siska, P.E.: J. Chem. Phys. **74**, 1078–1092 (1981)
31. Brunetti, B., Candori, P., Falcinelli, S., Pirani, F., Vecchiocattivi, F.: J. Chem. Phys. **139**(16), 164305 (2013)

32. Falcinelli, S., Candori, P., Pirani, F., Vecchiocattivi, F.: Phys. Chem. Chem. Phys. **19**(10), 6933–6944 (2017)
33. Falcinelli, S., Vecchiocattivi, F., Pirani, F.: J. Phys. Chem. A **125**(7), 1461–1467 (2021)
34. Falcinelli, S., Vecchiocattivi, F., Pirani, F.: J. Chem. Phys. **150**(4), 044305 (2019)
35. Nakamura, H.: J. Chem. Phys. Jpn. **26**, 1473–1479 (1969)
36. Miller, W.H.: J. Chem. Phys. **52**, 3563–3572 (1970)
37. Falcinelli, S., Bartocci, A., Cavalli, S., Pirani, F., Vecchiocattivi, F.: Chem. Eur. J. **22**(2), 764–771 (2016)
38. Falcinelli, S., Rosi, M., Cavalli, S., Pirani, F., Vecchiocattivi, F.: Chem. Eur. J. **22**(35), 12518–12526 (2016)
39. Biondini, F., Brunetti, B.G., Candori, P., De Angelis, F., et al.: J. Chem. Phys. **122**(16), 164308 (2005)
40. Brunetti, B., Candori, P., Falcinelli, S., Lescop, B., et al.: Eur. Phys. J. D **38**(1), 21–27 (2006)
41. Brunetti, B.G., Candori, P., Cappelletti, D., Falcinelli, S., et al.: Chem. Phys. Lett. **539–540**, 19–23 (2012)
42. Falcinelli, S., Vecchiocattivi, F., Pirani, F.: Sci. Rep. **11**, 19105 (2021)
43. Falcinelli, S., Vecchiocattivi, F., Farrar, J.M., Brunetti, B.G., Cavalli, S., Pirani, F.: Chem. Phys. Lett. **778**, 138813 (2021)
44. Falcinelli, S., Vecchiocattivi, F., Farrar, J.M., Pirani, F.: J. Phys. Chem. A **125**(16), 3307–3315 (2021)
45. Falcinelli, S., et al.: RSC Adv. **12**, 7587–7593 (2022)
46. Skouteris, D., Balucani, N., Ceccarelli, C., Faginas Lago, N., et al.: MNRAS **482**(3), 3567–3575 (2019)
47. Bartocci, A., Belpassi, L., Cappelletti, D., Falcinelli, S., et al.: J. Chem. Phys. **142**(18), 184304 (2015)
48. Alagia, M., Candori, P., Falcinelli, S., Mundim, M.S.P., Pirani, F., et al.: J. Chem. Phys. **135**(14), 144304 (2011)
49. Alagia, M., Biondini, F., Brunetti, B.G., Candori, P., et al.: J. Chem. Phys. **121**(21), 10508–10512 (2004)
50. Brunetti, B., Candori, P., De Andres, J., Pirani, F., M. Rosi, et al.: J. Phys. Chem. A, **101**(41), 7505–7512 (1997)

An Ab Initio Computational Study of Binding Energies of Interstellar Complex Organic Molecules on Crystalline Water Ice Surface Models

Harjasnoor Kakkar$^{(\boxtimes)}$ (iD), Berta Martínez-Bachs (iD), and Albert Rimola (iD)

Departament de Química, Universitat Autònoma de Barcelona,
08193 Bellaterra, Catalonia, Spain
{harjasnoor.kakkar,berta.martinez,albert.rimola}@uab.cat

Abstract. The interstellar medium is extremely heterogeneous in terms of physical environments and chemical composition. Spectroscopic observations in the recent decades have revealed the presence of gaseous material and dust grains covered in ices predominantly of water in interstellar clouds, the interplay of which may elucidate the existence of more than 250 molecular species. Of these species of varied complexity, several terrestrial carbon-containing compounds have been discovered, known as interstellar complex organic molecules (iCOMs) in the astrochemical argot. In order to investigate the formation of iCOMs, it is crucial to explore gas-grain chemistry and in this regard, one of the fundamental parameters is the binding energy (BE), which is an essential input in astrochemical models. In this work, the BEs of 13 iCOMs on a crystalline H_2O-ice surface have been computed by means of quantum chemical periodic calculations. The hybrid B3LYP-D3 DFT method was used for the geometry optimizations of the adsorbate/ice systems and for computing the BEs. Furthermore, to refine the BE values, an ONIOM2-like approximation has been employed to obtain them at CCSD(T), which correlate well with those obtained at B3LYP-D3. Additionally, aiming to lower the computational cost, structural optimizations were carried out using the HF-3c level of theory, followed by single point energy calculations at B3LYP-D3 in order to obtain BE values comparable to the full DFT treatment.

Keywords: Interstellar medium · Complex organic molecules · Binding energy

1 Introduction

The interstellar medium (ISM) is the region between the stars consisting of matter and energy. The ISM presents a vast range of physical conditions, with temperatures spanning from 10 K to 10^6 K, and densities varying from 10^{-4} to 10^8 cm^{-3} [1]. The interstellar matter is composed of gaseous material (such as atoms, molecules, and ions) and solid constituents (like submicron carbonaceous and silicate dust grains) [2]. In the cold (10 K) and dense (10^4 cm^{-3}) environments of prestellar cores [3], also known as interstellar dark molecular clouds, volatile species such as H_2O, CO, CO_2, NH_3, CH_3OH, and CH_4

© The Author(s) 2022
O. Gervasi et al. (Eds.): ICCSA 2022 Workshops, LNCS 13378, pp. 281–292, 2022.
https://doi.org/10.1007/978-3-031-10562-3_21

freeze-out onto the dust grain surfaces forming ice mantles, predominantly of amorphous H_2O ices and apolar CO ices [4].

As the gravitational collapse of the cold cloud proceeds, the interstellar gaseous material emits IR radiation, resulting in warm (100 K) inner envelopes around the low-mass protostar known as hot corinos, which provide viable conditions for rich grain surface chemistry, such as the formation of interstellar complex organic molecules (iCOMs), i.e., chemical compounds containing 6–13 atoms, of which at least one is C [5]. Although the complex species present in the cold regions are generally unsaturated and exotic, the warm cores in young stellar objects contain saturated (hydrogen-rich) complex organic molecules [6]. Despite being formed on the icy dust grains, as the temperature in the hot corinos reaches the mantle sublimation temperature, the species frozen in the ice mantles are released into the gas phase, where they can be detected and observed via their rotational spectra [7].

The molecules require energy to desorb from the grain surfaces. If the chemical, thermal, or radiative energy acquired by the molecule is higher than its interaction energy (i.e., binding energy, BE) with the icy surface, the species can evaporate into the gas phase. In case the acquired energy is a fraction of the BE, the molecules diffuse on the grain surface which is a critical step for chemical reactions to occur [7]. Therefore, obtaining highly accurate BE values of chemical compounds on dust grains is of utmost importance. Moreover, they are important input parameters in gas-grain astrochemical modelling for the characterization of the chemical composition and evolution of the ISM. The molecular species can interact with the icy grain surfaces by either forming chemical bonds (chemisorption), or via weaker (e.g., hydrogen bonding or dispersion) interactions (physisorption). This work focuses on the physical adsorption of iCOMs on a crystalline water ice surface model.

While techniques such as temperature programmed desorption (TPD) are used to estimate BEs experimentally, they are unable to accurately reproduce the extreme and diverse interstellar conditions in the laboratory setup. Hence, quantum chemical calculations provide a robust alternative to determine accurate BEs. This work, therefore, aims to add to the existing network of BEs those of some iCOMs recurrently found in astronomical observations, some of which, furthermore, might be regarded as potential prebiotic precursors.

2   Methods

In this work, the BEs of 13 closed-shell species have been calculated at various levels of theory adopting a periodic approach. Structural optimizations for the adsorption of iCOMs on the crystalline ice surface were carried out using the density functional theory (DFT), with the CRYSTAL17 program package [8]. The hybrid B3LYP functional [9–11] was utilized, along with Grimme's DFT-D3 correction with the Becke-Johnson damping [12, 13] for the treatment of London-dispersion interactions, and an Ahlrichs triple zeta valence (TZV) basis set with polarization functions (henceforth referred to as B3LYP-D3(BJ)/Ahlrichs-TVZ*). Due to the use of finite basis set, the basis set superposition error (BSSE) was encountered, which has been treated using the counterpoise method (CP) [14] as follows:

$$\Delta E^{CP} = \Delta E - BSSE \tag{1}$$

where ΔE is the interaction energy and ΔE^{CP} is the BSSE-corrected interaction energy using the CP method. The calculated binding energies are defined as the inverse of ΔE^{CP}:

$$BE = -\Delta E^{CP} \tag{2}$$

In addition to the computationally costly B3LYP-D3(BJ)/Ahlrichs-TVZ* calculations, the significantly cheaper semi-empirical HF-3c treatment has also been applied for the geometry optimization of the adsorbate/ice complexes. This method uses the MINIX basis set, along with three a posteriori corrections pertaining to London dispersion interactions, BSSE, and short-range deficiencies arising due to the minimal basis set [15]. In order to further improve the HF-3c BEs, single-point energy calculations at B3LYP-D3(BJ)/Ahlrichs-TVZ* on the HF-3c optimized geometries were performed. This relatively accurate yet computationally cost-effective method (hereafter referred to as DFT//HF-3c) can be further employed to calculate BEs on more realistic (i.e., larger and amorphous) ice models, where full DFT treatment is unfeasibly costly.

In order to check the accuracy of the DFT calculated BEs, the CCSD(T) method has been employed, with the Gaussian16 software package [16], along with correlation consistent basis sets extrapolated to the complete basis set limit (CBS). Due to the exceedingly high computational cost of the CCSD(T)/CBS method, an ONIOM2-like approach has been applied, in which the molecules involved in the binding (the model zone, i.e., the iCOMs and three/four water molecules, depending on the size of the iCOM) are treated at CCSD(T), while computing at DFT for the whole system [17]. By following this ONIOM2-like scheme, the BEs are given as:

$$BE(CCSD(T), system) = BE(CCSD(T), model)$$
$$+BE(DFT, system) - BE(DFT, model) \tag{3}$$

The water ice surface has been modelled using the bulk of the proton ordered crystalline structure of P-ice. A 2D-periodic slab model has been defined by cutting along the (010) surface [18–20]. The slab has a null electric dipole moment along the non-periodic z-axis. The slab model consists of twelve atomic layers involving 24 water molecules per unit cell, with cell parameters $|a| = 8.980$ Å and $|b| = 7.081$ Å. Due to the larger size of CH_3CH_2CN, a 2×1 supercell was used to ensure that the intermolecular interactions with the ice slab are predominantly considered, and spurious lateral interactions between adjacent unit cells are avoided.

3 Results

In this work, the BEs of 13 iCOMs (shown in Fig. 1) have been computed on the crystalline (010) water ice surface model. All the compounds are closed-shell species and, except HCOOH, comprise of 7–11 atoms. They contain different functional groups that determine their adsorption properties.

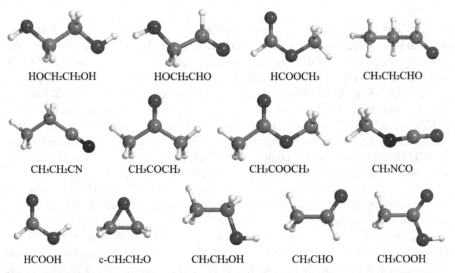

HOCH₂CH₂OH   HOCH₂CHO   HCOOCH₃   CH₃CH₂CHO

CH₃CH₂CN   CH₃COCH₃   CH₃COOCH₃   CH₃NCO

HCOOH   c-CH₂CH₂O   CH₃CH₂OH   CH₃CHO   CH₃COOH

Fig. 1. Set of iCOMs used in this study for the calculation of their BEs on the crystalline water ice surface model.

Based on likely hydrogen bond (H-bond) patterns, the iCOMs can be divided into two different categories: (i) iCOMs containing both H-bond donors and acceptors, due to the presence of hydrogen atoms chemically bonded to electronegative constituents, and (ii) iCOMs containing H-bond acceptors only, due to the presence of electronegative atoms such as O or N.

3.1 BEs of iCOMs at DFT

The B3LYP-D3(BJ)/Ahlrichs-TVZ* optimized geometries of the adsorbed species on the water ice surface are given in Fig. 2, while the calculated BEs and dispersion contribution are listed in Table 1. H-bonding and dispersion interactions are the two major binding driving forces. The general trend observed is that the species acting as both H-bond acceptors and donors have considerably higher BEs (around 80 kJ mol⁻¹) than the molecules that can only act as H-bond acceptors (ranging 50–65 kJ mol⁻¹). The formation of multiple H-bonds by different atoms of the adsorbate (namely, with one H-bond donor and at least one H-bond acceptor) leads to H-bond cooperativity, which strengthens the H-bond network in the system and, in turn, binds the iCOM more efficiently to the ice surface. The number of H-bonds significantly affects the values of the BEs, since for the complexes presenting three H-bonds, they are about 30 kJ mol^{-1} higher than those forming two H-bonds with the ice surface (e.g., CH_3COOH and CH_3CH_2OH respectively). In this work, all the H-bonds between the adsorbate and the surface lie in the range of 1.65–2.16 Å.

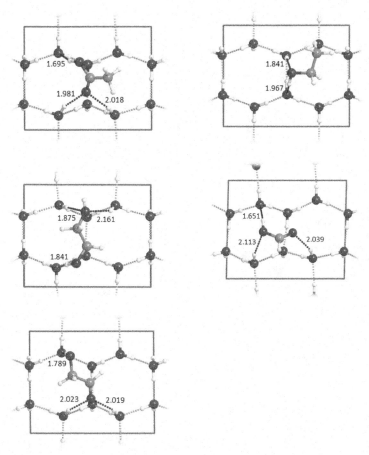

(i) Top-view of B3LYP-D3(BJ)/Ahlrichs-VTZ* optimized geometries of CH_3COOH, CH_3CH_2OH, $HOCH_2CH_2OH$, $HCOOH$, and $HOCH_2CHO$ (hydrogen bond donors and acceptors).

Fig. 2. Optimized geometries of complex organic molecules adsorbed on crystalline (010) ice slab. The hydrogen bond lengths are given in angstrom. Color legend: Red, O; White, H; Grey, C; Blue, N (Color figures online). CH_3CH_2CN uses a 2×1 supercell due to the large system size.

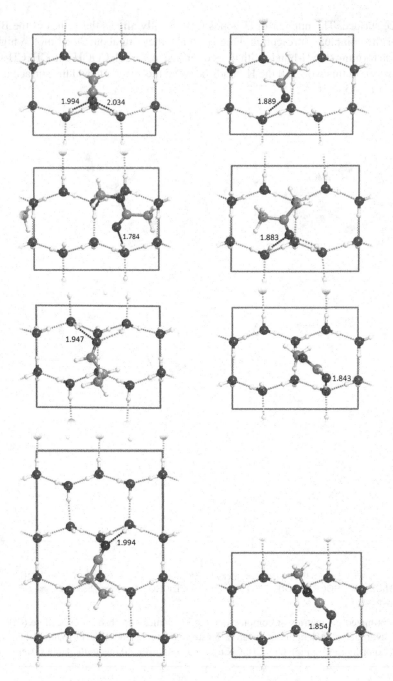

(ii) Top-view of B3LYP-D3(BJ)/Ahlrichs-VTZ* optimized geometries of c-CH₂CH₂O, CH₃CHO, CH₃COOCH₃, CH₃COCH₃, CH₃CH₂CHO, HCOOCH₃, CH₃CH₂CN, and CH₃NCO (hydrogen bond acceptors).

Fig. 2. continued

Table 1. Calculated BEs of iCOMs obtained at B3LYP-D3(BJ)/Ahlrichs-VTZ* level of theory. The contributions of dispersion and non-dispersion interactions are listed, along with the fraction of dispersion contribution to the total BE. The computed BEs are compared with BE values taken from the literature. The values are given in kJ mol^{-1}.

Species	Total BE	No dispersion	Only dispersion (% disp)	Literature values
CH_3COOH	84.8	51.0	33.8 (40%)	67.8[a],
CH_3CH_2OH	51.5	28.9	22.6 (44%)	56.5[a], 34.9[b], 45.7[b]
$HOCH_2CH_2OH$	80.3	44.3	36.0 (45%)	
$HCOOH$	72.0	45.6	26.4 (37%)	
$HOCH_2CHO$	80.8	51.4	29.4 (36%)	
$c\text{-}CH_2CH_2O$	53.0	29.0	24.0 (45%)	
CH_3CHO	57.2	29.0	28.2 (49%)	44.9[b]
CH_3COCH_3	64.7	32.0	32.7 (51%)	
CH_3CH_2CN	57.4	29.7	27.7 (48%)	
CH_3COOCH_3	57.4	24.5	32.9 (57%)	
$HCOOCH_3$	48.2	24.8	23.4 (48%)	38.5[a],
CH_3NCO	47.9	23.4	24.5 (51%)	39.1[b]
CH_3CH_2CHO	57.3	26.9	30.4 (53%)	37.4[b]

[a] Lattelais et al. 2011 [21], [b] Wakelam et al. 2017 [22]

It is also worth noting that, in the species containing only H-bond acceptor groups, the contribution of dispersion interactions in the total BE is higher than those having acceptor and donor H-bond groups, which can form multiple H-bonds. Species such as HCOOH and HOCH$_2$CHO have significantly lower extent of dispersion (37% and 36% of the total respective BEs), whereas dispersion plays the main role (more than the non-dispersive interactions) in molecules such as CH$_3$CH$_2$CHO and CH$_3$COOCH$_3$ (53% and 57% of the total respective BEs). The species containing only H-bond acceptor groups exhibit a more non-polar character than the species containing both H-bond donor and acceptor groups, because the former are CH-containing compounds with heteroatoms. Accordingly, the dispersive contribution to the binding energy of this set of compounds is higher than the H-bond donor and acceptor groups, while for the species in the latter group, the main contributor to the binding energy is H-bonding.

Our computed BE values have been compared with data available in the literature. There are some discrepancies between the previously reported and our computed values. They can mainly be attributed to the difference of the water ice surface models employed. In this work, we have employed a periodic crystalline ice slab consisting of 24 H$_2$O molecules per unit cell. A similar periodic strategy was used in Lattelais et al. [21]. However, they used a different quantum mechanical methodology: calculations at GGA PW91 level of theory using planewaves as basis set. The main difference between the two methods is that Lattelais' calculations lack the inclusion of dispersive forces, which have been seen in this work to be crucial. Consequently, in most of the cases, their values are lower than ours. In Wakelam et al. [22], M06-2X was employed, which is a meta-hybrid DFT method and includes dispersive forces in their definition. However, the main difference is in the water ice surface model: they used a single water molecule. Accordingly, long-range effects such as H-bond cooperativity and extended, actual iCOM-surface dispersion interactions are neglected. Thus, Wakelam's values are also lower than those computed by us.

3.2 Comparison with HF-3c Method

Since most of the water ice in the ISM exists in the amorphous form [4], it is imperative to compute the BEs of the species of interest on the amorphous water ice system. Due to the lack of definite long-range order, an amorphous water ice model would require a much larger unit cell to accurately simulate the structural variability of the amorphous ice mantle. In such cases, methods like DFT would be unfeasible and, accordingly, comparatively cheaper methods such as HF-3c are more practical. Thus, we test the accuracy of the BEs provided by the DFT//HF-3c scheme against those from B3LYP-D3(BJ)/Ahlrichs-TVZ* (here referred to as DFT//DFT). Computed BEs are listed in Table 2.

Table 2. Computed BEs of iCOMs at different levels of theory. The energies are given in kJ mol^{-1}. The values marked with asterisks present BEs on a different HF-3c optimized structure (more details in the text).

Species	DFT//HF-3c	DFT//DFT	CCSD(T)//DFT
CH_3COOH	76.5	84.8	83.3
CH_3CH_2OH	50.8/40.3*	51.5	49.0
$HOCH_2CH_2OH$	79.2/66.4*	80.3	77.6
$HCOOH$	66.3	72.0	68.3
$HOCH_2CHO$	74.6/70.1*	80.8	79.3
$c-CH_2CH_2O$	46.3/44.3*	53.0	53.9
CH_3CHO	58.9/43.6*	57.2	57.8
CH_3COCH_3	59.7	64.7	64.3
CH_3CH_2CN	54.0	57.4	57.9
CH_3COOCH_3	55.3/43.9*	57.4	55.7
$HCOOCH_3$	46.9	48.2	50.1
CH_3NCO	50.0	47.9	49.9
CH_3CH_2CHO	57.5/48.5*	57.3	57.4

Interestingly, in some cases, HF-3c optimized structures differed from the DFT optimized ones, hence resulting in inconsistencies in the computed BEs with both methods. For these cases, we used the DFT optimized structures as the initial guess structure for a new geometry optimization at HF-3c, thus obtaining another BE value that correlates better with the DFT one. Table 2 presents both BEs, with the different geometries marked with asterisks.

3.3 CCSD(T) Refinement

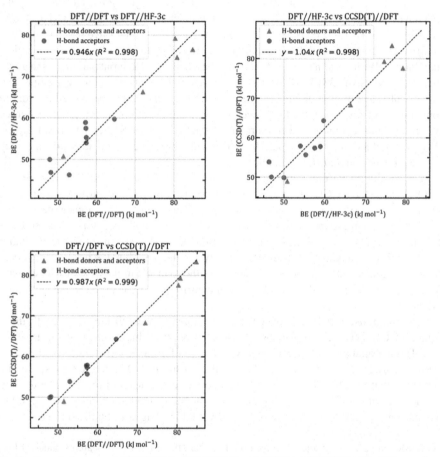

Fig. 3. Correlation among DFT//HF-3c, DFT//DFT, and CCSD(T)//DFT BEs. The BEs are given in kJ mol^{-1}.

Table 2 also presents the BEs computed with the ONIOM2-like CCSD(T)/CBS scheme (referred to as CCSD(T)//DFT) to check the accuracy of the DFT//DFT BEs. Our ONIOM2-like approach includes the effects of the whole system at B3LYP-D3 level of theory and provides energies comparable to a full CCSD(T) treatment. Figure 3 shows the correlation among the computed BE values with the different methodologies: DFT//DFT, DFT//HF-3c, and ONIOM2-like CCSD(T)//DFT. Results indicate that the CCSD(T)/CBS computed BEs are in an excellent correlation with the DFT//DFT values, and the cost-effective DFT//HF-3c BEs are in good correlation with DFT//DFT and

CCSD(T)//DFT too. Hence, it turns out that DFT//HF-3c is a reliable method for the calculation of BEs of iCOMs on large amorphous water ice surfaces.

4   Conclusion

In this work, the BEs of 13 iCOMs (i.e., CH_3COOH, CH_3CH_2OH, $HOCH_2CH_2OH$, HCOOH, $HOCH_2CHO$, c-CH_2CH_2O, CH_3CHO, CH_3COCH_3, CH_3CH_2CN, CH_3COOCH_3, $HCOOCH_3$, CH_3NCO, and CH_3CH_2CHO) on a P-ice based crystalline (010) water ice surface model have been computed by means of periodic quantum chemical calculations. With the aim to provide accurate BEs to be used as input parameters in astrochemical models, the B3LYP-D3(BJ)/Ahlrichs-TVZ* level of theory has been employed. Additionally, a computationally cheaper DFT//HF-3c scheme has been used in order to compare its accuracy with the DFT level for further application in amorphous water ice systems. BE refinements have been carried out at CCSD(T)/CBS level of theory using the ONIOM2 methodology to compare the DFT results with the "gold-standard" accuracy provided by the coupled cluster method.

The results reveal that some iCOMs (CH_3COOH, CH_3CH_2OH, $HOCH_2CH_2OH$, HCOOH, and $HOCH_2CHO$) can act as both H-bond donors and acceptors when adsorbed on the water ice surface. The BEs of such species are significantly higher (nearly 25–30 kJ mol^{-1}) than the molecules that can only act as H-bond acceptors. This indicates that the former species are much strongly bonded to the ice surface than the latter counterparts due to larger H-bond cooperativity. On the other hand, it has been observed that dispersion accounts for a major fraction of the total BE in the case of iCOMs with H-bond acceptors only, while the dispersion contribution is lower in species with high BE values.

The comparison of BEs between those obtained at the full DFT level with those refined at CCSDT//DFT shows an excellent correlation, indicating that the B3LYP-D3(BJ) functional yields accurate energy estimation for the adsorbate/ice systems. Furthermore, DFT//HF-3c results correlate well with the full DFT treated ones, thereby indicating that the DFT//HF-3c scheme is a cost-effective level of theory to be utilized for the calculation of BEs on amorphous ice systems, which require a larger unit cell to account for the lack of long-range order and varied binding sites on the surface.

Acknowledgment. This project received funding from the European Research Council (ERC) under the European Union's Horizon 2020 research and innovation programme grant agreement No. 865657 for the project "Quantum Chemistry on Interstellar Grains" (QUANTUMGRAIN), and from the European Union's Horizon 2020 research and innovation programme under the Marie Skłodowska-Curie grant agreement No. 811312 for the project "Astro-Chemical Origins" (ACO). A.R. is indebted to "Ramón y Cajal" program. CSUC supercomputing center is acknowledged for the allowance of computer resources.

References

1. Van Dishoeck, E.F.: Astrochemistry of dust, ice and gas: introduction and overview. Faraday Discuss. **168**, 9–47 (2014)
2. Williams, D.A., Herbst, E.: It's a dusty universe: surface science in space. Surf. Sci. **500**, 823–837 (2002)
3. Herbst, E., Yates, J.T.: Introduction: astrochemistry. Chem. Rev. **113**(12), 8707–8709 (2013)
4. Boogert, A.A., Gerakines, P.A., Whittet, D.C.: Observations of the Icy universe. Ann. Rev. Astron. Astrophys. **53**(1), 541–581 (2015)
5. Herbst, E., van Dishoeck, E.F.: Complex organic interstellar molecules. Annu. Rev. Astron. Astrophys. **47**(1), 427–480 (2009)
6. Herbst, E.: Chemistry in the interstellar medium. Annu. Rev. Phys. Chem. **46**, 27–53 (1995)
7. Caselli, P., Ceccarelli, C.: Our astrochemical heritage. Astron. Astrophys. Rev. **20**(1), 56 (2012)
8. Dovesi, R., et al.: Quantum-mechanical condensed matter simulations with crystal. WIREs Comput. Mol. Sci. **8**(4), e1360 (2018)
9. Becke, A.D.: Density-functional exchange-energy approximation with correct asymptotic behavior. Phys. Rev. A **38**, 3098–3100 (1988)
10. Becke, A.D.: Density-functional thermochemistry. iii. The role of exact exchange. J. Chem. Phys. **98**(7), 5648–5652 (1993)
11. Lee, C., Yang, W., Parr, R.G.: Development of the Colle-Salvetti correlation-energy formula into a functional of the electron density. Phys. Rev. B **37**, 785–789 (1988)
12. Grimme, S., Antony, J., Ehrlich, S., Krieg, H.: A consistent and accurate ab initio parametrization of density functional dispersion correction (DFT-D) for the 94 elements H-Pu. J. Chem. Phys. **132**(15), 154104 (2010)
13. Grimme, S., Ehrlich, S., Goerigk, L.: Effect of the damping function in dispersion corrected density functional theory. J. Comput. Chem. **32**(7), 1456–1465 (2011)
14. Boys, S.F., Bernardi, F.: The calculation of small molecular interactions by the differences of separate total energies. Some procedures with reduced errors. Mol. Phys. **19**(4), 553–566 (1970)
15. Sure, R., Grimme, S.: Corrected small basis set Hartree-Fock method for large systems. J. Comput. Chem. **34**(19), 1672–1685 (2013)
16. Frisch, M.J., et al.: Gaussian16 Revision C.01. Gaussian Inc., Wallingford CT (2016)
17. Dapprich, S., Komáromi, I., Byun, K., Morokuma, K., Frisch, M.: A new ONIOM implementation in gaussian98. Part I. The calculation of energies, gradients, vibrational frequencies and electric field derivatives. Comput. Theor. Chem. **461–462**, 1–21 (1999)
18. Ferrero, S., Zamirri, L., Ceccarelli, C., Witzel, A., Rimola, A., Ugliengo, P.: Binding energies of interstellar molecules on crystalline and amorphous models of water ice by ab initio calculations. Astrophys. J. **904**(1), 11 (2020)
19. Pisani, C., Casassa, S., Ugliengo, P.: Proton-ordered ice structures at zero pressure. A quantum-mechanical investigation. Chem. Phys. Lett. **253**(3–4), 201–208 (1996)
20. Zamirri, L., Casassa, S., Rimola, A., Segado-Centellas, M., Ceccarelli, C., Ugliengo, P.: IR spectral fingerprint of carbon monoxide in interstellar water-ice models. Mon. Not. R. Astron. Soc. **480**(2), 1427–1444 (2018)
21. Lattelais, M., Bertin, M., Mokrane, H., Romanzin, C., Michaut, X., et al.: Differential adsorption of complex organic molecules isomers at interstellar ice surfaces. Astron. Astrophys. **532**, A12 (2011)
22. Wakelam, V., Loison, J.-C., Mereau, R., Ruaud, M.: Binding energies: new values and impact on the efficiency of chemical desorption. Mol. Astrophys. **6**, 22–35 (2017)

International Workshop
on Computational Methods for Porous
Geo-materials (CompPor 2022)

Optimization of the Training Dataset for Numerical Dispersion Mitigation Neural Network

Kirill Gadylshin[1(✉)], Vadim Lisitsa[1] ⓘ, Kseniia Gadylshina[2],
and Dmitry Vishnevsky[1]

[1] Institute of Petroleum Geology and Geophysics SB RAS,
3 Koptug ave., Novosibirsk 630090, Russia
{GadylshinKG,lisitsavv}@ipgg.sbras.ru
[2] Sobolev Institute of Mathematics SB RAS,
4 Koptug ave., Novosibirsk 630090, Russia

Abstract. We present an approach to construct the training dataset for
the numerical dispersion mitigation network (NDM-net). The network is
designed to suppress numerical error in the simulated seismic wavefield.
The training dataset is the wavefield simulated using a fine grid, thus
almost free from the numerical dispersion. Generation of the training
dataset is the most computationally intense part of the algorithm, thus
it is important to reduce the number of seismograms used in the training
dataset to improve the efficiency of the NDM-net. In this work, we introduce the discrepancy between seismograms and construct the dataset, so
that the discrepancy between the dataset and any seismogram is below
the prescribed level.

Keywords: Deep learning · Seismic modelling · Numerical dispersion

1 Introduction

Seismic modeling is widely used to study elastic wave propagation in complex
Earth models. In particular, numerical simulation allows understanding peculiarities of the wavefields in models with small-scale heterogeneities [7], anisotropic
[17], viscoelastic [2], and poroelastic [13] media, in models with complex free-
surface topology [10,19]. However, seismic modeling is a computationally intense
procedure that requires the use of high-performance computations. In particular,
simulation of the wavefield corresponding to one source may take up to several

---

V.L. developed the algorithm of optimal dataset construction under the support of
RSF grant no. 22-11-00004. D.V. performed seismic modeling using NKS-30T cluster
of the Siberian Supercomputer Center under the support of RSF grant no. 22-21-00738.
Kseniia Gadylshina performed numerical experiments on NDM-net training under the
support of the RSF grant no. 19-77-20004. Kirill Gadylshin optimized the NDM-net
hyperparameters under the support of the grant for young scientists MK-3947.2021.1.5.

O. Gervasi et al. (Eds.): ICCSA 2022 Workshops, LNCS 13378, pp. 295–309, 2022.
https://doi.org/10.1007/978-3-031-10562-3_22

thousand core hours, whereas thousands of shot positions should be simulated. A standard way to reduce the computational error, which is mainly appeared as the numerical dispersion due to the use of symmetric stencils for approximation, is to increase the grid step (reduce the number of grid points). However, it leads to a rapid increase in numerical dispersion. There are various ways to reduce numerical dispersion using a coarse mesh, including dispersion-suppression schemes [12], use of high order finite element, discontinuous Galerkin, and spectral element methods [1,4,9,11,15]. However, it does not necessarily lead to a reduction of the computational cost of the algorithm, because the number of floating-point operations per degree of freedom (per grid point) increases with the increase of the formal order of approximation.

The other approach to deal with the numerical dispersion is a pre- and post-processing of the emitted and recorded signals [6,14]. However, the numerical dispersion in the recorded signal depends on the wave's ray path and can hardly be formalized. Thus, it may be treated by the Machine Learning methods. In particular, application of the neural networks to suppress numerical dispersion at the post-processing stage was suggested in [3,5,18]. In our previous research, we suggested the approach called the Numerical Dispersion Mitigation network (NDM-net) which is designed to suppress numerical dispersion in already simulated wavefields, recorded at the free surface. We suggest using the peculiarity of the seismic modeling problem; which is the simulation of the wavefield for a high number of right-hand sides (source positions), assuming that the seismograms corresponding to neighboring sources are similar. In this case, the true solution (solution computed on a very fine mesh) corresponding to a relatively small number of source positions can be used as the training dataset. In [3] we illustrated the applicability of the approach to realistic 2D problems. We used as few as 10% of sources equidistantly distributed. However, we have not tried to study the effect of the training dataset on the accuracy of the NDM-net results. In this study, we consider two possible strategies to construct the training dataset. First, we try the different number of equidistantly distributed sources. Second, we construct the training datasets preserving maximal differences between all seismograms and the training dataset.

The remainder of the paper has the following structure. In Sect. 2 we remind the basic concepts of seismic modeling and NDM-net. In Sect. 3 we provide the analysis of the seismograms and introduce the measure in the seismograms space. Different strategies to training dataset construction is presented in Sect. 4.

2   Preliminaries

2.1   Seismic Modelling

Consider a typical statement of seismic modeling problem, where the elastic wave equation is solved in a half-space for a series of right-hand sides. In a short form, the problem can be presented as:

$$L[\boldsymbol{u}] = \boldsymbol{f}(t)\delta(\boldsymbol{x} - \boldsymbol{x}_j^s), \tag{1}$$

where $\boldsymbol{f}(t)$ is the time-dependent right-hand side, either external forces or seismic moments, δ is the delta-function, $\boldsymbol{x}_j^s$ is the location of j-th source. Operator L represents the linear differential operator corresponding to the elastic wave equation with appropriate initial and boundary conditions. The solution of the problem is considered at a free surface, which can be assumed flat for simplicity, for example, $x_3 = 0$. Thus, the solution of a single problem can be represented as

$$\boldsymbol{u}(\boldsymbol{x}_j^s, \boldsymbol{x}^r, t),$$

where $\boldsymbol{x}^r$ is the receiver positions. Note, that the regular acquisition system follows the source position, which is $\boldsymbol{x}^r = \boldsymbol{x}^r(\boldsymbol{x}_j^s)$. It is convenient to consider an independent parameter, called offset $\boldsymbol{x}^o = \boldsymbol{x}_j^s - \boldsymbol{x}^r(\boldsymbol{x}_j^s)$. This parameter varies in the same limits for all source positions; thus, two seismogramms can be directly compared as functions of $(\boldsymbol{x}^o, t)$.

If the wavefield is simulated using a numerical method, in this study, we focus on finite-differences with the fourth-order of approximation in space and second-order in time [8], it can be written as

$$L_h[\boldsymbol{u}_h] = \boldsymbol{f}_h(t)\delta(\boldsymbol{x} - \boldsymbol{x}_j^s), \tag{2}$$

where L_h is the finite-difference approximation of the original differential operator L, $\boldsymbol{f}_h$ is the approximation of the right-hand side, and $\boldsymbol{u}_h$ is the finite-difference solution corresponding to the grid with the step h. Due to the convergence of the finite-difference solution to that of the differential problem one gets the estimate:

$$\|\boldsymbol{u}(\boldsymbol{x}_j^s, \boldsymbol{x}^o, t) - \boldsymbol{u}_h(\boldsymbol{x}_j^s, \boldsymbol{x}^o, t)\| = \varepsilon_h \le C_1 h^4 + C_2 \tau^2 \le C h^2, \tag{3}$$

we assume that $\tau \approx C_0 h$ due to the Courant stability criterion. Parameters C_0, C_1, C_2, and C are constants independent of grid steps.

2.2   NDM-Net

The error estimate (3) means that the finer the grid step the lower the error ε_h; which is

$$\varepsilon_{h_1} \le \varepsilon_{h_2} \quad if \quad h_1 \le h_2.$$

However, the reduction of the grid step leads to a significant increase in the computational resources demand and the computational intensity of the algorithm. We suggested recently [3] using machine learning to map coarse-grid solution to the fine-grid solution:

$$\mathcal{N}[\boldsymbol{u}_{h_2}(\boldsymbol{x}_j^s, \boldsymbol{x}^o, t)] = \tilde{\boldsymbol{u}}_{h_2}(\boldsymbol{x}_j^s, \boldsymbol{x}^o, t),$$

so that

$$\|\tilde{\boldsymbol{u}}_{h_2}(\boldsymbol{x}_j^s, \boldsymbol{x}^o, t) - \boldsymbol{u}_{h_1}(\boldsymbol{x}_j^s, \boldsymbol{x}^o, t)\| \le \varepsilon_{21} << \varepsilon_{h_2}$$

for all source positions x_j^s. If we manage to construct the map, which ensures that ε_{21} is small enough, we get

$$
\begin{aligned}
\|\tilde{u}_{h_2}(x_j^s, x^o, t) &- u(x_j^s, x^o, t)\| \\
\leq \|\tilde{u}_{h_2}(x_j^s, x^o, t) - u_{h_1}(x_j^s, x^o, t)\| &+ \|u_{h_1}(x_j^s, x^o, t) - u(x_j^s, x^o, t)\| \\
\leq \varepsilon_{21} + \varepsilon_{h_1} &< \varepsilon_{h_2}.
\end{aligned}
$$

In particular, we use the special case of Convolutional Neural Network (CNN) - a U-Net [16], however other types of neural networks, for example, generative adversarial networks (GANs) have also been applied to suppress the numerical dispersion [5,18]. The main problem in the NDM-net implementation is the training dataset construction. In our previous study [3] we suggested that due to low model variability in the horizontal direction, the seismograms corresponding to neighboring sources are similar. Thus, we may compute a small number of seismograms corresponding to a small number of sources using a fine enough grid to use them as the training dataset. However, we provided no quantitative analysis of the assumption and effect of wavefield similarity on the NDM-net accuracy.

3 Analysis of Seismogramms

In this section, we provide the study of the seismogram's similarities in dependence on the distance between the sources. Consider a standards 2D acquisition system with sources placed in points x_j^s, $j = 1, ..., J_s$. In this case, the entire set of seismograms can be represented as The entire dataset is a union of the solutions corresponding to all source positions

$$
U = \bigcup_{j=1,...,J_s} u(x_j^s, x^o, t) = \bigcup_{j=1,...,J_s} u(x_j^s).
$$

we further omit the variables t and x^o assuming that they get the same values for all seismograms. To compare the seismograms and measure their similarity we suggest using the repeatable measure - normalized root mean square (NRMS). The NRMS between two traces a_t and b_t at point t_0 using a window size dt is the RMS of the difference divided by the average RMS of the inputs, and expressed as a percentage:

$$
NRMS(a_t, b_t, t_0) = \frac{200 \times RMS(a_t - b_t)}{RMS(a_t) + RMS(b_t)}
$$

where the RMS is defined as:

$$
RMS(x_t) = \sqrt{\frac{\sum_{t_0 - dt}^{t_0 + dt} x_t^2}{N}}
$$

and N is the number of samples in the interval $[t_0 - dt, t_0 + dt]$. We introduce the distance $d(u(x_j^s), u(x_k^s))$ as an average NRMS between $u(x_j^s)$ and $u(x_k^s)$ seismograms.

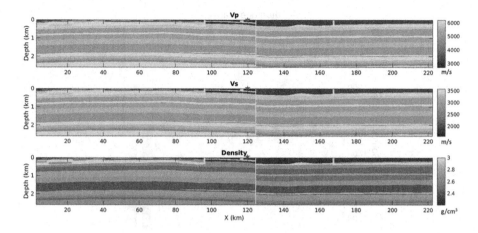

Fig. 1. Vanavar model

We constructed the distance matrix for the entire dataset, computed for the Vanavar model. The model is shown in Fig. 1. The size of the model was 220 km by 2.6 km. The acquisition included 1901 sources with a distance of 100 m. Similarly, we recorded the wavefield by 512 receivers for each shot with maximal source-receiver offsets of 6.4 km. The distance between the receivers was 25 m. The source wavelet was a Ricker pulse with a central frequency 30 Hz. Perfectly matched layers, including the top layer, were used at all boundaries.

We simulated wavefields using grids with steps 2.5 and 1.25 m. After that, for each set of seismograms, we computed the distance matrices, the one corresponding to the simulations with the step 2.5 m is presented in Fig. 2. The values of the distances are low within a narrow band near the main diagonal, which means that the distance between the seismograms is low if the sources are close enough, but it grows rapidly until reaching the value of about 100% of NRMS. To illustrate the relation between the NRMS of two seismograms with respect to the distance between the sources, we provide plots of several columns of the distance matrix in Fig. 3. Each line in this plot represents the NRMS-distance between the given seismogram and all the others. The distance is equal to zero if the seismogram is compared with itself. If the seismogram is compared with those corresponding to the nearby sources, the distance grows almost linearly, from some starting value to a limiting value. After that, the NRMS-distance is almost independent of the distance between the source position. Thus, for each source position x_j^s exist two numbers k_j^+ and k_j^- and value e_j, so that for all $k < k_j^-$ and $k > k_j^+$ the following statement holds $d(u(x_j^s), u(x_k^s)) \approx e_j$. However, these values are individual for each source position. To analyze the boundaries $k_j^{\pm}$ and error value e_j, we study the averaged values of the distance. For each seismogram, we compute the symmetric distance as:

$$d_j(\Delta j) = \frac{1}{2}(d(u(x_j^s), u(x_{j+\Delta j}^s)) + d(u(x_j^s), u(x_{j-\Delta j}^s)))$$

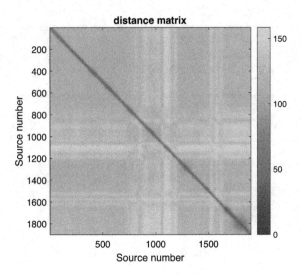

Fig. 2. Vanavar model. The distance matrix

If $j + \Delta j > J_s$ or $j < \Delta j < 1$ we assume that the corresponding distance is equal to symmetric one. After that we compute the mean and standard deviation of the with respect to j obtaining the functions of Δj:

$$M_d(\Delta j) = \frac{1}{J_s} \sum_{j=1}^{J_s} d_j(\Delta j), \tag{4}$$

$$\Sigma_d(\Delta j) = \frac{1}{J_s} \sum_{j=1}^{J} {}_s(d_j(\Delta j) - M_d(\Delta j))^2. \tag{5}$$

The plots of $M_d(\Delta j)$, $M_d(\Delta j) \pm \Sigma_d(\Delta j)$, and $M_d(\Delta j) \pm 3\Sigma_d(\Delta j)$ are presented in Fig. 4. It illustrates that the discrepancy increases if $j < 30$, after that the error stabilizes in average at the value of approximately 120%. The standard deviation starts from 10% for nearby sources to 20% for long-distance sources. It follows from the plot, that if we use equidistantly distributed sources to construct the training dataset we can not use fewer than each thirties, otherwise, we will lose the representativity of the dataset. This analysis allows restricting the lowest number of sources that are reasonable to use if equidistantly distributed (in this particular case it should be more than 3%). On the other hand, it allows us to choose the seismograms to keep the prescribed discrepancy level within the training dataset and this level. In particular, it is worth considering discrepancies between 60 and 90 %.

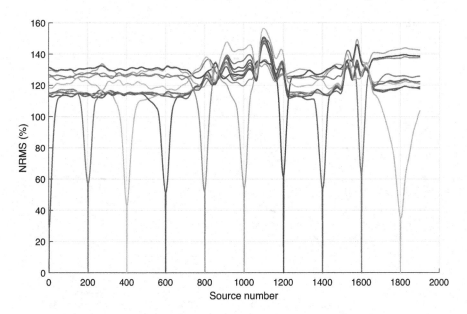

Fig. 3. Vanavar model. The distances for 19 seismogramms.

4   Training Dataset Construction Algorithms

Before describing particular algorithms of dataset construction let us introduce a measure, that characterizes the representativity of a dataset. Assume we have chosen a dataset

$$D_t = \bigcup_{j \in J_t} u(x_j^s),$$

where J_t is a set of training dataset sources indices, so that $J_t \subset \{1, ..., J_s\}$. Thus the training dataset is also a subset of the entire dataset $D_t \subset U$. Let us define the distance from a single seismogram to the dataset as

$$b(x_k^s, D_t) = \min_{j \in J_t} d(u(x_j^s), u(x_k^s)).$$

This function indicates the distance from a given seismogram to the closest one from the training dataset. It is clear that for $k \in J_t$ the distance will be zero. After that, the distance between the datasets can be introduced as

$$B_{\mathrm{inf}} = \max_{k \in \{1, ..., J_s\}} b(x_k^s, D_t) = \max_{k \in \{1, ..., J_s\}} \min_{j \in J_t} d(u(x_j^s), u(x_k^s)).$$

In our further considerations we will use both function $b(x_k^s, D_t)$ and the distance between the datasets B_{inf}.

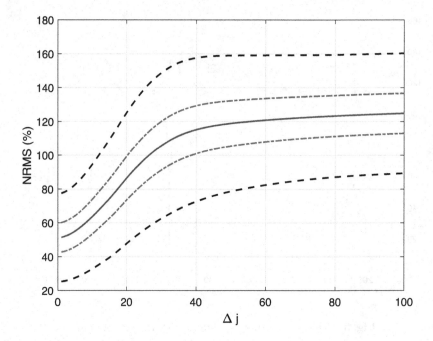

Fig. 4. Vanavar model. Mean NRMS-distance with respect to distance between the sources (solid line), dash-dotted lines correspond to $M_d \pm \Sigma_d$, and dashed lines correspond to $M_d \pm 3\Sigma_d$.

4.1   Equidistantly Distributed Dataset

The first and the simplest algorithm to construct the training dataset is to take the seismograms corresponding to the equidistantly distributed sources. In particular, we considered the datasets composed of 5, 10, and 20% of the entire seismograms denoting them as $D_{5\%}$, $D_{10\%}$, and $D_{20\%}$, respectively. We constructed the functions $b(\boldsymbol{x}_k^s, D_t)$ for the three datasets, as presented in Fig. 5. It is clear, that if a source position belongs to the training dataset the distance $b(\boldsymbol{x}_k^s, D_t)$ is equal to zero, so we excluded these points from the plots. Note, that functions $b(\boldsymbol{x}_k^s, D_{10\%})$ and $b(\boldsymbol{x}_k^s, D_{20\%})$ are almost indistinguishable, whereas variation of $b(\boldsymbol{x}_k^s, D_{5\%})$ is much stronger. This means that dataset $D_{5\%}$ may be under-representative providing poor data for NDM-net. On the contrary, increasing the number of sources in the dataset above 10% does not provide new valuable information.

We used these three datasets to train the NDM-net to map solution computed on a grid with the step of 2.5 m to that simulated using grid with the step of 1.25 m. Consequently, we estimate the accuracy of the NDM-net by introducing the measure

$$q(\boldsymbol{x}_j^s, D_t) = d(\boldsymbol{u}_{h_1}(\boldsymbol{x}_j^s), \tilde{\boldsymbol{u}}_{h_2}(\boldsymbol{x}_j^s)) = d\left(\boldsymbol{u}_{h_1}(\boldsymbol{x}_j^s), \mathcal{N}\left(\boldsymbol{u}_{h_2}(\boldsymbol{x}_j^s)\right)\right), \tag{6}$$

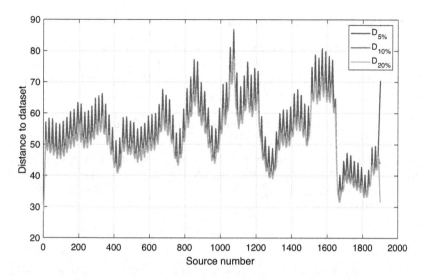

Fig. 5. Distances between the seismograms and the training datasets $b(x_k^s)$ for different dataset of equidistantly distributed source $D_{5\%}$, $D_{10\%}$, and $D_{20\%}$.

where $h_1 < h_2$. That is a source-by-source seismogramms comparison (in this study we simulated the wavefield using the fine mesh for all source positions to be able to validate the NDM-net action). We computed the mean value of

$$M_q = \frac{1}{J_s} \sum_{j=1}^{J_s} q(x_j^s, D_t)$$

overall source positions. The results mean NRMS between the fine grid solution and NDM-net action for three considered training datasets are presented in Table 1. Note, that the error is relatively high for the case of $D_{5\%}$, however, the cases of $D_{10\%}$ and $D_{20\%}$ provide approximately the same accuracy of the NDM-net prediction. Thus, the dataset $D_{10\%}$ can be considered as an optimal one among the datasets of equidistantly distributed sources.

Table 1. Datasets of equidistantly distributed sources.

Dataset	Number of sources	Average NRMS
$D_{5\%}$	283	44.28%
$D_{10\%}$	191	31.91%
$D_{20\%}$	86	29.41%

4.2   Distance-Preserving Datasets

We assume that the entire dataset computed on a coarse mesh is available. Thus, we can compute the distances between all seismograms and chose the source positions to compute the training dataset. We suggest constructing the training dataset by solving the max-min problem:

$$\max_{k \in \{1,...,J_s\}} b(\boldsymbol{x}_k^s, D_t) = \max_{k \in \{1,...,J_s\}} \min_{j \in J_t} d(\boldsymbol{u}(\boldsymbol{x}_j^s), \boldsymbol{u}(\boldsymbol{x}_k^s)) \le Q,$$

where Q is the desired error level. We considered several values of Q starting from 60% to 100% (according to the mean NRMS distances presented in Fig. 11). We considered five datasets $D_{60\%}^{NRMS}$, $D_{70\%}^{NRMS}$,...,$D_{100\%}^{NRMS}$, so that $D_{m\%}^{NRMS}$ correspond to the case of

$$\max_{k \in \{1,...,J_s\}} b(\boldsymbol{x}_k^s, D_t) = \le m,$$

In Figs. 8, 9 and 10 we provide the functions $b(\boldsymbol{x}_k^s, D_{m\%}^{NRMS})$. We kept the sources that belong to the training dataset to visualize the number of sources in the dataset. In particular, the distance is equal to zero if the source belongs to the dataset. For example, the dataset $D_{60\%}^{NRMS}$ contains all sources with numbers from 1500 to 1650. This means that even for the two neighboring sources in this range the NRMS between the seismograms exceeds 60%. Increasing the level of the acceptable NRMS one may reduce the number of sources in the training dataset accelerating the NDM-net. However, it may lead to significant accuracy degradation.

Next, we consider the one-to-one NRMS between the coarse- and fine-mesh solutions $q(\boldsymbol{x}_j^s, D_t)$ as defined by formula (6) for the considered datasets in Fig. 11. We also considered the averaged values of $q(\boldsymbol{x}_j^s, D_t)$ over the source position. The results are provided in Table 2. According to the plots in Fig. 11, all datasets provide similar accuracy in the leftmost part of the model (source numbers up to 800), where the model was relatively simple but original NRMS between fine- and coarse-mesh solutions was about 70%. The main difference between the datasets is associated with source numbers 800 to 1500, where the results of NDM-net applications are different for different datasets. For the source numbers 1500 to 1650, where $D_{60\%}^{NRMS}$ includes all the sources, its accuracy is higher than that of any other dataset. However, in the simplest part of the model (source numbers 1650–1900), all adaptive datasets include very sparse sources distribution,

Table 2. Distance-preserving datasets

Dataset	Number of sources	Average NRMS
$D_{60\%}^{NRMS}$	414	30.28%
$D_{70\%}^{NRMS}$	109	34.69%
$D_{80\%}^{NRMS}$	56	35.11%
$D_{90\%}^{NRMS}$	43	35.68%
$D_{100\%}^{NRMS}$	34	36.26%
$D_{10\%}$	191	31.91%

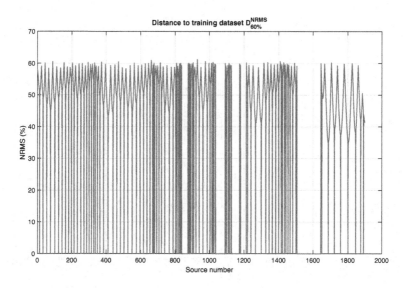

Fig. 6. Distances between the seismograms and the training dataset $b(\boldsymbol{x}_k^s, D_{60\%}^{NRMS})$.

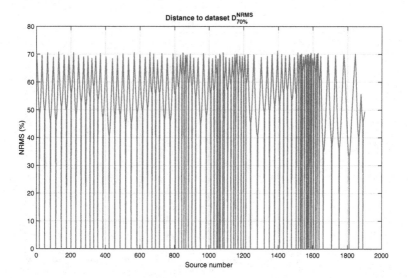

Fig. 7. Distances between the seismograms and the training dataset $b(\boldsymbol{x}_k^s, D_{70\%}^{NRMS})$.

which leads to an increase of the NRMS after the NDM-net application. On average, dataset $D_{60\%}^{NRMS}$ provides the highest accuracy of the NDM-net. However, the number of sources in $D_{60\%}^{NRMS}$ is 414 which is twice as many as in $D_{10\%}$. Dataset $D_{70\%}^{NRMS}$ includes half of the sources of $D_{10\%}$ but it caused a significant increase of the NRMS. According to the presented results, it is reasonable to construct the datasets with adaptive NRMS levels (Figs. 6 and 7).

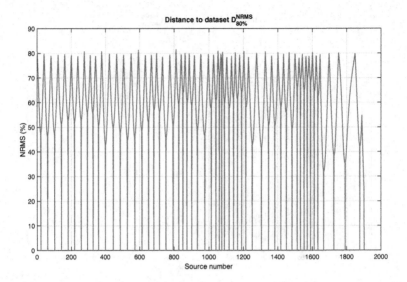

Fig. 8. Distances between the seismograms and the training dataset $b(\boldsymbol{x}_k^s, D_{80\%}^{NRMS})$.

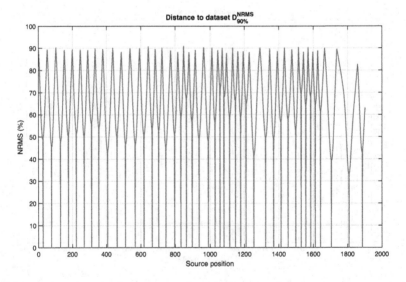

Fig. 9. Distances between the seismograms and the training dataset $b(\boldsymbol{x}_k^s, D_{90\%}^{NRMS})$.

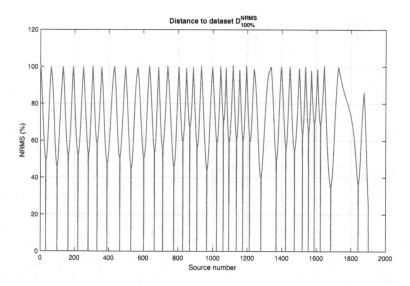

Fig. 10. Distances between the seismograms and the training dataset $b(x_k^s, D_{100\%}^{NRMS})$.

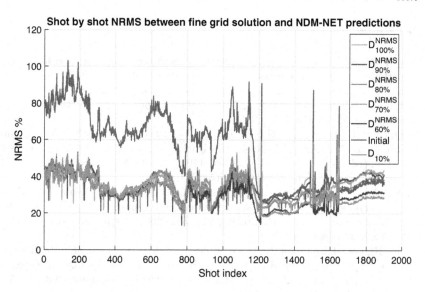

Fig. 11. Shot-by-shot distances between the fine-grid solution and NDM-net corrected solutions for different training datasets.

5   Conclusions

In this study, we consider two possible ways to construct the training datasets for the Numerical Dispersion Mitigation network or NDM-net. The network was designed to suppress numerical error in the seismic modeling results. It is constructed to map a noisy solution computed using a coarse mesh to that computed by a fine mesh. The training dataset is composed of the wavefields corresponding to a small number of sources from the considered acquisition system. Thus, the smaller the number of seismograms in the training dataset the more efficient the approach is. We considered two ways to construct the training datasets. The first one is based on the equidistantly distributed sources. In this case, the optimal set of sources to generated the training dataset is 10% of the entire number of the sources. Reduction of the number of sources leads to rapid error increase. The use of a denser system of sources requires higher computational time to generate the training dataset without significant accuracy improvement. The second way to construct the training dataset is the requirement, that the NRMS-based distance from the entire dataset and the training dataset does not exceed a prescribed level. In this case, the error can be reduced, however, an extremely dense source system may be needed, which may significantly increase the number of source positions in the training dataset. Moreover, due to the variation of the NRMS distance between the seismograms depending on the source position, it seems reasonable to consider an adaptive choice of the NRMS-based distance level to construct the training dataset.

References

1. Ainsworth, M.: Dispersive and dissipative behaviour of high order discontinuous Galerkin finite element methods. J. Comput. Phys. **198**(1), 106–130 (2004)
2. Blanch, J., Robertsson, J., Symes, W.: Modeling of a constant Q: methodology and algorithm for an efficient and optimally inexpensive viscoelastic technique. Geophysiscs **60**(1), 176–184 (1995)
3. Gadylshin, K., Lisitsa, V., Gadylshina, K., Vishnevsky, D., Novikov, M.: Machine learning-based numerical dispersion mitigation in seismic modelling. In: Gervasi, O., et al. (eds.) ICCSA 2021. LNCS, vol. 12949, pp. 34–47. Springer, Cham (2021). https://doi.org/10.1007/978-3-030-86653-2_3
4. Kaser, M., Dumbser, M., Puente, J.D.l., Igel, H.: An arbitrary high-order discontinuous Galerkin method for elastic waves on unstructured meshes III. Viscoelastic attenuation. Geophys. J. Int. **168**(1), 224–242 (2007). https://doi.org/10.1111/j.1365-246X.2006.03193.x
5. Kaur, H., Fomel, S., Pham, N.: Overcoming numerical dispersion of finite-difference wave extrapolation using deep learning. In: SEG Technical Program Expanded Abstracts, pp. 2318–2322 (2019). https://doi.org/10.1190/segam2019-3207486.1
6. Koene, E.F.M., Robertsson, J.O.A., Broggini, F., Andersson, F.: Eliminating time dispersion from seismic wave modeling. Geophys. J. Int. **213**(1), 169–180 (2017)
7. Kostin, V., Lisitsa, V., Reshetova, G., Tcheverda, V.: Local time-space mesh refinement for simulation of elastic wave propagation in multi-scale media. J. Comput. Phys. **281**, 669–689 (2015)
8. Levander, A.R.: Fourth-order finite-difference P-SV seismograms. Geophysics **53**(11), 1425–1436 (1988)
9. Lisitsa, V.: Dispersion analysis of discontinuous Galerkin method on triangular mesh for elastic wave equation. Appl. Math. Model. **40**, 5077–5095 (2016). https://doi.org/10.1016/j.apm.2015.12.039

10. Lisitsa, V., Kolyukhin, D., Tcheverda, V.: Statistical analysis of free-surface vari-
    ability's impact on seismic wavefield. Soil Dyn. Earthq. Eng. **116**, 86–95 (2019)
11. Lisitsa, V., Tcheverda, V., Botter, C.: Combination of the discontinuous Galerkin
    method with finite differences for simulation of seismic wave propagation. J. Com-
    put. Phys. **311**, 142–157 (2016)
12. Liu, Y.: Optimal staggered-grid finite-difference schemes based on least-squares for
    wave equation modelling. Geophys. J. Int. **197**(2), 1033–1047 (2014)
13. Masson, Y.J., Pride, S.R.: Finite-difference modeling of Biot's poroelastic equations
    across all frequencies. Geophysics **75**(2), N33–N41 (2010)
14. Mittet, R.: Second-order time integration of the wave equation with dispersion
    correction procedures. Geophysics **84**(4), T221–T235 (2019)
15. Pleshkevich, A., Vishnevskiy, D., Lisitsa, V.: Sixth-order accurate pseudo-spectral
    method for solving one-way wave equation. Appl. Math. Comput. **359**, 34–51
    (2019)
16. Ronneberger, O., Fischer, P., Brox, T.: U-Net: convolutional networks for biomed-
    ical image segmentation. In: Navab, N., Hornegger, J., Wells, W.M., Frangi, A.F.
    (eds.) MICCAI 2015. LNCS, vol. 9351, pp. 234–241. Springer, Cham (2015).
    https://doi.org/10.1007/978-3-319-24574-4_28
17. Saenger, E.H., Gold, N., Shapiro, S.A.: Modeling the propagation of the elastic
    waves using a modified finite-difference grid. Wave Motion **31**, 77–92 (2000)
18. Siahkoohi, A., Louboutin, M., Herrmann, F.J.: The importance of transfer learning
    in seismic modeling and imaging. Geophysics **84**, A47–A52 (2019). https://doi.org/
    10.1190/geo2019-0056.1
19. Tarrass, I., Giraud, L., Thore, P.: New curvilinear scheme for elastic wave propaga-
    tion in presence of curved topography. Geophys. Prospect. **59**(5), 889–906 (2011).
    https://doi.org/10.1111/j.1365-2478.2011.00972.x

Numerical Solution of Anisotropic Biot Equations in Quasi-static State

Sergey Solovyev[1], Mikhail Novikov[2]([✉]), and Vadim Lisitsa[1]📵

[1] Institute of Mathematics SB RAS, Koptug ave. 4, Novosibirsk 630090, Russia
lisitsavv@ipgg.sbras.ru
[2] Institute of Petroleum Geology and Geophysics SB RAS, Koptug ave. 3,
Novosibirsk 630090, Russia
novikovma@ipgg.sbras.ru

Abstract. Frequency-dependent seismic attenuation can be applied to indicate transport properties of fractured media and fluid mobility within it. In particular, wave-induced fluid flow (WIFF) appears during seismic wave propagation between fractures and background as well as within interconnected fractures, and causes intensive attenuation. We present effective algorithm for numerical upscaling to estimate attenuation in anisotropic fractured porous fluid-saturated media. Algorithm is based on numerical solution of quasi-static Biot equations using finite-difference approximation. Presented algorithm is used to estimate seismic attenuation in fractured media with high fracture connectivity. Results of numerical experiments demonstrate the influence of physical properties and microscale anisotropy of fracture-filling material on seismic attenuation.

Keywords: Poroelaticity · Wave-induced fluid flow · Quasi-static state · Finite differences · Direct methods for SLAE

1 Introduction

Modern challenges in geophysics for seismic monitoring methods (4D seismic) are motivated by the developing of such fields as CO_2 sequestration [8,13] and geothermal energy exploration [11,18]. In particular, transport properties of fractured reservoir, fluid mobility should be estimated properly using obtained seismic data. One popular seismic attribute in modern research is seismic attenuation. In fluid-saturated media seismic wave generates fluid pressure gradients on the interfaces and induces fluid flows, significantly affected by mesoscale geometry structure of reservoir as well as microscale properties of rock (porosity, permeability etc.) and fluid properties. Low-frequency wave (having large wave cycle)

S.S. developed the algorithm and V.L. designed the experiments under the support of RSCF grant no. 19-77-20004. M.N. performed the analysis of the results. Numerical simulations were performed using the supercomputer facilities of Siberian Supercomputer Center (Cluster NKS-30T).

O. Gervasi et al. (Eds.): ICCSA 2022 Workshops, LNCS 13378, pp. 310–327, 2022.
https://doi.org/10.1007/978-3-031-10562-3_23

propagation provide enough time for fracture-to-background flow (FB-WIFF) to form. FB-WIFF is most intensive at low frequencies and depends mostly on the contrast of physical properties between background rock and fracture-filling material [10,16]. Intensive wave-induced fluid flow also appear during high frequency wave propagation between intersecting fractures (fracture-to-fracture WIFF, FF-WIFF). This flow is defined by fracture-filling material properties and fracture connectivity [7,10,16]. Theoretical study of FF-WIFF effect consider relatively simple fractured media structure, and fracture connectivity is presented by intersecting pairs of differently oriented fractures [10]. FF-WIFF phenomena is also studied numerically, but recent studies are often restricted by the same fracture connectivity criteria [10,16]. One exception is study [9] where authors apply discrete fracture network of more complex structure and then stronger connectivity. Also, studies are mostly limited by isotropic case [9,19], when significant anisotropy of elastic and transport properties is typical for rocks, for example, affected by partial dissolution by CO_2. For finite-difference or finite-element approach to simulate seismic wave propagation or creeping test for fractured porous media grid step depends on fracture width, and computation domain size depends on representative volume [1,15] to consider typical percolation length and fracture connectivity [14,21]. Moreover, to obtain effective attenuation estimations numerical upscaling requires a series of experiments for statistically equivalent model of the fracture system. Thus, an efficient numerical upscaling algorithm should be able to solve series on large-scale problems including the microscale anisotropy of the media.

In this paper, we continue and extend our previous studies in numerical solution of quasi-static Biot equations [19] to anisotropic case and present an algorithm for numerical upscaling to estimate effective properties of fractured-porous anisotropic media in low-frequency range. To do so, we develop the algorithm to solve anisotropic Biot equations in quasi-static state. Finite-difference approximation of Biot anisotropic equations is applied to solve boundary value problem to calculate stress-strain behaviour in anisotropic fractured porous fluid-saturated sample. Set of stress-strain state problems form poroelastic media upscaling problem to obtain frequency-dependent stiffness tensor, corresponding to the visco-elastic media. We consider details of the numerical solution of the SLAE resulting from Biot equations approximation. We perform a set of numerical experiments using our algorithm demonstrating the effect of fracture-filling material properties and anisotropy on frequency-dependent attenuation caused by wave-induced fluid flow within highly connected fracture system.

2   Stress-Strain Behaviour in Poroelastic Fluid-Saturated Problem Set Up

2.1   Anisotropic Biot Equations in Quasi-static State

To calculate the stress-strain state in porous fluid-saturated media we consider the anisotropic quasi-static Biot equations in frequency domain governing the diffusion processes in fluid-filled poroelastic anisotropic media in a low-frequency regimes [2,3]. In Cartesian coordinates and 2D case the equations can be written as follows:

$$\frac{\partial}{\partial x}\left[C_{11}(\omega)\frac{\partial u_x}{\partial x} + C_{13}(\omega)\frac{\partial u_z}{\partial z} + \alpha_1 M\left(\frac{\partial w_x}{\partial x} + \frac{\partial w_z}{\partial z}\right)\right]$$
$$+ \frac{\partial}{\partial z}\left[C_{55}(\omega)\left(\frac{\partial u_x}{\partial z} + \frac{\partial u_z}{\partial x}\right)\right] = 0,$$
$$\frac{\partial}{\partial x}\left[C_{55}(\omega)\left(\frac{\partial u_x}{\partial z} + \frac{\partial u_z}{\partial x}\right)\right]$$
$$+ \frac{\partial}{\partial z}\left[C_{13}(\omega)\frac{\partial u_x}{\partial x} + C_{33}(\omega)\frac{\partial u_z}{\partial z} + \alpha_2 M\left(\frac{\partial w_x}{\partial x} + \frac{\partial w_z}{\partial z}\right)\right] = 0, \quad (1)$$
$$\frac{\partial}{\partial x}\left[M\left(\alpha_1\frac{\partial u_x}{\partial x} + \alpha_2\frac{\partial u_z}{\partial z}\right) + M\left(\frac{\partial w_x}{\partial x} + \frac{\partial w_z}{\partial z}\right)\right] = i\omega\frac{\eta}{k_1}w_x,$$
$$\frac{\partial}{\partial z}\left[M\left(\alpha_1\frac{\partial u_x}{\partial x} + \alpha_2\frac{\partial u_z}{\partial z}\right) + M\left(\frac{\partial w_x}{\partial x} + \frac{\partial w_z}{\partial z}\right)\right] = i\omega\frac{\eta}{k_2}w_z,$$

where $u = (u_x, u_z)^T$ is the solid matrix displacement vector, $w = (w_x, w_z)^T$ is the vector of the relative fluid displacement with respect to the solid matrix, C_{ij} is the frequency-dependent stiffness tensor C components of fluid-saturated material, α_1, α_2 are Biot-Willis parameters in axes directions, M is the fluid storage coefficient, η is the fluid dynamic viscosity, κ_1, κ_2 are the absolute permeability of the rock in axes directions, and ω is the temporal frequency. Parameters λ_u, M, and α_1, α_2 are defined by the drained material stiffness tensor and fluid and solid bulk moduli [6,12].

The strain tensor ε and its components are defined as

$$\varepsilon = \begin{pmatrix} \varepsilon_{xx} & \varepsilon_{xz} \\ \varepsilon_{xz} & \varepsilon_{zz} \end{pmatrix} \quad \varepsilon_{xx} = \frac{\partial u_x}{\partial x}, \quad \varepsilon_{zz} = \frac{\partial u_z}{\partial z}, \quad \varepsilon_{xz} = \frac{1}{2}\left(\frac{\partial u_z}{\partial x} + \frac{\partial u_x}{\partial z}\right). \quad (2)$$

The following boundary conditions provide an unique solution of system (1):

$$\sigma_{xx} = C_{11}(\omega)\frac{\partial v_x}{\partial x} + C_{13}(\omega)\frac{\partial v_z}{\partial z}\bigg|_{x = L_1^x} = \phi_x \quad (3)$$
$$w \cdot n|_{\partial D} = 0.$$

Here the first condition is the normal compression at the boundary, and second one is the torsion of domain and the third – no-flow boundary condition at all boundaries.

2.2   Effective Viscoelastic Media

We aim to reconstruct the effective viscoelastic media, so that for any stresses σ_0 applied to a unit volume the average strains in the reconstructed anisotropic

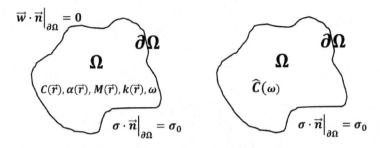

Fig. 1. Anisotropic media: original fluid-saturated poroelastic media (left); reconstructed effective frequency-dependent viscoelastic media (right). $C(r), \alpha(r), M(r), k((r))$ are the space dependent parameters of Eq. (1), n is the outer normal.

(Fig. 1, right) and the original fluid-saturated poroelastic media (Fig. 1, left) should coincide. By reconstruction we mean retrieving the frequency-dependent effective stiffness tensor $\hat{C}$, such that

$$\frac{\partial}{\partial x}\left[\hat{C}_{11}(\omega)\frac{\partial v_x}{\partial x} + \hat{C}_{13}(\omega)\frac{\partial v_z}{\partial z}\right] + \frac{\partial}{\partial z}\left[\hat{C}_{55}(\omega)(\frac{\partial v_x}{\partial z} + \frac{\partial v_z}{\partial x})\right] = 0,$$

$$\frac{\partial}{\partial x}\left[\hat{C}_{55}(\omega)(\frac{\partial v_x}{\partial z} + \frac{\partial v_z}{\partial x})\right] + \frac{\partial}{\partial z}\left[\hat{C}_{13}(\omega)\frac{\partial v_x}{\partial x} + \hat{C}_{33}(\omega)\frac{\partial v_z}{\partial z}\right] = 0,$$

(4)

where $v = (v_x, v_z)^2$ is the displacement vector in viscoelastic media.

Therefore, we need to compute reconstructed tensor $\hat{C}$ (which is constant in Ω) by the space-dependent parameters $C, \alpha_x, \alpha_z, M, k_1, k_2$ in the same geometry in terms of average strains $\int_\Omega \varepsilon dv = \int_\Omega \hat{\varepsilon} dv = I_\varepsilon$ ($\varepsilon \in \{\varepsilon_{xx}, \varepsilon_{zz}, \varepsilon_{xz}\}$) and same stress σ_0 at the boundary $\partial\Omega$. The strains ε distributions in original medium satisfy the solution of the Eq. (1) with boundary conditions (Fig. 1, left). Thus for $\sigma \in \{\sigma_{xx}, \sigma_{zz}, \sigma_{xz}\}$ (components of stress tensor in porous fluid-saturated media) $\int_\Omega \sigma dv = \int_\Omega \hat{\sigma} dv = I_\sigma$.

The strains ε distributions in original medium satisfy the solution of the Eqs. (1) with boundary conditions (Fig. 1, left).

In reconstructed medium $I_\sigma = \int_\Omega \hat{\sigma} dv = \int_\Omega \hat{C}\hat{\varepsilon} dv = \hat{C} \int_\Omega \hat{\varepsilon} dv = \hat{C}I_\varepsilon$

We obtain the SLAE $I_\sigma = \hat{C}I_\varepsilon$ with three equations and nine unknown tensor $\hat{C}$ components:

$$I_{\sigma xx} = \hat{C}_{11}I_{\varepsilon xx} + \hat{C}_{13}I_{\varepsilon zz} + \hat{C}_{15}I_{\varepsilon xz},$$
$$I_{\sigma zz} = \hat{C}_{31}I_{\varepsilon xx} + \hat{C}_{33}I_{\varepsilon zz} + \hat{C}_{35}I_{\varepsilon xz},$$
$$I_{\sigma xz} = \hat{C}_{51}I_{\varepsilon xx} + \hat{C}_{53}I_{\varepsilon zz} + \hat{C}_{55}I_{\varepsilon xz}.$$

(5)

As the solution for $\hat{C}$ is not unique, we need to perform at least three experiments with different sets of boundary conditions to determine the system (5). The strain tensors $I_\varepsilon^{(1)}, I_\varepsilon^{(2)}, I_\varepsilon^{(3)}$ and stress tensors $I_\sigma^{(1)}, I_\sigma^{(2)}, I_\sigma^{(3)}$ obtained as solutions of the system (1) with boundary conditions $\sigma_0^{(1)}, \sigma_0^{(2)}$ and $\sigma_0^{(3)}$, correspondingly, are used to construct SLAE with unique solution:

$$\begin{aligned}
I_{\sigma xx}^{(1)} &= \hat{C}_{11}I_{\varepsilon xx}^{(1)} + \hat{C}_{13}I_{\varepsilon zz}^{(1)} + \hat{C}_{15}I_{\varepsilon xz}^{(1)}, \\
I_{\sigma zz}^{(1)} &= \hat{C}_{31}I_{\varepsilon xx}^{(1)} + \hat{C}_{33}I_{\varepsilon zz}^{(1)} + \hat{C}_{35}I_{\varepsilon xz}^{(1)}, \\
I_{\sigma xz}^{(1)} &= \hat{C}_{51}I_{\varepsilon xx}^{(1)} + \hat{C}_{53}I_{\varepsilon zz}^{(1)} + \hat{C}_{55}I_{\varepsilon xz}^{(1)}, \\
I_{\sigma xx}^{(2)} &= \hat{C}_{11}I_{\varepsilon xx}^{(2)} + \hat{C}_{13}I_{\varepsilon zz}^{(2)} + \hat{C}_{15}I_{\varepsilon xz}^{(2)}, \\
I_{\sigma zz}^{(2)} &= \hat{C}_{31}I_{\varepsilon xx}^{(2)} + \hat{C}_{33}I_{\varepsilon zz}^{(2)} + \hat{C}_{35}I_{\varepsilon xz}^{(2)}, \\
I_{\sigma xz}^{(2)} &= \hat{C}_{51}I_{\varepsilon xx}^{(2)} + \hat{C}_{53}I_{\varepsilon zz}^{(2)} + \hat{C}_{55}I_{\varepsilon xz}^{(2)}, \\
I_{\sigma xx}^{(3)} &= \hat{C}_{11}I_{\varepsilon xx}^{(3)} + \hat{C}_{13}I_{\varepsilon zz}^{(3)} + \hat{C}_{15}I_{\varepsilon xz}^{(3)}, \\
I_{\sigma zz}^{(3)} &= \hat{C}_{31}I_{\varepsilon xx}^{(3)} + \hat{C}_{33}I_{\varepsilon zz}^{(3)} + \hat{C}_{35}I_{\varepsilon xz}^{(3)}, \\
I_{\sigma xz}^{(3)} &= \hat{C}_{51}I_{\varepsilon xx}^{(3)} + \hat{C}_{53}I_{\varepsilon zz}^{(3)} + \hat{C}_{55}I_{\varepsilon xz}^{(3)},
\end{aligned} \tag{6}$$

2.3 Upscaling Procedure

To reconstruct the effective viscoelastic tensor we consider a rectangular domain $D = [L_1^x, L_2^x] \times [L_1^z, L_2^z]$ of anisotropic poroelastic fluid-saturated media with boundary condition $\sigma \cdot \boldsymbol{n} = \sigma_0 = (\sigma_{01}, \sigma_{02})^t$ in following form:

$$\begin{aligned}
(\sigma_{xx}, \sigma_{xz})^t|x = L_1^x, x = L_2^x = (\sigma_{01}, \sigma_{02})^t = (\phi_x, \psi), \\
(\sigma_{xz}, \sigma_{zz})^t|z = L_1^z, z = L_2^z = (\sigma_{01}, \sigma_{02})^t = (\psi, \phi_z),
\end{aligned} \tag{7}$$

where ϕ_x is the load σ_{xx} at the left and right faces; ϕ_z is the load σ_{zz} at the down and top faces; ψ – the value of σ_{xz} at each face.

The boundary conditions for the solid matrix velocity vector $(u_x, u_z)^T$ and for the relative fluid velocity $(w_x, w_z)^T$ vector are set by applying the equations [6].

$$\begin{aligned}
\sigma_{xx} &= C_{11}\frac{\partial u_x}{\partial x} + C_{13}\frac{\partial u_z}{\partial z} + \alpha_1 M\left(\frac{\partial w_x}{\partial x} + \frac{\partial w_z}{\partial z}\right) \\
\sigma_{zz} &= C_{13}\frac{\partial u_x}{\partial x} + C_{33}\frac{\partial u_z}{\partial z} + \alpha_2 M\left(\frac{\partial w_x}{\partial x} + \frac{\partial w_z}{\partial z}\right) \\
\sigma_{xz} &= C_{55}\left(\frac{\partial u_x}{\partial z} + \frac{\partial u_z}{\partial x}\right).
\end{aligned} \tag{8}$$

Note that in the corners of domain both conditions of (7) must be taken into account. Therefore the full set of boundary conditions for Eqs. (1) in terms of velocity are the following:

1. Left and right faces, normal load:

$$\sigma_{xx} = C_{11}\frac{\partial u_x}{\partial x} + C_{13}\frac{\partial u_z}{\partial z} + \alpha_1 M\left(\frac{\partial w_x}{\partial x} + \frac{\partial w_z}{\partial z}\right) = \phi_x$$

2. Down and top faces, normal load:

$$\sigma_{zz} = C_{13}\frac{\partial u_x}{\partial x} + C_{33}\frac{\partial u_z}{\partial z} + \alpha_2 M\left(\frac{\partial w_x}{\partial x} + \frac{\partial w_z}{\partial z}\right) = \phi_z$$

3. All faces, shear load:

$$\sigma_{xz} = C_{55}\left(\frac{\partial u_x}{\partial z} + \frac{\partial u_z}{\partial x}\right) = \psi$$

4. Corners of the left and right faces, normal load:

$$\sigma_{xx} - \frac{C_{13}}{C_{33}}\sigma_{zz} = \left(C_{11} - \frac{C_{13}^2}{C_{33}}\right)\frac{\partial u_x}{\partial x} + M\left(\alpha_1 - \alpha_2\frac{C_{13}}{C_{33}}\right)\frac{\partial w_x}{\partial x}$$

$$+ M\left(\alpha_1 - \alpha_2\frac{C_{13}}{C_{33}}\right)\frac{\partial w_z}{\partial z} = \phi_x - \frac{C_{13}}{C_{33}}\phi_z$$

5. Corners of the down and top faces, normal load:

$$\sigma_{zz} - \frac{C_{13}}{C_{11}}\sigma_{xx} = \left(C_{33} - \frac{C_{13}^2}{C_{11}}\right)\frac{\partial u_z}{\partial z} + M\left(\alpha_2 - \alpha_1\frac{C_{13}}{C_{11}}\right)\frac{\partial w_x}{\partial x}$$

$$+ M\left(\alpha_2 - \alpha_1\frac{C_{13}}{C_{11}}\right)\frac{\partial w_z}{\partial z} = \phi_z - \frac{C_{13}}{C_{11}}\phi_x$$

6. Left and right faces, no-flow condition:

$$w_x = 0$$

7. Down and top faces, no-flow condition:

$$w_z = 0$$

b Schematic picture of domain (Fig. 2) demonstrate the distribution of applied boundary conditions at domain boundaries.

In particular, we consider three sets of boundary conditions (7) with basic loads:

– x-direction uniaxial compression:

$$\phi_x = 1, \quad \phi_z = 0, \quad \psi = 0; \tag{9}$$

– z-direction uniaxial compression:

$$\phi_x = 0, \quad \phi_z = 1, \quad \psi = 0; \tag{10}$$

– xz-torsion:

$$\phi_x = 0, \quad \phi_z = 0, \quad \psi = 1; \tag{11}$$

Details of the numerical solution of Biot equations (1) with presented boundary conditions (Sect. 2.3) are described below.

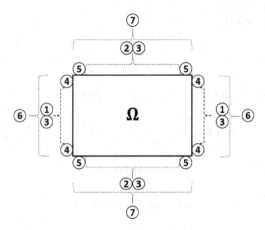

Fig. 2. Positions at the domain of the seven kinds boundary conditions.

3   Numerical Solution of Biot Equations

For numerical solution of Biot equations (1) we apply domain discretization by the rectangular grid $N_x \times N_z$. The domain boundaries cross the mesh at the centers of mesh edges as shown on Fig. 3.

Displacement x-components u_x and w_x are defined in the centers of vertical edges (z-edges), and z-components u_z and w_z are defined in the centers of horizontal edges (x-edges) (Fig. 4). The physical properties of the media $C_{11}, C_{13}, C_{33}, C_{55}, \alpha_1, \alpha_2, M$ are located in the middle nodes of the cells; η, k_1, k_2 – in the centers of the edges.

The first order partial derivatives $\frac{\partial}{\partial x}$ and $\frac{\partial}{\partial z}$ in Eq. (1) are approximated by the finite-differences. Solution components localization and derivatives discretization lead to the second order approximation of the first and the third equations of (1) in the centers of vertical edges, and the second and fourth equations in the centers of horizontal edges.

3.1   Approximation Stencils

The finite-difference approximation of Biot equations (1) leads to the four different stencils depending of the grid node. Each of them uses various number of unknowns (Fig. 5). Therefore, first and second equations of (1) use 16 components of total solution vector, third and fourth – 14 components. Boundary conditions (Sect. 2.3) approximation stencils are presented on the Fig. 6.

3.2   SLAE Construction

The Biot equations are approximated inside the computational domain (Fig. 3) and at its boundary. So $N_x \times N_z$ are performed $(N_x - 1)N_z$ approximations of the first and third equations and $N_x(N_z - 1)$ approximations of the second

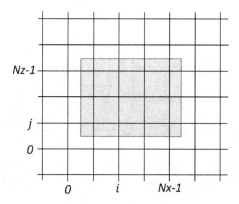

Fig. 3. Discretization of the computational domain.

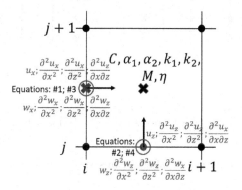

Fig. 4. Solution components, derivatives physical parameters location on computational grid, and nodes, where equations are approximated.

and fourth equations of (1) are performed on computational grid (approximated positions marked as "gray" circles at the Fig. 5). Total number of the linear equations is

$$N_{eq} = 2 * (N_x - 1)N_z + 2 * N_x(N_z - 1).$$

The stencils of such approximations cover the all internal points and points located out of the computational domain (see Fig. 7). So the number of unknown solution components in SLAE more than number of equations:

$$N_{un} = N_{eq} + 4 * N_x + 2 * (N_x - 1) + 4 * N_z + 2 * (N_z - 1).$$

Additional unknowns are excluded by boundary conditions. The external points are removed by boundary conditions (Sect. 2.3) with stencils (Fig. 6). Numbering of these points is demonstrated on the Fig. 7.

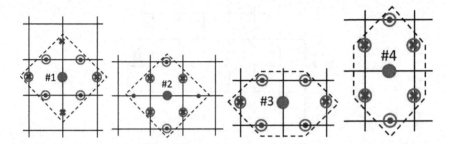

Fig. 5. Biot equations (1) stencils. Blue dots and crosses are solid matrix displacement components, red circles – fluid displacement components. (Color figure online)

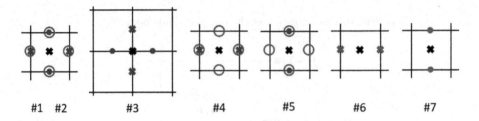

Fig. 6. Boundary condition stencils. Blue dots and crosses are solid matrix displacement components, red circles – fluid displacement components. (Color figure online)

By described approximation of equations and boundary conditions we obtain the consistent SLAE

$$Ax = b, \quad A \in \mathbb{C}^{N \times N}, \quad b \in \mathbb{C}^N,$$
$$N = N_{un} = N_{eq} = 2 * (N_x - 1)N_z + 2 * N_x(N_z - 1). \tag{12}$$

3.3 Numerical SLAE Solution

The SLAE (12) is non-symmetrical with complex values, so the approach to solve it is the special topic which is out of scope of this paper. Moreover, the SLAE is ill-conditioned because of 2nd order boundary conditions.

To resolve this SLAE the robust commercial product Intel MKL PARDISO solver is used. It is based on direct methods and has appropriate efficiency and memory consumption to solve 2D problems.

First the optimal computational parameters were found to run the set of numerical experiments to reconstruct the effective viscoelastic media. The choice of major parameters (grid size and the number of frequencies) is done corresponding to used hardware (Intel(R) Xeon(R) CPU E5-2690 v2 @ 3.00 GHz with 10 cores and 256G RAM). In these performance experiments we used a grid with 1000 nodes in both spatial directions. The Biot equation is resolved for the set of 48 frequencies from 0 to $6.4 * 10^9$ Hz. The SLAE for each frequency is resolved for 3 different right hand sides raised from boundary conditions (9)–(11).

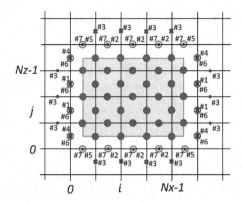

Fig. 7. Nodes, where Biot equations are approximated (gray circles) and unknowns are defined (gray, red circles, blue points and crosses). Numbering correspond to boundary conditions numerations. (Color figure online)

A single launch of the algorithm (one model, one frequency) include the following steps:

1. Approximate problem to get matrix A and right the hand sides b;
2. PARDISO Reordering step. It's a preliminary step to decrease memory consumption and time of the next factorization step;
3. PARDISO Factorization step. It perform LU decomposition of matrix A;
4. PARDISO Solve step. Solve the system to get solution x;
5. Postprocessing: Construct the stiffness tensor C.

Note that if the frequency is changed then only the main diagonal of the matrix A should be corrected. Thus, step (1) can be done for zero-frequency, but only diagonal entries are changed at further loops of the algorithm. The reordering step (2) depends only on position non-zero elements A, and can be done only once and applied for all ω. As a result, the entire algorithm (steps 1–5) is applied only to the first frequency in the raw, whereas the reduced version of the algorithm (steps 3–5) can be applied to all other frequencies.

The computational time for one model and all 48 frequencies is $\approx$3250 s. The timing profiling of the 48 runs is:

1. Approximate problem: $\approx$13 s.
2. PARDISO Reordering step: $\approx$20 s.
3. PARDISO Factorization step. 48$\times$ ($\approx$57 s.)
4. PARDISO Solve step. 48$\times$ ($\approx$7 s.)
5. Postprocessing: 48$\times$ ($\approx$3 s.)

Execution of the algorithm for the problem size of 1000^2 requires about 18Gb RAM, thus it can be done on any desktop machine, or ported to a standard GP-GPU to improve performance.

4   Numerical Reconstruction of Effective Viscoelastic Media

To reconstruct effective viscoelastic stiffness tensor $\hat{C}$ we numerically compute the strain $(\varepsilon(i,j))$ and the stress tensor components $(\sigma(i,j))$ in the centers of the each cell:

$$
\begin{aligned}
\varepsilon_{xx}(i,j) &= \frac{u_x(i+1,j) - u_x(i,j)}{h_x} \\
\varepsilon_{zz}(i,j) &= \frac{u_z(i,j+1) - u_z(i,j)}{h_z} \\
\varepsilon_{xz}(i,j) &= \frac{1}{2}(\frac{u_x(i,j) - u_x(i-1,j)}{h_z} + \frac{u_z(i,j) - u_z(i,j-1)}{h_z}) \\
Dw(i,j) &= \frac{w_x(i+1,j) - w_x(i,j)}{h_x} + \frac{w_z(i,j+1) - w_z(i,j)}{h_z} \\
\sigma_{xx}(i,j) &= C_{11}(i,j)\varepsilon_{xx}(i,j) + C_{13}(i,j)\varepsilon_{xx}(i,j) + \alpha_1(i,j)M(i,j)Dw(i,j) \\
\sigma_{zz}(i,j) &= C_{13}(i,j)\varepsilon_{xx}(i,j) + C_{33}(i,j)\varepsilon_{zz}(i,j) + \alpha_2(i,j)M(i,j)Dw(i,j) \\
\sigma_{xz}(i,j) &= C_{55}(i,j)\varepsilon_{xz}(i,j)
\end{aligned}
\tag{13}
$$

$$
\begin{aligned}
I_{\varepsilon xx} &= \sum_\Omega \varepsilon_{xx}(i,j) \quad I_{\sigma xx} = \sum_\Omega \sigma_{xx}(i,j) \\
I_{\varepsilon zz} &= \sum_\Omega \varepsilon_{zz}(i,j) \quad I_{\sigma zz} = \sum_\Omega \sigma_{zz}(i,j) \\
I_{\varepsilon xz} &= \sum_\Omega \varepsilon_{xz}(i,j) \quad I_{\sigma xz} = \sum_\Omega \sigma_{xz}(i,j)
\end{aligned}
\tag{14}
$$

To reconstruct the tensor $\hat{C}$, the SLAE 9×9 (6), with coefficients (14) should be solved.

It is convenient to apply seismic characteristics comparison except of effective stiffness tensor itself. So in all the numerical experiments we follow [4,20] to estimate the velocities and quality factors from complex frequency-dependent stiffness tensor components:

$$
\begin{aligned}
V_{px} &= \Re\sqrt{\frac{\hat{C}_{11}}{\rho}}, \; V_{pz} = \Re\sqrt{\frac{\hat{C}_{33}}{\rho}}, \; V_s = \Re\sqrt{\frac{\hat{C}_{55}}{\rho}}, \\
Q_{px} &= \frac{\Re\hat{C}_{11}}{\Im\hat{C}_{11}}, \quad Q_{pz} = \frac{\Re\hat{C}_{33}}{\Im\hat{C}_{33}}, \quad Q_s = \frac{\Re\hat{C}_{55}}{\Im\hat{C}_{55}}.
\end{aligned}
\tag{15}
$$

5   Numerical Experiments

We perform sets of numerical experiments to check quality of our approach of recovering effective viscoelastic anisotropic media in three aspects:

1. Approximation quality of the differential statement of Biot equation and boundary conditions. In particular, we apply comparison with analytic

velocity and attenuation (inverse quality factor $1/Q$) estimations for isotropic layered model. Such experiments help to estimate the quality of approximation for various number of layers and mesh points per one layer.

2. Quality of solving SLAE (12).
   The SLAE is solved via external package Intel MKL PARDISO, and we control the relative residual (usually about 10^{-11}) for each right hand side.
3. Correctness of the formulation of the problem by stress-strain state computation for anisotropic poroelastic fractured fluid-saturated media and reconstruction of the effective viscoelastic media properties, resulting in velocity and attenuation estimation. Analysis of obtained results is provided.

5.1 Approximation Quality

Realistic material properties of isotropic layered model (see Table 1.) are taken from the paper [17].

Mesh Refinement on Fixed Geometry. The geometry of model is square wits size 0.8×0.8 m with 8 layers: four internal Material #2 layers, three internal Material #1 layers and one external half-covered at the top and down of model (see Fig. 8). To check the approximation quality we decrease the mesh step in both (x) and (z) directions: $hx = hz = 0.05, 0.05/2, 0.05/4, 0.05/8, 0.05/16, 0.05/32$ (Fig. 8). Therefore, the number of points per layer are 2, 4, 8, 16, 32, 64.

The results (Fig. 9) show the convergence of presented approximation scheme by the convergence of frequency-dependent estimations of inverse quality factor.

Extending the Layered Model Along Layers. Other experiments involve the stretching of the layered model along the layers. The "square", "rectangle" and "thin" geometry is considered (see Fig. 10). The number of layers is 32, the number of points per layer is 16. Three corresponding grids have sizes 512×512, 1024×512 and 2048×512.

Figure 11 shows the quality factor $\frac{1}{Q}$ and P-wave phase velocity V_p (normal to layers) estimations compared with analytical estimations [5]. The peak position of attenuation estimation is at the same frequency, and the values of $\frac{1}{Q}$ and P-wave V_p are well agreed in general. The attenuation along layers is decreased to zero (Fig. 12) with model stretching as expected. Figure 13 demonstrates P-wave and S-wave velocity dependence on incidence angle. Estimations behaviour is similar to results demonstrated in [17].

Table 1. Material properties of isotropic layered model.

	Material #1	Material #2
Grain bulk modulus, K_s (GPa)	37	37
Grain density, ρ_s (g/cc)	2.65	2.65
Porosity, ϕ	0.1	0.37
Dry rock bulk modulus, K_m (GPa)	26	2.68
Dry rock shear modulus, μ_m (GPa)	31	0.857
Permeability, k (D)	10^{-3}	1
Brine: Builk modulus, K_f (GPa)	2.25	2.25
Brine: Density, ρ_f (g/cc)	1.09	1.09
Brine: Viscosity, η(P)	0.01	0.01
Biot-Willis parameter α	0.297297297	0.927567567
$\dfrac{Fluid\ dynamic\ viscosity}{Absolute\ permeability\ of\ the\ rock}, \dfrac{\eta}{k_0}$	$1 * 10^{12}$	$1 * 10^9$
Fluid storage coefficient M	$20.0896787868906 * 10^9$	$5.57060118727366 * 10^9$

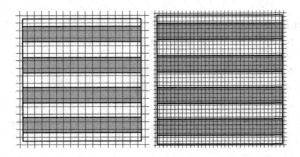

Fig. 8. Layered model geometry and mesh refinement. White layers have material #1 properties, black ones with material #2 properties from the Table 1.

5.2 Attenuation Estimation of Seismic Attenuation in Media with High Fracture Connectivity

We provide an example of described algorithm application to attenuation estimation in fractured porous fluid-saturated media. To study the influence of fracture-filling material physical properties and its anisotropy on seismic attenuation due to wave-induced fluid flow, we perform three sets of numerical experiments. We consider 8 realisations of fractured media model (example is shown on Fig. 14), containing two sets of vertical and horizontal fractures. Fractures are highly connected and form percolating chains through all domain. Details of corresponding fracture networks generation are described in [14]. Physical properties of isotropic background material are the same for all numerical experiments (Table 2, second column). We consider three fracture-filling materials, two isotropic – "stiff" (Table 2, third column) and "soft" (Table 2, fourth column), and one anisotropic, having different properties in horizontal and vertical fractures (Table 2, fifth and sixth columns, correspondingly).

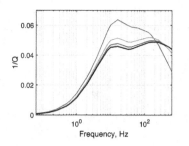
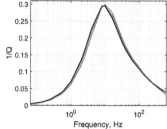

Fig. 9. Attenuation along layers (left) and normal to layers (right) for different grid step. Colors correspond to grid step decrease in the following order: red, green, blue, cyan, magenta, black. (Color figure online)

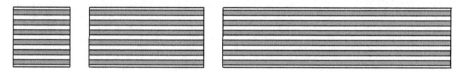

Fig. 10. The "square", "rectangle" and "thin" geometry.

Table 2. Physical properties of fractured media: background material (second column), isotropic fracture-filling materials (third and fourth columns), and anisotropic fracture-filling material for vertical and horizontal fractures (fifth and sixth columns).

	Background	"Stiff" fractures	"Soft" fractures	Horizontal frac	Vertical frac
C_{11}	$6.910 * 10^{10}$	$4.643 * 10^{10}$	$3.896 * 10^{10}$	$3.896 * 10^{10}$	$4.643 * 10^{10}$
C_{13}	$7.159 * 10^{10}$	$2.381 * 10^{10}$	$1.634 * 10^{10}$	$1.981 * 10^{10}$	$1.981 * 10^{10}$
C_{33}	$6.910 * 10^{10}$	$4.643 * 10^{10}$	$3.896 * 10^{10}$	$4.643 * 10^{10}$	$3.896 * 10^{10}$
C_{55}	$3.097 * 10^{10}$	$1.131 * 10^{10}$	$1.131 * 10^{10}$	$1.131 * 10^{10}$	$1.131 * 10^{10}$
M	$2.010 * 10^{10}$	$9.488 * 10^{9}$	$9.330 * 10^{9}$	$9.429 * 10^{9}$	$9.429 * 10^{9}$
$\alpha_1 M$	$5.953e * 10^{9}$	$5.767 * 10^{9}$	$6.854 * 10^{9}$	$6.502 * 10^{9}$	$6.051 * 10^{9}$
$\alpha_2 M$	$5.953e * 10^{9}$	$5.767 * 10^{9}$	$6.854 * 10^{9}$	$6.051 * 10^{9}$	$6.502 * 10^{9}$
$\frac{\eta}{k_1}$	$1 * 10^{12}$	$1.887 * 10^{9}$	$7.072 * 10^{6}$	$7.072 * 10^{6}$	$1.887 * 10^{9}$
$\frac{\eta}{k_2}$	$1 * 10^{12}$	$1.887 * 10^{9}$	$7.072 * 10^{6}$	$1.887 * 10^{9}$	$7.072 * 10^{6}$

Resulting averaged (with respect to realizations) attenuation estimations for P-wave (Fig. 15, left) and S-wave (Fig. 15, right) are demonstrated on for "stiff", "soft" and anisotropic fracture-filling materials. One can clearly see two peaks in "soft" material case, as well as in anisotropic material case for P-wave ans S-wave. Low frequency peak corresponds to FB-WIFF attenuation, and high-frequency peak corresponds to FF-WIFF attenuation mechanisms. High fracture connectivity through whole model results in high FF-WIFF peak attenuation in comparison with FB-WIFF peak for "soft" material case. "Stiff" material attenuation estimation for P-wave demonstrate only one peak, and that may be explained by

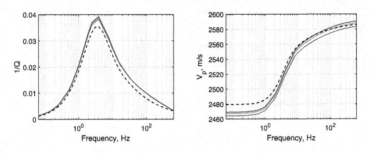

Fig. 11. Attenuation (left) and P-wave phase velocity (right) normal to layers. Colored curves represent numerical results (red for "square", green - "rectangular", blue - "thin" geometry). Black dashed lines represent analytical estimations from White's 1D model. (Color figure online)

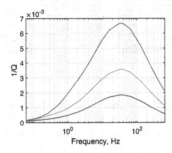

Fig. 12. Attenuation in direction along layers. Colored curves represent results for different domains: red for "square", green - "rectangular", blue - "thin" geometry. (Color figure online)

no significant fracture-to-fracture flow provided due to low permeability within the fractures. As the contrast between background and "stiff" fracture material permeabilities is still quite high, we still can note significant P-wave attenuation caused by FB-WIFF. Most interesting result is P-wave attenuation in anisotropic material case. Anisotropic case attenuation peak at low frequencies is flattened and lower than both other cases peaks, but in comparison with "stiff" material case, we can clearly observe the presence of the high frequency peak. This peak shows that in anisotropic case FF-WIFF is still significant, although it is much less than FF-WIFF peak in "soft" material case, as expected. For S-wave attenuation, we see much less attenuation at high frequencies for "stiff" case, and only low frequency FB-WIFF attenuation peak is clearly visible. FF-WIFF peak is also may be shifted in frequency is "stiff" material case, as we see relative small perturbation near 1 MHz. Indistinct of high frequency peak for S-wave attenuation in "soft" material case may be caused by the number of realizations with considered frequencies used in experiments.

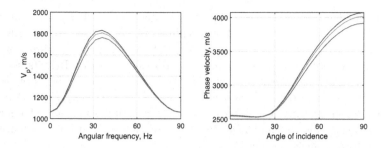

Fig. 13. P-wave (left) and S-wave (right) phase velocity as functions of incidence angle in horizontally layered media. The frequency 10 Hz. Colored curves represent results for different domains: red for "square", green - "rectangular", blue - "thin" geometry. (Color figure online)

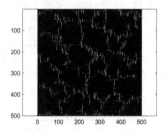

Fig. 14. Fractured media model. Black color represents background, yellow color represents fractures. (Color figure online)

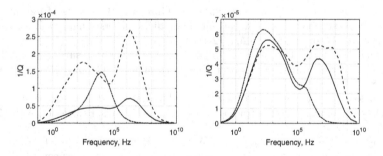

Fig. 15. P-wave (left) and S-wave (right) attenuation estimations for "stiff" (dash-dotted lines), "soft" (dashed lines) and anisotropic (solid lines) fracture-filling material.

6 Conclusion

Algorithm of numerical upscaling is presented to reconstruct effective viscoelastic stiffness tensor and estimate frequency-dependent attenuation in anisotropic fractured porous fluid-saturated media. Finite-difference approximation of quasi-static Biot equations is presented and applied to solve basic load test problems and recover effective viscoelastic properties related to poroelastic fluid-saturated sample. Resulting effective viscoelastic stiffness tensor is used to estimate both seismic attenuation and phase velocity. Algorithm is applied to estimate seismic attenuation in highly connected fractured poroelastic media. Results show the effect of fracture-filling material properties and its microscale anisotropy on seismic attenuation caused by FB-WIFF and FF-WIFF. However, one of dominant attenuation mechanisms at high frequencies is scattering, so presented algorithm can be used combined with algorithms of attenuation estimation, which are able to take dynamic effects into account.

References

1. Bazaikin, Y., et al.: Effect of CT image size and resolution on the accuracy of rock property estimates. J. Geophys. Res. Solid Earth **122**(5), 3635–3647 (2017)
2. Biot, M.A.: Theory of propagation of elastic waves in a fluid-saturated porous solid. II. Higher frequency range. J. Acoust. Soc. Am. **28**, 179–191 (1956)
3. Biot, M.A.: Theory of propagation of elastic waves in fluid-saturated porous solid. I. Low-frequency range. J. Acoust. Soc. Am. **28**, 168–178 (1956)
4. Carcione, J.M., Cavallini, F.: A rheological model for anelastic anisotropic media with applications to seismic wave propagation. Geophys. J. Int. **119**, 338–348 (1994)
5. Carcione, J., Picotti, S.: P-wave seismic attenuation by slow-wave diffusion: effects of inhomogeneous rock properties. Geophysics **7**, O1–O8 (2006). https://doi.org/10.1190/1.2194512
6. Cheng, A.H.D.: Material coefficients of anisotropic poroelasticity. Int. J. Rock Mech. Min. Sci. **34**(2), 199–205 (1997)
7. Guo, J., Rubino, J.G., Glubokovskikh, S., Gurevich, B.: Effects of fracture intersections on seismic dispersion: theoretical predictions versus numerical simulations. Geophys. Prospect. **65**(5), 1264–1276 (2017)
8. Huang, F., et al.: The first post-injection seismic monitor survey at the Ketzin pilot CO_2 storage site: results from time-lapse analysis. Geophys. Prospect. **66**(1), 62–84 (2018)
9. Hunziker, J., et al.: Seismic attenuation and stiffness modulus dispersion in porous rocks containing stochastic fracture networks. J. Geophys. Res. Solid Earth **123**(1), 125–143 (2018)
10. Kong, L., Gurevich, B., Zhang, Y., Wang, Y.: Effect of fracture fill on frequency-dependent anisotropy of fractured porous rocks. Geophys. Prospect. **65**(6), 1649–1661 (2017)
11. Marty, N.C.M., Hamm, V., Castillo, C., Thiéry, D., Kervévan, C.: Modelling water-rock interactions due to long-term cooled-brine reinjection in the Dogger carbonate aquifer (Paris basin) based on in-situ geothermal well data. Geothermics **88**, 101899 (2020)
12. Masson, Y.J., Pride, S.R., Nihei, K.T.: Finite difference modeling of Biot's poroelastic equations at seismic frequencies. J. Geophy. Res. Solid Earth **111**(B10), 305 (2006)
13. Menke, H.P., Reynolds, C.A., Andrew, M.G., Pereira Nunes, J.P., Bijeljic, B., Blunt, M.J.: 4D multi-scale imaging of reactive flow in carbonates: assessing the impact of heterogeneity on dissolution regimes using streamlines at multiple length scales. Chem. Geol. **481**, 27–37 (2018)

14. Novikov, M.A., Lisitsa, V.V., Bazaikin, Y.V.: Wave propagation in fractured-porous media with different percolation length of fracture systems. Lobachevskii J. Math. **41**(8), 1533–1544 (2020). https://doi.org/10.1134/S1995080220080144
15. Ovaysi, S., Wheeler, M., Balhoff, M.: Quantifying the representative size in porous media. Transp. Porous Media **104**(2), 349–362 (2014)
16. Rubino, J.G., Muller, T.M., Guarracino, L., Milani, M., Holliger, K.: Seismoacoustic signatures of fracture connectivity. J. Geophys. Res. Solid Earth **119**(3), 2252–2271 (2014)
17. Rubino, J.G., Caspari, E., Müller, T.M., Milani, M., Barbosa, N.D., Holliger, K.: Numerical upscaling in 2-D heterogeneous poroelastic rocks: anisotropic attenuation and dispersion of seismic waves. J. Geophys. Res. Solid Earth **121**(9), 6698–6721 (2016)
18. Salaun, N., et al.: High-resolution 3D seismic imaging and refined velocity model building improve the image of a deep geothermal reservoir in the Upper Rhine Graben. Lead. Edge **39**(12), 857–863 (2020)
19. Solovyev, S., Novikov, M., Kopylova, A., Lisitsa, V.: Numerical solution of Biot equations in quasi-static state. In: Gervasi, O., et al. (eds.) ICCSA 2021. LNCS, vol. 12949, pp. 519–531. Springer, Cham (2021). https://doi.org/10.1007/978-3-030-86653-2_38
20. Vavrycuk, V.: Velocity, attenuation, and quality factor in anisotropic viscoelastic media: a perturbation approach. Geophysics **73**(5), D63–D73 (2008)
21. Xu, C., Dowd, P.A., Mardia, K.V., Fowell, R.J.: A connectivity index for discrete fracture networks. Math. Geol. **38**(5), 611–634 (2006)

Effect of the Interface Roughness
on the Elastic Moduli

Tatyana Khachkova[1]([✉]) [iD], Vadim Lisitsa[2] [iD], and Dmitry Kolyukhin[2]

[1] Institute of Petroleum Geology and Geophysics SB RAS,
Koptug Avenue 3, Novosibirsk 630090, Russia
khachkovats@ipgg.sbras.ru
[2] Institute of Mathematics SB RAS,
Koptug Avenue 4, Novosibirsk 630090, Russia
lisitsavv@ipgg.sbras.ru

Abstract. In this paper, we study the effect of the interface roughness on the elastic parameters of layered media. We consider the three-
dimensional models of a layered medium with two different elastic materials inside and outside the layer. We generate the first class of models, where the interfaces between the layers are rough, and the elastic
parameters of the inner layers are fixed. Then, the numerical upscaling
technique is applied to estimate the effective stiffness tensor. Next, we
downscale the stiffness tensor to reconstruct the new elastic parameters
of the inner layer for the model of second class with flat interfaces; that
is the uncertainty of the model geometry is mapped to the uncertainty
of the stiffness tensor component for a fixed geometry of the model.
After that, we propose an algorithm for extending the results of restor-
ing the elastic tensors for arbitrary parameters of uncertainty applying
the bilinear regression with respect to interface rough parameters and
bilinear interpolation using two nearest points with respect to the physical parameters of the inner layers. Verification of the algorithm shows
that the errors in the recovering covariance matrix do not exceed 7%;
that is, it can be used to statistically simulate models of the second class
with a flat interface by arbitrary values of the interface roughness and
the physical parameters of the layers in the first class of models.

Keywords: Rough interfaces · Static elastic loading · Upscaling ·
Statistical analysis

1 Introduction

The construction of continuum mathematical models of physical processes is
always based on the consideration of a representative volume in which the change

V. Lisitsa performed numerical simulation using the Supercomputer of the Saint-
Petersburg Polytechnic University under the support of Russian Science Foundation
grant no. 21-71-20003. D. Kolyukhin generated the models with rough interfaces under
the support of Russian Science Foundation grant no. 19-77-20004. T. Khachkova did
the statistical analysis of the results.

O. Gervasi et al. (Eds.): ICCSA 2022 Workshops, LNCS 13378, pp. 328–339, 2022.
https://doi.org/10.1007/978-3-031-10562-3_24

in the considered physical parameters is assumed to be negligible, i.e. volume can be considered homogeneous. Relations between physical parameters on the microscale are replaced by the equations of state implicitly accounting for the macroscopic quantities. To establish these relationships, either the data of laboratory experiments are used, or the problem is solved at the micro level with a subsequent transition to a larger scale (upscaling). As the result, fundamentally different mathematical models can be considered on different scales. An example of problems for which the same mathematical models are used on the micro- and macroscales are the problems of "averaging" or the homogenization of medium models. For example, a rock on the scale of centimeters can be considered as a perfectly elastic body, which, however, is extremely inhomogeneous. At the same time, the characteristic scale of the studied seismic processes is from tens of meters in seismic exploration to tens of kilometers in seismology, which is determined by the characteristic wavelength. In this case, no significant scattering occurs at such small inhomogeneities, and the medium can be considered as a homogeneous ideally elastic medium, but with complex anisotropy. To construct homogenizations for elastic media, both analytical methods [2,3,10] for relatively simple media (periodic structures) and numerical methods based on solving a set of problems in the static theory of elasticity [1,8,9] can be used.

Analytical homogenization methods for layered media have become widespread in exploration seismic [2,4,10], since in the structure of the earth's crust in the absence of tectonic disturbances predominantly there is a weak variability of the parameters of the environment in the horizontal direction in comparison with the vertical one. Therefore, one of the most applicable models in seismic exploration is a model of a vertical transversely isotropic medium as a result of averaging a layered shared with isotropic layers. However, these models are obtained under the assumption that the interfaces between the layers are flat [2,10], which is not always true. The presence of irregularities at the interfaces leads to errors in determining the parameters of effective models, which, in turn, can affect quality of processing and interpretation of seismic data. In our work, we provide a numerical study of the effect of interfaces roughness on the effective elastic parameters of the averaged models. Moreover, the algorithm allows for given roughness parameters to reconstruct models with flat layer boundaries, but allowing fluctuations of elastic parameters in the layer, such that the effective parameters for the original and restored models are the same. Establishing the relationship between the interfaces roughness, which is always outside the resolution seismic methods, and uncertainties in the effective model will significantly increase reliability of seismic data interpretation.

In paper [5] we presented the results of the two-dimensional experiments and the corresponding conclusions. This work considers the three-dimensional case. Since the calculations for such models are much more resource-intensive, the series of experiments were performed in a smaller volume compared to the two-dimensional case, but the results obtained confirm the earlier conclusions.

The paper has the following structure. We formulate the problem in Sect. 2. Numerical experiments are provided in Sect. 3. Statistical modeling the tensors of elastic moduli for the inner layer with using the covariance matrices is presented in Sect. 4.

2   Statement of the Problem

Let's consider static elastic equation stated in a rectangular representative volume $\Omega = [0, X_1] \times [0, X_2] \times [0, X_3]$ written in Cartesian coordinates

$$
\frac{\partial}{\partial x_1}\left(c_{11}\frac{\partial u_1}{\partial x_1} + c_{12}\frac{\partial u_2}{\partial x_2} + c_{13}\frac{\partial u_3}{\partial x_3}\right) + \frac{\partial}{\partial x_2}\left(c_{66}\frac{\partial u_1}{\partial x_2} + c_{66}\frac{\partial u_2}{\partial x_1}\right) + \frac{\partial}{\partial x_3}\left(c_{55}\frac{\partial u_1}{\partial x_3} + c_{55}\frac{\partial u_3}{\partial x_1}\right) = 0,
$$

$$
\frac{\partial}{\partial x_1}\left(c_{66}\frac{\partial u_2}{\partial x_1} + c_{66}\frac{\partial u_1}{\partial x_2}\right) + \frac{\partial}{\partial x_2}\left(c_{12}\frac{\partial u_1}{\partial x_1} + c_{22}\frac{\partial u_2}{\partial x_2} + c_{23}\frac{\partial u_3}{\partial x_3}\right) + \frac{\partial}{\partial x_3}\left(c_{44}\frac{\partial u_2}{\partial x_3} + c_{44}\frac{\partial u_3}{\partial x_2}\right) = 0,
$$

$$
\frac{\partial}{\partial x_1}\left(c_{55}\frac{\partial u_3}{\partial x_1} + c_{55}\frac{\partial u_1}{\partial x_3}\right) + \frac{\partial}{\partial x_2}\left(c_{44}\frac{\partial u_3}{\partial x_2} + c_{44}\frac{\partial u_2}{\partial x_3}\right) + \frac{\partial}{\partial x_3}\left(c_{13}\frac{\partial u_1}{\partial x_1} + c_{23}\frac{\partial u_2}{\partial x_2} + c_{33}\frac{\partial u_3}{\partial x_3}\right) = 0
$$

$$(1)$$

with boundary conditions:

$$
\begin{aligned}
\sigma_{11}(0, x_2, x_3) &= \sigma_{11}^1, \ \sigma_{11}(X_1, x_2, x_3) = \sigma_{11}^1, \\
\sigma_{12}(0, x_2, x_3) &= \sigma_{12}^1, \ \sigma_{12}(X_1, x_2, x_3) = \sigma_{12}^1, \\
\sigma_{13}(0, x_2, x_3) &= \sigma_{13}^1, \ \sigma_{13}(X_1, x_2, x_3) = \sigma_{13}^1, \\
\sigma_{22}(x_1, 0, x_3) &= \sigma_{22}^2, \ \sigma_{22}(x_1, X_2, x_3) = \sigma_{22}^2, \\
\sigma_{12}(x_1, 0, x_3) &= \sigma_{12}^2, \ \sigma_{12}(x_1, X_2, x_3) = \sigma_{12}^2, \\
\sigma_{23}(x_1, 0, x_3) &= \sigma_{23}^2, \ \sigma_{23}(x_1, X_2, x_3) = \sigma_{23}^2, \\
\sigma_{33}(x_1, x_2, 0) &= \sigma_{33}^3, \ \sigma_{33}(x_1, x_2, X_3) = \sigma_{33}^3, \\
\sigma_{13}(x_3, x_2, 0) &= \sigma_{13}^3, \ \sigma_{13}(x_1, x_2, X_3) = \sigma_{13}^3, \\
\sigma_{23}(x_3, x_2, 0) &= \sigma_{23}^3, \ \sigma_{23}(x_1, x_2, X_3) = \sigma_{23}^3,
\end{aligned}
$$

$$(2)$$

where u_1, u_2 and u_3 are the displacement vector components, σ_{ij} are the stress tensor components and c_{ij} are the components of the stiffness tensor. The stress-stain relation can be written with using Voight notations:

$$
\begin{pmatrix} \sigma_{11} \\ \sigma_{22} \\ \sigma_{33} \\ \sigma_{23} \\ \sigma_{13} \\ \sigma_{12} \end{pmatrix} = \begin{pmatrix} c_{11} & c_{12} & c_{13} & & & \\ c_{12} & c_{22} & c_{23} & & 0 & \\ c_{13} & c_{23} & c_{33} & & & \\ & & & c_{44} & & \\ & 0 & & & c_{55} & \\ & & & & & c_{66} \end{pmatrix} \begin{pmatrix} \varepsilon_{11} \\ \varepsilon_{22} \\ \varepsilon_{33} \\ 2\varepsilon_{23} \\ 2\varepsilon_{13} \\ 2\varepsilon_{12} \end{pmatrix},
$$

$$(3)$$

where the strain tensor components are defined as

$$
\varepsilon_{ij} = \frac{1}{2}\left(\frac{\partial u_j}{\partial x_i} + \frac{\partial u_i}{\partial x_j}\right).
$$

Here we consider orthotropic media with symmetry axes coinciding with the coordinate axes; that is, the stiffness tensor C is assumed to have a block-diagonal structure, as shown in the Eq. (3).

As before [5] for the stiffness tensor C, which generally depends on spatial coordinates, we determine the constant effective stiffness tensor $\tilde{C}$ relating the volume-averaged stress and strain:

$$< \sigma >_V = \tilde{C} < \varepsilon >_V, \tag{4}$$

where $<>_V$ denote the volume-averaging:

$$< f >_V = V^{-1} \int_V f(x_1, x_2, x_3) dv. \tag{5}$$

The solution of this problem is described in detail by us in paper [5].

Further, for two models with stiffness tensors $C^{(1)}(\vec{x})$ and $C^{(2)}(\vec{x})$, the concept of equivalence is introduced, which means that their effective stiffness tensors coincide: $\tilde{C}^{(1)} = \tilde{C}^{(2)}$.

And for two sets of models, statistical equivalence means the coincidence of the first two moments of the solutions to the upscaling problems for them:

$$< (\tilde{C}^{(1)})_m >_E = < (\tilde{C}^{(2)})_m >_E,$$
$$cov\left((\tilde{C}^{(1)})_m\right) = cov\left((\tilde{C}^{(2)})_m\right). \tag{6}$$

Here $<>_E$ denotes the average value over the statistical realizations ensemble; for tensors and vectors it's considered component-wise mean. And cov is the covariance matrix of the stiffness tensor components.

Let's consider a model of a layered medium with two different elastic materials inside and outside the layer. We suggest that the outer material is fixed, these are quartz grains, while the parameters of the inner material can be changed. After that, two sets of models are considered.

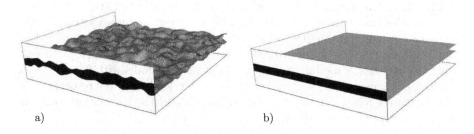

a)   b)

Fig. 1. Examples of 3D-models of the layered medium from two types of sets: a) with rough interfaces and constant stiffness tensor of the inner layer; b) with flat interfaces and variable stiffness tensor of the one.

First one are models with rough interfaces and constant elastic moduli of the inner layer (Fig. 1a):

$$C(x_1, x_2, x_3) = \begin{cases} C^f, & x_3 < b_u(x_1, x_2), \\ C^s, & b_u(x_1, x_2) \leq x_3 \leq b_d(x_1, x_2), \\ C^f, & x_3 > b_d(x_1, x_2), \end{cases} \tag{7}$$

where C^f, C^s are stiffness tensors of first and second layers and interfaces $b_d(x_1, x_2)$ and $b_u(x_1, x_2)$ are random functions:

$$b_d(x_1, x_2) = \langle b_d \rangle + b'_d(x_1, x_2),$$
$$b_u(x_1, x_2) = \langle b_u \rangle + b'_u(x_1, x_2),$$

where mean values $\langle b_u \rangle$, $\langle b_d \rangle$ and inner layer thickness are constant.

Perturbations b'_u and b'_d are random-valued functions with Gaussian probability distribution with zero mean value and covariance function $Q(r)$ defined by the standard deviation φ and correlation length I:

$$Q(r) = \varphi^2 exp(-\frac{\pi r^2}{4I^2}),$$

where r is a separation distance along the interface. For simulation of b'_u and b'_d it's used the method described in [7].

Second set are models with flat interfaces and variable stiffness tensor of the layer (Fig. 1b):

$$\hat{C} = \begin{cases} C^f, & x_3 < \langle b_u \rangle, \\ \hat{C}^s, & \langle b_u \rangle \leq x_3 \leq \langle b_d \rangle, \\ C^f, & x_3 > \langle b_d \rangle. \end{cases} \qquad (8)$$

Here $\hat{C}^s = \left\langle \hat{C}^s \right\rangle + (C^s)'$, where $(C^s)'$ is a random tensor-function with a given covariance matrix.

The main idea of the method is mapping the uncertainty of the model geometry to the uncertainty of stiffness tensor component values for a fixed geometry of the model; that is recovering perturbations of the tensor $\hat{C}^s$ as a function of the roughness of b'_u and b'_d so that two models were statistically equivalent.

The algorithm of equivalent model construction is described in detail in paper [5]. It includes two steps. First, for each model with rough interfaces we apply the numerical upscaling technique to estimate the effective stiffness tensor $\tilde{C}$. And then, we downscale the stiffness tensor using formulae Schoenberg [10] to reconstruct the new elastic moduli of the inner layer for the model with flat interfaces [5]. The main question is: how the covariance of the elements of the reconstructed stiffness matrix related to the standard deviation and the correlation length of the interfaces in the original set?

3   Experiments

First, we generate the set of models with rough interfaces (Fig. 1a) using the next parameters: the interlayer thickness is L, the size of model along the layer is $20L$ and the size across the layer is $5L$. The discretization step of the model $h = 0.02L$ was chosen equal in both spatial directions, therefore the discretization of the computational domain was 1000 points along and 250 points across the layer.

As materials outside and inside the layer, we consider an isotropic medium for which the following relations hold $c_{11} = c_{22} = c_{33}$, $c_{44} = c_{55} = c_{66}$ and $c_{12} = c_{13} = c_{23} = c_{11} - 2c_{55}$. Therefore, we define only c_{55} and c_{11} components or c_{11}/c_{55} ratio. For material outside the layer (C^f in Eq. 7) it's used the quartz grains parameters ($c_{11} = 95.4$ GPa, $c_{55} = 44$ GPa), and the different clay parameters $c_{55} \in [6, 16.8, 33]$ GPa, $c_{11}/c_{55} \in [2.17, 3.06, 4]$ GPa are determined for inner layer (C^s in Eq. 7). To simulate rough interfaces at the top and bottom of the inner layer the different values of next two parameters are defined: the correlation length $I \in [0.1L, 0.2L, 0.4L, 0.8L, 1.6L]$ and the standard deviation $\varphi \in [0.02L, 0.04L, 0.1L, 0.2L]$. Specified interfaces are generated independently.

For each of 180 statistical model $[I_i, \varphi_j, (c_{55})_n, (c_{11}/c_{55})_m]$ we generate a series of 10 realizations, after that upscaling and downscaling technique are performed for each such implementation. Finally, the sets of reconstructed stiffness tensors are obtained for interlayers of the equivalent models with flat interfaces (Fig. 1b).

The first characteristic of the obtained tensors, which we investigate, is the component mean calculated over the ensemble of realizations for each statistical models. Since the average layer thickness in the original models is fixed and we keep the same thickness for models of second type, then it's expected that the mean values of the reconstructed tensors coincide with the initial tensors for inner layer. Having calculated the mean values of the tensor components and their confidence intervals, we estimated the relative error between the obtained mean and the tensor, which corresponds to the original model, using the Frobenius norm. Comparison of the obtained relative differences with the norms of relative confidence intervals, which are calculated using the t-distribution, confirms the expected coincidence (Fig. 2). The confidence interval in this case has the form

$$c_{kl} = \bar{c}_{kl} \pm t_{\alpha, N-1} \frac{S}{\sqrt{N}}, \quad S^2 = \frac{\sum_{i=1}^{N}(c_{kl})_i^2 - N\bar{c}_{kl}^2}{N - 1}, \tag{9}$$

where $\bar{c}_{kl}$ is mean, S is the variance, N is the size of realizations sample, $t_{\alpha, N-1}$ is a quantile of Student's distribution of level $1 - \alpha/2$ with $N - 1$ degree of freedom, the significance level is chosen as $\alpha = 0.05$.

Then, to check out that the components of the reconstructed stiffness tensors are normally distributed, the Royston's test [6] was used, which confirmed the multivariate normal distribution of the tensor components. Therefore, this distribution is completely determined by the mean vector and the covariance matrix.

So, we have obtained the values of the restored elastic moduli at the points of the regular sparse mesh in four-dimensional parametric space. Further, we propose a method for extending the results to the entire domain of interest, which makes it possible to construct tensors of elastic moduli of the inner layer for arbitrary values of the specified parameters using the covariance matrices.

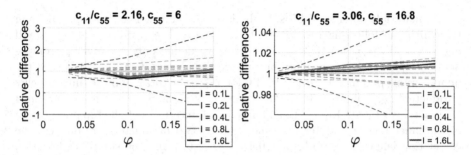

Fig. 2. The relative differences (solid lines) between the original model and mean tensors depending on the standard deviation φ and the correlation length I of rough interface, the dashed lines represent the norms of relative confidence intervals.

4   Interpolation of the Covariance Matrices

The covariance matrices are symmetric and strictly positive definite. Since their size in 3D case is 9×9, then we can apply any explicit method of linear algebra. Let's consider the following interpolation:

$$\Sigma(t) = \exp\left[(1-t)\ln(\Sigma_1) + t\ln(\Sigma_2)\right], \tag{10}$$

where Σ_1 and Σ_2 are the covariance matrices in the point of mesh, and exp and ln are the matrix functions that can be defined as

$$f(\Sigma) = Uf(D)U^*,$$

where f is the function (exp or ln), U is the eigenvectors matrix, D is the diagonal matrix whose main diagonal consists of the eigenvalues of Σ, and U^* is conjugate of U. Since the matrix Σ is symmetric, then D is diagonal and real, U is orthogonal and $f(\Sigma)$ is symmetric. And since the functions of the logarithm and exponent are monotonic, then the interpolated matrix is positive.

In order to consider the element-wise representation of the $\Psi = \ln(\Sigma)$ matrices, we fix three of the four parameters and examine the components of Ψ as a function of the remaining parameters (Fig. 3 and 4). We have obtained that the coefficients of the matrices almost linearly depend on $\ln(\varphi)$ and $\ln(I)$, in while the dependence on the physical parameters of the inner layer is more complex.

Next, we calculated a component-wise linear regression of the matrices for each parameter, while the other three parameters were fixed, and considered the relative error of such regression, which is the solution of the equations system by the least-squares method:

$$A_0 y_j + A_1 = \Psi(y_j).$$

Here y is the parameter by which the linear regression is constructed, y_j are its known values, matrices A_0 and A_1 are the coefficients of regression that don't

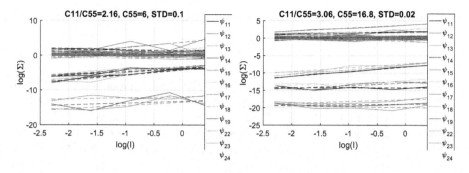

Fig. 3. The components of the matrices $\ln(\Sigma)$ depending on the correlation length I of rough interface for different values of c_{11}, c_{55} and φ. Solid lines correspond to the matrix components; dashed lines are the linear regression of the corresponding components.

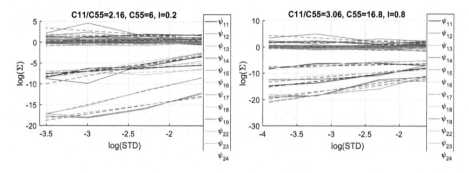

Fig. 4. The components of the matrices $\ln(\Sigma)$ depending on the standard deviation φ of rough interface for different values of c_{11}, c_{55} and I. Solid lines correspond to the matrix components; dashed lines are the linear regression of the corresponding components.

dependent on y. It should be noted that the problem is overdetermined, and it is possible to find a least-squares solution that minimizes the residual:

$$\sum_{ij} \sum_{k} ((A_0)_{ij} y_k + (A_1)_{ij} - \psi_{ij}(y_k))^2 = \sum_{k} \|A_0 y_k + A_1 - \Psi(y_k)\|_F^2 \to \min.$$

Here $\|\ \|_F$ is a Frobenius norm

$$\|\Psi\|_F^2 = \sum_{ij} \psi_{ij}^2.$$

For the fixed physical parameters of the inner layer $(c_{11}/c_{55})_m, (c_{55})_n$, we compute the linear regression with respect to I for each φ_j:

$$\hat{\Psi}_{ijmn} = \hat{\Psi}((c_{11}/c_{55})_m, (c_{55})_n, \varphi_j, I)$$
$$= A_0((c_{11}/c_{55})_m, (c_{55})_n, \varphi_j)I + A_1((c_{11}/c_{55})_m, (c_{55})_n, \varphi),$$

then calculate its value at the nodes $I = I_i$ and estimate the relative error as:

$$R_{ijmn} = \frac{\|\hat{\Psi}_{ijmn} - \Psi_{ijmn}\|_F}{\|\Psi_{ijmn}\|_F}.$$

Here Ψ_{ijmn} is the available matrices $\ln(\Sigma)$ at the quadruplets nodes. In Fig. 5 we represent the calculated relative errors depending on the experiment number $M = 5(j - 1) + i$, where $j = 1,...4$, and $i = 1,...5$. In contrast to the two-dimensional case, where the relative error with respect to the correlation length I did not exceed 0.05, in 3D case the relative errors are larger, reaching 0.15–0.18.

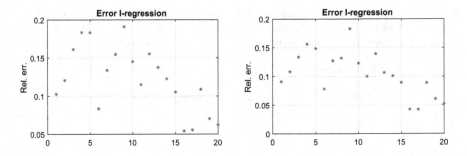

Fig. 5. Relative errors of the linear regression with respect to the correlation length depending on the experiment number at fixed values of the physical parameters $c_{11}/c_{55} = 2.16, c_{55} = 6$ (left) and $c_{11}/c_{55} = 3.06, c_{55} = 16.8$ (right).

Then, similarly, we calculated the relative errors of the linear regression with respect to the standard deviation φ. The results are represented in Fig. 6.

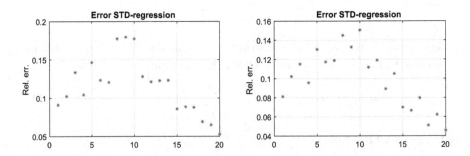

Fig. 6. Relative errors of the linear regression with respect to the standard deviation depending on the experiment number at fixed values of the physical parameters $c_{11}/c_{55} = 2.16, c_{55} = 6$ (left) and $c_{11}/c_{55} = 3.06, c_{55} = 16.8$ (right).

It should be noted that in the Figs. 5 and 6 every five experiments correspond to a fixed standard deviation $\varphi \in [0.2L, 0.1L, 0.04L, 0.02L]$. The first five correspond to $\varphi = 0.2L$ and so on. It can be seen that the relative error decreases with the standard deviation decrease. And within each five experiments, the correlation length grows, $I \in [0.1L, 0.2L, 0.4L, 0.8L, 1.6L]$. These graphs (Figs. 5 and 6) confirm that the matrix $(\Psi = \ln(\Sigma))$ components are linearly dependent on the functions $\ln(\varphi)$ and $\ln(I)$.

Accordingly, for each pair $((c_{11}/c_{55})_m, (c_{55})_n)$ it can be calculated the bilinear regression; that is, the approximation of Ψ can be founded as:

$$
\begin{aligned}
\tilde{\Psi}_{mn} &= \tilde{\Psi}((c_{11}/c_{55})_m, (c_{55})_n, \varphi, I) \\
&= B_{mn}^{11} \varphi I + B_{mn}^{10} \varphi + B_{mn}^{01} I + B_{mn}^{00},
\end{aligned}
\tag{11}
$$

where matrices $B_{mn}^{11}, ..., B_{mn}^{00}$ depend on the parameters $((c_{11}/c_{55})_m, (c_{55})_n)$.

Using the technique described above, we calculate the relative errors for bilinear regression (Fig. 7), which, just like in the case of linear regression, decrease along with a decrease in the standard deviation.

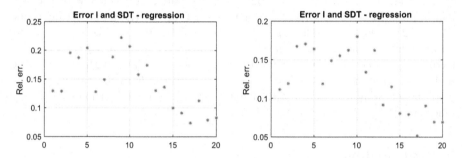

Fig. 7. Relative errors of the bilinear regression depending on the experiment number at fixed values of the physical parameters $c_{11}/c_{55} = 2.16, c_{55} = 6$ (left) and $c_{11}/c_{55} = 3.06, c_{55} = 16.8$ (right).

Since the dependence on the physical parameters of the inner layer is more complex, another interpolation method is needed to restore the values between them. We use standard linear component-wise interpolation with using two nearest points, that is described in detail by us in paper [5].

So, applying the bilinear regression with respect to $\ln(I)$ and $\ln(\varphi)$ and bilinear interpolation using two nearest points with respect to the physical parameters, we reconstruct the covariance matrix at arbitrary point of four-dimensional parametric space.

To test the algorithm for recovering the covariance matrix, we created 18 data sets, each containing 10 realizations, for random quadruples. After calculating the covariance matrices for these datasets, we compared them with the values predicted by the algorithm. In the Fig. 8, we represent the relative errors of the reconstructed matrices Ψ compared to those obtained from numerical experiments using the Frobenius norm. The left plot show the errors depending on the experiment number, the right are the errors depending on the position of input parameters (I, φ).

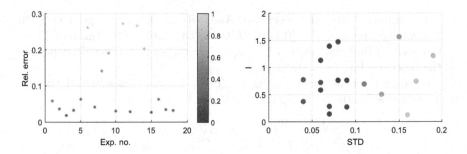

Fig. 8. Relative errors of the recovered matrices Ψ for random quadruplets. Markers colour correspond to the value of error. (Color figure online)

It can be seen that the matrix reconstruction error increases with the growth of the standard deviation φ of rough interface and is independent of the correlation length I. In addition, it can be seen that for the standard deviation $\varphi \in [0.02L, 0.04L, 0.1L, 0.2L]$ only the interval $\varphi \in [0.1L, 0.2L]$ leads to large errors, while for standard deviation values from 0.02L to 0.1L, the errors in recovering the covariance matrix do not exceed 0.07. Thus, with the specified limitation on standard deviation values, the recovery algorithm can be used to statistically simulate models with a flat interface and random material in the inner layer.

5   Conclusions

In this paper, we research the effect of uncertainty of the model geometry on the elastic parameters of the layered media. We consider the three-dimensional models of a layered medium with two different elastic materials inside and outside the layer. Initially, we generate the first class of models, where the interfaces between the layers are rough, and the elastic parameters of the inner layers are fixed. Such models are constructed for different values of parameters at the points of the regular sparse mesh in four-dimensional parametric space. The varied parameters are standard deviation and correlation length of interface roughness and two physical parameters (components of stiffness tensor) of the inner layer. Then, for each model the numerical upscaling technique is applied to estimate the effective stiffness tensor. After that, we downscale the stiffness tensor to reconstruct the new elastic parameters of the inner layer for the model of second class with flat interfaces that is equivalent to original; that is their effective stiffness tensors are coincide.

The next step is investigation of the characteristic of the obtained stiffness tensors, that confirm the coincidence of the mean values of the reconstructed tensor (over the ensemble of realizations) and the initial tensors for inner layer in first models. Since Royston's test confirmed the multivariate normal distribution of the tensor components, and this distribution is completely determined by the mean vector and the covariance matrix, we use this matrix to extend the results to the entire domain of four-dimensional parametric space; that is we propose the algorithm to reconstruct the stiffness tensors of the inner layer with flat interfaces for arbitrary values of the interface roughness and the physical parameters of the inner layer in the first class of models.

We apply the bilinear regression with respect to interface rough parameters and bilinear interpolation using two nearest points with respect to the physical parameters of the inner layers. So, we reconstruct the covariance matrix at arbitrary point of four-dimensional parametric space. Verification of the algorithm shows that the errors in the recovering covariance matrix do not exceed 7%. Thus, with the specified limitation on standard deviation values, the proposed algorithm can be used to statistically simulate models of the second class with a flat interface by given parameters of uncertainty for the inner layer in the first class of models.

References

1. Andra, H., et al.: Digital rock physics benchmarks - part II: computing effective properties. Comput. Geosci. **50**, 33–43 (2013)
2. Backus, G.E.: Long-wave elastic anisotropy produced by horizontal layering. J. Geophys. Res. **67**(11), 4427–4440 (1962)
3. Capdeville, Y., Guillot, L., Marigo, J.J.: 2-D non-periodic homogenization to upscale elastic media for P–SV-waves. Geophys. J. Int. **182**(2), 903–922 (2010). https://doi.org/10.1111/j.1365-246X.2010.04636.x
4. Henriques, J.P., et al.: Experimental verification of effective anisotropic crack theories in variable crack aspect ratio medium. Geophys. Prospect. **66**(1), 141–156 (2018)
5. Khachkova, T., Lisitsa, V., Kolyukhin, D., Reshetova, G.: Influence of interfaces roughness on elastic properties of layered media. Probab. Eng. Mech. **66**, 103170 (2021)
6. Royston, J.P.: Some techniques for assessing multivarate normality based on the Shapiro-Wilk w. J. R. Stat. Soc. Ser. C (Appl. Stat.) **32**(2), 121–133 (1983)
7. Sabelfeld, K.K.: Monte Carlo Methods in Boundary Value Problems. Springer, Cham (1991)
8. Saenger, E.H.: Numerical methods to determine effective elastic properties. Int. J. Eng. Sci. **46**(6), 598–605 (2008)
9. Saenger, E.H., et al.: Analysis of high-resolution X-ray computed tomography images of Bentheim sandstone under elevated confining pressures. Geophys. Prospect. **64**(4), 848–859 (2016)
10. Schoenberg, M., Muir, F.: A calculus for finely layered anisotropic media. Geophysics **54**(5), 581–589 (1989)

International Workshop
on Computational Science and HPC
(CSHPC 2022)

Design and Implementation of an Efficient Priority Queue Data Structure

James Rhodes$^{(\boxtimes)}$ and Elise de Doncker

Western Michigan University, Kalamazoo, MI 49008, USA
{james.rhodes,elise.dedoncker}@wmich.edu

Abstract. Priority queues are among the most useful of all data structures. Existing priority queues have a vast amount of overhead associated with them. There is a need to have a simple data structure that can be used as a priority queue with low overhead. The data structure should have the operation where the data item with the minimum/maximum value is the next item to be deleted. The data structure should also support the function of a calendar queue where elements with the same or similar priority have the same key. For example, all of today's appointments will have today's date as their key. To that end, a bucket data structure has been developed that has both of these features. We address the functionality and efficiency of the data structure for the applications of adaptive multivariate integration and the 15-puzzle. In adaptive multivariate integration, the key is an error estimate, which is a floating-point number. The data for this application has many items with similar keys and the maximum is the next item to be deleted. The key for the 15-puzzle is generated by a heuristic cost function. The data for this application has many items with the same key and the minimum is the next item to be deleted. This paper presents an implementation of the bucket priority queue and discusses its performance.

Keywords: Priority queue · Bucket data structure · Heap · Performance

1 Introduction and Background

Various software has been developed to perform multivariate integration of difficult functions using adaptive region partitioning [1,2,5,7,8,13,14,18]. The idea is to reduce the error estimate below a tolerated error. Recent research has been conducted on how to perform multivariate integration efficiently using GPUs [9–12,16,17]. The software utilizes a max heap to hold regions for possible further subdivision. The region with the maximum error is deleted from the max heap and subdivided into smaller regions. The smaller regions are inserted into the max heap, and this process continues until either the total error is sufficiently reduced or a maximum number of function evaluations have been performed. We developed a more efficient data structure to replace the max heap as a priority queue that improves the performance of the application.

© The Author(s), under exclusive license to Springer Nature Switzerland AG 2022
O. Gervasi et al. (Eds.): ICCSA 2022 Workshops, LNCS 13378, pp. 343–357, 2022.
https://doi.org/10.1007/978-3-031-10562-3_25

Heaps are a great data structure but they require constant data manipulation in order to maintain the heap property. The heap property guarantees that the children keys of any node are not less or greater than the parent key for a min or max heap, respectively. In order for heaps to satisfy this property, maintenance is required when both inserting and deleting data. We refer to this process as "heapifying". The objective of our data structure is to improve the CPU time by eliminating the maintenance associated with min/max heaps.

1.1 Problem Statement

A min/max heap has insert and delete methods that both have $\mathcal{O}(\log n)$ time complexity. For most applications this is acceptable. However, for some applications where the number of elements on the heap is huge, the performance of the application can be adversely affected. Our novel priority queue data structure lowers the time complexity for target applications, while maintaining reasonable accuracy.

In particular, task partitioning methods support algorithms maintaining task pools, such as some branch-and-bound methods and adaptive numerical integration. Adaptive region partitioning is also applied to produce fine meshes delineating areas of lighting and sharp shadows for image processing such as in ray tracing and radiosity methods, and likewise for intensive partitioning in the vicinity of difficult function behavior in adaptive integration. Our focus is on global adaptive integration methods where important (high error) regions are selected, subdivided and their children regions processed at each step, until the global error estimate falls below the tolerated error or the maximum number of function evaluations is achieved. Previous methods used a heap or linked list priority queue and deleted one region at a time, which is inefficient as the heap may become extremely large. We have developed a bucket-based data structure that improves efficiency without significantly affecting the accuracy by selecting moderately important regions for a parallel evaluation at each step.

1.2 Background

A review of data structures and algorithms was conducted as a preliminary step of our development of a priority queue data structure. The most common data structures associated with priority queues and a wide variety of applications were reviewed. Fast implementations of priority queues are important to many high-performance computing (HPC) applications, and hence optimizing their performance has been a subject of intense research [20]. Although priority queues can be implemented with data structures such as simple arrays and linked lists, heaps are most commonly used. After search trees, heaps are the second most studied data structure [3]. Heaps support the insert, find-min (find-max), and delete-min (delete-max) operations. Heaps that support the minimum key operations are called min-heaps, and heaps that support the maximum key operations are called max-heaps.

The heap data structure was invented by J. W. J. Williams in 1964 for a sorting application [21]. This connection is important since sorting n elements has a complexity of $\mathcal{O}(n \log n)$. Building a min heap with n elements can be done in $\mathcal{O}(n)$ time. Then, n delete-min operations can be performed to sort the elements in non-decreasing order. A delete-min operation can get and delete an element in $\mathcal{O}(\log n)$ time, thus "heapsort" has $\mathcal{O}(n \log n)$ time complexity. Heaps have a tree-based structure. They have the property where the key of a parent node is not smaller (greater) than the key of any of its children nodes for a min-heap (max-heap) [3]. There are different types of heaps with various operations and complexities. The "standard" heap is known as a binary heap based on a binary tree. Our data structure improves the efficiency of this type of heap for applications of interest.

The 15-puzzle application is a special case of the $(n^2 - 1)$-puzzle problem, which has applications to practical relocation problems [6,19]. In this case, $n = 4$ ($4^2 - 1 = 15$). It consists of a 4×4 matrix with 15 tiles numbered 1 to 15 and an empty space. In the initial state of the matrix, the tiles are not in numerical order. The idea is to arrange the tiles in numerical order from 1 to 15 starting with the top left corner and going left to right and top to bottom. This state is considered the solution or goal state. A tile can be moved up, down, left, or right depending on its position in the matrix. The tiles are moved in such a way as to work towards the goal state. Some applications utilize branch-and-bound or genetic algorithms.

The term branch-and-bound refers to all state space search methods in which all children of the node being expanded (E-node) are generated before any other node can become the E-node. There are two basic search strategies, BFS and D-search [15]. A BFS state space search is a FIFO (First In First Out) search as the list of live nodes is a first-in-first-out list (or queue). In D-search, the exploration of a new node cannot begin until the current E-node is fully explored. A D-search state space search is a LIFO (Last In First Out) search as the list of live nodes is a last-in-first-out list (or stack). In both FIFO and LIFO branch-and-bound, the selection rule for the next E-node is rigid and blind. It does not give any priority to a node that has the best chance of finding the solution node quickly. We will apply an LC (Least Cost) state space search for the 15-puzzle.

An LC state space search uses a cost function to give priority to a node with the least cost. A node with the least cost becomes the E-node. The idea is that the solution node will be found faster if the node with the least cost is expanded next. Since the node with the least cost value has highest priority, a min-heap is used for the 15-puzzle application. The state space is represented by a tree. For example, Fig. 1 displays part of a state space tree (generated breadth-first) for a 15-puzzle instance.

The cost function [15] for the 15-puzzle application is $c(x) = f(x) + g(x)$ where $f(x) =$ the length of the path from the root of the state space tree to node x and $g(x) =$ the number of tiles that are not in the correct position in the matrix for the state of node x.

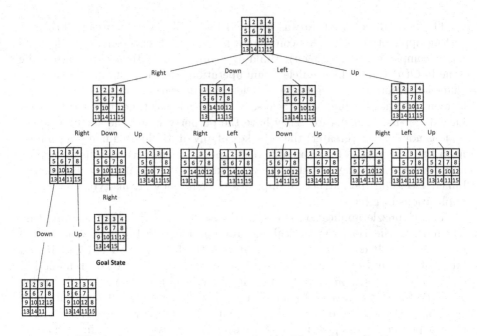

Fig. 1. A partial state space tree for the 15-puzzle application

Subsequently, this paper covers methods and implementation in Sect. 2, detailed results in Sect. 3, and gives conclusions in Sect. 4.

2   Methodology and Implementation

In order to improve on the efficiency of heaps, the "heapify" process must be optimized or eliminated. Many applications that utilize a heap insert an enormous number of items with the same or similar key value. Each of these items is stored in a separate node. So, it is possible for many nodes with the same or similar key values to be stored on a heap. When a node is deleted from a heap, the "heapify" process rearranges the heap by first moving the last entry (rightmost leaf) of the heap to the root position. The "heapify" function is called recursively and moves the node down through the heap until every node in the heap satisfies the heap property. A similar process occurs when a node is inserted into a heap. If the heap has millions of nodes, this could take up a great deal of CPU time. The smaller the number of nodes on the heap, the less time it takes to "heapify". This leads to the question of whether nodes with the same or similar key values on a heap could be coalesced to make the application more efficient.

We investigated the answer to this question by using buckets. We start this discussion by associating buckets with nodes on a heap; therefore, it is necessary to make a distinction between a data item (or list item) and a heap item. A list item is an individual node that is a member of a linked list. A heap item is the head node of a linked list, corresponding to a heap node that can be inserted into or deleted from the heap, thus each node on the heap corresponds to a bucket of nodes with the same or similar key value. The bucket may have many nodes in it, organized as a linked list, but only the head node of the linked list will reside on the heap. More precisely, the node on the heap will be a pointer to the head node of a linked list. This means that the node collection could have millions of items in it but less than a percentage of those nodes would have to be manipulated whenever the "heapify" process is called. The nodes in a bucket will be moved as if they were one node.

Also, with buckets, not only is the time spent to "heapify" lowered but the "heapify" process is called less frequently. If there is more than one node on the linked list in the root bucket on a heap, and a list item is deleted from it, at least one node will remain in the linked list, and a call to the "heapify" process will be avoided. Likewise, if a node that is being inserted into the heap has a key value that already exists in a bucket on the heap, a "heapify" process can be avoided by simply placing the new node at the end of the linked list, provided that the correct bucket can be determined.

By using a hashing function on the buckets, this can be done in $\mathcal{O}(1)$ time. This led us to develop a priority queue data structure in the form of a hash table that gets rid of the heap altogether. This not only reduces the number of nodes to manipulate, but eliminates the "heapify" process entirely.

Note that there is a trade-off between the size of the buckets and a possible reduction in accuracy. If, in the multivariate integration application, the regions are not selected in the order of decreasing error estimate, this may degrade the accuracy. For a parallel application it will further be of interest to enable deletion and insertion of multiple regions at a time.

2.1   Buckets Implementation

The buckets to store data items will form an array of pointers to the heads of linked lists. The array is dynamic and can grow and shrink as required. If the last node of a bucket is deleted, to find the next (minimum or maximum) bucket, the array is searched linearly to find the next non-null bucket. If the bucket for a new item that is being inserted is empty, there is not any additional data manipulation required.

Multivariate Integration. In certain applications, the key values on the linked list will not be required to be exactly the same. For these applications, it is sufficient for the values to be similar or within a range. Multivariate integration is such an application. Previous implementations with a standard heap store an integer index of a region on the heap. The index is used to get a value that represents the error estimate of that region from an array [17]. The error estimate referenced by the index is used as the priority to position the index on the heap. Since the region with the largest error estimate is deleted first, the application utilizes a max heap. Our new data structure stores a node according to its error estimate as well. However, the log_{10} and the mantissa of the value are used to create a bucket. The first bucket value is at the head of the linked list and a pointer to the head is inserted into the array. Then, when another value is inserted into a bucket with the same log_{10} value and mantissa as a head node of a linked list already in the array, it is placed at the end of the linked list for which that node is the head. If a node with the log_{10} and the mantissa of the value is not in the array, then the value is associated with a new head node and a pointer to the new head node is inserted into the array. The precision (length) of the mantissa is configurable in the data structure. In essence, the bucket structure for the multivariate integration application is a 2-dimensional array of pointers that point to head nodes of linked lists. A hashing function used to find the correct bucket is based on a simple function $g(log_{10}E)$ and the mantissa of the error estimate E, for example, BUCKETS[$g(log_{10}E)$][mantissa].

The buckets for the data structure are represented by an array of pointers to the head nodes of the linked lists for the priorities of an application. The priority (error estimate) for the (double precision) multivariate integration application has a non-negative double data type. The desired mantissa length has to be established for bucket purposes. Let's assume the priority queue considers key values with 1 to 6 digits of precision. As mentioned, the BUCKETS array for the multivariate integration application will have two dimensions.

The first dimension represents log_{10} of the error estimate. For example, with the dimension size for the log_{10} index set to 601, the log_{10} value ranges between -300 and 299, inclusively. An offset of 301 is added to the log_{10} value of the error estimate so that the index values will be 1 to 600, inclusively. The index 0 element is used for the value of 0. The second dimension represents the mantissa of the error estimate. The dimension size for the mantissa can be a power of 10 between 10 and 1,000,000, inclusively, depending on the desired mantissa length. If 1 digit of precision is desired, then the dimension size is set to 10. That would allow the index to be in the range from 0 to 9. If 6 digits of precision are desired, then the dimension size is set to 1,000,000. That would allow the index to be a value from 0 to 999,999. So, if 3 digits of precision are desired, then the BUCKETS array is declared as BUCKETS[601][1000]. The size and number of the dimensions may be configured differently depending on the application.

The values of the indexes for the BUCKETS array will be determined by taking the base 10 logarithm of the error estimate, and an offset is added to eliminate negative numbers. This will be the index for the first dimension. The error estimate is multiplied by 10 to the power of $(-)$ its base 10 logarithm plus the base 10 logarithm of the selected precision -1. The integer part of this value will provide the mantissa, which is the index for the second dimension. Using e_a to denote the error estimate, and m_{size} for the required mantissa size, the exponent and the mantissa are calculated as follows:

$$\text{exponent} = \lfloor \log_{10}(e_a) \rfloor + \text{offset (to eliminate negative logarithms)}$$

$$\text{mantissa} = \lfloor e_a * 10^{-\lfloor \log_{10}(e_a) \rfloor + \log_{10}(m_{size}) - 1} \rfloor = \lfloor e_a * m_{size} * 10^{-\lfloor \log_{10}(e_a) \rfloor - 1} \rfloor$$

For example, if the error estimate is $e_a = 0.002316$, then the exponent is $\lfloor \log_{10}(0.002316) \rfloor + 301 = -3 + 301 = 298$, and the mantissa with 1 digit of precision is $\lfloor 0.002316 * 10 * 10^{-\lfloor \log_{10}(0.002316) \rfloor - 1} \rfloor = \lfloor 0.002316 * 10 * 10^{-\lfloor -2.635261 \rfloor - 1} \rfloor = \lfloor 0.002316 * 10 * 10^2 \rfloor = \lfloor 2.316 \rfloor = 2$. The mantissa with 3 digits of precision is $\lfloor 0.002316 * 10^3 * 10^{-\lfloor \log_{10}(0.002316) \rfloor - 1} \rfloor = \lfloor 0.002316 * 10^3 * 10^{-\lfloor -2.635261 \rfloor - 1} \rfloor = \lfloor 0.002316 * 10^3 * 10^2 \rfloor = \lfloor 231.6 \rfloor = 231$.

The truncation of the error estimate according to the desired length of precision is only for determining the bucket in which to put the data item. The actual values are stored in the node and used in the application. Thus, the actual error estimates are not affected.

15-Puzzle. The 15-puzzle is another heap application that we implemented with our bucket data structure. All key values on the linked list within a bucket will be exactly the same for this application. The (integer) key value represents the cost of a state of the puzzle, which is used as the priority. Since the state with the smallest cost is deleted first, the priority queue implemented as a heap would use a min heap. For the bucket data structure, the cost of a node will be used to create a bucket. The cost of a node equals the number of tiles that are not in their proper place plus the length of the path from the root of the state space tree to the node. The first occurrence of a particular cost becomes the head node of the linked list and a pointer to the head node is inserted into the appropriate bucket. Then, when another cost is inserted with the same value as a head node of a linked list already in the bucket array, it is placed at the end of the linked list for which that node is the head. If no node with the cost value is on the heap, then the value is associated with a new head node and a pointer to the head node is inserted into the array. The buckets for the 15-puzzle application form a 1-dimensional array of pointers that point to the head node of a linked list for a particular cost of a state. A hashing function that is used to find the correct bucket is based on the cost of the node's state, for example, BUCKETS[cost]. Once the hashing function is established, our data structure will operate in the same manner for any application.

The priority (cost) for the 15-puzzle application has an integer data type, is non-negative and the cost is assumed to be less than 100 in the implementation. As mentioned, the BUCKETS array for this application will have one dimension. The dimension size is set to 100, assuming that the cost can be a value between 0 and 99, inclusively. The index for the dimension will also be a value from 0 to 99. The BUCKETS array would be declared as BUCKETS[100].

2.2   Further Specifications

There was some consideration of using data structures in the C++ Standard Template Library (STL) to develop the data structure. However, this would force the result to be programming language dependent, which is undesirable. Therefore we used data types common to various programming languages, thus allowing the data structure to be used with any of the popular programming languages.

Our data structure implementation uses an array with one or more dimensions. The data items in the array are pointers to nodes, which are heads of linked lists of nodes. When a data item is inserted, if a node with a certain bucket value is not already present, then the node becomes the head node for that bucket and a pointer to the head node is inserted into the array. If a pointer to a head node with a certain bucket value is already in the array, then the node is added to end of the linked list and it becomes the tail node for that bucket. When a data item is deleted from a linked list, if it is the only node of the linked list, then the pointer to that bucket is set to NULL after deleting the node. If the linked list has more than one node, then this operation is not required.

Our implementation has several methods allowing it to be a self-contained and fully functional data structure that can perform all the operations necessary to accommodate an application. This includes performing its own garbage collection when applicable.

2.3 Priority

The priority of a data item that is inserted or deleted by an application is used to determine the bucket in which to put the data item. The priority is also used for comparison purposes to determine the min or max value. This will be different for every application. Our data structure has a MIN pointer that points to the bucket with the minimum value or a MAX pointer that points to the bucket with the maximum value dependent on the application. The priority is the first most important update that has to be made when incorporating the data structure into an application. For the multivariate integration application, the priority is the value of a region's error estimate. The largest value has the highest priority. For the 15-puzzle, the priority is the value of the cost of the state associated with a node. The smallest value has the highest priority. For other applications, someone at the domain or subject matter expert level will have to determine the correct priority.

2.4 Hashing

The hashing function to determine the proper bucket for a given priority will be different for every application. This is the second most important update that has to be made when incorporating the data structure into an application. The hashing function has to return a unique index for each priority value. For the multivariate integration application, the hashing function determines the bucket according to the log_{10} and the mantissa of the error estimate with the desired number of digits. For the 15-puzzle application, the hashing function produces the cost of a node's state. For other applications, someone at the domain or subject matter expert level will have to implement a suitable hashing function.

2.5 Configuration

The data structure can be configured to function like a min heap or a max heap. It can also be configured to delete one or more data items at a time when the delete method is called. Some of the other configurable features include settings for the maximum number of data items, the bucket size and precision of the keys in the buckets.

2.6   Adjustments

The size of the buckets array can be adjusted according to the number of data items being stored in the array, using methods described in [4] − by allowing it "to oscillate between the top and the bottom of a larger array". For example, "shrinking" can be accomplished by manipulating a subarray. The structure starts out with a default size. Each time the current size is reached, the size of the array will be doubled. If ever the size drops significantly, the current size will be cut in half. The point at which the size of the array is cut in half will be when the size of the data falls below the current size divided by four. However, the array is never reduced to a size lower than the original size.

2.7   Required Changes for Different Applications

Our data structure was developed so that the code modifications required for one application or another are minimal. The first step to update the data structure for a particular application is to decide on the configuration settings discussed in the Configuration Sect. 2.5. This includes determining if the application will function as a min or a max heap; if it will delete one or more data items at a time; setting the maximum number of data items, and the size and precision of the buckets. The next step is to implement the "node" class, which encapsulates the type of information that will be inserted and deleted. This will be unique and designed specifically for each application. The only code changes required for the available methods concern the logic necessary for comparing items for priority purposes, the logic for the hash function to determine the bucket in which each data item will be placed, and the actions to take when there is an overflow or underflow. The logic for these code changes will affect the insert, delete, and contains methods. The logic for comparison and the hash function for buckets is isolated in the get_priority and get_bucket methods, respectively. It was our goal to avoid having to change any of the other methods.

3   Testing and Results

For this paper, two applications were implemented, multivariate integration and the 15-puzzle. For both of these applications, the behavior of the heap was observed. Normally, the multivariate integration application uses a max heap, and the 15-puzzle uses a min heap. The basic operations of a heap are insert and delete. Whenever an item is inserted into a heap, the position to place the newly added data item has to be determined. This requires a certain amount of CPU time. Likewise, whenever a data item is deleted from a heap, the remaining items have to be rearranged so that the heap property is maintained, which requires a certain amount of CPU time as well. This "heapify" process is executed recursively. If the application is heap-intensive, meaning millions of data items are inserted into and deleted from the heap, the total time spent inserting and deleting may be substantial.

Testing was conducted on the two applications discussed in this paper, for the 15-puzzle and multivariate integration. The original programs using a heap priority queue for these applications were run. A counter was incremented to keep track of the number of "heapify" calls in the course of each execution.

The programs for these applications were rewritten and the min and max heaps were replaced with the bucket data structure. The programs incorporating the new data structure for the applications were run and the results were obtained. The total time spent in insertions and deletions were recorded and compared to the corresponding times incurred in the heap code.

3.1   15-Puzzle

The 15-puzzle application was executed with the initial state shown in Fig. 2, on a x86_64 machine running a Ubuntu version of Linux.

2		8	3
1	5	6	12
13	15	9	7
14	11	10	4

Fig. 2. The initial state used for the 15-puzzle instance

The 15-puzzle heap implementation created a heap with 17,413,988 individual nodes. For example, it executed 8,231,156 direct "heapify" calls when deleting nodes from the heap, which further made 141,475,164 recursive calls for a total 149,706,320 calls. The buckets implementation had the 17,413,988 nodes distributed among 21 buckets.

- For the total deletion time, the heap code took 5.82 s (seconds); the buckets code took 0.71 s for a 87.9% improvement.
- For the total insertion time, the heap code took 2.97 s; the buckets code took 1.23 s, a 58.8% improvement.
- The total insertion and deletion time was 8.79 s for the heap code, and 1.93 s for the buckets code, a 78.0% improvement.
- The total execution time for the puzzle instance was 22.10 s for the heap code, and 13.82 s for the buckets code, a 37.5% improvement.

The number of "heapify" calls is thus extremely large (in the original 15-puzzle application code). The bucket data structure eliminates this type of data manipulation, which results in a substantial reduction in the amount of CPU utilization.

As an added bonus, the original program for the 15-puzzle application found the solution in 8,231,156 iterations. The program rewritten to incorporate the bucket data structure found the solution in 7,174,751 iterations, which is 1,056,405 less iterations than the original program.

For the multivariate integration application, we used adaptive integration code from [17], which performs its region subdivisions sequentially, as well as its heap manipulations, but evaluates the integration rules on the GPU (for two or four regions at a time). A new program was generated by replacing the max heap in the original code by the new bucket data structure.

3.2   Multivariate Integration

The multivariate integration application was executed for the test cases in [17], including an integral of a function with discontinuous derivatives, one with a singular integrand, and a linear model function. We observed a similar behavior for the different test cases, and thus report on one of these, shown as the integral I_1 below, where d denotes the dimension (set to $d = 10$), and $\mathcal{C}_d$ is the d-dimensional unit cube. The exact result of the integral is 1.

$$I_1 = \int_{\mathcal{C}_d} f(\mathbf{x})\ d\mathbf{x} = \frac{6}{5d} \int_{\mathcal{C}_d} \sum_{i=1}^{d} |3x_i - 1|\ d\mathbf{x}. \tag{1}$$

The error tolerance was set to 0, and the maximum number of function evaluations was 5,000,000,000. The original (CUDA) program creates a heap (on the host) with 3,920,800 individual nodes. The integrand evaluations are performed on the GPU. The program incorporating the bucket data structure for the application had the 3,920,800 nodes distributed among 124 buckets, and uses one digit of precision for the length of the mantissa to determine the proper bucket for insertion. The program runs are done on a system with dual 8C Intel Xeon E5-2670 @2.6 GHz as the host, with 128 GB of RAM, and Kepler 20(m) GPU. The Kepler 20(m) GPU has 2496 CUDA cores, 4.8 GB global memory and 956.8 GFLOPS double precision theoretical performance. This is an old GPU with low performance compared to, e.g., the GV100 at 8.33 TFLOPS for FP64 (double).

The original multivariate integration program had 980,199 direct "heapify" calls when deleting nodes from the heap, and 18,468,590 recursive calls for a total of 19,448,789 calls. Albeit a large number, it is only 13.0% of the corresponding number for the 15-puzzle application.

For a time comparison between the heap and buckets versions, we have the following:

Table 1. I_2, I_3 parameters α, integral approximation Q, total absolute error E_a, processing time improvement (%), and total processing time (inserting and deleting) in seconds, for the heap and buckets data structures

Int.	α	Q	E_a		% Improv.	Time [s]	
			Heap	Buckets		Heap	Buckets
I_2	2	4.483234514 e−02	1.44 e−08	1.44 e−08	36.1	6.28	4.01
	4	2.42941120 e−03	1.41 e−08	1.41 e−08	40.8	6.83	4.04
	8	1.94867 e−05	5.70 e−09	5.58 e−09	38.1	6.72	4.16
	9	1.0053 e−05	7.80 e−09	7.65 e−09	35.4	6.31	4.08
	9.2	9.6524 e−06	9.41 e−09	9.24 e−09	36.0	6.60	4.22
	9.4	9.91012 e−06	1.28 e−08	1.26 e−08	43.1	7.43	4.23
	9.5	1.04454 e−05	1.57 e−08	1.54 e−08	38.9	6.90	4.21
I_3	0.1	2.8660 e 00	1.07 e−02	1.09 e−02	40.8	7.08	4.20
	0.2	9.276 e 00	1.23 e−01	1.23 e−01	37.3	6.69	4.19
	0.3	34.87 e 00	1.04 e 00	1.05 e 00	43.8	7.13	4.01

- The total deletion time was 3.80 s for the heap code, and 0.88s for the buckets version, a 76.9% improvement.
- The total insertion time was 3.44 s for the heap code, vs. 3.25 s for the buckets version, a 5.4% improvement.
- Thus, the total insertion and deletion time was 7.24 s for the heap, and 4.13 s for the buckets code, a 42.9% improvement.
- In this case, however, the total execution time was heavily dominated by the function evaluation time which was 85.0% of the total execution time and amounted to 142.30 s using the heap, vs. 138.13 s using buckets, 2.9% improvement. The data structure performance would be more significant if, for example, a faster GPU would be used effectively.

The test results further demonstrate that the programs rewritten to incorporate the bucket data structure maintained the accuracy for both applications. The original integration program returns the value 1.0000564607886810 e+00 as the total integral approximation, i.e., the sum of the results of all subregions on the heap, with total absolute error estimate 8.334 e−05 as the sum of the absolute errors of all subregions on the heap. These values are very close to the total integral approximation of the program rewritten to incorporate the bucket data structure, 1.0000561894650961 e+00, which is the sum of the results of all subregions, with 8.294 e−05 as the sum of the absolute errors of all subregions.

Thus, the accuracy of the multivariate integration application was not affected.

As additional test problems, we consider the integrals

$$I_2 = \int_{\mathcal{C}_d} \frac{1}{\left(\sum_{i=1}^{d} x_i\right)^\alpha} \, d\mathbf{x}, \quad \alpha < 10 \tag{2}$$

Table 2. I_2, I_3 parameters α, the number of "heapify" calls on insert, on delete (directly, recursively, and the total), and the number of buckets required for the heap and buckets data structures

Int.	α	Heapify/Ins.	Heapify/Del.	Rec. Heapify	Tot. Heapify/Del.	Buckets
I_2	2	3,920,800	980,199	18,677,644	19,657,843	85
	4	3,920,800	980,199	18,655,439	19,635,638	90
	8	3,920,800	980,199	18,728,412	19,708,611	110
	9	3,920,800	980,199	18,744,088	19,724,287	164
	9.2	3,920,800	980,199	18,767,907	19,748,106	191
	9.4	3,920,800	980,199	18,770,462	19,750,661	241
	9.5	3,920,800	980,199	18,781,981	19,762,180	287
I_3	0.1	3,920,800	980,199	18,641,266	19,621,465	82
	0.2	3,920,800	980,199	18,686,978	19,667,177	86
	0.3	3,920,800	980,199	18,896,512	19,876,711	83

and

$$I_3 = \int_{\mathcal{C}_d} \frac{1}{\prod_{i=1}^{d} x_i^\alpha} \, d\mathbf{x}, \quad \alpha < 1, \tag{3}$$

both for $d = 10$ dimensions. I_2 has an integrand singularity at the origin, and the integrand function of I_3 is singular at $x_i = 0$ for $1 \le i \le d$, which will cause excessive subdivisions by the adaptive strategy near the singularities.

Table 1 displays total loads (in terms of error), the percentage of processing time improvement that the buckets data structure provided, and processing times (inserting and deleting) for the heap and buckets data structures.

The efficiency of the buckets data structure also emerges from the Table 2 data, showing the number of "heapify" calls when inserting, deleting (directly, recursively, and the total) in the heap code, and the number of buckets that were required to accommodate all the nodes that were inserted and deleted in the buckets code. The number of "heapify" calls when inserting (3,920,800) corresponds to the number of regions generated, and the number of "heapify" calls when deleting directly (980,199) corresponds to the number of regions evaluated. The number of regions generated is four times the number of regions evaluated since, for every region evaluated, four subregions are generated.

4   Conclusions

We presented a priority queue data structure based on buckets to group the same or similar data items together, which makes the data structure more efficient than a heap for the applications at hand. A heap inserts these data items as separate nodes, while using the "heapify" process to maintain the heap property. In the bucket data structure, data items with the same or similar values reside on linked lists within the buckets and can be inserted or deleted without relying on "heapify" calls. In essence, this eliminates the need for a heap in the applications for multivariate integration and the $(n^2 - 1)$-puzzle/relocation problem considered. The only cost is the time to find the bucket in which to put a data item, an operation that has a $\mathcal{O}(1)$ time complexity using a hashing function. The actual values of the data items are stored, despite the fact that a truncated value may be used in the hashing function to determine the bucket. This allows the data structure to be more efficient than a min/max heap without affecting the accuracy of the results.

Our data structure can be used with any application that uses a min/max heap. It can be reconfigured for a different application with minimal effort. The most difficult part of configuring the data structure for a different application is determining the priority of the data items and the hashing function for the buckets. This will have to be implemented by someone who is familiar with the data at the domain or subject matter expert level. In future work we will continue to test with other applications, as priority queues constitute one of the most utilized data structures in Computer and Computational Sciences.

Acknowledgements. The authors would like to thank Dr. J. Kapenga and Dr. D. Zeitler for their valuable comments, as well as the reviewers of this paper.

References

1. Berntsen, J., Espelid, T.O., Genz, A.: An adaptive algorithm for the approximate calculation of multiple integrals. ACM Trans. Math. Softw. **17**, 437–451 (1991)
2. Berntsen, J., Espelid, T.O., Genz, A.: Algorithm 698: DCUHRE-an adaptive multi-dimensional integration routine for a vector of integrals. ACM Trans. Math. Softw. **17**, 452–456 (1991)
3. Brass, P.: Advanced Data Structures. Cambridge University Press, Cambridge (2008)
4. Brown, R.: Calendar queues: a fast o(1) priority queue implementation for the simulation event set problem. Commun. ACM **31**(10), 1220–1227 (1988)
5. Cools, R., Haegemans, A.: CUBPACK: progress report. In: Espelid, T.O., Genz, A.C. (eds.) Numerical Integration, Recent Developments, Software and Applications, pp. 305–315. NATO ASI Series C: Mathematical and Physical Sciences (1992)
6. Demaine, E.D., Rudoy, M.: A simple proof that the $(n^2 - 1)$-puzzle is hard (2017)
7. de Doncker, E., Genz, A., Gupta, A., Zanny, R.: Tools for distributed adaptive multivariate integration on NOW's: PARINT1.0 release. In: Supercomputing'98 (1998)
8. de Doncker, E., Kaugars, K., Cucos, L., Zanny, R.: Current status of the ParInt package for parallel multivariate integration. In: Proceedings of Computational Particle Physics Symposium (CPP 2001), pp. 110–119 (2001)
9. de Doncker, E., Yuasa, F.: Self-energy Feynman diagrams with four loops and 11 internal lines. In: Gervasi, O., et al. (eds.) ICCSA 2021. LNCS, vol. 12953, pp. 160–175. Springer, Cham (2021). https://doi.org/10.1007/978-3-030-86976-2_11
10. de Doncker, E., Yuasa, F., Almulihi, A.: Efficient GPU Integration for Multi-loop Feynman Diagrams with Massless Internal Lines. In: Okada, H., Atluri, S.N. (eds.) ICCES 2019. MMS, vol. 75, pp. 737–747. Springer, Cham (2019). https://doi.org/10.1007/978-3-030-27053-7_62

11. de Doncker, E., Yuasa, F., Almulihi, A., Nakasato, N., Daisaka, H., Ishikawa, T.: Numerical multi-loop integration on heterogeneous many-core processors. In: The Journal of Physics: Conference Series (JPCS), vol. 1525, no. 012002 (2019). https://doi.org/10.1088/1742-6596/1525/1/012002

12. de Doncker, E., Yuasa, F., Olagbemi, O., Ishikawa, T.: Large scale automatic computations for Feynman diagrams with up to five loops. In: Gervasi, O., et al. (eds.) ICCSA 2020. LNCS, vol. 12253, pp. 145–162. Springer, Cham (2020). https://doi.org/10.1007/978-3-030-58814-4_11

13. Genz, A., Malik, A.: An adaptive algorithm for numerical integration over an n-dimensional rectangular region. J. Comput. Appl. Math. **6**, 295–302 (1980)

14. Hahn, T.: Cuba − a library for multidimensional numerical integration. Comput. Phys. Commun. **176**, 712–713 (2007)

15. Horowitz, E., Sahni, S., Rajasekaran, B.: Computer Algorithms/C++. Computer Science Press (1997)

16. Jarząbek, Ł, Czarnul, P.: Performance evaluation of unified memory and dynamic parallelism for selected parallel CUDA applications. J. Supercomput. **73**(12), 5378–5401 (2017). https://doi.org/10.1007/s11227-017-2091-x

17. Olagbemi, O.E., de Doncker, E.: Scalable algorithms for multivariate integration with ParAdapt and CUDA. In: Proceedings of the 2019 International Conference on Computer Science and Computational Intelligence. IEEE Computer Society (2019)

18. Piessens, R., de Doncker, E., Überhuber, C.W., Kahaner, D.K.: QUADPACK, A Subroutine Package for Automatic Integration, Springer Series in Computational Mathematics, vol. 1. Springer, Cham (1983). https://doi.org/10.1007/978-3-642-61786-7

19. Ratner, D., Warmuth, M.: The $(n^2 - 1)$-puzzle and related relocation problems. J. Symb. Comput. **10**(2), 111–137 (1990)

20. Ros-Giralt, J., Commike, A., Cullen, P., Lucovsky, J., Madathil, D., Lethin, R.: Multiresolution priority queues (2017)

21. Williams, J.W.J.: Algorithm 232: heapsort. Commun. ACM **7**(6), 374–378 (1964)

Acceleration of Multiple Precision Solver for Ill-Conditioned Algebraic Equations with Lower Precision Eigensolver

Tomonori Kouya[✉] [iD]

Shizuoka Institute of Science and Technology,
2200-2 Toyosawa, Fukuroi 437-8555, Japan
kouya.tomonori@sist.ac.jp

Abstract. There are some types of ill-conditioned algebraic equations that have difficulty in obtaining accurate roots and coefficients that must be expressed with a multiple precision floating-point number. When all their roots are simple, the problem solved via eigensolver (eigenvalue method) is well-conditioned if the corresponding companion matrix has its small condition number. However, directly solving them with Newton or simultaneous iteration methods (direct iterative methods) should be considered as ill-conditioned because of increasing density of its root distribution. Although a greater number of mantissa of floating-point arithmetic is necessary in the direct iterative method than eigenvalue method, the total computational costs cannot obviously be determined. In this study, we target Wilkinson's example and Chebyshev quadrature problem as examples of ill-conditioned algebraic equations, and demonstrate some concrete numerical results to prove that the direct iterative method can perform better than standard eigensolver.

Keywords: Multiple precision floating-point arithmetic · Algebraic equation · Eigensolver · Parallelization

1 Introduction

Currently, multiple precision floating-point (MPF) arithmetic are executed using reliable and highly performed de-facto standard libraries such as QD [2] and MPFR [14]. These have been developed since the end of the 20th century, and have been in use for over two decades. More convenient and efficient multiple precision numerical computation libraries such as MPLAPACK [10] and our BNCpack [6], are constructed on that of MPF libraries. Therefore, we can easily deal with various types of ill-conditioned problems in normal hardware and software environment for consumers.

We have published a paper [7] about deriving highly accurate abscissas of Gauss-type integration formulas using both "eigenvalue method" via symmetric tridiagonal matrix originated by Golub and Welsch [4] and "direct iterative

O. Gervasi et al. (Eds.): ICCSA 2022 Workshops, LNCS 13378, pp. 358–372, 2022.
https://doi.org/10.1007/978-3-031-10562-3_26

method" to directly calculate zeros of orthogonal polynomial using Newton iteration [17–19]. Our previous task with direct iterative method was to calculate the approximation of abscissas with user-required accuracy by combining binary64 eigensolver of LAPACK [8] and MPFR Newton iteration.

The performance of the eigenvalue method is slow but robust for effect by round-off error; hence, we can obtain good approximations even with binary64 arithmetic. However, the direct iterative method is well-performed with multiple precision arithmetic, but good initial guesses must be employed to guarantee its convergence. In our previous paper, we concluded that, to shorten total computational time, it is best to combine binary64 LAPACK eigensolver to derive initial guesses, and multiple precision Newton iteration to exploit a higher accuracy. From the current point of view, our proposed combination employs the "mixed precision" method for solving algebraic equations.

From these experiences, we have tried to combine multiple precision eigensolver of MPLAPACK, and the second- and third-order simultaneous iterative methods of BNCpack, such as the Durand-Kerner (DK) method, to solve two types of ill-conditioned algebraic equations with real and complex roots. One is well-known Wilkinson's problem [16], and another is calculation of abscissas of Chebyshev quadrature formula (Chebyshev quadrature problem for short), which has been experimented by Harumi Ono [12,13]. The algebraic equation of Chebyshev quadrature problem has already been studied in several published papers, but those papers are not known except in Japan because most of them are written only in Japanese. The only English non-referred paper in English can be found in Kokyoroku series of RIMS in Kyoto University [5]. Here, we briefly explain the previous results on the Chebyshev quadrature problem.

The algebraic equations derived from the Chebyshev quadrature formula have difficulties not only in calculating the accurate coefficients, but also in solving the equations with multiple precision floating-point arithmetic. Harumi Ono published papers on how to solve the 1024-th degree polynomial in 1979, and 2048-th degree in 1981 using the DK method with her original multiple precision arithmetic on Cray. Additionally, Masumoto et al. reported the numerical property of 20480-th degree coefficients when deriving them with MPF arithmetic [9] but did not achieve the obtained roots of those. We will describe the mathematical expression of the Chebyshev quadrature problem in Sect. 4.

In this study, we will relay the results obtained by combining DD Rgeev of MPLAPACK as a generator of initial guesses and the MPFR Durand-Kerner method from the point of view for mixed precision approach of acceleration.

The following two computational environments, EPYC and Xeon, are used in the rest of this study. MPLAPACK and our BNCpack, including QD and MPFR/GNU MP, are natively compiled with Intel Compiler.

EPYC AMD EPYC 7402P 24 cores, Ubuntu 18.04.6 LTS, GCC 7.5.0, Intel Compiler version 2021.4.0, MPLAPACK 1.0.1, BNCpack 0.8, MPFR 4.1.0

Xeon Intel Xeon W-2295 3.0 GHz 18 cores, Ubuntu 20.04.3 LTS, GCC 9.3.0, Intel Compiler version 2021.5.0, MPLAPACK 1.0.1, BNCpack 0.8, MPFR 4.1.0

OpenMP is applied for parallelization of DK method with the following compile option as `icpc -03 -qopenmp`.

2   Brief Introduction of Current Multiple Precision Arithmetic and Performance of Current MPFR GEMM

The need for computational processing, mainly using floating-point arithmetic, will increase and not decrease in the foreseeable future, not only for scientific computing, but also for deep learning and other applications. The performance of hardware such as CPUs and GPUs is mainly improved in the following two ways:

- Parallel processing capabilities such as SIMD instructions and an increase in the number of cores in CPUs and GPUs,
- Increasing or decreasing the length of mantissa and exponent in floating-point numbers according to the necessity of user's computation, such as the introduction of half-precision and single-precision floating-point operations.

In similar ways, the whole process of computation is now accelerated by enhancing parallelization techniques. While taking advantage of these hardware performance improvement and employing parallelization techniques such as SIMD instruction, OpenMP, and MPI, mixed precision techniques combining IEEE binary32, binary64, and half precision arithmetic, which are standard in hardware, are also used. Regarding MPF operations using software libraries, which are adopted in adverse conditions where binary64 arithmetic lacks computational accuracy, the heavy MPF processing requires the active use of similar software performance improvement techniques.

Currently, the mainstream of MPF arithmetic falls into two types of implementations: a multi-component method that combines multiple binary32 and binary64, using error-free transformation (EFT) techniques to extend the mantissa length, and an integer-based many-digit method. The QD library by Bailey et al. [2] is well-known for those based on the multi-component method, and the MPFR library using the arbitrary-length natural number kernel (MPN) of GNU MP (GMP) has a significant number of users because of its superior speed and reliability. MPLAPACK by Maho Nakata is a multiple precision linear computation library based on the C++ converted from the original Fortran code of LAPACK/BLAS, which was developed from these two multiple precision floating-point arithmetic libraries, and the latest version (Version 1.0.1) as of February 2022, provides the parallelized BLAS code with OpenMP. In addition, the main driver and calculation routines of LAPACK are available in multiple precision. However, we will wait to gain all driver routines benefit from this high-performance MPBLAS. In addition, SIMD instructions, the use of CUDA, and the introduction of the Ozaki scheme have not officially improved performance as faster multiple precision ATLAS, OpenBLAS, and Intel Math Kernel.

The current driver routine Rgeev of MPLAPACK used in this paper is not parallelized. In addition, the arbitrary-precision routines have been determined to be slower than our native C implementation of the basic linear subprogram, probably because of the use of MPREAL, a wrapper C++ class library for MPFR adopted in the arbitrary-precision calculations. In fact, the computational time of the MPFR block matrix multiplication supported by our BNCmatmul library is illustrated in Fig. 1 compared to the time of the MPFR Rgemm of MPBLAS. The results of Strassen matrix multiplication are also included for comparison.

Evidently, both 212 bits (the same bits of QD precision) and 1024 bits are 2.0–2.4 times (212 bits) and 1.4–1.5 times (1024 bits) larger than MPBLAS (Rgemm) by block matrix multiplication (matmul_mpfmatrix_block) in both EPYC and Xeon environments. Thus, the smaller the number of MPF mantissa, the faster the C native MPFR direct call (BNCmatmul) is than MPLAPACK/MPBLAS (MPREAL).

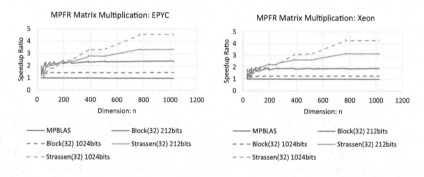

Fig. 1. Speedup ratio of MPFR matrix multiplication against MPBLAS (Rgemm)

The eigenvalue driver routine (Rgeev) for real matrices in MPLAPACK are adopted as eigensolvers, while the simultaneous iterative method for MPFR algebraic equations, prepared as a competitor, employs a direct call to MPFR for its implementation. Therefore, it is expected to perform better than the implementation using MPREAL.

3 Mixed Precision Approach for Solving Algebraic Equations

We target the following algebraic equations

$$p_n(x) = 0, \tag{1}$$

where the following real coefficient polynomial $p_n(x)$ is adopted as the left term,

$$p_n(x) = \sum_{i=0}^{n} a_i x^i \ (a_i \in \mathbb{R}, a_n \neq 0). \tag{2}$$

It is well-known that the n-th degree algebraic equation (1) definitely has n roots $\alpha_i \in \mathbb{C}$, $i = 1, 2, ..., n$ at most. In this time, suppose that we have no prerequisite for roots of the Eq. (1).

For convenience, the corresponding monic polynomial $q_n(x)$ derived from $p_n(x)$

$$q_n(x) = x^n + \sum_{i=0}^{n-1} c_i x^i \ (c_i = a_i/a_n), \tag{3}$$

is also prepared.

3.1   Eigenvalue Method for Roots of Algebraic Equation

As explained in several number of textbooks of linear algebra, it is well-known that our targeted algebraic equation (1) can be expressed by the following companion matrix C_n with the same eigenvalues as the roots of $q_n(x)$.

$$C_n = \begin{bmatrix} 0 & 1 & 0 & \cdots & 0 \\ \vdots & \ddots & \ddots & \ddots & \vdots \\ 0 & \cdots & 0 & 1 & 0 \\ 0 & \cdots & \cdots & 0 & 1 \\ -c_0 & -c_1 & \cdots & -c_{n-2} & -c_{n-1} \end{bmatrix}. \tag{4}$$

As the polynomial (2) has only real coefficients, we can exploit xGEEV driver routines in LAPACK [8] as an eigensolver to obtain all the eigenvalues of C_n. If C_n has a negligible condition number, is diagonalizable, and does not have multiple eigenvalues, we do not require additional digits of mantissa of MPF. We call this approach the "eigenvalue method" for solving the algebraic equation (1).

3.2   Simultaneous Iteration Method for Directly Solving Algebraic Equation

The standard direct iterative method for obtaining solutions to algebraic equations of degree five or higher (1) is the Newton method and its related simultaneous methods. In recent years, several higher-order direct iterative methods have been proposed, and Petkovic has compactly summarized the results up to 2012 [11]. However, as described below, the higher the order of the direct iterative method, the more computational complexity per iteration increase, and global convergence is not guaranteed. Therefore, we employ only the second- and third-order methods [1]. The comparison of the initial guess setting methods will be discussed later.

For both second- and third-order DK methods, by expressing the approximation at k times iteration as

$$\mathbf{z}_k = [z_1^{(k)} \ z_2^{(k)} \ ... \ z_n^{(k)}]^T \in \mathbb{C}^n,$$

their iteration formulas are described with monic polynomial (3) as follows:

Second Order DK Method

$$z_i^{(k+1)} := z_i^{(k)} - \frac{q_n(z_i^{(k)})}{\prod_{j=1, j\neq i}^n (z_i^{(k)} - z_j^{(k)})} \tag{5}$$

Third Order DK Method

$$z_i^{(k+1)} := z_i^{(k)} - \frac{\dfrac{p_n(z_i^{(k)})}{p_n'(z_i^{(k)})}}{1 - \dfrac{p_n(z_i^{(k)})}{p_n'(z_i^{(k)})} \displaystyle\sum_{\substack{j=1 \\ j\neq i}}^n (z_i^{(k)} - z_j^{(k)})^{-1}} \tag{6}$$

To compare with our initial guess approach, we adopt Aberth's initial guesses [1] as follows:

$$z_i^{(0)} := -\frac{c_{n-1}}{n} + r \exp\left\{ \left(\frac{2(i-1)\pi}{n} + \frac{3}{2n} \right) i \right\}. \tag{7}$$

We adopt r in Aberth's initial guess (7) as follows:

$$r := \max_{0 \le i \le (n-1)} |n_{\mathrm{nz}} c_i|^{1/(n-i)},$$

where $n_{\mathrm{nz}} \le n$ is the number of non-zero coefficients in (2).

4   Examples of Ill-Conditioned Algebraic Equations

Unlike the eigenvalue problem, when solving the equation, the higher the density of roots, the larger the error in the values of the polynomial $p_n(x)$ and $q_n(x)$. Therefore, if we employ a method such as the Danilevsky method [3], where the eigenvalue problem of a matrix is transformed into a companion matrix and coefficients of the eigenequation are obtained directly, despite the good conditions for an eigenvalue problem, it becomes a bad problem to solve the algebraic equation. This is why this method via eigenequation is not recommended currently because fast eigenvalue solving methods such as the QR method, which uses a shift of the origin in the Householder transformation to obtain stable and accurate eigenvalues exist.

However, regarding bad algebraic equations that require longer MPF numbers than binary64 for the coefficients, MPF operations are essential in the process of solving, and there is a possibility that the solution of algebraic equations with less computational complexity can be performed faster than matrix eigenvalue routines. In our previous study on the Gauss-type integral formulas for the quantile calculations [7], there were several cases where the computation time was reduced using the Newton method with the approximate eigenvalues obtained by binary64 as the initial guess, rather than computing all eigenvalues using the MPF eigensolver for real symmetric matrices. As an extension of this result, it is expected that similar mixed-precision techniques will be effective in reducing computation time for solving arbitrary real coefficient algebraic equations using a real asymmetric companion matrix to derive useful initial guesses.

Here, we consider two examples: Wilkinson's example (with all real roots) and an algebraic equation whose solution is the quantile of the Chebyshev quadrature formula (with almost all complex roots) to prove that such examples exist in the benchmark test.

Wilkinson's Example

This well-known example is explained in detail in Wilkinson's book [16].

The polynomial (2) is provided as $p_n(x) = \prod_{i=1}^{n}(x - i)$, with $\alpha_i = i$ as its roots. The absolute value of coefficients is glowing up on $O(n!)$. For $n = 20$, the coefficients are

$$a_0 = 2432902008176640000$$
$$a_1 = -8752948036761600000$$
$$\vdots$$
$$a_{19} = -210$$
$$a_{20} = 1.$$

In the case of $n = 128$ adopted in our experiments, $a_0 = 3.8562\cdots \times 10^{215}$ means that coefficients at over $n = 128$ may not be treated in binary64, DD (double-double, 106 bits), and QD (quad-double, 212 bits) arithmetic. Figure 2 illustrates how the approximation of the roots with DD, QD, and MPREAL 512 bits Rgeev are distributed on Gauss plane. Although DD and QD Rgeev produce different but results located near each other on the Gauss plane, it is obvious that DD and QD Rgeev cannot provide an accurate approximation of the roots.

However, as indicated later, DD approximations are useful as initial guesses to accelerate MPF direct iterative methods.

Chebyshev Quadrature Problem

As previously described, this example is solely popular in Japan. In this subsection, we mathematically explain the Chebyshev quadrature problem [15].

The Chebyschev quadrature formula is not categorized in the Gaussian quadrature formula. When $[-1, 1]$ is provided as the integration interval, we can obtain the following discrete quadrature formula as follows:

$$\int_{-1}^{1} w(x)f(x)dx \approx \sum_{k=1}^{n} w_k(x_k)f(x_i). \tag{8}$$

In the above formula (8), we suppose that $w(x) = 1$ and $w_k(x_k) = 2/n$ is fixed. In this case, the "Chebyshev quadrature formula" is determined by choosing the appropriate abscissas.

Wilkinson n=128, DD, QD, MPREAL(512bits) Rgeev

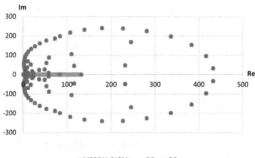

Fig. 2. Approximation of DD, QD, and MPREAL (512 bits) Rgeev (MPLAPACK) on Gauss plane: Wilkinson's example $n = 128$

The coefficients of polynomial (2) with roots as their abscissas are derived, starting with $a_n = 1$, as follows:

$$\begin{cases} a_{n-(2k-1)} := 0 \\ a_{n-2k} := -\sum_{2j+1}^{k} a_{n-2(k-j)} \end{cases}, \tag{9}$$

where $k = 1, 2, ..., \lfloor n/2 \rfloor$. The odd numbered terms are zero, and the even numbered terms are derived from (9).

The algebraic equation (1) with the above coefficients has real roots only in the case of $n = 1, 2, ..., 7, 9$, and almost all conjugate complex pairs of roots in the case of $n = 8$ or $n \geq 10$. According to Moriguchi and Iri's prediction [5], regarding $n \rightarrow \infty$, all roots are nearly distributed on the curve and expressed as the following functions of z,

$$|(z + 1)^{(z+1)/2}(z - 1)^{-(z-1)/2}| = 2. \tag{10}$$

The prediction has been precisely confirmed in numerical experiments through Ono's works [12, 13].

In addition, we should handle the process of calculating the coefficients using the formula (9). With an increase in the order of the polynomial, the digits decrease significantly, and longer MPF operations are required to obtain accurate coefficients. According to studies of Masumoto et al., we need over 215 decimal digits to guarantee the over six decimal significant digits of a_0 in case of $n = 1024$. As illustrated in Fig. 3, DD and QD precision arithmetic are not sufficient to obtain accurate roots even in the case of $n = 256$.

5   Benchmark Tests

Our mixed precision approach for targeted algebraic equation (1) is set up as follows:

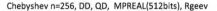

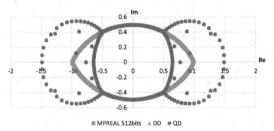

Fig. 3. The abscissas of Chebyshev quadrature problem ($n = 256$) obtained by DD, QD, and MPREAL (512 bits) Rgeev on Gauss plane

1. Calculate all coefficients a_i, $i = 0, 1, ..., n$ of the algebraic equation with L bits MPFR arithmetic.
2. Transfer MPFR a_i to the DD ones and then calculate only eigenvalues $\lambda_i^{(DD)} \in \mathbb{C}$ of the corresponding companion matrix C using DD Rgeev of MPLAPACK.
3. Set the above DD eigenvalues $\lambda_i^{(DD)}$ as initial guesses $\mathbf{z}_0 \in \mathbb{C}^n$ for second- or third- order DK methods. Converged values $\mathbf{z}_{end} \in \mathbb{C}^n$ are adopted as final approximations of roots α_i.

Although the value of $|p_n(x)|$ is normally used for the stopping rule of iteration, we use the difference of $z_i^{(k)}$ and $z_i^{(k+1)}$, relative tolerance ε_{rel}, and absolute tolerance ε_{abs} as follows:

$$|z_i^{(k+1)} - z_i^{(k)}| \leq \varepsilon_{rel}|z_i^{(k)}| + \varepsilon_{abs}. \tag{11}$$

We recognize that the $(k+1)$-th iterated approximation is converged when the condition of (11) is satisfied. The relative errors as illustrated in latter figures are calculated by comparing with the eigenvalues using MPREAL 2048 bits Rgeev.

5.1   Wilkinson's Example

First, we relay the results for Wilkinson's example ($n = 128$) on EPYC and Xeon.

As shown in Fig. 4, the 512 bit MPREAL Rgeev can obtain an accurate approximation of the roots. In contrast, to obtain an approximate solution with sufficient accuracy using the DK method, the 1024 bits are necessary, and $\varepsilon_{rel} := 7.5 \times 10^{-145}$ and $\varepsilon_{abs} := 1.0 \times 10^{-300}$ are adopted to determine convergence. Accordingly, we can confirm that the approximations of both ways can reach the same level of relative errors as illustrated by Fig. 4.

The entire results via benchmark tests are presented in Table 1, where "DKA2" (second order DK) and "DKA3" (third order DK) indicate the computational time (in seconds) and number of iterations for each number of threads

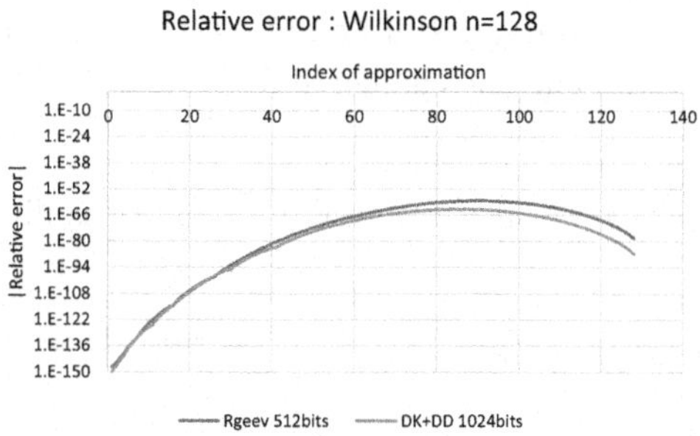

Fig. 4. Relative errors of Rgeev (512 bits) and DK-DD (1024 bits) for Wilkinson's example: $n = 128$.

when the initial guess of Aberth is employed, and "DK2+DD" and "DK3+DD" indicate the computational time and number of iterations when DD Rgeev is adopted to derive the initial guess. Currently, the Rgeev of MPLAPACK is not parallelized; hence, it cannot be accelerated with over two threads. Underlined computational time means being faster than those of MPFR Rgeev.

Table 1. Wilkinson's example:$n = 128$, Rgeev 512 bits, DK methods 1024 bits

EPYC	Rgeev (Second)		DKA2 (1024 bits)		DKA3 (1024 bits)		DK2 (1024 bits)+DD		DK3 (1024 bits)+DD	
#Thr.	DD	MPFR 512 bits	Second	#Iter.	Second	#Iter.	Second	#Iter.	Second	#Iter.
1	0.106	5.1	79.7	1374	71.1	691	30.8	538	51.5	512
2			39.8	1374	36.6	691	15.4	535	26.1	509
4			20	1374	18.7	691	8.14	562	14	512
8			10.1	1374	8.96	691	1.43	195	1.42	99
16			5.15	1374	4.55	691	2.07	557	3.32	511
24			3.88	1374	3.74	691	1.52	541	2.69	511
Xeon	Rgeev (Second)		DKA2 (1024 bits)		DKA3 (1024 bits)		DK2 (1024 bits)+DD		DK3 (1024 bits)+DD	
#Thr.	DD	MPFR 512 bits	Second	#Iter.	Second	#Iter.	Second	#Iter.	Second	#Iter.
1	0.09	3.5	58.6	1374	50.8	691	22.3	538	37.1	512
2			29.2	1374	26.2	691	11.3	535	18.8	509
4			15.3	1374	13.7	691	6.18	562	9.93	512
8			7.78	1374	6.86	691	1.1	195	0.996	99
16			4.73	1374	4.16	691	1.86	557	2.97	511
18			4.88	1374	4.37	691	1.9	557	3.05	511

We can observe the following results from Table 1.

1. When Aberth's initial values are adopted, the number of iterations for DK3 is approximately half that of DK2. However, the computation time has not decreased much.
2. When using the eigenvalues of DD Rgeev as initial values, the number of iterations is reduced compared to when using Aberth's initial values, and the

computation time is also reduced. In addition, the number of iteration varies from thread to thread, and the change in $\lambda_i^{(DD)}$ has a significant effect on the iterative process.

The speedup ratio by parallelization with OpenMP is illustrated in Fig. 5.

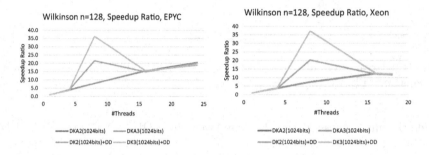

Fig. 5. Speedup ratio of Wilkinson's problem

These figures illustrate that parallelization is efficient for direct iterative methods, and that the decrease of iterative times pull up the speedup ratio at DK2+DD and DK3+DD using eight threads.

5.2 Chebyshev Quadrature Problem

Here, we explain the results of solving the Chebyshev quadrature formula quantile problem ($n = 256$ and $n = 512$).

First we present the case of $n = 256$. MPREAL (256 bits) is adopted to derive the coefficients based on the (9) formula, and the DK methods employs a 512 bits calculation and $\varepsilon_{rel} := 8.6 \times 10^{-68}$ and $\varepsilon_{abs} := 1.0 \times 10^{-300}$ for convergence determination. As illustrated in Fig. 6, this results in accurate approximations of roots of approximately 30 decimal digits.

Table 2 presents all the results of the benchmark tests. MPREAL (256 bits) Rgeev took 57.6 s on EPYC and 38.4 on Xeon, respectively. The DK methods employ a 512 bits MPF arithmetic, $\varepsilon_{rel} := 8.6 \times 10^{-68}$ and $\varepsilon_{abs} := 1.0 \times 10^{-300}$ for convergence determination.

The common numerical properties and trend of computational times is presented in Table 2. In addition, we observe that the serial DK2+DD is faster than MPREAL Rgeev.

According to the results presented in Table 2 and Fig. 7 illustrates the speedup ratio. We can confirm that the second- and third-order DK methods have approximately achieved an ideal speedup ratio. As seen in Wilkinson's example, a decrease in the number of iterations in DK methods increases the speedup ratio.

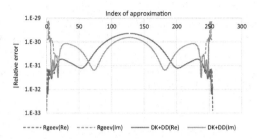

Fig. 6. Relative errors of Rgeev (256 bits) and DK-DD (512 bits) for the Chebyshev quadrature problem: $n = 256$.

Table 2. Chebyshev quadrature: $n = 256$, Rgeev 256 bits, DK methods 512 bits

EPYC	Rgeev (Second)		DKA2 (512 bits)		DKA3 (512 bits)		DK2 (512 bits)+DD		DK3 (512 bits)+DD	
#Thr.	DD	MPFR 256 bits	Second	#Iter.	Second	#Iter.	Second	#Iter.	Second	#Iter.
1	1.29	57.6	203.0	1344	183	671	<u>47</u>	316	70.3	264
2			100.0	1344	93.8	671	<u>7.11</u>	96	<u>10</u>	73
4			<u>51.1</u>	1344	<u>46.3</u>	671	<u>3.62</u>	96	<u>5.08</u>	71
8			<u>25.3</u>	1344	<u>23.1</u>	671	<u>5.77</u>	308	<u>8.92</u>	264
16			<u>13</u>	1344	<u>11.7</u>	671	<u>0.918</u>	95	<u>1.22</u>	71
24			<u>8.96</u>	1344	<u>8.1</u>	671	<u>2.02</u>	263	<u>3.1</u>	263
Xeon	Rgeev (Second)		DKA2 (512 bits)		DKA3 (512 bits)		DK2 (512 bits)+DD		DK3(512 bits)+DD	
#Thr.	DD	MPFR 256 bits	Second	#Iter.	Second	#Iter.	Second	#Iter.	Second	#Iter.
1	1.12	38.4	146.0	1344	127	671	<u>32.1</u>	316	49.2	264
2			69.0	1344	64.8	671	<u>4.95</u>	96	<u>6.91</u>	73
4			<u>36.3</u>	1344	<u>34.2</u>	671	<u>2.61</u>	96	<u>3.59</u>	71
8			<u>18.6</u>	1344	<u>17.7</u>	671	<u>4.2</u>	671	<u>6.63</u>	264
16			<u>10.9</u>	1344	<u>10</u>	1344	<u>0.779</u>	95	<u>1.07</u>	71
18			<u>10.8</u>	1344	<u>9.79</u>	671	<u>2.32</u>	293	<u>3.62</u>	250

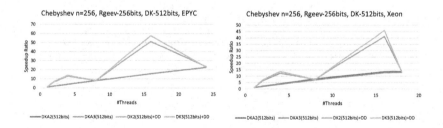

Fig. 7. Speedup Ratio of Chebyshev quadrature: $n = 256$

Second, the results of solving the Chebyshev quadrature problem in the case of $n = 512$ are indicated. MPREAL (512 bits) was adopted to derive the coefficients based on the (9) formula, and the MPREAL 512 bits Rgeev took 636.0 s and 441.0 s on EPYC and Xeon, respectively. The DK method employs a 1024 bits calculation, $\varepsilon_{rel} := 7.5 \times 10^{-145}$, and $\varepsilon_{abs} := 1.0 \times 10^{-300}$ for the convergence decision. This results in correct approximate solutions from 58 to 62 decimal digits, as illustrated in Fig. 8.

Chebyshev n=512, 512bits-Rgeev, 1024bits-DK

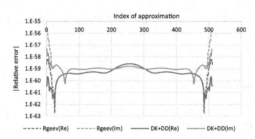

Fig. 8. Relative errors of Rgeev (512 bits) and DK-DD (1024 bits) for the Chebyshev quadrature problem: $n = 512$.

All results obtained through the benchmark test are presented in Table 3.

Table 3. Chebyshev quadrature: $n = 512$, Rgeev 512 bits, DK methods 1024 bits

EPYC	Rgeev (Second)		DKA2 (1024 bits)		DKA3 (1024 bits)		DK2 (1024 bits)+DD		DK3 (1024 bits)+DD	
#Thr.	DD	MPFR 512 bits	Second	#Iter.	Second	#Iter.	Second	#Iter.	Second	#Iter.
1	9.84	636.0	2850.0	3032	2460	1510	500	552	577	357
2			1370.0	3032	1210	1510	90.9	200	67.8	82
4			689	3032	614	1510	46.1	202	34.5	83
8			344	3032	307	1510	62.8	550	77.3	363
16			173	3032	174	1510	11.8	205	8.59	84
24			120	3032	106	1510	8.47	213	6.57	83
Xeon	Rgeev (Second)		DKA2 (1024 bits)		DKA3 (1024 bits)		DK2 (1024 bits)+DD		DK3 (1024 bits)+DD	
#Thr.	DD	MPFR 512 bits	Second	#Iter.	Second	#Iter.	Second	#Iter.	Second	#Iter.
1	9.25	441.0	2030.0	3032	1740	1510	362	552	407	357
2			1020.0	3032	937	1510	67.3	200	48.2	82
4			528	3032	471	1510	35.3	202	25.5	83
8			267	3032	238	1510	48.5	550	56.5	363
16			159	3032	142	1510	10.8	205	7.85	84
18			151	3032	132	1510	26.9	539	31.7	365

We confirm that more cases of DK2+DD and DK3+DD are faster than MPFR Rgeev.

The speedup ratio of parallelized DK methods is illustrated in Fig. 9.

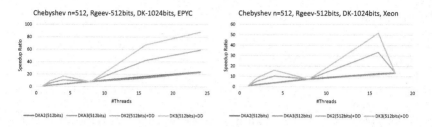

Fig. 9. Speedup Ratio of Chebyshev quadrature: $n = 512$

These two examples indicate that even if the low precision eigensolver cannot obtain good approximations, there are examples where speedup can be easily achieved using low-precision results as an initial guess for direct iterative methods with higher precision. Of course, it is not useful for all ill-conditioned algebraic equations, but the easy-to-implement and highly parallelizable direct iterative method may have a reasonably wide range of applications.

6   Conclusion and Future Works

For two types of ill-conditioned algebraic equations, initial guesses were obtained using a low-precision eigensolver, and it was indicated that a highly accurate and efficient direct iterative method can be accelerated by parallel computing. Considering the stability of the iterative computing process, it is possible to speed up the eigensolver, for example, by using the approximate eigenvalues from low-precision computation as the origin shift in high-precision computation, but the direct iterative method for algebraic equations with its computational complexity, parallelism, and ease of implementation, is also useful in long precision computing environments. In several cases, it can be stated that our mixed precision approach is useful in solving high degree algebraic equations with long precision floating-point arithmetic.

For future studies, we will implement and confirm the effectiveness of our approach as follows :

1. Application to higher degree ill-conditioned algebraic equations,
2. Implementation of other third-order (or above) direct iteration methods and comparison among them,
3. Application to general linear equations with Chebyshev Proxy Rootfinder.

In addition, we accelerate the above applications using multi-component MPF arithmetic.

Acknowledgment. This study was supported by JSPS KAKENHI, Grant Number JP20K11843, and Shizuoka Institute of Science and Technology. We acknowledge all the organizations that are continuously encouraging our study. We also appreciate the unknown referees who provided concrete and useful suggestions for revising our paper.

References

1. Aberth, O.: Iteration methods for finding all zeros of a polynomial simultaneously. Math. Comp. **27**, 339–344 (1973)
2. Bailey, D.: QD. https://www.davidhbailey.com/dhbsoftware/
3. Faddeev, D.K., Faddeeva, V.N.: An Introduction to Numerical Linear Algebra. Dover (1959)
4. Golub, G.H., Welsch, J.H.: Calculation of Gauss quadrature rules. Math. Comput. **23**, 221–230 (1969)
5. Iri, M., Yamashita, H., Terano, T., Ono, H.: An algebraic-equation solver with global convergence property. RIMS Kokuroku **339**, 43–69 (1978)
6. Kouya, T.: BNCpack. https://na-inet.jp/na/bnc/
7. Kouya, T.: Practical proposition of empirical error estimation and its application to calculation of absissas of Gauss-type integration. Trans. IPSJ **48**(SIG18(ACS20)), 1–11 (2007). (in Japanese)
8. LAPACK. http://www.netlib.org/lapack/
9. Masumoto, H., Fujino, S., Ono, H., Kojima, A.: Parallelisim of multi-precision arithmetic for computation of coefficients of algebraic equation of 20480 degrees. Trans. IPSJ **40**(12), 4159–4168 (1999). (in Japanese)
10. MPLAPACK/MPBLAS: Multiple precision arithmetic LAPACK and BLAS. https://github.com/nakatamaho/mplapack
11. Petkovic, M.S., Neta, B., Petkovic, L.D., Dzunic, J.: Multipoint Methods for Solving Nonlinear Equations. Elsevier (2013)
12. Ono, H.: On numerical computation of a high degree polynomial equation by the methods of Durand - Kerner and Aberth. Trans. IPSJ **20**(5), 399–404 (1979). (in Japanese)
13. Ono, H.: On numerical computation of a high degree polynomial equation using the method of Durand, Kerner and Aberth. Trans. IPSJ **22**(2), 165–168 (1981). (in Japanese)
14. The MPFR library. https://www.mpfr.org/
15. Weisstein, E.W.: Chebyshev Quadrature. MathWorld-A Wolfram Web Resource. https://mathworld.wolfram.com/ChebyshevQuadrature.html
16. Wilkinson, J.H.: Rounding Errors in Algebraic Process (Reprint edition). Dover (1994)
17. Yamashita, S.I.: Computation of the abscissas and weight coefficients for Gaussian quadrature formulae. IPSJ **5**, 206–215 (1964). (in Japanese)
18. Yamashita, S.I., Satake, S.: Computation of the abscissas and weight coefficients for the Hermite-Gaussian quadrature formulae. IPSJ **5**, 266–270 (1965). (in Japanese)
19. Yamashita, S.I., Satake, S.: Computation of the abscissas and weight coefficients for the Laguerre-Gaussian quadrature formulae. IPSJ **4**, 216–220 (1965). (in Japanese)

Study of Galaxy Collisions and Thermodynamic Evolution of Gas Using the Exact Integration Scheme

Koki Otaki[1,2]($\boxtimes$) (ID) and Masao Mori[3] (ID)

[1] Degree Programs in Pure and Applied Sciences, Graduate School of Science and Technology, University of Tsukuba, Tennodai 1-1-1, Tsukuba, Ibaraki 305-8577, Japan
otaki@ccs.tsukuba.ac.jp

[2] Degree Programs in Systems and Information Engineering, Graduate School of Science and Technology, University of Tsukuba, Tennodai 1-1-1, Tsukuba, Ibaraki 305-8577, Japan

[3] Center for Computational Sciences, University of Tsukuba, Tennodai 1-1-1, Tsukuba, Ibaraki 305-8577, Japan
mmori@ccs.tsukuba.ac.jp

Abstract. Radiative cooling of the interstellar medium plays a vital role in the context of galaxy formation and evolution. On the other hand, the cooling time in the high-density regions involving star formation is much shorter than the dynamical time of the gas. In numerical simulations, it is challenging to solve physical phenomena coexisting on significantly different timescales, and it is known as the overcooling problem in the study of galaxy formation. Townsend (2009) has developed the Exact Integration (EI) scheme that provides a stable solution for the cooling term in the energy equation of astrophysical fluid dynamics, regardless of the size of the simulation time step. We apply the EI scheme to define the effective cooling time that accounts for the temperature dependence of the cooling rate and investigate the thermodynamic evolution of gas in colliding dark matter subhalos. The results show that the conventional cooling time always indicates a shorter than the effective cooling time derived by the EI scheme because it does not include the dependence of the cooling rate on temperature. Furthermore, we run three-dimensional galaxy collision simulations to examine the difference in thermodynamic evolution between the EI scheme and the conventional Crank–Nicholson method for solving the cooling equation. Comparing the results of the two simulations, we find that the EI scheme suppresses the rapid temperature decrease after galaxy collisions. Thus, the EI scheme indicates considerable potential for solving the overcooling problem in the study of galaxy formation.

Keywords: Galaxies · Evolution · Galaxies · Interactions · Methods · Numerical

O. Gervasi et al. (Eds.): ICCSA 2022 Workshops, LNCS 13378, pp. 373–387, 2022.
https://doi.org/10.1007/978-3-031-10562-3_27

1 Introduction

In the standard galaxy formation model, it is believed that cold dark matter (CDM) drives a hierarchical structure formation in the universe. Based on the CDM model, small structures grow up to larger structures through repeated collisions and merges. The gas falls into the deep gravitational potential well created by the dark matter halo, where stars and galaxies are formed. Radiative cooling is an essential physical process in galaxy formation since stars form in dense regions of gravitational contraction triggered by gas energy loss. The cooling time in the high-density regions where star formation occurs is much shorter than the dynamical time. It is a hard problem to solve physical phenomena with significantly different timescales. Supernova explosion significantly affects the evolution of the galaxy and suppresses star formation since it ejects a large amount of energy into the surrounding gaseous medium. However, in simulations of galaxy formation, it is difficult to capture the effects of supernova energy feedback because most of the energy is lost through radiative cooling. The problem of too short a cooling time to simulate physical phenomena in the simulations is known as the overcooling problem. Typically, the implicit integration method is used to handle time steps longer than the cooling time in the simulations. However, this method may estimate inadequate solutions due to multiple roots.

Townsend [19] has developed the Exact Integration (EI) to solve the cooling term of the energy equation stably regardless of the size of the simulation time step. This scheme provides temperatures by using a temporal evolution function given a cooling rate fitted by a piecewise power law, which accounts for the temperature dependence of the cooling rate. Zhu et al. [22] have applied this scheme to cosmological simulations and showed that it is closer to the equilibrium state where radiative cooling balances the heating effect of the ultraviolet background compared to previous simulations. This study defines the effective cooling time by adopting the EI scheme. We then estimate and compare gas timescales using an analytical model of dark matter subhalo collisions in Sect. 2. In Sect. 3, we run three-dimensional simulations of dark matter subhalo collision and discuss the differences of the thermodynamic evolution in the EI scheme and the Crank–Nicholson scheme, respectively.

2 Simple Model for Colliding Homogeneous Gas

Galaxy collision events are different from the standard evolution of isolated galaxies as the shock waves generated at the collision surface have a significant impact on the thermodynamic evolution. We focus on the head-on collisions between two dark matter subhalos (DMSHs) in the massive dark matter halo to evaluate the thermodynamic evolution of gas and timescales of physical processes. A DMSH consists of gas and dark matter components and is assumed to have a dark matter mass to gas mass ratio of 5.36. We consider a collision between two gas clouds in each DMSH as a simple one-dimensional (1D) model and assume that two colliding clouds both have mass M_{gas}, uniform density ρ_0

within scale radius $r_{\rm s}$. For simplicity, initial pressure and specific internal energy are zero, $p_0 = u_0 = 0$. The centres of the cloud are initially at $x = \pm r_{\rm s}$, and the bulk-velocities are $\mp v_{\rm col}$, respectively (Fig. 1). The subscript 0 indicates the physical quantities before the collision of the DMSHs, and the subscript 1 indicates the physical quantities after the shock wave generated by the collision has passed through.

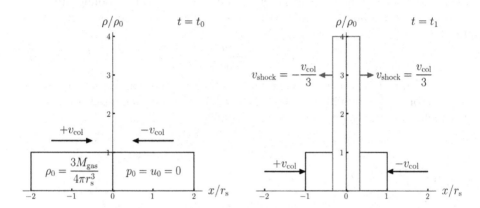

Fig. 1. 1D analytical model of hydrodynamics. Left panel: the gas density before the collision. Right panel: the gas density after the collision.

The density of the shocked clouds according to the Rankine-Hugoniot (RH) condition for strong adiabatic shocks are

$$\rho_1 = \frac{\gamma + 1}{\gamma - 1}\rho_0, \tag{1}$$

where γ is the specific heat ratio. We have used $\gamma = 5/3$. Assuming that the system's kinetic energy is converted into thermal energy as $u_1 = v_{\rm col}^2/2$, we can derive the shock velocity using RH condition for the conservation of mass,

$$v_{\rm shock} = \frac{1}{3}v_{\rm col}. \tag{2}$$

When the shock waves reach their cloud surface, shock-breakout ejects most of the gas from the system. The shock-crossing time is given by

$$t_{\rm cross} = \frac{2r_{\rm s}}{v_{\rm col} + v_{\rm shock}}. \tag{3}$$

We compare it with other timescales such as the free-fall time and cooling time to evaluate the thermodynamic evolution. The gravitational free-fall time in the shocked gaseous medium is

$$t_{\rm ff} = \sqrt{\frac{3\pi}{32G\rho_1}}, \tag{4}$$

where G is the gravitational constant. The conventional cooling time in the shocked gaseous medium is

$$t_{\text{cool,conv}} = \frac{k_B m_p \mu_1 T_1}{(\gamma - 1)\rho_1 \Lambda_1},$$

(5)

where k_B is the Boltzmann constant, m_p is the atomic mass, T is the temperature, $\mu = \mu(T)$ is the molecular weight and $\Lambda = \Lambda(T)$ is the cooling rate. The internal energy u and the temperature T are related to the equation of state of ideal gas:

$$T = \frac{(\gamma - 1)\mu m_p u}{k_B}.$$

(6)

We assume that the gas has the solar metal abundance and is under the collisional ionization equilibrium. MAPPINGS V [17,18] code is adopted to calculate the molecular weight and the cooling rate (Fig. 2).

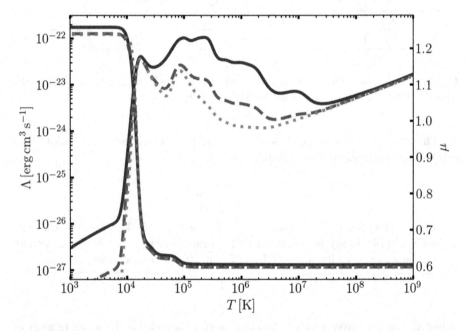

Fig. 2. Cooling rate Λ (blue) and mean molecular weight μ (red) for gas assuming the collisional ionisation equilibrium. The solid, dashed and dotted lines correspond to the gas metallicity of $1\,Z_\odot$, $0.1\,Z_\odot$ and $0.01\,Z_\odot$, respectively. The metallicity refers to the abundance of elements heavier than helium, and $Z_\odot$ represents the metal abundance in the Sun. (Color figure online)

Here, in order to consider the thermodynamic evolution of the gas in DMSH collision, we then define the effective cooling time based on the feature that the EI scheme provides the exact temperature change. The cooling term of the energy equation is expressed as

$$\frac{dT}{dt} = -\frac{(\gamma - 1)\rho \Lambda}{k_B m_p \mu}, \tag{7}$$

and to integrate it, the temporal evolution function (TEF) is defined as

$$Y_T(T) \equiv \frac{\Lambda_{ref}}{\mu_{ref} T_{ref}} \int_T^{T_{ref}} \frac{\mu}{\Lambda} dT, \tag{8}$$

where the subscript "ref" means an arbitrary reference value in the EI scheme. We also take into account the time evolution of the mean molecular weight. $Y_T(T)$ can be calculated analytically as a table by fitting the cooling function Λ with a piecewise power-law. Using the TEF, we integrate Eq. (7) from the temperature T_{start} to cooled temperature T_{end} and define the effective cooling time,

$$t_{cool,\,eff}(T_{start} \rightarrow T_{end}) \equiv [Y_T(T_{end}) - Y_T(T_{start})]\frac{k_B m_p}{(\gamma - 1)\rho}\frac{\mu_{ref} T_{ref}}{\Lambda_{ref}}. \tag{9}$$

If we set $T_{ref} = T_{start}$, the effective cooling time is expressed as

$$t_{cool,\,eff}(T_{start} \rightarrow T_{end}) = Y_T(T_{end}) \times t_{cool,\,conv}(T_{start}). \tag{10}$$

Hence, it can be understood that Y_T acts as a correction factor in the relationship between the effective cooling time and the conventional cooling time.

The timescales of each physical process as a function of collision velocity are shown in Fig. 3. Assuming that the kinetic energy of the collision velocity is fully converted to internal energy, the timescales are represented as a function of the temperature of the gas after the collision. In the DMSHs collision with a collision velocity of $100\,km\,s^{-1}$, the cooling times are shorter than the shock-crossing time and the free-fall time. The conventional cooling time for 2.4×10^5 K corresponding to the collision velocity of $100\,km\,s^{-1}$ is about 0.02 Myr. On the other hand, the effective cooling time for decreasing from 2.4×10^5 K to 10^4 K is about 0.07 Myr, but the effective cooling time for $T_{end} = 10^3$ K is 10 Myr. The conventional definition estimates a shorter cooling time than the new definition because of the temperature dependence of the cooling rate. Thus, it is found that there is a decisive difference between the conventional method and the present method and that it is quite effective to employ the present method in order to track accurate radiative cooling in the study of galaxy formation.

3   Three-Dimensional Model for Colliding Gas Spheres

In the previous section, we presented a physical insight into the collision process of DMSHs using an analytical model. However, this analytical model contains several uncertain assumptions. For instance, it is easy to foresee that the uniformity of the system is severely violated in an actual situation, and the inhomogeneity of the density field makes further nonlinear evolution of the gaseous system. In this section, we perform three-dimensional collision simulations of DMSHs to study the evolution of gas.

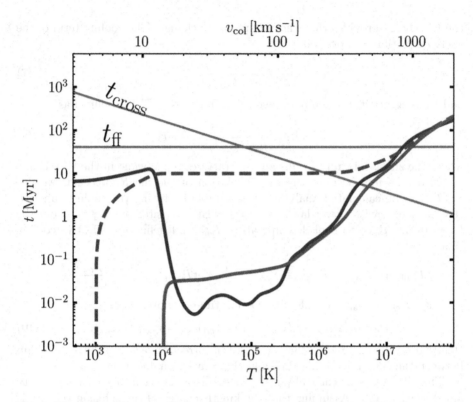

Fig. 3. Timescales after gas collisions. The temperature is calculated using the assumption that the kinetic energy is fully converted to internal energy. The blue line is the conventional cooling time $t_{\text{cool, conv}}(T)$. The red lines are the effective cooling time $t_{\text{cool, eff}}(T \rightarrow T_{\text{end}})$, with the solid line corresponding to the timescale for $T_{\text{end}} = 10^4 \, \text{K}$, and the dashed line corresponding to the timescale for $T_{\text{end}} = 10^3 \, \text{K}$. The two gray lines are the shock-crossing time t_{cross} and free-fall time t_{ff}, respectively. (Color figure online)

3.1   Model Description

The simulations of DMSHs collision are performed with the N-body/Smoothed Particle Hydrodynamics (SPH) [6,8] code implemented for this study. This code is parallelized by the Framework for Developing Particle Simulators (FDPS) [7,11]. It includes functions such as domain decomposition, redistribution of particles, and gathering of particle information for interaction calculation. The FDPS libraries are implemented using OpenMP for intra-node parallelism and MPI for inter-node parallelism. Using these libraries, users can easily implement parallelized programs by writing sequential code for interaction calculations. We use C++ bindings of FDPS. The gravitational forces are calculated with a tree algorithm [2,3] and the tree-opening angle is 0.7.

In the SPH formulation, the gas density of one particle at position $\boldsymbol{r}_i$ is given by

$$\rho_i = \sum_{j=1}^{N_{\text{neigh}}} m_j W(r_{ij}, h_i), \tag{11}$$

where m_j is the mass of gas particle at $\boldsymbol{r}_j$, $r_{ij} = |\boldsymbol{r}_i - \boldsymbol{r}_j|$, $W(r, h)$ is the smoothing kernel, h is the smoothing length and $N_{\text{neigh}} = 200$ is the number of neighbour particles. For the kernel function, we adopt the Wendland C^4 function [20],

$$W(r, h) = \frac{495}{32\pi h^3} \begin{cases} (1 - q)^6 (1 + 6q + \frac{35}{3}q^2), & q \leq 1, \\ 0, & q > 1, \end{cases} \tag{12}$$

where $q = r/h$, to avoid the clumping instability [4,21]. We followed the formulation of SPH introduced by [16]. The smoothing length h_i of each particle is determined by

$$\frac{4\pi}{3} h_i^3 \rho_i = \overline{m} N_{\text{neigh}}, \tag{13}$$

where $\overline{m}$ is an average mass of gas particles. These Eqs. (11) and (13) need to be solved implicitly for ρ_i and h_i. However, the minimum value of smoothing length h is set to gravitational softening ϵ in order to match the spatial resolution.

The momentum equations for the gas particles are given by

$$\frac{d\boldsymbol{v}_i}{dt} = -\sum_j m_j \left[f_i \frac{p_i}{\rho_i^2} \nabla_i W(r_{ij}, h_i) + f_j \frac{p_j}{\rho_j^2} \nabla_i W(r_{ij}, h_j) + \Pi_{ij} \nabla_i \overline{W}_{ij} \right] \tag{14}$$

where $\boldsymbol{v}_i$, and p_i are the velocity and pressure of gas particle respectively, f_i is defined by

$$f_i = \left(1 + \frac{h_i}{3\rho_i} \frac{\partial \rho_i}{\partial h_i} \right)^{-1}, \tag{15}$$

and $\overline{W}_{ij}$ is a symmetrized kernel,

$$\overline{W}_{ij} = \frac{1}{2} \left[W(r_{ij}, h_i) + W(r_{ij}, h_j) \right]. \tag{16}$$

The artificial viscosity Π_{ij} needs to treat shocks. We adopt Monaghan's [10] artificial viscosity combined with Balsara's [1] switch (see Otaki & Mori [14]). The entropy $A_i = p_i/\rho_i^\gamma$ is conserved in adiabatic flow, but it is generated by artificial viscosity via shocks. The entropy equations are given by

$$\frac{dA_i}{dt} = \frac{1}{2} \frac{\gamma - 1}{\rho_i^{\gamma-1}} \sum_j m_j \Pi_{ij} \boldsymbol{v}_{ij} \cdot \nabla_i \overline{W}_{ij}, \tag{17}$$

where $A_i = p_i/\rho_i^\gamma$ is the entropic function for specific entropy.

In the dissipationless case, we solve only Eq. (17) for the thermal evolution, but since radiative cooling is important in galaxy formation, it is needed to add a dissipation term to the entropy equation:

$$\left(\frac{dA}{dt}\right)_{\text{cool}} = \frac{\gamma - 1}{\rho^{\gamma}}\left(\frac{du}{dt}\right)_{\text{cool}} \tag{18}$$

where

$$\left(\frac{du}{dt}\right)_{\text{cool}} = -\frac{n^2 \Lambda(u, Z)}{\rho} = -\frac{\rho \Lambda(u, Z)}{\mu^2 m_{\text{p}}^2}, \tag{19}$$

where n, μ and m_{p} are the number density of the gas, the mean molecular weight and the proton mass, respectively. Λ is the cooling function of the specific internal energy u and the metallicity Z. To compare the results of different methods of solving the cooling equation (19), we implement the EI scheme [19] and the Crank–Nicholson scheme.

For EI scheme, to integrate from time t^n to $t^{n+1} = t^n + \Delta t$ for each gas particle i, the Eq. (19) becomes

$$\int_{u_i^n}^{u_i^{n+1}} \frac{\mu(u)^2}{\Lambda(u, Z)} du = -\frac{\rho_i}{m_{\text{p}}^2}\Delta t, \tag{20}$$

where $u_i^n = u_i(t^n)$ and $u_i^{n+1} = u_i(t^{n+1})$. Then, using the temporal evolution function defined by

$$Y_u(u) = \frac{\Lambda_{\text{ref}}}{\mu_{\text{ref}}^2 u_{\text{ref}}} \int_u^{u_{\text{ref}}} \frac{\mu(u)^2}{\Lambda(u, Z)} du, \tag{21}$$

the specific internal energy at the next time step is expressed by

$$u_i^{n+1} = Y_u^{-1}\left[Y_u(u_i^n) + \frac{\Lambda_{\text{ref}}}{\mu_{\text{ref}}^2 u_{\text{ref}}} \frac{\rho_i}{m_{\text{p}}^2}\Delta t\right]. \tag{22}$$

$Y_u(u)$ is the function corresponding to $Y_T(T)$, which is the result of the transformation using Eq. (6). It can be obtained as a table by fitting the cooling function Λ with a piecewise power-law. By using this integrated function $Y(u)$, the time evolution equation can be solved taking into account the temperature dependence of the cooling rate and is not sensitive to the size of the time step.

On the other hand, to solve the Eq. (19) with the Crank–Nicholson method,

$$u_i^{n+1} = u_i^n - \frac{1}{2}\left(\frac{\Lambda(u^{n+1})}{\mu(u^{n+1})^2} + \frac{\Lambda(u^n)}{\mu(u^n)^2}\right)\frac{\rho_i}{m_{\text{p}}^2}\Delta t, \tag{23}$$

we calculate it using the bisection method.

The simulation time steps Δt share the same value throughout the system. It is determined by the CFL conditions:

$$\Delta t = \min_i(\Delta t_{i,\text{grav}}, \Delta t_{i,\text{hydro}}), \tag{24}$$

$$\Delta t_{i,\text{grav}} = C_{\text{CFL}}\sqrt{\frac{\epsilon}{|d\boldsymbol{v}_i/dt|}}, \tag{25}$$

$$\Delta t_{i,\text{hydro}} = C_{\text{CFL}}\frac{h_i}{\max_j(v_{ij}^{\text{sig}})}, \tag{26}$$

where C_{CFL} is the CFL constant and we set $C_{\text{CFL}} = 0.3$. We adopted the second-order Runge–Kutta method for the time integration.

3.2   Thermodynamic Evolution of Colliding Gas

In order to study the physics of DMSH collisions, we have set up an ideal situation for a head-on collision. The two colliding DMSHs both have mass $M_{\text{tot}} = M_{\text{DM}} + M_{\text{gas}} = 10^9\,M_\odot$, the mass ratio between dark matter and gas is 5.36. A DMSH contains no stellar components.

The DMSH centres are initially at

$$(x, y, z) = (\pm 5, 0, 0)\,\text{kpc}, \tag{27}$$

and the initial bulk velocities of the DMSHs are

$$(v_x, v_y, v_z) = (\mp 100, 0, 0)\,\text{km\,s}^{-1}, \tag{28}$$

respectively. The density distribution of dark matter is the NFW profile [12, 13]. The gas is assumed under the hydrostatic equilibrium in the gravitational potential of the dark matter halo,

$$\rho_{\text{gas}}(r) = \rho_{\text{gas},0}\exp\left[-\frac{\mu m_{\text{p}}}{k_{\text{B}}T_{\text{vir}}}\Phi_{\text{NFW}}(r)\right], \tag{29}$$

where Φ_{NFW} is the gravitational potential of NFW profile and T_{vir} is the virial temperature of DMSH defined as

$$T_{\text{vir}} = \frac{c(c^2 + 2c - 2(1 + c)\ln(1 + c))}{2((1 + c)\ln(1 + c) - c)^2}\frac{GM_{\text{tot}}\mu m_{\text{p}}}{3k_{\text{B}}R_{200}}, \tag{30}$$

where $c = R_{200}/r_{\text{s}}$ is a concentration parameter [15] and R_{200} is the radius that the average density of the DMSH is 200 times the critical density of the universe. To generate the initial conditions of a DMSH, we use the MAGI [9]. After generating particle distributions using MAGI, we calculated a DMSH for several hundred million years in an isolated system of adiabatic processes to suppress density fluctuations and reach dynamical equilibrium. We use 1.36 million particles in each simulation. The mass resolution of N-body particles and SPH particles is $\sim 10^3\,M_\odot$, and all particles have the same mass. The spatial resolution is 100 pc.

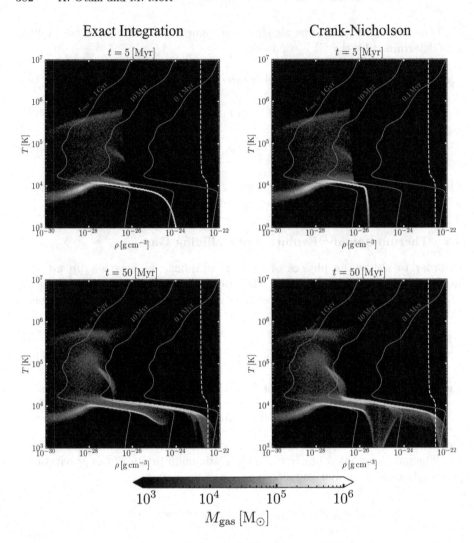

Fig. 4. Density-Temperature phase diagram for gas in the simulation box. The two left panels show the result of a simulation using the EI scheme at 5 Myr (upper panel) and 50 Myr (lower panel). The two right panels show the result of a simulation using the Crank–Nicholson scheme at 5 Myr (upper panel) and 50 Myr (lower panel). The white dashed line is a number density of $10 \, \mathrm{cm}^{-3}$. The white solid lines represent the cooling time calculated by the conventional definition.

The cooling rate for gas with a metallicity of $1 \, \mathrm{Z_\odot}$ is calculated by MAPPINGS V [17,18] assuming the collisional ionisation equilibrium.

Figure 4 shows the resultant gas distributions on the density-temperature plane for collision simulations using the EI scheme and the Crank–Nicholson scheme, respectively. The upper left and lower left panels in Fig. 4 show the thermodynamic evolution using the EI scheme, at 5 Myr and 50 Myr from the initial condition, respectively. On the other hand, the two panels on the right in Fig. 4 indicate the simulation results solved using the Crank–Nicholson scheme. The white solid lines are the conventional cooling time and the white dashed line indicate a number density of $10\,\mathrm{cm}^{-3}$.

At 5 Myr, the temperature increases to $\sim 10^{5-6}\,\mathrm{K}$ in the low-density region due to collisions in between the outside of the DMSHs. At the same time, in the high-density region at the centre of the DMSHs, the temperature suddenly decreases due to radiative cooling. Comparing the two schemes, the EI scheme has gas with $10^4\,\mathrm{K}$ and $10^{-25}\,\mathrm{g\,cm}^{-3}$, whereas the Crank–Nicholson scheme indicates that most of the gas denser than about $10^{-26}\,\mathrm{g\,cm}^{-3}$ have the temperature lower than $10^4\,\mathrm{K}$. The conventional cooling time for the gas with $\sim 10^{-25}\,\mathrm{g\,cm}^{-3}$ and $\sim 10^4\,\mathrm{K}$ is enough longer then 5 Myr, suggesting that the Crank–Nicholson scheme estimates a larger amount of energy loss.

The temperature distribution in the high-density regions is completely different between the two schemes at 50 Myr, when the centre of the two DMSHs collide. The Crank–Nicholson scheme overestimates the cooling efficiency. Therefore, the temperature drops rapidly, and the pressure-gradient force against self-gravity is weakened, eventually increasing the gas density to above $10\,\mathrm{cm}^{-3}$. On the other hand, the modest change of the temperature seen in the result of the EI scheme does not cause a significant increase in gas density than that of the Crank–Nicholson scheme. Several recipes are proposed to model star formation in simulations of galaxy formation. Commonly, stellar particles are produced as collision-less particles in high-density regions beyond the gas density of the star formation threshold n_{SF}. The result of our simulation shows that the total gas masses for typical star formation thresholds $n_{\mathrm{SF}} > 10\,\mathrm{cm}^{-3}$ and temperature $10^3\,\mathrm{K} < T < 10^4\,\mathrm{K}$ are $1.1 \times 10^7\,\mathrm{M}_\odot$ and $6.2 \times 10^6\,\mathrm{M}_\odot$ using the Crank–Nicholson scheme and the EI scheme, respectively. Therefore, this result clearly shows that the overestimation of cooling efficiency in our galaxy collision simulations using the Crank–Nicholson scheme critically affects the formation of galaxies. The numerical accuracy of the radiative cooling plays a significant role in studying galaxy formation precisely.

4 Performance Analysis

4.1 Algorithm Efficiency

We compare the algorithm efficiency of the two schemes. The cooling rate required to calculate the cooling equation is a complicated function of temperature and metallicity and is usually expressed as a numerical table. It is calculated to obtain the cooling rate for any given temperature and metallicity using linear interpolation. The Crank–Nicholson scheme is a second-order implicit scheme, and we use the bisection method for root finding. This method calls for repeated linear interpolation of the cooling rate until a root is found. On the other hand, in the EI scheme, the temporal evolution function is pre-computed before the simulation and is used as a numerical table. Instead of linearly interpolating the

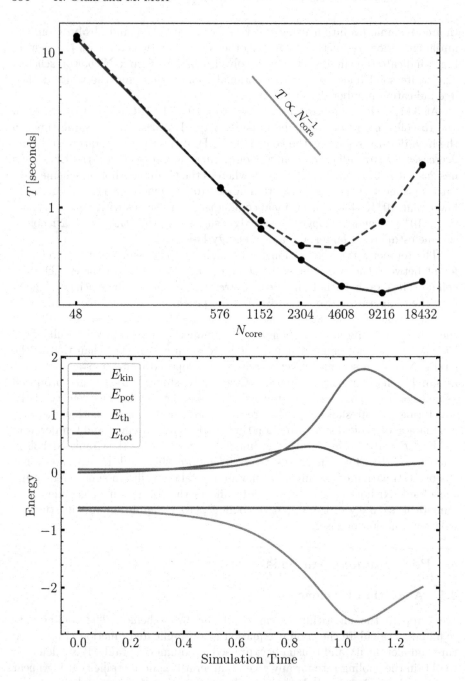

Fig. 5. Top panel: strong scaling for Evrard Collapse Test. The solid black line indicates the scalability with 4 MPI processes and 12 threads per node, and the dashed black line indicates the result with 24 MPI processes and 2 threads per node. The solid gray line represents $T \propto N_{\mathrm{core}}^{-1}$. We note that the A64FX processor has 48 cores. Bottom panel: energy evolution in the simulation. Blue, orange, green and red lines are kinetic, potential, thermal, and total energy, respectively. (Color figure online)

cooling rate, the temporal evolution function is calculated using linear interpolation for any given temperature and metallicity. Next, following Eq. (22), the temperature is linearly interpolated for the temporal evolution function after the time step. Therefore, it is clear that the EI scheme is more computationally efficient than the Crank–Nicholson scheme without any loss of accuracy. In addition, parallelization of the EI scheme is straightforward, and no additional difficulties exist because of no communications to other threads.

4.2 Scalability

We investigate the performance of our simulation code using the library of FDPS [7,11]. This study is conducted using the FUJITSU Supercomputer Wisteria/BDEC-01 at the University of Tokyo. This supercomputer consists of 7,680 nodes, with an A64FX processor with 48 cores per node. The topology of the network interconnect named TofuD is a 6-D mesh torus with 12 nodes. As a benchmark test, we ran the Evrard Collapse, a gravitational collapse simulation of self-gravitating gas [5]. The initial density distribution of the gas is

$$\rho(r) = \frac{M}{2\pi R^2}\frac{1}{r}\Theta(R - r), \tag{31}$$

where M is the total mass of gas and R is the radius of the gas sphere, and $\Theta(r)$ is the step function. We set $M = R = 1$. The initial velocity is 0 and the initial internal energy is $0.05\,GM/R$. We measure the elapsed time per time step of each simulation with the number of particles set to 1,099,136. The gravitational softening length is 10^{-5}. The top panel in Fig. 5 shows the strong scaling of this problem using the Wisteria/BDEC-01. We measured the scalability for two types of parallelization, one is the 4 MPI processes and 12 threads per node (solid black line), and the other is 24 MPI processes and 2 threads per node (dashed black line). The bottom panel in Fig. 5 shows kinetic, potential, thermal and total energies as a function of simulation time in the Evrard Collapse. The final energy error in the simulation is about 0.1%.

5 Summary

In this paper, we have investigated the thermodynamic evolution of gas in head-on collisions between dark matter subhalos using one-dimensional analytical models and three-dimensional simulations based on the Exact Integration (EI) scheme by Townsend [19]. We have applied this scheme to define the effective cooling time including temperature dependence of cooling rate. In one-dimensional model, the conventional cooling time is estimated to be shorter than the effective cooling time. We implement the Crank–Nicholson scheme and EI scheme in the gravitational N-body and hydrodynamic simulation code and run the three-dimensional galaxy collision simulations. The results show that the Crank–Nicholson scheme overestimates the cooling efficiency and critically affects the formation of galaxies. Therefore, to discuss the thermodynamic evolution of the gas properly, it is necessary to take into account the effective cooling time. The scheme impacts the evaluation of radiative cooling in the dense region. Using this scheme has the potential capability to solve the overcooling problem. Finally, we estimate the computational efficiency of the two schemes. The Crank–Nicholson scheme needs iterative calculations to solve the cooling equation, thus linear interpolations must be computed each time. However, the EI scheme requires two linear interpolations to obtain the temperature after a time step. The EI scheme is more computationally efficient than the Crank–Nicholson scheme.

Acknowledgements. Numerical computations were performed with computational resources provided by the Multidisciplinary Cooperative Research Program in the Center for Computational Sciences, University of Tsukuba, Oakforest-PACS operated by the Joint Center for Advanced High-Performance Computing (JCAHPC) and FUJITSU Server PRIMERGY GX2570 (Wisteria/BDEC-01) at the Information Technology Center, The University of Tokyo. This work was supported by JSPS KAKENHI Grant Numbers JP21J21888, JP20K04022.

References

1. Balsara, D.S.: Von Neumann stability analysis of smoothed particle hydrodynamics–suggestions for optimal algorithms. J. Comput. Phys. **121**(2), 357–372 (1995). https://doi.org/10.1016/S0021-9991(95)90221-X
2. Barnes, J., Hut, P.: A hierarchical O(N log N) force-calculation algorithm. Nature **324**(6096), 446–449 (1986). https://doi.org/10.1038/324446a0
3. Barnes, J.E.: A modified tree code: don't laugh; it runs. J. Comput. Phys. **87**(1), 161–170 (1990). https://doi.org/10.1016/0021-9991(90)90232-P
4. Dehnen, W., Aly, H.: Improving convergence in smoothed particle hydrodynamics simulations without pairing instability. Mon. Not. R. Astron. Soc. **425**(2), 1068–1082 (2012). https://doi.org/10.1111/j.1365-2966.2012.21439.x
5. Evrard, A.E.: Beyond N-body: 3D cosmological gas dynamics. Mon. Not. R. Astron. Soc. **235**(3), 911–934 (1988). https://doi.org/10.1093/mnras/235.3.911
6. Gingold, R.A., Monaghan, J.J.: Smoothed particle hydrodynamics: theory and application to non-spherical stars. Mon. Not. R. Astron. Soc. **181**(3), 375–389 (1977). https://doi.org/10.1093/mnras/181.3.375
7. Iwasawa, M., Tanikawa, A., Hosono, N., Nitadori, K., Muranushi, T., Makino, J.: Implementation and performance of FDPS: a framework for developing parallel particle simulation codes. Publ. Astron. Soc. Jpn. **68**(4), 54-1–54-22 (2016). https://doi.org/10.1093/pasj/psw053
8. Lucy, L.B.: A numerical approach to the testing of the fission hypothesis. Astron. J. **82**, 1013–1024 (1977). https://doi.org/10.1086/112164
9. Miki, Y., Umemura, M.: MAGI: many-component galaxy initializer. Mon. Not. R. Astron. Soc. **475**(2), 2269–2281 (2018). https://doi.org/10.1093/mnras/stx3327
10. Monaghan, J.J.: SPH and Riemann solvers. J. Comput. Phys. **136**(2), 298–307 (1997). https://doi.org/10.1006/jcph.1997.5732
11. Namekata, D., et al.: Fortran interface layer of the framework for developing particle simulator FDPS. Publ. Astron. Soc. Jpn. **70**(70) (2018). https://doi.org/10.1093/pasj/psy062

12. Navarro, J.F., Frenk, C.S., White, S.D.M.: The structure of cold dark matter halos. Astrophys. J. **462**, 563 (1996). https://doi.org/10.1086/177173
13. Navarro, J.F., Frenk, C.S., White, S.D.M.: A universal density profile from hierarchical clustering. Astrophys. J. **490**(2), 493 (1997). https://doi.org/10.1086/304888
14. Otaki, K., Mori, M.: The formation of dark-matter-deficient galaxies through galaxy collisions. J. Phys. Conf. Ser. **2207**(1), 012049 (2022). https://doi.org/10.1088/1742-6596/2207/1/012049
15. Prada, F., Klypin, A.A., Cuesta, A.J., Betancort-Rijo, J.E., Primack, J.: Halo concentrations in the standard Λ cold dark matter cosmology. Mon. Not. R. Astron. Soc. **423**(4), 3018–3030 (2012). https://doi.org/10.1111/j.1365-2966.2012.21007.x
16. Springel, V., Hernquist, L.: Cosmological smoothed particle hydrodynamics simulations: the entropy equation. Mon. Not. R. Astron. Soc. **333**(3), 649–664 (2002). https://doi.org/10.1046/j.1365-8711.2002.05445.x
17. Sutherland, R., Dopita, M., Binette, L., Groves, B.: MAPPINGS V: Astrophysical plasma modeling code. Astrophysics Source Code Library p. ascl:1807.005 (July 2018)
18. Sutherland, R.S., Dopita, M.A.: Effects of preionization in radiative shocks. I. Self-consistent models. Astrophys. J. Suppl. Ser. **229**(2), 34 (2017). https://doi.org/10.3847/1538-4365/aa6541
19. Townsend, R.H.D.: An exact integration scheme for radiative cooling in hydrodynamical simulations. Astrophys. J. Suppl. Ser. **181**(2), 391–397 (2009). https://doi.org/10.1088/0067-0049/181/2/391
20. Wendland, H.: Piecewise polynomial, positive definite and compactly supported radial functions of minimal degree. Adv. Comput. Math. **4**(1), 389–396 (1995). https://doi.org/10.1007/BF02123482
21. Zhu, Q., Hernquist, L., Li, Y.: Numerical convergence in smoothed particle hydrodynamics. Astrophys. J. **800**(1), 6 (2015). https://doi.org/10.1088/0004-637X/800/1/6
22. Zhu, Q., Smith, B., Hernquist, L.: Gas cooling in hydrodynamic simulations with an exact time integration scheme. Mon. Not. R. Astron. Soc. **470**(1), 1017–1025 (2017). https://doi.org/10.1093/mnras/stx1346

Regularization of Feynman 4-Loop Integrals with Numerical Integration and Extrapolation

E. de Doncker[1(✉)] and F. Yuasa[2]

[1] Western Michigan University, Kalamazoo, MI 49008, USA
elise.dedoncker@wmich.edu
[2] High Energy Accelerator Research Organization (KEK),
1-1 OHO Tsukuba, Ibaraki 305-0801, Japan
fukuko.yuasa@kek.jp
http://www.cs.wmich.edu/elise

Abstract. In this paper we continue our recent work on evaluating numerical approximations for a set of 4-loop self-energy integrals required in the computation of higher orders in perturbation theory. The results are given by a Laurent expansion in the dimensional regularization parameter, ε, where ε is related to the space-time dimension as $\nu = 4 - 2\varepsilon$. Although the leading-order coefficients for the diagrams with massless internal lines are given in analytic form in the literature, we obtain them using a numerical approach and with modern computational techniques. In a similar manner, we derive results for diagrams with massive lines. The SIMD (Single Instruction, Multiple Data) nature of the computation of loop integrals based on composite lattice rules lends itself to an efficient GPU implementation. We further apply double exponential numerical integration layered over the message passing interface (MPI) parallel platform. Limits of integral sequences (as the regularization parameter tends to zero) are implemented numerically using linear and nonlinear extrapolation procedures. Numerical results are given to illustrate the versatility of the methods and show some robustness with regard to the selection of sequence parameters.

Keywords: Feynman loop diagrams · Dimensional regularization · Lattice rules · High performance computing · GPUs

1 Introduction

As elementary particle physics experiments are becoming more and more precise, it is an important task to provide theoretical predictions with commensurate accuracy. To achieve this, the study of radiative corrections based on the perturbative method in Quantum Field Theory is inevitable. In higher order corrections, Feynman loop integrals appear and their accurate evaluation is a fundamental ingredient to obtain physically meaningful results. For the calculation of loop integrals, the most important issue is how to deal with divergences due to thresholds and/or ultra-violet or infrared singularities, and many excellent methods have been developed following analytical (e.g., [2, 17]), semi-numerical or fully numerical schemes (e.g., [3, 16]).

© The Author(s), under exclusive license to Springer Nature Switzerland AG 2022
O. Gervasi et al. (Eds.): ICCSA 2022 Workshops, LNCS 13378, pp. 388–405, 2022.
https://doi.org/10.1007/978-3-031-10562-3_28

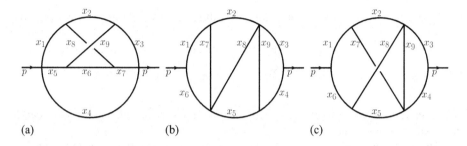

Fig. 1. 4-loop $N = 9$ diagrams [2] (a) M_{43}; (b) M_{44}; (c) M_{45}

Our work on the development of a fully numerical method for the evaluation of Feynman loop integrals emphasizes two distinct components, an extrapolation with respect to regularization parameters incorporated in the integrands to treat singularities and an efficient multivariate integration method. This versatile approach can be applied to multi-loop integrals with arbitrary masses occurring in quantum electrodynamics (QED), quantum chromodynamics (QCD) and the electroweak theory. We have presented our achievements so far for 1-, 2-, 3- and some 4-loop integrals in our past work (e.g., [5,6,8,9,26]). In this paper, we apply our methods to three other 4-loop diagrams, which are among the first thirteen complicated diagrams shown in Fig. 2 of [2]. Of these three, we presented some numerical results for two in the massless cases [8].

The 4-loop self-energy diagrams with $N = 9$ internal lines represented in Fig. 1 are referenced as M_{43}, M_{44} (*lala*) and M_{45} (*nono*) in [2,17]. The associated Feynman integral with L loops and N internal lines satisfies $\mathcal{F} = \mathcal{I}/(4\pi)^{\nu L/2}$ with

$$\mathcal{I} = \Gamma(N - \frac{\nu L}{2})(-1)^N \int_{\mathcal{C}_N} \prod_{r=1}^{N} dx_r\ \delta(1 - \sum x_r)\ \frac{C^{N-\nu(L+1)/2}}{(D - i\varrho C)^{N-\nu L/2}}$$

$$= \Gamma(N - \frac{\nu L}{2})(-1)^N \int_{\mathcal{S}_{N-1}} \prod_{r=1}^{N-1} dx_r\ \frac{C^{N-\nu(L+1)/2}}{(D - i\varrho C)^{N-\nu L/2}}. \tag{1}$$

C and D are polynomials in Feynman parameters x_r of degree L and $L+1$, respectively, determined by the topology of the diagram and physical parameters; $\nu = 4 - 2\varepsilon$ is the space-time dimension (i.e., where $\nu = 4$ is modified in terms of the regularization parameter ε), and ϱ in the denominator is zero unless D vanishes within the domain. Eliminating the δ-function in the integral over the N-dimensional unit hypercube $\mathcal{C}_N$ yields the integral (1) over the d-dimensional unit simplex with $d = (N - 1)$, $\mathcal{S}_d = \{\mathbf{x} \in \mathcal{C}_d \mid \sum_{j=1}^{d} x_j \leq 1\}$.

Substituting $L = 4$ and $N = 9$ in (1), the integrand function is

$$f(\mathbf{x}) = \frac{C^{-1+5\varepsilon}}{D^{1+4\varepsilon}}. \tag{2}$$

Furthermore, the loop integral (1) is multiplied with a normalization factor,

$$\eta(\varepsilon)^L = \left(\frac{\Gamma(2 - 2\varepsilon)}{\Gamma(1 + \varepsilon)\Gamma(1 - \varepsilon)^2}\right)^L, \tag{3}$$

in [2], which we apply in the computation for the massless diagrams (for comparison purposes).

The integral $\mathcal{I} = \mathcal{I}(\varepsilon)$ is expanded asymptotically as $\varepsilon \to 0$ in integer powers of ε,

$$\mathcal{I}(\varepsilon) \sim \sum_{k \geq \kappa} C_k \, \varepsilon^k \qquad \text{as } \varepsilon \to 0. \tag{4}$$

The expansion is satisfied with $\kappa = 0$ for the diagrams considered, and it is our goal to approximate C_0, C_1 and C_2.

The numerical methods are outlined in Sect. 2 of this paper. Results are presented in Sect. 3, with Conclusions in Sect. 4. We include the C and D functions for the integrand definitions in Appendix A.

2   Methods

As the diagrams of Fig. 1 give rise to an expansion of the form (4) with $\kappa = 0$, an approximation to C_0 can be obtained as an integral approximation to (1) for $\varepsilon = 0$. The integrand function (2) is then $f(\mathbf{x}) = 1/(CD)$.

For calculating leading term coefficients, a sequence of integral approximations $A(\varepsilon) \approx \mathcal{I}(\varepsilon)$ for a decreasing sequence of ε can be used in an extrapolation as shown below.

2.1   Integration and Transformations

Transformations. The integral (1) is first transformed to the $(N - 1)$-dimensional unit cube by the transformation $x_j = (1 - \sum_{k=1}^{j-1} x_k) \, t_j$, $1 \leq j \leq d$, with Jacobian $\prod_{j=1}^{d}(1 - \sum_{k=1}^{j-1} x_k)$, for $d = N - 1$.

We apply Sidi's Ψ_6 transformation [20] to the resulting integral,

$$\Psi_6(t) = t - (45\sin(2\pi t) - 9\sin(4\pi t) + \sin(6\pi t))/(60\pi), \quad \Psi_6'(t) = \frac{16}{5}\sin^6(\pi t),$$

which helps alleviate boundary singularities.

Lattice Rules. We perform Quasi-Monte Carlo (QMC) integration based on lattice rules [21]. For the rules with $100M \approx 100,000,007$ points, $200M \approx 200,000,033$ points, and $400M \approx 400,000,009$ points, we previously computed the generators (see, e.g., [1]) using the component by component (CBC) algorithm [13,14]. We also use an 8-d projection of a rule with 5^{13} points, originally derived for dimension $d = 15$ (see [7]) via Lattice Builder [10].

For increased accuracy, we apply m^d-copy lattice rules $Q^{(m)}$ of a basic rank-1 lattice rule Q [21] with $m = 2$ or 3, which corresponds to subdividing the $[0, 1]$-range into m equal parts in each coordinate direction and scaling the basic rule to each subcube. An error estimate is also given.

In cases where an analytic result is not available, we compare our results with the pySecDec 1.4.5 [3] package, which uses randomized (shifted) lattice rules [21] obtained by adding a random vector to all integration points, with an error estimate given in terms of the standard deviation.

Double Exponential Integration. The double-exponential method (DE) [22,23] transforms the one-dimensional integral $\int_0^1 f(x)\,dx = \int_{-\infty}^{\infty} f\left(\phi(t)\right)\phi'(t)\,dt$ using

$$x = \phi(t) = \frac{1}{2}\left(\tanh\left(\frac{\pi}{2}\sinh(t)\right) + 1\right),$$

with

$$\phi'(t) = \frac{\pi\cosh(t)}{4\cosh^2\left(\frac{\pi}{2}\sinh(t)\right)}.$$

The DE technique involves a truncation of the infinite range and application of the trapezoidal rule,

$$I_h^{N_{eval}} = \sum_{k=-N_-}^{N_+} f(\phi(kh))\,\phi'(kh)$$

with mesh size h and $N_{eval} = N_- + N_+ + 1$ function evaluations in each direction.

2.2   Numerical Extrapolation

Nonlinear Extrapolation. We apply the ϵ-algorithm [18,25] for nonlinear extrapolation, with an implementation based on the code of *qext()* from the QuadPack package [15]. The computation produces a triangular table, where the entry (input) sequence of integral approximations $A(\varepsilon)$ for a decreasing geometric sequence of ε forms the first column, and a new lower diagonal of the table is generated when a new input element is available. A table element *"new"* relative to preceding e_0, e_1, e_2, e_3 is obtained as

$$\begin{array}{l} e_0 \\ e_3\ e_1\ new = e_1 + 1/(1/(e_1 - e_3) + 1/(e_2 - e_1) - 1/(e_1 - e_0)), \\ e_2 \end{array}$$

and its *distance* measure from neighboring elements is set to

$$\Delta = |e_2 - e_1| + |e_1 - e_0| + |e_2 - new|. \tag{5}$$

The new lower diagonal element with the smallest value of the distance estimate is returned as the extrapolated result.

Linear Extrapolation. Whereas the ϵ-algorithm does not utilize information regarding the underlying asymptotic expansion, linear extrapolation generates a solution to a linear system determined by the orders (e.g., powers in ε) of the leading terms. The expansion (4) is truncated after $2, 3, \ldots, K$ terms in the summation, to form a linear system in the first K variables C_k (with $k = \kappa,\ \kappa + 1,\ \ldots,\ \kappa + K - 1$, and κ is the index of the first term in (4)). This procedure is a generalized form of Richardson extrapolation [4,12,19].

3   Results

The computations were performed in double precision on an x86_64 host machine with Intel(R) Xeon(R) Gold (dual) 6230 CPU@2.10 GHz, 20 cores (40 HT—hyper-threaded), under GNU/ Linux and using CUDA C, accelerated by an NVIDIA Quadro GV100 (Volta) GPU with compute capability 7.0, 5120 cores and 640 tensor cores, and 32 GB HBM2 global memory, (boost) clock rate of 1627 MHz, and memory clock rate 848 MHz [24]. The GPU performs at (a theoretical peak of) 8.33 TFLOPS for FP64 (double).

The main program ran sequentially on the host, and called the CUDA kernel for the lattice rule computation. The number of threads per block was set to the 1024 maximum for one dimension, except for the M_{44} computations where the number of threads per block was 512. Whereas the maximum number of blocks in a grid is $2^{31} - 1$, it is suggested in [11] to have the number of CUDA blocks close to the number of threads per block. For the results obtained subsequently, the number of blocks was set to 1024 unless otherwise noted. Timing was done using *cudaEvent* timing functions in the main program for each iteration of the while loop over successive values of the ε-parameter.

In Sects. 3.1–3.3 below, we report results for the diagrams of Fig. 1 with massive internal lines where all masses $= 1$, and $s = p^2 = 1$, and for massless lines with all masses $= 0$.

3.1 M_{43}

Table 1 lists the integral approximations $Q^{(m)} = Q^{(m)}(\varepsilon)$ for the massive M_{43} diagram with $\varepsilon = 0$ and $m = 1, 2$ and 3, for lattice rules with $100M$, $200M$, $400M$ and 5^{13} points. Also given are the estimated absolute error $|Est_a|$, and the actual absolute error $|Err_a|$ with respect to the result from pySecDec 1.4.5 [3] considered as the "exact" value. The digits underlined in the results correspond to the accuracy estimated by pySecDec.

The pySecDec code was run on the same system with the GV100 GPU, and further used 64 threads on the CPU host, yielding estimated accuracies for C_0, C_1 and C_2 as listed in Table 4 by performing 3,328,627,552 evaluations in a total execution time of 34,238 s. A pySecDec result obtained with 337,930,896 function evaluations in 3,471 s estimated its error for the integral at $\pm 3.07\,e{-}09$.

For the lattice rules, it emerges in Table 1 that the computation time increases proportionally with the number of points used, as expected since the work per function evaluation is more or less fixed. The accuracy improves with higher m, and (with some exceptions) for higher numbers of evaluations and the same m, but appears to reach a limiting accuracy level at $m = 3$. Note that an estimated error is not available for $m = 1$.

Table 1. M_{43}, massive lines, $Q^{(m)}(\varepsilon)$ with $\varepsilon = 0$

m	Rule # points	$Q^{(m)}(\varepsilon = 0)$	$|Est_a|$	$|Err_a|$	Time [s]
1	$100M$	-1.0315282429238		$1.58\,\mathrm{e}{-04}$	0.035
	$200M$	-1.0314118459476		$4.16\,\mathrm{e}{-05}$	0.069
	$400M$	-1.0313596638223		$1.06\,\mathrm{e}{-05}$	0.136
	5^{13}	-1.0313649504581		$5.33\,\mathrm{e}{-06}$	0.384
2	$100M$	-1.0313702568810	$2.53\,\mathrm{e}{-07}$	$2.76\,\mathrm{e}{-08}$	7.03
	$200M$	-1.0313702864127	$5.92\,\mathrm{e}{-08}$	$1.97\,\mathrm{e}{-09}$	14.0
	$400M$	-1.0313702892440	$5.92\,\mathrm{e}{-09}$	$4.80\,\mathrm{e}{-09}$	28.1
	5^{13}	-1.0313702831956	$2.19\,\mathrm{e}{-09}$	$1.25\,\mathrm{e}{-09}$	89.4
3	$100M$	-1.0313702845002	$2.97\,\mathrm{e}{-08}$	$5.57\,\mathrm{e}{-11}$	218.2
	$200M$	-1.0313702844244	$2.61\,\mathrm{e}{-09}$	$2.01\,\mathrm{e}{-11}$	452.2
	$400M$	-1.0313702844655	$5.86\,\mathrm{e}{-10}$	$2.10\,\mathrm{e}{-11}$	909.7
pySecDec:		-1.0313702844445			
Accuracy:		$\pm 8.10\,\mathrm{e}{-12}$			

Table 2. Nonlinear extrapolation for M_{43}, massive lines, $Q^{(m)}(\varepsilon)$ with 5^{13}-point rule Q and $m = 2, \varepsilon = 1.2^{-\ell}, \ell = 25, 26, \ldots$

ℓ	$Q^{(2)}(\varepsilon)$	$|Est_a|$	Time [s]	Extrapolated	Distance Δ
25	-0.949721884198076	$1.77\,\mathrm{e}{-09}$	108.5		
26	-0.962670060436187	$1.83\,\mathrm{e}{-09}$	117.1		
27	-0.973654888967547	$1.89\,\mathrm{e}{-09}$	118.5	-1.0351144345	
28	-0.982946959438472	$1.93\,\mathrm{e}{-09}$	118.8	-1.0339539968	$1.16\,\mathrm{e}{-03}$
29	-0.990787906493011	$1.97\,\mathrm{e}{-09}$	119.4	-1.0312791735	$2.67\,\mathrm{e}{-03}$
30	-0.997390772578446	$2.01\,\mathrm{e}{-09}$	119.4	-1.0313177886	$3.86\,\mathrm{e}{-05}$
31	-1.002941486113812	$2.04\,\mathrm{e}{-09}$	119.4	-1.0313716995	$5.39\,\mathrm{e}{-05}$
32	-1.007600983997484	$2.06\,\mathrm{e}{-09}$	119.9	-1.0313709638	$7.36\,\mathrm{e}{-07}$
33	-1.011507644305459	$2.08\,\mathrm{e}{-09}$	119.9	-1.0313702688	$6.95\,\mathrm{e}{-07}$
34	-1.014779802141804	$2.10\,\mathrm{e}{-09}$	119.9	-1.0313702772	$8.39\,\mathrm{e}{-09}$
35	-1.017518202051021	$2.12\,\mathrm{e}{-09}$	119.8	-1.0313702847	$7.57\,\mathrm{e}{-09}$
36	-1.019808298726949	$2.13\,\mathrm{e}{-09}$	119.8	-1.0313702823	$2.45\,\mathrm{e}{-09}$
pySecDec Korobov-3:				-1.0313702844	
pySecDec accuracy:				$\pm 8.10\,\mathrm{e}{-12}$	

Table 3. Nonlinear extrapolation for M_{43}, massive lines, $Q^{(m)}(\varepsilon)$ with 5^{13}-point rule Q and $m = 2$, $\varepsilon = 1.2^{-\ell}$, $\ell = 30, 31, \ldots$

ℓ	$Q^{(2)}(\varepsilon)$	$\|Est_a\|$	Time [s]	Extrapolated	Distance Δ
30	-0.997390772578446	$2.01\,\mathrm{e}{-09}$	119.4		
31	-1.002941486113812	$2.04\,\mathrm{e}{-09}$	119.4		
32	-1.007600983997484	$2.06\,\mathrm{e}{-09}$	119.9	-1.0319620020	
33	-1.011507644305459	$2.08\,\mathrm{e}{-09}$	119.9	-1.0317802705	$1.82\,\mathrm{e}{-04}$
34	-1.014779802141804	$2.10\,\mathrm{e}{-09}$	119.9	-1.0313644627	$4.16\,\mathrm{e}{-04}$
35	-1.017518202051021	$2.12\,\mathrm{e}{-09}$	119.8	-1.0313669208	$2.46\,\mathrm{e}{-06}$
36	-1.019808298726949	$2.13\,\mathrm{e}{-09}$	119.8	-1.0313703196	$3.40\,\mathrm{e}{-06}$
37	-1.021722358649551	$2.14\,\mathrm{e}{-09}$	119.5	-1.0313703008	$1.89\,\mathrm{e}{-08}$
38	-1.023321342972643	$2.15\,\mathrm{e}{-09}$	119.5	-1.0313702829	$1.79\,\mathrm{e}{-08}$
39	-1.024656569986113	$2.15\,\mathrm{e}{-09}$	119.6	-1.0313702833	$4.41\,\mathrm{e}{-10}$
	pySecDec Korobov-3:			-1.0313702844	
	pySecDec accuracy:			$\pm 8.10\,\mathrm{e}{-12}$	

Table 4. Linear extrapolation for M_{43}, massive lines, $Q^{(m)}(\varepsilon)$ with 5^{13}-point rule Q and $m = 2$, $\varepsilon = 1.2^{-\ell}$, $\ell = 25, 26, \ldots$

ℓ	$Q^{(2)}(\varepsilon)$	$\|Est_a\|$	Time [s]	C_0	C_1	C_2
25	-0.9497218841980715	$1.77\,\mathrm{e}{-09}$	126.7			
26	-0.9626700604361829	$1.83\,\mathrm{e}{-09}$	132.3	-1.0274109416	7.411242153	
27	-0.9736548889675424	$1.89\,\mathrm{e}{-09}$	133.6	-1.0312337816	8.213547992	-41.74742269
28	-0.9829469594384685	$1.93\,\mathrm{e}{-09}$	134.1	-1.0313666693	8.259692219	-47.02980432
29	-0.9907879064930063	$1.97\,\mathrm{e}{-09}$	134.8	-1.0313702070	8.261503803	-47.37292710
30	-0.9973907725784428	$2.01\,\mathrm{e}{-09}$	135.1	-1.0313702819	8.261556977	-47.38777961
31	-1.0029414861138075	$2.04\,\mathrm{e}{-09}$	135.5	-1.0313702832	8.261558197	-47.38825191
32	-1.0076009839974789	$2.06\,\mathrm{e}{-09}$	135.5	-1.0313702832	8.261558221	-47.38826436
33	-1.0115076443054538	$2.08\,\mathrm{e}{-09}$	135.6	-1.0313702832	8.261558220	-47.38826400
	pySecDec Korobov-3:		34,238	-1.0313702844	8.261558253	-47.38826443
	pySecDec accuracy:			$\pm 8.10\,\mathrm{e}{-12}$	$\pm 5.29\,\mathrm{e}{-11}$	$\pm 2.13\,\mathrm{e}{-10}$

As another comparison with $\varepsilon = 0$, DE/MPI returned a result of -1.031370286859 with an absolute error of $2.41\,\mathrm{e}{-09}$ in $4473.0\,\mathrm{s}$ using 121 threads.

Tables 2 and 3 show nonlinear extrapolations with the ϵ-algorithm [18, 25], for a sequence of integral approximations $A(\varepsilon) = Q^{(2)}(\varepsilon)$, with the 5^{13}-point lattice rule Q, and $\varepsilon = 1.2^{-\ell}$, $\ell \geq \ell_0$. Different starting points, $\ell_0 = 25$ and 30, are used in Tables 2 and 3, respectively, resulting in similar accuracy. $|Est_a|$ is the estimated absolute accuracy for the integration result $Q^{(2)}(\varepsilon)$, whereas Δ refers to the distance estimate (5) for the extrapolation, used to assess convergence in the extrapolation table. It should be noted that the error in the input sequence $A(\varepsilon)$ affects the accuracy of the extrapolated results.

A linear extrapolation for M_{43} with massive lines achieves excellent agreement with pySecDec in Table 4, for $Q^{(2)}(\varepsilon)$, the 5^{13}-point lattice rule Q and $\varepsilon = 1.2^{-\ell}$, $\ell \geq \ell_0 = 25$. This computation was performed using 64 CUDA blocks, resulting in slightly higher execution times than those of Tables 2 and 3 with 1024 blocks.

Computation for the massless case is in progress and will be presented at the next opportunity.

3.2 M_{44}

For the M_{44} diagram with massive lines, Table 5 displays results $Q^{(m)}(\varepsilon)$, $m = 2, 3$, and $\varepsilon = 0$ for lattice rules Q with $100M$, $400M$ and 5^{13} points. Also given are $|Est_a|$, $|Err_a|$ and the time in seconds Time[s]. The number of blocks was set to 1024, which delivered results in slightly shorter times than those incurred using 512 blocks.

According to the estimated error, the accuracy appears to plateau at $m = 3$. The estimated accuracy of pySecDec is reported at $\pm 4.75\,\mathrm{e}{-}11$.

Table 5. M_{44}, massive lines, $Q^{(m)}(\varepsilon)$ with $\varepsilon = 0$

m	Rule # points	$Q^{(m)}(\varepsilon = 0)$	$|Est_a|$	$|Err_a|$	Time [s]
2	$100M$	-5.19282488199	$9.18\,\mathrm{e}{-}07$	$1.33\,\mathrm{e}{-}06$	7.17
	$400M$	-5.19282330726	$8.05\,\mathrm{e}{-}07$	$2.47\,\mathrm{e}{-}07$	28.7
	5^{13}	-5.19282357097	$1.03\,\mathrm{e}{-}06$	$1.72\,\mathrm{e}{-}08$	90.1
3	$100M$	-5.19282354549	$2.12\,\mathrm{e}{-}07$	$8.29\,\mathrm{e}{-}09$	223.3
	$400M$	-5.19282354800	$2.50\,\mathrm{e}{-}08$	$5.77\,\mathrm{e}{-}09$	935
	5^{13}	-5.19282355028	$1.69\,\mathrm{e}{-}08$	$3.50\,\mathrm{e}{-}09$	2880
pySecDec:		-5.19282355049			
Accuracy:		$\pm 4.75\,\mathrm{e}{-}11$			

Table 6. Linear extrapolation for M_{44}, massive lines, $Q^{(m)}(\varepsilon)$ with 5^{13}-point rule Q and $m = 2$, $\varepsilon = 1.2^{-\ell}, \ell = 20, 21, \ldots$

ℓ	$Q^{(2)}(\varepsilon)$	$|Est_a|$	Time [s]	C_0	C_1	C_2
20	-4.0601894422804277	5.32 e$-$07	109.7			
21	-4.2238374065452442	5.93 e$-$07	118.3	-5.0420772279	37.643221095	
22	-4.3671653932011543	6.49 e$-$07	119.9	-5.1786419142	49.161458176	-240.8626718
23	-4.4917175874154758	7.00 e$-$07	120.7	-5.1918120637	50.999337564	-325.4145335
24	-4.5992353524890035	7.46 e$-$07	120.8	-5.1927668799	51.195835153	-340.3714509
25	-4.6915260313418741	7.86 e$-$07	121.0	-5.1928210212	51.211281283	-342.1053175
26	-4.7703690293820973	8.22 e$-$07	120.6	-5.1928234777	51.212216470	-342.2508460
27	-4.8374532649071398	8.52 e$-$07	121.0	-5.1928235682	51.212261249	-342.2601444
28	-4.8943393898615151	8.79 e$-$07	121.2	-5.1928235709	51.212262985	-342.2606132
	pySecDec Korobov-3:		5573	-5.1928235505	51.212262947	-342.2606447
	pySecDec accuracy:			±4.75 e$-$11	±5.89 e$-$10	±7.84 e$-$09

Table 7. Linear extrapolation for M_{44}, massive lines, $Q^{(m)}(\varepsilon)$ with 5^{13}-point rule Q and $m = 2$, $\varepsilon = 1.2^{-\ell}, \ell = 25, 26, \ldots$

ℓ	$Q^{(2)}(\varepsilon)$	$|Est_a|$	Time[s]	C_0	C_1	C_2
25	-4.6915260313418741	7.86 e$-$07	121.0			
26	-4.7703690293820973	8.22 e$-$07	120.6	-5.1645840196	45.127942331	
27	-4.8374532649071398	8.52 e$-$07	121.0	-5.1917163129	50.822242214	-296.2989083
28	-4.8943393898615151	8.79 e$-$07	121.2	-5.1927909230	51.195392228	-339.0154278
29	-4.9424405485681984	9.02 e$-$07	121.4	-5.1928228185	51.211725494	-342.1090275
30	-4.9830163208244791	9.21 e$-$07	121.5	-5.1928235571	51.212249839	-342.2554878
31	-5.0171755408111256	9.38 e$-$07	121.4	-5.1928235708	51.212262778	-342.2604972
	pySecDec Korobov-3:		5573	-5.1928235505	51.212262947	-342.2606447
	pySecDec accuracy:			±4.75 e$-$11	±5.89 e$-$10	±7.84 e$-$09

Tables 6 and 7 show linear extrapolation results based on $Q^{(2)}(\varepsilon)$ with the 5^{13}-point lattice rule Q, $\varepsilon = 1.2^{-\ell}, \ell \geq \ell_0 = 20$ or 25, respectively. Linear extrapolation results are further given in Table 8, for $Q^{(3)}(\varepsilon)$, the $100M$-point lattice rule Q, and a sequence of $\varepsilon = 1.15^{-\ell}, \ell \geq \ell_0 = 30$. Note that $|Est_a|$ applies to $Q^{(m)}(\varepsilon)$. Here pySecDec supplies $C_0 \pm 4.75$ e$-$11, $C_1 \pm 5.89$ e$-$10, and $C_2 \pm 7.84$ e$-$09, using 3,328,627,552 integrand evaluations in 5573 s total.

Table 8. Linear extrapolation for M_{44}, massive lines, $Q^{(m)}(\varepsilon)$ with $100M$-point rule Q and $m = 3, \varepsilon = 1.15^{-\ell}, \ell = 30, 31, \ldots$

ℓ	$Q^{(3)}(\varepsilon)$	$|Est_a|$	Time [s]	C_0	C_1	C_2
30	-4.4913762732977878	1.21 e−07	264.7			
31	-4.5752619125386005	1.30 e−07	277.2	-5.1344995075	42.582328922	
32	-4.6499486706965047	1.38 e−07	278.5	-5.1892894903	50.381973890	-276.2290982
33	-4.7162480921373637	1.46 e−07	279.0	-5.1926552224	51.155824291	-335.1528054
34	-4.7749487570631164	1.53 e−07	278.9	-5.1928169988	51.209310838	-341.7306166
35	-4.8268027974846168	1.60 e−07	279.0	-5.1928233329	51.212138553	-342.2306794
36	-4.8725170366719457	1.66 e−07	279.1	-5.1928235396	51.212258325	-342.2592792
37	-4.9127477956401648	1.71 e−07	279.1	-5.1928235456	51.212262716	-342.2606404
	pySecDec Korobov-3:		5573	-5.1928235505	51.212262947	-342.2606447
	pySecDec accuracy:			± 4.75 e−11	± 5.89 e−10	± 7.84 e−09

Table 9. M_{44}, massless lines, $Q^{(m)}(\varepsilon)$ with $\varepsilon = 0$

m	Rule # points	$Q^{(m)}(\varepsilon = 0)$	$|Est_a|$	$|Err_a|$	Time [s]
2	$100M$	55.586159640	7.92 e−04	9.06 e−04	7.19
	$400M$	55.585470422	1.21 e−04	2.17 e−04	28.7
	5^{13}	55.585200215	2.23 e−04	5.37 e−05	90.6
3	$100M$	55.585320218	8.05 e−05	6.63 e−05	223.3
	$400M$	55.585267930	2.80 e−05	1.40 e−05	930
	5^{13}	55.585249864	2.55 e−05	4.05 e−06	2861
Baikov & Chetyrkin:		55.585253916			
$= 441\,\zeta_7/8$					

M_{44} with massless lines is covered in [2], and in [17] as *lala*, yielding the coefficients C_0, C_1 and C_2 of the expansion (4) as

$$C_0 = \frac{441}{8}\zeta_7 \approx 55.585253915678496$$

$$C_1 = -\frac{81}{5}\zeta_{5,3} + \frac{18567}{80}\zeta_8 - \frac{1323}{4}\zeta_7 - 135\,\zeta_3\,\zeta_5 \approx -268.8399321507655$$

$$C_2 = \frac{486}{5}\zeta_{5,3} + \frac{4583}{2}\zeta_9 - \frac{55701}{40}\zeta_8 + \frac{1323}{2}\zeta_7 - 81\,\zeta_4\,\zeta_5 - \frac{675}{2}\zeta_3\,\zeta_6 + 810\,\zeta_3\,\zeta_5 - 267\,\zeta_3^3$$
$$\approx 1607.707392047691$$

Table 9 lists values of $Q^{(m)}(\varepsilon)$ for $\varepsilon = 0$, $m = 2, 3$, and lattice rules Q with $100M$, $400M$ and 5^{13} points. Table 10 shows an approximation of C_0 by a nonlinear extrapolation with the ϵ-algorithm [18,25] on a sequence of $Q^{(3)}(\varepsilon)$ for $\varepsilon = 1.2^{-\ell}, \ell = 25, 26, \ldots$, and $400M$-point lattice rule Q. The results improve significantly on our integral approximation reported in [8], obtained with ParInt using adaptive multivariate integration (which was 55.585150 with absolute error estimate 1.80 e−04).

Table 10. Nonlinear extrapolation for M_{44}, massless lines, $Q^{(m)}(\varepsilon)$ with $400M$-point rule Q and $m = 3$, $\varepsilon = 1.2^{-\ell}$, $\ell = 25, 26, \ldots$

ℓ	$Q^{(3)}(\varepsilon)$	$|Est_a|$	Time [s]	Extrapolated	Distance Δ
25	56.2134036983660	3.01 e−05	1094		
26	56.0899314726530	2.97 e−05	1095		
27	55.9927679820992	2.94 e−05	1112	55.63392352	
28	55.9157800203784	2.92 e−05	1112	55.62200104	1.19 e−02
29	55.8543905428634	2.90 e−05	1093	55.58542148	3.66 e−02
30	55.8051558873379	2.88 e−05	1109	55.58531681	1.05 e−04
31	55.7654636982214	2.86 e−05	1112	55.58525139	6.54 e−05
32	55.7333158567362	2.85 e−05	1111	55.58525674	5.35 e−06
	Baikov & Chetyrkin: $= 441\,\zeta_7/8$			55.58525392	

Table 11. Linear extrapolation for M_{44}, massless lines, $A(\varepsilon) = \eta(\varepsilon)\,Q^{(m)}(\varepsilon)$ with $400M$-point rule Q and $m = 3$, $\varepsilon = 1.15^{-\ell}$, $\ell = 20, 21, \ldots$.

ℓ	$A(\varepsilon)$	$|Est_a|$	Time[s]	C_0	C_1	C_2
20	43.826387228315269	3.23 e−05	1094			
21	44.930346690376020	3.14 e−05	1114	52.2900764	−138.5213	
22	45.977317510031128	3.08 e−05	1097	55.0254843	−234.7750	842.623
23	46.958581053438841	3.02 e−05	1114	55.4904447	−261.1200	1339.98
24	47.869144536903718	2.98 e−05	1114	55.5728826	−267.9371	1544.79
25	48.706947067994378	2.95 e−05	1100	55.5837878	−269.1405	1597.39
26	49.472182119424062	2.92 e−05	1116	55.5851203	−269.3314	1608.66
27	50.166723337484783	2.90 e−05	1117	55.5852549	−269.3558	1610.52
	Ruijl, Ueda & Vermaseren:			55.5852539	−268.8399	1607.71

Linear extrapolation results for M_{44} with massless lines are given in Table 11 for $A(\varepsilon) = \eta(\varepsilon)\,Q^{(3)}(\varepsilon)$, $\varepsilon = 1.15^{-\ell}$ with $\ell = 25, 26, \ldots$, and $400M$-point lattice rule Q. The normalization factor $\eta(\varepsilon)$ of (3) is used so the coefficients C_k of the expansion (4) would correspond to the published expressions in [2, 17]. Without the normalization factor, C_0 remains the same but the coefficients differ from C_1 on, as can be seen by multiplying the series expansions of $\eta(\varepsilon)$ and $Q^{(m)}(\varepsilon)$. Good agreement was achieved for C_0, while C_1 and C_2 are in the range of the analytic results in [17]. In view of the integrand structure, the numerical evaluation for these massless diagrams is more difficult than for the corresponding diagrams with massive lines, which also emerges from the larger estimated errors for the massless cases.

We further obtained a corresponding expansion with DE as

$$\eta(\varepsilon)\, Q^{(3)}(\varepsilon) \sim 55.5852571 - 269.3577\,\varepsilon + 1610.69\,\varepsilon^2 + \ldots$$

Table 12. M_{45}, massive lines, $Q^{(m)}(\varepsilon)$ with $\varepsilon = 0$

m	Rule # points	$Q^{(m)}(\varepsilon = 0)$	$|Est_a|$	$|Err_a|$	Time [s]
2	$100M$	-4.69481370796	2.00e−06	2.73e−06	6.97
	$400M$	-4.69481095482	1.78e−06	2.42e−08	28.07
	5^{13}	-4.69481104268	2.52e−07	6.37e−08	90.0
3	$100M$	-4.69481102162	6.70e−07	4.26e−08	217.6
	$400M$	-4.69481098486	1.56e−07	5.87e−09	909
	5^{13}	-4.69481098180	2.97e−08	2.80e−09	2791
pySecDec:		-4.69481098172			
Accuracy:		±4.14e−11			

3.3 M_{45}

Approximations with $\varepsilon = 0$ for the constant coefficient C_0 in the expansion (4) of the massive M_{45} diagram are displayed in Table 12. Table 13 gives a nonlinear extrapolation that derives C_0. In Table 12, $m = 2$ and 3 are used for the lattice rules with $100M$,

Table 13. Nonlinear extrapolation for M_{45}, massive lines, $Q^{(m)}(\varepsilon)$ with $100M$-point rule Q and $m = 3$, $\varepsilon = 1.15^{-\ell}$, $\ell = 35, 36, \ldots$

ℓ	$\mathcal{Q}^{(3)}(\varepsilon)$	$|Est_a|$	Time [s]	Extrapolated	Distance Δ
35	-4.3616359607327739	5.10e−07	258.0		
36	-4.4032370844791213	5.28e−07	270.5		
37	-4.4398507092408011	5.45e−07	272.2	-4.7086342263	
38	-4.4720250633753444	5.60e−07	257.4	-4.7052140602	3.42e−03
39	-4.5002602654438046	5.73e−07	270.6	-4.6944830070	1.07e−02
40	-4.5250094047667391	5.85e−07	273.3	-4.6945961688	1.13e−04
41	-4.5466805811799018	5.95e−07	258.0	-4.6948163050	2.20e−04
42	-4.5656395641499143	6.05e−07	275.5	-4.6948140308	2.27e−06
43	-4.5822128112051645	6.13e−07	272.1	-4.6948109712	3.06e−06
44	-4.5966906515695101	6.20e−07	273.0	-4.6948109828	1.15e−08
		pySecDec:		-4.6948109817	
		Accuracy:		±4.14e−11	

Table 14. Linear extrapolation for M_{45}, massive lines, $Q^{(m)}(\varepsilon)$ with $400M$-point rule Q and $m = 3, \varepsilon = 1.15^{-\ell}, \ell = 35, 36, \ldots$

ℓ	$Q^{(3)}(\varepsilon)$	$\lvert Est_a \rvert$	Time [s]	C_0	C_1	C_2
35	-4.3616359324804321	$1.22\,e{-}07$	1217			
36	-4.4032370552356630	$1.26\,e{-}07$	1238	-4.6805778736	42.475259950	
37	-4.4398506791081358	$1.29\,e{-}07$	1293	-4.6943713695	46.424715423	-281.3332185
38	-4.4720250324481521	$1.32\,e{-}07$	1221	-4.6948003601	46.623102971	-311.7166389
39	-4.5002602338092768	$1.35\,e{-}07$	1240	-4.6948107758	46.630029426	-313.4299532
40	-4.5250093725044307	$1.38\,e{-}07$	1242	-4.6948109814	46.630213989	-313.4956014
41	-4.5466805483618833	$1.40\,e{-}07$	1220	-4.6948109849	46.630218037	-313.4975457
	pySecDec Korobov-3:		7718	-4.6948109817	46.630217902	-313.4975487
	pySecDec accuracy:			$\pm4.14\,e{-}11$	$\pm6.27\,e{-}10$	$\pm5.19\,e{-}09$

Table 15. M_{45}, massless lines, $Q^{(m)}(\varepsilon)$ with $\varepsilon = 0$

m	Rule # points	$Q^{(m)}(\varepsilon = 0)$	$\lvert Est_a \rvert$	$\lvert Err_a \rvert$	Time [s]
2	$100M$	52.0184973113	$7.91\,e{-}04$	$6.29\,e{-}04$	6.96
	$400M$	52.0178434626	$4.99\,e{-}05$	$2.53\,e{-}05$	28.1
	5^{13}	52.0178700693	$6.73\,e{-}05$	$1.33\,e{-}06$	88.3
3	$100M$	52.0179169807	$5.87\,e{-}05$	$4.82\,e{-}05$	217.6
	$400M$	52.0178734009	$8.27\,e{-}06$	$4.66\,e{-}06$	907
	5^{13}	52.0178680744	$6.66\,e{-}06$	$6.69\,e{-}07$	2791
	Exact:	52.0178687436			
	$= 36\,\zeta_3^2$				

$400M$ and 5^{13} points. Table 13 applies $m = 3$, a sequence of the regularization parameter $\varepsilon = 1.15^{-\ell}, \ell = 35, 36, \ldots$, and lattice rule Q with $100M$ points. Both tables agree with the pySecDec value, which is supplied within $\pm4.14\,e{-}11$ accuracy.

Excellent agreement with the pySecDec results is also observed in Table 14, which presents a linear extrapolation to approximate C_0, C_1 and C_2 in the series expansion for the massive M_{45} diagram.

Subsequently, we address the M_{45} diagram with massless lines. The expansion is given in [2], and as the *nono* diagram in [17], with coefficients

$$C_0 = 36\,\zeta_3^2 \approx 52.01786874361083$$

$$C_1 = 108\,\zeta_3\,\zeta_4 - 378\,\zeta_7 \approx -240.64650248085806$$

$$C_2 = \frac{3024}{5}\zeta_{5,3} - \frac{26901}{10}\zeta_8 + 2844\,\zeta_3\,\zeta_5 + 432\,\zeta_3^2$$

$$\approx 1471.472731850490$$

Table 16. Nonlinear extrapolation for M_{45}, massless lines, $A(\varepsilon) = \eta(\varepsilon)\, Q^{(m)}(\varepsilon)$ with 5^{13}-point rule Q and $m = 2$, $\varepsilon = 1.2^{-\ell}$, $\ell = 25, 26, \ldots$

ℓ	$A(\varepsilon)$	$|Est_a|$	Time [s]	Extrapolated	Distance Δ
25	49.651577093927614	7.28 e−05	125.7		
26	50.025083811898384	7.18 e−05	131.5		
27	50.342512562887130	7.10 e−05	133.1	52.13931472	
28	50.611407572808432	7.04 e−05	134.2	52.10118612	3.81 e−02
29	50.838572116548526	6.98 e−05	134.9	52.01527075	8.59 e−02
30	51.030048013982416	6.94 e−05	134.5	52.01636014	1.09 e−03
31	51.191137039502031	6.90 e−05	134.7	52.01791539	1.56 e−03
32	51.326447659351238	6.87 e−05	135.2	52.01789174	2.36 e−05
33	51.439955375016694	6.84 e−05	135.4	52.01786959	2.21 e−05
34	51.535068675662060	6.82 e−05	135.7	52.01786990	3.09 e−07
		Exact:		52.01786874	
		$= 36\,\zeta_3^2$			

Table 17. Linear extrapolation for M_{45}, massless lines, $A(\varepsilon) = \eta(\varepsilon)\, Q^{(m)}(\varepsilon)$ with 5^{13}-point rule Q and $m = 2$, $\varepsilon = 1.2^{-\ell}$, $\ell = 20, 21, \ldots$.

ℓ	$A(\varepsilon)$	$|Est_a|$	Time [s]	C_0	C_1	C_2
20	46.647094292297766	8.22 e−05	125.9			
21	47.427685634869917	7.94 e−05	132.4	51.3306423	−179.555992	
22	48.110521177339116	7.72 e−05	134.1	51.9657365	−233.121558	1120.1320
23	48.703053492158347	7.54 e−05	134.4	52.0143244	−239.901941	1432.0643
24	49.213792353197562	7.40 e−05	134.8	52.0176864	−240.593844	1484.7303
25	49.651577093927614	7.28 e−05	125.7	52.0178620	−240.643944	1490.3534
26	50.025083811898384	7.18 e−05	131.5	52.0178698	−240.646890	1490.8126
	Ruijl, Ueda & Vermaseren:			52.0178687	−240.646503	1471.4727

Table 15 provides results for C_0 computed with $\varepsilon = 0$. $Q^{(m)}(\varepsilon = 0)$ is listed for $m = 2, 3$ and lattice rules Q with $100M$, $400M$ and 5^{13} points, together with their estimated absolute error $|Est_a|$, actual absolute error $|Err_a|$, and execution time in seconds. $|Est_a|$ is a slight upper bound of $|Err_a|$ in these cases.

Tables 14, 16, and 17 were computed using 64 CUDA blocks. Table 16 gives a nonlinear extrapolation for the massless M_{45} diagram, using the 5^{13}-point lattice rule Q, $m = 2$, and the regularization sequence $\varepsilon = 1.2^{-\ell}$ for $\ell = 25, 26, \ldots$. The extrapolated results exhibit good convergence to within a relative accuracy of 2.2×10^{-8}.

Linear extrapolation results are shown in Table 17 for $A(\varepsilon) = \eta(\varepsilon)\, Q^{(2)}(\varepsilon)$ where $\eta(\varepsilon)$ is the normalization factor (3) from Baikov and Chetyrkin [2], with $\varepsilon = 1.2^{-\ell}$, $\ell = 20, 21, \ldots$ and lattice rule Q with 5^{13} points.

4 Conclusions

We handled three self-energy diagrams with four loops and 9 internal lines, for the computation of the associated loop integrals and leading-order coefficients in their asymptotic expansions with respect to the regularization parameter.

We obtained results by various solution methods, using composite lattice rule approximations and double exponential integration, as well as linear and nonlinear extrapolation on the sequences of integral approximations for different sequences of the regularization parameter or with different starting points.

For the massless diagrams, comparisons could be made with expansions given in the literature. We improved on our previously published results, and hope to make further improvements in future work. New numerical results are provided for the massive versions of the diagrams, where all masses $= 1$, and $s = p^2 = 1$. Good agreement is found with computations by the pySecDec package.

For these diagrams, it has thus been shown that a numerical treatment is possible, only relying on a definition of the integrand function in the loop integral.

Acknowledgments. We acknowledge the support of the Grant-in-Aid for Scientific Research (JP20K03941) from JSPS KAKENHI, and the National Science Foundation Award Number 1126438 that funded work on multivariate integration. The authors of this paper sincerely appreciate the reviewers' comments, which have helped us improve the paper considerably.

Appendix

A C and D Functions

In the code of the C and D functions below for the diagrams of Fig. 1, we denote $x_{i_1...i_k} = \sum_{j=i_1}^{i_k} x_j$.

A.1 M_{43} Massive Diagram

```
C = (x129*x4 + x1234*x5)*(x6*x7 + x67*x8) +
    x39*(x2*x47*x56 + x1*x6*x7 + x4*x5*x8 +
    x4567*(x1*x2 + x12*x8)) + x3*(x18*x47 + x1478*x56)*x9
    + (x1*x45 + x4*x5)*(x3*x6 + x2*x67 + x7*x9);

D = x12356789*(x29*x3*x5*x7 + (x1259*x3 + x2*x5)*x6*x7 +
    x3*x67*(x18*x29 + x15*x8) +
    (x15*x4 + x1*x5)*(x37*x6 + x3679*x8) + x12*x56*x7*x9 +
    x67*(x1*x28 + x2*x8)*x9 + x1568*x4*(x29*x37 + x2*x9) +
    x18*x5*(x2*x3679 + x3*x9) + x4*x6*(x158*x2 + x8*x9)) +
    x4*x4*(x1259*x37*x6 + x29*x367*x8 + x1568*x2*x9 +
    x15*(x28*x367 + x37*x9 + x8*x9));
```

A.2 M_{44}

For M_{44} (and M_{45}) we use the expression for W from [8], and further

$$U = C$$

$$V = M^2 - \frac{W}{U}$$

$$D = UV = UM^2 - W$$

$$M^2 = \sum_{r=1}^{N} m_r^2 x_r$$

Thus $M^2 = 1$ for massive diagrams with all masses $m_r = 1$, and $D = U - W = C - W$. For massless diagrams, $M^2 = 0$, so $D = -W$.

```
C =  x7*x8*x9*x123456 + x7*x8*x1256*x34 + x7*x9*x126*x345
     + x8*x9*x16*x2345 + x7*x5*x126*x34 +x8*x16*x25*x34 +
     x9*x2*x16*x345 + x2*x5*x16*x34;

W = x7*x8*x9*x123*x456 + x7*x8*(x12*x3*x456 + x123*x4*x56)
    + x7*x9*(x12*x3*x456 + x123*x45*x6)
    + x8*x9*(x1*x23*x456 + x123*x45*x6) + x7*x5*(x12*x3*x46
    + x123*x4*x6) + x8*(x12*x34*x5*x6 + x1*x2*x34*x56 +
    x16*x3*x4*x25) + x9*x2*(x13*x45*x6 + x1*x3*x456) +
    x2*x5*(x1*x3*x46 + x13*x4*x6);

D = C - W; // massive diagram
// D = -W; // massless diagram
```

A.3 M_{45}

```
C =  x7*x8*x9*x123456 + x7*x8*(x12*x34 + x34*x56) +
     x7*x9*x126*x345 + x8*x9*x156*x234 + x7*x5*x126*x34
     + x8*x2*x156*x34 + x9*(x16*x2*x345 + x126*x34*x5)
     + x2*x5*x16*x34;

W = x7*x8*x9*x123*x456 + x7*x8*(x12*x4*x356 + x124*x3*x56)
    + x7*x9*(x12*x3*x456 + x123*x45*x6) + x8*x9*(x1*x23*x456
    + x123*x4*x56) + x7*x5*(x12*x3*x46 + x123*x4*x6) +
    x8*x2*(x13*x4*x56 + x1*x3*x456) + x9*(x1*x2*x36*x45 +
    x12*x36*x4*x5 + x14*x25*x3*x6) + x2*x5*(x1*x3*x46
    + x13*x4*x6);

D = C - W; // massive diagram
// D = -W; // massless diagram
```

References

1. Almulihi, A., de Doncker, E.: Accelerating high-dimensional integration using lattice rules on GPUs. In: Proceedings of the 2017 International Conference on Computational Science and Computational Intelligence, CSCI 2017. CPS IEEE (2017)
2. Baikov, B.A., Chetyrkin, K.G.: Four loop massless propagators: an algebraic evaluation of all master integrals. Nucl. Phys. B **837**, 186–220 (2010)
3. Borowka, S., Heinrich, G., Jahn, S., Jones, S.P., Kerner, M., Schlenk, J.: A GPU compatible quasi-Monte Carlo integrator interfaced to pySecDec. Comput. Phys. Commun. **240**, 120–137 (2019). Preprint: arXiv:1811.11720v1 [hep-ph]. https://arxiv.org/abs/1811.11720. https://doi.org/10.1016/j.cpc.2019.02.015
4. Brezinski, C.: A general extrapolation algorithm. Numer. Math. **35**, 175–187 (1980)
5. de Doncker, E., Shimizu, Y., Fujimoto, J., Yuasa, F.: Computation of loop integrals using extrapolation. Comput. Phys. Commun. **159**, 145–156 (2004)
6. de Doncker, E., Yuasa, F.: Self-energy Feynman diagrams with four loops and 11 internal lines. In: Gervasi, O., et al. (eds.) ICCSA 2021. LNCS, vol. 12953, pp. 160–175. Springer, Cham (2021). https://doi.org/10.1007/978-3-030-86976-2_11
7. de Doncker, E., Yuasa, F., Almulihi, A., Nakasato, N., Daisaka, H., Ishikawa, T.: Numerical multi-loop integration on heterogeneous many-core processors. J. Phys. Conf. Ser. (JPCS) **1525**(012002) (2019). https://doi.org/10.1088/1742-6596/1525/1/012002
8. de Doncker, E., Yuasa, F., Kato, K., Ishikawa, T., Kapenga, J., Olagbemi, O.: Regularization with numerical extrapolation for finite and UV-divergent multi-loop integrals. Comput. Phys. Commun. **224**, 164–185 (2018). https://doi.org/10.1016/j.cpc.2017.11.001
9. de Doncker, E., Yuasa, F., Olagbemi, O., Ishikawa, T.: Large scale automatic computations for Feynman diagrams with up to five loops. In: Gervasi, O., et al. (eds.) ICCSA 2020. LNCS, vol. 12253, pp. 145–162. Springer, Cham (2020). https://doi.org/10.1007/978-3-030-58814-4_11
10. L' Equyer, P., Munger, D.: Algorithm 958: lattice builder: a general software tool for constructing rank-1 lattice rules. ACM Trans. Math. Softw. **42**(2), 15:1–15:30 (2016)
11. Lyman, J.: Optimizing CUDA for GPU architecture – choosing the right dimensions. http://selkie.macalester.edu/csinparallel/modules/CUDAArchitecture/build/html/index.html
12. Lyness, J.N.: Applications of extrapolation techniques to multidimensional quadrature of some integrand functions with a singularity. J. Comput. Phys. **20**, 346–364 (1976)
13. Nuyens, D., Cools, R.: Fast algorithms for component-by-component construction of rank-1 lattice rules in shift-invariant reproducing kernel Hilbert spaces. Math. Comp. **75**, 903–920 (2006)
14. Nuyens, D., Cools, R.: Fast component-by-component construction of rank-1 lattice rules with a non-prime number of points. J. Complex. **22**, 4–28 (2006)
15. Piessens, R., de Doncker, E., Überhuber, C.W., Kahaner, D.K.: QUADPACK, A Subroutine Package for Automatic Integration, Springer Series in Computational Mathematics, vol. 1. Springer-Verlag, Heidelberg (1983). https://doi.org/10.1007/978-3-642-61786-7
16. Pittau, R., Webber, B.: Direct numerical evaluation of multi-loop integrals without contour deformation. Eur. Phys. J. C **82**(1), 1–22 (2022). https://doi.org/10.1140/epjc/s10052-022-10008-6
17. Ruijl, B., Ueda, T., Vermaseren, J.A.M.: Forcer, a Form program for the parametric reduction of four-loop massless propagator diagrams. Comput. Phys. Commun. **253**, 107198 (2020)
18. Shanks, D.: Non-linear transformations of divergent and slowly convergent sequences. J. Math. Phys. **34**, 1–42 (1955)
19. Sidi, A.: Practical Extrapolation Methods - Theory and Applications. Cambridge University Press (2003). ISBN 0-521-66159-5

20. Sidi, A.: Extension of a class of periodizing transformations for numerical integration. Math. Comp. **75**(253), 327–343 (2005)
21. Sloan, I., Joe, S.: Lattice Methods for Multiple Integration. Oxford University Press (1994)
22. Sugihara, M.: Optimality of the double exponential formula - functional analysis approach. Numer. Math. **75**(3), 379–395 (1997)
23. Takahasi, H., Mori, M.: Double exponential formulas for numerical integration. Publ. Res. Inst. Math. Sci. **9**(3), 721–741 (1974)
24. Techpowerup. https://www.techpowerup.com/gpu-specs/quadro-gv100.c3066
25. Wynn, P.: On a device for computing the $e_m(s_n)$ transformation. Math. Tables Aids Comput. **10**, 91–96 (1956)
26. Yuasa, F., et al.: Numerical computation of two-loop box diagrams with masses. J. Comput. Phys. Commun. **183**, 2136–2144 (2012)

Acceleration of Matrix Multiplication Based on Triple-Double (TD), and Triple-Single (TS) Precision Arithmetic

Taiga Utsugiri and Tomonori Kouya$^{(\boxtimes)}$

Shizuoka Institute of Science and Technology, 2200-2 Toyosawa, Fukuroi 437-8555, Japan
`2121002.ut@sist.ac.jp`

Abstract. In this study, we present the results obtained from the acceleration of multi-component-type multiple precision matrix multiplication using the Ozaki Scheme and OpenMP. We aim for triple-double (TD) and triple-single (TS) precision matrix multiplication on CPU and GPU using TD and TS arithmetic based on error-free transformation (EFT) arithmetic. Owing to these combined techniques, on CPU, our implemented multiple precision matrix multiplications were more than seven times faster than Strassen matrix multiplication with AVX2. Similar to using Ozaki scheme, on GPU, our implemented multiple precision matrix multiplications could not be accelerated. Furthermore, we report that our accelerated matrix multiplication can modify parallelization performance with OpenMP on CPU.

Keywords: Triple-double (TD) precision arithmetic · Triple-single (TS) precision arithmetic · Matrix multiplication · Ozaki scheme

1 Introduction

It is well-known that the accuracy of numerical computation with floating-point arithmetic is limited owing to the effect of round-off errors. The hardware-supported IEEE754 floating-point numbers are binary32 (24-bit fraction) and binary64 (53-bit fraction). Although binary128 was officially proposed by IEEE754-2008, there is no implementation that normal users can use other than the float128 type supported by GCC.

The implementation of multi-precision floating-point (MPF) arithmetic is divided into two categories: multi-digit-type based on integer arithmetic and multi-component-type (or multi-term), which is constructed by several binary32 or binary64 numbers using error-free transformation (EFT) [1]. EFT is a method to calculate floating point arithmetic without loss of errors. The QD library proposed by Bailey et al. [2] is a de-facto standard, one of the multi-component-type MPF libraries, which consists of double-double (DD) and quadruple-double (QD) data types for the CPU. Lu et al. implemented the GQD library on GPU, which is based on QD library [3].

O. Gervasi et al. (Eds.): ICCSA 2022 Workshops, LNCS 13378, pp. 406–423, 2022.
https://doi.org/10.1007/978-3-031-10562-3_29

While DD and QD precision arithmetic have been recently prepared, triple-double (TD) or triple-single (TS) arithmetic between DD and QD have been proposed in some studies. TD and TS calculations are expected to be comparable in performance and accuracy between DD and QD; thus, they can achieve the best results not quiet with DD but in shorter computational time than QD.

As an optimized three-word operation method, Fabiano et al. [4] proposed optimized triple-word operation. Although we cannot confirm its good performance, we selected it to implement TD by omitting the forth and least element of QD value.

The Strassen algorithm [5] is known as a high-speed matrix multiplication algorithm that uses divide-and-conquer approach to reduce the number of operations. Kouya [6] implemented and accelerated TD Strassen matrix multiplication with AVX2. On the contrary, the Ozaki scheme developed by Ozaki et al. [7, 8] is a highly accurate and well-performed matrix multiplication method that takes advantage of highly performed hard-wired precision matrix multiplication, such as DGEMM or SGEMM, provided by Intel Math Kernel and cuBLAS. Mukunoki et al. [9] used the Ozaki scheme to implement MPF matrix multiplication. Utsugiri et al. [10] reported some implementations on GPU.

As mentioned above, various optimization techniques for multi-component-type multiple precision matrix products have already been exploited, however, which technique is the most efficient is unclear. In particular, we should know the effectiveness of the Ozaki scheme compared with other optimized matrix products in TD and TS arithmetic. The main purpose of our study is to show the performance of the Ozaki scheme on CPU and GPU environments.

In this paper, we describe the implementation of the Ozaki scheme for TD and TS matrix multiplication on CPU and GPU and discuss the performance and accuracy. The programming languages used are C++ with an Intel compiler in the CPU and C++ with CUDA in the GPU.

Furthermore, we parallelize the inner process of the Ozaki scheme using OpenMP on CPU. As a result of our benchmark test, we can reveal that the Ozaki scheme implemented on CPU is faster than Strassen + AVX2 matrix multiplication. However, on the GPU, the Ozaki scheme is slower than the simple matrix multiplication. In addition, we can observe that the parallelized TD matrix multiplication on CPU is faster in a small matrix size than without OpenMP, but not in a large matrix size.

2 Overview of Triple-Double (TD), Triple-Single (TS)

Figures 1 and 2 show the schematics of the TD and TS precisions, respectively. TD precision consists of three double precisions (binary64) and TS precision consists of three single precisions (binary32). Our implementation uses the "td_real" data type as the TD floating-point number, and "ts_real" as the TS number; therefore, both "td_real a" and "ts_real a" consist of a[0] as the most significant component, a[1] in the middle, and a[2] as the least significant component. The precision of floating-point numbers is determined by the length of the mantissa part. The double-precision mantissa part has 53 bits (including one hidden bit), with a precision of approximately 16 decimal digits, and the single-precision mantissa part has 24 bits, with a precision of approximately seven decimal digits. The mantissa part of the TD precision is 159 bits, with a precision of

408   T. Utsugiri and T. Kouya

approximately 48 decimal digits, and the mantissa part of the TS precision is 72 bits, with
a precision of approximately 21 decimal digits. On the contrary, IEEE754 binary128 has
a 15-bit exponent portion; TD precision has 11 bits and TS precision has only 8 bits.
Therefore, the valid range of these TD and TS values are smaller than binary128.

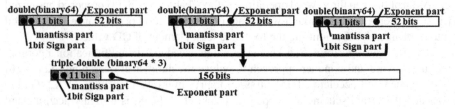

Fig. 1. Schematic of triple-double (TD) precision

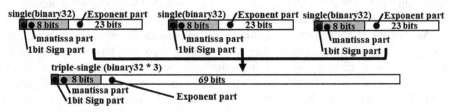

Fig. 2. Schematic of triple-single (TS) precision

2.1 Implementation of Triple-Double (TD) Precision on the CPU

Here, we describe that implementation of TD precision arithmetic. EFT, TD multiplication and TD addition are necessary to implement TD precision matrix multiplication.
They have been implemented as inline functions to approve fast call of TD arithmetic
functions. The implemented codes are shown in Algorithm 1–Algorithm 7.

Two_Sum (Algorithm 1) is addition that considers double precision rounding error.
In *Two_Sum*, the result of $a+b$ operation with binary64 is stored in s, and the correspoind
rounding error is stored in *err*. *Quick_Two_Sum* (Algorithm 2) is also addition that considers double precision rounding error as Two_Sum, however it can be used only when
$|a| \geq |b|$. *Three_Sum* (Algorithm 3) is implemented by combining multiple *Two_Sum*
and adding three floating-point numbers. *Three_Sum* reconstructs the upper part as a,
the middle part as b, and the lower part as c, while exchanging the values of a, b, and c.
Two_Sum and *Three_Sum* can be used to calculate the number of floating-point numbers.
TD_Add (Algorithm 5) is implemented by combining *Two_Sum* and *Three_Sum* for TD
precision addition. Renorm (Algorithm 4) is a function to normalize TD number.

The conventional QD library as default, does not use the FMA operation; therefore, in our TD arithmetic, *Two_Prod_FMA* (Algorithm 6) is prepared as a default Two_Prod function with "fma" utilized in the standard C library, which is a multiplication that considers double precision rounding errors. *Two_Prod_FMA* stores the $a \times b$ result in p and the rounding errors that occur in err. Because of these EFT functions, *TD_Prod* (Algorithm 7) performs TD precision multiplication.

Algorithm 1. inline double Two_sum(double a, double b, double &err)

```
inline double Two_sum(double a, double b, double &err){
    double s, bb;
    s = a + b;
    bb = s - a;
    err = (a - (s - bb)) + (b - bb);
    return s;
}
```

Algorithm 2. inline double Quick_Two_Sum(double a, double b, double &err)

```
inline double Quick_Two_Sum(double a, double b, double &err){
    double s;
    s = a + b;
    err = b - (s - a);
    return s;
}
```

Algorithm 3. inline void Three_sum(double &a, double &b, double &c)

```
inline void Three_Sum(double &a, double &b, double &c){
    double t1, t2, t3;
    t1 = Two_Sum(a, b, t2);
    a = Two_Sum(c, t1, t3);
    b = Two_Sum(t2, t3, c);
}
```

Algorithm 4. inline void Renorm(double &c0, double &c1, double &c2, double &3)

```
inline void Renorm(double &c0, double &c1, double &c2, double &c3){
  double s0, s1, s2 = 0.0;
  s0 = Quick_Two_Sum(c2, c3, c3);
  s0 = Quick_Two_Sum(c1, s0, c2);
  c0 = Quick_Two_Sum(c0, s0, c1);
  s0 = c0;
  s1 = c1;
  if(s1 != 0.0){
    s1 = Quick_Two_Sum(s1, c2, s2);
    if(s2 != 0.0)
      s2 = s2 + c3;
    else
      s1 = Quick_Two_Sum(s1, c3, s2);
  }else{
    s0 = Quick_Two_Sum(s0, c2, s1);
    if(s1 != 0.0)
      s1 = Quick_Two_Sum(s1, c3, s2);
    else
      s0 = Quick_Two_Sum(s0, c3, s1);
  }
  c0 = s0;
  c1 = s1;
  c2 = s2;
}
```

Algorithm 5. inline td_real td_real::TD_Add(const td_real &a, const td_real &b)

```
inline td_real td_real :: TD_Add(const td_real &a, const td_real &b){
  double s0, s1, s2, t0, t1, t2;
  double v0, v1, v2, u0, u1, u2;
  double w0, w1, w2;
  s0 = a[0] + b[0]; s1 = a[1] + b[1]; s2 = a[2] + b[2];
  v0 = s0 - a[0]; v1 = s1 - a[1]; v2 = s2 - a[2];
  u0 = s0 - v0; u1 = s1 - v1; u2 = s2 - v2;
  w0 = a[0] - u0; w1 = a[1] - u1; w2 = a[2] - u2;
  u0 = b[0] - v0; u1 = b[1] - v1; u2 = b[2] - v2;
  t0 = w0 + u0; t1 = w1 + u1; t2 = w2 + u2;
  s1 = Two_Sum(s1, t0, t0);
  Three_Sum(s2, t0, t1);
```

```
t0 = Two_Sum(t2, t0, t2);
t0 = t0 + t1;
Renorm(s0, s1, s2, t0);
return td_real(s0, s1, s2);
}
```

Algorithm 6. inline double Two_Prod_FMA(double a, double b, double &err)

```
inline double Two_Prod_FMA(double a, double b, double &err){
    s = a × b
    err = fma(a, b, −p);
    return s;
}
```

Algorithm 7. inline td_real td_real:: TD_Prod(const td_real &a, const td_real &b)

```
inline ts_real td_real::TD_Prod(const td_real &a, const td_real &b) {
    double p0, p1, p2, q0, q1, q2, s0;
    p0 = Two_Prod_FMA(a[0], b[0], q0);
    p1 = Two_Prod_FMA(a[0], b[1], q1);
    p2 = Two_Prod_FMA(a[1], b[0], q2);
    Three_Sum(p1, p2, q0);
    Three_Sum(p2, q1, q2);
    s0 = a[0] * b[2] + a[2] × b[0] + a[1] × b[1] + q0 + q1 + q2;
    Renorm(p0, p1, p2, s0);
    return td_real(p0, p1, p2);
}
```

2.2 Implementation of Triple-Single Precision (TS) on GPU

This section describes the implementation of TS precision on GPU. The GPUs for consumers, such as NVIDIA GTX and RTX series, are designed to accelerate the processing graphics of games and deep learning using binary16 and binary32, rather than to outperform binary64. We confirmed performance of binary64 and TD arithmetic by comparing it to binary32 and TS arithmetic, and as the result, we conclude that TS matrix multiplication should be made available to users of consumer GPUs.

Algorithm 8–Algorithm 14, which contain parts of TS matrix multiplication, are almost similar to TD and related arithmetic as shown in Sect. 2.1. We can observe from these algorithms, that all data types used are "float" (binary32) rather than "double" (binary64).

Algorithm 8. __device__ __forceinline__ float Two_Sum(float a, float b, float &err)

```
_device_ _forceinline_ float Two_Sum(float a, float b, float &err){
    float s, bb;
    s = a + b;
    bb = s - a;
    err = (a - (s - bb)) + (b - bb);
    return s;
}
```

Algorithm 9. __device__ __forceinline__ float Quick_Two_Sum(float a, float b, float &err)

```
_device_ _forceinline_ float Quick_Two_Sum(float a, float b, float &err){
    float s;
    s = a + b;
    err = b - (s - a);
    return s;
}
```

Algorithm 10. __device__ __forceinline__ void Three_Sum(float &a, float &b, float &c)

```
_device_ _forceinline_ void Three_Sum(float &a, float &b, float &c) {
    float t1, t2, t3;
    t1 = two_sum(a, b, t2);
    a = two_sum(c, t1, t3);
    b = two_sum(t2, t3, c);
}
```

Algorithm 11. __device__ __forceinline__ void Renorm(float &$c0$, float &$c1$, float &$c2$, float &$c3$)

```
_device_ _forceinline_ void Renorm(float &c0, float &c1, float &c2,
float &c3) {
  float s0, s1, s2 = 0.0;
  s0 = Quick_Two_Sum(c2, c3, c3);
  s0 = Quick_Two_Sum(c1, s0, c2);
  c0 = Quick_Two_Sum(c0, s0, c1);
  s0 = c0;   s1 = c1;
  if (s1 != 0.0) {
    s1 = Quick_Two_Sum(s1, c2, s2);
  if (s2 != 0.0)
    s2 = s2 + c3;
  else
    s1 = Quick_Two_Sum(s1, c3, s2);
  } else {
    s0 = Quick_Two_Sum(s0, c2, s1);
    if (s1 != 0.0)
      s1 = Quick_Two_Sum(s1, c3, s2);
    else
      s0 = Quick_Two_Sum(s0, c3, s1);
  }
  c0 = s0;   c1 = s1;   c2 = s2;
}
```

Algorithm 12. __device__ __forceinline__ ts_real ts_real::TS_Add(const ts_real &a, const ts_real &b)

```
_device_ _forceinline_ ts_real ts_real :: TS_Add(const ts_real &a,
const ts_real &b) {
  float s0, s1, s2, t0, t1, t2;
  float v0, v1, v2, u0, u1, u2, w0, w1, w2;
  s0 = a[0] + b[0]; s1 = a[1] + b[1]; s2 = a[2] + b[2];
  v0 = s0 - a[0]; v1 = s1 - a[1]; v2 = s2 - a[2];
  u0 = s0 - v0; u1 = s1 - v1; u2 = s2 - v2;
  w0 = a[0] - u0; w1 = a[1] - u1; w2 = a[2] - u2;
  u0 = b[0] - v0; u1 = b[1] - v1; u2 = b[2] - v2;
  t0 = w0 + u0; t1 = w1 + u1; t2 = w2 + u2;
```

$s1 = Two\_Sum(s1, t0, t0);$
$Three\_Sum(s2, t0, t1);$
$t0 = Two\_Sum(t2, t0, t2);$
$t0 = t0 + t1;$
$Renorm(s0, s1, s2, t0);$
$return\ ts\_real(s0, s1, s2);$
}

Algorithm 13. __device__ __forceinline__ float Two_Prod_FMA(float a, float b, float &err)

$\_device\_\_forceinline\_ float\ Two\_Prod\_FMA(float\ a, float\ b, float\ \&err)\{$
$s = a \times b$
$err = fmaf(a, b, -p);$
$return\ s;$
}

Algorithm 14. __device__ __forceinline__ ts_real ts_real::TS_Prod(const ts_real &a, const ts_real &b)

$\_device\_\_forceinline\_ ts\_real\ ts\_real :: TS\_Prod(const\ ts\_real\ \&a,$
$const\ ts\_real\ \&b)\ \{$
$float\ p0, p1, p2, q0, q1, q2, s0;$
$p0 = Two\_Prod\_FMA(a[0], b[0], q0);$
$p1 = Two\_Prod\_FMA(a[0], b[1], q1);$
$p2 = Two\_Prod\_FMA(a[1], b[0], q2);$
$Three\_Sum(p1, p2, q0);$
$Three\_Sum(p2, q1, q2);$
$s0 = a[0] * b[2] + a[2] \times b[0] + a[1] \times b[1] + q0 + q1 + q2;$
$Renorm(p0, p1, p2, s0);$
$return\ ts\_real(p0, p1, p2);$
}

3   Ozaki Scheme for Triple-Double (TD) Precision on CPU

The Ozaki scheme [7, 8] has been developed to implement accurate and fast matrix multiplication by accumulating binary64 or binary32 matrix products using highly optimized DGEMM or SGEMM. Here, we describe its usage for TD and TS matrix multiplications.

First, we explain the Ozaki scheme for TD matrix multiplication. To begin, we split the original TD matrices into binary64 matrices. The divided binary64 matrices were multiplied using the DGEMM of the BLAS library implemented in the Intel Math Kernel Library (intel MKL) [11]. Algorithm 15 shows how to split three binary64 matrices from the original matrices A and B. However, as described later, the number of splits should be limited to a maximum of 10.

In the 1st line, the first binary64 value of TD elements of A and B are copied to $A^{(D)}$ and $B^{(D)}$. In the 4th and 5th lines, the maximum value of each row of $A^{(D)}$ and each column of $B^{(D)}$ are obtained and stored in M_A and M_B, respectively. In the 6th to 7th lines, T_A and T_B are obtained using M_A and M_B, respectively and 53 bits which is a double precision mantissa. In the 8th to 9th lines, T_A and T_B, and the translocation vector E are used to find S_A and S_B. The reason for finding S_A and S_B is to make a matrix that does not cause overflow when A_1 and B_1, which include elements near the $\pm$Inf of binary64 value, perform matrix multiplication. In the 10th and 11th lines, A_1 and B_1 are obtained so that overflow does not occur. In the 12th and 13th lines, A_1 and B_1 are subtracted from A and B, and the elements of A and B are updated. At this time, because it is divided into three, A_3 and B_3 are obtained on the 16th and 17th lines after repeating this operation twice. By performing this operation from producing A_α and B_α even with binary64 arithmetic, round-off errors owing to digit overflow does not create and then can gain precise numerical results with TD arithmetic. From the 18th to 23rd lines, after DGEMM run for obtaining each C_α, TD addition is performed and then each element is stored in the TD matrix C.

Algorithm 15. Ozaki scheme algorithm used for the created TD precision matrix multiplication (in the case of three splits, n × n matrix)

Input: A, B : A and B are TD precision square matrices
Output: C : C is a TD type square matrix
1: $A^{(D)} = A$, $B^{(D)} = B$
               : $A^{(D)}$ and $B^{(D)}$ are double precision $n \times n$ square matrices.
2: $\alpha = 1$
3: $While(\ \alpha < 3)$
4:   $M_A(i)_{1 \leq i \leq n} = max_{1 \leq k \leq n} |A^{(D)}{}_{i,k}|$: M_A is an nth − order vector
5:   $M_B(j)_{1 \leq j \leq n} = max_{1 \leq p \leq n} |B^{(D)}{}_{p,j}|$: M_B is an nth − order vector
6:   $T_A(i)_{1 \leq i \leq n} = 2^\wedge(ceil\left(log2(M_A(i))\right) + ceil\left(\dfrac{53 + log2(n)}{2}\right))$
       : T_A is an nth − order vector
7:   $T_B(j)_{1 \leq j \leq n} = 2^\wedge(ceil\left(log2(M_B(j))\right) + ceil\left(\dfrac{53 + log2(n)}{2}\right))$
       : T_B is an nth − order vector
8:   $S_A = T_A \cdot E^t$: $E = (1, 1, 1, ..., 1)^t$ is an nth − order vector
9:   $S_B = E \cdot T_B{}^t$
10:   $A_\alpha = \left(A^{(D)} + S_A\right) - S_A$
11:   $B_\alpha = \left(B^{(D)} + S_B\right) - S_B$
12:   $A = A - A_\alpha$, $B = B - B_\alpha$: Calculated by TD calculation
13:   $A^{(D)} = A$, $B^{(D)} = B$
14:   $\alpha = \alpha + 1$
15: $End\ While$
16: $A_\alpha = A^{(D)} - A_{\alpha-1}$
17: $B_\alpha = B^{(D)} - B_{\alpha-1}$
18: $For(\alpha = 1; \alpha < 4; \alpha + +)$
19:   $For(\beta = 1; \beta < 5 - \alpha; \beta + +)$
20:     $C_\beta = A_\alpha \times B_\beta$: Calculated by Dgemm of intel MKL
21:   $End\ For$
22:   $C += \sum_{k=1}^{4-\alpha} C_k$: Calculated by TD addition
23: $End\ For$

---

4   Ozaki Scheme for Triple-Single (TS) Precision on GPU

The concrete implementation of the TS precision Ozaki scheme is described as shown in Algorithm 16.

Primarily, it is similar to Algorithm 15 for TD Ozaki scheme, however there are different points caused by change of base precision from binary64 to binary32. For example, A and B are TS matrices constructed from elements of three binary32 numbers, and $A^{(S)}$, $B^{(S)}$, S_A, S_B M_A, M_B, A_α, and B_α are also binary32 matrices. In addition, SGEMM for $A_\alpha \times B_\beta$ is performed using CUDA's cuBLAS Sgemm [12].

Algorithm 16. Ozaki scheme algorithm used for the created TS precision matrix multiplication (in the case of three splits, n × n matrix)

$Input: A, B$: A and B are TS precision square matrices
$Output: C$: C is a TS precision square matrix

1: $A^{(S)} = A, \ B^{(S)} = B$
                 : $A^{(S)}$ and $B^{(S)}$ are single precision $n \times n$ square matrices.

2: $\alpha = 1$

3: $While(\ \alpha < 3)$

4:    $M_A(i)_{1 \le i \le n} = max_{1 \le k \le n} |A^{(S)}{}_{i,k}|$: M_A is an nth − order vector

5:    $M_B(j)_{1 \le j \le n} = max_{1 \le p \le n} |B^{(S)}{}_{p,j}|$: M_B is an nth − order vector

6:    $T_A(i)_{1 \le i \le n} = 2^\wedge (ceil\left(log2(M_A(i))\right) + ceil\left(\dfrac{24 + log2(n)}{2}\right))$
     : T_A is an nth − order vector

7:    $T_B(j)_{1 \le j \le n} = 2^\wedge (ceil\left(log2(M_B(j))\right) + ceil\left(\dfrac{24 + log2(n)}{2}\right))$
     : T_B is an nth − order vector

8:    $S_A = T_A \cdot E^t$: $E = (1,1,1,\dots,1)^t$ is an nth − order vector

9:    $S_B = E \cdot T_B{}^t$

10:   $A_\alpha = \left(A^{(S)} + S_A\right) - S_A$

11:   $B_\alpha = \left(B^{(S)} + S_B\right) - S_B$

12:   $A = A - A_\alpha, \ B = B - B_\alpha$: Calculated by TS calculation

13:   $A^{(S)} = A, \ B^{(S)} = B$

14:   $\alpha = \alpha + 1$

15: $End\ While$

16: $A_\alpha = A^{(S)} - A_{\alpha-1}$

17: $B_\alpha = B^{(S)} - B_{\alpha-1}$

18: $For(\alpha = 1; \ \alpha < 4; \ \alpha++)$

19:   $For(\beta = 1; \ \beta < 5 - \alpha; \ \beta++)$

20:    $C_\beta = A_\alpha \times B_\beta$: Calculated by Sgemm of cuBLAS

21:   $End\ For$

22:   $C += \sum_{k=1}^{4-\alpha} C_k$: Calculated by TS addition

23: $End\ For$

5 Performance Evaluation by TD and TS Matrix Multiplications

We show the results of the benchmark tests performed on both CPUs and GPUs, whose execution environments are show in Table 1. TD matrix multiplication is performed on CPUs, and TS matrix multiplication on GPUs. We use intel Core i7 6700k and intel Core i7 11700 as CPUs, and NVIDIA GeForce GTX 1080 and NVIDIA GeForce RTX 3070 as GPUs. Intel MKLs with multi-thread facilities are also utilized to achieve the best performance of the Ozaki scheme.

The elements of matrices used for the benchmark test are random numbers calculated by $(ru - 0.5) \times exp(1.0 \times rn)$, where ru is a uniform random number of $[0, 1]$, and rn is a random number associated with the standard normal distribution. Our benchmark tests are categorized into the following four types:

1. Execution time and relative error comparison of matrix multiplication using TD precision 12-split Ozaki scheme and matrix multiplication using Strassen + AVX2 on CPU.
2. Comparison of execution time using TD precision 12-split Ozaki scheme between the serial matrix multiplication and parallelized one using OpenMP on CPU.
3. Profiling the execution time among matrix size of matrix multiplication split, multiplication, and addition using TD 12-split Ozaki scheme + OpenMP on CPU.
4. Execution time comparison of matrix multiplication using TS precision 7 to 12-split Ozaki scheme and matrix multiplication using shared memory on GPU.

The maximum relative error is calculated by (1), where $E_{i,j}^*$ represents the operation result of the QD type matrix multiplication, and $E_{i,j}$ represent the elements of the result of the matrix multiplication of the Ozaki scheme.

$$
\max_{1 \leq i,j \leq n} \frac{\left| E_{i,j}^* - E_{i,j} \right|}{\left| E_{i,j}^* \right|} \tag{1}
$$

Table 1. Execution environment of i7 6700K (left) and i7 11700 (right)

CPU	Intel Core i7 6700K	CPU	Intel Core i7 11700
Memory	16GB	Memory	32GB
GPU	NVIDIA GeForce GTX 1080	GPU	NVIDIA GeForce RTX 3070
OS	Ubuntu20.04.2 LTS	OS	Ubuntu20.04.2 LTS
gcc/g++	9.3.0	gcc/g++	9.3.0
icc/icpc	2021.5.0	icc/icpc	2021.5.0
CUDA	10.1	CUDA	10.1

5.1 Comparison with Matrix Multiplication Execution Time of TD 12-Split Ozaki Scheme and TD Strassen + AVX2 on CPU

The comparison of the execution time of the matrix multiplication using the 12-division Ozaki scheme of TD precision and matrix multiplication using Strassen + AVX2 of TD precision is shown in Fig. 3. The matrix size (N) was 500 to 5000, and the performance was evaluated at 500 pitch of matrix size. For performance comparison, we used the matrix multiplication using Strassen + AVX2 as counterpart.

The left part of Fig. 4. uses the i7 6700k, and the right part uses the i7 11700 for performance evaluation. On i7 6700K, we can observe that the matrix product using the TD Ozaki scheme is up to approximately 4.5 times faster than the matrix multiplication using Strassen + AVX2. On the right of Fig. 3, the i7 11700 shows that the matrix product using the TD Ozaki scheme is approximately 4.5 times faster than the matrix product using Strassen + AVX2. At N = 4500, Strassen + AVX2 is much slower and the speed difference with the Ozaki scheme is significant.

Figure 4 shows the comparison with their relative error of up to 5000 as matrix size (N). The number of split of matrices in the Ozaki scheme has trade-off relation with the accuracy of the final TD matrix C. While we can gain accurate C by increasing the number of splits, the performance becomes slower owing to more number of DGEMM and TD additions.

We can observe that the Ozaki scheme has better accuracy than Strassen + AVX2 when the number of splits exceeds 10. As the matrix size increases, the accuracy of the Ozaki scheme gradually deteriorates until the matrix size (N) is up to 5000, however if

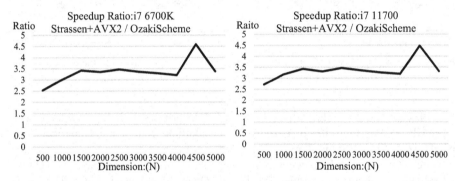

Fig. 3. Speedup ratio of matrix multiplication using TD precision 12-split Ozaki scheme and Strassen + AVX2 in Core i7 6700K (left) and Core i7 11700 (right)

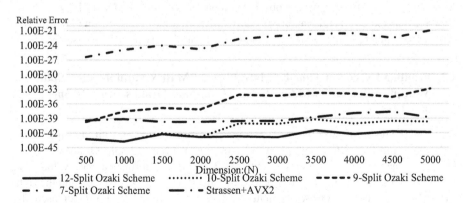

Fig. 4. TD type Ozaki scheme 7–12 division and relative error comparison of matrix product using Strassen + AVX2

the matrix size is larger than 5000, it is possible that more splits are required. Up to N = 5000, the maximum accuracy is achieved with 12 splits.

5.2 Comparison of Execution Time Between Matrix Multiplication Using the TD 12-Split Ozaki Scheme and Parallelized One with OpenMP on CPUs

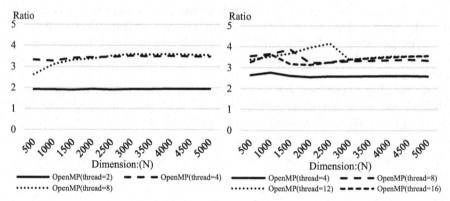

Fig. 5. It shows how fast it is for each number of threads used compared to the execution time when OpenMP is not used in Core i7 6700K (left) and Core i7 11700 (right).

Subsequently, we evaluated parallelization of the process of the Ozaki scheme with OpenMP. At this time, the parts of split and addition from original matrices were parallelized because of the shortening development time.

The comparison of the execution time with matrix multiplication using the TD type 12-split Ozaki scheme and parallelized one with OpenMP is shown in Fig. 5. The matrix size (N) was 500 to 5000, and the performance was evaluated in 500 pitch as matrix size. The left part of Fig. 5 is the performance on i7 6700K, and right part on i7 11700. According to these figures, we can observe that parallelization with OpenMP is not more effective than expected and that the speedup ratio by parallelization in small matrix size is better than in larger matrix size. The analysis of parallelization will be described later.

5.3 Profiling Execution Time by Matrix Size of Matrix Multiplication Split, Multiplication, and Addition Using TD 12-Split Ozaki scheme + OpenMP on CPU

We show the result of profiling the execution time constructed by split, multiplication, and addition in the matrix multiplication using the TD type 12-split Ozaki scheme + OpenMP. The matrix size (N) was 500 to 5000, and the performance was evaluated at 500 as matrix size. The left part of Fig. 6 shows the profile on the i7 6700k and the right part of Fig. 6 on the i7 1700. As the matrix size of i7 6700K and i7 11700 increased, the multiplication ratio in total execution time increased.

As shown in Fig. 6, we understand that the parallelization performance causes the multiplication ratio in the Ozaki scheme to occupy the major part of the computation time.

Because multiplication is calculated using the DGEMM of intel MKL, the parallelization of OpenMP is considered inapplicable to the entire process of our implementation. As previously described, the parallelization with OpenMP is performed only in the part of the split and addition from the original matrices; therefore we believe that acceleration can be achieved by parallelizing the usage of multiple MKL DGEMM processes with OpenMP.

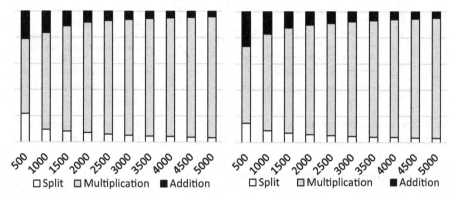

Fig. 6. Ratio of execution time by matrix size of matrix multiplication split, multiplication, addition using TD 12-split Ozaki scheme in Core i7 6700K (left) and Core i7 11700 (right)

5.4 Speed Comparison of Matrix Multiplication Using TS Precision 7 to 12-Split Ozaki Scheme and Blocked Using Shared Memory on GPU

The execution times of the 7 to 12-split TS Ozaki scheme and the TS blocked matrix product are shown in Fig. 7. The matrix size (N) was 512 to 4096, and the performance was evaluated in 512 as matrix size pitch. The matrix multiplication was blocked by 32 × 32. "tsmatmul" represents the execution time of a TS simple matrix multiplication. As environment of GPU, Fig. 7 (left) uses GeForce GTX 1080 and Fig. 7 (right) uses GeForce RTX 3070.

It is observed that the TD precision Ozaki scheme is inefficient for GPUs without support of special hardware, such as TensorCore, and that the TS simple matrix multiplication using shared memory is faster than using the TS Ozaki scheme. Furthermore, the matrix multiplication using the TS Ozaki scheme is faster on GTX 1080 than on RTX 3070. TS simple matrix multiplication on RTX3070 is faster than on GTX1080, because memory access of Ozaki scheme and unused Tensor core may be the cause of delay.

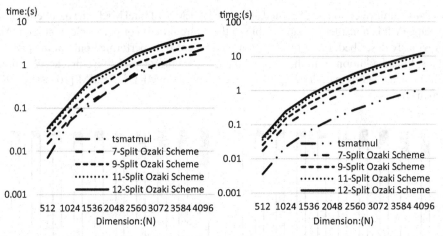

Fig. 7. Speed comparison of matrix multiplication using TS type 7 to 12 split Ozaki scheme and matrix multiplication using shared memory in GTX 1080 (left) and RTX 3070 (right)

6 Conclusion and Future Work

In this study, we explain the implementation of the matrix multiplication using the TD and TS Ozaki scheme and then evaluate their performances on CPUs and GPUs, respectively. We can conclude that on CPU, the TD Ozaki scheme is faster than the matrix multiplication using TD Strassen + AVX2. However, on GPU, the TS simple matrix multiplication using shared memory is faster than using the TS Ozaki scheme. We believe that lack of special hardware causes slow TS matrix multiplication with Ozaki scheme.

For our future work, we will solve the following issues:

1. Confirm how the usage the Tensor core on the GPUs are effective.
2. Review, on CPUs, the multiplication program to accelerate our current TD matrix multiplication that uses Intel MKL.
3. Implement and evaluate quadruple-double (QD) and quadruple-single (QS) precision matrix multiplication with the Ozaki scheme, OpenMP on CPUs and GPUs, respectively.

Acknowledgement. This study was supported by JSPS KAKENHI, Grant Number JP20K11843, and the Shizuoka Institute of Science and Technology. We acknowledge all organizations that continuously encourage our study and appreciate unknown referees who provide sharp and fruitful advice for polishing this paper.

References

1. Dekker, T.J.: A floating-point technique for extending the available precision. Numer. Math. **18**, 224–242 (1971). https://doi.org/10.1007/BF01397083

 2. Bailey, D.: QD. https://www.davidhbailey.com/dhbsoftware/
 3. Lu, M., et al: supporting extended precision on graphics processors. In: Proceedings of the Sixth International Workshop on Data Management on New Hardware, pp. 19–26, June 2010
 4. Fabiano, N., Muller, J.M., Picot, J.: Algorithms for triple words arithmetic. IEEE Trans. Comput. **68**, 1573–1583 (2019)
 5. Strassen, V.: Gaussian elimination is not optimal. Numer. Math. **13**, 354–356 (1969)
 6. Kouya, T.: Performance evaluation of multiple precision matrix multiplications using parallelized Strassen and Winograd algorithms. JSIAM Lett. **8**, 21–24 (2016). https://doi.org/10.14495/jsiaml.8.21
 7. Ozaki, K., Ogita, T., Oishi, S.I., Rump, S.M.: Error free transformations of matrix multiplication by using fast routines of matrix multiplication and its applications. Numer. Algorithms **59**, 95–118 (2012)
 8. Ichimura, S., Katagiri, T., Ozaki, K., Ogita, T., Nagai, T.: Threaded accurate matrix-matrix multiplications with sparse matrix-vector multiplications. In: IRRR International Parallel and Distributed processing Symposium Workshop (2018)
 9. Mukunoki, D., Ozaki, K., Ogita, T., Imamura, T.: Accurate matrix multiplication on Binary128 format accelerated by Ozaki scheme (2021)
10. Utsugiri, Kouya: konsyumamuke GPU wo motiita 3baiseido (Triple Single) gyouretsuseki no seinouhyouka, HPC-182, pp. 1–8 (2021). (in Japanese)
11. Intel Corp.: Intel® oneAPI Math Kernel Library https://www.intel.com/content/www/us/en/developer/tools/oneapi/onemkl.html
12. Nvidia Corp.: Cuda toolkit documentation. https://docs.nvidia.com/cuda/cublas/index.html

International Workshop on Cities, Technologies and Planning (CTP 2022)

Fragile Territories Around Cities: Analysis on Small Municipalities Within Functional Urban Areas

Chiara Di Dato [✉] [iD] and Alessandro Marucci [iD]

DICEAA, University of L'Aquila, Via Giovanni Gronchi 18, 67100 L'Aquila, Italy
chiara.didato@graduate.univaq.it

Abstract. Although many disadvantaged areas are remote and isolated, the others are close to dense urban areas. Then, these peri-urban areas are both peripheral and fragile and, often, they are fringes from urban areas, where the well-known urban-rural mix occur.

In recent years, the most fragile urban areas have experienced growing attention. Through the years policymakers have tried to face territorial disadvantages with the help of funds.

Law 158/2017 is one example. The Law proposes twelve criteria to identify and then select small municipalities to support. In 2020 a research proposed to convert Law's criteria into indicators. Combinations of indicators can identify different conditions of fragility. These combinations are useful starting point for further studies on small municipalities.

The present work uses data from the mentioned research for focusing on peri-urban areas. The case study is composed of small municipalities within three Functional Urban Areas (FUA) in Central Italy: L'Aquila, Perugia and Terni. Then, the paper points out the relationship between indicators of fragility and peculiar features of these areas.

Keywords: Peri-urban · Urban Atlas · Computational analysis

1 Introduction

Nowadays, researchers in urban studies became familiar with considering even less valid the urban-rural dualism as a territorial feature. In fact, a great variety of factors have contributed to expand borders of urban areas. Among the others, there are demographic and economic dynamics. Often, uncontrolled expansion has generated the well-known dispersed settlements of which structural problems are still under investigation [1–3].

Those complex urban areas that are not compact cities neither rural areas can be defined 'territories-in-between' [4] and they are well-known as peri-urban areas. Peri-urban is defined as the area between urban settlements and their rural hinterland. Peri-urban can include town and villages within urban agglomeration, and they are characterised by complex pattern of land use and landscape [5].

O. Gervasi et al. (Eds.): ICCSA 2022 Workshops, LNCS 13378, pp. 427–438, 2022.
https://doi.org/10.1007/978-3-031-10562-3_30

The unsustainability of uncontrolled expansion is already evident [6]. In addition, as UN-Habitat advocates, peri-urbanization is an on-going process. Frequently, it implies growing land consumption unbound from real population growing [7, 8]. This is a main question in urban studies and researchers have implemented various methods for studying causes and consequences. As every process which involves land, peri-urbanization is a site-specific issue, and it varies depending on characteristics of geographical context [9–11].

However various approaches are valid for studying peri-urban areas, the attention should go to the power of urban policies in giving adequate support for territorial issues [12, 13]. Proper urban policies are necessary for recovering urban disease caused by dispersion.

Some examples from Italy may be useful. In recent years, various funds have addressed disadvantaged areas. Inner Areas (52% of Italian municipalities) are example of very specific and fragile territories to which the National Strategy for Inner Areas (SNAI) is addressed to [14]. Also, specific funds from National Recovery and Resilience Plan (PNRR) are allocated to 21 selected villages [15] to support sustainable development in less-favoured municipalities up to 5000 inhabitants with a recognisable historical centre.

Similarly, Law 158/2017 was born to give "Support measures for promotion of small municipalities and regulations for redevelopment and restoration of their historical centres". In Italy small municipalities (up to 5000 inhabitants) occupy around the seventy percent of the territory. A series of problems threaten these areas, such as the impoverish of natural resources and the abandonment of estates and artistic heritages.

The law aims to address most disadvantaged areas. In addition, it gives parameters that are interesting to analyze. Marucci et al. [16] have based research on Law 158/2017 for studying disadvantaged characteristics that conduct small municipalities to territorial fragility. The present work elaborates results of the research on a limited area. Then, it aims to add further consideration on typologies of urban settlement that characterize peri-urban areas. The study area offers diverse examples of urban-rural gradient. Then, the paper aims to start a discussion on results about fragility of peri-urban areas and urban settlements.

2 Case Study

The choice of the study area is linked to the LIFE IMAGINE UMBRIA Integrated Project. This LIFE Integrated Project aims to sustain integrated strategies for management of Nature2000 Network in Umbria Region [17]. Although the project refers basically to Umbria Region, the case study is here extended.

In particular, the paper aims to compare three areas from central Italy: Perugia, Terni, and L'Aquila. Central Apennines is the geographical context for the three cities. Perugia and Terni are in Umbria Region and L'Aquila is in Abruzzo Region. In recent years, this part of the Country has drawn attention because of the seismic events happened. Earthquakes of 2009 and 2016 have stroked this geographical area causing differential grades of damage, and then revealing high local level of fragility.

2.1 Functional Urban Areas

The study areas are Functional Urban Areas (FUA) in which the cities are included (Fig. 1). FUA are result of classification methodology produced by European Commission and the OECD (Organization for Economic Co-operation and Development). The data is available at Urban Atlas database [18], which provides pan-European comparable land cover and land use data for FUA. Elements in one FUA are one city (urban centre with at least 50000 inhabitants) and its communing zone (at least 15% of workers, who live in the commuting zone and work inside the city). Evidently, FUA is based on functional links between the city and suburban areas. In addition, it represents useful territorial unit for studying phenomena regarding urban centres.

2.2 FUA of Perugia, Terni and L'Aquila

FUA of Perugia extends for 1310,59 kmq with 281305 inhabitants, FUA of Terni has 1023,65 kmq for 171877 residents and FUA of L'Aquila extends for 1402,55 kmq with 98186 inhabitants. FUA of Perugia is the densest populated, whereas FUA of L'Aquila is the biggest in dimensions. FUA of Terni includes also three municipalities belonging to Rieti, in Lazio Region.

Within the three FUA mentioned, the analysis is conduct on municipalities up to 5000 inhabitants.

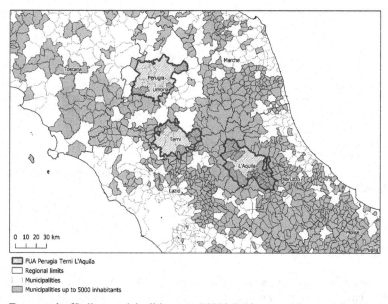

Fig. 1. Framework of Italian municipalities up to 5 000 inhabitants and the municipalities included in the FUA of Perugia, Terni and L'Aquila. The figure shows the extensive presence of small municipalities in the area considered (Central Italy).

The study area comprehends 39 municipalities (Fig. 2). Among these, 3 municipalities are part of FUA of Perugia (Valfabbrica, Bettona, Piegaro); 10 are part of FUA of Terni (Polino, Arrone, Stroncone, Ferentillo, Montefranco, Acquasparta, Avigliano Umbro in Umbria Region and then Configni, Labro, Colli sul Velino which are in Lazio Region) and 26 are part of L'Aquila FUA (San Demetrio Ne' Vestini, Prata D'Ansidonia, Pizzoli, Poggio Picenze, Castelvecchio Calvisio, Montereale, Navelli, Ocre, Barete, Barisciano, San Pio delle Camere, Sant'Eusanio Forconese, Scoppito, Lucoli, Acciano, Villa Sant'Angelo, Tornimparte, Tione degli Abruzzi, Caporciano, Carapelle Calvisio, Capitignano, Cagnano Amiterno, Calascio, Fagnano Alto, Fontecchio, Fossa).

The total area analysed occupies about 1550 kmq. Perugia's small municipalities extend for 236,19 kmq. In the case of Terni, they occupy 391,23 kmq and for L'Aquila they are 929,78 kmq.

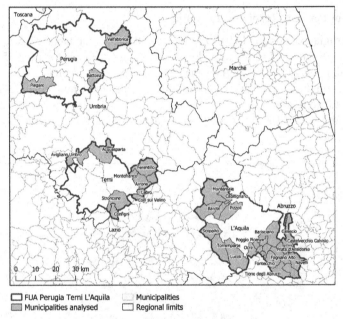

Fig. 2. Case study: Central Italy municipalities up to 5 000 inhabitants. The figure shows that the three FUA also comprehends municipalities not involved in the analysis.

3   Material and Methods

For the present work, analysis is conducted on the 39 municipalities mentioned. The analysis is based on results of research "Marginality assessment: Computational Applications on Italian municipalities" [16]. As mentioned, the research uses the parameters of Law 158/2017, which provides 12 criteria to identify areas affected by a disadvantaged economy, depopulation, and environmental issues.

The analysis has been conducted on 8 criteria and proper data are evaluated for each criterion. This selection has been considered acceptable also because the Law itself gives the same relevance to every criterion. Then, if every one of the criteria is equally valid for accessing the fund, also selection of them can be representative of the areas' fragility. The list of criteria used in the research:

1. Hydrogeological instability (IndA)
2. Economic backwardness (IndB)
3. Strong population decrease (IndC)
4. Residential density low than 80 inhabitants per kmq (IndG)
5. Municipalities that are part of a mountain community or municipalities that implement together the basic functions (IndI)
6. Municipalities partially or totally included in a national park, a regional park, or other protected areas (IndL)
7. Municipalities created by fusion (IndM)
8. Municipalities included in peripheral or ultra-peripheral areas, as defined by National Strategy for Inner Areas (IndN)

Criteria analyses is done with open-source software (QGis). The base is data set of Italian municipalities (updated in 2017, the same year of the Law considered). It is integrated with the additional spatial or tabular data mentioned. Through this database every municipality receives one value per each criterion. The value is 1 if the municipality meets the reference criterion or value can be 1 if there is correspondence. The result are the sums of values each municipality got.

Then, the analysis of results obtained is processed using MATLAB software. The analysis permit to find 255 classes of possible combinations among the 8 criteria [16].

4   Results

Results are resumed in two sections. The first section highlights the most shared criteria among the municipalities analysed. Frequency is useful to detect the relevance of each criterion for the specific area. The second section highlights combinations of criteria. These sequences give the general picture of disadvantage condition in the study area.

4.1   Frequencies of Criteria

As evidenced by numerical correspondence of indicators (Fig. 3), there is a correspondence with criteria related to low residential density (IndG) and hydrogeological instability (IndA) in 33 municipalities out 39. Moreover, more than half of municipalities are affected by strong economic backwardness (IndB). Presence of half of the municipalities inside a national park or other protected areas (IndL) attests environmental relevance of the area considered. Also, population decrease (IndC) affects almost half of the municipalities. The remaining indicators are less significant.

As expected, correspondence with peripheral and ultra-peripheral areas (IndN) affected only few municipalities out 39 (the 3 municipalities Acciano, Tione degli Abruzzi, Calascio, all within the FUA of L'Aquila). In this sense, the proximity of study areas to the cities is determinant. In fact, the municipalities analysed are peri-urban areas in the fringe between high urbanised centres and more peripheral areas.

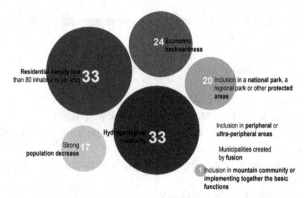

Fig. 3. Bubble chart of frequencies of each criterion. The chart shows how many times each criterion appears within the case study. Bigger dimensions and darkness colours suggest the prominence of the criteria in the area.

Focusing on singular municipalities, only one municipality (Bettona) results in lack of correspondence to the criteria analysed (Fig. 4). Indeed, the FUA of Perugia has only three cases of small municipalities. The other two, Piegaro and Valfabbrica, have respectively 2 and 3 correspondences to the criteria.

FUA of Terni presents a mixed result, in which municipalities go from 1 (Labro) to 5 (Polino) criteria met.

Instead, municipalities in the FUA of L'Aquila proved to be the most affected by the criteria of disadvantages analysed. 14 municipalities out 26 have correspondence to more than 4 criteria (Acciano, Tione degli Abruzzi, Calascio, Castelvecchio Calvisio, Montereale, Navelli, Barete, Barisciano, Lucoli, Carapelli Calvisio, Capitignano, Cagnano Amiterno, Fognano Alto, Fontecchio). In the worst cases, Calascio and Tione degli Abruzzi have respectively 7 and 6 criteria.

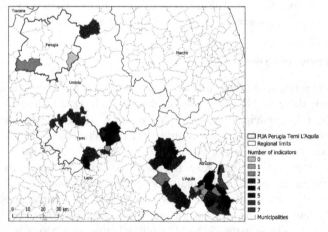

Fig. 4. Numbers of criteria met for each municipality

4.2 Sequences of Criteria

Various combinations occur considering specificity of criteria involved (Fig. 5). Among all the possible combinations, 19 sequences characterize the 38 municipalities (Bettona excluded). However, the conditions are heterogeneous because about half of combinations are isolated cases (11 combinations are valid only for one municipality). In other words, a quarter of the total municipalities observed has a specific condition of disadvantage generated by a singular combination of criteria.

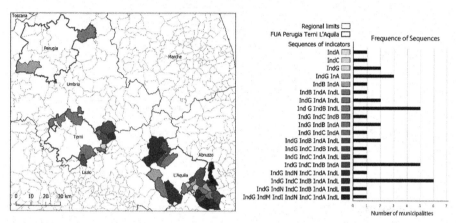

Fig. 5. Sequences of criteria. Map shows sequences for each municipality, going from less disadvantaged conditions (1 criterion, yellow coloured) to the most disadvantaged condition (7 criteria, dark red coloured). Bar graph shows the frequency of each sequence within the case study. Frequencies goes from 1 municipality to 6 municipalities involved. (Color figure online)

The sequence which has the highest incidence is IndG IndC IndB IndA IndL (Table 1). The 6 cases are almost in FUA of L'Aquila (5 municipalities: Castelvecchio Calvisio, Montereale, Lucoli, Cagnano Amiterno, Fagnano Alto) and then 1 case is in FUA of Perugia (Polino).

The following sequences with high incidence are IndG IndB IndA IndL and IndG IndB IndA. In detail, for IndG IndB IndA IndL more than half are in FUA of L'Aquila (Fontecchio, Barisciano, Barete), whereas the other two are in FUA of Perugia (Ferentillo, Arrone).

For the sequence IndG IndB IndA most cases are in FUA of Terni (Stroncone, Avigliano Umbro, Acquasparta), one case is in FUA of Perugia (Valfabbrica) and one other in FUA of L'Aquila (Tornimparte).

In particular, the three criteria IndG IndB IndA are present in the three most shared combination. Low residential density, economic backwardness, and hydrogeological instability are the main causes of disadvantage in the area considered.

Table 1. Detail of the 3 most frequent sequences

Sequences		Detail of Sequences	Incidence
IndG IndC IndB IndA IndL	IndG	Municipalities characterised by a residential density low than 80 inhabitants per kmq	6
	IndC	Municipalities affected by strong population decrease compared to the general census of 1981	
	IndB	Municipalities affected by strong economic backwardness	
	IndA	Municipalities included in areas affected by hydrogeological instability	
	IndL	Municipalities partially or totally included in a national park, a regional park or other protected areas	
IndG IndB IndA IndL	IndG	Municipalities characterised by a residential density low than 80 inhabitants per kmq	5
	IndB	Municipalities affected by strong economic backwardness	
	IndA	Municipalities included in areas affected by hydrogeological instability	
	IndL	Municipalities partially or totally included in a national park, a regional park or other protected areas	
IndG IndB IndA	IndG	Municipalities characterised by a residential density low than 80 inhabitants per kmq	5
	IndB	Municipalities affected by strong economic backwardness	
	IndA	Municipalities included in areas affected by hydrogeological instability	

5 Discussion

5.1 Local Disadvantaged Conditions

Although FUA share the same geographical context, some differential conditions can explain the results obtained. As seen, small municipalities in FUA of L'Aquila are the most numerous and the most fragmented. In these areas, frequency of criteria met is higher. Disadvantages relate both to loss of social opportunities and to fragile territories. Disconnection and isolation of the area could be one main reason. Indeed, some observations emerge in comparison with the other two cases analysed.

Firstly, Umbria's FUA are more connected to the nearest Regions, such as Lazio and Toscana. They are in a dense road network that crosses Central Italy from north to south. Between the two, Perugia is the highest connected. Moreover, its territory is the most densely urbanized among the three compared. Instead, Terni represents a middle ground. It has direct infrastructural connection between Umbria, Abruzzo, and Lazio Regions, but it has also environmental quality's areas in its surroundings. Differently, natural areas are the main feature of FUA of L'Aquila. Its infrastructural network is composed mainly by one axis (A24), which connects to Lazio Region at west and to coastal cities at east.

In fact, FUA of L'Aquila is immersed in a mountain system. This morphological condition deeply impacts on disadvantages, as the analysis has shown. Difficult connections and absence of diffuse services on territory can cause lack of economic development. Through the years, these aspects have led to concentrate resources in the city centre of L'Aquila at the expense of an even less developed surroundings. Comparing to the other two, FUA of L'Aquila has the highest level of fragility.

In this framework, further interesting data are about seismic risk. Some municipalities are part of the highly damaged areas in the recent earthquakes in Central Italy and they are mainly concentrated in L'Aquila or in its proximity. In detail, 9 of them are part of the so-called Seismic Crater: 5 municipalities are in FUA of L'Aquila (Barete, Cagnano Amiterno, Capitignano, Montereale, Pizzoli) and 4 are in FUA of Terni (Arrone, Ferentillo, Montefranco, Polino). This further criterion of disadvantage weighs on the overall fragile condition of these areas. This means also that these municipalities have received extra funds for recovering and that they are involved in plans for reconstructions. Evidently, specific and different funds are welcome.

However, diverse funds focalize on diverse territorial disadvantages, whereas functional planning strategies require organic and long-term vision. A structural assess of all fragilities and then a comprehensive and coherent territorial policy would be preferable.

5.2   From Disadvantaged Areas to Urban Settlements

The current work presents some characteristics related to fragility and disadvantages. It has been possible through the previous research on disadvantages in small municipalities. In fact, the reference analysis settles a framework in which each municipality fits. Disadvantages analysed cover mainly territorial aspects, economical background, and population dynamics. These results represent a coarse-grain layer of information. In fact, this categorization gives the overall picture of local urban conditions (Fig. 6).

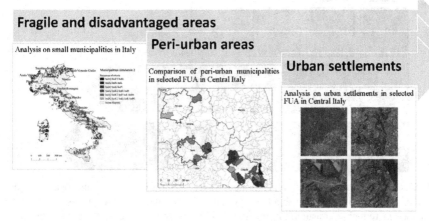

Fig. 6. The diagram shows phases in the research's workflow. It has started from the analysis on disadvantaged areas at national scale. The intermediate step is represented by the analysis on FUA of Perugia, Terni, L'Aquila. The third part on urban settlements is going to be developed.

Other types of analysis are needed for deepening further aspects. Indeed, peri-urban areas comprehend a series of elements readable through a more detailed scale. As instance, main differences in classification of peri-urban settlement regard context, dimensions, shape, and structure.

In future steps of the research, a first approach will consist in setting a window of defined dimensions for catching urban settlements at the same scale.

For this purpose, Urban Atlas will be used as base for finding recurring peri-urban settlement and then to classify them in categories. Of course, this tool is not sufficient, and it will be implemented, as instance with satellite images. Then, a fine-grained analysis on aspects at a detailed scale will be needed.

The on-going research is part of LIFE IMAGINE UMBRIA Project. Purposely, extracting models of peri-urbanization is one of the project's goals. Although in preliminary phase, the reasoning about connections of different scales and approaches seemed interesting to develop at this point.

One other aspect of the reasoning is linked to sustainability related to urban context. Indeed, among the other causes, urban structure and dispersion deeply impacts on land management [19]. This research opens also to comparisons in measuring settlements' performances. In perspective, this may be useful for evaluating sustainability of settlements (Fig. 7).

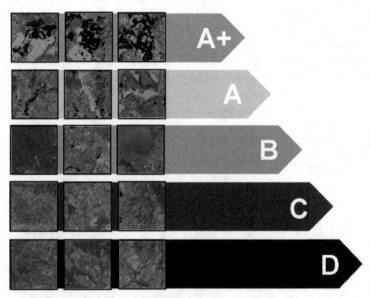

Fig. 7. Comparison among diverse types of land transformation in terms of sustainability. The figure refers to the typical scheme about energy class of buildings. It goes from cases of higher-performance in reducing consumption of land (A+) to the worst case in terms of dispersion (D).

6 Conclusion

As shown, some conditions of peri-urban areas are comprehensible considering limits of commuting zones as classified from Urban Atlas. The unit of FUA has been functional to investigate similar territorial conditions for different urban areas. Moreover, the case study represents an interesting subject of further analysis. Then, considering a larger part of the related geographical area (Central Italy) may be desirable.

The case study has highlighted that local gradient of fragility can be combined with other analyses even if different in approach. In fact, the work is a first-step analysis for approaching multi-spectral problems of peri-urban areas. Forthcoming research will focus mainly on assets of urban settlements. In this perspective, the use of FUA is also functional for extending the comparisons on European territory, by virtue of urban information available in the Urban Atlas database.

Acknowledgments. The analysis for characterizing urban and peri-urban areas thought the use of Functional Urban Areas is a work developed within Integrated Project LIFE IMAGINE UMBRIA (LIFE19 IPE/IT/000015 - Integrated MAnagement and Grant Investments for the N2000 NEtwork in Umbria).

References

1. Cattivelli, V.: Methods for the identification of urban, rural and peri-urban areas in Europe: an overview. J. Urban Regen. Renew. **14**(3), 240–246 (2021)
2. Shaw, B.J., van Vliet, J., Verburg, P.H.: The peri-urbanization of Europe: a systematic review of a multifaceted process. Landsc. Urban Plan. **196** (2020). https://doi.org/10.1016/j.landurbplan.2019.103733
3. Romano, B., Zullo, F., Fiorini, L., Ciabò, S., Marucci, A.: Sprinkling: an approach to describe urbanization dynamics in Italy. Sustain. (Switz.) **9**(1) (2017). https://doi.org/10.3390/su9010097
4. Wandl, A., Rooij, R., Rocco, R.: Towards sustainable territories-in-between: a multidimensional typology of open spaces in Europe. Plan. Pract. Res. **32**(1), 55–84 (2017). https://doi.org/10.1080/02697459.2016.1187978
5. Piorr A., Ravetz J., Tosics I.. Peri-urbanisation in Europe: Towards a European Policy to sustain Urban-Rural Futures. University of Copenhagen/Academic Books Life Sciences (2011). 144 p. ISBN: 978-87-7903-534-8
6. Geneletti, D., la Rosa, D., Spyra, M., Cortinovis, C.: A review of approaches and challenges for sustainable planning in urban peripheries. Landsc. Urban Plan. **165**, 231–243 (2017). https://doi.org/10.1016/j.landurbplan.2017.01.013
7. UN-Habitat: State of the World's Cities 2012/2013. Prosperity of Cities. Routledge (2013). 208 pages. ISBN: 978-04-1-583888-7
8. UN-Habitat: World Cities Report 2020 (2020). ISBN: 978-92-1-132872-1
9. Gottero, E., Cassatella, C., Larcher, F.: Planning peri-urban open spaces: methods and tools for interpretation and classification. Land **10**(8) (2021). https://doi.org/10.3390/land10080802
10. Inostroza, L., Hamstead, Z., Spyra, M., Qhreshi, S.: Beyond urban–rural dichotomies: measuring urbanisation degrees in central European landscapes using the technomass as an explicit indicator. Ecol. Ind. **96**, 466–476 (2019). https://doi.org/10.1016/j.ecolind.2018.09.028
11. Serra, P., Vera, A., Tulla, A.F., Salvati, L.: Beyond urban-rural dichotomy: exploring socioeconomic and land-use processes of change in Spain (1991–2011). Appl. Geogr. **55**, 71–81 (2014). https://doi.org/10.1016/j.apgeog.2014.09.005
12. Romano, B., Fiorini, L.: Abbandoni, costi pubblici, dispersione. Alla ricerca di risposte migliori. Sentieri Urbani **26**, 66–73 (2018). ISSN: 2036-3109

13. Fiorini, L., Zullo, F., Marucci, A., Di Dato, C., Romano, B.: Planning Tool Mosaic (PTM): a platform for Italy, a country without a strategic framework. Land **10**(3) (2021). https://doi.org/10.3390/land10030279

14. National Strategy for Inner Areas (SNAI). https://www.agenziacoesione.gov.it/strategia-nazionale-aree-interne/la-selezione-delle-aree/. Accessed 11 Apr 2022

15. Ministero della Cultura, Investimento 2.1 "Attratività dei borghi". https://cultura.gov.it/borghi. Accessed 11 May 2022

16. Marucci, A., Fiorini, L., Di Dato, C., Zullo, F.: Marginality assessment: computational applications on Italian municipalities. Sustain. (Switz.) **12**(8) (2020). https://doi.org/10.3390/SU12083250

17. Integrated Project LIFE IMAGINE UMBRIA (LIFE19 IPE/IT/000015 - Integrated MAnagement and Grant Investments for the N2000 NEtwork in Umbria). https://www.lifeimagine.eu/. Accessed 11 May 2022

18. Copernicus, Urban Atlas. https://land.copernicus.eu/local/urban-atlas. Accessed 11 Apr 2022

19. United Nations, the 2030 Agenda for Sustainable Development. With reference to "Goal 11: Make cities and human settlements inclusive, safe, resilient and sustainable" and to "Goal 14: Conserve and sustainably use the oceans, seas and marine resources for sustainable development". https://sdgs.un.org/goals. Accessed 11 May 2022

Assessing Coastal Urban Sprawl and the "Linear City Model" in the Mediterranean – The Corinthian Bay Example

Apostolos Lagarias(✉) 🄳, Ioannis Zacharakis, and Anastasia Stratigea🄳

Department of Geography and Regional Planning, School of Rural, Surveying and Geoinformatics Engineering, National Technical University of Athens, Athens, Greece
lagarias@iacm.forth.gr, stratige@central.ntua.gr

Abstract. Urban sprawl and tourism urbanization, as prevailing trends in the Mediterranean coast, result in particular forms of 'linear' urban development, stretching with deployment of low-density urban fabric along extended areas near the shoreline. The evolving 'linear city model' accounts for: land and marine environmental degradation; higher vulnerability to climate change; and unsustainable future pathways of coastal urban constellations. Assessment of this model of urban development in coastal zones can reveal distinct spatial and functional irregularities and properly guide policy remediation action. This work elaborates on the exploration and identification of the "linear city" concept in the Mediterranean by use of a methodological approach that integrates: high-resolution multi-temporal data for built-up areas and their GIS-enabled elaboration; spatial metrics for quantifying morphological as well as spatial peculiarities and qualities of this linear city type; Principal Component and Cluster Analysis, unveiling built environment typologies; and correlation analysis, illuminating functionality weaknesses by linking these typologies with urban variables, e.g. population density, accessibility to transport and urban facilities. Implementation of this approach on a Greek example – Corinthian Bay, Northern Peloponnese – witnesses the discrete spatial typologies and highlights the fragmented and rather unsustainable, in spatial and functional terms, linear urban pattern of this coastal urban area.

Keywords: Coastal urban sprawl · Linear city model · Spatial metrics · Spatial planning and policy · Mediterranean coast

1 Introduction

The Mediterranean Region is perceived as a distinct part of the world, mainly due to its exceptional natural peculiarities, among which fall the extraordinary coastal and insular ecosystems and complexes as well as the 46,000 km of coastline, endowed with fabulous land and seascapes [1–4]. Concurrently, this Region is characterized by its distinguishable cultural diversity and wealth, being the result of its location at the crossroad of three continents; and the intense, back to ancient world, cultural and commercial interaction of eastern and western civilizations through the Mediterranean Sea routes [1, 2].

© The Author(s), under exclusive license to Springer Nature Switzerland AG 2022
O. Gervasi et al. (Eds.): ICCSA 2022 Workshops, LNCS 13378, pp. 439–456, 2022.
https://doi.org/10.1007/978-3-031-10562-3_31

At the heart of the Mediterranean historical trajectory lays its *coast*. Thus Mediterranean coastal regions are, through centuries, grasped as the most flourishing, prosperous, and preponderant places; and an appealing home to large population shares [1]. This, in turn, is reflected in the escalating population trends witnessed in Mediterranean regions [5]. In fact population in the Mediterranean Region has increased from approximately 419 million inhabitants in 2000 to 475 million in 2010 (+13,4%) and 529 million in 2020 (+11.4%) [6, 7]; while is predicted to reach the 572 million inhabitants by 2030 (https://www.medqsr.org/population-and-development). Additionally, a certain population concentration in the Mediterranean coastal part and especially coastal cities is noticeable [8, 9], since almost one third of the population of the states surrounding the Mediterranean Sea resides in the coastal zone and more than 70% of it in coastal cities [8].

Currently, coastal areas in the Mediterranean are highly rated, extremely attractive and globally acknowledged *tourism destinations* due to their mild climate, remarkable natural/cultural assets and the warm hospitality/spirit of the Mediterranean people. In fact, Mediterranean coastal regions host almost one third of the international tourism (32%), rising from 58 million of international tourist arrivals in 1970 to 314 million in 2014, a number that is expected to reach the 500 million tourists in 2030 (an increase of 59%, compared to 2014). The overarching coastal mass tourism model in the Mediterranean shoreline, however, coupled with the steadily increasing pressure from the intense habitation and economic activity pattern, leads to an immense demand for urban land in coastal zones, largely affecting the land- and sea-scape state [9, 10]; and severely endangering sustainability performance of relevant coastal territories. The growing urbanization due to habitation and economic activity (tourism included) [9], coupled with the repercussions of Climate Change (CC) in coastal areas, create an explosive mixture that renders coastal urban areas quite *vulnerable and fragile*. In fact, Mediterranean region as a whole is currently perceived as a *hotspot* in many respects [2] – all touching one way or another future developments of coastal urban constellations – namely a CC hotspot [11, 12]; a biodiversity hotspot [13]; an urbanization hotspot [2]; a sea transportation hotspot [11]; and a tourism hotspot [14], to name but a few.

Within such a decision environment, the study of urban development and/or sprawl patterns in Mediterranean coastal zones calls for articulating proactive and evidence-based *policy responses* to diverse risks, e.g. overpopulation, overtourism, biodiversity loss of land and marine environment, CC. That said, the *aim* of this research is to explore and illuminate aspects of the form/structure of the *linear city model* in Mediterranean coastal areas as the output of, among others, coastal intense urbanization processes. Therefore, a *geospatial methodology* is developed for quantifying density, dispersion, connectivity and accessibility of coastal urban contexts that bear attributes of a distinct 'linear city' model. The proposed methodology is implemented in Corinthian Bay in Greece, an area that is marked by a tough coastal urban sprawl pattern, displays a coastal-hinterland urban divide and is categorized as a typical "linear city" model. Urban sprawl patterns in the Corinthian coastal zone are analyzed at a disaggregated level, using high-resolution Global Human Settlement Layer data for 2017–2018 (GHS-BUILT-S2); and spatial metrics are applied to identify urban pattern typologies.

The *structure* of the paper has as follows: in Sect. 2 the emergence of the 'linear city' model as the prevailing urban form in the Mediterranean coast is shortly discussed;

Sect. 3 summarizes the proposed methodological approach, data sets, spatial metrics and methods used in order for the coastal urban form/development to be explored; Sect. 4 implements this approach in the specific case study of Corinthian Bay and comments on results obtained; while in Sect. 5 some key conclusions are drawn.

2 Evolving Linear City Patterns in the Mediterranean Coast

Apart from the physical characteristics as well as the social, environmental and economic processes, the *urban form* constitutes a key determinant in pursuing sustainability objectives in respective areas [15]. This is not only related to the repercussions of the urban form on mobility patterns [16], but also to the fact that shape, size, density and settlements' configuration as well as land use patterns have direct environmental, economic and social repercussions [17]. In urban planning theory, specific prototypes of the urban form – e.g. *compact, polycentric, linear and sprawl urban forms* – have been assessed and studied so far [15], with the urban sprawl model being generally regarded as an unsustainable one.

As a means to restrain urban sprawl, the *compact city model* has been proposed by various researchers. This model is delineated by higher built-up and population densities, land use diversity and sustainable mobility that favors public transport, walking and cycling [18]. 'Compactness' is also defined as high-density or monocentric development [19]; while Anderson et al. [20] suggest that both monocentric and polycentric forms can be grasped as compact ones. Compact cities are currently perceived as sustainable urban forms, as opposed to the urban sprawl model. This is due to their potential to contain soil sealing and loss of surrounding natural and agricultural land as well as reduce the amount of travel, car dependency, energy consumption and related air pollution [15]. However, to their detriment fall the high population densities that can lead to traffic congestion, poor environmental quality and lack of adequate open space.

Among the urban form typologies, the *linear city model* constitutes a particular and interesting case. From Soria y Mata's conceptualization of the linear city in the 19th century [21] to Gauthier's long mega-structures [22], linear cities have often been regarded as an 'ideal' type of urbanization, providing direct access to natural landscape and favoring a transit-based human settlement around major transport infrastructures. Linear urban development patterns are largely observed in coastal urban areas, taking the form of ribbon sprawl along *major roads/transit infrastructure* [23] such as the case of *coastal tourism destinations* [10, 24]. This form, however, usually defined as *coastalization* [25], mostly ascribes a linear and largely unsustainable urban development type, encroaching along the coastline. It is generally associated with uncoordinated forms of low-density sprawl, consuming agricultural and natural land; while is the source of severe environmental degradation. It is mostly driven by increasing population density in coastal zones; suburbanization; mass-tourism installations' investments, coupled with the spreading of vacation houses; and deployment of transport infrastructure in coastal zones.

In the Mediterranean Region, *coastalization* is identified as an increasingly alerting phenomenon that is heavily affecting its vulnerable land and marine ecosystems [25]. Historically, the dense concentration of urban settlements along the coastal zone is a prevalent feature of the Region's landscape, taking the form of port towns as hubs of socio-economic activity; and serving intense economic/commercial transactions through the well-established ancient Mediterranean Sea routes. Recently, however, urban growth in coastal zones is no longer restricted around the main coastal cities. Additionally, the lack of a consistent and regulatory spatial planning framework, coupled with the massive urbanization investments of the past few decades, have led to extensive urban sprawl in the Mediterranean coastal zone [10, 26]. This, in turn, poses severe threats to coastal and marine ecosystems, consumes resources in a rather unsustainable way and increases vulnerability to CC [27, 28].

Coastalization in the Mediterranean Region is, among others, the outcome of the quite notable large-scale *mass tourism development*, calling for the deployment of tourist accommodation, leisure facilities and transport infrastructure, to name a few. Such a tourism-related developmental trajectory is further stressed by the increase of second homes/vacation houses [24, 29]; the new suburban style, observed in compact Mediterranean cities since late '80s [30]; and the so-called residential tourism phenomenon, i.e. mass movement of Northern Europe retirees, migrating to the sunny and hospitable European south [31]. The largely unplanned or poorly regulated character of settlements' expansion, accompanied by the large-scale infrastructure deployment in coastal and periurban areas [32] have so far been the key drivers of the linear urban expansion and growth of the Mediterranean economies [30]. Quite exemplar cases of this trend constitute the highly rated tourism destinations in the Mediterranean, e.g. in Spain, Italy, Greece and Turkey. Distinct examples are the *linear city model* along the Spanish coast [33]; the *Italian Adriatic Coast*, described as the longest – more than 1470 km – urban stretch in southern Europe [34]; the *Crete linear pattern* of urban development as the prevailing one in the northern, heavily based on tourism, coastline of the island [10]; to name but a few. In the subsequent sections, a methodology is articulated in order for the (linear) urban sprawl patterns at a disaggregated level to be quantified and assessed for properly guiding planning and policy action.

3   Data and Methods

The steps of the proposed methodological approach for assessing urban sprawl patterns and built environment typologies in coastal regions are shown in Fig. 1.

More specifically, urban sprawl patterns in the coastal zone are studied by use of high-resolution data (10 m resolution), based on Sentinel-2 images for the period 2017–2018 (GHS-BUILT-S2). Thus, the structural and morphological attributes of urban spatial patterns can be assessed. GHS built-up grid (GHS-BUILT-S2 R2020A) corresponds to a global map of built-up areas, expressed in terms of a probability grid at 10 m spatial resolution. Built-up probability values are rescaled in the range 0–100 (0 for probability $= 0$ and 100 for probability $= 1$). Dataset is delivered as tiff raster files, while each tile inherits the projection of the UTM grid zone to which it belongs to. A binary version of the raster is constructed by setting a 0.2 threshold point to binarize the probabilistic output (built-up probability values). A cell is defined as "built-up" if raster value > 20, while otherwise is defined as "non built-up".

Density, dispersion, connectivity and accessibility of urban compartments are quantified and evaluated at a disaggregated/local level by use of *spatial metrics*. The latter are

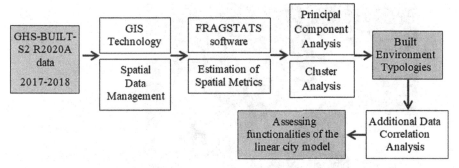

Fig. 1. Steps of the proposed geospatial methodological approach.

estimated at the urban patches level, i.e. discrete areas that are homogenous as to land use, habitat type, etc.; and are identified based on a binary raster grid, using a developed/non-developed land dichotomy approach. GHS-BUILT data are processed in the ArcGIS environment, reclassified and exported as GeoTIFF grids in FRAGSTATS software. Area/density/edge, shape, proximity, contagion/diversity and connectivity metrics are estimated by use of FRAGSTATS. Uniform tiles, with a side-length of 1 km, are used to calculate spatial metrics at a disaggregated/local level. Using the above steps, areas mostly affected by urban sprawl (low density, disconnected development etc.), can be identified; and the morphological and spatial peculiarities of this linear urban sprawl type can be assessed.

Following the estimation of the aforementioned spatial metrics, Principal Component Analysis (PCA) is used as a means to end up with a set of *explanatory factors*; and is followed by a Cluster analysis for assessing different built environment typologies. At a final step, the previously ascertained spatial patterns and typologies are linked to data regarding urban facilities, accessibility to transportation networks and major urban towns/settlements, population density and geomorphology. Additional data sources are used, including *Open Street Map*, the Greek official *geodata portal* (geodata.gov.gr) and data from the European Environmental Agency; while *correlation analysis* is applied for grasping functionality aspects of the linear city model.

The application of the above described methodological steps in the Corinthia coastal zone is presented in the following.

4   The Corinthian Case Study

4.1   The Study Region

The study region – coastal front of Corinthia Regional Unit (RU) (Fig. 2) – is located in the north-eastern part of Peloponnese, at a distance of 80 km from the Athens metropolitan area. In 2011, the RU of Corinthia counts for 145,082 habitants, while covers a surface of 2,297 km². Corinthus is the administrative center and a major urban area, located along the national road axis that connects Athens to Patras. Corinthia's coastal front is surrounded by two significant gulfs – Corinthian and Saronic – connected through the Isthmus canal.

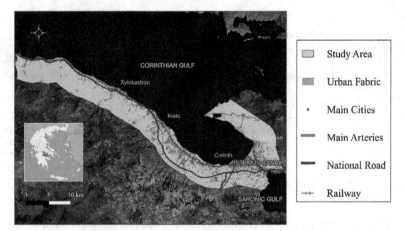

Fig. 2. The study area – a 5 km coastal zone in Corinthia RU.

The region's geomorphology demonstrates a great variety, ranging from a rough terrain, endowed by distinct mountainous ecosystems in the western part; to a smoother one along the eastern part of the Isthmus canal area. Cultural resources in the study area are abundant; same holds for natural ones, featured by four protected NATURA 2000 areas – both in land and sea – among which falls the Corinthian Gulf as a whole [35]. The proximity of the study area to the national road and railway networks – part of them unfolds across its coastal zone – is critical for the RU's geography. Tourism infrastructure includes hotels and Airbnb rooms/apartments, spreading along the coastal zone. The coastal front is marked as an Airbnb hotspot due to the extended offer of relevant accommodations [36]. Internal tourist flows prevail, with the tourism sector playing a significant role in the local economy and employment. Heavy industry facilities (Seveso areas) are also located in the eastern part of the Isthmus Canal.

In the study region, the concept of the *'linear city model'* is explored by elaborating on a 5 km zone adjacent to the shoreline (Fig. 2). This decision is justified by the fact that in the gradient of built-up density along distance from coast the curve is practically flattening beyond this distance [37].

4.2 Built Environment Typologies and 'Linear City' Pattern in Corinthian Bay

The previously presented methodological approach is applied in the Corinthian Bay case study. Towards this end and based on the GHS-BUILT-S2 R2020A data for 2017–2018 [38], a set of spatial metrics is firstly estimated, in order for built-up areas along the Corinthian Bay to be analyzed. The *spatial metrics* used are as follows:

- COV (Coverage): the sum of the areas (in ha) of all patches expressed as percentage (%) of total landscape area.
- PD (Patch Density): the number of patches divided by the total landscape area.
- ED (Edge Density): The total length (m) of patch edges, divided by total landscape area.

- MPS (Mean Patch Size): The mean area of patches.
- LPI (Largest Patch Index): The area of the largest patch, divided by the total landscape area.
- LSI (Landscape Shape Index): The sum of the landscape boundary and all edge segments (m) of patches within the landscape boundary, divided by the square root of the total landscape area (m^2), adjusted by a constant for square standard (raster).
- GYRATE (Gyration): The mean distance (m) between each cell in the patch and the patch centroid. This metric is estimated as "mean" value for all parches (mn) and "area-weighted" (am), based on the area of each patch.
- ENN (Euclidean Nearest neighbor, mn): The mean Euclidean distance to the nearest neighboring patch, calculated as the shortest edge-to-edge distance.
- CONNECT (Connectance): A measure of connection between patches, divided by the total number of possible connections among them. In this study, threshold distance is set to 200 m.
- PROX (Proximity): The sum of patch area, divided by the nearest edge-to-edge squared distance between the patch and the focal patch of all patches within a threshold distance (in this study set to 1000 m) of the focal patch. PROX is area-weighted (am) in this study, so that larger patches in proximity to be accounted for.

Uniform tiles of 1 × 1 km grid are applied; tiles mainly covered by sea (over 75%) are excluded. Overall, *408 valid tiles* are processed and summary statistics for spatial metrics are reported in Table 1.

Table 1. Summary statistics for spatial metrics based on tile analysis (1 × 1 km grid)

Spatial metric	Mean	Std. deviation	Minimum	Maximum
COV	7.04	10.73	0.01	78.3
PD	45.05	35.56	1.0	157.0
ED	87.88	93.74	0.40	441.4
LSI	7.24	3.96	1.0	17.1
LPI	3.45	9.19	0.01	76.9
MPS	0.17	0.55	0.01	8.7
GYRATE (mn)	11.06	5.51	5.0	69.9
GYRATE (am)	45.04	58.08	5.0	367.3
ENN (mn)	55.05	71.56	0.0	1052.6
CONNECT	28.11	21.95	0.0	100.0
PROX (am)	11.84	24.89	0.0	242.0

Results obtained demonstrate considerable variations in the spatial distribution of built-up areas along the coastal zone of Corinthia Bay. COV and ED high values are concentrated in the proximity of Corinthus town (from Isthmus to Kiato area – Fig. 2) and are spanning along a thin strip in the coastline zone; while high PD values spread towards the hinterland. In areas further inside from the coast, COV is less than 15, indicating development of very low-density (Fig. 3). LPI is maximized in the central zone of Corinthus and Kiato towns and is relatively low (<10) in all other tiles. A different pattern is observed in the distribution of GYRATE (mn) values, namely high values are spread in a larger inland zone and in proximity to major transportation infrastructure, where large elongated clusters tend to form. GYRATE metric is generally estimated higher for large irregular/elongated patches.

PCA analysis is used to reduce a number of variables to a minimum set of factors and obtain synthetic indicators, taking into account the variables assessed by spatial metrics. Towards this end, Varimax rotation with Kaiser Normalization is used in this work, while the KMO (Kaiser–Meyer–Olkin) and communalities indexes are also estimated. The indexes used for the evaluation of the PCA analysis show a good performance. KMO was estimated equal to 0.773, while communalities passed the test of 0.5. Bartlett's test of sphericity presents a small value of significance level (less than 0.01), therefore the hypothesis that correlation matrix is an identity matrix is rejected.

A total of three factors – Factor 1 (F1), Factor 2 (F2) and Factor 3 (F3) – are obtained (Table 2), with eigenvalues >1, accounting for 85.2% of the total variance of the initial variables. In F1, the COV, MPS, LPI, GYRATE (am) and GYRATE (mn) metrics are loaded on the positive axis; therefore, F1 is related to *high levels of patch aggregation, high density/coverage and complex/elongated patch shapes*. In F2, the PD, ED, LSI, PROX (am) are loaded on the positive axis; and CONNECT on the negative axis. Therefore, F2 is related to *high patch density with complex edge formation, high proximity and low connectivity*. In F3, ENN is loaded on the negative axis; therefore, F3 is related to *low levels of patch dispersion*.

To explore the "linear city" model and further analyze built-up typologies in the case study area, *Cluster Analysis* is performed using the three factors identified by the PCA (F1, F2 and F3). Hierarchical cluster, using Wards' method and squared Euclidean distance, is applied. A range of options regarding the number of clusters (5 to 7) is examined; and 7 clusters are selected as the optimal choice. Summary statistics for the spatial metrics per cluster are tabulated in Table 3.

Mean Factor values (F1, F2 and F3) per cluster group are displayed in Table 4. Inspection of results show that *Clusters 1* and *3* are the most important typologies identified, counting for 202 and 142 tiles respectively. *Cluster 1* is characterized by small (negative) average F1 values, therefore these tiles present low coverage, low patch aggregation and relatively simple patch shapes; while F3 is negative, a fact related to relatively high distances between built-up patches. *Cluster 3* is characterized by very high (positive) average F2 values (Table 4), therefore these tiles present high patch and edge complexity and low connectivity as well as patch aggregation and coverage close to the mean; while F3 is positive, a fact related to short nearest-neighbor distances between built-up patches.

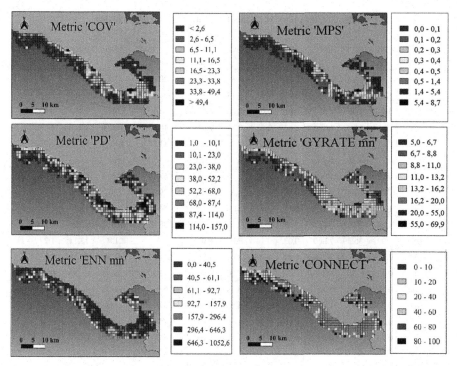

Fig. 3. Maps of COV, PD and ENN (mn) metrics and distribution of respective values along the 1 × 1 km tiles

Table 2. PCA results (rotated component matrix)

	F1	F2	F3		F1	F2	F3
LPI	0.961			LSI		0.898	
COV	0.908	0.366		ED	0.669	0.683	
MPS	0.885			CONNECT		−0.655	0.433
GYRATE (am)	0.880	0.308		PROX (am)	0.479	0.518	
GYRATE (mn)	0.856			ENN (mn)			−0.893
PD		0.904					

The next three important typologies are *Cluster 2* (26 tiles), *Cluster 6* (20 tiles), and *Cluster 4* (10 tiles). *Cluster 2* is characterized by negative F1 and F2 values, presenting low coverage, patch density and mean patch size; and very high (positive) F3 values. *Cluster 6* scores high on F1 and very high on F2 values, therefore patch and edge density in these areas is high, combined with high coverage and patch aggregation. In *Cluster 4*, all Factor values are negative, with F3 scoring very low on the average, a fact related to very high nearest-neighbor distances between built-up patches. *Cluster 7*

Table 3. Summary statistics for spatial metrics per cluster

Cluster		COV	PD	ED	MPS	LPI	LSI	GYRATE (mn)	GYRATE (am)	ENN	PROX	CONNECT
1	Mean	1.70	24.61	28.73	0.07	0.61	4.89	9.0	20.6	57.2	27.91	1.83
N = 202	S.D.	1.75	16.70	22.66	0.07	1.07	2.18	3.1	21.2	33.0	16.66	2.44
	Min	0.01	1.00	0.40	0.01	0.01	1.00	5.0	5.0	0.0	0.00	0.00
	Max	10.24	65.00	112.40	0.42	9.02	9.37	23.1	188.4	189.4	100.00	15.35
2	Mean	0.31	6.31	6.10	0.04	0.15	2.39	7.8	11.2	41.0	93.03	0.90
N = 26	S.D.	0.37	4.36	5.73	0.03	0.17	0.87	2.4	6.0	14.6	11.73	1.20
	Min	0.03	2.00	1.00	0.01	0.02	1.25	5.0	5.0	20.0	61.54	0.04
	Max	1.40	15.71	23.21	0.10	0.58	4.38	14.2	29.0	73.3	100.00	4.51
3	Mean	10.49	81.49	150.38	0.14	3.17	11.02	12.8	54.7	34.7	18.20	17.43
N = 142	S.D.	4.94	26.65	54.17	0.07	2.41	2.11	2.6	28.3	5.2	4.82	15.66
	Min	2.99	36.39	72.60	0.03	0.14	7.07	6.9	9.6	21.9	12.11	1.27
	Max	26.73	157	304.60	0.47	10.93	17.11	19.6	144.2	49.1	34.19	76.42
4	Mean	0.19	4.76	4.29	0.05	0.12	2.00	8.7	11.3	303.0	15.33	0.12
N = 10	S.D.	0.12	2.90	2.36	0.04	0.09	0.42	4.2	6.5	134.9	13.90	0.31
	Min	0.04	2.25	1.40	0.01	0.02	1.50	5.0	5.0	214.0	0.00	0.00
	Max	0.36	12.23	9.17	0.16	0.30	2.60	19.1	25.6	646.3	33.33	1.01
5	Mean	0.02	2.0	0.80	0.01	0.01	1.33	5.0	5.0	1052.6	0.0	0.0
N = 1	S.D.	–	–	–	–	–	–	–	–	–	–	–
6	Mean	29.20	72.58	285.53	0.45	19.86	12.35	16.5	195.7	31.6	19.45	86.76
N = 20	S.D.	7.59	21.99	68.01	0.18	9.37	2.61	3.1	54.3	7.2	7.14	60.26
	Min	19.00	25.63	102.91	0.21	6.84	5.70	13.3	79.2	23.8	13.57	2.37
	Max	48.00	104.0	398.81	0.76	41.41	15.79	25.7	299.7	54.6	46.44	242.03
7	Mean	63.39	24.34	397.72	3.47	61.56	9.85	37.3	304.7	26.2	40.38	32.57
N = 7	S.D.	9.11	9.55	26.90	2.66	9.99	1.66	17.9	38.6	4.4	20.12	21.17
	Min	49.44	9.00	366.71	1.45	45.69	6.28	23.6	251.5	22.2	20.31	1.96
	Max	78.34	34.11	441.37	8.70	76.88	10.94	69.9	367.3	34.7	83.33	64.03
Total	Mean	7.04	45.05	87.88	0.17	3.45	7.24	11.1	45.0	55.1	28.11	11.84
N = 408	S.D.	10.73	35.56	93.74	0.55	9.19	3.96	5.5	58.1	71.6	21.95	24.89
	Min	0.01	1.00	0.40	0.01	0.01	1.00	5.0	5.0	0.0	0.0	0.0
	Max	78.34	157.0	441.37	8.70	76.88	17.11	69.9	367.3	1052.6	100.0	242.03

(7 tiles) displays the highest mean F1 values and negative F2 values, corresponding to very compact urban areas. *Cluster 5* contains only 1 tile (a mine site), therefore is not considered as a typology, but rather as a unique case.

Clusters along the 5 km coastal zone of Corinthian Bay are mapped in Fig. 4. In this figure, *Cluster 3* is identified as 'urban sprawl pattern' forming a continuum along the coast and incorporating some pre-existing settlements. These areas are characterized by similar built-up patterns, with low-density development (mean COV 10.49%), very high PD and ED, low ENN and relatively high connectivity (CONNECT) at the 100 m threshold. High GYRATE values are related to geometrically complex built-up patches,

Table 4. Mean Factor values per cluster group

| | Mean Factor values | | |
	F1	F2	F3
Cluster 1	−0.33	−0.48	−0.11
Cluster 2	−0.17	−1.83	1.42
Cluster 3	−0.03	0.93	0.22
Cluster 4	−0.16	−0.66	−3.39
Cluster 5	0.27	−0.18	−12.55
Cluster 6	1.71	1.33	0.03
Cluster 7	6.03	−1.22	0.04

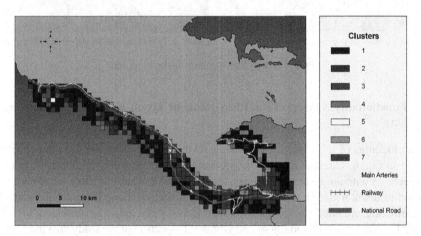

Fig. 4. Pattern of clusters/typologies along the Corinthian Bay.

usually with elongated shapes along the coast and/or the main road network. Taking into consideration the aforementioned results, Cluster 3 tiles are perceived as those shaping the *'linear city'* along Corinthian Bay.

Cluster 1 tiles are mostly identified in hinterland areas and semi-mountainous zones, characterized by low PD, LSI, ED values and Coverage between 1–10%. In most cases these refer to old rural settlements not affected by coastalization processes or isolated structures. *Cluster 2* tiles are similar to *Cluster 1*, with very low coverage areas, located far from the coast and next to the southern part of the national transportation network, presenting mostly isolated structures. *Cluster 6* tiles are suburban areas along the coast with a mean coverage of 29.20%; high PD, ED, MPS, LPI values; very high connectivity (CONNECT); and low ENN. *Cluster 7* tiles are compactly built-up areas with a mean coverage of 63.39%. Typical tiles of the above typologies, overlaid on satellite base map, are depicted in Fig. 5.

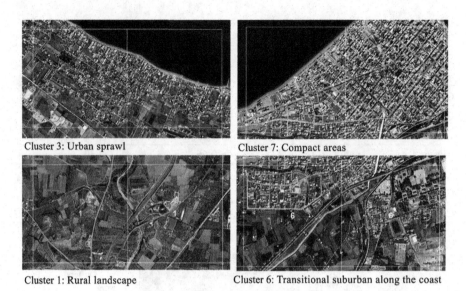

Cluster 3: Urban sprawl   Cluster 7: Compact areas

Cluster 1: Rural landscape   Cluster 6: Transitional suburban along the coast

Fig. 5. Typical tiles of main cluster typologies.

4.3   Functionality Aspects of Built Environment Typologies in Corinthia 'Linear City'

Urban facilities of the study area and their relation to the transportation networks are shown in Fig. 6. Commercial uses are concentrated in the major settlements, in close connectivity with the main road arteries. Most cultural facilities are in close proximity to Corinthus town, while recreational facilities are dispersed along the coast. Educational infrastructure and public buildings are more evenly distributed in the settlement network, located in coastal and inland towns and villages. Inspection of the study area leads to the conclusion that this, although spatially connected, seems to be *functionally disconnected*. Indeed, urban facilities are mainly concentrated around major settlements; while the largest part of sprawling patterns is almost solely equipped by recreational and tourism facilities, and is marked by the lack of public space, a clear neighborhood structure and a complete urban facilities' network. Highway and railway networks act more like boundaries, separating the coastline part from inland zones.

In this linear city model, the *spatial relationship* between built-up typologies on the one hand and urban facilities – e.g. commerce, recreation, education – as well as transportation networks on the other, is examined by means of correlation analysis at the tile level. Towards this end, the following *additional variables* are defined:

- *UrbFacil:* The number of urban facilities, including commerce, education, public buildings, culture and other services, located within each tile.
- *Dis_Highway:* The mean distance from the nearest highway transportation node.
- *Dis_RoadNet:* The mean distance from the nearest point of the main road network (major roads connecting settlements).
- *Dis_RailNet:* The mean distance from the nearest railway transportation node.

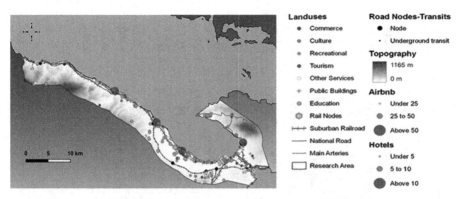

Fig. 6. Attributes of the Corinthian study region - urban facilities and networks

- *Dis_Settl:* The mean distance from the nearest major settlement.
- *Pop_den:* The mean population density, as estimated at the 250 m level by GHSL 2015 population grid (https://ghsl.jrc.ec.europa.eu/ghs_pop.php).
- *Elev:* The mean elevation of each tile (as estimated by the EU-*DEM* v1.1).

Results obtained from correlation analysis (Pearson correlation) at the tile level, incorporating F1, F2, F3 factors and the additional variables, are tabulated in Table 5. Among the additional variables, *Pop_den* is highly correlated to the distribution of urban facilities (*UrbFacil*); and negatively correlated to distance from transport infrastructure and major settlements, since population is only partly concentrated around major towns and settlements. *UrbFacil* is also negatively correlated to distance from transport infrastructure (*Dis_RoadNet, Dis_RailNe* and *Dis_Highway)* and elevation (Elev), but with very low Pearson's r values (close to 0.1).

F1 is positively and highly correlated to density of urban facilities (*UrbFacil*) and population density (*Pop_Den*); and negatively correlated to elevation (*Elev*) and distance from transportation infrastructure (mainly *Dis_RoadNet* and *Dis_RailNet*). F2 is negatively correlated (r values < 0.5, yet statistically significant at the 0.05 level) to all variables, apart from population density. This shows that while urban sprawl areas (Cluster 3, high F2) are in close proximity to the transportation network, they lack of positive relationship with urban facilities. Additionally, urban sprawl areas are generally areas with low elevation, as they are closer to the coast. This fact explains the negative correlation of *Elev* variable with F2. On the contrary, in rural/agricultural areas (Cluster 1), where both F1 and F2 are low, urban facilities are scarce and both accessibility and population density are lower. The distribution of F3 values (related to dispersion of built-up patches) is not correlated to any of the variables used in this analysis.

Finally, distance from highway seems to affect to a lesser degree the formation of urban sprawl typology (Cluster 3), as Pearson's r between F2 and *Dis_Highway* is significantly lower than the corresponding value for *Dis_RailNet*. This can be explained by the fact that the national road network acts more like a physical boundary, splitting up the area into two different compartments; it is only recently constructed; and in the eastern part of the study area is relatively far from the coast. Settlement areas are mostly

Table 5. Pearson correlation and statistical significance levels among main variables.

		F1	F2	F3	UrbFacil	Dis_Highway	Dis_RoadNet	Dis_RailNet	Dis_Settl	Pop_den	Elev
F1	r	1	0.0	0.00	0.712	−0.100	−0.272	−0.198	−0.198	0.796	−0.249
	Sig.		1.00	1.00	0.00	0.04	0.00	0.00	0.00	0.00	0.00
F2	r		1	0.00	−0.116	−0.274	−0.427	−0.369	−0.317	0.320	−0.474
	Sig.			1.00	0.02	0.00	0.00	0.00	0.00	0.00	0.00
F3	r			1	−0.009	−0.069	−0.058	−0.030	−0.041	0.035	−0.047
	Sig.				0.80	0.16	0.25	0.54	0.406	0.48	0.34
UrbFacil	r				1	−0.026	−0.116	−0.101	−0.100	0.625	−0.104
	Sig.					0.60	0.02	0.04	0.04	0.00	0.04
Dis_Highway	r					1	0.034	0.029	−0.083	−0.164	0.325
	Sig.						0.49	0.57	0.092	0.00	0.00
Dis_RoadNet	r						1	0.195	0.238	−0.317	0.575
	Sig.							0.00	0.00	0.00	0.00
Dis_RailNet	r							1	0.905	−0.310	0.48
	Sig.								0.00	0.00	0.00
Dis_Settl	r								1	−0.254	−0.100
	Sig.									0.00	0.043
Pop_den	r									1	−0.31
	Sig.										0.00
Elev	r										1
	Sig.										

connected by the local road network that spans close to coastline and links residential areas to the sea.

5 Discussion and Conclusions

In the current millennium, escalating urbanization trends within a complex and uncertain, rapidly changing decision environment constitute a huge challenge for planners and policy makers. This holds even truer in case of *coastal areas*, taking into consideration the vulnerability of such areas to contemporary challenges, e.g. climate change, rising sea level, overtourism; the concern as to the integrated coastal zone management for ensuring environmental health and qualities of both land and marine ecosystems; but also the attractiveness of such areas for both habitation and economic activity.

In the Mediterranean Region in particular, *coastal urbanization* constitutes a historically witnessed trend that comes from the distant past; and continues to attract the interest of, among others, the residential, coastal tourism, maritime commerce and transport sectors, perceived as key drivers of change in coastal urban environments. However, the rapidly rising pattern of this trend is today grasped as an alarming, mostly beyond the carrying capacity of land and marine ecosystems, phenomenon that needs to be assessed and properly handled or restrained. Moreover, in many Mediterranean cases, coastal urban development displays a linear pattern, resulting in a city form that is marked by extremely thin low-density ribbon features along the coast; and is lacking the functional dimensions and qualities of typical cities and neighbourhoods of inland urban contexts. Additionally, such a linear urban form seems to hamper sustainability and resilience objectives and expose respective areas and their population to various (un)predictable risks.

Planning and policy formulation for restraining coastal urban sprawl and expanded linear forms of urban development as well as handling potential risks from such urban models presupposes a deep insight into each specific urban pattern and the distinct attributes of its various zones. This implies a data-intensive approach and extraction of knowledge out of it that is capable of guiding more informed *policy decisions*, e.g. zone regulation and protection, building rules and restrictions, design of adequate public space as well as provision of efficient services and infrastructures, to name but a few. The geospatial methodological approach proposed in this work can support these requests by indulging into novel and contemporary spatial data with regard to the built environment. Proper handling of these data by means of GIS technology; use of carefully selected and meaningful spatial metrics for assessing the nature and intensity of built environment in the various coastal city's compartments; and utilization of mature statistical methods (PCA and Cluster Analysis) for shifting data outcomes to planning-related spatial information, can support efforts of planners and policy makers towards identifying bottlenecks in that kind of coastal cities and designing specific zone-related policy remediation measures.

The implementation of this approach in the case of Corinthian Bay proves its value to explore and pinpoint morphological and structural attributes of the 'linear' urban development pattern; and classify zones that are mostly affected by urban sprawl. Additionally, this knowledge and its correlation to current city structural elements – e.g.

type and distribution of services, transport networks – unveils the inadequacies of this linearly-expanded coastal urban model as a functional space that are mainly due to the unplanned and incohesive urban sprawl pattern. Actually, overall evaluation of results can generally classify Corinthian region as a typical example of unsustainable linear coastal development that is featured by uncoordinated urban sprawl, a heavily burdened coastal part and severe deficiencies in terms of functionality aspects.

Acknowledgements. The present research work is part of the postdoctoral research co-financed through the Greek State and EU (European Social Fund) accomplished under the framework of the Program "Workforce development, Education and continuing education", within the act "Financial assistance for Postdoc researchers – Cycle B" (MIS 5033021), led by IKY (State Scholarships Foundation of Greece).

Operational Programme
Human Resources Development,
Education and Lifelong Learning

Co-financed by Greece and the European Union

Ευρωπαϊκή Ένωση
European Social Fund

ΕΣΠΑ
2014-2020

References

1. Leka, A., Lagarias, A., Panagiotopoulou, M., Stratigea, A.: Development of a Tourism Carrying Capacity Index (TCCI) for sustainable management of coastal areas in Mediterranean Islands – case study Naxos, Greece. Ocean Coast. Manag. **216** (2022). https://doi.org/10.1016/j.ocecoaman.2021.105978
2. Stratigea, A., Leka, A., Nicolaides, C.: Small and medium-sized cities and insular communities in the Mediterranean: coping with sustainability challenges in the smart city context. In: Stratigea, A., Kyriakides, E., Nicolaides, C. (eds.) Smart Cities in the Mediterranean. PI, pp. 3–29. Springer, Cham (2017). https://doi.org/10.1007/978-3-319-54558-5_1. ISBN 987-3-319-54557-8
3. Koutsi, D., Stratigea, A.: Unburying hidden land and maritime cultural potential of small islands in the Mediterranean for tracking heritage-led local development paths. Heritage **2**(1), 938–966 (2019). https://doi.org/10.3390/heritage2010062
4. Koutsi, D., Stratigea, A.: Releasing cultural tourism potential of less-privileged island communities in the Mediterranean: an ICT-enabled, strategic, and integrated participatory planning approach. In: Marques, R.P., Melo, A.I., Natário, M.M., Biscaia, R. (eds.) The Impact of Tourist Activities on Low-Density Territories. THEM, pp. 63–93. Springer, Cham (2021). https://doi.org/10.1007/978-3-030-65524-2_4. ISBN 978-3-030-65523-5
5. García-Nieto, A.P., Geijzendorffer, I., Baró, F., Roche, P., Bondeau, A., et al.: Impacts of urbanization around Mediterranean cities: changes in ecosystem service supply. Ecol. Ind. **91**, 589–606 (2018). ffhal-01744748ff
6. United Nations Environment Programme/Mediterranean Action Plan and Plan Bleu (UNEP/MED-Plan Bleu): SoED 2020 - Summary for Decision Makers. Job No: DEP/2298/NA (2020). ISBN 978-92-807-3800-1
7. Ambrosetti, E.: Demographic challenges in the Mediterranean. IEMed Mediterranean Yearbook 2020 (2020). https://www.iemed.org/publication/demographic-challenges-in-the-mediterranean/. Accessed 10 Feb 2022

8. Plan Bleu: Demographic trends and outlook in the Mediterranean, Plan Bleu Notes #38 (2020). https://planbleu.org/wp-content/uploads/2020/10/Note38_-English-version.pdf. Accessed 12 Feb 2022
9. Battarra, R., Mazzeo, G.: Challenges of Mediterranean metropolitan systems: smart planning and mobility. Transp. Res. Procedia **60**(6), 92–99 (2022). https://doi.org/10.1016/j.trpro.2021.12.013
10. Lagarias, A., Stratigea, A.: High-resolution spatial data analysis for monitoring urban sprawl in coastal zones: a case study in Crete Island. In: Gervasi, O., et al. (eds.) ICCSA 2021. LNCS, vol. 12958, pp. 75–90. Springer, Cham (2021). https://doi.org/10.1007/978-3-030-87016-4_6. ISBN 978-3-030-87015-7
11. EEA (European Environment Agency): The European Environment - State and Outlook. Publications Office of the European Union, Luxembourg (2015)
12. Lionello, P., Platon, S., Rodo, X.: Preface: trends and climate change in the Mediterranean region. Glob. Planet. Change **63**, 87–89 (2008). https://doi.org/10.1016/j.gloplacha.2008.06.004
13. Cuttelod, A., García, N., Abdul Malak, D., Temple, H., Katariya, V.: The Mediterranean: a biodiversity hotspot under threat. In: Vié, J.-C., Hilton-Taylor, C., Stuart, S.N. (eds.). The 2008 Review of the IUCN Red List of Threatened Species. IUCN Gland, Gland (2008). https://www.researchgate.net/publication/285086595_The_Mediterranean_a_biodiversity_hotspot_under_threat. Accessed 12 Jan 2022)
14. Simpson, M.C., Gössling, S., Scott, D., Hall, C.M., Gladin, E.: Climate Change Adaptation and Mitigation in the Tourism Sector: Frameworks, Tools and Practices. UNEP, University of Oxford, UNWTO, WMO, Paris (2008). ISBN 978-92-807-2921-5
15. Coppola, P., Papa, E., Angiello, G., Carpentieri, G.: Urban form and sustainability: the case study of Rome. XI Congreso de Ingenieria del Transporte (CIT 2014). Procedia Soc. Behav. Sci. **160**, 557–566 (2014). https://doi.org/10.1016/j.sbspro.2014.12.169
16. Stojanovski, T.: Urban form and mobility choices: informing about sustainable travel alternatives, carbon emissions and energy use from transportation in Swedish neighbourhoods. Sustainability **11**(2), 548 (2019). https://doi.org/10.3390/su11020548
17. Boarnet, M.G., Crane, C.R.: Travel by design the influence of urban form on travel. Oxford Scholarship Online: November 2020 (2001). https://doi.org/10.1093/oso/9780195123951.001.0001
18. Saleh, A., Biswajeet, P., Shattri, M., Abdul, R.: GIS-based modeling for the spatial measurement and evaluation of mixed land use development for a compact city. GISci. Remote Sens. **52**(1), 18–39 (2015). https://doi.org/10.1080/15481603.2014.993854
19. Gordon, P., Richardson, H.W.: Are compact cities a desirable planning goal? J. Am. Plann. Assoc. **63**(1), 95–106 (1997)
20. Anderson, W.P., Kanaroglou, P.S., Miller, E.J.: Urban form, energy and the environment: a review of issues, evidence and policy. Urban Stud. **33**(1), 7–35 (1996)
21. Antyufeev, F., Antyufeeva, O.A.: Linear cities: controversies, challenges and prospects. IOP Conf. Ser. Mater. Sci. Eng. **687**, 055025 (2019). https://doi.org/10.1088/1757-899X/687/5/055025. International Conference on Construction, Architecture and Technosphere Safety
22. Cathcart, R.B.: Gauthier's 'Linear City.' Environ. Conserv. **23**(4), 286 (1996). https://doi.org/10.1017/S0376892900039114
23. Verbeek, T., Boussauwa, K., Pisman, A.: Presence and trends of linear sprawl: explaining ribbon development in the north of Belgium. Landsc. Urban Plan. **128**, 48–59 (2014)
24. Soto, M.T.R., Clavé, S.A.: Second homes and urban landscape patterns in Mediterranean coastal tourism destinations. Land Use Policy **68**, 117–132 (2017). https://doi.org/10.1016/j.landusepol.2017.07.018. Accessed 18 Feb 2022

25. Lagarias, A., Sayas, J.: Urban sprawl in the Mediterranean: evidence from coastal medium-sized cities. Reg. Sci. Inq. **0**(3), 15–32 (2018). http://www.rsijournal.eu/ARTICLES/December_2018/1.pdf. Accessed 19 May 2021

26. Valdunciel, J.: The anatomy of urban sprawl in the Mediterranean region: the case of the Girona districts, 1979–2006. Urban and Environmental Analysis Group, University of Girona. In: O'Donoghue, D. (ed.) Urban Transformations: Centres, Peripheries and Systems, Chap. 1. Routledge, London (2014). ISBN 9781315548685

27. Sterzel, T., Lüdeke, M.K.B., Walther, C., Kok, M.T., Sietz, D., Lucas, P.L.: Typology of coastal urban vulnerability under rapid urbanization. PLoS ONE **15**(1), e0220936 (2020). https://doi.org/10.1371/journal.pone.0220936

28. Anfuso, G., Postacchini, M., Di Luccio, D., Benassai, G.: Coastal sensitivity/vulnerability characterization and adaptation strategies: a review. J. Mar. Sci. Eng. **9**(1), 72 (2021). https://doi.org/10.3390/jmse9010072

29. EEA (European Environment Agency): The Changing Faces of Europe's Coastal Areas. EEA Report No 6. Publications Office of the European Union, Luxembourg (2006). ISBN 92-9167-842-2

30. Munoz, F.: Lock living: urban sprawl in Mediterranean. Cities **20**(6), 381–385 (2003). https://doi.org/10.1016/j.cities.2003.08.003

31. Membrado, J.C., Huete, R., Mantecón, A.: Urban sprawl and northern European residential tourism in the Spanish Mediterranean coast. Via Tour. Rev. **10** (2016). https://doi.org/10.4000/viatourism.1426. https://journals.openedition.org/viatourism/1426

32. Salvati, L., Smiraglia, D., Bajocco, S., Munafò, M.: Land use changes in two Mediterranean coastal regions: do urban areas matter? Int. J. Environ. Ecol. Eng. **8**(9), 562–566 (2014). https://doi.org/10.5281/zenodo.1337441

33. Andrés, M., Barragán, J.M., Sanabria, J.G.: Relationships between coastal urbanization and ecosystems in Spain. Cities **68**, 8–17 (2017). https://doi.org/10.1016/j.cities.2017.05.004

34. Romano, B., Zullo, F.: The urban transformation of Italy's Adriatic coastal strip: fifty years of unsustainability. Land Use Policy **38**, 26–36 (2014). https://doi.org/10.1016/j.landusepol.2013.10.001

35. Natura 2000 network. https://natura2000.eea.europa.eu/. Accessed 2 Mar 2022

36. Airbnb platform. https://www.airbnb.gr/. Accessed 1 Mar 2022

37. Lagarias, A., Zacharakis, J., Stratigea, A.: Assessing the coastal vs hinterland divide by use of multitemporal data: case study in Corinthia, Greece. Eur. J. Geograp. **13**, 1–26 (2022)

38. GHS-BUILT, multitemporal data. https://ghsl.jrc.ec.europa.eu/ghs_bu2019.php. Accessed 5 Jan 2022

I Wish You Were Here. Designing a Geostorytelling Ecosystem for Enhancing the Small Heritages' Experience

Letizia Bollini[1](✉) ⓘ and Chiara Facchini[2]

[1] Free University of Bozen-Bolzano, Piazza Università 1, 39100 Bolzano, Italy
letizia.bollini@unibz.it
[2] Politecnico di Milano, Campus Bovisa, 20158 Milan, Italy

Abstract. Slow tourism is a different, more conscious, and sustainable way of travelling, often targeting places secondary to cities of art and mainstream locations but rich in lesser-known social, intangible, and cultural heritage. While digital technologies have provided tools to make cultural places, institutions and GLAMs accessible, top-down and informative communication no longer seem sufficient to intercept both the more sophisticated public and the new generations whose information consumption is increasingly mediated by smartphones and the experience shared with peers through social networks. The paper proposes and discusses an exploration and a design prototype of a mobile geostorytelling ecosystem related to the "Prosecco hills" – Valdobbiadene e Conegliano in Treviso province, in northern Italy, recently designated a UNESCO heritage site – based on a bi-directional communication platform whose aim is, on the one hand, to allow minor realities – the "small heritages" – to be present efficiently and effectively on online and in-situ thanks to shared communication tools, on the other hand, to allow people to create georeferenced points of interest for other users through mechanisms of storytelling and emotional connection with the territory and the visitors.

Keywords: Cultural and intangible heritage · Geo-storytelling · Experience and interfaces design · Emotional design · Small heritages

1 Slow Tourism, *Small* Heritages, and Emotional Design

In recent decades we have witnessed the profound transformation of the tourism industry and its impact on urban and territorial transformations in the broadest sense. A phenomenon that emerged in western cultures mainly as a result of the economic boom and prosperity after World War II, mass tourism has also become a significant and strategic component of local economies [1]. Especially in Italy, the phenomenon has generated entire economic chains and completely transformed cities of art [2], coastal areas [3, 4], or mountain territories [5]. The evolution, however, increasingly seems to be showing the structural limits of a now uncontrolled growth that, in fact, is turning it into *over*-tourism [6] in a dynamic of consumption, gentrification with partial cancellation or deterioration of cultures, habitats and communities that are being 'transfigured' to accommodate and

© The Author(s), under exclusive license to Springer Nature Switzerland AG 2022
O. Gervasi et al. (Eds.): ICCSA 2022 Workshops, LNCS 13378, pp. 457–472, 2022.
https://doi.org/10.1007/978-3-031-10562-3_32

adapt. Moreover, this evolution seems to be always in the balance between the needs for mutation of cities and territories – dictated by the needs of inhabitants in social, economic, and political terms – and the needs for preservation and conservation of historical, cultural, and intangible heritage. We can in fact see how art cities become simulacra of themselves in a sort of Las Vegas-like transformation [7]. The tourists' experience is no longer a discovery and a contamination with the local culture, but rather a superficial contact with the *other*, the *different*, the *picturesque*, the *famous*. Places and traditions merge and flatten into a tour guided knowledge well represented by the iconography of kitsch souvenirs and photographs, where "to be there" is the most important than the human and cultural enrichment. Paraphrasing Augé [8] we could say that the cultural heritage becomes – paradoxically – the *non-lieux* of culture by definition. Or a postmodern nightmare [9] where we became witness of the *disneylandisée* phenomenon occurring to city such as Paris, Mont Saint-Michel, Venice, Rome, Florence – just to mention a few – where the real city, with its inhabitants, its history, social and symbolic becomes a mere scenic backdrop irrelevant to the tourist's visit. If not even a place of predation in which the mutual exchange is replaced by abuse or lack of respect for culture, resources, and local living conditions. As mentioned in a previous study conducted back in 2014 on the city of Milan about sustainable tourism: "This kind of globalization brings tourists to experience a place regardless of its specific genius loci, but rather in a sort of mash-up that unifies and blends multiple main-stream cultures in a sort of massive international style. Everything is similar and the differences are hardly mentioned in a sort of vernacular *taste*, and nothing is sufficiently different, strange, abnormal to involve and maybe even offend the sensibilities of those who looks thinking of seeing. Remaining on the surface that emphasizes what's equal rather than the other risk to misunderstand the concept of trip itself and its aim: find the strange to acquire consciousness and enrich knowledge by differences" [10].

In these imbalances exacerbated also by the phenomena of sharing and gig-economy – for instance AirB&B [11] – and by the difficulty of managing and regulating them at a political and collective level in a sort of 'cultural lag' [12], digital technologies and large market players have a crucial role. The cultural lag, in fact, implies that it is technology that dictates the agenda for changes [13], both in terms of timing and objectives, as has happened in many sectors, from the music industry to commerce, from transport to urban transformation. In this *mature* Anthropocene – the contemporary epoch where even the natural, biological etc. characteristics of the planet are strongly conditioned by the effects of human action, represents, according to Green, a *spatial turn*, a sort of leap of discontinuity that also implies new interpretative tools and parameters [14]. Crystallised in an unreal and "postcard" image, the *actual* city does not find its own space of representation, unknown to tourists, alien to those who live there. So hyper-realistic as to become *hyper-lieux* [15].

However, these phenomena have already been partially addressed starting from very specific domains – e.g., local food patrimony, viticulture – with its associated heritage of culture and traditions. Back in the 1980s, in Bra (Cuneo), a small city in the northwest of Italy – and Langhe region the "slow food" movement originated [16], both as a form of protection and promotion, in political and industrial terms, of local agricultural and wine production, and, more generally, as a reaction to the phenomena of mondialisation symbolic represented by McDonald and the *fast* food, alien to the typical Italian nutrition culture. This approach was quickly transferred and reproduced in other areas, moving from the material dimension alone to that of the territory and cities and even to tourism, which is 'nourished' by these cultures and places or a consumption factor [17].

The *slow city* movement – which originated ten years later in 1999 Italy and especially in the Umbria-Tuscany area – also explicitly forwarded the concept of size. The resident population, in order to join the network, must be less than 50,000 inhabitants. Its philosophy also implies adherence to the six fundamental pillars: 1) environment policies, 2) infrastructural policies, 3) technologies and facilities for urban quality, 4) safeguarding autochthonous productions, 5) Hospitality and 6) Awareness. The last two explicitly address the relationship between inhabitants and tourists. *Hospitality* is explicitly aimed to "supporting conviviality through local cultural events and the establishment of convivia, increase local gastronomic traditions", and *Awareness* underlines the need to involve "bot locals and visitors" in education programmes. The Slow Movement becomes an approach to the contemporary epochal transformations we are facing, which seeks to offer a *possible, desirable* and alternative vision to globalisation and its impact especially at the community level. The vision is holistic and inclusive of phenomena at different scales, which nevertheless place the *micro* – the local – at the centre. "The same principles and philosophy can be easily applied to tourism. Central to the meaning of *Slow Tourism* is the shift in focus from a achieving a quantity and volume of experience, while on holidays towards the quality of (generally fewer) experiences. It Is a form of tourism that respects local cultures, history and environment and values social responsibility while celebrating diversity and connecting people (tourists whit other tourists and whit host communities)" according to Heitmann, Robinson and Povey [18]. The concept of *slowness* becomes transversal and encompasses many aspects of experience, including from the point of view of design and the way in which these phenomena are addressed, understood, and communicated [19].

In this critical context that brings into question many of the founding aspects of tourism dynamics and the related industry, the concept of cultural and immaterial assets is also being transformed in line with new sensitivities according to the UNESCO definition of intangible heritage that involves people, space and their intertwingled relationships, according to which ICH is *community-based*: "intangible cultural heritage can only be heritage when it is recognized as such by the communities, groups or individuals that create, maintain and transmit it – without their recognition, nobody else can decide for them that a given expression or practice is their heritage" [20]. Local communities thus become the driver of the process, on the one hand of cultural production, but above all its interpretation and evolution. If in the Slow movement concept, it is the small towns – the boroughs according to the approach that is emerging as a reaction to the COVID emergency and the urban and climate crisis – that are the nodes of the network, in the cultural sphere the scale changes. From the major art cities, the focus shifts to the peripheral and smaller realities the *small* heritages [21]. Of course, it is not just a matter of scale or size. Rather, the issue of the need for a paradigm shift lies in the re-evaluation of artistic and historical aspects that have been neglected in favour of an authorial and *exceptional* conception of the culture construction. At the same time, this implies reconsidering two ethical and antithetic aspects. Firstly, what impact tourism could have on these smaller towns and villages, given what has already happened in the big cities, to avoid the same devastating gentrification and social *dispossession*. Secondly, how to help these peripheral localities to be attractive with the resources available, by networking or creating shared tools for collective service instead of continuing to replicate communication and promotion platforms unsustainable in terms of resources and maintenance. Digital technologies can play a fundamental role in this regard, if properly conceived, designed, and adopted within a strategic vision of development and conservation of these territories, their communities, and their specificities. The relationship between the internet and the world's leading museums is as "old as the web" – Louvre and Uffizi were *online* since the very beginning back in the 1990s for example – and has evolved over the years, opening up spaces for the experimentation of languages and offers increasingly oriented towards involving visitors as co-authors/actors in participative and performative experiences of access to knowledge. Besides, mobile devices have further expanded these possibilities by borrowing dynamics of engagement and communication from other areas of interaction: gamification, ubiquity, narrative approaches, blended and cross-reality, social sharing and user-generated content are some of the modalities that can be adopted by small heritages to create touchpoints and connections with new potential visitors and tourists.

The idea is therefore to design a "technologically-mediated" environment – a sort of enhanced layer – that enables people to access and explore the space both actual as well as digital as Andy Budd states: "environments in which experiences happen" [22] or as mentioned by Mark Wiser – initiator together with John Seely Brown of the liminal movement *Calm Technology* back in the 1990s – referring to ubiquitous technologies "where virtual reality puts people inside a computer generated world UC forces the computer to live out here in the world with people" as in the case of a mobile device and a georeferenced information [23]. However, user involvement can be approached on several levels. First of all, a mobile application that can be used in the explored territory offers deferred and contextualised information at the same time creating an articulated communication framework. But involvement can take place at a deeper level, as in the case of interacting, albeit through mediation, with other people who have visited a place before and who can offer us a glimpse, a fragment of the experience, of the interest, of the emotion that it has aroused in them. It is not just a matter of designing experiences that act or use the visceral language of emotion, but rather creating a frame, or perhaps a container, that allows these emotions to settle and be shared. A sort of word of mouth – emotionally connected with the place and then shared among visitors – that involves people directly in a virtuous handover of discovery and testimony. In fact, the engagement mechanism becomes crucial compared to the functionality offered by other platforms. People who embrace the slow philosophy are unlikely to follow mainstream routes or top-down generated by mass tourism and "consumerist" strategies. Personal experience and its sharing are elements that generate curiosity, trust and at the basis of exploratory mechanisms that intercept unexpected places or deeper keys to interpretation. The spontaneous generation of content, personal access and exploration, the circularity of the discovery and sharing process are factors that bring people closer to small heritage by placing them at the centre of a different relationship with visitors.

2 *Pssst! Not on Everybody's Lips:* A Case Study

The proposed study is based on an analysis of the communication methods employed by cultural institutions, their relationship with digital technologies and visitors. Particular attention is paid to what will be defined in this project as "small heritages", that is, private and municipal museums, small cultural and artistic sites, monuments and historical buildings and their difficulties in keeping pace with technological innovation.

The aim of the project is to provide an accurate analysis of the context in which small heritages operate and the needs of the users who visit them. In this way, the paper proposes new narrative keys and digital transmedia tools that connect cultural institutions with their visitors. The formers were investigated by checking their presence on certain digital platforms, while the latter were investigated through reports on tourism in the area in which the case study was conducted. Thanks to this research work, it was possible to analyse the limits and dynamics of small realities located in non-iconic areas and with limited resources.

2.1 The Alta Marca Trevigiana, the Prosecco Hills

In order to make the project's concept more understandable and concrete, it was decided to set it in a context located in a well-defined area. This made it possible to base the research on real data and to better grasp the realities in which these small heritages usually are immersed. An area in the Veneto region, the Alta Marca Trevigiana, also known as "Hills of Prosecco Conegliano-Valdobbiadene", was chosen as it proved to have optimal characteristics for the purposes of the project. The opportunity lies primarily in the many small cultural heritages, deriving from the area's rich history. Furthermore, it is only in recent years that the characteristic aspects of tourism in the area have begun to emerge. On the institutional websites, the wine and food aspect is now the protagonist, followed by experiential excursion proposals, as can be seen on visitproseccohills.it, one of the first to appear on search engines. The willingness to promote what is defined as 'slow tourism' [24] with activities such as walks, and bike rides may prove to be fertile ground for encouraging exploration. But building up a food and wine narrative, risks swallowing up the attention that could be paid to cultural heritage. It was also preferable for the project to be contextualised in a place that was not subject to mass tourism. The ideal audience is indeed curious people, looking for unusual experiences, exploration, and contemplation.

- One of the aims of the project is to provide a digital space for small heritages, giving them greater visibility and improving interaction with the visitor. From the point of view of digitalisation, there is a constellation of websites managed by municipal and regional authorities, or even by pro loco or groups of enthusiasts, which try to tell the peculiar story of the area. But the problems of these portals are manifold, and their main criticalities are reported below:
- Lack of information architecture: information is divided into categories and sub-categories that lack a logical and functional criterion, preventing easy navigation of the site.

- Obsolescence of information and long-winded texts: the text setup does not take into account the needs of today's users, whose attention threshold is usually very low, as is their willingness to listen.
- Paraphrasing of visual information works of art and architecture are described by means of long texts, which lack a direct link to the image, and force the reader to continually search for the part of the image to which they refer.
- Description of the location of places of interest: here again, the text replaces information that could be conveyed more effectively. The use of the GPS position would in fact also solve problems related to places that are difficult to identify by their address.
- Uninviting graphics, poor image quality
- No mobile version, non-responsive website
- Institutions in the area are not unaware of the potential of digitalisation, but rather focus their attention on the many wineries in the area: annual reports include survey data on the digital marketing activities of individual wineries.[2]
- The management of these websites is entrusted to very different bodies, which are in charge of reporting the information of well delimited areas. This prevents the creation of synergies between heritages dealing with similar topics. In this regard, in the report of the Alta Marca Trevigiana one of the area's weaknesses is the "current insufficient capacity to propose an integrated and aggregated tourist offer". *[xx]*
- Small heritages have more space in local periodicals, such as "Visit Conegliano Valdobbiadene". Sometimes there are even whole articles about them, but the main problem with this type of communication lies in its ephemeral nature. These magazines are distributed periodically within the accommodation facilities or tourist offices and the visitor only has access to the information contained in the issue published at that time. There is therefore a lack of information aggregation by thematic criteria and free from time limits.

A similar problem is faced when dealing with social networks. In the area there is only one Instagram page where there is clearly an investment in terms of social media management: @collinepatrimoniounesco. This page is committed to creating content that tells also about the cultural-historical aspect and traditions of the area. The quality of the content is excellent, but unfortunately the problem of time is persistent: the videos and photographs are lost in the feed over time, and it is even worse when they are uploaded on the platform as stories.

Although these heritages have major gaps in digital communication, this does not mean that they are completely lacking in communication tools. The most common approach is to set up information panels on site. However, these too have problems which a digital approach would solve:

- the panels often have long texts, and the user rarely finishes reading the whole content;
- if these panels are located outdoors, exposed to the weather, they risk deteriorating quickly, especially if not enough time was invested in the design phase to find materials and inks suitable for outdoor use;

- this type of communication always requires a more or less substantial financial investment and therefore local institutions have to assess whether the heritage is worth the investment. Therefore, sometimes the user reaches a place where there is clearly something that could be explored, but finds no information about it;

In addition to the economic investment, the acquisition of these communication systems often requires long bureaucratic procedures, which further raises the barriers of access to the production of information panels. A clear representation of the dynamics characterising the communication system of small heritages is provided in Fig. 1.

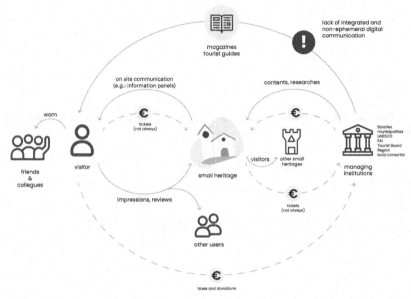

Fig. 1. Representation of the dynamics characterising the communication system of small heritages.

In order to have a clearer representation of the types of users who might use the service some proto-personas were developed. In particular, they are based on the Annual Economic Reports of the Conegliano Valdobbiadene Prosecco DOCG District for the years 2018, 2019 and 2020 together with a Context Analysis elaborated by the Local Development Programme of the G.A.L. of 'Alta Marca Trevigiana published in 2015. Although this source is less recent than the others, it provides a different perspective on the territorial context. It is therefore possible to compare it with more recent sources and to develop hypotheses on future trends. The reports were fundamental in understanding the reasons for travelling, the duration and location of the stay (Figs. 2 and 3).

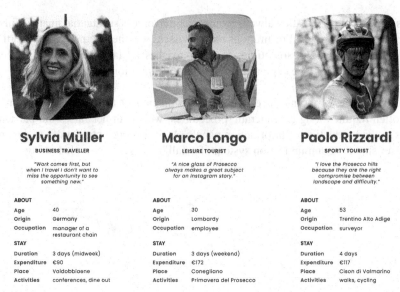

Fig. 2. Proto-personas displaying reasons for travelling, age, average expenditure, duration and location of the stay.

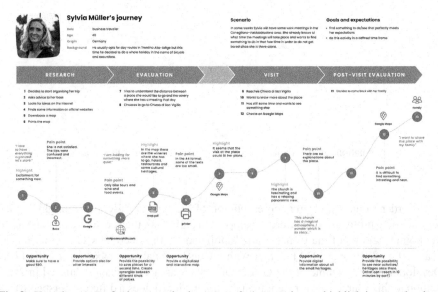

Fig. 3. User journey referring to a business traveler's experience, highlighting touchpoints, scenario and possible problems during the exploration of the area.

2.2 Site-Specific Geostorytelling

One of the main focuses of the research was the geo-referencing of the content. This mode of conveying information is extremely effective when dealing with experiences linked to a territory. We cannot imagine a trail guide without a map. Although this approach is widely used in printed media, in the digital world the potential of georeferencing is still largely unexplored. Applications using this type of content visualisation tend to limit the design to mere functionality. Instead, it is much more common to find printed maps with a narrative connotation, as in the example of "La Rimini di Fellini", which focuses on locations of the city of Rimini where the films of the famous Italian director were shot. What is often missing in the digital world is therefore what is called 'geostorytelling', which consists of drawing the users into a narrative world that can emotionally involve them and which "is based on the use of mobile devices and on forms of digital anchoring to the territory, such as GPS or other forms of geolocation." [25]. As in the previous example, one of the most common approaches consists in the development of a thematic route. In this way, it is easy to draw the attention of very specific publics, which are interested in the topic. Moreover, this kind of storytelling is strictly related to the territory and gives the possibility to enhance its peculiarities. In the case of the area of Conegliano Valdobbiadene four main routes were defined, starting from the identification and clusterization by topic of the many points of interest in the hills. The first one is dedicated to the many artistic works of the 16th century scattered in the area, including artists such as Tiziano and Cima da Conegliano. Another itinerary is dedicated to the places where the detective film "Finché c'è Prosecco c'è speranza" was shot. The area was the setting for many events of the First World War, and for this reason a dedicated itinerary has been created. Finally, an itinerary was proposed that retraced the 2020 Giro d'Italia "Conegliano Valdobbiadene Wine stage". It is therefore clear that the area presents a very diversified offer, potentially attracting very different tourists and visitors, both in terms of psychographics and age (Fig. 4).

Each heritage site involved in the itineraries was then subjected to a careful analysis aimed at defining its digital presence. From these data it was then decided to choose three heritages that were able to represent different levels of digitisation. This made it possible, as with the Personas, to facilitate the development of the project by packing a wide range of information into just three exemplary heritages (see Fig. 5).

In the charts (see Fig. 6) we can see that the most difficult process is to keep the owned digital means of communication updated with events and last news. Also, social networks are not widely used. Only two heritages have a virtual tour. This is probably due to the bigger economic investment required to produce this kind of media. No heritage is present on all digital channels analyzed, but it is important to note that there are no heritages that completely lack of digital communication. There is always some information about them on the web, but it is often third-party and disconnected from other heritages.

2.3 Far-Reaching Geostorytelling

One of the main goals of the project was not only to give a greater visibility to small heritages, but also to get more people to visit them. The analysis carried out on tourism in the area showed how varied the motivations of the visitors are. It is impossible to imagine attracting a sports tourist to a small heritage site by basing the narrative solely on the cultural aspect. Thus, there was a need to find a storytelling that would allow users to consider a deviation from their thematic route. Something that connects the different

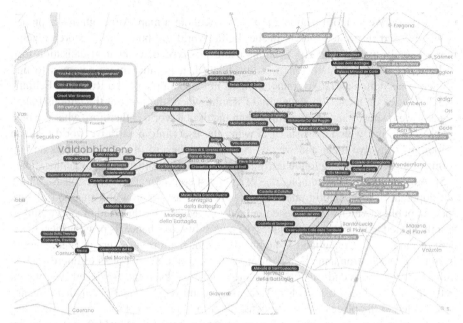

Fig. 4. Map showing the thematic routes identified in the area. These paths aim at enhancing the historic and cultural peculiarities of the specific territory.

Fig. 5. Characteristics of three exemplary heritages that emerged from the analysis on their communication means and digital presence.

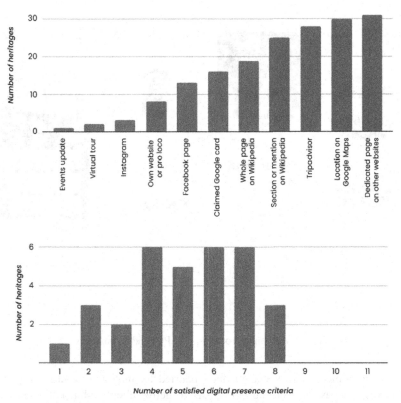

Fig. 6. Charts showing the results of the analysis of the small heritages' digital presence. On top: how many of them were present in the analyzed platforms, while bottom: displayed how many heritages satisfied a certain number of criteria.

narrative levels of the thematic routes. In addition, this narrative had to be wide-ranging, detached from the peculiarities of the territory and the boundary defined by institutions. In fact, the project is also designed for application outside of the case study. For this reason, it was decided to develop a system that would allow visitors to co-create the value of the application through their emotionality, which is applicable in all places and is by definition common to all human beings (Fig. 7).

Once they have reached a point of interest, users will then have the opportunity to send a virtual "postcard" to an acquaintance or someone close to them. The postcard will feature a sentence inviting people to write. This expedient allows a standardisation of the form of the thoughts written by the sers, which are then used to enhance the place from which the postcard was sent. On each site there will be a virtual "postbox", which will make it possible to know how many people have decided to send a postcard from that particular place, thus measuring how much that place has left a mark on their hearts. This feature allows users to choose the next place to visit based on how many postcards have been sent there. The importance of co-creation of content in the digital age is also highlighted in "The future of museums" [26]:

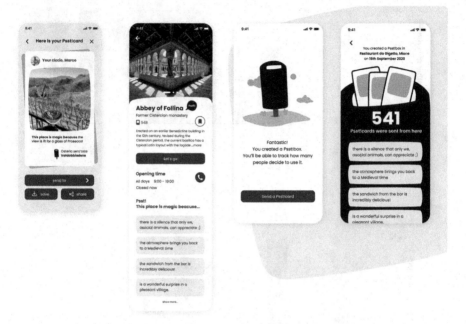

Fig. 7. Mechanism for creating a POI and a "postcard"

- "Using multimedia and social media-supported technologies, visitors have transformed from passive learning consumers to active clients who participate in co-authoring their visits."[4]
- Whereas the need to bring cultural storytelling (which in our case is focused on small heritages) closer to tourism storytelling (with the postcard theme) is discussed in "Digital storytelling nel marketing culturale e turistico":
- "In today's communication field, cultural storytelling has become increasingly similar to tourism storytelling, especially in terms of language. Both have the specific aim of attracting the cultural user/visitor with stories that make the cultural heritage, the museum, the work of art, the territory with its typical features and so on, attractive: telling a place through little stories, anecdotes, curiosities linked to the characters who wrote the history, the events that happened, are all strategic topics in both cultural and tourist communication, as they allow an emotional approach able to arouse interest in the public." [27] (Fig. 8).

2.4 Geo-Referenced User-Generated Contents and POIs

The application allows the managers of small heritages, or their communication managers, to create and share with their visitors' content that enhances interaction. The aim is to generate situated interactions. When visitors reach a point of interest, they "unlock" the contents related to it. The aim is to make them feel "informed inc et nunc" [28] through the use of notifications and geolocalisation. Moreover, the contents will be of different kinds: video, audio, visual in order to create a transmedia narrative.

Fig. 8. Final mockup of the application *Pssst! Not on everybody's lips*

The analysis of the digital presence of these heritages allowed to understand their needs in terms of communication, but also their digital skills. This aspect was fundamental to develop peculiar features for the content creator part of the App. For instance, a minimalist approach was applied to the content creation process, trying to avoid making the users deal with a screen full of information requests. Those who interact with this part of the application dedicated to small heritages will be supported in the creation of content thanks to tips provided by the application to avoid the most common mistakes.

2.5 Content to Connect and Engage

Descriptive Photos. One of the most popular technologies used by large, cutting-edge museums is augmented reality, which allows visitors to get more information about what they see in front of them. The development of this technology almost always requires the intervention of a specialist and a substantial financial investment. Augmented reality, in many cases, simply adds textual information to the world around us, moving with it (or rather, with the section we frame with our device's camera). Although the so-called "WOW effect" is strong, this continuous movement of information can worsen the accessibility of content, especially for an audience less familiar with digital technologies. Therefore, it was decided to give small heritages the possibility to achieve a similar result in terms of communicative power in a very simple way: they will be able to take a picture of the interested subject and add the content on the areas of the picture they refer to. The visitors can then zoom in on the selected area.

The user is also given some tips on how to take better photographs, assuming that it is not always possible to have professional photographs available, and that even with a smartphone you can achieve good results in terms of image quality if you make certain adjustments.

Embed Links. The analysis of the digital presence of the heritages in the area shows that some of them have content present on other platforms or social networks. Such as Youtube and Instagram. The application provides them the possibility of aggregating content from different platforms into their own tab, so that the visitor can easily access it as soon as they reach the site. The advantage of this operation also lies in linking this content to the geographical location of the heritage sites. As mentioned above, videos uploaded on YouTube, Facebook or Instagram are strictly linked to the temporal dimension, and over the years they lose their visibility because they end up at the bottom of the feed, and no user will scan their finger on the screen for hours just to see a piece of content they can't even imagine exists.

The second advantage of allowing the integration of the API of platforms for the production of audio guides such as izi.travel, or YouTube Creator Academy by hooking up to them for the production and editing of files.

Curiosities. Curiosities are a type of content inspired by typical blog formats. The idea is to provide the visitor with content 'in a nutshell': short sentences to which images can be attached.

The small heritage will be limited in the number of characters that can be inserted in, to avoid long-winded and ineffective texts.

Short Stories. Unlike the curiosities, the short stories have their own continuity but there will always be a character limit and the possibility of inserting suggestive images.

Before inserting the text, some advice will be given on how to structure the content, emphasising the importance of creating a climax and giving a proper end to the narration. These last two formats also present the possibility of communicating intangible heritage to visitors.

Events. Perhaps one of the biggest difficulties for small heritages is communicating events and keeping possible visitors up to date, so the application provides the possibility to enter events which can then be grouped and filtered according to tags (Fig. 9).

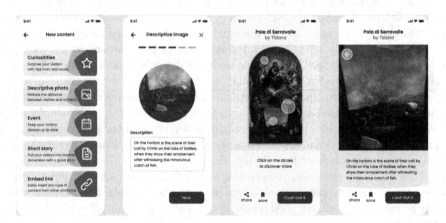

Fig. 9. Content insertion interface for small heritages

3 Conclusions

Another way of experiencing tourism is not only desirable, but above all possible. A path of discovery of minor territories, local communities, cultures and knowledge otherwise obliterated by mass consumption that instead of enhancing them, erases diversity. The protagonists of this transformation are the "users", people who can play an active role in building relationships both with the territory – site specific – and with an open, multiform and digitally interconnected community. Connected with the places, but above all emotionally with the experience made and shared that creates links with other potential visitors who, through these traces, can reconstruct an emotional itinerary even more than a local one. In order to generate and manage this change of scale and increased awareness, it is strategic that small heritages, the realities on the ground that guarantee their vitality, have access to scalable tools and resources to relate to visitors with new sensitivities and interests. The case study presented, although referring to a specific territory, represents a possible approach that includes both extremes of this interaction between a place and the people who explore it in a progressive process of knowledge. Furthermore, the research, in its subsequent developments, envisages that the interactive prototype will be tested with subjects belonging to the identified personas and in situ in order to assess and implement criticalities and potentialities with people for a future implementation on a local scale and as a transversal and scalable model to other similar realities.

Acknowledgements. Although the paper is a result of the joint work of all the authors, Letizia Bollini is in particular author of paragraphs 1 and 3 and Chiara Facchini is author of paragraphs 2. The authors wish to express their thanks the colleague Prof. Pietro Corraini.

References

1. Sezgin, E., Medet, Y.: Golden age of mass tourism: its history and development. In: Kasimoglu, M. (ed.) Vision for Global Tourism Industry: Creating and Sustaining Competitive Strategies, pp. 73–90. InTech, Rijeka (2012)
2. Zanini, S.: Tourism pressures and depopulation in Cannaregio: effects of mass tourism on Venetian cultural heritage. J. Cult. Herit. Manag. Sustain. Dev. **7**(2), 164–178 (2017)
3. Bonomi, A.: Il distretto del piacere. Bollati Boringhieri, Torino (2000)
4. Bollini, L.: Territorio e rappresentazione. Paesaggi urbani. Paesaggi Sociali. Paesaggi Digitali. Rimini e l'altro Mediterraneo. In: Giovannini, M., Prampolini, P. (eds.) Spazi e culture del Mediterraneo, pp. 28–42. Edizioni Centro Stampa di Ateneo, Reggio Calabria (2011)
5. Franco, C., Maumi, C.: The construction of a territory in the Alps. Infrastructure for mass tourism. TECHNE J. Technol. Archit. Environ., 172–179 (2016)
6. Dodds, R., Butler, R.: The phenomena of overtourism: a review. Int. J. Tour. Cities **5**(4), 519–528 (2019)
7. Bollini, L.: The urban landscape and its social representation. A cognitive research approach to rethinking historical cultural identities. In: Amoruso, G. (ed.) INTBAU 2017. Lecture Notes in Civil Engineering, vol. 3, pp. 834–842. Springer, Cham (2018). https://doi.org/10.1007/978-3-319-57937-5_86
8. Augé, M.: Non-Lieux. Introduction à une anthropologie de la surmodernité. Editions du Seuil, Parigi (1992)

9. Augé, M.: L'impossible voyage: le tourisme et ses images. Éditions Payot & Rivages, Parigi (1997)
10. Bollini, L., Pietra, R.: Mobile apps for a sustainable cultural-tourism experience. "The Betrothed 3.0" an Italian case study. In: Amoêda, R., Lira, S., Pinheiro, C. (eds.) Heritage 2014 – Proceedings of the 4th International Conference on Heritage and Sustainable Development, pp. 1207–1212. Green Lines Institute, Barcelos (2014)
11. Dolnicar, S., Zare, S.: COVID19 and Airbnb-disrupting the disruptor. Ann. Tour. Res. **83**, 102961 (2020)
12. Bucchi, M.: Io & Tech. Piccoli esercizi di tecnologia, Bompiani (2020)
13. Mattern, S.: A city is not a computer. In: The Routledge Companion to Smart Cities, pp. 17–28. Routledge (2020)
14. Green, J.: The Anthropocene Reviewed: Essays on a Human-Centered Planet. Dutton Penguin, New York (2021)
15. Lussault, M.: Hyper-lieux. Les nouvelles géographies de la mondialisation. Média Diffusion (2017)
16. Petrini, C.: Slow Food: The Case for Taste. Columbia University Press, New York (2003)
17. Pink, S.: Sense and sustainability: the case of the Slow City movement. Local Environ. **13**(2), 95–106 (2008)
18. Heitmann, S., Peter R., Ghislain, P.: Slow food, slow cities and slow tourism. In: Research Themes for Tourism. CAB International (2011)
19. Ceccarelli, N.: Slow complexity. In Ceccarelli, N., Jiménez-Martínez, C. (eds.) 2CO Communicating Complexity 2017, pp. XI–XXV. Universidad de La Laguna, Santa Cruz de Tenerife (2017)
20. What is Intangible Cultural Heritage? UNESCO Infokit (2011). https://ich.unesco.org/en/what-is-intangible-heritage-00003
21. Harvey, D.C.: The history of heritage. In: Graham, B., Howard, P. (eds.) The Ashgate Research Companion to Heritage and Identity, pp. 19–36. Routledge, London (2016)
22. Budd, A. (2019). https://twitter.com/andybudd/status/1142582773984763904
23. Weiser, M., Seely Brown, J.: Designing Calm Technology. Xerox Park (1995). https://calmtech.com/papers/designing-calm-technology.html
24. Boatto, V., Pomarici, E., Barisan, L.: Il distretto del Conegliano Valdobbiadene Prosecco DOCG - Centro studi. Rapporto Economico 2020 - Offerta e struttura delle imprese della DOCG Conegliano Valdobbiadene Prosecco nel 2020. Grafiche Antiga spa, December 2020
25. Boatto, V., Pomarici, E., Barisan, L.: Il distretto del Conegliano Valdobbiadene Prosecco DOCG - Centro studi. Rapporto Economico 2019 - Offerta e struttura delle imprese della DOCG Conegliano Valdobbiadene Prosecco nel 2019. Grafiche Antiga spa, December 2019
26. Bonacini, E.: Digital storytelling nel marketing culturale e turistico. Manuale pratico con esempi applicativi. Flaccovio Dario (2021)
27. Bast, G., Carayannis, E.G., Campbell, D.F.J. (eds.): The Future of Museums. Springer, Cham (2018). https://doi.org/10.1007/978-3-319-93955-1
28. Boatto, V., Pomarici, E., Barisan, L.: Distretto del Conegliano Valdobbiadene Prosecco DOCG - Centro Studi. Rapporto economico 2019. Grafiche Antiga spa, December 2019
29. Programma di sviluppo locale del G.A.L dell'Alta Marca Trevigiana. Analisi di contesto del PSL 2014–2020 del G.A.L. dell'Alta Marca Trevigiana (2014)

Smart City and Industry 4.0

New Opportunities for Mobility Innovation

Ginevra Balletto[1(✉)], Giuseppe Borruso[2], Mara Ladu[1], Alessandra Milesi[1],
Davide Tagliapietra[3], and Luca Carboni[4]

[1] DICAAR – Department of Civil and Environmental Engineering and Architecture, Università
degli studi di Cagliari, University of Cagliari, Via Marengo 2, 09123 Cagliari, Italy
{balletto,mara.ladu}@unica.it

[2] DEAMS – Department of Economic, Business, Mathematic and Statistical Sciences "Bruno
de Finetti", University of Trieste, Via A. Valerio, 4/1, 34127 Trieste, Italy
giuseppe.borruso@deams.units.it

[3] America's Cup Team American Magic, New York, USA
davide@styacht.com

[4] Atena - Associazione Italiana di tecnica navale, Diten, Genova, Italy
sardegna@atenanazionale.it

Abstract. The manufacturing industry is undergoing profound changes, so much
so that it recognizes a new phase, called Industry 4.0 both for the shape and structure of the supply chain, production of goods and energy transition. An example
of this epochal change is the auto and boat motive sector; a sector that sees the
increasingly marked electrification of its products, characterized in parallel by
a percentage modification of materials and production methods (new materials,
3D print). The development of the smart city is closely connected with the phenomenon of Industry 4.0, both in terms of mutual capacity for innovation, integration and digital transition, which makes its effects felt as well as in production
lines, supply chains, with tangible effects in the urban spatial distribution of goods
and services. The goal of the paper is to investigate the effects of industry 4.0 on
slow and ecological mobility and in new environments: protected areas (natural
and semi-natural parks) and historical and cultural areas. The case studies are: Natural Park of Molentargius and Archaeological Park ok Nora in the metropolitan
city of Cagliari (Sardinia, Italy) that represent a significant case of contamination
lab (Luna Rossa team, Atena and Dicaar of University of Cagliari).

Keywords: Smart city · Industry 4.0 · Sustainable city · Slow mobility

1 Introduction

From the first industrial revolution we are proceeding into the fourth industrial revolution,
the digital one, which is not only an advancement of internal processes, but an absorption
and improvement of all emerging technologies [1].

O. Gervasi et al. (Eds.): ICCSA 2022 Workshops, LNCS 13378, pp. 473–484, 2022.
https://doi.org/10.1007/978-3-031-10562-3_33

Industry 4.0 optimizes the overall traditional production cycle, making all phases and processes interact and represents a radical change in industrial production with profound implications for the city and its community [2]. The focus is not only on large numbers, those approved by consumers, but also on the individual buyer who, with his choices, can change the dynamics of production (3D print). Industry 4.0 is based on digital manufacturing technologies: Internet of Things (IoT), Big Data, Cloud Computing and Robotics.

Industry 4.0 today also finds concrete application in Smart Cities, or cities based on sustainable development in order to ensure a high quality of life for their inhabitants. In this case, the Internet of Things (IoT) is applied to logistics and intelligent mobility services (IoS) and the use of natural resources (IoE) [3, 4].

In this synthetic framework, the city is not only a reality of stone [5], but above all a place of highly expressive cultural conceptualities with multiple meanings between past, present and future. In fact, in the city all our aspirations are possible, which manifest through the progression of "innovation" [6], capable of reaching new cultural and technological goals, also through a new image of the city [7, 8]. Since the 19th century, the interaction between industry and technologies has guided the spatial distribution of both cities and transport and the community and related services [9].

Industrial development is in fact the engine of urban development. The 'industrial cities' grew rapidly and from that moment in Europe all urban development models contemplated industry, in all its phases, up to industry 4.0, which in the Smart City finds application through internal/external connections in all network directions, both physical and digital [10].

In fact, the Smart City, characterized by an essentially instrumental system, is aimed at regulating all urban circuits in a faster and more comprehensive way, through communication networks inside and outside the city, which is well suited to industrial processes 4.0 [11].

However, even the Smart City itself has evolved and manifests itself through 'digital ecosystems' without apparent limits and with the aim of global "Urban Quality". Industry 4.0, in short, favors the optimization of resources and products in the complex global fluidity of urban markets, between digital sociality (social network), personalization (3D print, just on time) and process and result quality (digital ecosystem) [12].

In this synthetic framework, the objective of the paper is to evaluate the opportunities that can be generated by recent innovations in mobility to promote sustainable accessibility in areas of environmental protection (natural and semi-natural parks) and/or historical and cultural areas (archaeological parks).

After the introduction (Sect. 1), the paper develops: Sect. 2 - Industry 4.0 and sustainable mobility (Sect. 2.1 Sustainable mobility and opportunities for territorial enhancement; Sect. 2.2 Sustainable mobility and circular economy; Sect. 2.3 Sustainable mobility and smart city); Sect. 3 - Materials: GoGo, innovation in slow mobility (Sect. 3.1 GOGO: main characteristics); Sect. 4 - Case study: Natural Park of Molentargius and Archaeological Park of Nora in the metropolitan city of Cagliari (Sardinia, Italy) and Sect. 5 - Discussions and conclusions.

2   Industry 4.0 and Sustainable Mobility

Digital, artificial intelligence and electrification have long since become part of the great theme of urban mobility. In the next few years, urban vehicles (land and naval) will have a profile and functions that are very different from the current ones, becoming real concentrates of innovation [13]. Technological development combined with the need to reduce energy consumption, together with the digital transition, has a decisive influence both in production processes, on the creation of increasingly smart models, and on consumer habits, more oriented towards sharing than possession [14].

2.1   Sustainable Mobility and Opportunities for Territorial Enhancement

Electric mobility, which derives from industry 4.0, offers greater opportunities for accessibility and sustainable use of places to the entire population [15], not only in urban areas but also in contexts of protected natural and cultural value, where accessibility is often denied or severely limited. Furthermore, the growing innovations in this research field introduces new contents in urban mobility planning, which concern both the management of potential impacts [35] and, particularly, the design of roads and waterways, which become a fundamental component of more comprehensive urban regeneration policies and projects [16]. This is even more evident in the cases of waterways which nowadays do not properly constitute communication routes. Here, not only energy supply systems but also the entire organizational and functional structure of the immediate context should be planned, especially when this is not yet adequately served due to scarce use. It means that the introduction of electric vehicles and electric boats may significantly affect the processes of enhancement and use of urban and suburban settings characterized by a different degree of naturalness, opening new scenarios also in terms of tourism development [17] (Fig. 1). More precisely, when we consider the ancient core of the city and its consolidated parts, it is expected that the existing road network will constitute the main network infrastructure on which to organize the electric charging points and related services (parking, car and bike sharing, e-mobility, refreshment points) in contexts not yet equipped. The introduction of electric cars, e-micro mobility and electric public transport makes it possible to promote a sustainable use of the ancient core, which is essential for ensuring an integrated conservation of cultural heritage. Even in the rural context, the existing roads will constitute the network infrastructure for electric cars, electric farm machinery, and e-micro mobility, which is strongly related to slow tourism. In the latter case, the growing use of new forms of sustainable mobility supports an overall regeneration and enhancement of the territorial capital (tangible and intangible) in response to the demand for slow tourism [18, 19]. Moreover, electric vehicles and boats may represent an extraordinary opportunity to rethink the entire system of accessibility and use of protected natural areas, both terrestrial and marine, characterized by different degrees of conservation. In Italy, Natural Parks constitute significant components of the natural heritage, both in quantitative and qualitative terms. In the Sardinia Region, the extension of the Natural Parks and their morphological features [20] make the introduction of electric vehicles (electric car, e-micro mobility) essential to allow sustainable accessibility.

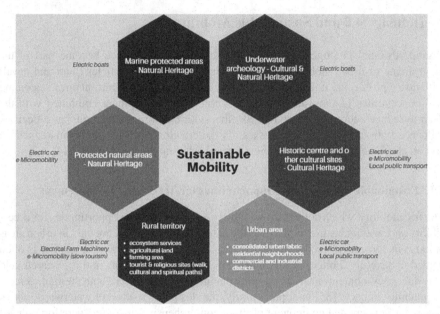

Fig. 1. Sustainable mobility in the metropolitan city (Authors: M Ladu and G Balletto, 2022)

Also in this case, the need to ensure an efficient energy supply system requires the selection of the most suitable areas for creating charging stations for electric vehicles, car parks, interchange nodes and, where necessary, refreshment points. Finally, in the case of marine protected areas, the main effects of the introduction of electric boats may concern the landing and mooring points of the boats, coinciding with the starting and ending points of the planned routes. As a matter of fact, the design schemes of these new hubs should include the construction of piers equipped with the necessary infrastructure, shading and refreshment points. Similar design schemes may also concern underwater archeology sites, which cannot always be visited before the introduction of the recent electric boats.

2.2 Sustainable Mobility and Circular Economy

In the paradigm of sustainable mobility, sharing, innovation and sustainability are the cornerstones to which are added: modularity, longevity and recyclability of products and services and consumption models based on the logic of sharing economy and "product as a service", abundantly referred to in the National Resilience Recovery Plan (PNRR) [21, 22]. In this sense, the fundamental factors that identify and guarantee efficiency in terms of environmental sustainability and reduction of waste of resources are represented by:

– The use of renewable energy and recyclable raw materials;
– The product is offered to the customer no longer as an owned asset but as a service to be used for a certain time and with specific methods;
– The development of platforms for sharing goods and services, to encourage savings and the efficiency of resources.

– Design and manufacture of goods with a long-term life cycle;
– The recovery and recycling of waste for the creation of new production cycles.

Furthermore, the right to mobility, precisely because it is recognized in the Constitution of the major international countries, has contributed to the renewal of the corresponding economic sector. Furthermore, this sector, in line with the 2030 objectives, innovates towards the circular economy of: sharing, slow and e-mobility, which represent new mobility opportunities also for areas of environmental/cultural protection. In addition, the thrust of the system of increasingly stringent laws in terms of emissions (Green Deal) and management of the vehicle end-of-life phase has led to a consequent remodeling of the products offered on the market. In particular, the electrification of vehicles combined with reuse/recycling deriving from design for disassembly continues to emerge. However, the problems associated with the batteries of electric vehicles are not negligible [23, 24]. In this synthetic picture, it emerges that the circular economy constitutes a fundamental dimension of sustainable mobility.

2.3 Sustainable Mobility and Smart City

The binomial sustainable mobility in the smart city represents the set of networks (both transport and technological) in support of economic progress and the improvement of citizens' quality standards, managing to ensure safe travel with a minimum environmental impact. The goal that combines sustainable mobility and the smart city is ecological and economic sustainability of travel within the urban network, protecting health and developing a renewed sense of community and belonging to places. Furthermore, the progressive innovation of digital technologies strengthened the pillar of Smart Mobility, sometimes making it the characterizing pillar of the smart city [25]. In other words, Smart Mobility intervenes on two levels, that is, it makes "intelligent infrastructures", developing applications such as Telepass Pay, easy Park, etc.; it makes intelligent vehicles" by developing sharing associated with electric motors on cars, bikes, micro-mobility and boats. In particular, boat sharing has all the characteristics to become the new strategic frontier in urban mobility of waterfront cities, similar to Uber and BlaBlaCar for cars [26].

3 Materials: GoGo, Innovation in Slow Mobility

Strategic urban metropolitan planning represents the tool for harmoniously governing urban innovation in all its forms, trying to identify shared scenarios of municipal urban planning. On the occasion of the drafting of the Strategic Plan of the Metropolitan City of Cagliari (2021) in the participation phase with the local community, the proposals from the community were collected. In particular, innovative mobility proposals were collected which are also capable of enhancing protected environmental and cultural areas. Among the proposals received, the one called GoGo was consistent on two planning levels: the Metropolitan Strategic Urban Plan of Cagliari and the Sustainable Mobility Plan of the Metropolitan City of Cagliari. In particular, the GoGo is an electric boat suitable for navigation, and has the aim of becoming part of the growing offer of sustainable mobility. Furthermore, the GoGo project (poject of ing. D. Tagliapietra) is part of the stay in Cagliari of the America's Cup Luna Rossa Prada Pirelli team in which Tagliapietra actively participated, thus being able to evaluate the main geographical and environmental characteristics of the navigable contexts. The contamination between the strategic policies of the metropolitan city of Cagliari, the Luna Rossa team, and in particular with the engineer Tagliapietra made it possible to systematize the GoGo project, which in 2022 was also included both in the Urban Plan of Cagliari and in the seaside part of the archaeological park of Nora.

3.1   GOGO: Main Characteristics

The GOGO is an electrified boat, a trimaran of 3.85 m in length and 2 m in width, having the ability to carry from four to six passengers depending on the variants and the ability to recharge the batteries through photovoltaic panels that also perform the dual function of protecting the passengers themselves from solar radiation. GOGO is a boat particularly suitable for use in inland waters and protected marine areas, guaranteeing environmental protection and passenger safety. The GOGO is built in thermoplastic material - multilayer polyethylene - entirely recycled, allowing energy savings - and raw material cost - during construction, as the Table 1:

Table 1. Operation during construction

Operation	MJ/Kg
Plastic recycling	22.00
New plastic production	86.00
Energy saving with recycled plastic	64.0
Energy recovered with burning plastics	48.00
Electric energy converted from thermal energy	24.00

In addition, the Solar Roof allows you to perform multiple functions (Fig. 2):

Battery charger and passenger screen function creation of a possible hub for energy exchange between different boats.

Fig. 2. Energy interchange scheme - joystick for direction and speed control and max speed 4 knots – mooring (source: D. Tagliapietra and L. Carboni)

The GOGO also allows access to protected marine areas, maximum environmental and underwater archaeological protection:

– absence of environmental pollution
– no wave motion formation
– no use of fuel and consequent emissions into the atmosphere and water

Finally, the living deck configuration allows a wide view and passenger involvement. The recycled multilayer polyethylene structure ensures sturdiness to the hull, which does not require maintenance [27].

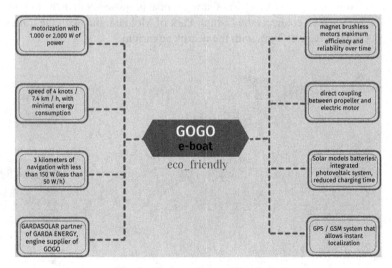

Fig. 3. Characteristics of the GoGo

In summary, the GOGO is an extremely innovative, eco-friendly, customizable boat, made with entirely recycled thermoplastic material, to the benefit of production costs, whose main technical characteristics are summarized as in Fig. 3.

4   Case Study: Natural Park of Molentargius and the Archaeological Park of Nora (Metropolitan City of Cagliari Sardinia, Italy)

The Metropolitan City of Cagliari was established in 2016 and consists of 17 Munici-palities: covers a territory of about 1250 square km with a resident population of about 414,370 inhabitants. The Strategic Plan of the city and the Sustainable Urban Mobil-ity Plan (SUMP) represent the main forms of government of the Metropolitan City of Cagliari in the aftermath of the health crisis [28] and in the light of the urgent energy transition. In this sense, the health crisis and the energy crisis have confirmed how to give priority to the design of open spaces and their external mobility (pedestrian, cycling, boat and micro-mobility) as well as the simultaneous containment of consumption and production of automatic energy and circular economy [29–31]. Open spaces include natural parks and archaeological areas that are an integral and substantial part of the ecosystem services of metropolitan cities and the surrounding area [32, 33]. Moreover, the strengthening of outdoor activities such as street sports have implicitly confirmed the latent urban and territorial walkability [34]. In this sense, the Strategic Plan of the metropolitan city of Cagliari has understood as sustainable mobility, energy efficiency, improvement of the quality of life of people and reduction of air pollution, are decisive goals for the near future, a change in people's lifestyle that passes for the health and well-being of the community. Below is the combined framework between the stress lev-els of cycling and sharing e-boat proposed both in the city of Cagliari, which connects the Molentargius natural park and the train station. Below is the combined framework between the stress levels of cycling and sharing e-boat proposed both in the metropolitan city of Cagliari, which connects the Natural Park of Molentargius with the train station and Archeological Park of Nora with the nearby aquarium.

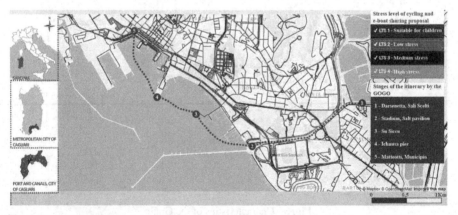

Fig. 4. E_Boat sharing proposal for the Molentargius natural park, in the metropolitan city of Cagliari (Authors: M. Ladu and G. Balletto, 2022; Source: https://bicistressatedaltraffico.it/#13. 55/38.99162/9.01987).

The result is shown in Figs. 4 and 5.

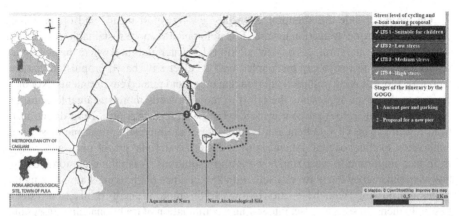

Fig. 5. E_Boat sharing proposal for the Nora archaeological park, in the Metropolitan city of Cagliari (Authors: M. Ladu and G. Balletto, 2022, Source: https://bicistressatedaltraffico.it/#13. 55/38.99162/9.01987)

5 Discussions and Conclusions

Innovation in mobility are key factors for fostering the ecological and energy transitions, with reference to how cities and territories will cope with the challenges they are - and will be - facing in the present and in the years to come. The topic examined here address different aspects of the planning process in transition (ecological, energetic and circular). What is examined here in fact deals with two aspects tightly connected: the network side and the means side. On one side mobility at urban level is still mostly referred to private cars/mopeds and bikes, with a modal share. Observing data from the Plans for Sustainable Urban Mobility (SUMP) the municipality of Cagliari, touches 60%, with bigger figures for other municipalities part of the metropolitan city - i.e., Capoterra reaches 79% of private transport share over the entire traffic composition. There is a need in a such sense to keep moving the traffic share towards more sustainable transport means, including walking, biking, local public transport and other forms of mobility. These are generally referred to landforms of transport and means, that can be integrated, in cities and settlements located in proximity of sea, rivers and canals and other water bodies, by waterways and related means that can contribute to moving away people's movement from land, decongesting it and therefore reducing the pressure on the urban environment. This process passes through the design of new routes on the water, changing the lines and orientation of communications and using and reinterpreting connecting lines, in many cases not used for long periods of time. In the case of Cagliari, the presence of the sea is related to the organization of the same city following capes, gulfs and inlets and, in parallel, the presence of other water bodies as lagoons, lakes and canals. These can host alternative routes to other more traditional ones, in line with the most recent trends of modal shift towards more sustainable transport means, by covering in parallel similar destinations as those touched by traditional transport means, and also allowing a re-discovery of locations that can be more easily reached by water transport instead of land-based ones. In such a sense, a mixed of integration and decongestion of traffic can be reached.

On the other side the attention is on the new transport means and opportunities for fostering a proper transition towards less pollutant and more sustainable vehicles. In such a sense, our attention was on new solutions for water mobility, where the new water routes and lines as above can be coupled with an electricity-based propulsion that can therefore allow moving pollution out from the urban and related environments. Not only electricity, however, represents the effort to move towards more sustainable transport means, but a full re-design of transport vehicles represents a promising direction to head towards. The GOGO boat can be considered an interesting example of such an application of this concept. Attention over design, materials, power units and weight is, in fact, important for realizing new transport solutions. A similar attention must be, however, put also on the network itself, with the presence of charging stations along the routes and, possibly, with the creation of low-carbon emission energy production systems to supply them. In such a sense, integrating a combination of traditional charging points and the implementation of micro-gridding systems could highly benefit a modal shift and a more sustainable mobility.

Acknowledgments. The authors Balletto G.; Ladu M. and Milesi A. took part in the following research activities: Strategic Plan of the Metropolitan City of Cagliari, commissioned to the Temporary Business Association (ATI), constituted by Lattanzio Advisory and Lattanzio Communication in 2019: Luigi Mundula (PI) University of Cagliari, "Investigating in the relationships between knowledge-building and design and decision-making in spatial planning with geodesign—CUP F74I19001040007—financed by Sardinia Foundation—Annuity 2018". Professor Michele Campagna (PI) University of Cagliari. Accessibility Lab, interdepartmental center of the University of Cagliari (Sardinia, Italy).

References

1. Matheri, A.N., Ngila, J.C., Njenga, C.K., Belaid, M., Van Rensburg, N.J.: Smart city energy trend transformation in the fourth industrial revolution digital disruption. In: 2019 IEEE International Conference on Industrial Engineering and Engineering Management (IEEM), pp. 978–984. IEEE (2019)
2. Allam, Z.: Cities and the Digital Revolution: Aligning Technology and Humanity. Springer, Cham (2019). https://doi.org/10.1007/978-3-030-29800-5
3. Lom, M., Pribyl, O., Svitek, M.: Industry 4.0 as a part of smart cities. In: 2016 Smart Cities Symposium Prague (SCSP), pp. 1–6. IEEE (2016)
4. Ghodmare, S.D., Khode, B.V., Ladekar, S.M.: The role of artificial intelligence in Industry 4.0 and smart city development. In: Gupta, L.M., Ray, M.R., Labhasetwar, P.K. (eds.) Advances in Civil Engineering and Infrastructural Development. LNCE, vol. 87, pp. 591–604. Springer, Singapore (2021). https://doi.org/10.1007/978-981-15-6463-5_58
5. Balletto, G.: Stone in the city. Publica (2018)
6. Florida, R., Adler, P., Mellander, C.: The city as innovation machine. Reg. Stud. **51**(1), 86–96 (2017)
7. Borruso, G., Balletto, G.: The image of the smart city: new challenges. Urban Sci. **6**(1), 5 (2022)
8. Balletto, G., Borruso, G., Donato, C.: City dashboards and the Achilles' heel of smart cities: putting governance in action and in space. In: Gervasi, O., et al. (eds.) ICCSA 2018. LNCS, vol. 10962, pp. 654–668. Springer, Cham (2018). https://doi.org/10.1007/978-3-319-95168-3_44

9. Zimmermann, C. (ed.): Industrial Cities: History and Future, vol. 2. Campus Verlag, Frankfurt (2013)

10. Safiullin, A., Krasnyuk, L., Kapelyuk, Z.: Integration of Industry 4.0 technologies for "smart cities" development. In: IOP Conference Series: Materials Science and Engineering 2019, vol. 497, no. 1, p. 012089. IOP Publishing, March 2019

11. Kashef, M., Visvizi, A., Troisi, O.: Smart city as a smart service system: human-computer interaction and smart city surveillance systems. Comput. Hum. Behav. **124**, 106923 (2021)

12. Ruohomaa, H.J., Salminen, V.K.: Mobility as a service in smart cities-new concept for smart mobility in Industry 4.0 framework. In: ISPIM Conference Proceedings, pp. 1–12. The International Society for Professional Innovation Management (ISPIM) (2019)

13. Ceder, A.: Urban mobility and public transport: future perspectives and review. Int. J. Urban Sci. **25**(4), 455–479 (2021)

14. Kon, F., Ferreira, É.C., de Souza, H.A., Duarte, F., Santi, P., Ratti, C.: Abstracting mobility flows from bike-sharing systems. Public Trans., 1–37 (2021). https://doi.org/10.1007/s12469-020-00259-5

15. Zaffagnini, T., Lelli, G., Fabbri, I., Negri, M.: Innovative street furniture supporting electric micro-mobility for active aging. In: Scataglini, S., Imbesi, S., Marques, G. (eds.) Internet of Things for Human-Centered Design, pp. 313–327. Springer, Singapore (2022). https://doi.org/10.1007/978-981-16-8488-3_15

16. Huston, S.: Smart Urban Regeneration: Visions, Institutions and Mechanisms, 15 April 2016. SSRN: https://ssrn.com/abstract=2765454 or https://doi.org/10.2139/ssrn.2765454

17. Ray, S.: A pragmatic approach of interaction between technology and tourism-hospitality. In: Hassan, A., Sharma, A. (eds.) The Emerald Handbook of ICT in Tourism and Hospitality, pp. 19–29. Emerald Publishing Limited, Bingley (2020). https://doi.org/10.1108/978-1-83982-688-720201002

18. Balletto, G., Milesi, A., Ladu, M., Borruso, G.: A dashboard for supporting slow tourism in green infrastructures. A methodological proposal in Sardinia (Italy). Sustainability **12**(9), 3579 (2020)

19. Balletto, G., Borruso, G., Ladu, M., Milesi, A.: Smart and slow tourism: evaluation and challenges in Sardinia (Italy). In: La Rosa, D., Privitera, R. (eds.) INPUT 2021. Springer, Cham (2022). https://doi.org/10.1007/978-3-030-96985-1_20

20. Ladu, M., Marras, M.: Nature protection and local development: a methodological study implemented with reference to a natural park located in Sardinia (Italy) (2021)

21. Franceschini, D., Cirimele, V., Longo, M.: Analysis of possible scenarios on the future development of electric mobility: focus on the Italian context. In: 2021 AEIT International Annual Conference 2021 (AEIT), pp. 1–6. IEEE, October 2021

22. Balletto, G., Borruso, G., Murgante, B., Milesi, A., Ladu, M.: Resistance and resilience. a methodological approach for cities and territories in Italy. In: Gervasi, O., et al. (eds.) ICCSA 2021. LNCS, vol. 12952, pp. 218–229. Springer, Cham (2021). https://doi.org/10.1007/978-3-030-86973-1_15

23. Fleischer, J., Gerlitz, E., Rieß, S., Coutandin, S., Hofmann, J.: Concepts and requirements for flexible disassembly systems for drive train components of electric vehicles. Procedia CIRP **98**, 577–582 (2021)

24. Qiao, D., Wang, G., Gao, T., Wen, B., Dai, T.: Potential impact of the end-of-life batteries recycling of electric vehicles on lithium demand in China: 2010–2050. Sci. Total Environ. **764**, 142835 (2021)

25. Khamis, A.: Smart Mobility: Exploring Foundational Technologies and Wider Impacts. Apress, Berkeley (2021)

26. Warmington-Lundström, J., Laurenti, R.: Reviewing circular economy rebound effects: the case of online peer-to-peer boat sharing. Resour. Conserv. Recycl. X **5**, 100028 (2020)

27. Lucertini, G., Musco, F.: Circular city: urban and territorial perspectives. In: Amenta, L., Russo, M., van Timmeren, A. (eds.) Regenerative Territories, pp. 123–134. Springer, Cham (2022). https://doi.org/10.1007/978-3-030-78536-9_7

28. Dettori, M., et al.: Air pollutants and risk of death due to COVID-19 in Italy. Environ. Res. **192**, 110459 (2021)

29. Trevisan, A.H., Zacharias, I.S., Liu, Q., Yang, M., Mascarenhas, J.: Circular economy and digital technologies: a review of the current research streams. Proc. Des. Soc. **1**, 621–630 (2021)

30. Balletto, G., Ladu, M., Milesi, A., Borruso, G.: A methodological approach on disused public properties in the 15-minute city perspective. Sustainability **13**(2), 593 (2021)

31. Mundula, L., Ladu, M., Balletto, G., Milesi, A.: Smart marinas. The case of metropolitan city of Cagliari. In: Gervasi, O., et al. (eds.) ICCSA 2020. LNCS, vol. 12255, pp. 51–66. Springer, Cham (2020). https://doi.org/10.1007/978-3-030-58820-5_5

32. Palumbo, M.E., Mundula, L., Balletto, G., Bazzato, E., Marignani, M.: Environmental dimension into strategic planning. The case of metropolitan city of Cagliari. In: Gervasi, O., et al. (eds.) ICCSA 2020. LNCS, vol. 12255, pp. 456–471. Springer, Cham (2020). https://doi.org/10.1007/978-3-030-58820-5_34

33. Battino, S., Balletto, G., Borruso, G., Donato, C.: Internal areas and smart tourism. Promoting territories in Sardinia Island. In: Gervasi, O., et al. (eds.) ICCSA 2018. LNCS, vol. 10964, pp. 44–57. Springer, Cham (2018). https://doi.org/10.1007/978-3-319-95174-4_4

34. Ladu, M., Balletto, G., Borruso, G.: Sport and smart communities. Assessing the sporting attractiveness and community perceptions of Cagliari (Sardinia, Italy). In: Misra, S., et al. (eds.) ICCSA 2019. LNCS, vol. 11624, pp. 200–215. Springer, Cham (2019). https://doi.org/10.1007/978-3-030-24311-1_14

35. Acampa, G., Grasso, M., Ticali, D.: MCDA to evaluate alternative paths for urban electric micromobility. In: AIP Conference Proceedings, vol. 2343, no. 1, p. 090009. AIP Publishing LLC, March 2021

Digital Ecosystem and Landscape Design. The Stadium City of Cagliari, Sardinia (Italy)

Ginevra Balletto[1]([⊠]), Giuseppe Borruso[2], Giulia Tanda[1], and Roberto Mura[3]

[1] Department of Civil and Environmental Engineering and Architecture, University of Cagliari,
Via Marengo 2, 09123 Cagliari, Italy
balletto@unica.it

[2] Department of Economics, Business, Mathematics and Statistics "Bruno de Finetti",
University of Trieste, Via Tigor 22, 34127 Trieste, Italy
giuseppe.borruso@deams.units.it

[3] Department of Culture of the Ca' Tron Project, IUAV University of Venice, Santa Croce,
1957-30135 Venice, Italy
rmura@iuav.it

Abstract. Increasingly articulated and sensitive is the approach to issues of conservation and/or balanced exploitation of resources because of the value of environmental quality at the basis of fundamental rights. This is demonstrated by the successive regulatory instruments: Environmental Impact Assessment (EIA, 1969), Strategic Environmental Assessment (SEA, 2001), Single Environmental Text (2006 containing the rules on environmental protection and waste management) and Agenda 2030 (2015, 17 for Sustainable Development), aimed at avoiding-containing situations of degradation and risk to the quality of life, including future generations. Among the different methodological approaches, we must also consider the Landscape Assessment to be carried out before and, in an evolutionary continuum, after the realization of a work. This in order to identify interference and impacts on the territory (with descriptive-quantitative investigation of the elements characterizing its structure from a naturalistic, anthropic and historical-cultural point of view, and perceptive investigation of the visual impact) its relations, qualities and balances. Certainly, responding to the purposes of a complete and innovative landscape assessment appears to be the use, although not yet fully widespread, such as that of Open Source Tools, available without a commercial license, therefore within the reach of all: designers and evaluators. In this framework, the aim of the work is to evaluate the role of Open-Source Tools in the assessment of the visual, ante and post operam impact of important urban works that require continuous information-collective participation and therefore capable of fostering the creation of shared values.

Keywords: Landscape assessment · Visual impact · Open source tools

1 Introduction

The Digital Ecosystem takes the concept from biology, Natural Ecosystem, with which it identifies a set of living organisms and non-living matter that interact with each other in

O. Gervasi et al. (Eds.): ICCSA 2022 Workshops, LNCS 13378, pp. 485–493, 2022.
https://doi.org/10.1007/978-3-031-10562-3_34

a given environment, constituting a self-sufficient system. Within this environment there are interactions and exchanges of information, all in an ever-changing dynamic balance. The concept of Digital Ecosystem, on the other hand, contains a link of interconnection and interoperability between software of various kinds and hardware. The concept of Digital Ecosystem, applied to the current context of the case study of environmental impact assessment (EIA), in particular referring to the Stadio Sant'Elia of Cagliari, and its demolition and construction of the new structure, It benefits from the research and applications developed, in particular during the 90s of the past century, and related to Spatial Decision Support Systems (Systems-SDSS) in turn, the result of the integration between Decision Support Systems (DSS) and Geographic Information Systems (GIS). These systems, which form the basis of the application of ICT technologies to territorial problems, concern (and mainly concerned) the application of interactive systems based on IT media, designed to assist final decision-makers on semi-spatial problems.

The Environmental Impact Assessment (EIA) is a technical-administrative procedure preparatory to the implementation of a work and identifies and describes the effects on the environment, health and human well-being. In addition, the EIA identifies measures to prevent, eliminate or reduce negative impacts on the environment. The EIA was thus created (in the late 1960s) to find new decision-making methods, since the traditional ones were no longer adequate to an increasingly complex social reality and to an ever-increasing demand for participation by the various Stakeholders and the introduction of environmental factors between the design components [1, 2].

In relation to the Environmental factors, we must emphasize one of the fundamental aspects of the VIA, the context analysis, better defined as the scenic landscape, within which falls the work to be protected/created. The aesthetic aspect of the landscape, unlike the environment and the territory, is the generalized and social perception, determined by assessable scenic-spatial connotations (height, shape, etc.), characteristic and expressive of a landscape [3].

In particular, as reported in the Guidelines of MiBACT [4], the analysis of the scenic landscape, which correlates the observer with the observed reality, is based on the identification of landscape observation points, such as landscape observation points (panoramic points, public viewpoints with filtered access, equipped belvedere), routes with good or high panoramic views, perspective axes and routes (connecting and crossing, interior and exploration, slow to use) of landscape-environmental interest; on the identification of overall and detailed panoramic beauties, such as visual fulcrums (of the built environment, natural, on a local scale, supra-local and isolated), landscape profiles (ridges, skylines), other elements that contribute to the identity of the landscape (urban and natural fronts, trees, rows, hedges) and visual relationships (intervisibility, focal views, focal views on off-paper elements and frontal elements, wide panorama, visual gates); on the identification of areas characterized by critical elements and with visual detractions, such as punctual, linear and areal critical factors and qualified visual detraction situations based on the type of alteration in progress (perceptual degradation, de-connotation, intrusion, obstruction) [5, 6]. The complex assessment of the landscape context is more and more often carried out with the aid of Open-Source software, thus allowing maximum sharing of the results and the related process.

In this synthetic framework, the objective of the work is to select an open-source digital ecosystem to support the EIA for large urban projects. In this synthetic framework, the goal of the work is to select an open-source digital ecosystem to support the EIA for large urban projects, to identify levels of danger, risk on the landscape, so as to guide from the early planning stages and avoid subsequent mitigation actions [7].

The case study is the project of the new Cagliari Stadium developed on an urban brownfield, in collaboration also with the DICAAR (Department of Environmental Civil Engineering and Architecture, University of Cagliari) for landscape, environmental and circular economy themes.

After the first section (Introduction), the rest of the paper is organized as follows, Sect. 2: Methodology and data: the analysis of digital ecosystems is developed that can support the assessment of the visual impact of the new Cagliari stadium; Sect. 3: case study of Cagliari stadium (Sect. 3); Sect. 4: Results and discussions; and Sect. 5: Conclusions.

2   Methodology and Data

After developing the context analysis (geographic, urban and environmental and landscape), the analysis of digital ecosystems is developed that can support the assessment of the visual impact of the new Cagliari stadium. The analysis focused on 'Open-Source Tools', such as Google MyMaps and Google Earth Pro, which made it possible to identify and evaluate possible critical issues and visual interference on the landscape. In particular: Google Earth Pro is a geospatial software application that offers the ability to analyze and acquire geographic data (as does Google Earth but with additional functions). Google Earth Pro allows you to print high resolution images (the resolution is lower with Google Earth), to automatically geolocate images of the GIS geographic information system (Google Earth requires manual geolocation), to overlay images (Google Earth can only import image files), create animated movies, map points simultaneously, and access demographic and traffic data layers. Google MyMaps is a geospatial software that offers updated and complete information of urban activities and functions and since mobile searches are correlated to the position, it is a real traffic driver. Through the use of Google MyMaps, it is possible to identify the most significant observation points of the city from which the work and (ante operam and post operam), Table 1.

Table 1. Main observation points and routes.

Case study: New Stadium of Cagliari	Points and routes of observation
Points in Proximity	Ranges from: 30–50–70–100–200 m distant big buildings
Points and routes panoramic	Variable spatial distribution

3   Case Study and Data

The Cagliari stadium constitutes the case study, which represents a real symbol for the city, but today it constitutes a Brownfield that requires urgent interventions in the context of an urban regeneration aimed at enhancing the landscape. The city Cagliari, capital of the metropolitan city (population = 421.488 - Istat 2021) with its landscape and naturalistic beauties, dynamic for its international dimension (ICT, economic and industrial hub, sport,..) with the stadium included in an urban project intends to redevelop and integrate parts of cities, some of which have been on the fringes of urban life for too long [8–11].

The work includes a Stadium Building (with multiple sports activities and those connected to them such as museums, bars, commercial activities) and a Hotel Building (with some views of the rooms directly on the playing field). A stadium responding to the needs of landscape-architectural and urban-functional balance, thanks to its circular shape, characterized by an open structure free of barriers, which gives it a symbolic value of social cohesion, generating the idea of a shared collective landscape, also thanks to the presence of a large Mediterranean garden [12–14]. A stadium that will be found to support an improvement of the same city, also passing through the recovery of the suburbs, thanks to a fair distribution of volumes in space, the creation of a system of quality public spaces, the introduction of services in the neighborhoods that today lack them (in a polycentric perspective), the limitation of the environmental impact in the places of commerce, reducing the voids of parking areas, and thanks to the improvement of vehicular accessibility (public and private) and soft mobility, with the creation of a new road network that connects the different functional areas [15–17] (Fig. 1).

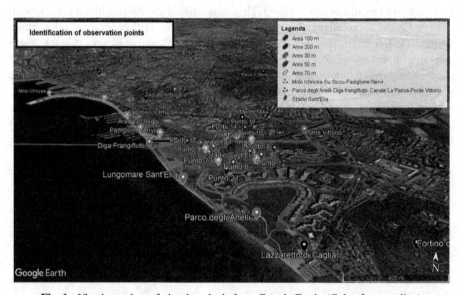

Fig. 1. Viewing points of visual analysis from Google Earth. (Color figure online)

Through the use of Google Earth Pro, the visual analyzes relating to the points previously reported were carried out, first considering the old Stadium (Fig. 2 and Fig. 4) and then the new one (Fig. 3 and Fig. 5), so as to then being able to highlight the differences and compare the data collected.

It can be seen from the following photos that what is visible from the identified point is graphically colored in green, while the absence of green indicates the total absence of visibility of the old and new Stadium.

Displayed equations are centered and set on a separate line.

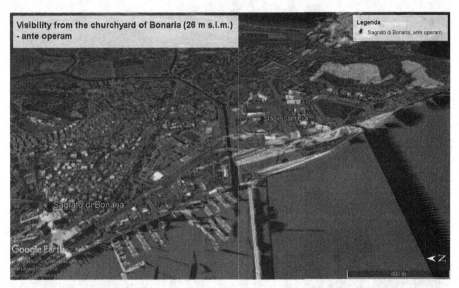

Fig. 2. Visibility analysis from the panoramic point of the Basilica of Bonaria, ante operam. (Color figure online)

The visual analysis carried out with the Open Source Tools revealed significant differences between the old and the new structure. In particular, the ante operam analysis confirms the condition of Brownfield of Stadio Sant'Elia from all the points identified, while from the post operam one can see how the Stadium fits into the landscape without creating discontinuity and fragmentation.

Fig. 3. Visibility analysis from the panoramic point of the Basilica of Bonaria, post operam. (Color figure online)

Fig. 4. Visibility analysis from the panoramic point of the Poetto Tower, ante operam. (Color figure online)

Fig. 5. Visibility analysis from the panoramic point of the Poetto Tower, post operam. (Color figure online)

4   Results and Discussions

The paper is part of a broader research on the demolition and construction of the new Cagliari stadium and has shown above all the usability of open geographic information tools, in particular those based on online platforms.

First of all, through the visual analysis on Google Earth, it was possible to identify the starting points from which to carry out the visibility analysis, and, more generally, the environment within which to integrate various aspects of a territorial nature, relating to the object of study analyzed. In particular, the tool served as a preliminary analysis for the correct localization of the observation points, combining the experience of the research group with the potential offered by the tool (e.g. geolocation, also three-dimensional, calculation of distances as the crow flies and on the road, visibility analysis, etc.).In particular, the work environment of Google Earth Pro made it possible to familiarize with the territory at a scenario level, also observing, thanks also to the presence of the 3-dimensional simulations of the buildings, as well as the profile of the land, the relationships existing between the structure in question (Stadium of Cagliari) in the two moments, ante operam and post operam.

The work also included the insertion of an object identifying the new stadium within the Google Earth platform, in such a way as to more correctly simulate the shape and visual dimensions of the structure, thus facilitating the ante and post operam. The visibility analysis, in particular, proved to be useful and interesting, thanks to the possibility of a correct georeferencing of the observation points, also taking into account the altimetric component, and therefore managing to simulate the point of view even at a height of 'man', fixing the observation point at 2 m above ground level and, therefore, highlighting the areas that are actually visible.

5 Conclusions

This research has dealt with the visual impact of the old and new Sant'Elia Stadium (Cagliari); in relation to this aspect, we limited ourselves to that deriving from the use of "Open-Source Tools", such as Google MyMaps and Google Earth Pro, which made it possible to identify and evaluate possible critical issues and visual interference on the landscape. From the analysis carried out with these tools, it emerged that, at present, from the points considered (panoramic, proximity and along the promenade), no significant visual interference is identified in the post-construction project, unlike the existing structure. The use of such software therefore made it possible to understand whether the new construction project could fit harmoniously into the context, without creating interference, discontinuity, and fragmentation of the landscape. The work carried out also made it possible to test on a practical case the usability of geographic tools, freely available, applied to the evaluation processes (Environmental Impact Assessment and Strategic Environmental Assessment), showing their potential and communicative effectiveness, even to a non-public necessarily expert.

References

1. Morgan, R.K.: Environmental impact assessment: the state of the art. Impact Assessm. Project Appraisal **30**(1), 5–14 (2012)
2. Wood, C.: Environmental Impact Assessment: A Comparative Review. Routledge (2014)
3. van Heck, S., Valks, B., Den Heijer, A.: The added value of smart stadiums: a case study at Johan Cruijff Arena. J. Corp. Real Estate (2021)
4. Ministero dei Beni e delle Attività Culturali e del Turismo (MiBACT) Direzione Regionale per i beni culturali e paesaggistici del Piemonte Regione Piemonte Direzione Programmazione strategica, politiche territoriali ed edilizia Dipartimento Interateneo di Scienze, Progetto e Politiche del Territorio (DIST), Politecnico e Università di Torino: Linee guida per l'analisi, la tutela e la valorizzazione degli aspetti scenico-percettivi del paesaggio (2014)
5. Wartmann, F.M., Stride, C.B., Kienast, F., Hunziker, M.: Relating landscape ecological metrics with public survey data on perceived landscape quality and place attachment. Landsc. Ecol. **36**(8), 2367–2393 (2021)
6. Spielhofer, R., Hunziker, M., Kienast, F., Hayek, U.W., Grêt-Regamey, A.: Does rated visual landscape quality match visual features? An analysis for renewable energy landscapes. Landsc. Urban Plan. **209**, 104000 (2021)
7. Shan, P., Sun, W.: Research on 3D urban landscape design and evaluation based on geographic information system. Environ. Earth Sci. **80**(17), 1–15 (2021)
8. Judd, D.R., Smith, J.M.: The new ecology of urban governance: special-purpose authorities and urban development. In: Governing Cities in a Global Era, pp. 151–160. Palgrave Macmillan, New York (2007)
9. Bale, J.: Stadium and the City. Edinburgh University Press (2019)
10. Elgammal, Y., Abdel-Razek, N.M.: Architecture design of stadium facilities between ancient times and today. Egypt. Int. J. Eng. Sci. Technol. **38**(1), 26–40 (2022)
11. Hyun, D.: Proud of, but too close: the negative externalities of a new sports stadium in an urban residential area. Ann. Reg. Sci. (2021). https://doi.org/10.1007/s00168-021-01095-6
12. Ahlfeldt, G., Maennig, W.: Stadium architecture and urban development from the perspective of urban economics. Int. J. Urban Reg. Res. **34**(3), 629–646 (2010)

13. Balletto, G., Giuseppe, B.: Sport and City. The case of Cagliari's new stadium (Sardinia–Italy). In: 22nd International Scientific Conference on Mind Scenery in the Landscape-Cultural Mosaic. Palimpsests, Networks, Participation. Aversa/Caserta (Italy), July 2nd–3rd, 2018. ITA (2018)
14. Ladu, M., Balletto, G., Borruso, G.: Sport and the city, between urban regeneration and sustainable development. TeMA-J. Land Use Mobil. Environ. **12**(2), 157–164 (2019)
15. Ladu, M., Balletto, G., Borruso, G.: Sport and smart communities. Assessing the sporting attractiveness and community perceptions of Cagliari (Sardinia, Italy). In: Misra, S., et al. (eds.) ICCSA 2019. LNCS, vol. 11624, pp. 200–215. Springer, Cham (2019). https://doi.org/10.1007/978-3-030-24311-1_14
16. Tjønndal, A., Nilssen, M.: Innovative sport and leisure approaches to quality of life in the smart city. World Leisure J. **61**(3), 228–240 (2019)
17. Balletto, G., Borruso, G., Tajani, F., Torre, C.M.: Gentrification and sport. Football stadiums and changes in the urban rent. In: Gervasi, O., et al. (eds.) ICCSA 2018. LNCS, vol. 10964, pp. 58–74. Springer, Cham (2018). https://doi.org/10.1007/978-3-319-95174-4_5

Towards an Augmented Reality Application to Support Civil Defense in Visualizing the Susceptibility of Flooding Risk in Brazilian Urban Areas

Gustavo Vargas de Andrade[1]([✉]) [iD], Victor Luis Padilha[2] [iD],
Adilson Vahldick[1] [iD], and Francisco Henrique de Oliveira[2] [iD]

[1] Universidade do Estado de Santa Catarina (UDESC), Rua. Dr. Getúlio Vargas,
2822. Bela Vista, Ibirama, SC, Brazil
gustavo.andrade@edu.udesc.br, adilson.vahldick@udesc.br
[2] Universidade do Estado de Santa Catarina (UDESC), Av. Me. Benvenuta,
2007. Itacorubi, Florianópolis, SC, Brazil
victor.padilha@edu.udesc.br, francisco.oliveira@udesc.br

Abstract. This paper presents an augmented reality application for visualizing floods in urban environments, aiming to raise awareness people about occupied areas susceptible to flooding, based on historical occurrences. Environmental disasters by flooding have occurred frequently in different regions of Brazil, where urban areas grow fast with low Master Plan control and integrated use of flood risk maps. At the same time, Civil Defense have been challenged on strategies of how to convince the population to abandon their homes in an iminent flood disaster. In order to help and support these strategies, mainly to avoid occupations in susceptible areas that may be irregular or not, it was developed a mobile application named "Neocartografia". The main goal of the application is to show ordinary people the potential effect of flooding presented in real time and interacting with their own buildings. The idea of integrating the real image of the environment with the simulation of a certain depth of flooding, being set manually or related with hydrological and hydraulic modelling. The Civil Defense agents point the smartphone to the citizen's home, through augmented reality and simulates in the real world the depth of the water according to some parameters (speed, depth and extent of the flood waves). The results earned with the first version of Neocartografia Application have been sufficiently good to reach the developers first goals, but still must be improved.

Supported by organization CNPq through: a) project 426911/2018-0 by the call MCTIC/CNPq Nº 28/2018 - Universal; b) project 307153/2019-3 by the call CNPq Nº 06/2019 - Bolsa de Produtividade e Pesquisa. Special thanks to FAPESC by support the research through the call FAPESC Nº 27/2020 - FAPESC/TO - 2021TR857, and also to Santa Catarina Civil Defense for the partnership. This research received financial support from the Coordination of Superior Level Staff Improvement - CAPES - Brazil (PROAP/AUXPE).

O. Gervasi et al. (Eds.): ICCSA 2022 Workshops, LNCS 13378, pp. 494–506, 2022.
https://doi.org/10.1007/978-3-031-10562-3_35

Keywords: Flooding visualization · Augmented reality · Augmented reality geovisualization · Flood awareness

1   Introduction

Brazil is characterized by being a country that has a territorial extension of continental size and therefore several and different planning actions in its 5.570 municipalities. Among the forms of territorial occupation associated with the land use policies, there are regions with specific profiles, which present specific constructive characteristics in buildings and which are distinctly related to land uses and occupation. The national territorial extension of the Brazilian coastline is approximately 7.491 mi. In addition to the Brazilian hydrography characteristics, it occupies a prominent place among the natural elements.

Despite its national hydrographic wealth, Brazil has historically shown, since its discovery, a significant importance in the formation of urban centers in coastal regions or in inland cities where the banks of rivers have become natural attractions and a basic parameter for the formation of the urban nucleus.

Santa Catarina was a state in southern Brazil that suffered an intense process of rural exodus and therefore many urban centers were impacted by occupation and increase of population density. In this scenario, new buildings appeared in areas not recommended by the master plan, or by occupation actions in irregular areas. As a side effect, the ability of a city (or locality) to become resilient to natural flood events has gradually received more attention from managers/planners, as well as technological advances. Even aware of the susceptibility and vulnerability to which the citizen is exposed, as well as the urgent demand for risk reduction, genuine mitigation actions often only occur after the occurrence of flood events and their respective harmful effects [15].

Therefore, knowing, respecting and especially understanding the meteorological and hydrodynamic condition associated with the risk of flooding or floodable areas around rivers (especially in urban areas) and ocean coast is essential to exercise the resilience of cities and urban centers, as well as to define adequate urban parameters and restrictions on occupation. The concept of population resilience in the context of flood risk management relates to when the urban water system faces floods [15]: (a) the ability to withstand flood events, or the ability to protect against floods; (b) resilience when events occur, or the ability to minimize damage and support a fast recovery; and (c) the self-organization, learning, and adaptability of the system.

The impacts of floods are recognized by the scientific community as being of a complex nature and therefore need to be properly identified and monitored [14]. Whatever the severity of a flood, the impacts to those who are affected can be complex and intangible. Tangible flood impacts are those that are easy to estimate in monetary terms, such as the cost of reintegration or replacement cost of damaged items [13]. While intangible impacts are those that cannot be immediately evaluated, they are often described in qualitative or quantitative terms, such as loss of irreplaceable items or items of sentimental value, health impacts, and psychological effects of flooding.

Residential resilience can be understood as the ability of a property/building to recover from the effects of floods without causing serious damage to the property/building structure [6]. Therefore, to achieve resilience to floods in homes [12], different procedures can be taken: for example, raising items such as electrical outlets above the expected flood levels; or using resilient materials that do not deform/disintegrate on contact with water, such as cementitious materials. In all cases, it is important to ensure that water can be expelled from the building quickly and that an adequate air circulation is maintained around the exposed elements for a reasonably quick drying. For any building, therefore, there may be several possible strategic solutions to contain the ingress of water [12].

On the other hand, it is also important to consider and act in the development of technologies that help citizens to recognize the susceptibility associated with the location of their property, making public and disseminating technology applied to the recognition of potential areas subject to flooding is fundamental, since the climate change is increasingly showing the recurrence of disastrous and intense effects in urban centers related to flooding. In this context, Brazil is still at an embryonic stage in technological development and therefore needs to exercise the power of mitigation or scalable solution of the climatic events effects of intense and concentrated rains that cause floods in urban areas.

Aware of the Brazilian reality and especially the susceptibility of Santa Catarina cities to the harmful effects of floods and the need of the Santa Catarina Civil Defense to act with technological resources accessible to the citizen, aiming to recognize the vulnerability of their housing, the "Neocartografia" application was developed. This application has the purpose of identifying the building and by computational processing using Augmented Reality resources, to project potential different levels of flooding on the image that represents the building.

2   Related Work

At the beginning of the research, the idea was to develop an application that would be useful to society and that would also support actions to raise awareness of society for the Santa Catarina Civil Defense in urban occupations of areas that are susceptible to flooding. In this way, as a starting point for the development of the application, a bibliographic survey was carried out to support the technical, scientific and technological development of the application, taking as a reference other products, prototypes or similar applications - therefore, based on other ideas, the Neocartografia application proposal was conceived and built by the team of researchers. The initial aim was to carry out a scientific exploration of applications for smartphones that could combine scientific and technological concepts related to the interface between cartography and augmented reality.

Based on this principle, the study developed "Innovative Applications of VR: Flash-flood control and monitoring" presents a technical solution that combines virtual reality with augmented reality [8], aiming at the control and monitoring of flash floods.

Another study considered as a reference in the Neocartografia application development research was the one entitled "An efficient flood dynamic visualization approach based on 3D printing and augmented reality" [16], which presents a technical-scientific and technological proposal that combines the reality increased with the creation of a 3D world at different scales, aiming to simulate the occurrence of floods.

Another application entitled "Disaster Scope" [5] was identified in the scientific databases. This application has similar characteristics to the Neocartografia, and its main purpose is to make citizens aware of local susceptibilities and risks related to flooding processes. It was developed in Japan based on the earthquakes that occurred in 2011, as well as the heavy rains of 2018. These natural events hit the population and caused severe consequences, suffering and loss, since people were aware of the size of the risk they were facing and the degree of susceptibility involved.

The"Disaster Scope" application provides local flood simulation, as well as the Neocartografia application, where the augmented reality visualization is calculated in real time and can have its parameters changed in real time. However, the "Disaster Scope" flood simulation requires the use of virtual reality glasses so that the user can immerse himself in the scenario even more.

Another relevant scientific research that was highlighted in the databases, whose projected parameters are similar to those developed in the Neocartografia project, was verified in the Mobile Augmented Reality for Flood Visualization [2] project. The research presents a differentiated application, since the device, instead of performing the detection of plans automatically, demands from the users to touch the screen in order to create anchor points that later will be used for occlusion on the flood simulation.

The studies developed in [11] considering "3D Virtual Environment for Storm Surge Flooding Animation" were focused on the 3D representation of the virtual world where floods occur. Based on the same concept, it appears that the company Bentley Systems also develops similar software solutions and presents"OpenFlows" to the market. Thus, Bentley software produced a digital twin model of the city's water supply, wastewater, stormwater, and bathing water systems, to forecast flooding and water quality issues, thereby improving city response and resilience. But especially the FLOOD software from Bentley allows modeling and the simulation of extreme events using accurate and reliable flood risk analysis data. FLOOD is a flood modeling software for analyzing and mitigating flood risk in urban, riverine, and coastal areas. Using spatially distributed numerical models, users can quickly simulate all hydrological and hydraulic processes to support emergency planning and green-initiative design. Apply a multi-scale 1D/2D approach to support flood early warning systems (FEWS).

It is also important to consider the promising scientific work presented by [3] entitled "In-situ Flood Visualization Using Mobile AR", which uses a computed method of interactive triangulation by correspondences of natural and augmented point.

Finally, it is possible to assume that Virtual Reality (VR) and Augmented Reality (AR) are new techniques for flood visualization that could provide a stereoscopic 3D visualization environment and a user-friendly interactive interface through natural body posture. In fact, VR flood visualizations are more intuitive and realistic than are screen-based 3D visualizations and involve more natural human-computer interactions [1,7,9]. However, VR creates a totally immersive virtual environment isolating users from the real-world circumstances, which also impedes interactions with other users. AR flood visualization superimposes virtual flood scenes over the real environment; therefore, virtual floods and real environments could be observed synchronously, which could address the issue that exists in current VR flood visualizations.

3 Floods in Santa Catarina and Civil Defense

Between 1980 and 2010, there were 1.344 episodes of gradual flooding in Santa Catarina, which left thousands of people homeless and caused serious socioeconomic impacts in the municipalities that were affected. It is noteworthy that the gradual floods correspond to 19,6% of the total natural disasters that devastated the State in the period between 1980 and 2010 [4]. In the months of May and July of 1983 and August of 1984, Santa Catarina faced the most catastrophic floods of the 20th century, which mainly affected the Itajaí Valley mesoregion. During the floods of July 1983 in the State, 197.770 homeless and 49 deaths were recorded, and the most damaged cities were located in the Itajaí-Açu River basin, highlighting the city of Blumenau with 50.000 homeless and 8 dead, which represented 29.3% of the total population of the municipality [4,10].

The year of 2008 was also defined as exceptional, due to the rainfall that occurred during the months of September to November, and were continuous, from moderate to strong intensity, and between the 21st and 23rd of November 2008, from moderate to strong intensity, in a constant way, causing a fast rise in the level of rivers with overflow and rapidly expanding the flooded areas. In fact, the disasters occurred in different cities between the 21st and the 25th of November, whose records of concentrated rainfall during those 5 d were greater than 550 mm.

The graph on Fig. 1, the importance of developing the "Neocartografia" application is concretely justified, since the repeatability of the occurrence of the flood phenomenon recorded by the official state agency (Civil Defense) associated with the harmful effect of the event (flood) that was reported between the years 2013 to 2021. The harmful effect of floods on the state's economy is significant and undoubtedly much more marked in people's lives.

As a clear example of the occurrence of floods, Figs. 2 and 3 present two affected urban areas, the first representing affected low-income buildings and the second a more economically wealthy place. The two examples demonstrate that in the urban planning of Brazilian cities, there is a lack of care when occupying places susceptible to the event of flooding and the lack of control of the public authorities to stop constructive advances - since the real estate market has great power.

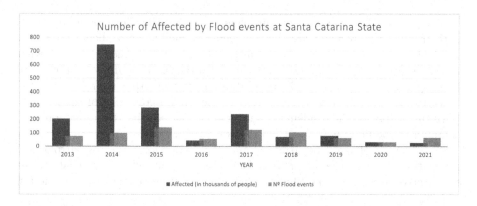

Fig. 1. Graph shows the relation between floods in Santa Catarina State and people affected by it.

Fig. 2. Frequent example of an urban area flooded after a heavy and fast rain. Source: Town hall of São Bento do Sul city/Santa Catarina (2021)

Fig. 3. Effects of flash flood on wealthy building underground garage at Florianópolis city. Source: the authors (2020)

The developed device will allow the user to download the application and, having available the historical data of the occurrence of floods in the place, will automatically have available on the screen of his smartphone the projection of the water feature (virtual) applied to the real image - facade of the building.

The visual impression will help the user to decide to stay in the place of potential risk, recognize and identify the dynamics of the phenomenon and the impact on buildings and movable assets.

The Civil Defense will have one more instrument to support management and raise awareness of citizens which, through the visualization device of the "Neocartografia", will place themselves in a new environment of interaction and spatial representation of the object and phenomenon.

For people who experience the problem and suffer the effects of flood events, in their respective recurrences, they somehow learned to be resilient and Figs. 4 and 5 show it in fact. In the first situation (Fig. 4) it is verified that the owner raised the level of the public sidewalk and also the internal area of the patio of his property. In Fig. 5 - from the opposite point of view, that is, from inside the house to the outside view, it is possible to see the lowered level of the kitchen. With this behavior, the resident of the property created a barrier and a potential damming if the flood is higher than the concreted area.

The history of flood occurrences and the projections by "Neocartografia" application will help the owner to make decisions about which technical actions are effective to mitigate the effects of the flood, or improve resilience techniques.

Fig. 4. Main entrance to the house with a concrete elevation on the sidewalk to avoid the damaging effects of flooding. Source: authors (2019)

Fig. 5. House with backyard elevation that causes a barrier and the kitchen at same street level, resilience that needs to be improved. Source: authors (2019)

Currently, the Santa Catarina Civil Defense has a weather forecasting system as technology, which, when identifying a potential flood event, automatically sends an alert SMS (Short Message Service) to residents (registered on the alert platform).

Therefore, the Neocartografia application is configured as another instrument to help prior to the occurrence of the flooding phenomenon that will instruct the users about the susceptibility to the problem, since occupying these places in which their houses are located, they are subjected to the harmful effects of its occurrence.

4 Neocartografia: An AR Application to Simulate Flood Events

The main goal of the Neocartografia application is to be used as a tool to demonstrate real-world flood events in an augmented reality situation. It can be used by civil defense agents to predict the water depth situation in the residents' homes, and then convince them to withdraw and secure their lives.

This software was developed using the Unity game engine, which has several libraries and documentation about augmented reality applications that facilitate the development process of this solution. The first version of Neocartografia was made using the ARCore framework. It's a development platform for creating augmented reality applications developed by Google. It had the limitation of only being able to create android AR applications. However, ARKit is a set of tools created by Apple to aid developers in creating augmented reality applications for iOS devices. ARKit had a similar limitation as ARcore, it only allowed to create IOS applications. Finally, AR Foundation is a package that allows building cross-platform AR applications in Unity. AR Foundation can use features from both ARCore and ARKit frameworks in the same project. So, it works with horizontal and vertical plane detections on Android devices. Then we decided to migrate from ARCore to the AR Foundation framework.

It's well known and established that iOS devices can run augmented reality softwares better in major categories. Devices like iPad Pro 2021 and iPhone 13 Pro have a LiDAR scanner next to the camera that increases the overall potential of identified areas in augmented reality. Most Android devices can run AR applications up to 30 Frames Per Second (FPS) where iOS devices can run up to 60 FPS, giving a more fast and fluid simulation. There are some Android devices like most Pixel phones and the Xiaomi MI 11 Ultra that support the same as iPhone. The choice was obvious to develop Neocartografia for iOS devices. However, the sale prices of Apple devices in Brazil make it infeasible to purchase them. Consequently, we focus development and testing on Android devices.

AR Foundation uses a point cloud manager that creates sets of feature points. A feature point is a specific point in the point cloud which the device uses to determine its location in the world. The plane detection is done from a point cloud. Figure 6 illustrates the algorithm of real-time extraction of planes developed in Neocartografia. First, it identifies the planes and assigns an anchor to some specific points. Then, it identifies the lowest point on the Y axis, this will be the main anchor until a lower anchor is identified. So this main anchor is used to generate the water plane in the X axis and place it in a Y axis + Depth (informed by the user). This plane has all the water attributes to make a more real simulation, such as wave size and speed, and it uses a mesh renderer with water texture. The waves simulation is made by displacing parts of the mesh, giving to the texturized plane a water-like movement. AR Foundation gives an extra layer of immersion with an occlusion script that interacts directly with the water plane, merging real life objects with the simulation. This feature is illustrated in Fig. 7 where the water plane is occluded by the walls.

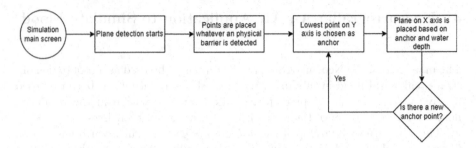

Fig. 6. A macro view of Neocartografia's system using a flowchart.

Neocartografia's interface was designed to make interaction with the application simple, without major compromises. The user simply points his or her smartphone in some direction, and the planes are detected. Then she/he can adjust the simulation parameters: water depth, waves speed and size. Figure 7 demonstrates this interface.

Fig. 7. Neocartografia simulation screen with settings menu open.

The National Water and Sanitation Agency (ANA) collects all kinds of data about rivers and lakes and uses it to ensure sustainable use of these river resources. It provides a free API to query a specific water station on a series of dates. The use of the API and some of the results are exemplified in Fig. 8.

http://telemetriaws1.ana.gov.br/ServiceANA.asmx/DadosHidrometeorologicos?CodEstacao=83800002&DataInicio=07/08/2021&DataFim=08/08/2021

```
<DadosHidrometereologicos diffgr:id="DadosHidrometereologicos1" msdata:rowOrder="0">
    <CodEstacao>83800002</CodEstacao>
    <DataHora>2021-08-08 23:45:00 </DataHora>
    <Vazao>9.97</Vazao>
    <Nivel>25.00</Nivel>
    <Chuva>0.00</Chuva>
</DadosHidrometereologicos>
```

Fig. 8. Example of communication with ANA's API.

In order to give an opportunity to investigate data from historic flood events, Neocartografia allows simulates flood events, for instance, that happened ten years ago. This feature enables civil defense agents to get information about areas that are at risk from a similar event.

Figure 9 demonstrates on (A) how this integration works in Neocartografia. In the search screen, the user fills the station ID and the time period. Figure 9 (B) shows the search result. There is a combobox component with the series of measurements identified by the date and time stamp. User selects an item from the combobox and clicks on it to reveal what was the flood situation on that specified date and time.

Based on our tests and with the limitations in mind we came up with a formula of how to improve floor identification; Fig. 10 represents on (A) the main formula, where the effective view and effective simulation distance lean on the angle of A and total height of Y, simplified by the function:

$$E = |(A/180)| - Y \tag{1}$$

Where E =<0 better

Still considering the same figure, example B shows the simple scheme where by the formula, the effective view and effective simulation distance lean on the function:

$$E = |Y| - |A| \tag{2}$$

Where E =<0 better

The two formulas presented above aim to obtain a more precise identification of the X plane and an ideal height of the device for this purpose is between 1 and 8 m.

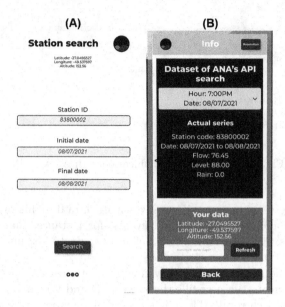

Fig. 9. ANA API screens on Neocartografia being: A) Search screen of ANA historic event; B) Result screen.

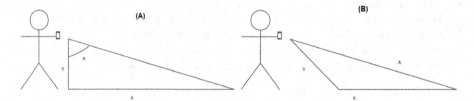

Fig. 10. User and device triangulation on (A) and device view triangulation on (B).

5   Discussion and Conclusions

The general perspective of AR flood visualization involves superimposing the scaled virtual flood scenario onto the real environment (urban), which could provide a better visual perspective for flood mitigation or decision making.

As described in the Related Work section, significant efforts are being made to simulate flooding and visualize it in real urban environments. The present work presents a low-cost solution that can be affordable for the population. Although the focus of the development is on an application to support Civil Defense, the direction of this work is focused on moving forward so that any citizen can perform the simulations at home.

The tests performed by the Neocartografia application developer team showed that the device is useful and has a valuable application for the reality of irregular occupations in urban centers of Brazilian cities, presenting an excellent performance as a system for visualizing the susceptibility to risk in which families are exposed. The visualization of AR projected on the buildings brings a real approximation regarding the risk and danger to which the residents are susceptible, reducing the need for technical arguments about moving from that place.

The application also proved to be important for integrating databases, such as the National Water Agency (ANA), which keeps historical data on the processes and harmful effects of urban floods, as well as other technical parameters.

Finally, the system described on the present study indicates the assessment of 3D hydrodynamic simulation of flooding, allowing to assist the state, through control and monitoring, by showing people what a expected amount of rain is capable to reach in terms of flooded areas instantaneously visualized, based in the specific topography established. In conclusion, the Neocartografia is an application useful for assisting participants, civil defense and decision makers in understanding the flood hazard at different urban areas and also provide a more intuitive and realistic visual experience, based on AR technology.

From the observed results, it can be concluded that the Neocartografia application reached its purpose considering the first version developed. Furthermore, similar technologies related to this research must be supported by national agencies and developed by experts from academia to mitigate the effects from natural disasters in Brazil.

6   Challenges and Recommendations for Future Work

Therefore, still some issues need to be solved. First, hardware limitations restricted some test cases. Besides not being able to use iOS devices with LiDAR technology, we noticed that in a wide outdoor environment, the camera in our smartphones (Samsung Galaxy Note and S 2018 series) could not correctly identify distant planes, thus not placing the water plane in places not close to the device's camera.

Test suites in partnership with the experts that work at Santa Catarina Civil Defense will be the next steps, from which we can improve the application features.

For future work we intend to make the Neocartografia searches on ANA's API even better to query and visualize on the application. An IOS version could carry out the actual potential of simulation to a higher level and also adopt the smoothness of data input using the LIDAR technology, which would make the distance measurements more trustful for collecting data and also for visualization.

References

1. Chen, M., Lin, H.: Virtual geographic environments (vges): originating from or beyond virtual reality (vr)? Int. J. Digital Earth **11**(4), 329–333 (2018). https://doi.org/10.1080/17538947.2017.1419452, https://www.tandfonline.com/doi/abs/10.1080/17538947.2017.1419452

2. Haynes, P., Hehl-Lange, S., Lange, E.: Mobile augmented reality for flood visualisation. Environ. Model. Soft. **109**, 380–389 (2018). https://doi.org/10.1016/j.envsoft.2018.05.012, https://www.sciencedirect.com/science/article/pii/S1364815217302529

3. Haynes, P.S., Lange, E.: In-situ flood visualisation using mobile ar. In: 2016 IEEE Symposium on 3D User Interfaces (3DUI), pp. 243–244 (2016). https://doi.org/10.1109/3DUI.2016.7460061

4. Hermann, M.d.P.: Atlas de desastres naturais do estado de santa catarina: período de 1980 a 2010. IHGSC, Florianopolis (2014)

5. Itamiya, T., Tohara, H., Nasuda, Y.: Augmented reality floods and smoke smartphone app disaster scope utilizing real-time occlusion. In: 2019 IEEE Conference on Virtual Reality and 3D User Interfaces (VR), pp. 1397–1397 (2019). https://doi.org/10.1109/VR.2019.8798269

6. Joseph, R., Proverbs, D., Lamond, J.: Assessing the value of intangible benefits of property level flood risk adaptation (PLFRA) measures. Nat. Hazards **79**(2), 1275–1297 (2015). https://doi.org/10.1007/s11069-015-1905-5

7. Massaâbi, M., Layouni, O., Oueslati, W., Alahmari, F.: An immersive system for 3d floods visualization and analysis. In: iLRN (2018)

8. Padilha, V.L., de Oliveira, F.H., Proverbs, D., Füchter, S.K.: Innovative applications of vr: Flash-flood control and monitoring. In: 2019 IEEE International Symposium on Measurement and Control in Robotics (ISMCR), pp. A2-4-1–A2-4-3 (2019). https://doi.org/10.1109/ISMCR47492.2019.8955726

9. Philips, A., et al.: Immersive 3d geovisualization in higher education. J. Geogr. High. Educ. **39**(3), 437–449 (2015). https://doi.org/10.1080/03098265.2015.1066314, https://doi.org/10.1080/03098265.2015.1066314

10. Pompêo, C.A.: Drenagem urbana sustentável. Revista Brasileira de Recursos Hídricos **5**(1), 15–23 (2000)

11. Presa Reyes, M.E., Chen, S.C.: A 3d virtual environment for storm surge flooding animation. In: 2017 IEEE Third International Conference on Multimedia Big Data (BigMM), pp. 244–245 (2017). https://doi.org/10.1109/BigMM.2017.54

12. Rose, C., Lamond, J., Dhonau, M., Joseph, R., Proverbs, D.: Improving the uptake of flood resilience at the individual property level. Flood Risk Manag. Response **6**(3), 153–162 (2016)

13. Tapsell, S., Penning-Rowsell, E., Tunstall, S., Wilson, T.: Vulnerability to flooding: Health and social dimensions. Philos. Trans. Ser. A, Math. Phys. Eng. Sci. **360**, 1511–25 (2002). https://doi.org/10.1098/rsta.2002.1013

14. Werritty, A., Houston, D., Ball, T., Tavendale, A., Black, A.: Exploring the social impacts of flood risk and flooding in scotland, January 2007

15. Xu, W., Cong, J., Proverbs, D., Zhang, L.: An evaluation of urban resilience to flooding. Water **13**(15), 2022 (2021)

16. Zhang, G., et al.: An efficient flood dynamic visualization approach based on 3d printing and augmented reality. Int. J. Digital Earth **13**(11), 1302–1320 (2020). https://doi.org/10.1080/17538947.2019.1711210, https://doi.org/10.1080/17538947.2019.1711210

Framework Proposal of Smart City Development in Developing Country, A Case Study - Vietnam

Tu Anh Trinh[1]([⊠]), Thi Hanh An Le[2], Le Phuc Tam Do[1], Nguyen Hoai Pham[1], and Thi Bich Nguyet Phan[2]

[1] Institute of Smart City and Management, University of Economics Ho Chi Minh City, 700000 Ho Chi Minh City, Vietnam
{trinhtuanh,tamdlp,hoaipm}@ueh.edu.vn
[2] University of Economics Ho Chi Minh City, 700000 Ho Chi Minh City, Vietnam
{anlth,nguyettcdn}@ueh.edu.vn

Abstract. Smart City has been determined as the key solution for urbanization which is considered as the inevitable trend of urban areas. There are many cities aim for the goals to become Smart City, but mostly failed because of different challenges. There is a desire for Smart City strategy of developing countries which having cities that begin to significantly face with the difficulties of urbanization process. This study offered a revolutionary smart city approach and framework for implementation in developing countries considering Vietnam cities as the case studies. Based on the revision for critical factors and journey of Smart City development of successful smart cities globally, the study proposed groups of solutions and framework for building Smart City effectively towards the sustainability in developing countries.

Keywords: Smart City · Framework · Urbanization · Urban planning · Sustainable development

1 Introduction

Urbanization is currently taking place strongly around the world, in four different directions which are natural population growth mainly in Asia and Africa; rural-to-urban migration; regional reclassification between rural and urban areas [1] and international migration from poorer, politically unstable countries to wealthier countries. The population living in cities currently accounts for nearly 55% of the total population and is forecast to grow to 70% by 2050, accounting for 60% of global GDP [2]. It brings positive impacts to urban life such as economic benefits, social communication, job opportunities, innovation, life utilities including health care, education, entertainment, and transportation [3, 4]. On the other hand, urbanization has negative impacts on the natural environment, public health, and social equity [5].

Urbanization leads to land degradation. It is estimated that 10% of land with agricultural potential gets concreted every year, which reduced crop and pasture productivity,

O. Gervasi et al. (Eds.): ICCSA 2022 Workshops, LNCS 13378, pp. 507–519, 2022.
https://doi.org/10.1007/978-3-031-10562-3_36

poor vegetation, and biodiversity loss, which has the reverse impact of hastening soil erosion and degradation [6, 7]. Ninety-seven percent of the earth's areas has changed significantly due to human impacts on construction, population density, electricity infrastructure, productivities of crops and pastures, and transportation infrastructure, resulting in a biodiversity crisis [8]. Urbanization also has a negative effect on the environment. Industrial exhaust gas, vehicles are the main sources of air pollution. Wastewater from industrial parks, business households, and even from household activities caused surface and ground water pollution and soil pollution [3]. Urbanization contributes to global warming due to the heating of buildings, concrete structures, as well as the use of fossil fuels such as coal, gasoline, and diesel in industrial parks, leading to the greenhouse effect which negatively affects citizens' health, labor productivity, and leisure activities [9]. As cities have expanded in terms of space, both breadth and height, population density have increased, and the infrastructure becomes overburdened, resulting in higher transaction costs and opportunity costs. More especially, urban land capacity for living is shrinking. It was reported that about 25% of Vietnamese people cannot afford housing and 20% of accommodation is inadequate for living [10]. The process of urbanization brings both positive and negative effects to all cities globally in the same trend, just different in the matter of level and the reaction of them.

Considering solving the above urban issues, government and different stakeholders have implemented different ways and aim for different objectives. In fact, scholars have found, applied and developed many city models in urban development. Every type of city like eco-city, livable city, green city, or smart city was established and developed aiming to solve urban problems that citizens have to face and fulfill the demand of the citizens [11]. With the ongoing Industrial Revolution 4.0, Smart City has established as an inevitable trend, highlighted with the transformation of the real world into the digital world through new technologies such as the Internet of Things (IoT), Artificial Intelligence (AI), Virtual Reality (VR), social media, cloud computing, mobility, big data [12]. Although the concept of Smart City has only been widely known since 2015 [13]. Smart City was mentioned in the 1990s when the most important factor in developing Smart City was information and communication technology [14].

Nowadays, when researchers looking for the definition of Smart City, there can be many approaches. From an academic point of view, a city is "smart" when the investment is mainly for human resources, society, traditional and modern communication infrastructure and promotes sustainable economic growth and improves quality of life, with wise management of natural resources, through participatory governance [15, 15]. From the government point of view, Smart City is a high-tech-intensive and advanced city that connects people, information, and elements of the city to create more sustainable green cities, competitive trade, innovation, and improves the quality of life with the city's simple management and maintenance system [17]; in Smart City, information technology is combined with infrastructure, architecture, appliances, and even our bodies to solve economic, social, and environmental problems [12]. Therefore, the development strategies of Smart City are various in different cities leading or inspiring by many different organizations. This paper is focusing on generating the framework of Smart City implementation according to the revision of both successful and fail cases.

2 Success Development Factors

According to OECD experts, about 80% of current Smart City projects are not adequate because they just can provide solution for one issue of cities, but not focusing on upgrade the living quality of residents [18]. In fact, cities are ranked in top Smart City lists of many countries around the world proves that building and developing Smart City is a long journey. Reality has proved that every Smart City takes years to accomplish the development. In order to figure out the suitable framework for Smart City implementation, authors have conducted a revision of successful Smart City cases and collected the similarities among them which are believed to be the main factors to contribute to their success.

2.1 Smart City Image

The first common action which can be observed clearly from the best Smart Cities in the world that every period, cities will create a city brand or city image which usually based on cities characteristic or intense problem that needed to be solved, to become the vision for city development, thus government can set up mutual goals for every stakeholder. As the goals and vision are achieved in the set period through effective application of new technologies and strategy, Smart Cities continue to choose a new urban identity, as well as the vision of city authority which support the orientation [19]. It can be clearly observed from the top 10 smartest city development in 2018 which are New York, London, Paris, Tokyo, Reykjavi, Singapore, Seoul, Toronto, Hong Kong and Amsterdam [20].

There are two main methodology which have been applied by numbers of places. Cities with unique characteristics usually take their strengths or features in terms of culture, economy and society as the urban image. Has been well known as the European tourist capital, Paris decided to take its image as "Attractive destination" focusing on green and friendly environment, high-speed Wi-Fi in crowded tourist spots, and densely populated places; solutions for clean energy and renewable energy; public transport and promotion of non-motorized traffic [20]. Somerville in USA, which has the smartest government in the Commonwealth and pioneer in the Smart City development trend, selected "Smart livable community" as urban image. This has clearly reflexed the strengths of Somerville when it was 3 times nominated as All- America city (The award honors communities whose citizens work together to identify and solve common challenges) and one of the 100 Best Communities for Young People. On the other hand, many cities don't own any special characteristic to compete with other region, thus they follow the second methodology that concerning their current urban problems while selecting the city identity. For example, with the slowing down of economic growth after the financial crisis, New York (NYC) decided to build an urban identity as an "Innovative Economic City". All areas are developed solely for the purpose of promoting economic activity for all groups [20]. London, one of the most congested cities in the world, decided to solve this problem by developing a more efficient public transport system. And London takes the development of an efficient public transport system as its urban image [20]. And most importantly, the vision of city authority in this period needs to be very clear, and consistent with city identity. Smart vision also approach target stakeholders in the city (authorities, businesses, investors, tourists, residents) and will be the key point to mobilize the participation of all sectors, contributing to the success of each project [21].

2.2 Prioritization of Goals

Each stage of Smart City development should focus on the main prioritized objectives according to the selected above city identity, the other aspects of the city will be integrated into the set of action plan when solving the main problems. After all, all urban problems are related to each other and should not be clearly separated in city vision. But with the application of different priority focus, city authority and developers will be able to take advance of their resources effectively and avoid wasting of investment on various issues [22, 23]. In the Smart City development strategy from 2016–2018 of Seoul taking Smart Transportation as the city image [20], authority raised up two problem which are traffic congestion (too much use of personal transport) and environmental pollution (air, waste discharge, lack of electricity) affecting the quality of life for solution. Then the city decide to gain the prioritization for the following objectives: (i) smart mobility, (ii) waste management and smart energy; (iii) smart citizens combined with smart management, (iv) smart public space [20, 24]. In the Smart Environment strategy of Singapore Smart City, its government has separated it into 3 clear phases which are: (i) Aiming to create more green spaces in the city (gardens in the city); (ii) linking the green spaces with each other to create large green areas, occupying the main space in the city (city in the garden); (iii) establishing a friendly living environment with the existing green spaces, so that urban life can be closer to the natural ecosystem (City in nature). With this effort, although Singapore's population density is among the largest in the world, the tree cover density ranks among the highest in the world [20, 25].

2.3 Smart Government and Smart Community

The foundation of successful Smart City cases is not the well implementation of information technology platforms, but the initiative to change the mindset and approach of the government regarding of the transparency, willingness to share resources, data, management methods, decision-making for all target groups in society. Residents and other stakeholders (enterprises, agencies, institute, university, etc.) participation for initiatives, smart solutions, decision making will support government in the process of deploying smart economy, smart mobility, smart environment and smart life for smart city development. That is called smart government and smart community [20, 26–28]. For example, Seoul opened a call center called Dasan with the dial number 120. This is the government's phone complaint handling system, directing all inquiries and questions from the residents to one integrated call center, responding to people's daily complaints more quickly and conveniently based on a one-to-one consultation. The system was operating 24/7 and received about 22,000 calls per day in 2016 [20, 29]. Government leading all the cooperation activities with transparent information, objectives and action plan is the highlight strategy of Smart City development in Singapore. This city authority also mobilized all the social resources regarding of public private partnership projects, including the participation of private enterprises, universities, research institutes, residents [30].

The Somerville government vision in urban planning and design is "access – dialogue – decision – implementation" aiming to encourage residents to collaborate them to look at the future of the city, towards a sustainable quality of life. New ideas are tested and determined by gathering input and feedback from the community through the open data platforms Dropbox, which allows uploading and downloading; online survey platforms such as Survey Monkey, Mind Mixer; crowdsourcing social networking platforms like Twitter and Facebook and a network of community connecters - able to reach and chat with anyone in the area [31].

2.4 Integrated Platform

There is an essential requirement of an integrated platform to advance the connection, communicate understanding, mobilize participation, effective cooperation with different stakeholders like authorities, universities, research institutes, organizations, private enterprises, residents in the development process, urban problems solving [19, 32]. Seoul in its period of smart transportation promotion, applied the advance technology including Smart Transportation application, Bus Management System and GPS has reduced the waiting time of public means of transport and encouraged citizens to stop using private vehicles. As result, MRT and Bus users' proportion was about 70% while the percentage of people driving private car was just 30% in 2013 [18, 33]. Smart Nation is an important platform in the Smart City Development journey of Singapore, which include 5 smaller projects: national electronic identification system; smart sensor system across the country; intelligent transportation platform; electronic payment gateway and moments of Life (mobile application that helps the Government deliver the right services quickly to residents) [34]. Regarding the aspect of smart urban planning and management, Somerville government has published an application called ResiSTAT running on the SomerSTAT data system for create connection between government and citizens. The city leaders can supervise all the activities and performances of each department based on the system and have in-time decisions [35].

2.5 Detail Plan for Once at a Time

Last but not least, it is necessary to generate a clear plan with only one or two main goals for each stage as it has been confirmed that a smart city is a journey, not a destination, thus no one can foresee what a Smart City will look like at the end [36, 37]. Seoul is considered one of the leading cities of Smart City development trend when it is continuously in the list of the top 10 smartest cities globally through the assessment of measurement organizations on global Smart Indexes. In fact, Seoul has had a long way to develop Smart City with urban images generated in each stage from urban solving problems or enhancing the satisfaction of citizens. Looking back the journey of Seoul, it can be seen that 2001 marked the first Smart City image as "The world's leading E-government", in 2011–2015 as "Smart Seoul", and 2016–2018 as "Smart Technology in Transportation". From 2019 onwards, Seoul authority has considered this city as the "Data Capital" [38–40].

3 Analysis of Smart City Developments in Viet Nam

In order to generate the suitable solution in the context of developing countries, there must be a revision and analysis of existing cases. Facing the urban challenges, Vietnam cities have huge opportunity when the government has been paying a huge attention on Smart City as the chosen solution along with the industry 4.0 and digital transformation emergence. Since 2015, government of Vietnam and the related departments have released many policies to favor and encourage the development of Smart City, especially the publish of E-government to support Smart Economy, Smart Tourism, Smart Education or Smart Healthcare [41, 42].

In general, many cities in Vietnam have shifted their focus on Smart City Development, even before government encouragement. According to MIC, there are nearly a half of Vietnamese provinces and cities develop services relating to Smart City; 12 out of them have deployed their Smart City management center; 13 provinces and cities have published Smart services which 10 of them have transportation e-service [43]. Looking back along the journey of Smart City development in Vietnam, authors have realized the critical change in the strategy of Vietnamese cities leader's approach.

At the starting point when developing Smart City, most of Vietnamese cities believe that post and telecommunications will be the main foundation, so it can be seen that 30 provinces and cities have signed agreements to develop Smart City or Smart City features with big post and telecommunications companies in Vietnam like VNPT, Viettel. In order to strengthen the services of telecommunication and internet application, cities like Da Nang, Phu Quoc improved their accessibility for internet connection by constructing the free Wi-Fi, smart Wi-Fi system and infrastructure [44, 45].

After a while developing Smart City, city leaders in Vietnam have shifted their focus to technology, thus many localities believe that new technology is a prerequisite for smart city development. The establishment of executive boards in many Vietnamese cities for Smart City Development has usually included technology firms' representatives. Regarding of information technology, Ha Noi, Ho Chi Minh and Binh Duong are the three big cities in Vietnam deployed government and economic e-service with internet platform. The localities believe that the transparency and responsiveness will enhance the connection between authority and other private stake holders and citizens [46–48]. On the other hand, focusing on technology implementation for solving urban problems or infrastructure management, Da Nang and Phu Quoc invested in many projects of transportation, environment management [49, 50]. However, learning from successful cases of Smart City globally, the developed post and telecommunication system as well as technology are just supportive tools in the journey of Smart City development. For successful and effective development, it is necessary to have executive representatives from many different groups in society having experience in the field of Smart City development.

In the following period, Smart City development strategy aims for too many goals and too many development directions, thus many projects do not even account for the characteristics of the city or the urban problems which needed to be solved. There are no prioritization in order of action plan while the participation among related stake holders nearly didn't exist. Despite the success in Smart City technology application. Da Nang authority hasn't really had a prioritized list of issues to be resolved first, but rather spread out all issues. The objectives are in all different directions while there is limitation in resources. A few solutions have been proposed and implemented, but there has not been an appropriate priority in the context of limited resources. Technology incentives have been mainly focused so solve urban problems according to authority's vision, but not from citizens perspective while the city image is livable city [51].

Learning from the world when Smart City has become the global trend for urban area. Many Vietnamese cities have connected or corporate to be consulted by foreign companies with experience in Smart City development. Most important when they have now generated their own cities images according to their characteristics or prioritized problems. To develop smart city, Binh Duong consulted the experience of implementing the Eindhoven Smart City model (Kingdom of the Netherlands) and set its most important goals is attracting high skill labor force for economic development. City authority signed a memorandum of understanding with Eindhoven city (2015) to cooperate and implement the economic development model based on the cooperative society (state, entrepreneurs, scientists/research institutes/universities) [52]. Da Nang has also promoted its urban image as "Livable Smart City" after referencing many Smart City cases in Barcelona, Valencia, Amdalusia, Yokohama, Keihanna, Kitatyushu, Kashiwa-no-ha Seoul, Songdo, Pangyo, Anyang, Busan, Dongtan, London, Dubai, Singapore [53]. The Smart City project of Can Tho was implemented starting in 2020 with the consultation of the Asian Development Bank (ADB). ADB proposes Can Tho to become a smart and energy-saving city (identified as the city image). The focus of the project is on developing LED lights infrastructure and smart technology for public lighting and buildings [54].

4   Framework Proposal for Smart City Implementation

According to the analysis of previous experience for Smart City development, this paper aims to develop a framework for the implementation of Smart City for cities in developing countries facing the difficulties of urbanization. In fact, there is no one universal formula for all cities. Each city is a separate entity with many different factors to consider, thus the authors only propose a framework of development steps (see Fig. 1).

When the developers begin to study further for specific practice, each step will have to be implemented in detail. It is necessary to go into in-depth analysis and reality of each locality to have a smart city development program most suitable to the context of each locality and most effective.

4.1   Identification of City Image and Government Vision

The first and most important thing is that the government must determine the characteristics or image of the city to build and develop in each stage of the Smart City development strategy. The selection of the city image can be based on the inherent characteristics of the city in terms of culture, economy and society. When determining the characteristics of each locality or city, developers should take in mind that they should consider the unique image or strength of their place in the whole region. However, there will be many cities don't own any specialty, thus authors suggest that developers should identify the most critical issues to generate the solution and keep it as their city's images.

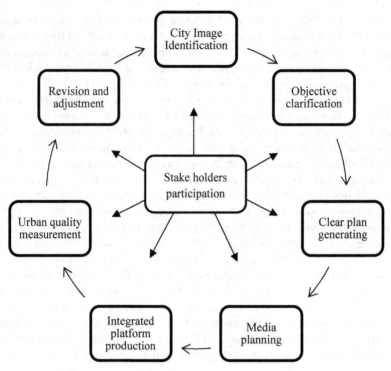

Fig. 1. The proposal framework for Smart City implementation

For instance, 2001 marked the first image of Seoul as "the world's leading e-government", in 2011–2015 was "Smart Seoul", and 2016–2018 was "smart technology in transportation", from 2019 onwards, it has become the "City data" [38–40]. This is a proof that smart city development is a continuous process aimed at solving various urban problems and enhance the quality of life, not just stopping with the destination as a title about smart city. City authority usually use the city image to express their vision and clearly describe their long-term orientation in the future. In particular, a smart vision also needs to show the cohesion of target groups in the city from authorities, businesses, investors, tourists and residents together to achieve the goal which is quality of life.

4.2 Objective Clarification

It is necessary to define clear and specific goals in each development stage corresponding to the selected city image with reasonable resources. However, the important thing is that only 2 to 3 goals should be selected in one stage, thus the implementation is more focused, and solutions are provided effectively.

The chosen objectives should directly connect to the city image made by urban unique characteristic and the prioritized problems. In this stage, developers and city authority must define the potential stake holders, in order to mention them and create a sense of responsibility for them. The other aspects of the city will also be integrated into the set of objectives at the same time to solve other urban problems, more over the mutual solutions for the mutual problems will give more opportunities for stake holders to find a win-win solution and save the resources.

4.3 Clear Plan Generating

There should be a clear strategy to achieve each objective regarding of methodology, testing, monitoring, solution steps proposal, and how to mobilize human resources must also be given, Stakeholders should also identify their roles and how they will involve in each phase.

Technology is an important factor in planning for Smart City implementation, but in this stage, a detail plan including clear responsibility and action plan for each stakeholder is essentially required.

4.4 Media Planning

This is an important stage to attract the participation of the entire society in the development and implementation of Smart City projects. City branding not only helps city authority to address weaknesses, as well as exploit strengths of the city, but also to increase the competitiveness globally. Moreover, the branding project will spread the information to internal stake holders (local community, local enterprises, university and institute) and external groups (investors, tourists, immigrants), which helps them understand the possible benefits and values of the city's overall strategy, along with guidelines and policies that facilitate and reduce the barriers. The city branding strategy should take place throughout the implementation of smart city projects with messages focused on the City Image. Thus, the efficiency of smart city development will be enormous, and it will be easier to call for capital in all aspects as well as increase the responsibilities of stakeholders. At these integrated platforms (it can be social media, websites, apps, or new technology programs, etc.) businesses, university, research institutes, communities can together propose ideas, test, monitor, plan, manage, feedback. With this integrated platform widely sharing information to every stakeholders, the transparent, the commitment and responsibility, roles and impact of all parties will clearly be observed.

4.5 Integrated Platform Production

The government needs to create an integrated platform where stakeholders can directly interact to each other in different projects in order to achieve mutual objectives. On this integrated platform which can be social sites, websites/apps, new technology programs, etc., businesses, universities, research institutes, communities can propose ideas, test, monitor, plan, manage, and give feedback. With this way, the information will be transparent to all stake holders and the responsibility of all parties will be clearly expressed.

It is necessary to realize that the role of universities and research institutes is not only producing qualified human resource and fulfill the social needs, but also to apply research and assist in solving real urban problems, decision making. They can as well verify the implementation and sustainable development of projects to ensure Smart City development in the right direction and efficiency.

4.6 Urban Quality Measurement

Building smart city is a continuous and long journey, thus a set of indicators is needed to monitor and measure the development process. According to this research, smart city is a developed towards the ultimate goal of quality of life, so the index to evaluate smart cities should focus on the factors that assess the urban living quality.

This paper propose to generate an evaluation on 7 main groups of factors which are the quality of the natural environment, the quality of the physical environment, the quality of the traffic environment, the quality of the social environment, the quality of the economic environment, the quality of the political environment and the quality of the spiritual environment (discussed in detail in Sect. 3). However, the evaluation specific methodology will vary depend on the local condition of different cities.

4.7 Revision and Adjustment

This is the final stage in developing a Smart City towards the quality of urban life. If the process has been done well, city authority just have to inherit and keep developing, otherwise, it needs to be adjusted for the next round of Smart City development.

In conclusion, in order to successfully develop an effective smart city, in addition to implementing the proposed framework, three extremely important complementary factors need to be also considered area (1) the transparency about information and activities of the government, (2) mobilization to create a foundation (mechanisms, policies, regulations) for stakeholders to actively participate in the smart city development process in the future in every stages, (3) it is necessary to have a consultant from local and international universities, research institutes with enough experience and capacity to lead and support the process.

5 Conclusion

From the analysis of urbanization process and its problems globally, it can be realized that every city has its own characteristics in terms of people - culture - natural conditions but the effect of the urbanization process is inevitable. Smart city development is a necessary movement for them to effectively solve internal problems towards sustainability. Although Vietnam is not a pioneering country in Smart City development, thus city leaders should learn from experience of successful smart cities in the world and extract the suitable knowledge for local implementation. Smart city is a journey; thus no one knows what the destination would look like. Therefore, smart cities can be developed with their own characteristics solving their local problems. Reality has proven that the policies and mechanisms proposed by Vietnamese government show the support for Smart City development. However, the previous implementation strategies haven't showed any critical result in satisfying the demand of citizens, which is due to the lack of thorough understanding, subjectivity, unwillingness to be transparent about activities in government representatives and organizations, poor cooperation among stakeholders, lack of strict control. All smart cities in the world have acknowledged that the prerequisite for success to develop smart cities is how people manage, interact and react with the development instead of just advance technology. Therefore, to effectively deploy the Smart City project in any country, city leaders must have a clear framework from city identity selection to urban quality measurement.

References

1. ESCAP, U.: Housing the Poor in Asian Cities, No 5: Housing Finance: Ways to Help the Poor Pay for Housing (2008)
2. DESA, U.J.U.N.: World Urbanization Prospects—Population Division: The 2018 Revision. Department of Economic and N. Social Affairs, New York (2018)
3. Rasoolimanesh, S.M., Badarulzaman, N., Jaafar, M.J.J.o.S.D.: Achievement to sustainable urban development using city development strategies: a comparison between cities alliance and the World Bank definitions. 4(5), 151 (2011)
4. Nguyễn, T.L.: Nhận diện vấn đề đô thị và quản lý phát triển đô thị khi đất nước dần trở thành nước công nghiệp theo hướng hiện đại, in Tạp chí Cộng sản. Tạp chí Cộng sản, Hanoi (2021)
5. Gu, C.J.S.B.: Urbanization: positive and negative effects. 64, 281–283 (2019)
6. Gomiero, T.J.S.: Soil degradation, land scarcity and food security: reviewing a complex challenge. 8(3), 281 (2016)
7. Chiến, M.: Cả nước có 1,3 triệu ha đất bị suy thoái, in Nông nghiệp Việt Nam. Báo Nông nghiệp, Việt Nam (2019)
8. Venter, O., et al.: Global terrestrial human footprint maps for 1993 and 2009. 3(1), 1–10 (2016)
9. Dasgupta, S., et al.: The impact of sea level rise on developing countries: a comparative analysis. 93(3), 379–388 (2009)
10. Ngô, V.: Cảnh báo đô thị hóa thiếu chiến lược. In: VnEconomy. VnEconomy, Hanoi (2007)
11. Anh, T.T.T., Mai, T.T.T.Q.: Tác động của chuyển đổi số trong phát triển đô thị: cơ hội và thách thức cho Việt Nam, in Kinh tế Việt Nam trên con đường chuyển đổi số. Trường Đại học Kinh tế Tp, Hồ Chí Minh (2021)
12. Townsend, A.M.: Smart cities: big data, civic hackers, and the quest for a new utopia. WW Norton & Company (2013)
13. Fu, Y., Zhang, X.J.C.: Trajectory of urban sustainability concepts: a 35-year bibliometric analysis. 60, 113–123 (2017)
14. Albino, V., Berardi, U., Dangelico, R.M.J.J.o.u.t.: Smart Cities: Definitions, Dimensions, Performance, and Initiatives. 22(1), 3–21 (2015)
15. Caragliu, A., Del Bo, C., Nijkamp, P.J.J.o.u.t.: Smart cities in Europe. 18(2), 65–82 (2011)
16. Song, H., et al.: Sustainability in smart cities: balancing social, economic, environmental, and institutional aspects of urban life (2017)
17. Capdevila, I., Zarlenga, M.I.J.J.o.S.: Management, Smart city or smart citizens? The Barcelona case (2015)
18. OECD. Smart cities and inclusive growth. In: Kamal-Chaoui, L. (ed.) Smart Cities and Inclusive Growth. OECD, Paris (2019)
19. Angelidou, M.J.J.o.U.T.: The role of smart city characteristics in the plans of fifteen cities. 24(4), 3–28 (2017)
20. IESE. IESE cities in motion index. In: Cities in Motion. IESE Business (2018)
21. Bendel, P.R.: Branding New York City — The Saga of 'I Love New York.' In: Dinnie, K. (ed.) City Branding, pp. 179–183. Palgrave Macmillan UK, London (2011). https://doi.org/10.1057/9780230294790_24

22. Toli, A.M., Murtagh, N.J.F.i.B.E.: The concept of sustainability in smart city definitions. **6**, 77 (2020)
23. Dameri, R.P.J.P.i.I.: Smart City Implementation. Springer, Genoa (2017)
24. Lee, J.H., et al.: Towards an effective framework for building smart cities: lessons from Seoul and San Francisco. **89**, 80–99 (2014)
25. Yeo, M.T.: From garden city to city in a garden and beyond. In: Dense and Green Building Typologies. SADT, pp. 21–25. Springer, Singapore (2019). https://doi.org/10.1007/978-981-13-0713-3_5
26. Gascó-Hernandez, M.J.C.o.t.A.: Building a smart city: lessons from Barcelona. **61**(4), 50–57 (2018)
27. Kassim, N., et al.: A conceptual paper of the smart city and smart community. 39–47 (2019)
28. Nanni, S., Mazzini, G.J.J.o.C.S.: Systems, From the Smart City to the Smart Community, Model and Architecture of a Real Project: SensorNet. **10**(3), 188–194 (2014)
29. Im, T., et al.: ICT as a buffer to change: a case study of the Seoul Metropolitan Government's Dasan Call Center. **36**(3), 436–455 (2013)
30. Wang, J.J.R.P.: Innovation and government intervention: a comparison of Singapore and Hong Kong. **47**(2), 399–412 (2018)
31. Okner, T., Preston, R.J.S.c.F.: Principles, and applications. In: Smart Cities and the Symbiotic Relationship Between Smart Governance and Citizen Engagement, pp. 343–372 (2017)
32. Nesi, P., et al.: An integrated smart city platform. In: Semanitic Keyword-based Search on Structured Data Sources. Springer (2017)
33. Kang, M.J.W.B.B.R.M.: How is Seoul, Korea transforming into a smart city. **21**, 2020 (2020)
34. Ho, E.J.U.S.: Smart Subjects for a Smart Nation? Governing (Smart) Mentalities in Singapore. **54**(13), 3101–3118 (2017)
35. Mcgrath, M.: Data points create position of mutual understanding and knowledge in Somerville, MA. In: National Civic League, America (2017)
36. Khan, H.H., et al.: Challenges for sustainable smart city development: a conceptual framework. **28**(5), 1507–1518 (2020)
37. Angelidou, M.: Strategic planning for the development of smart cities. Αριστοτέλειο Πανεπιστήμιο Θεσσαλονίκης (ΑΠΘ). Σχολή Πολυτεχνική. Τμήμα ... (2015)
38. Chung, C.-S.J.G.e.m.p.: The introduction of e-Government in Korea: development journey, outcomes and future. **3**(2), 107–122 (2015)
39. Russell, P.J.J.o.E.-G.: Smart Seoul 2015 vision announced. **35**(1), 3 (2012)
40. Kang, M.: How is Seoul, Korea transforming into a smart city? In: World Bank Blog (2020)
41. Thanh, H.: Vietnamese Information and Communication Ministry. In: Thông tin và Truyền thông, Bộ thông tin và truyền thông, Vietnam (2021)
42. Nga, Q.: Việt Nam đang là quốc gia tích cực xây dựng thành phố thông minh. In: Công Thương. Bộ Công Thương, Vietnam (2020)
43. M.T.: Đã có 19 địa phương thí điểm các dịch vụ đô thị thông minh. In: itcnews. Vietnamnet, Hanoi (2020)
44. Vnpt. Những thay đổi tích cực ở đô thị thông minh Phú Quốc. In: Tin tức, Vietnam (2019). https://vnpt.com.vn/tin-tuc/nhung-thay-doi-tich-cuc-o-do-thi-thong-minh-phu-quoc.html
45. Nguyên, Đ.: Wi-Fi miễn phí ở Đà Nẵng được truy cập 60 phút/lần. In: Zing News. Vietnam (2016)
46. Hà, L.: Xây dựng Hà Nội thành Smart City vào năm 2030. In: VnEconomy. Hội Khoa học Kinh tế Việt Nam, Vietnam (2018)
47. Anh, P.: TP HCM sắp có 4 trung tâm thông minh. In: Người Lao Động, Vietnam (2018)
48. Lê, P.: Thành phố thông minh Bình Dương với mô hình "ba nhà". Báo Bình Dương (2018)
49. Linh, H.: Phát triển đô thị thông minh, Đà Nẵng đột phá trong giai đoạn mới. In: Thông tin và Truyền thông. Bộ thông tin và truyền thông, Vietnam (2021)

50. BBT. Thành phố Phú Quốc: Gˊăn kết chặt chẽ vˊơi chuyển đổi số và phát triển đô thị thông minh. In: Cổng thông tin điện tử tỉnh Kiên Giang. UBND Tỉnh Kiên Giang, Vietnam (2021). https://kiengiang.gov.vn/trang/TinTuc/286/25701/TP.-PHU-QUOC--Gan-ket-chat-che-voi-chuyen-doi-so-va-phat-trien-do-thi-thong-minh.html

51. Tâm, C.: Phát triển Đà Nˉăng theo hưˊơng đô thị nén thông minh, sáng tạo và bền vˉưng. In: Cổng thông tin điện tử thành phố Đà Nˉăng. UBND Thành phố Đà Nˉăng, Vietnam (2021)

52. Thẩm, L.: Bình Dương hợp tác vˊơi TP Eindhoven (Hà Lan) về lĩnh vực công nghiệp và đô thị. In: Nhân Dân. Báo Nhân Dân, Vietnam (2015)

53. Thành, Đ.T.: Thế nào là một Đô thị thông minh? Và một số cách tiếp cận. In: ICTPress. Liên chi hội Nhà báo Thông tin và Truyền thông, Vietnam (2017)

54. Liêm, T.: ADB đề xuất thực hiện dự án thành phố thông minh tại Cần Thơ. In: BNews. Thông tấn xã Việt Nam, Vietnam (2021)

Studying Urban Space from Textual Data: Toward a Methodological Protocol to Extract Geographic Knowledge from Real Estate Ads

Alicia Blanchi[1(✉)], Giovanni Fusco[1], Karine Emsellem[1], and Lucie Cadorel[2]

[1] Université Côte d'Azur, CNRS, ESPACE, Nice, France
`alicia.blanchi@etu.univ-cotedazur.fr`, {`giovanni.fusco,`
`karine.emsellem`}`@univ-cotedazur.fr`
[2] Université Côte d'Azur, INRIA, CNRS, I3S, Sophia-Antipolis, France
`lucie.cadorel@inria.fr`

Abstract. Real estate ads are a rich source of information when studying social representation of residential space. However, extracting knowledge from them poses some methodological challenges namely in terms its spatial content. The use of techniques from artificial intelligence to find and extract knowledge and relationships from textual data improves the classical approaches of Natural Language Processing (NLP). This paper will first conceptualize what kind of information on urban space can be targeted in real estate ads. It will then propose an automated protocol based on artificial intelligence to extract named entities and relationships among them. The extracted information will finally be modeled as RDF graphs and queried through GeoSPARQL. First results will be proposed from the case study of real estate ads on the French Riviera, with a focus on toponymy. Perspectives of quantitative spatial analysis of the geolocated RDF models of real-estate ads will also be highlighted.

Keywords: Geographic information · Real estate ads · Social representation of space · Toponyms · French Riviera

1 Introduction

The retrieval and treatment of geographic information from textual data are a challenge in urban geography. New data sources are booming since the emergence of social networks, online platforms, etc. that offer the possibility to collect massive digital corpora. Indeed, *"a large number of statements that humans produce account for spatial phenomena"* [1]. This is the case of real estate ads. They are short texts presenting properties inscribed in urban space and are intended for a target population. They describe what are thought to be key characteristics of a piece of real estate, but they indirectly show how potential buyers/renters can project themselves in it, hinting also at the urban environment around it. They speak indirectly of individuals, social groups, places, urban forms, objects in space as they are represented by its authors who are trying to talk the symbolic language

O. Gervasi et al. (Eds.): ICCSA 2022 Workshops, LNCS 13378, pp. 520–537, 2022.
https://doi.org/10.1007/978-3-031-10562-3_37

of their targeted readers. Real estate listings are not spatial reality, but a transformation of it for a particular purpose.

Thus, most of the real estate ad deals with the description of the property, anchored by observed and sometimes even measured elements (housing type, surface, number of rooms, quality of materials, energy efficiency, financial charges, price, etc.). When it comes to the location of the piece of real estate, its close or further urban environment, we wouldn't expect realtors to carry out any specific analysis: they will more or less consciously reflect the prevailing social representation of the neighborhood, eventually filtering negative aspects following conventional selling strategies. If we can extract the few elements concerning the spatial context of the property and if the ads can be geolocated, we have the possibility to project in space dominant aspects of shared social representations of different subspaces in the city. More generally, the analysis of a large corpus of geolocated real estate ads gives us the possibility to produce a geography of the social representation of urban spaces, and even to understand to what degree fragments of urban space become recognized and socially characterized places in the city [2]. Understanding real estate advertisements also makes it possible to grasp the strategies and behaviors of the various urban actors in space: inhabitants, real estate actors, developers, local authorities. The study of urban space through real estate ads requires the creation of an effective analysis protocol to be able to extract automatically and organize in a formal structure the socio-spatial phenomena included in the texts. There are many approaches and a variety of applications that allow processing and extracting information from textual data including geographic information. Various studies have already been initiated in this direction [3–6]. However, rare are those who ventured into the analysis of real estate ads and their peculiar characteristics.

The goal of this paper is thus to present a new methodological protocol, allowing to carry out studies on urban space from a corpus of geolocated real estate ads. The main challenge lies in the recognition, extraction, classification, and analysis of geographic information from the corpora of real estate ads.

The case study to which our protocol will be applied is a corpus of real estate ads from the housing market of the French Riviera (from 2019 and 2021). The proposed protocol will thus include a few specific steps linked to the syntax and the semantics of the French language. It is understood that every natural language will need a specific conceptualization of the way in which the valuable geographic information is encoded in the text of the ad. However, the overall structure of the protocol applies to any natural language and can open new perspectives in the use of real estate ads to understand the perceived geography of vast metropolitan areas.

The remainder of the paper will be organized as follows. In the next section we will highlight what kind of geographic information can be found and targeted in real estate ads. Section 3 will present a methodology to extract this geographic content, linking information extraction, relationship extraction and information modeling through RDF graphs. Section 4 will present first results from the analysis of real estate ads from the French Riviera and perspectives of future research.

2 Geographic Information in Real Estate Ads

Real estate ads are massive punctual data that contain descriptive texts partly focused on urban space. Real estate ads talk about space in different ways: they can describe physical features that are observable around the advertised property; they can describe the relational properties of the location by highlighting proximities and travel connections; they can give specific perceived attributes to the area, eventually using known toponyms; they can indirectly suggest what kind of populations are to be found in this spatial context. Of course, their readers are not geographers and planners. They will thus hardly use analytical knowledge or standardized representations of space (XY coordinates, census tracts, planning perimeters, location quotients, etc.). Even their use of isochrones (travel time) has to be understood as qualitative and evocative of lifestyles (what travel modes are implied? What destinations?) more than a precise description of urban space-time. They use more the evocative language of places than the abstract representation of space. These speeches of these places are to enhance the residential space and/or location in a particular urban context [3, 4]. Indeed, the literature indicates that location, or better socially perceived location, is a preponderant criterion in the acquisition of real estate: by choosing a piece of real estate, the buyer (or renter) chooses an address within a place, with all its implications in terms of aspired social status, lifestyle, and perceived needs within the life cycle of the household [5–7]. If social status, lifestyle and position within the life cycle of the household are the most important explanatory factors of residential mobility (together with ethnicity, as shown in [5]), we can assume that social representations of space carried by real estate ads will at least include these dimensions. We will thus expect to be able to identify places associated with higher vs lower social status, urban vs suburban lifestyles, needs for the young, the seniors, families with children, etc. But these associations will be rarely dealt with directly. Much more often they will have to be inferred by the semantic associations implied in the text.

Toponyms, i.e., place names, are specific issue of place analysis in geographic space through real estate ads. What we are interested in is the capacity of toponyms to structure the perception of places in the city. What subspaces are identified as places through the coherent use of a given toponym? What is the spatial extent of the use of this toponym? Are there areas where the use of toponyms is erratic or even absent? Can the use of a toponym beyond the area of its established perimeter be an indicator of the particularly strong social representation of the designated place? The use of toponyms within real estate ads will have to deal with specific disambiguation problems. If we don't want to pollute our use of real estate ads as a corpus of social representation of space in use, we have to consider toponyms, beyond their administrative definitions.

In order to take up these knowledge challenges, we must identify a few specificities of the description of space and place within real estate ads. These will call for specific methodological choices within our extraction and structuring protocol which we will present in the rest of the text.

First, as in any descriptive text for a generic public, the authors of real estate ads use common geographic entities to talk about places: city-center, neighborhood, village, sector, street, square, park, streetcar, train station, highway, school, campus, new project, etc. are all common terms, well understood by people, beyond any formal definition by geographers or planners. It is also through the combination of these natural or man-made amenities, linear, punctual or surface-like features, that the place characteristics within which a property is located are evoked to the reader.

As we have seen, another fundamental category of geographic entities are the named entities, i.e., toponyms. These can come or not in association with common geographic entities. We can distinguish three situations:

- Geographic entity + associated toponym ("City of Nice", "Cimiez district", "Salis Beach"). These combined occurrences are particularly informative, because they directly associate the toponym to a more general category, indicating what the toponym means for the potential readers. And different geographic terms are differently evocative of social representations: identifying a given toponym as a *quartier*[neighborhood] evokes an urban life at a pedestrian scale or an older part of town, which is not the case if it is identified as a *secteur*[sector, district], which is a much more neutral subdivision of space.
- Unnamed geographic entity ("city", "neighborhood", "beach"). In this case the place is characterized as corresponding to a specific feature, but it appears nameless, hinting at a weakness of the placename and, possibly, at the whole place-making as a unique and specific location in urban space.
- Toponym alone without a corresponding geographic entity ("Nice", "Cimiez", "Salis"). This situation is the most enigmatic. We have evidence of the strength of a place name, but we don't exactly know what geographic feature is meant by that name.

Whenever placenames are used (cases 1 and 3), we can project their occurrences in space and study the spatial patterns resulting from them, eventually as a function of the associated geographic features.

Attributes associated to the geographic entities are also particularly informative of social representations. Indeed, the authors use the specific characteristics of places that give a context to the proposed property. Example: "famous district", "dynamic sector", "lively street", etc. Of course, the absence of any specific attribute could also be indicative of shared negative representations of a given place. We will thus never find descriptions like "crime-ridden district", "rough neighborhood", "dwindling village", etc.

Second, even if real estate ads mainly talk about places, they also use specific and relatively simple concepts of spatial relations: inclusion, proximity, concentration, connection, etc. Sometimes, even the multi-scale nature of property location within a neighborhood, a city and a whole metropolitan area is addressed. Spatial relations play a central role in the understanding of urban space (or spatial context of valorization). They make it possible to realize the relationships that the different geographic objects (the property, toponyms and/or common geographic entities) have in space. Examples: "In the heart of Cimiez", "Between the industrial zone and city-center", "Close to shops and transit", etc.

The words that express a spatial relationship are specific to the language of the real estate ads. As for French, two different ways of dealing with spatial relations arise:

- Simple spatial prepositions can encode the spatial relationship between two entities (E.g., in English, "in", "on", "at").

– Compound prepositions that combine simple spatial prepositions with other words (nouns, adjectives, verbs) or prepositional locutions [1] that we call here "Space-time entities" (E.g., in English, "at 3 min from", "near", "close to").

Beyond the specific linguistic encoding, there are four main types of spatial relationships in the texts of real estate ads:

– Spatial relationship of **situation**: We are talking here about location within a given space and at a given scale. Examples are: "*au coeur* [in the heart of]" represents location within the most central part of an entity, "*dans*[in]", it represents insideness within a designated spatial extent, whether named or unnamed; "*à*[at]" can also be used to represent location in a place, as if it were observed at a wider scale, where the place becomes a point on a map; "*sur*[on]" and locates objects on topographic and hydrographic features. Sometimes, real estate ads omit the prepositions and the spatial relation with the entity and/or the toponym remains understood.
– Spatial relationship of **proximity**:

  • Temporal or metric distances that are quantities, real or perceived (E.g., "10 min from", "100 m from").
  • Qualitative relationship of proximity (E.g., "close to", "near").

– Spatial relationship of **accessibility/connexion**: They are often complex since most accessibility relationships are between the places and not directly between the property and the places (E.g.: "Apartment near the station that **serves** Monaco", "Neighborhood served by the line 4 of the subway").
– Spatial relationship of **visibility**: this spatial relationship involves sight instead of movement. It completes our understanding of the spatiality of the geographic entities mentioned in real estate ads. Hence the interest in studying visibility relationships at the same level of more physical spatial relationships.

It is these different relationships that make it possible to understand the spatial context and the links between places. For example, the fact that the property is "in" the neighborhood of Cimiez or that the property is "close" to the neighborhood of Cimiez have a different meaning in terms of place-making. In the first case, we just remark the use of a given toponym to identify a place, which is already evidence of the toponym's strength. In the second case, the strength of the toponym Cimiez is even higher: a piece of real estate which is out of its perceived boundary refers to the proximity to it in order to have a meaningful location for the reader. Conversely, the sense of place of the subspace where the property lies, whose toponym is not even mentioned in the ad, could be much lower.

In addition, the fact that the property is located "close" to certain features is informative of the targeted population and, therefore, of the kind of population associated with that space, in terms of all the relevant factors: social status, lifestyle, life cycle, ethnicity. Proximity to commerce and schools is not the same as proximity to a golf course, proximity to bars and nightlife is not the same as proximity to a park or to health-care facilities. What is to be considered here is the putative social status of the different facilities (which could vary in different cultural contexts), the opposition between natural vs. urban amenities as well as the specific social groups (or just age groups) that a given facility or service caters to.

Third, when advertisers talk about features near the property, they also emphasize the presence of travel modes that allow readers to better understand the spatiality of the places and to project themselves into a wider urban space. However, the spatial relationships mentioned in the ads do not always allow assessing the precise distance between places. Two difficulties arise: The assessment of distance differs among people [8] especially in the context of real estate ads [9], and spatial relationships are often qualitative. What is more interesting is often the travel modes indicated (or just assumed) in the text, as it relates indirectly to lifestyles and/or to constraints in the household agenda. "Ten minutes from the motorway exit" is not just important for its quantification of the time-distance, but for its understanding of car mobility to access metropolitan space. "10-min walk from schools and commerce", notwithstanding its truthfulness, hints at a pedestrian city-life or village-life. Indeed, according to the location factors cited, the spatial relationships expressed by the travel modes could also hint at different spatial scales and spatial contexts for a given place. For example, the fact that a property is at a particular walking distance from some geographic features contributes to the neighborhood characterization of the place, whereas its transit or car distance from higher-end services or centers could contribute to characterize a place as a cell within a larger urban and metropolitan space.

« **Nice**, appartement avec vue mer, au cœur du quartier calme de **Cimiez**, proche du centre-ville et à 2 minutes à pied des commerces. »
« Nice, apartment with sea view at the heart of the quiet neighborhood of cimiez, near the city-center and at 2 minutes walk from shops».

Fig. 1. Geographic information in the real estate ad

Figure 1 shows the main targeted features that must be extracted from a real estate ad to understand the perceived qualities of places in the city, with the subtleties of the semantic and spatial relations among them. In what follows we will see which protocols can be used to extract these features and to assign them the correct label while keeping the structure of the semantic and spatial relations.

3   Methods to Extract Geographic Information and Relationships

This section will deal with the specificities of automated processing of real estate ads to analyze social perceptions of urban space. To be able to use real estate ads, it is essential to transform raw ads into analyzable data which go beyond the usual pre-treatments applied to texts such as data standardization, removal of special characters, lemmatization, and others. Usual techniques of lexicometry are of little use: pertinent information of social representation of urban space is minority within the real estate ad and can be scattered within it. Thus, analysis of associations based on word co-occurrences could even be misleading [10]. We need to know what terms are used to describe the piece of property and which ones are used to characterize its spatial setting. An understanding of the language syntax is thus necessary prior to lexicometry.

The challenge is therefore based on the creation of an adapted protocol to deal with the socio-spatial phenomena and geospatial information in the texts. There are different methods and techniques that can be used to process texts, including geographic textual information [11, 12]. To detect and extract the targeted geographic knowledge in the real estate ads, we decided to implement a particular protocol based on NLP (Natural Language Processing) techniques. This protocol can only be realized with a hindsight and conceptualization of the knowledge present in the texts and the phenomena to be described. It is therefore necessary to know beforehand a set of indicators that provide information as presented in the previous section.

The extraction of information from texts can be established on specific methods using artificial intelligence techniques. The chosen method is based on the extraction of all the information and its relationships in two steps: The named entities recognition (NER) and dependency parsing.

3.1 Information Extraction (IE)

The Information Extraction (IE) from text is made possible, among others, by the named entities recognition (NER) approach. NER is based on a natural language processing (NLP) technique that makes it possible to extract valuable elements from a text in particular by "*the automatic identification of the entities present in a text to the classification of these entities into different categories*" [13]. This method is among the most widely used methods for extracting a mass of information automatically from text [14, 15]. It is based on machine learning approaches. The real challenge here is to be able to extract all the geographic information automatically in the rich corpora of real estate ads.

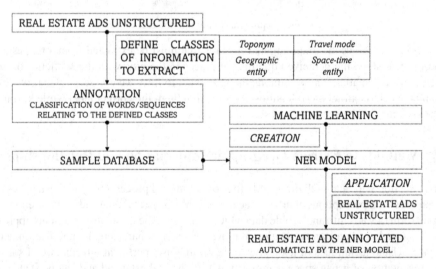

Fig. 2. Steps to extract information from the real estate ads by using NER model.

Figure 2 shows the different steps of the method that will be detailed below:

The **first step** is focused on the identification of the targeted information, i.e., the geographic information previously presented that echoes the environment of the property: geographic entities of several kinds, toponyms, travel modes, and the words used to create a spatial relationship (space-time entities).

The **second step** creates a sample database of real estate ads annotated according to the previously defined classes (Fig. 3). The goal is to have a training dataset for the creation of the NER learning model. For this, it was necessary to classify manually the words of each category identified in each real estate ad. The annotation is a very time-consuming task that must be performed on a significant number of real estate ads for the training dataset to be effective. In this research, we annotated more than 1000 ads of the real estate market of the French Riviera. They were written by professional real estate agents in 2021 and concern all types of real estate (sale and rental; apartments, houses and garages). This sample is representative of the real estate market of the French Riviera in its diversity (coastal cities, suburban areas, hinterland, etc.).

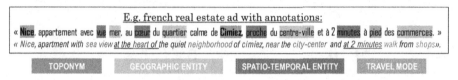

Fig. 3. Annotation of real estate ads according to the information selected.

The **third step** is based on the training of a NER learning model. There are many NER learning models [16, 17]. SPACY and FLAIR are two leading models for performing the named entity recognition task. These models are different in terms of neural network architecture and algorithms, requiring different computing power. In this study, we have tested these different NER models on our annotated French real estate ads. The results and performances have been detailed in a previous paper [18]. Here we will focus on FLAIR which is based on the BiLSTM Language Model. The advantages of this model are the consideration of the context of labeled words and the fact that it is more easily optimized [19].

At the end of the information extraction task, we obtained annotated ads according to the defined classes and a list of words (all named entities) for each ad. The extracted elements of information have no links among them, and the geographic information is devoid of context. In order to reconstruct context and relations, we need to go through automatic syntactic analysis, as shown in the next section.

3.2   Relationship Extraction (RE)

Methods of Relationship Extraction (RE) pay attention to the semantic structure of the text and are based on syntactic analysis of sentences or groups of words that compose the text [20]. We selected the dependency parsing method deriving the links between words from the lexical relationships that words keep in the text. The objective here is to extract, from the grammatical dependencies, all the links between geographic objects and their socio-spatial properties through the lexical and sometimes specialized relationships that animate them. There are different libraries to study the grammatical relationships of texts. We chose the STANZA library developed by Stanford NLP Group [21] and which adapts to French texts. This library allows to highlight the grammatical structures between the words in the text by *"creating a tree of words from an input sentence that represents the syntactic dependency relationships between words"* [21] (Fig. 4). The dependency tree was created from the structure of sentences and the nature of words (called Part-of-Speech (POS)). Stanza's dependency tree of texts makes it possible to understand the grammatical relationships between each word or sequence of words for each sentence of each real estate ads (Fig. 4).

The careful analysis of grammatical dependencies of real estate ads, more particularly by focusing on the lexical links among named entities make it possible to highlight four main types of relationship: association, spatial, modal, and attribute relationship, as detailed below.

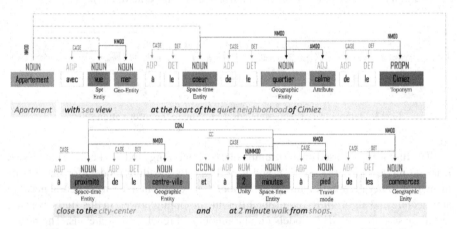

Fig. 4. Grammatical dependencies of STANZA on the real estate ad (Table 1).

Table 1. List of Part-Of-Speech (POS) of the tokens in the ad (Fig. 4).

Part-of-Speech (POS)	Noun (NOUN), Adposition (ADP), Determiner (DET), Proper noun (PROPN), Adjective (ADJ), Coordinating conjunction (CCONJ)

Association relationships focus on the dependency between two kinds of named entities (toponyms and geographic entities) to recompose the geographic objects. This makes it possible to understand exactly which toponym is related to which geographic entity in the texts. The relationship between toponyms and a geographic entity is usually a direct nominal relationship (NMOD) as can be seen in this example (Table .2). E.g., Relation between the geographic entity *"quartier*[Neighborhood]" and the toponym "Cimiez" (Fig. 4).

Spatial Relationships express the spatial context of the geographic objects, the spatial scale of valorization, the surroundings of the property, the position that a geographic

Table 2. STANZA legend of the dependency tree in the ad (Fig. 4).

Type	Legend of dependencies [21]
[NMOD] Nominal modifier	*«Is used for nominal modifiers of nouns or noun phrases»*
[AMOD] Adjective modifier	*«Is any adjectival phrase that serves to modify the meaning of the nominal head»*
[NUMMOD] Numeric modifier	*«Is any number phrase that serves to modify the meaning of the noun with a quantity»*
[CONJ] Conjunction	*«Relation between coordinated elements. The head of the relation is the first conjunct and other conjuncts depend on it»*
[CC] Coordinating conjunction	*«Relation between the head conjunct of a coordinate structure and any of the coordinating conjunction involved in the structure»*
[CASE] Case	*«Is used for any preposition introducing a nominal construction in French. Prepositions are treated as dependents of the noun ...»*
[DET] Determiner	*«Relation between the head of a nominal phrase and its determiner»*

object takes in space. There are different spatial relationships in these texts between the property and the geographic objects. The grammatical relationships as spatial relationships are rarely direct between the property and geographic objects. Spatial relationships are expressed by prepositional locutions based on annotated space-time entities (e.g., "Apartment close to the city-center", "Apartment at two minutes from shops"). The links are only made through a space-time entity (*proche*[near], *minutes*[minutes], *proximité*[close] (Fig. 4)). The space-time entities that are at the center of these spatial relationships are between the property and geographic objects. In this case, the property has a direct nominal modifier relationship (NMOD) with a space-time entity and the latter has a second grammatical relationship with a geographic object (Fig. 4). Exceptions exist when geographic objects are simply cited or when the spatial relationship is based on spatial prepositions considered as an adposition (ADP) (e.g., "Apartment *dans*[in] Nice", "Apartment *sur*[on] the mountain", "Nice"). The spatial relationship based on the spatial preposition is directly recognized in the grammatical dependencies because the property has a direct grammatical link with a geographic object as a nominal relationship (NMOD).

Modal Relationships concern the travel modes when they express a distance between the geographic objects (either between the property and the geographic object or between two geographic objects). Generally, the travel mode has a nominal modifier relationship (NMOD) with a named entity. In this example (Fig. 4), the travel mode "*pied*[walk]" has a grammatical relationship with a space-time entity "*minutes*[minutes]" making the spatial relationship between the property (subject) and the geographic entity "*commerces*[shops]".

Attribute relationships provide a quality of the geographic objects mentioned in the texts (E.g., "Dynamic sector", "Quiet district"). These characteristics are an additional key to understanding spaces in their social dimension. Generally, the named entities have a direct adjective modifier relationship (AMOD) with their characteristics because these are most often adjectives (E.g., relationship between "*quartier*[neighborhood]" and "*calme*[quiet]" (Fig. 4)). In addition, spatial relationships can also have attributes (E.g., qualitative attribute "very close to" or quantitative attribute "2 min from" where "2" have a numeric modifier relationship (NUMMOD) with "minutes" (Fig. 4)).

The extraction of these relationships is founded on a rule-based approach according to recurring identified structures expressed by the grammatical relationships. Many parameters have been added to integrate the specificities of each relationship (number of words between tokens, type of part-of-speech of token, type of relationship between tokens, etc.). This step makes it possible to link all the pieces of geographic information together and give them meaning and context.

To conclude, the extracted information and relationships can be modeled in the form of a grammatical dependency graph. A dependency graph is a set of dependency relationships between tokens in a sentence [22], where the tokens can be represented by a node and the dependencies by arcs between nodes. In our case, the nodes are the pieces of extracted geographic information (space-time entity, toponym, geographic entity, travel mode and attribute) and the arcs are the grammatical relationships extracted among them (spatial relationship, association relationship, attribute relationship, modal relationship) (Fig. 5).

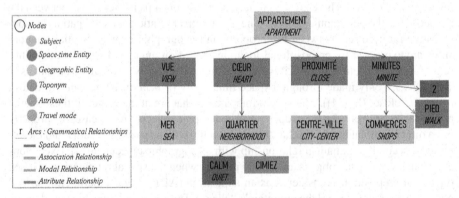

Fig. 5. Conceptual grammatical dependency graph.

3.3 Structuring Information Extracted

The analysis of the extracted geographic information is only possible after the transformation of the real estate ads into analyzable data. The objective is to structure the extracted entities and their relationships in a common graph structure. Several modeling choices allow us to produce a common graph structure for all the ads.

First, we include white/empty nodes in the model to indicate the absence of some key pieces of information (Fig. 6). In fact, the real estate ads are not constituted in the same way: some do not refer to the geographic area where the property is situated, others do not mention any toponym, some mention just the geographic entities without associated toponyms. The presence of empty nodes allows all real estate ads to keep the same structure but also to model the absence of information.

Second, we link all information related to the geographic object to toponyms, whether they are present or not in the ad (Fig. 6). A toponym can be considered as a unique identifier even if it is linked to several types of geographic entities. We will therefore link the geographic entities to the toponym. The possibility of linking an attribute node to the toponym (and, through the latter, to the geographic entity) is always foreseen. These nodes can, of course, be empty (Fig. 6).

Third, we link the property to the toponym in order to represent the spatial relationship between them. All the information relating to the kind of spatial relationship that has been extracted are modeled as parameters of the link (presence or not of space-time entity, travel mode, and attribute) (Fig. 6).

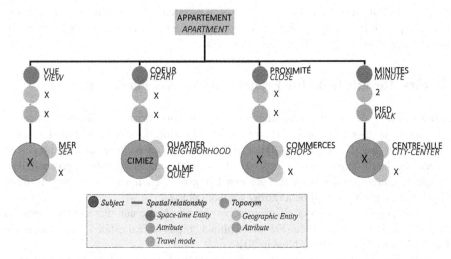

Fig. 6. Ad structured according to the extracted geographic information and relationships

Figure 6 gives an example of the resulting graph. In this ad, a piece of property (here an apartment) has a spatial relationship with four geographic objects:

- The first spatial relationship is with an "empty" toponym concerning the "*mer*[sea]" as geographic entity through a visibility relationship according to the space-time entity "*vue*[view]". In many ads of the French Riviera, the toponym "*Méditerranée*[Mediterranean]" is understood for the geographic entity "*mer*[sea]".
- The second spatial relationship is of situation kind, according to the space-time entity "*coeur*[heart]" with the toponym "Cimiez" which is considered in this ad as a "*quartier*[neighborhood]" that is "*calme*[quiet]" according to its attribute.
- The third spatial relationship is one of distance between the apartment and the "*centre-ville*[city-center]" through the space-time entity "*proximité*[close]". We do not have more precision for this spatial relationship because of the absence of information at the level of the parameters "travel mode" and "unity" or "attribute";
- The last spatial relationship is also of distance between the apartment and the "*commerces*[shops]" according to the space-time entity "*minutes*[minute]". We can see that its attribute is "2" and its travel mode is "*pied*[walk]".

Structuring ads, first conceptually and then practically in a graph, is of paramount importance to make their exploitation possible. The transformation of real estate ads also allows to store them properly, for example, in the form of an RDF graph. The RDF graph is an appropriate formalization for transformed real estate ads since relationships are grammatical dependencies and follow the same logic as the grammatical dependency graphs that we created. The basic components of RDF data modeling are a triple of elements: a first node (the subject), a second node (the object) and a relation between the two (the predicate). This intrinsic graph-based framework can also be easily queried. Its GeoSPARQL vocabulary is particularly well adapted to geographic information and makes it possible to make spatial queries, opening the field of possible applications for the analysis of urban space from real estate ads.

4   First Results from the French Riviera and Perspectives

Whenever they contain precise location (XY coordinates or street address), real estate ads become punctual data and therefore give the opportunity, once transformed, to carry out different spatial analyses to understand urban spaces in their socially perceived dimension. In this paper, we will limit ourselves to different cartographic representations of specific query results from the presented protocol. Their spatial analysis through appropriate algorithms will be a further research phase.

Application 1: The spatial distribution of the real estate ads that use the toponym "Gambetta" in the city of Nice (France) within a corpus of real estate ads geolocated at the address within the city of Nice in 2019.

Many real estate ads mention the toponym "Gambetta" which makes "Gambetta" a frequently used toponym in the marketing of real estate ads in our study area. There are different geographic entities related to the toponym "Gambetta". In the real estate ads in Nice, the toponym "Gambetta" can be of the "Boulevard" type which makes it an odonym but can also be used to designate a "neighborhood" and a "sector" (Fig. 7). The widespread use of the placename Gambetta makes it a prominent place within the social representation of the city of Nice, insisting on its urban connotation: a main street

Real estate ads of Nice geolocated at the address in 2019 that use the toponym « Gambetta ».

Ads that use the toponym 'Gambetta':
Type of spatial relation :
■ Spatial relation of Situation
● Spatial relation of Distance
Type of Geo-entity :
● Boulevard
● Quartier
○ Secteur

+ Ads that do not use the toponym 'Gambetta'

-- Boulevard Gambetta

0   100 m

Background MAP : OSM Standard

Fig. 7. Real estate ads of Nice geolocated in 2019 that use the toponym "Gambetta".

(boulevard) and a neighborhood, the latter being used mainly for its southern section. The more generic and less urban word "sector" is seldom used, and our relatively limited sample doesn't contain other designations of the street denoting its role of transportation artery. Further analyses of this point pattern are needed on a larger corpus of data, and in conjunction with the attributes associated to the boulevard/neighborhood/sector, as well as with the other geographic features included in the RDF graphs, the associated transport modes, etc.

Application 2: The spatial distribution of the real estate ads that use the toponym "Promenade des Anglais" in the city of Nice (France) within a corpus of real estate ads geolocated at the address within the city of Nice in 2019.

The Promenade des Anglais is one of the most mentioned places in real estate ads in Nice in 2019. Some ads locate the property by the proximity to the Promenade des Anglais using a quantitative or a qualitative distance while others use the notion of visibility (Fig. 8). This toponym has thus an area of influence beyond the perimeter of properties situated on it (at least according to the ads). Even the "view" of the Promenade des Anglais is a very rewarding criterion. Its view from the eastern hills is a "canonical view" within the imagery of the French Riviera as painted or photographed by artists and then used by tourist guides and other marketing supports for almost two centuries [23]. Interestingly enough, the Promenade des Anglais has an area of influence well beyond the coastal strip, but is not used to characterize the situation of real estate ads in its westernmost section. We observe here a clear discrepancy between the official odonym and the social representation of the place Promenade des Anglais. For the users of real estate ads, the Promenade des Anglais is the famous seafront promenade of the city of Nice. Some ads develop this dimension of social representation by the presence of attributes such as "*célèbre*[renowned]" or "*fameuse*[famous]". When the street departs

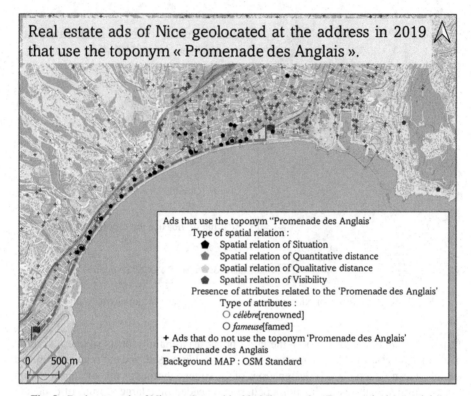

Real estate ads of Nice geolocated at the address in 2019
that use the toponym « Promenade des Anglais ».

Fig. 8. Real estate ads of Nice geolocated in 2019 that use the "Promenade des Anglais".

from the seaside and is bordered by the wastewater treatment plant and the airport, its surrounding areas are no longer considered part of this well-established urban place.

Application 3: The spatial distribution of the real estate ads that use the toponym "Montfleury" in the city of Cannes (France) within a corpus of real estate ads geolocated at the address within the city of Cannes in 2019.

According to the number of real estate ads located at the address in 2019 within the city of Cannes, the real estate market in "Montfleury" was very dynamic (Fig. 9). We would therefore like to know from the real estate ads what are the characteristics attached to the toponym "Montfleury". In fact, for more than half of the real estate ads mentioning the toponym "MontFleury"/"Montfleury", this is considered as "Residential" and only residential (no other attributes are used within the ads). Residential areas are less generally considered neighborhoods in France and Montfleury is not an exception. The term *"secteur"*[sector] is thus much more frequent than *"quartier"*[neighborhood], in contrast with what was observed for the toponym Gambetta in Nice.

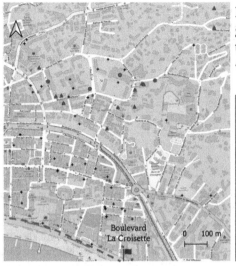

Real estate ads of Cannes geolocated at the address in 2019 that use the toponym « Montfleury ».

Ads that use the Toponym "Montfleury":
Type of geo-entity :
● quartier
▲ secteur
Attributes related to the toponym:
● résidentiel[residential]
● *None*

+ Ads that do not use the Toponym 'Montfleury'

Background MAP : OSM Standard

Fig. 9. Real estate ads of Cannes geolocated in 2019 that use the toponym "Montfleury".

Beyond these three cartographic analyses, we can easily set the goal of conducting quantitative spatial analysis on this data to provide an additional key to understanding the social representation of urban spaces through real estate ads. Different statistical methods and spatial analysis techniques of point patterns analysis (PPA) could be applied to the geolocated ads and their associated RDF graphs. For example, it would be very interesting to work around the question of toponymy (dominant and most used toponyms within a given study area) while integrating space (intensity of use of toponyms in space, spaces that have a hesitant toponymy when there are different toponyms to designate the same space, spaces that have a low of presence of toponyms, etc.). These spatial distributions of the punctual data can be measured by applying methods based on spatial density (kernel density, DBSCAN) and by using methods around the fuzzy logics, to model core and support space for a given toponym. Area of influence of toponyms could also be studied beyond the relationship of spatial situation, as shown in the example of the Promenade des Anglais in Nice.

We could also study the factors of location used within the real estate ads located in a given space (type of geographic object mentioned, multiplicity of the factors of location or not and their copresence in an ad and in all ads, preponderant characteristics mentioned, as well as absence of the characteristics). We can show the consensus of information if it exists through the study of convergences and divergences within the same space, and infer the target population of the ads (if they are not explicitly mentioned). Linking target populations and mentioned places will be another important aspect of their social representation.

We can also study the question of the reputation of places by studying the characteristics associated to the places mentioned in the real estate ads (dominant attributes, multiplicity of attributes to qualify a given place, copresence of attributes, absence of attributes, sense of attributes).

These perspectives will be the object of a further research endeavor. We think that the main contribution of the present paper is to have proposed a new methodology for extracting spatial information from real estate ads, that opens the way of spatial quantitative analysis of one of the most qualitative aspects of urban research: social representation of places.

Acknowledgement. This research was carried out thanks to a research grant by KCityLabs, KINAXIA Group (CIFRE Agreement with UMR ESPACE).

References

1. Stosic, D.: 'par' et 'à travers' dans l'expression des relations spatiales: comparaison entre le français et le serbo-croate (2002). https://hal.archives-ouvertes.fr/tel-00272907/
2. Relph, E.: Place and placelessness (1976). https://doi.org/10.4135/9781446213742.n5
3. Alba, M., et al.: La publicité immobilière à l'assaut de l'environnement dans une grande ville du Sud, Mexico, 1950–2000. Ecol. Polit. **39**(1), 55 (2010). https://doi.org/10.3917/ecopo.039.0055
4. Blanchi, A., et al.: The real estate ads, a new data source to understand the social representation of urban space. In: ECTQG21 (2021)
5. Shearmur, R., et al.: From Chicago to L.A. and back again: a Chicago-inspired quantitative analysis of income distribution in Montreal. Prof. Geogr. **56**(1), 109–126 (2004). https://doi.org/10.1111/j.0033-0124.2004.05601016.x
6. Thomas, M.-P.: Les choix résidentiels: Une approche par les modes de vie, pp. 1–41 (2018)
7. Sigaud, T.: Accompagner les mobilités résidentielles des salariés: l'épreuve de l'entrée en territoire. Espaces et sociétés **162**, 129–145 (2015)
8. Bailly, A.: Ditances et espaces : vingt ans de géographie des représentations. Espac. géographique **14**(3), 197–205 (1985).https://doi.org/10.3406/spgeo.1985.4033
9. McKenzie, G., et al.: The 'nearby' exaggeration in real estate. In: Proceedings of the Cognitive Scales of Spatial Information, CoSSI 2017 (2017)
10. Lancia, F.: Word co-occurrence and similarity in meaning: some methodological issues. Mind Infin. Dimens., 1–39 (2007)
11. McKenzie, G., et al.: Identifying urban neighborhood names through user contributed online property listings. ISPRS Int. J. GeoInf. **7**(10), 388 (2018)
12. Hu, Y., et al.: A Semantic and sentiment analysis on online neighborhood reviews for understanding the perceptions of people toward their living environments. Ann. Am. Assoc. Geogr. **109**(4), 1052–1073 (2019)
13. Shrivarsheni: How to Train spaCy to Autodetect New Entities (NER) (2020). https://www.machinelearningplus.com/nlp/training-custom-ner-model-in-spacy/
14. Andrey from Prodigy Support: Former ensemble NER et extraction de relations (RE), pp. 3–5 (2021). www.support.prodi.gy/t/training-ner-and-relations-extraction-re-together/3911
15. Wang, J., et al.: NeuroTPR: a neuro-net toponym recognition model for extracting locations from social media messages. Trans. GIS **24**(3), 719–735 (2020). https://doi.org/10.1111/tgis.12627
16. Benesty, M.: NER algo benchmark: spaCy, Flair, m-BERT and camemBERT on anonymizing French commercial legal cases. Towards Data Science (2019). https://towardsdatascience.com/benchmark-ner-algorithm-d4ab01b2d4c3
17. Hu, Y., et al.: How do people describe locations during a natural disaster: an analysis of tweets from hurricane Harvey. In: Leibniz International Proceedings of Informatics, LIPIcs, vol. 177, no. 23, pp. 1–16 (2020)

18. Cadorel, L., et al.: Geospatial knowledge in housing advertisements: capturing and extracting spatial information from text (2021). HAL Id: hal-03518717
19. Duffy, S.: Is Flair a suitable alternative to SpaCy? (2020). https://medium.com/@sapphireduffy/is-flair-a-suitable-alternative-to-spacy-6f55192bfb01
20. Perera, N., Dehmer, M., Emmert-Streib, F.: Named entity recognition and relation detection for biomedical information extraction. Front. Cell Dev. Biol. **8**, 673 (2020)
21. Sanford NLP Group: Stanza - A Python NLP Library for Many Human Languages | Stanza. https://stanfordnlp.github.io/stanza/. https://universaldependencies.org/
22. Alfared, R.: Acquisition de grammaire catégorielle de dépendances de grande envergure (2013). HAL Id: tel-00822996
23. Hérault, M.: La Riviera, pays de l'éternel printemps: Imaginaire paysager et transferts culturels, à Nice et dans son territoire, du Grand Tour à nos jours, Thèse de Doctorat, Sorbonne Université, Paris (2021). https://www.theses.fr/2021SORUL022

International Workshop on Digital Sustainability and Circular Economy (DiSCE 2022)

Transforming DIGROW into a Multi-attribute Digital Maturity Model. Formalization and Implementation of the Proposal

Paolino Di Felice[1]([⊠])[ID], Gaetanino Paolone[2], Daniele Di Valerio[2], Francesco Pilotti[2], and Matteo Sciamanna[3]

[1] Department of Industrial and Information Engineering and Economics, University of L'Aquila, 67100 L'Aquila, Italy
paolino.difelice@univaq.it
[2] Gruppo SI S.c.a.r.l, 64100 Teramo, Italy
{g.paolone,d.divalerio,f.pilotti}@softwareindustriale.it
[3] Selfresh S.r.l., Via Alcide de Gasperi 9, Penna Sant'Andrea, 64039 Teramo, Italy
m.sciamanna@selfresh.it

Abstract. SMEs form the backbone of the economy of most countries all over the world. Increasing their performance is therefore essential to increase the well-being of peoples. The adoption of digital technologies is universally recognized an essential component to achieve this objective. Unfortunately, European SMEs are lagging behind in the transition to digital. The Digital Europe Programme has been started at the end of 2021 to speed up such a transition. As part of the Programme, a network of so-called European Digital Innovation Hubs (EDIHs) is under construction in order to lend a hand to SMEs in the process of increasing their digitalization awareness and capabilities. The availability of a digital maturity assessment tool would be of great help for EDIHs to accomplish their mission. This paper gives a contribution in such a direction. It starts from an already published framework suitable to assess the digital maturity level of SMEs, as well as the capabilities associated to each level that are necessary for promoting a digital enabled growth. Then, the framework is transformed into a qualitative multi-attribute digital maturity model and the latter is implemented by means of a freeware software. The proposed model returns transparent indications to SMEs about where they are on their digital transformation journey, together with eventual strengths and weaknesses. SMEs should take these indications into account before allocating the budget for future investments.

Keywords: Digital transformation · Digital maturity model · Digital maturity assessment · Multi-attribute model · SME · DEXi

This research was funded by Software Industriale.

1   Introduction

Small and medium-sized enterprises (SMEs) are the backbone of the economy of most countries all over the world. "They represent 99% of all businesses in the EU. They employ around 100 million people, account for more than half of Europe's GDP and play a key role in adding value in every sector of the economy."[1] "SMEs are made up of enterprises which employ fewer than 250 persons and which have an annual turnover not exceeding EUR 50 million, and/or an annual balance sheet total not exceeding EUR 43 million."[2] A 2021 report by Innovation Finance Advisory states that in Italy, for example, there are about 4.4 million SMEs which account for 99.9% of the total number of active companies (95% of which are micro-enterprises – firms with fewer than 10 employees and less than €2 million of turnover), 80% of employment and 70% of value added [1].

In the today world of emergent and continuous changes, SMEs, in order to be competitive, must equip themselves with IT technologies and adopt new information processes capable of favoring the sharing of resources and the engagement of all corporate stakeholders, including their customers [2].

[3] is a recent paper which analyzed 58 peer-reviewed studies published between 2001 and 2019, dealing with different aspects of digital transformation. From such a study, Nadkarni and Prügl concluded that digital transformation is an actor-driven organizational transformation (touching capabilities, structures, processes and business models of the firm) triggered by the adoption of suitable digital technology. It follows that, to exploit IT potentials, SMEs must embed "IT-capabilities" in themselves. Unfortunately, for most SMEs this is not true as stressed, for example, in a 2020 document of the European Commission: "SMEs often must [...] overcome structural barriers such as a lack of management and technical skills [...]".[3] At present, the level of digitalization of European SMEs is still around 20%, which is highly unsatisfactory [4]. The situation is even worse in Italy where productivity of SMEs is lower than the EU average and its closest peers (i.e., France and Germany). In light of the findings in [4], awareness and capabilities are the primary knowledge gaps that limit the ability of SMEs to adopt digital solutions. Drawing on the evidence coming from case studies carried out in several European states, authors of [5] confirm that making digital maturity assessment tools available to SMEs is a way to raise their awareness on the competitive advantages implied by a digital transformation.

Through the Digital Europe Programme (DIGITAL),[4] started at the end of 2021, and European Regional Development Funds, important amounts of funding are foreseen to strengthen a network of European Digital Innovation

---

[1] https://ec.europa.eu/growth/smes.en.

[2] Definition based on Article 2 of the Annex to Commission Recommendation 2003/361/EC; https://ec.europa.eu/docsroom/documents/42921.

[3] User Guide to the SME Definition. Publications Office of the European Union, Luxembourg 2020. https://op.europa.eu/en/publication-detail/-/publication/79c0ce87-f4dc-11e6-8a35-01aa75ed71a1 (Accessed on January 23, 2022).

[4] https://digital-strategy.ec.europa.eu/en/activities/digital-programme.

Hubs (EDIHs) that will geographically cover the whole territory of Europe. Each EDIH will be at working distance of their stakeholders, speak their language, and will help firms to become more performant by improving their business/production processes, products (and services) through digital technology. A EDIH is a regional multi-partner cooperation built around: Competence Centers; Technical universities and Research and technologies organizations; and Local Knowledge and Innovation Communities. In collaboration with: Vocational training institutes; Industry associations; Public administrations/National and regional authorities; Incubators, Accelerators, Investors; Clusters; Chambers of commerce; Enterprise Europe Network; Business development agencies and others. DIHs were funded for the first time within the H2020 EU Programme to support SMEs with experimentation and testing of advanced digital technologies. To distinguish them from other DIHs, the label European has been added.

EDIHs are called to provide the following four main categories of services to the local SMEs/public sector beneficiaries: Test before invest; Skills and training; Support to find investments; and Innovation ecosystem and networking. These categories are described in detail in [4], here we restrict our attention to the first one to remain within the scope of the present study. Services under the "Test before invest" category include: awareness raising and digital maturity assessment. In addition, EDIHs will act as European Regional Development Funds intermediary for providing SMEs with vouchers to be spent to progress in their internal digitalization [4] (p. 17).

Fletcher and Griffiths, talking about digital transformation during lockdown [6], stated that organizations must improve their digital maturity, because organizations with higher levels of digital maturity are generally more flexible, while less digitally mature organizations are more fragile. In this paper, we address the problem of digital maturity assessment of SMEs. Today many (digital) maturity models ((D)MMs) are available, however they weren't developed to fit the above mentioned needs of EDIHs. To fill the twofold gap, we propose assessing the digital maturity level of SMEs as a multi-attribute decision problem, where the characteristics of the SMEs are mapped into their digital maturity level. Among available methods, in this paper we adopt the so-called DEcision eXpert (DEX) method [7–9]. Modeling the (expert) knowledge about the firm's digital maturity using the DEX method offers the following advantages: it is easy to be used; it provides convincing explanations about the final score; it allows multiple interactive analyses; it is largely accepted by users; and finally it has been implemented in the DEXi freeware software [10]. DEXi supports two basic tasks: development of qualitative multi-attribute models; application of these models for the evaluation and analysis of decision options.

Very recently, North et al. [11] introduced the DIGROW framework as a conceptual tool useful to assess the digital maturity level of SMEs, as well as the capabilities associated to each level, that are necessary for promoting a digital enabled growth. DIGROW has a solid theoretical basis, but the underlying data analysis might result tricky in real contexts. For example, North et al. [11] carried out descriptive analysis using the IBM SPSS software whose technical documentation is quite complex to be mastered. To keep things as simple as possible on the EDIHs'/SMEs' side, we rearranged the DIGROW framework into a multi-attribute DMM, then the latter was evaluated by making recourse to the DEXi software tool. The theoretical soundness of the DIGROW framework combined with the advantages featured by the DEX method make our proposal appropriate for assisting the nascent EDIHs in the activities devoted to awareness raising and digital maturity assessment of SMEs (two services belonging to the "Test before invest" category [4]), as well as in the evaluation process for awarding vouchers to SMEs.

The remaining part of the paper is structured as follows. Section 2 (Materials and Methods) recalls the DIGROW framework. Section 3 (Results) discusses its mapping into a multi-attribute DMM based on the DEX method. The implementation of the model by means of the DEXi open-source tool [10] is part of this section, as well as a brief discussion of an example of usage of the model. Section 4 concerns the relevant related work, while Sect. 5 ends the paper.

2   Materials and Methods

As mentioned in Sect. 1, in this paper we adopt the DIGROW framework described in [11][5] which combines Teece's "sensing" (i.e., the ability to detect changes and to learn quickly), "seizing" (i.e., the ability to address new opportunities in the marketplace), and "transforming" triple [13], on one side, and Valdez-de-Leon [14] digital maturity model, on the other side.

Van Veldhoven and Vanthienen identified three types of frameworks: conceptual, transformation, and maturity frameworks [15]. The latter describe the different stages of digital maturity in organizations. This category of conceptual tools is useful to assess an SME's maturity, strengths, and weaknesses, or as a comparison tool between competitors. In light of such a classification, DIGROW is a maturity framework. In fact, DIGROW allows to assess the digital maturity level of SMEs and the capabilities associated to each level, that are necessary for promoting a digital enabled growth. In other words, the framework provides guidance to SMEs to detect and capture digitally enabled growth opportunities, as well as start a project based learning process to transform the firm in order to remain competitive in the global market. The framework by North et al. [11] overcomes the limits of most previously available frameworks regarding digital transformation, which lack a sound theoretical foundation and do not consider the specific features of SMEs. The theoretical underpinnings of the DIGROW framework come from dynamic capabilities theory and from digital transformation studies especially those referred to SMEs. Teece et al. [16] define dynamic capabilities as the firm's ability to integrate, build, and reconfigure internal and external competences to address rapidly changing environments. Dynamic capabilities are an established approach in the investigation of firms. For example, Yeow et al. [17] applied it to model the business-to-business company's aligning process to enact an Internet-based business-to-consumer digital strategy over

---

[5] [12] provides a preliminary description of DIGROW.

a five-year period; Inan and Bititci [18] postulated that dynamic and organizational capabilities are the pillars of the organizational capabilities of SMEs; while Canhoto et al. [19] adopted such a lens to examine digital strategy alignment in SMEs to identify the practices that allow them to take advantage of digital technologies.

The DIGROW framework is structured as a sequence of four "steps" (called "stages" in the following), each denoting a "challenge for" (/ "capability of") the firm:

- Stage 1: Sensing digitally enabled growth potentials;
- Stage 2: Developing a digitally enabled growth strategy and mindset;
- Stage 3: Seizing digitally enabled growth potentials;
- Stage 4: Managing resources for digital transformation.

In turn, each stage of DIGROW is defined in terms of four "capacities". Figure 1 depicts the stages (visualized as a "wheel of digitally enabled growth" in [11]) and the capacities. Each capacity is evaluated starting from level 0 up to level 5. In a pre-test with firms, the results of which are reported in [11], it was found that this number of levels allows a sufficient degree of differentiation among firms. Tables AI-AIV in the Appendix of [11] detail the DIGROW maturity model evaluation grid.

By browsing through the rows of the four stages of DIGROW, it emerges that such a DMM embeds in itself the seven key attributes that, according to a report published by the Digital Transformation Institute of Capgemini [20], are the constituent elements of the digital culture of a digital mature organization, namely: Customer Centricity; Innovation; Data-driven decision making; Open Culture; Digital-first mindset; Agility and Flexibility; and Collaboration.

DIGROW has been extensively validated in the reality, which represents the exception rather than the rule, as pointed out, for example, in [21]. In 2018, North et al. tested the applicability of their framework in two distinct pilot studies. The first study involved 52 sample SMEs [11], while the second one involved 427 SMEs [22]. In both cases, the firms were located in the Basque region of Spain. These studies have contributed to get to the final version of the questionnaire collected in the Appendix of [11].

As mentioned in Sect. 1, in this paper the DIGROW maturity framework is rearranged as a multi-attribute DMM by adopting the DEX qualitative multi-criteria decision analysis method, developed by a research team led by Bohanec, Bratko, and Rajkovič [7]. In DEX, all attributes are qualitative and can take values represented by simple words (e.g., "bad", "good" or "excellent"). Attributes are organized in a tree-like hierarchy. The *leaves* of the tree, which describe the options (also named the alternatives), are the *basic* attributes (in the context of this study, the leaves denote the levels corresponding to the four capacities for each of the four stages of DIGROW), while the internal attributes are the *aggregate* attributes. All attributes are assumed to be described by discrete and ordered values, so that a higher ordinal value represents a better preference.

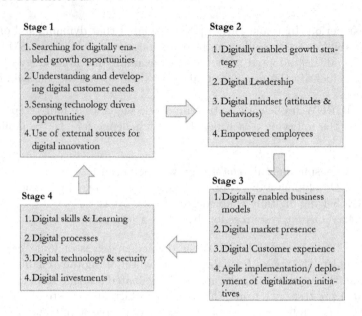

Stage 1

1. Searching for digitally enabled growth opportunities

2. Understanding and developing digital customer needs

3. Sensing technology driven
   opportunities

4. Use of external sources for
   digital innovation

Stage 2

1. Digitally enabled growth strategy

2. Digital Leadership

3. Digital mindset (attitudes &
   behaviors)

4. Empowered employees

Stage 4

1. Digital skills & Learning

2. Digital processes

3. Digital technology & security

4. Digital investments

Stage 3

1. Digitally enabled business
   models

2. Digital market presence

3. Digital Customer experience

4. Agile implementation/ deployment of digitalization initiatives

Fig. 1. The four stages of the DIGROW digital maturity framework.

The digital maturity level of a SME is evaluated through a bottom-up approach that aggregates the basic attribute values of the leaves up to a single root node value. The evaluation of decision alternatives is carried out by *utility functions*, which are represented in the form of "if-then" decision rules.

The DEX method can be formalized as follows. Let $A = a_1, a_2, \ldots, a_n$ be the finite set of attributes, and $O = o_1, o_2, o_3, \ldots$ be the (potentially infinite) set of options. The domain value of $a_i \in A$ belongs to the finite set of discrete values $D_i = v_{i_1}, v_{i_2}, \ldots, v_{i_j}$, with $v_{i_1} < v_{i_2} < \ldots < v_{i_j}$. An option ($o \in O$) is described by a vector of n values, where each value corresponds to an attribute in the set A, measured by the domain values from set D. Let $a_0 \in A$ be an aggregate attribute and $a_1, a_2, \ldots, a_n \in A$ be its children in the hierarchy, the utility function $f_0(a_1, a_2, \ldots, a_n)$ denotes the values of the aggregate attribute a_0, that is $a_0 = f_0(a_1, a_2, \ldots, a_n)$. A utility function maps *all* the combinations of the lower-level attribute values into the values of the direct aggregate attribute a_0. The mapping can be represented in a table, where each row gives the value of f for one combination of the lower-level attribute values. Rows are called *decision rules*, because each row can be interpreted as an if-then rule of the form: *if $a_1 = v_1$ and $a_2 = v_2$ and ... and $a_n = v_n$ then $a_0 = v_0$*, where $v_i \in D_i$, $i = 0, 1, \ldots, n$. Through utility functions, decision-makers can model their knowledge and preferences.

3   Results

By applying the DEX method to the DIGROW maturity framework, we obtained the multi-attribute DMM of Fig. 2. The model consists of four distinct trees corresponding to the stages of DIGROW. Each tree has a root and four leaves (the basic attributes of the hierarchy), one for each capacity of the corresponding stage. Overall, the multi-attribute model is composed of 16 basic attributes and 4 aggregate attributes (the four trees), totaling 20 attributes.

Domain Values
The domain values of all the basic attributes of the four trees belong to the ordered set of qualitative values: $0, 1, 2, 3, 4, 5$; while the domain values of the four aggregate attributes (i.e., the roots of the four trees) are the ordered set of qualitative values: Non-existent (0), Initial (1), Encouraged (2), Practiced (3), Managed (4), Optimized (5). The choice of these six values is inspired to the maturity models in [23] where these six degrees of maturity levels are adopted.

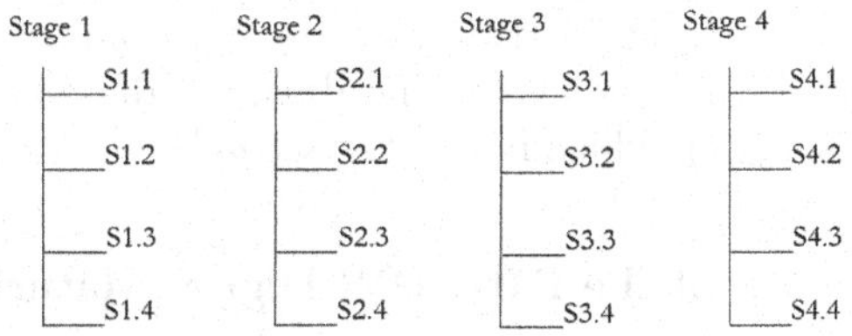

Stage 1	Stage 2	Stage 3	Stage 4
S1.1	S2.1	S3.1	S4.1
S1.2	S2.2	S3.2	S4.2
S1.3	S2.3	S3.3	S4.3
S1.4	S2.4	S3.4	S4.4

Fig. 2. The four trees defining the multi-attribute DMM based on the DIGROW framework, structured according to the DEX method.

Utility Functions
Column DMM in Table 1 collects summary data about the number of nodes and number of if-then rules for the four trees in Fig. 2.

Table 1. Summary data about two multi-attribute modeling of the DIGROW framework.

Tree levels	DMM		DMM restructured	
	Number of nodes	Number of if-then rules	Number of nodes	Number of if-then rules
h = 0 (the roots)	4	$6^4 \times 4$	4	$6^2 \times 4$
h = 1	4×4		2×4	$6^2 \times 8$
h = 2 (the leaves)			4×4	
		5184		432

This combinatorial explosion of the DEX model makes it unsuitable for actual usage. To overcome this problem, it was necessary to restructure the four hierarchies by introducing new aggregate nodes as it is shown in Fig. 3. For space reason, the figure shows only two restructured trees.

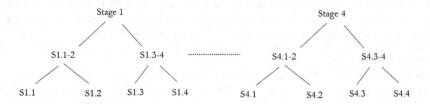

Fig. 3. Restructuring of the trees of the multi-attribute DMM of Fig. 2.

Table 1 collects summary data about the restructured trees. As we can see, the number of leaves (i.e., the number of basic attributes) is unchanged (16), while the number of aggregate attributes increased from 4 to 12. The positive side effect brought by this increase is the drastic reduction of the number of if-then rules (432 vs. 5184), which makes the second solution of practical utility.

3.1   Implementation of the DIGROW Digital Maturity Model

We have implemented the four restructured trees of the DIGROW DMM by means of the DEXi open source tool (Fig. 4).

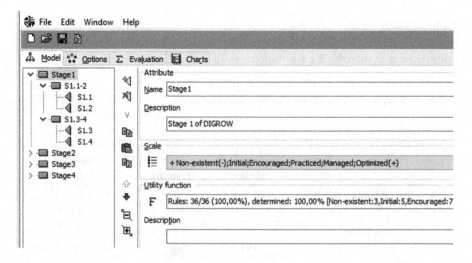

Fig. 4. The DEXi main screen showing the four trees of the multi-attribute DMM implementing the DIGROW framework.

Figure 5 shows the domain values of the basic and aggregate attributes (the latter both at height 1 and height 0 – the root of the tree) for Stage 1. The remaining three stages (namely, Stage 2, Stage 3, and Stage 4) of the multi-attribute model take the same values.

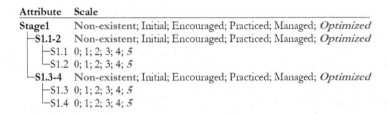

Fig. 5. The domain values with regard to Stage 1 of the multi-attribute proposed DMM.

Utility functions were defined for all the aggregate attributes belonging to the four hierarchies. Equation 1 defines the criterium adopted for the computation of the utility functions referring to the aggregate attributes at depth 1 (i.e., nodes S1.1-2, S1.3-4, ..., S4.1-2, S4.3-4 – Fig. 3), while Eq. 2 defines the criterium adopted for the computation of the utility functions referring to the aggregate attributes corresponding to the roots.

$$if \ a_1 = v_1 \ and \ a_2 = v_2 \ \ then \ \ \rightarrow \ \lfloor (v_1 + v_2)/2 \rfloor + 1 \tag{1}$$

$$if \ v_1 = v_2 \ \begin{cases} then \rightarrow v_1 \\ else \rightarrow max(v_1, v_2) \end{cases} \tag{2}$$

where $v_k (k = 1, 2) \in [0..5]$.

Figure 6 and Fig. 7 show the rows corresponding to the evaluation of the utility functions for, respectively, the aggregate attributes sons of the root and the root itself, with respect to Stage 1. In both cases, the utility functions are represented in terms of complex rules [10]. If expressed as simple rules, both these functions give rise to 36 rows (see Table 1). Converted to weights, Fig. 6 and Fig. 7 confirm that in the final assessment both aggregate attributes have the same weight (i.e., 50%). This result is an obvious consequence of the symmetry of the four trees that implement the DIGROW stages.

DEXi uses four types of weights: Local, Global, Loc.norm., and Glob.norm. Local weights refer to a single aggregate attribute and a single corresponding utility function, therefore the sum of weights of the attribute's immediate descendants is 100%. Global weights, on the other hand, take into account the structure of the tree and relative relevance of its sub-trees. A global weight of an attribute is calculated as a product of its local weight and the global weight of the attribute that lies one level above. From Fig. 8, we see that the global weight of the root attribute is 100%, while the global weight of Stage 1 is 25%: 50% (its local weight) × 50% (the global weight of S12), which gives 25%.

	S1.1	S1.2	S1.1-2
	50%	50%	
1	0	<=1	Non-existent
2	<=1	0	Non-existent
3	0	2:3	Initial
4	<=1	2	Initial
5	1	1:2	Initial
6	1:2	1	Initial
7	2	<=1	Initial
8	2:3	0	Initial
9	0	>=4	Encouraged
10	<=1	4	Encouraged
11	1	3:4	Encouraged
12	1:2	3	Encouraged
13	2	2:3	Encouraged
14	2:3	2	Encouraged
15	3	1:2	Encouraged
16	3:4	1	Encouraged
17	4	<=1	Encouraged
18	>=4	0	Encouraged
19	1:2	5	Practiced
20	2	>=4	Practiced
21	2:3	4	Practiced
22	3	3:4	Practiced
23	3:4	3	Practiced
24	4	2:3	Practiced
25	>=4	2	Practiced
26	5	1:2	Practiced
27	3:4	5	Managed
28	4	>=4	Managed
29	>=4	4	Managed
30	5	3:4	Managed
31	5	5	Optimized

	S1.1-2	S1.3-4	Stage1
	50%	50%	
1	Non-existent	<=Initial	Non-existent
2	<=Initial	Non-existent	Non-existent
3	<=Initial	Encouraged	Initial
4	Initial	Initial:Encouraged	Initial
5	Initial:Encouraged	Initial	Initial
6	Encouraged	<=Initial	Initial
7	<=Encouraged	Practiced	Encouraged
8	Encouraged	Encouraged:Practiced	Encouraged
9	Encouraged:Practiced	Encouraged	Encouraged
10	Practiced	<=Encouraged	Encouraged
11	<=Practiced	Managed	Practiced
12	Practiced	Practiced:Managed	Practiced
13	Practiced:Managed	Practiced	Practiced
14	Managed	<=Practiced	Practiced
15	<=Managed	*Optimized*	Managed
16	Managed	>=Managed	Managed
17	>=Managed	Managed	Managed
18	*Optimized*	<=Managed	Managed
19	*Optimized*	*Optimized*	*Optimized*

Fig. 6. Utility function at the root level.

Fig. 7. Utility function at the level of the sons of the root.

3.2   Examples

Aim of this section is to make use of the proposed multi-attribute DMM to show how simple is its usage and, at the same time, how simple to be interpreted are the outcomes.

The first example concerns the assessment of the digital maturity level of a single SME (called SME1). Figure 9 reports the numerical evaluation values; while Fig. 10 shows, pictorially, the corresponding attribute values. This test can be carried out either by a referent of a EDIH or by the manager of the SME. It is sufficient a rapid look at the radar chart of Fig. 10 to understand where the SME is, what its weaknesses are, and what should be done next in the digital transformation journey.

DEXi allows to compare several SMEs. As second example, Fig. 11 reports the numerical evaluation values of four SMEs; while Fig. 12 shows, pictorially, the corresponding attribute values. To save space, only Stage 1 is shown. We can think of the need to compare the digital maturity level of a set of SMEs as the preliminary step in the process of evaluation awarding vouchers to SMEs to support their digital transformation. Figure 13 provides the full picture about

Attribute	Local	Global	Loc.norm.	Glob.norm.
Stage1				
├─**S1.1-2**	50	50	50	50
│ ├─S1.1	50	25	50	25
│ └─S1.2	50	25	50	25
└─**S1.3-4**	50	50	50	50
├─S1.3	50	25	50	25
└─S1.4	50	25	50	25

Fig. 8. The average weights of aggregate attributes belonging to Stage 1.

the digital maturity level of each of the four SMEs taken into account. From such a figure it is easy to rank the SMEs.

4 Related Work

This work builds a multi-attribute DMM by transforming the DIGROW framework. Since the latter is already published, it is not necessary to defend its validity. Vice versa, it is necessary to compare the transformed model with potential competitors. Therefore, this section is organized as follows. We limit ourselves to recalling recent review articles about MMs, in particular those concerning the context of SMEs to which the reader is referred; then we compare our proposal with a very recent multi-attribute DMM referring to SMEs.

Section 2 in [11] classifies MMs for digital transformation according to the three application-specific purposes introduced in [24]: descriptive, prescriptive and comparative.

[25] is an up-to-date review about business MMs for SMEs covering the period 2001–2020 (first trimester). Only articles published in peer-reviewed journals and stored into the Scopus and the Web of Science databases were taken into account. Initially, 160 articles were selected. The classification process introduced in [26] was used to limit the article group to the articles that proposed a new MM in the SME context. At the end of the classification process, 20 articles were study in depth. Authors carried out a systematic literature review which allowed them to give an unbiased and more balanced summary of the literature compared to traditional reviews. They found that most of the articles used five maturity levels in their MMs; while in the remaining ones either three, four or six maturity levels were used. As said in Sect. 2, the DIGROW MM is structured in four stages, each of them has six levels that we mapped into the values Non-existent (0), Initial (1), Encouraged (2), Practiced (3), Managed (4), Optimized (5) of the proposed multi-attribute DMM (Sect. 3).

Previous papers reporting on the state of the art about (Digital) MMs are [27] and [28]. Below, few sentences about these studies are given to better emphasize the merits of the proposed multi-attribute DMM.

Evaluation results

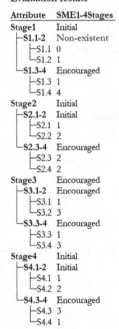

Attribute	SME1-4Stages
Stage1	Initial
├─**S1.1-2**	Non-existent
│ ├─S1.1	0
│ └─S1.2	1
└─**S1.3-4**	Encouraged
│ ├─S1.3	1
│ └─S1.4	4
Stage2	Initial
├─**S2.1-2**	Initial
│ ├─S2.1	1
│ └─S2.2	2
└─**S2.3-4**	Encouraged
│ ├─S2.3	2
│ └─S2.4	2
Stage3	Encouraged
├─**S3.1-2**	Encouraged
│ ├─S3.1	1
│ └─S3.2	3
└─**S3.3-4**	Encouraged
│ ├─S3.3	1
│ └─S3.4	3
Stage4	Initial
├─**S4.1-2**	Initial
│ ├─S4.1	1
│ └─S4.2	2
└─**S4.3-4**	Encouraged
│ ├─S4.3	3
│ └─S4.4	1

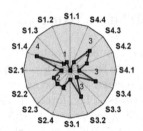

Fig. 9. Evaluation results for SME1.

Fig. 10. The radar chart displaying the evaluation results in Fig. 9.

[27] reports on a systematic literature review on DMMs for SMEs. This study confirms the lack of validation and specific proposals for SMEs. Moreover, the authors discovered six basic DMM dimensions for SMEs: Strategy; Products/Services; Technology; People and Culture; Management; and Processes. The most significant dimension is People and Culture. Stages 2 and 4 of the multi-attribute DMM proposed in our paper assign great relevance to "People and Culture". [28] reveals that the majority of existing DMMs addresses the manufacturing sector. Other domains like service are clearly under-represented. The proposed multi-attribute DMM is general.

In [29], Borštnar and Pucihar designed a DMM to be used to assess the different levels of digital maturity of SMEs. In their study, the DEX method was adopted to design a multi-attribute model. Their model takes inspiration from the findings in [3], where authors identified "technology" and "actor" as the two aggregate dimensions of digital transformation. In [29], these two dimensions are called "Digital capability" and "Organizational capability". We summarize the DMM in [29] in the following. The DMM is composed of two subtrees, one for each dimension. The tree is structured in terms of 34 basic attributes and 17 aggregated attributes. The Digital capability dimension consists of four groups of attributes: Use of Technology, Role of Informatics, Digital Business Model, and Strategy; while the Organizational capability consists of three groups of

Evaluation results

Attribute	SME1-Stage1	SME2-Stage1	SME3-Stage1	SME4-Stage1
Stage1	Initial	Encouraged	Practiced	Managed
└S1.1-2	Non-existent	Encouraged	Practiced	Managed
├S1.1	0	2	3	4
└S1.2	1	3	4	5
└S1.3-4	Encouraged	Practiced	Practiced	Managed
├S1.3	1	2	3	5
└S1.4	4	4	4	4

Fig. 11. Evaluation results for four SMEs.

attributes: Human Resources, Organizational Culture, and Management. For each attribute in the tree, the corresponding domain value was defined. The domain values were ordered, discrete, and qualitative. The DMM in [29] is in a pilot stage. So far, after a preliminary validation by an expert group, a test involving seven Slovenian SMEs from different sectors has been carried out. Authors plan to survey a larger number of SMEs, then another round of model validation will be conducted and the model will adjusted accordingly if needed.

It is quite obvious to verify that both MMs take into account the same concepts; that is, it is easy to relate the (sixteen) capacities describing the four stages of DIGROW to the seven (aggregate) attributes of the DMM in [29]. However, there is a relevant difference between the two models which comes from the abstraction level at which they capture the firms' characteristics. The multi-attribute MM based on DIGROW is much more abstract than the other one. A positive side effect of abstraction is "model stability", that is a desirable feature of a multi-attribute MM referring to the fact that the number of basic attributes on which it is built remains stable over time. To make things clear, let us refer to the DMM in [29]. As already said, "Digital technology" is one of the items of such a model. It is specialized in terms of three attributes: Advanced technologies, SMACIT, and Basic business technologies. In turn, the theme "Advanced technologies" include: Blockchain; Industry 4.0; and Data analytics. A pertinent question is: Why Internet of Things (IoT) is not included in the list? Zhu et al. [30] say that IoT is one of the five most representative technologies. For sure, IoT is a disruptive technology that has the potential to transform the world's future. In [29] (Sect. 4. Discussion), authors admit that they faced criticalities in the model building process, in connection with the phases of constructing the tree, assigning the domain values to attributes, and then defining the utility functions.

In summary, our work as that by Borštnar and Pucihar [29] are meant to be used by DIHs as well as by managers of the SMEs to assess the firm's digital maturity level. Moreover, both studies adopt the DEX method. Hence, the two proposals must be considered as alternatives. It would be interesting to validate both models on the same set of SMEs to compare their effectiveness in real settings.

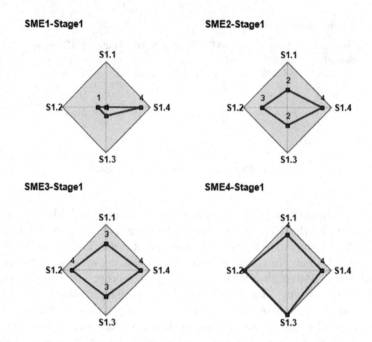

Fig. 12. The radar charts displaying the evaluation results in Fig. 11.

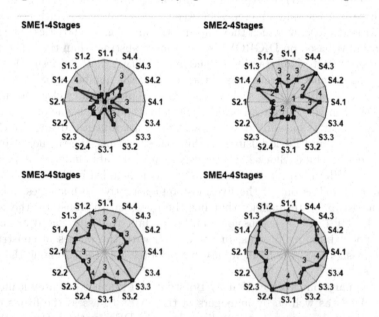

Fig. 13. The radar charts displaying the evaluation results about the four stages of the proposed multi-attribute DMM.

5   Conclusions

In February 2021, the SMB Group surveyed 761 North America Small and Medium Business (1–1,000 employee) decision-makers and mid-market (1,000–2,500 employee) organizations [31]. From the survey it emerged that, in the wake of COVID-19: (a) around 50% of American SMEs have established a digital transformation strategy and are executing on it; (b) 31% of their spending for technology solutions will increase in the 2022. Both these findings are good news, but it is important to recall that digital transformation does not imply better performance unless it goes with an adequate firm's digital maturity level, as it has been stressed, for instance, by Kane [32].

In this paper, we addressed the problem of digital maturity assessment of SMEs. Our proposal is based on a previously published DMM which we have transformed into a qualitative multi-attribute DMM in order to make it more simple to be understood and used. The model has been implemented as a DEXi model. Although many DMMs have been developed in the past, models that fit the needs of the EDIHs are rare. As said in Sect. 1, EDIHs are called to implement the "Test before Invest" service (which includes awareness raising and digital maturity assessment of SMEs), as well as to take charge of the awarding evaluation process for the assignment of vouchers to SMEs to be spent to progress in their internal digitalization. To achieve these goals the final score coming from the digital maturity assessment tool must be transparent in order to facilitate the interactions between the EDIH and the SME. The qualitative nature of our DMM ensures such a feature.

References

1. European Investment Bank: The digitalisation of small and medium-sized enterprises in Italy. Models for financing digital projects. Summary report, 2021. Prepared for COTEC Italia and the European Investment Advisory Hub By Innovation Finance Advisory (IFA) within EIB Advisory Services (2021)
2. Eller, R., Alford, P., Kallmünzer, A., Peters, M.: Antecedents, consequences, and challenges of small and medium-sized enterprise digitalization. J. Bus. Res. **112**, 11–127 (2020). https://doi.org/10.1016/j.jbusres.2020.03.004
3. Nadkarni, S., Prügl, R.: Digital transformation: a review, synthesis and opportunities for future research. Manag. Rev. Q. **71**, 233–341 (2021)
4. Kalpaka, A., Sörvik, J., Tasigiorgou, A.: Digital Innovation Hubs as policy instruments to boost digitalization of SMEs. In: Kalpaka, A., Rissola, G. (Eds.) EUR 30337 EN, Publications Office of the European Union, Luxembourg (2020). Accessed 02 Feb 2022 https://doi.org/10.2760/085193. ISBN 978-92-76-21405-2, JRC121604
5. Cavallini, S., Soldi, R.: The state of digital transformation at regional level and COVID-19 induced changes to economy and business models, and their consequences for regions. Report of the Commission for Economic Policy (2021). https://doi.org/10.2863/37402
6. Fletcher, G., Griffiths, M.: Digital transformation during a lockdown. Int. J. Inf. Manag. **55**, 102185 (2020)

7. Bohanec, M., Rajkovič, V.: An expert system for decision making. Processes and tools for decision Support. In: Sol, H.G. (ed.) IFIP 1983, pp. 235–248. North-Holland (1983)

8. Bohanec, M., Rajkovič, V.: DEX: an expert system shell for decision support. Sistemica **1**, 145–147 (1990)

9. Bohanec, M., Rajkovič, V.: Multi-attribute decision modeling: industrial applications of DEX. Informatica **23**, 487–491 (1999)

10. Bohanec, M.: DEXi: Program for Multi-Attribute Decision Making User's Manual (Version 5.05), Institut "Jožef Stefan", Ljubljana, Slovenija, May 2021

11. North, K., Aramburu, N., Lorenzo, O.: Promoting digitally enabled growth in SMEs: a framework proposal. J. Enterp. Inf. Manag. **33**(1), 238–262 (2020). https://doi.org/10.1108/JEIM-04-2019-0103

12. North, K., Aramburu, N., Lorenzo, O.: Promoting digitally enabled growth in SMEs: a framework proposal. In: Proceedings of IFKAD Conference, Delft, Holland, 4–6 July 2018, pp. 197–214 (2019). https://www.ifkad.org/

13. Teece, D.J.: Explicating dynamic capabilities: the nature and microfoundations of (sustainable) enterprise performance. Strateg. Manag. J. **28**(4), 1319–1350 (2007)

14. Valdez-de-Leon, O.: A digital maturity model for telecommunications service providers. Telev. New Media **6**(8), 19–32 (2016)

15. Van Veldhoven, Z., Vanthienen, J.: Digital transformation as an interaction-driven perspective between business, society, and technology. Electron. Mark. (2021). https://doi.org/10.1007/s12525-021-00464-5

16. Teece, D.J., Pisano, G., Shuen, A.: Dynamic capabilities and strategic management. Strateg. Manag. J. **18**(7), 509–533 (1997)

17. Yeow, A., Soh, C., Hansen, R.: Aligning with new digital strategy: a dynamic capabilities approach. J. Strat. Inf. Syst. **27**(1), 43–58 (2018)

18. Inan, G.G., Bititci, U.S.: Understanding organizational capabilities and dynamic capabilities in the context of micro enterprises: a research agenda. Procedia - Soc. Behav. Sci. **210**, 310–319 (2015). Proceedings of 4th International Conference on Leadership, Technology, Innovation and Business Management

19. Canhoto, A.I., Quinton, S., Pera, R., Molinillo, S., Simkin, L.: Digital strategy aligning in SMEs: a dynamic capabilities perspective. J. Strateg. Inf. Syst. **30**, 101682 (2021)

20. Buvat, J., Crummernel, C., Kar, K., et al.: The digital culture challenge: closing the employee leadership gap. Capgemini Digital Transformation Institute Survey (2017)

21. Lasrado, L.A., Vatrapu, R., Andersen, K.N.: Maturity models development in is research: a literature review. In: Proceedings of the 38th Information Systems Research Seminar in Scandinavia, Oulu, Finland 9–12 August 2015

22. North, K., Aramburu, N., Lorenzo, O., Zubillaga Rego, A.: Digital maturity and growth of SMEs: a survey of firms in the Basque country (Spain). In: Finland of International Forum on Knowledge Assets Dynamics, Matera, Italy, 5–7 June 2019

23. IT Governance Institute 2007 COBIT 4.1. The IT Governance Institute. Accessed 3 Jan 2022

24. de Bruin, T., Rosemann, M., Freeze, R., Kulkarni, U.: Understanding the main phases of developing a maturity assessment model. In: Proceedings of 16th Australasian Conference on Information Systems (ACIS), Sydney, Australia, 29 November–2 December 2005

25. Virkkala, P., Saarela, M., Hänninen, K., Kujala, J., Simunaniemi, A.-M.: Business maturity models for small and medium-sized enterprises: a systematic literature review. Management **15**(2), 137–155 (2020)

26. Thorpe, R., Holt, R., Macpherson, A., Pittaway, L.: Using knowledge within small and medium-sized firms: a systematic review of the evidence. Int. J. Manag. Rev. **7**(4), 257–81 (2005)
27. Williams, C., Schallmo, D., Lang, K., Boardman, L.: Digital maturity models for small and medium-sized enterprises: a systematic literature review. In: Proceedings of The ISPIM Innovation Conference - Celebrating Innovation: 500 Years Since da Vinci, Florence, Italy, 16–19 June 2019. www.ispim.org
28. Teichert, R.: Digital transformation maturity: a systematic review of literature. Acta Universitatis Agriculturae et Silviculturae Mendelianae Brunensis **67**(6), 1673–1687 (2019)
29. Borštnar, M.K., Pucihar, A.: Multi-attribute assessment of digital maturity of SMEs. Electronics **10**, 885 (2021). https://doi.org/10.3390/electronics10080885
30. Zhu, X., Ge, S., Wang, N.: Digital transformation: a systematic literature review. Comput. Ind. Eng. **162**, 107774 (2021)
31. SMB Technology Directions for a Changing World SMB Group (2021). https://www.smb-gr.com/reports/smb-technology-directions-for-a-changing-world/. Accessed 3 Jan 2022
32. Kane, G.C.: Digital Maturity, Not Digital Transformation. MIT Sloan Management Review, 4 April 2017. http://sloanreview.mit.edu. Accessed 3 Jan 2022

International Workshop
on Econometrics and Multidimensional
Evaluation in Urban Environment
(EMEUE 2022)

A Methodological Approach for the Assessment of the Non-OSH Costs

Maria Rosaria Guarini[1] , Rossana Ranieri[1(✉)] , Francesco Tajani[1] ,
Pierluigi Morano[2] , and Francesco Sica[3]

[1] Department of Architecture and Design, "Sapienza" University of Rome, 00196 Rome, Italy
rossana.ranieri@uniroma1.it
[2] Department of Civil, Environmental, Land, Building Engineering and Chemistry
(DI-CATECh), Polytechnic University of Bari, 70126 Bari, Italy
[3] Department of Civil, Environmental and Mechanical Engineering, University of Trento,
38123 Trento, Italy

Abstract. Accidents at work represent, globally, a significant social and business cost for all production processes and working sectors. With a view to achieve the Sustainability Goals shared in the 2030 Agenda, it is necessary to design tools that can analyze and monitor the costs related to the accidents phenomenon, to support the Public Administration in defining effective strategies for the implementation of health and safety in the workplace and to reduce the costs related to the phenomenon of injuries. On the basis of an analysis of the literature referring to existing models used to calculate the costs of the accident phenomenon, this research proposes a methodological approach to calculate the costs incurred, both in case and in the absence of injury by the main stakeholders (worker, enterprise, state, society) with reference to four different scenarios. The methodological approach can be used to different product sectors and different geographical contexts and could be used to compare the effectiveness of different political strategies.

Keywords: Health and safety at work · OSH costs · Accidents at work · Sustainability · Social goals

1 Introduction

Sustainable Goals shared in Agenda 2030 set out objectives covering social, economic and environmental aspects. In particular, with reference to the framework of the present research, objectives 3 "Ensure healthy lives and promote well-being for all at all ages" and 8 "Promote sustained, inclusive and sustainable economic growth, full and productive employment and decent work for all" are to be explored in depth [1]. Those objectives are connected to the mission state in 1947 by the World Health Organization (WHO) to achieve "by all populations of the highest possible standard of health", defined as "a state of complete physical, mental and social well-being' and not simply "the absence of disease or infirmity" [2]. In this sense, it should be highlighted that a safe and healthy workplace contributes to labor productivity and promotes economic growth [3–5] and it is therefore consistent in the pursuit of sustainable objectives.

O. Gervasi et al. (Eds.): ICCSA 2022 Workshops, LNCS 13378, pp. 561–571, 2022.
https://doi.org/10.1007/978-3-031-10562-3_39

Furthermore, it is recognized that accidents at work and occupational diseases represent a significant social and business cost in all production processes. In fact, the International Labour Organization (ILO) estimated in 2012 that globally 4% of GDP is lost due to work accidents and diseases [6]. A recent project by the European Agency for Safety and Health at Work (EU-OSHA) observed that the inconvenience of work injuries and diseases is 3.9% of global GDP and 3.3% of European GDP [7].

In this sense, it is worth pointing out that is estimated that globally every year the death of 2.3 million women and men is caused by work accidents or diseases, corresponding to over 6000 deaths every single day [8]. It means that around 340 million occupational accidents and 160 million victims of work-related illnesses annually occur. To outline the severity of this phenomenon, it should be taken into account that the Covid 19 pandemic has caused over 6 million deaths in almost two years (3 million per year) [9].

With a view to monitoring and analyzing the accident phenomenon, the aim of this contribution is to define a methodological approach to calculate and evaluate the costs related to the management of health and safety in the workplace. In particular, the approach is intended to assess the cost for the stakeholders involved in production processes (entrepreneur, collectivity, workers etc.), both in the event and in the absence of an accident in scenarios with different modalities of Occupational Safety and Health (OSH) management.

The methodological approach allows to calculate the costs resulting from an accident event for each of the above-mentioned stakeholders according to different scenarios. In detail, the proposed evaluation approach is planned to highlight and identify the investment costs in prevention and OSH and also to estimate the costs of non-safety. This is carried out by analyzing the consequences that may fall on the worker, the entrepreneur, both in case and in the absence of an accident. The proposed methodology allows to verify whether, and to what extent, the costs related to investments, both to comply with regulatory requirements and to implement occupational health and safety management systems, are lower than those that would be incurred in the event of an accident.

The work is organized in three Sections. In Sect. 2, an analysis of the current literature on the work-related accident costs and the methods to assess them is reported. In Sect. 3, the methodological approach and its main phases are illustrated. In Sect. 4, the conclusions and possible further insights are drawn.

2 Current Literature

Although an established method is not shared, many contributions aimed at assessing the work-related accident costs can be found in the current literature [10–16].

Assessing the work-related accident costs could be a complex task because the costs are hard to identify and - often - underestimated [17, 18]. In fact, the total costs of occupational accidents and illnesses are often undervalued, both because some costs are external to the company and internal company costs are particularly difficult to be quantified and recognized. Moreover, a large part of the consequences of a work-related accident are often hidden or cannot be financially quantified [3].

In detail, the Human Capital Approach and the Opportunity Cost Method [19–22] are frequently borrowed for the calculation of occupational accident and/or illness costs. Ma et al. analyze the factors that may influence the planning of strategies to implement safety investments in China from the point of view of the opportunity cost, intended as the ratio between the costs due to the occurrence of accidents and the costs related to investments in prevention [23]. Furthermore, the Friction Cost Method is often used [24], in particular for those accidents that causes long term consequences to the workers and create vacancy to be covered by an unemployed person [25] and when it is needed to specifically consider productivity [26, 27].

The methods and the approaches are various and heterogeneous in terms of type of costs studied and perspective of stakeholders involved. It is also necessary to highlight that an exhaustive and shared list of the cost items to be considered has not been developed, as some of them are immediately evident and quantifiable in monetary terms, whereas others are harder to be monetized [20, 28–32].

The OSH costs are assessed by many other points of view, by investigating the relationship between accidents at work and climatic condition [33] and the benefits of implementing an OSH management system [34, 35]. Furthermore, many studies relating the assessment of non-OSH costs on the communities have been carried out [36–38].

The definition of the he methodological approach requires the identification of the stakeholders involved and the costs to be considered. EU-OSHA [39] defines a framework that individuates: the main stakeholders involved (Table 1); the four potential outcomes of an accident; five main cost categories summarizing the consequences of accidents in economic terms (Table 2); the appropriate methods for each cost category; the distribution of the specific costs within each of the five categories among four types of stakeholders bearing the effects of accidents.

Table 1. Stakeholder involved

Name	Definition
Worker (W)	Person affected by the accident or their family members
Enterprise (E)	Company or organization where the individual works
State (C)	Public authority competent for social security benefits
Society (So)	All stakeholders - effect on society at large

Elaboration on EU - OSHA 2014 [39].

For the specific objectives of this study, the *Society* stakeholder and the cost category related to *quality of life* are excluded.

Table 2. Cost categories

Category	Costs related to
Productivity (P)	Decreasing of yield or production
Health (S)	Both direct and indirect medical services
Administrative (Am)	Social security benefits
Insurance (As)	Insurance benefits
Quality of life (Q)	The decrease in quality of life

Elaboration on EU - OSHA 2014 [39].

3 Methodological Approach

By borrowing the EU-OSHA framework, the proposed methodological approach considers: three stakeholders (worker, entrepreneur, collectivity); four cost categories (productivity, healthcare, administrative, insurance); four types of an accident (death, permanent inability, temporary inability and no consequences).

In addition to these elements, it is necessary to define [40]:

- connotation of accident categories;
- scenarios and alternatives to assess;
- stakeholder characteristics;
- time horizons and reference periods;
- list of macro-cost items and their declination.

3.1 Connotation of Accident Categories

It should be highlighted that the methodological approach only takes into account the consequences related to accidents and does not consider the occupational disease. Referring to the consequences of a work-related accident, this classification has been introduced: the death of the victim (D); a permanent disability (IP); a temporary disability (IT). The case of non-occurrence of the accident (NA) is also considered.

In order to quantify the monetary cost of the accident's consequences, an analysis of the main accidents dynamics has to be carried out, by examining the values published by Social Security and Insurance Institutions of the reference context and by detecting the most frequent accidents connotation related to the nature of the injury (e.g. wound, broken bone, fracture) and the body part injured. Having identified the most frequent data for the sector under consideration, the prognosis range, the type of disability and the type of compensation provided by the health/insurance system must be defined.

3.2 Scenarios and Alternatives

Scenarios
It is essential to establish the four scenarios taken as reference. The characteristics to be considered relate to: the contractual regularity of the company's workers (whether

or not there are illegal or undocumented workers); the compliance or non-compliance with health and safety regulations with respect to the legislative apparatus in force; the compulsory registration by the company with social security institutions; the possession of a certification relating to the occupational health and safety management system with requirements in accordance with the ISO 45.001 "Occupational health and safety management systems — Requirements with guidance for use" standard [41].

From the combination of the occurrence or non-manifestation of the conditions mentioned above, the four scenarios could vary from Scenario 1 (S1) where all conditions are non-compliant, to Scenario 4 (S4) where the conditions are compliant.

Alternatives (An)

Combining the outlined four scenarios and the four accident's categories it is possible to obtain sixteen alternatives that allows to assess how specific costs could change (e.g. same accidents occurring in different scenarios or different accident in the same scenario).

3.3 Stakeholder Characteristics

To identify those factors whose variability could have the utmost impact on the assessment of costs, it is mandatory to consider "ordinary" characteristics of the stakeholders. The data should be taken from statistical institutes or pension funds for the production sector examined.

For the worker, it is necessary to consider: working condition, sex, familiar composition, classification by type of work, workplace, type of contract, theoretical annual working hours, average annual working hours worked.

For the enterprise, it is necessary to collect information about average number of workers, average annual revenue, number of accidents in the two years prior the accident occurred.

For the State, noteworthy characteristics are to be found in the State articulation, in particular with reference to the Social Security and Insurance Institutions that directly manage the costs related to the work-related accident like healthcare and welfare costs.

For the Society, the main indicators are those that refer to the quality of life of citizens and the well-being and how the work-related accident could affect it.

3.4 Time Periods

Significant time periods should be defined (Ya, Yb and Yc), by considering three milestones: the first one is the average age of the worker in the year of the accident (Ea), the second one is the age the retirement age under current legislation (Er), the third one is referred to the age relative to life expectancy for the country in question (Ed). Named Ya the first year after the accident, the reference periods Yb and Yc can be determined and assessed through Eqs. (1) and (2):

$$Yb = Er - Ea \tag{1}$$

$$Yc = Ed - Er \tag{2}$$

If the worker is involved in an accident with a fatal outcome, the costs relating to the survivor after the date of the worker's assumed life expectancy should not be considered because the variables - closely linked to family conditions - cannot be traced back to an ordinary condition.

3.5 List of Macro-cost Items and Their Declination

For each stakeholder considered, it is necessary to define a list of specific items useful for the calculation of the cash flows. In Tables 3, 4 and 5 a brief classification of costs is reported, based on the cost categories identified above. All the costs' items must be specifically tailored to the context in which the methodological approach is to be applied, by considering the obligations deriving from the current legislative framework of the Entities responsible for managing accident costs. E.g., for the wage charges the national collective bargaining agreement adopted in the context of reference and its parameters in terms of minimum wage and hours worked should be considered.

Table 3. Macro-cost items for the worker

Cost category	Macro-cost items	Breakdown of cost items
Productivity	Wage charges	Hourly wage base non-regular worker
		Hourly wage base regular worker
		Annual salary for non-regular workers (in S1)
		Annual salary for regular workers (in S2; S3; S4)
	Tax and social security charges	Taxable income
		Contribution for worker's spouse
		Contribution for worker's ki
		Burdens on the employee
		Severance indemnity
	Insurance deductions	Employee's cash account
Healthcare costs	Costs for disability reduction	Removal of architectural barriers Prosthesis supply Supply of specific equipment
Administrative costs	Fines	Labor contract violations
		OSH-related violations
	Fees for the lawyer	Civil Proceedings
		Penal Proceedings
	Funeral expenses	*(only in case of death of the worker)*

(continued)

Table 3. (*continued*)

Cost category	Macro-cost items	Breakdown of cost items
Insurance costs	Indemnity	Temporary daily allowance
		Permanent daily allowance
		Survivor's pension

Table 4. Macro-cost items for the enterprise

Cost category	Macro-cost items	Breakdown of cost items
Productivity	Salary	Annual salary for non-regular workers (in S1)
		Annual salary for regular workers (in S2; S3; S4)
	Tax and social security charges	Contribution fees
		Severance indemnity
	Basic personal protective equipment	Safety helmet
		Safety shoes
		Gloves
		Goggles
		Disposable overall
	Replacement worker salary	*(for the friction period considered)*
	Average Annual Revenue	
Healthcare costs	Cost of appointments by position of responsibility	Competent occupational doctor Health Monitoring Protocol
Administrative costs	Fines	
	Costs worker's training	
	Fees for the lawyer	Civil Proceedings
		Penal Proceedings
	Costs relating to the production of documentation required by law	Risk assessment documents
	Cost of health and safety management system according to ISO 45001	
Insurance costs	Indemnity	Temporary daily allowance
		Permanent daily allowance

Table 5. Macro-cost item for the Society

Cost category	Macro-cost items	Breakdown of cost items
Productivity	Tax and social security charges	
Healthcare costs	Costs for disability reduction	Removal of architectural barriers
	Costs related to treatment provided by the national health service	Prosthesis supply
		Supply of specific equipment
Administrative costs	Fines	Labor contract violations
		OSH-related violations
	Trials	Civil Proceedings
		Penal Proceedings
Insurance costs	Indemnity	Temporary daily allowance
		Permanent daily allowance
		Survivor's pension

3.6 Assessment of the Costs

Once the list of cost items has been defined, the cash flows relating to the three time periods (F_1 for the first period Ya, F_2 for the second Yb and F_3 for the third period Yc) are determined. At the end, the sum of the cash flows F_{tot} is calculated by Eq. (3) for each alternative An (considering n varying between 1 and 16).

$$F_{tot}(z)\forall An = F_{1(z;An)} + F_{2(z;An)} + F_{3(z;An)} \tag{3}$$

For each stakeholder (indicated generically with z) and for each alternative to evaluate (An) the costs K can be calculated by Eq. (4):

$$F_1 f(z; A_n) = \Sigma_{Ya} f(z; A_n)[K_{Ya(P)} + K_{Ya(S)} + K_{Ya(Am)} + K_{Ya(As)}] \tag{4}$$

Consequently, F_2 can be determined by Eq. (6), taking into account that K_r represents the recurrent costs, K_b is the basic costs, y_{ir} is the year immediately following that of the accident:

$$K_{ir} = K_r + K_b \tag{5}$$

$$F_2 f(z; A_n) = \{\Sigma_{yir} f(z; A_n)[K_{ir\,(P)} + K_{ir\,(S)} + K_{ir\,(Am)} + K_{ir\,(As)}]\} * Yb \tag{6}$$

Finally, F_3 can be obtained through Eq. (7), by adding up the cost items related to pension contribution (K_3):

$$F_3 f(z; A_n) = \{\Sigma_{Yc} f(z; A_n)[K_{3(P)} + K_{3(S)} + K_{3(Am)} + K_{3(As)}]\} * Yc \tag{7}$$

In order to be able to assess the convenience to invest in OSH, the output obtained for each alternative must be compared, by considering the "best" accident consequence categories, that corresponds to the maximum positive value.

4 Conclusions

The issue of occupational health and safety management is closely linked to the Sustainable Goals of the 2030 Agenda [1]. It is clear that the implementation of working conditions, in particular safety management measures and all related prevention activities, could generate benefits in terms of: quality of life and overall well-being of workers; cost savings and increased productivity for the company; reduction of health costs for the community. The methodological approach developed in this research allows to assess the degree of benefit for each stakeholder considered, in relation to the scenarios foreseen. The articulation of the methodological approach takes into account the most influential factors that could change the costs related to the management of health and safety in the workplace.

By modifying the origin of the input values, the method can be applied to multiple contexts, considering any sector of economic activity and any legislative apparatus. The construction of the methodological approach allows, if the boundary conditions may vary, or if the approach should be adapted to contexts that do not provide for the collection of some specific data, to be used even partially to calculate the costs of interest. The research aims to be a useful reference for the awareness of entrepreneurs in the field of health and safety, highlighting the onerousness of the consequences related to the non-application of mandatory legislation on the subject and the convenience of investing in prevention. Moreover, the method could be an effective tool to support the public entities involved in the reduction of the accident situations and the spread of the safety culture at work.

Future insights of the research foresee a systematic application of the methodology to national data in order to outline a picture of which sectors and types of accidents have the greatest impact on the State budget and to support policy-makers in a more effective strategic planning of State investments in prevention, health and training.

The issue dealt with has strong ethical implications, as it affects the health and well-being of workers and citizens [2]. By recalling the words of the President of the Italian Republic in his inauguration speech for his second term in office (on the 3rd of March 2022), "the safety of work, of every worker, concerns the value we attribute to human life" [42].

References

1. Sustainable Development Goals. https://sdgs.un.org/goals. Accessed 2 Apr 2022
2. World Health Organization. https://www.who.int. Accessed 2 Apr 2022
3. Mossink, J.C.M., de Greef, M.: Inventory of Socioeconomic Costs of Work Accidents. Office for Official Publications of the European Communities, Luxembourg (2002)
4. De Greef, M., Van Den Broek, K., Van Der Heyden, S., Kuhl, K., Schmitz-Felton, E.: Socioeconomic costs of accidents at work and work-related ill-health. Full study report. European Union (2011)
5. Tompa, E., et al.: Economic burden of work injuries and diseases: a framework and application in five European Union countries. BMC Public Health **21**(1), 1–10 (2021)
6. International Labour Organization. https://www.ilo.org/moscow/areas-of-work/occupational-safety-and-health/WCMS_249278/lang--en/index.htm#:~:text=The%20ILO%20estimates%20that%20some,of%20work%2Drelated%20illnesses%20annually. Accessed 2 Apr 2022

7. Elsler, D., Takala, J., Remes, J.: Work-related accidents in EU cost 476 billion Euro per year–Results of a recent project of EU-OSHA and ILO (2021)
8. International Labour Organization. https://www.ilo.org/global/topics/safety-and-health-at-work/lang--en/index.htm. Accessed 2 Apr 2022
9. WHO Coronavirus Dashboard. https://covid19.who.int/. Accessed 2 Apr 2022
10. Steel, J., Godderis, L., Luyten, J.: Productivity estimation in economic evaluations of occupational health and safety interventions: a systematic review. Scand. J. Work Environ. Health **44**(5), 458–474 (2018)
11. Reniers, G., Brijs, T. An overview of cost-benefit models/tools for investigating occupational accidents. Chem. Eng. Trans. **36**, 43–48 (2014)
12. Chountalas, P.T., Tepaskoualos, F.A.: Implementing an integrated health. Safety and environmental management system: the case of a construction company. Int. J. Qual. Res. **11**, 733–752 (2017)
13. Health and Safety Executive. Costs to Britain of workplace fatalities and self-reported injuries and ill health 2012/13, London (2013)
14. International Labour Organization. Safety and Health at Work: A Vision for Sustainable Prevention: XX World Congress on Safety and Health at Work 2014: Global Forum for Prevention, 24–27 August 2014, Frankfurt, Germany/International Labour Office, Geneva (2014)
15. Bräunig, D., Kohstall, T., Insurance, G.S.A.: Calculating the international return on prevention for companies: costs and benefits of investments in occupational safety and health. International Social Security Association, Final Report (2013)
16. International Social Security Association. Social security and a culture of prevention: a three-dimensional approach to safety and health at work, Geneva (2013)
17. Mazzetti, M., Cipolla, C.: Sicurezza e salute sul lavoro: quale cultura e quali prassi? Sicurezza e salute sul lavoro, pp. 1–223 (2015)
18. Waehrer, G.M., Dong, X.S., Miller, T., Haile, E., Men, Y.: Costs of occupational injuries in construction in the United States. Accid. Anal. Prev. **39**(6), 1258–1266 (2007)
19. Miller, P., Whynes, D., Reid, A.: An economic evaluation of occupational health. Occup. Med. **50**(3), 159–163 (2000)
20. Miller, T.R., Waehrer, G.M., Leigh, J.P., Lawrence, B.A., Sheppard, M.A.: Costs of Occupational Hazards: A Microdata Approach. National Institute of Occupational Safety and Health, Washington, DC (2002)
21. Nunez, I., Prieto, M.: The effect of human capital on occupational health and safety investment: an empirical analysis of Spanish firms. Hum. Resour. Manag. J. **29**(2), 131–146 (2019)
22. Nichols, T.: Health and safety at work. Work Employ Soc. **12**(2), 367–374 (1998)
23. Ma, Y., Zhao, Q., Xi, M.: Decision-makings in safety investment: an opportunity cost perspective. Saf. Sci. **83**, 31–39 (2016)
24. Jallon, R., Imbeau, D., de Marcellis-Warin, N.: Development of an indirect-cost calculation model suitable for workplace use. J. Safety Res. **42**(3), 149–164 (2011)
25. Brady, W., Bass, J., Moser Jr, R., Anstadt, G.W., Loeppke, R. R., Leopold, R.: Defining total corporate health and safety costs-significance and impact: review and recommendations. J. Occup. Environ. Med. **39**, 224–231 (1997)
26. Tompa, E., Verbeek, J., Van Tulder, M., de Boer, A.: Developing guidelines for good practice in the economic evaluation of occupational safety and health interventions. Scand. J. Work Environ. Health **36**, 313–318 (2010)
27. Tompa, E., Culyer, A.J., Dolinschi, R.: Economic Evaluation of Interventions for Occupational Health and Safety: Developing Good Practice. Oxford University Press, Oxford (2008)
28. Miller, P., Whynes, D.: An economic evaluation of occupational health. Occup. Med. **50**(3), 159–163 (2000)

29. Biddle, E.A., Keane, P.R.: The economic burden of occupational fatal injuries to civilian workers in the United States based on the Census of Fatal Occupational Injuries, 1992–2002 (2011)
30. Leigh, J.P., Miller, T.R.: Ranking occupations based upon the costs of job-related injuries and diseases. J. Occup. Environ. Med. **39**, 1170–1182 (1997)
31. Leigh, J.P., Waehrer, G., Miller, T.R., Keenan, C.: Costs of occupational injury and illness across industries. Scand. J. Work Environ. Health **30**, 199–205 (2004)
32. Leipziger, D.: The Definitive Guide to the New Social Standard. Financial Times Prentice Hall, Upper Saddle River (2002)
33. Ma, R., et al.: Estimation of work-related injury and economic burden attributable to heat stress in Guangzhou, China. Sci. Total Environ. **666**, 147–154 (2019)
34. Rzepecki, J.: Cost and benefits of implementing an occupational safety and health management system (OSH MS) in enterprises in Poland. Int. J. Occup. Saf. Ergon. **18**(2), 181–193 (2012)
35. Yiu, N.S., Chan, D.W., Shan, M., Sze, N.: Implementation of safety management system in managing construction projects: benefits and obstacles. Saf. Sci. **117**, 23–32 (2019)
36. Takala, J., et al.: Global estimates of the burden of injury and illness at work in 2012. J. Occup. Environ. Hyg. **11**(5), 326–337 (2014)
37. Waehrer, G., Dong, X.S., Miller, T., Haile, E., Men, Y.: Costs of Occupational Injuries in Construction in the United States (2004)
38. Zainal Abidin, M., Rusli, R., Khan, F., Mohd, S.: A Development of inherent safety benefits index to analyse the impact of inherent safety implementation. Process Saf. Environ. Protect. **117**, 454–472 (2018)
39. Weerd, M.D., Tierney, R., Duuren-Stuurman, B.V., Bertranou, E.: Estimating the cost of accidents and ill-health at work: a review of methodologies (2014)
40. Guarini, M.R., Ranieri, R.: Is investing in safety worthwhile? A methodology for assessing the costs and benefits of accidents in the construction sector. In: Morano, P., Oppio, A., Rosato, P., Sdino, L., Tajani, F. (eds.) Appraisal and Valuation. GET, pp. 269–288. Springer, Cham (2021). https://doi.org/10.1007/978-3-030-49579-4_18
41. International Standard Organization. https://www.iso.org/standard/63787.html. Accessed 2022
42. Presidency of the Republic. https://www.quirinale.it/elementi/62298. Accessed 12 Mar 2022

A Decision-Making Process for Circular Development of City-Port Ecosystem: The East Naples Case Study

Sabrina Sacco and Maria Cerreta[✉]

Department of Architecture, University of Naples Federico II, via Toledo 402,
80134 Naples, Italy
{sabrina.sacco,cerreta}@unina.it

Abstract. The synergy between the city and the port area represents an essential component for the sustenance of the economic system of the European Union. In this perspective, ports have been recognised as critical strategic hubs in economic competitiveness, job opportunities and investments. Nowadays, ports as strategic hubs suggest an increase of spaces dedicated to logistics and a high economic, social and environmental impact on cities. The need to define a new process of circular development of the city-port ecosystem considers environmental and social challenges, proposing a circular urban model capable of renewing the relationship between city and port. Starting from the Sustainable Development Goals (SDGs) and Agenda 2030, and the instances of circular development, the research defines an adaptive and multidimensional decision-making process capable of integrating evaluation approaches and tools to deal with complex interactions and conflicts that characterise the city-port ecosystem. The outcomes were focused on evaluating alternative scenarios for the city-port ecosystem of East Naples, in Italy, towards a circular model implementation. The decision-making process supports an accurate reflection on the possibilities of regenerative transformation of a complex context, considering the specific needs of stakeholders and the local community.

Keywords: Circular city-port · Multi-criteria analysis · Community-driven process

1 Introduction

The synergy between the city and the port area represents an important issue for the economic system of the European Union as a trigger capable of relaunching its growth, improving its efficiency and stimulating innovation and long-term competitiveness [1].

In this perspective, ports have been recognised as critical strategic hubs for the transport system and economic competitiveness and their high potential for job opportunities and investments [2]. However, today speaking of ports as strategic hubs means considering an increase in spaces dedicated to logistics and consequently of a high impact on cities in economic, social and environmental terms [3].

O. Gervasi et al. (Eds.): ICCSA 2022 Workshops, LNCS 13378, pp. 572–584, 2022.
https://doi.org/10.1007/978-3-031-10562-3_40

Pollution, noise and visual impact are some of the leading social and environmental effects stemming from port expansion [4]. The awareness of the negative impact of port expansion has led to a growing public demand for the transformation process of urban waterfront [5, 6]. It is evident the need to define a new growth process for a port-city that considers environmental and social challenges, proposing an urban model capable of overcoming the dichotomy between the two polarities. This model could allow proposing integrated and circular visions that, on the one hand, enhance the port as a strategic hub and, on the other, favour a mending of the urban fabric and a renewed relationship between city and sea.

The definition of a circular vision of the city [7] is strongly interconnected with the Sustainable Development Goals (SDGs) of the Agenda 2030, drawn up by the United Nations [8]. The approach of the SDGs is particularly significant to address the issue of a regenerative transformation process and to deduce a set of actions focused on the complex urban and maritime issues affecting the city-port ecosystem.

In this perspective, the circular models stimulate reflections on the architectural project moving toward the re-use of existing heritage [9], understood as a fundamental element for local identity and social cohesion [10]. The concept of reuse has been enriched over time with cultural, socio-economic and ecological objectives that underline its potential as a driver of new strategies for the regeneration of the context for the reduction of soil consumption [11–16].

In the specific context of the city-port, the landscapes are dotted with imposing architectures, now abandoned or underused, which have marked a glorious past in terms of production and transformation of raw materials. Starting from a circular perspective, it is central to rewrite the role of these great ghost architectures present in the coastal landscape and the relationships between the city and the port.

The research aims to define an adaptive and multidimensional decision-making process that integrates evaluation approaches and tools to deal with complex interactions and conflicts that characterise the city-port ecosystem. Specifically, the research finds its field of application in the context of the port city of Naples (Italy), in the district of San Giovanni a Teduccio.

The decision-making process was developed as part of the activities carried out during the Second level Master in "Sustainable planning and design of port areas", at the Department of Architecture (DiARC) of the University of Naples Federico II, in support of the Metropolitan City of Naples and the Central Tyrrhenian Port System Authority (AdSP) in identifying alternative strategies for the redevelopment of the East Naples port area.

This contribution is organised as follows: the first Section indicates the materials and methods adopted to define a multidimensional and multi-methodological decision-making process for the city-port ecosystem of East Naples; the second Section describes the case study; the third one shows the application of the methodological approach to the case study and report the results. Finally, the last Section points out preliminary outcomes, highlighting potential and critical aspects.

2 Materials and Methods

The research aims to elaborate an evaluation framework for defining a decision-making process based on a multidisciplinary and integrated methodology [17], which can help a circular growth strategy for the city-port ecosystem.

The complexity of the city-port ecosystem necessarily requires the coexistence of specific and diversified skills and to structure the decision-making problem starting from the main issues that are the subject of public debate. Therefore, the research, from a sustainable and circular development perspective, considers the following transformation factors:

- the complex relationship between port and city;
- the economic issues of the port authority;
- the social issues of the local community.

The methodological process proposed by the research adopts a place-based approach [18–20]. This approach requires a decision-making process involving a plurality of local actors (governments, communities, businesses, universities, etc.) to support initiatives of territorial regeneration and cultural and social innovation [21].

The proposed methodology is based on mixed evaluation methods, eliciting soft and hard knowledge domains, which were then assessed to develop and define alternative, inclusive and sustainable solutions [22] (see Fig. 1).

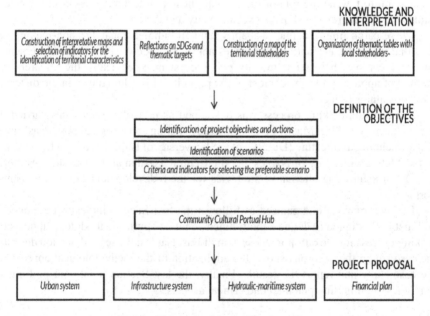

Fig. 1. The methodological framework.

The collection of hard data deriving from institutional sources was combined with a soft survey to bring out the needs and requirements of local communities. To this end, defining a map of the most significant stakeholders proved indispensable, orienting research towards a vision of balanced development and making it possible to assess compromise solutions (see Fig. 2).

Fig. 2. The stakeholders' map.

By organising thematic tables of discussion, a workflow was built focused on defining sustainable and inclusive design alternatives starting from choosing the most suitable SDGs objectives [8].

Each round table was organised by submitting to the various participants some of the SDGs objectives of Agenda 2030 regarding the subject of the meeting. For the study of the San Giovanni a Teduccio area, ten objectives (Objective 4 - Quality education; 7 - Affordable and clean energy; 8 - Decent work and economic growth; 9 - Industry, innovation and infrastructure; 11 - Sustainable cities and communities; 12 - Responsible consumption and production; 13 - Climate action; 14 - Life below water; 16 - Peace, justice and strong institutions; 17 - Partnerships for the goals) were selected, translated into project criteria and brought to the attention of stakeholders during the seminars.

During the discussion table, each participant was asked to express a score according to the Likert scale from 1 to 5, where 1 represents "strongly disagree" and 5 means "strongly agree" for each proposed SDG object to build together a set of macro objectives to be pursued. At the end of each seminar, sustainable goals and visions for the neighbourhood are extrapolated to help define project scenarios.

At the same time, a survey was submitted to associations and citizens of the neighbourhood to identify better moods regarding new transformation processes to be undertaken. The survey, constructed according to the CATWOE method [23], allows to highlight criticalities and potentialities of a possible transformation process and what are the advantages, disadvantages and traits of this transformation. The CATWOE method is structured as follows in Table 1.

Table 1. CATWOE method structure

Labels		Questions
C	Customers	Who could get advantages or disadvantages from the transformation?
A	Actors	Who could carry out the transformations? How could you help to build this vision?
T	Transformations	What objectives will be pursued to enhance San Giovanni a Teduccio district?
W	Weltanschauung	Which of these projects should occur in the area? What do you propose?
O	Owners	Who could oppose these transformations?
E	Environment	What environmental constraints could be conditioned?
+	Critical and potential	What are the significant criticalities and potentialities of the area?

Concerning the results of the discussion tables and the survey, the definition of several project alternatives made it possible to envisage different development scenarios. Thus, three design alternatives have been defined for a new spatial configuration of the city-port eco-system of the San Giovanni a Teduccio district which corresponds respectively to port-centred, community-centred and hybrid visions. Each scenario was evaluated using the Evamix method, implemented by the Definite 2.0 software [24]. A coalition dendrogram with the Naiade method [25] was also structured to highlight the conflicts and coalitions arising from each scenario. The preferable alternative is one capable of mediating more between the interests of the various stakeholders by proposing a new sustainable, circular and integrated configuration of the city-port eco-system of the San Giovanni a Teduccio district.

3   The Case Study: The East Naples City-Port

Located in Southern Italy, the district of San Giovanni a Teduccio lies on the east coastal zone of Naples city-port. With an original agricultural vocation, this area is nowadays squeezed between the administrative boundaries of the two systems: the Metropolitan City of Naples and the Central Tyrrhenian Port System Authority (AdSP). As in all those territories located in city-port interaction, the dichotomy between city and port has generated places in between, spaces waiting for a definition [26, 27]. Furthermore, communities who live in these contexts are marked by the uncomfortable conditions of the places themselves: communities that need essential services, in places where there are problems of welfare, illegality, and security (see Fig. 3).

Fig. 3. The focus area: the city-port of Naples, the district of San Giovanni a Teduccio.

This area of city-port interaction has been selected as a case study within the Master's Degree Course Level II in "Sustainable planning and design of port areas" of the Department of Architecture (DiARC), University of Naples Federico II.

The study and the analysis of the main urban planning tools have made it possible to have a broader and more solid knowledge of the complexities of the urban processes affecting the San Giovanni a Teduccio district. This operation has highlighted how the mending between the coastal strip and the rest of the urban fabric is indispensable and central to the planning of the territory. Another evidence is the quantity of stakeholders that affect this portion of territory, each with their requests, thus highlighting the need to build a collaborative decision-making process capable of finding a compromise between the various visions [28–30].

4   Results

The proactive involvement of the main stakeholders ensures the definition of a methodological approach based on support systems for collaborative decisions. The organisation of thematic discussion tables makes explicit the relation between SDGs and the decision-making processes of transformation of the study area. Representatives of the local municipal administration, urban regeneration experts, scholars, representatives of local associations and representatives of the Port Authority were involved in the thematic tables held between the spring and the summer of 2020 on the themes of the reuse of existing heritage, the right to the city and the sea, the opportunities related to SEZs the rearrangement of the road system and the water networks of East Naples.

During the first round table, entitled "The Ex-Corradini: industrial archaeology in East Naples", held on May 22, 2020, emerged the need to recover the existing architectural heritage, enhance the identity of the neighbourhood and its memory, to improve the quality of public space and create greater integration between port and city.

The second seminar is "The reconstruction of a shared vision for East Naples" on June 19, 2020. The emerging objectives concern the urgency of recovering the relationship between the community and the sea by stopping port expansion.

The third round table focused on the theme of "Special Economic Zones (SEZs): Opportunities and conflicts" on June 26, 2020, defined a new concept of cultural SEZs, tax-facilitated areas whose beneficiaries could be cultural and creative enterprises.

"Infrastructures and transport systems for East Naples" is the title of the fourth seminar held on 3 July 2020. This meeting made explicit the need to have a new port mobility system capable of respecting European standards and integrating with the urban context.

Finally, the last meeting entitled "Management of environmental resources and water networks in East Naples", held on 10 July 2020, focused on constructing a complete framework of knowledge about the situation of the district's water networks and the theme of reclamation of the coastal strip.

In parallel, a survey titled "Let's build together a new vision for East Naples" was submitted to associations and citizens of San Giovanni a Teduccio.

The main criticalities that emerged refer to the lack of direct access to the sea due to an almost total absence of overpasses or underpasses to overcome the physical barrier of the railway line. Another important criticality emerged is the urgency of reclamation works in the coastal area to allow its use for recreational purposes. In addition, there are inadequate housing conditions, the absence of job opportunities and a poor perception of safety. Instead, the proximity to the sea, the historic city, and the privileged view of the Gulf of Naples were considered the neighbourhood's potentialities to be enhanced and from which to start a new and more conscious city-port project.

These considerations confirm that the dichotomy between the two polarities of the city-port ecosystem has generated isolated and insecure places that urgently need structured and adaptive planning.

Furthermore, the survey shows the need to activate a process of regenerative transformation of the neighbourhood's city-port ecosystem that could aim to develop collaborative processes involving stakeholders and the local community and encourage greater physical and functional integration between the port and urban spaces.

The objectives and visions that emerged allowed the definition of three possible scenarios:

- "Portual International Hub" envisages the expansion of the Levante dock with consequent construction of the track bundle without any enhancement of the Ex-Corradini complex. In this scenario, the SEZs are intended as central retro-port areas to make San Giovanni a Teduccio an internationally competitive production hub.
- "Cultural Hub" gives voice to the community issues. The Ex-Corradini, the waterfront, the Pietrarsa railway museum, and the Apple Academy are configured as the neighbourhood's new centres of cultural and social development. In this proposal, the

SEZs are interpreted in the sense of cultural SEZs. Any expansion of the port spaces is renounced, allowing to think about a new urban waterfront configuration.

– "Community Cultural Portual Hub", is an equilibrium scenario. The expansion of the port is limited to the completion of the dock in its current configuration. This scenario favours a high development of the port towards the SEZs areas of the hinterland of East Naples. The neighbourhood would have the opportunity to reappropriate of the waterfront, recovering abandoned and degraded areas and converting some of the inland SEZs into SEZs with a cultural vocation. In this comprehensive system, the Ex-Corradini plant acts as a hinge between the issues addressed.

The three scenarios were compared using the software Definite 2.0 [24] and Evamix method to make a decision-making process operational and capable of combining the preferences of the various stakeholders and the hard data collected during the analysis and knowledge phase. The Evamix method, also pursuing the objective of tracing the preferable scenario, reasons by gradually attributing different weights to the various indicators, establishing an order of priority.

The three scenarios were assessed through the adoption of 7 criteria divided into further indicators, as showing in the following Table 2.

Table 2. Criteria and indicators used for the evaluation.

Criteria	Indicators
Capacity to accommodate maritime traffic	Goods handling
Commercial attractiveness of port activities	Surface of SEZs
Commercial attractiveness of cultural activities	Employment of shipping companies
	Construction/infrastructure companies
	Logistics companies
	Export companies
	Cultural companies
	Hi-tech companies
	Manufacturing companies
Alteration of the coastline	Size of the fill
City-port interaction	Presence of overpasses
	Presence of a promenade
	Beach reclamation
	Presence of public spaces along the coast
	Elimination of physical barriers
Enhancement of the archaeological and industrial heritage	Recovery of the Ex Corradini area
	Recovery of the area of the Ex Depuratore
	Enhancement of the "Forte di Vigliena" area

The criteria relating to the city-port interaction, the enhancement of the existing heritage and the alteration of the coastline are those to which greater weight has been attributed than the others because of their essentiality for a process of transformation of San Giovanni a Teduccio. The Evamix method highlights the preferability of the third scenario: Community Cultural Portual Hub (see Fig. 4).

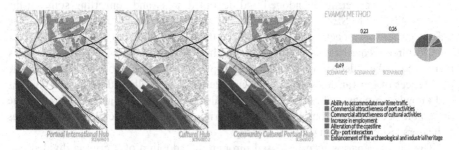

Fig. 4. The results of the alternative scenarios assessment with Evamix method.

Implementing the Naiade software [25], a coalition dendrogram was created that highlights the relationships between the proposed scenarios and the needs of the territorial stakeholders involved. The coalition dendrogram also identifies the third scenario as preferable. In particular, the dendrogram underlines how the time component is decisive for defining a compromise scenario capable of responding to the needs of the various stakeholders (see Fig. 5).

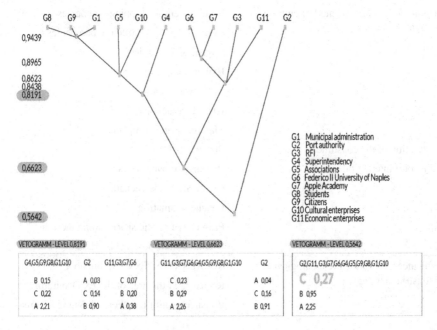

Fig. 5. The coalition dendrogram elaborated with Naiade method.

The Community Cultural Portual Hub scenario proposes an inclusive and partici-patory vision capable of explaining the needs of the various stakeholders involved and, therefore, defining possible mediation solutions. This scenario has been explored in this contribution by proposing a new coastline configuration through the design of a network of public spaces capable of reconnecting the environmental, urban and social resources that characterise the district of San Giovanni a Teduccio in an ecosystemic vision.

4.1 Community Cultural Portual Hub

The research defines a project proposal focused on recovering the relationship between the urban fabric of the neighborhood and its coastal strip (see Fig. 6).

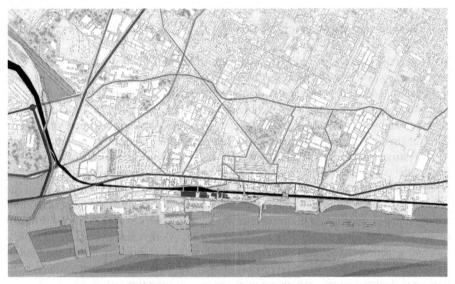

Fig. 6. A new spatial configuration of the city-port eco-system of the San Giovanni a Teduccio district.

The main purpose of the project proposal is to rethink the coastal strip in terms of a public space accessible from the city and capable of accommodating a new variety of activities and functions related to well-being, training and productivity [31].

The Ex-Corradini area acts as a hinge between the issues in this panorama of possible transformations. Parks, promenades, and beaches become the spaces of sociality.

The project concept focuses on the theme of public space as a connecting element in the city-port ecosystem and, therefore, capable of redefining the relationship between sea and neighborhood [32].

To foster this connection, it becomes necessary to think of a port infrastructure that is not expanding towards the coast but widespread inland, thanks to the presence of the SEZs. According to the concept project, SEZs are conceived as retro-port areas and areas dedicated to cultural productivity thanks to their economic potential. The proposal focuses on enhancing and strengthening a direct connection between the port and the Napoli Traccia station and between the port and the SEZs. According to the project, new tracks for access to the hub area are provided to meet the needs of accessibility and enhancement of the port logistics chain. At the same time, the research proposes separating the flows of urban mobility and port mobility by creating separate traffic systems.

To verify the project's economic sustainability, a financial analysis was carried out considering a time frame of twenty years.

The NPV (Net Present Value), calculated considering the conventional rate of return of 5%, was positive. The IRR (Internal Rate of Return) was higher than the discount rate chosen to calculate the NPV. Therefore, the investment can be considered profitable taking into account a public-private partnership.

5 Discussions and Conclusions

The research proposes an adaptive and multi-methodological decision-making process able to grasp the complexities of East Naples' city-port ecosystem and develop sustainable and circular strategies for new regenerative transformations of the study context. Adopting a circular approach makes it possible to interpret the complexity of the urban and socio-economic processes that characterise the city-port ecosystems and trace the drivers capable of activating new scenarios of regenerative transformation.

The structured decision-making process was helpful to support an accurate reflection on the possibilities of regenerative transformation of a complex context, focusing on the specific needs of stakeholders and the community. The main limitations of the proposed approach are to be found in the fact that the application of the process requires a long time as it focuses on the involvement and willingness to collaborate of different subjects. Furthermore, integrating hard and soft knowledge requires a careful collection and selection of the most appropriate data and a deep interaction with the various stakeholders. The methodological process chosen was structured in such a way as to be flexible and continuously contaminated by the new information collected.

The research aims are oriented to highlight the validity of a strategy applicable in cases of urban regeneration of complex areas affected by the influence of various stakeholders. From this point of view, the city-port ecosystem represents an enabling context capable of activating integration processes at multiple levels and accompanying the entire territory towards a circular and sustainable transition by stimulating territorial productivity, economic development and social cohesion.

References

1. Commission for Territorial Cohesion Policy and EU Budge: Opinion of the Regeneration of Port Cities and Port Areas. In: 121st Plenary Session, COTER-VI/018. European Committee of the Regions, Brussels (2017)
2. European Commission: Ports 2030, Gateways for the Trans European Transport Network. In: Communication from the Commission, COM (2013)295. European Commission, Brussels (2013)

3. Hein, C.: Port Cities: Dynamic Landscapes and Global Networks, 1st edn. Routledge, Milton Park (2011)
4. Saz-Salazar, S., García-Menéndez, L., Merk, O.: The Port and its environment: methodological approach for economic appraisal. Report 24. Technical report 24. OECD Regional Development Working Papers, Paris (2013)
5. Hoyle, B.: A rediscovered resource: comparative Canadian perceptions of waterfront redevelopment. J. Transp. Geogr. **2**, 19–29 (1994)
6. Olivier, D., Slack, B.: Rethinking the port. Environ. Plan. A **38**, 1409–1427 (2006)
7. Ellen MacArthur Foundation, Cities in the circular economy: An initial exploration. https://www.ellenmacarthurfoundation.org/. Accessed 30 Apr 2022
8. United Nations: Resolution adopted by the General Assembly on 6 July 2017, A/RES/71/313. https://unstats.un.org/sdgs/. Accessed 30 Apr 2022
9. Daldanise, G., Gravagnuolo, A., Oppido, S., Ragozzino, S., Cerreta, M., De Vita G.E.: Economie circolari per il patrimonio culturale: processi sinergici di riuso adattivo per la rigenerazione urbana. In: XXI Conferenza nazionale sui confini, movimenti, luoghi politiche e progetti per città e territori in transizione. Università degli studi di Firenze, Firenze (2018)
10. Van Balen, K., Vandesande, A.: Heritage Counts (Reflections on Cultural Heritage Theories and Practices), 1st edn. Garant Publishers, Chicago (2016)
11. Amit-Cohen, I.: Synergy between urban planning, conservation of the cultural built heritage and functional changes in the old urban center – the case of Tel Aviv. Land Use Policy **22**(4), 291–300 (2005)
12. Bullen, P.A., Love, P.E.D.: Adaptive reuse of heritage buildings. Struct. Surv. **29**(5), 411–421 (2011)
13. Cantell, S.: The Adaptive Reuse of Historic Industrial Buildings: Regulation Barriers, Best Practices, and Case Studies. VA Polytechnic Institute and State University (2005)
14. Conejos, S., Langston, C., Smith, J.: Improving the implementation of adaptive reuse strategies for historic buildings. In: International Forum of Studies SAVE HERITAGE: Safeguard Architectural, Visual, Environmental Heritage, Naples (2011)
15. Schipper, E.L.F., Langston, L.: A comparative overview of resilience measurement frameworks analysing indicators and approaches. Overseas Development Institute 422 (2015)
16. Yung, E.H.K., Chan, E.H.W.: Implementation challenges to the adaptive re-use of heritage buildings: towards the goals of sustainable, low carbon cities. Habitat Int. **36**, 352–361 (2012)
17. Fusco, G.L., Cerreta, M., De Toro, P.: Beyond Benefit Cost Analysis, 1st edn. Routledge, London (2004)
18. Barca, F.: An agenda for a reformed cohesion policy: a place-based approach to meeting European Union challenges and expectations. Independent report. European Commission, Brussels (2009)
19. Huggins, R., Clifton, N.: Competitiveness, creativity, and place-based development. Environ. Plan. A **43**(6), 1341–1362 (2011)
20. Pugalis, L., Bentley, G.: Place-based development strategies: possibilities, dilemmas and ongoing debates. Local Econ. **29**(4–5), 561–572 (2014)
21. Cerreta, M., Panaro, S., Poli, G.: A knowledge-based approach for the implementation of a SDSS in the Partenio Regional Park (Italy). In: Gervasi, O., et al. (eds.) ICCSA 2016. LNCS, vol. 9789, pp. 111–124. Springer, Cham (2016). https://doi.org/10.1007/978-3-319-42089-9_8
22. Ishizaka, A.: Multi-criteria Decision Analysis, 1st edn. Wiley, Chichester (2013)
23. Smyth, D.S., Checkland, P.B.: Using a systems approach: the structure of root definitions. J. Appl. Syst. Anal. **5**(1), 75–83 (1976)
24. Van Herwijnen, M., Janssen, R.: "Definite". In: Locket, A.G., Islei, G. (eds.) Improving Decision Making in Organizations. Springer, Cham (1988). https://doi.org/10.1007/978-3-642-49298-3_50

25. Munda, G.: Multicriteria Evaluation in a Fuzzy Environment: Theory and Applications in Ecological Economics, 1st edn. Physica-Verlag, Heidelberg (1995)
26. Koolhaas, R.: Junkspace. Per un ripensamento radicale dello spazio urbano, Quodlibet Editore, Macerata (2006)
27. Clement, G.: Manifesto del Terzo Paesaggio. Quodlibet Editore, Macerata (2016)
28. Cerreta, M., Poli, G., Regalbuto, S., Mazzarella, C.: A Multi-dimensional decision-making process for regenerative landscapes: a new harbour for Naples (Italy). In: Misra, S., et al. (eds.) Conference ICCSA 2019. LNCS, vol. 11622, pp. 156–170. Springer, Cham (2019)
29. Cerreta, M., Giovene di Girasole, E., Poli, G., Regalbuto, S.: Operationalizing the circular city model for Naples' City-Port: a hybrid development strategy. Sustainability 12, 2927 (2020)
30. Cerreta, M., Muccio, E., Poli, G., Regalbuto, S., Romano, F.: A multidimensional evaluation for regenerative strategies: towards a circular City-Port model implementation. In: Bevilacqua, C., Calabrò, F., Della Spina, L. (eds.) New Metropolitan Perspectives, NMP 2020. Smart Innovation, Systems and Technologies, vol. 178. Springer, Cham (2021). https://doi.org/10.1007/978-3-030-48279-4_100
31. Gausa, M.: Dispositivi geourbani. Area 79(166), 4–12 (2005)
32. Jacobs, J.: Città e libertà. Eleuthera (2020)

A Cost-Benefit Analysis for the Industrial Heritage Reuse: The Case of the Ex-Corradini Factory in Naples (Italy)

Marilisa Botte, Maria Cerreta[✉], Pasquale De Toro, Eugenio Muccio, Francesca Nocca, Giuliano Poli, and Sabrina Sacco

Department of Architecture, University of Naples Federico II, via Toledo 402, Naples, Italy
{marilisa.botte,maria.cerreta,pasquale.detoro,eugenio.muccio,
francesca.nocca,giuliano.poli,sabrina.sacco}@unina.it

Abstract. Cost-Benefit Analysis (CBA) is a method to evaluate a project or a policy, considering all costs and benefits to encourage a medium- and long-term vision. The CBA application fields are broad, counting as primary domains: transportation, environment, cultural heritage, energy, research and innovation, and information technology. This paper shows a Cost-Benefit Analysis for the reuse project of a former industrial plant in the city of Naples (Italy). The purpose is to identify the advantages and disadvantages of the cost-benefit approach in the case of industrial building reuse. The preliminary outcomes show that the following social externalities are likely to be subject to a certain risk degree: i) amount of CO_2 emissions saved, ii) benefits for businesses (enterprises and start-ups), iii) benefits for researchers, young professionals and students, iv) impacts on the real estate market, v) technological readiness of the involved innovative processes. The authors highlight that specialists should also search for ways to generate acceptable shadow prices for short-run consequences using empirical data from different sources.

Keywords: Cost-Benefit Analysis · Industrial heritage reuse · Multidimensional impact evaluation · Risk analysis

1 Introduction

Cost-Benefit Analysis (CBA) is a method to evaluate a project or a policy, considering all costs and benefits to encourage a medium- and long-term vision [1]. It can represent a powerful tool to assess the desirability of an investment project as a measure of its impacts on social well-being.

Approximately 60,603 open access publications mentioning CBA in the Scopus database appeared in the last decade, including articles, reviews, conference proceedings, editorials, and others. The CBA application fields are broad, counting as primary domains: transportation, environment, cultural heritage, energy, research and innovation, and information technology. Nevertheless, no publication in scientific databases makes explicit reference to the application of Cost-Benefit Analysis to industrial reuse

O. Gervasi et al. (Eds.): ICCSA 2022 Workshops, LNCS 13378, pp. 585–599, 2022.
https://doi.org/10.1007/978-3-031-10562-3_41

projects. In the industrial field, CBA is mainly adopted to assess the impacts of energy efficiency projects and waste management in working industrial buildings.

A primary principle linked to this tool is inherent to social opportunity cost as the likely gain from the optimal alternative within a set of mutually-exclusive options [2]. This rationale at the foundation of CBA addresses an effort to consider multiple stakeholders' values and interests, which converge into the notion of complex social value [3], trying to include intangible assets in Decision-Making [4]. The final goal of CBA aims to compare aggregated benefits with the costs to estimate the investment desirability in monetary terms [5].

This study aims to identify the advantages and disadvantages of applying a cost-benefit approach in the case of industrial heritage reuse. In particular, the presented CBA relates to the reuse project of a former industrial plant in the Eastern port area of Naples (Italy), referred to as Corradini Factory.

The article proceeds as follows: Sect. 2 focuses on an overview related to applications of CBA in different fields; Sect. 3 describes the proposed methodology; Sect. 4 introduces the case study; Sect. 5 shows the application results; finally, Sect. 6 identifies the potentials and criticalities of the approach and highlights the research follow-ups.

2   Cost-Benefit Analysis: An Overview of the Main Application Fields

As aforementioned, CBA is applied in different fields of knowledge and professions. In the environmental area, several applications of the CBA method to assess benefits and costs linked to ecological scenarios for Ecosystems Services (ES) conservation or restoration have been increasing in recent years [6]. These studies mainly use benefit transfer methods to estimate costs derived from biodiversity loss and degradation of ES due to anthropic land-use changes [7, 8]. Most frequently, at the ground of this research, the goal is to assess government projects at national or regional scales.

Conversely, urban studies focus mainly on costs related to city sprinkling and land consumption [9]. On the other hand, the real-estate market studies address estimating the benefits of the preservation of cultural heritage [10], adaptive buildings reuse [11, 12], or soil saving [13]. As an example, in Balena et al. 2013, the cost-benefit balance of soil conservation in the outer urban fringe of the cities has been estimated to increase the property market values [14].

Within the transportation sector, as widely shown by the literature [15–17], the cost-benefit analysis is, together with cost-effectiveness assessment and multicriteria methods, the main evaluation technique to be used for investigating the feasibility and viability degree of a project. The peculiarity of CBA in transport applications lies in the need of developing a simulation model, replacing demand and supply features, whose output represents the input of CBA. General details on how to model transportation systems can be found in [18]; however, each mobility mode has specific issues to be taken into account (see, for instance, [19–21]).

On one side, cost items to be considered are in common with other CBA applications, i.e., investment and construction costs, as well as operation and maintenance costs; on the other side, benefits to be computed present some noteworthy specific features. First of all, benefits for users, i.e., travelers who adopt the transport system under evaluation, can be divided into two main categories: perceived and not perceived benefits. The former is mainly related to the reduction in time and monetary cost items to be incurred to make a trip (e.g., highway tolls, waiting times at the bus stop, etc.). Specifically, as shown by [22], such benefits can be computed by evaluating the reduction in the user generalised costs or the increase in travelers' satisfaction in the design scenario with respect to the non-intervention strategy. On the other hand, some instances of non-perceived benefits are represented by the reduction in maintenance frequency or tyre wear associated with lower use of private cars. Such benefits, generally, occur when the choice mode model is involved in the evaluation and, therefore, the modal split is properly simulated. Finally, non-user benefits are related to who, although not directly involved in the use of the infrastructure or service under evaluation, is affected by the impacts of such a project (such as air and noise pollution, accident rate, congestion degree, etc.). An exhaustive list of such externalities can be found in [23].

Both linear infrastructures and transport nodes (i.e., terminals), across every mobility mode option, have been largely analysed by adopting CBA; among the other, new highways [24, 25], railway lines and services [26, 27] and air transport terminals and facilities [28, 29]. In the case of the maritime sector, freight transport represents a particularly crucial factor to be analysed, as shown by [30, 31]; while, concerning maritime terminals, i.e., ports, a specific discussion is provided below, with a particular focus on their interactions with the surrounding urban and suburban environment.

Specifically, as regards the areas of city-port interaction, a CBA process is particularly useful to consider the environmental and social impacts of the actions to be carried out. Furthermore, a CBA analysis has to consider the complex issues that characterize the areas of city-port interaction to encourage a synergic relation between these two polarities. The synergy between the city and the port area represents an essential component for the economic system of the European Union [32]. To operationalize this synergic relation, it is evident the need to define a new process of circular growth of the city-port ecosystem that takes into account the environmental and social challenges, overcoming the conflicts and proposing integrated visions.

Defining the main goals that move the project or the policy is the first step to starting a CBA able to promote and implement a circular growth of the city-port. The definition of the objectives is extremely relevant to CBA, which must demonstrate what kind of results can be achieved through the project implementation [2].

As regards to the areas of city-port interaction and their circular development, it is important to refer to the Sustainable Development Goals (SDGs), drawn up by United Nations [33]. The SDGs approach is particularly significant for addressing the issue of the regeneration of the city-port ecosystem for its capability to address the challenges raised by climate change and to reduce any form of inequality, ensuring the long-term economic, environmental and social sustainability of human communities.

In 2018, the Association Internationale Villes et Ports (AIVP) launched the AIVP 2030 Agenda, the first global initiative that adapts the 17 SDGs to the specific context of the city-port ecosystem. The Agenda is designed to guide the actions and projects of the city-port stakeholders to ensure sustainable relationships between the city and the port [34]. Starting from the SDGs and the AIVP 2030 Agenda, it is possible to deduce a set of actions focused on the complex urban and maritime transformations affecting the city-port ecosystem and also to consider the impacts of these actions from an environmental, economic and socio-cultural point of view.

Starting from the objectives and the effects generated, CBA measures the impact of a project or a policy on social well-being. In the case of the city-port interaction areas, there are many positive and negative impacts that must be taken into account.

The global dimension of today's traffic has led the maritime transport industry to new technological advances [35]. These changes have brought an expansion of the ports area, in order to meet the standards of ship size and hinterland connection [36].

Land, air and water pollution, noise and visual impact are the main social and environmental effects resulting from a port expansion [37]. It is essential to take these externalities into account in a CBA of port area expansion and transformation project.

CBA is mainly carried out for maritime projects (i.e. terminals and docks design). These are projects of relevant social and environmental impact because they involve both the occupation of areas that could be used for other community-oriented functions and the occupation of (often) polluted areas (that require water reclamation works). In such cases, the CBA pays great attention to the environment and energy since port areas play a key role in climate change issues and in the use of the sea as a common good. The CBA method insists on the definition of objectives and impacts linked to the context and is consistent with the European reference framework.

3   The Methodological Workflow

The research aims to identify advantages and disadvantages of the Cost-Benefit Analysis method applied to an industrial heritage reuse project (see Fig. 1). According to this aim, the proposed methodology consists of five main phases: 1. literature review; 2. testing; 3. indicators sampling; 4. results; 5. follow-ups.

A preliminary review of the CBA literature was conducted in four areas of interest (transportation, environment, port-areas, and industrial heritage), identifying industrial buildings as the main research field related to the selected case study. This led to the testing phase, which considers the analysis of the operational context, the selection of objectives in line with the Sustainable Development Goals (SDGs), and the program of the project functions. The latter was identified through the structuring of a Multi-Criteria Decision Analysis (MCDA) by means of the Evamix method [38], which has allowed the optimal combination of alternative design functions to be determined.

Subsequently, the phase concerning the demand analysis - which identifies the need for investment by assessing current and future demand - has provided the definition of a set of indicators, by prior identification of the selection criteria for them.

The CBA results showed as the output of the financial analysis a negative Net Present Value (FNPV), defined as the sum that results when the discounted expected investment and operating costs of the project are deducted from the discounted value of the expected revenues [2]. This provided the need to perform an economic analysis, which consists of the use of shadow prices to reflect the social opportunity cost of goods and services [2]. This analysis envisaged a series of market distortions adjustments such as correction

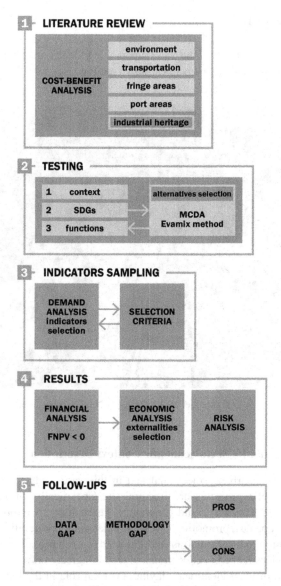

Fig. 1. The methodological workflow

for externalities, i.e. any cost or benefit that spills over from the project towards other parties without monetary compensation [2].

Therefore, a risk assessment has been conducted to deal with the uncertainty that permeates investment projects. In this way, it was possible to identify data and methodology gap in order to define the advantages and disadvantages of the Cost-Benefit Analysis of an industrial heritage reuse project, as envisaged by the overall goal.

4 The Case-Study

The study area is located within the San Giovanni a Teduccio district, which is part of
the VI Municipality of the city of Naples, Italy (see Fig. 2).

Fig. 2. The study area (source: Municipality of Naples, www.comune.napoli.it)

4.1 The Project and Its Socio-economic Context Description

The architectural and urban project analysed in this case study intends to activate -
through the reuse of a protected architectural site (specifically, an abandoned industrial building) - a process that could generate a cluster district equipped with research
structures and companies, promoting new economies and opportunities for social interaction, consumption and information and knowledge exchange. The project provides
with cultural services related to education and technological innovation, contributing to
the reformulation of local identities, playing a central role in the local economic system and redefining the character of lifestyles and urban consumption models of the San
Giovanni a Teduccio district and the city of Naples.

Over the years, the district has been equipped with an infrastructural connection
system that has made it one of the most suitable for the location of production plants,
most of which are now abandoned.

San Giovanni a Teduccio counts about 24.000 residents (with a slight decrease in
the last 20 years according to the Italian National Institute of Statistics census. It is one
of the least populous districts in the eastern area of Naples, even though with a relevant
population density (about 10,000 inhab/sq. km). The structure of the population shows
a strong presence of young people with a low level of education; the unemployment rate

(about 17%, higher than the city of Naples one which is about 15%) is also high. On the other hand, the current economic structure of the district is characterized by a marked vitality of the tertiary sector and the commercial one.

The ex-Corradini building was chosen for the project to provide highly knowledge-intensive structures in a marginalized context characterized by a recent de-industrialization, which in recent years has been experiencing the activation of technological, urban and social regeneration interventions.

For example, in VI Municipality, specifically in Ponticelli district, the Athena Future Technology District is located, whose Fuel Cell Lab (hydrogen) studies, manufactures and develops high-tech systems for mobile, stationary and propulsive applications, with which the project functions intend to activate collaboration and experimentation processes.

In addition, a business incubator is located in San Giovanni a Teduccio district. Its function is to encourage the birth of entrepreneurial projects, support and assist the development of creative and innovative companies, providing spaces and services. The activation of partnerships is foreseen: it will allow the incubator to be connected to the industrial and academic sectors of the city, enhancing professional experiences and know-how.

In this regard, it should be highlighted that the Campania region is among the top ones in Italy according to the number of innovative start-ups. In fact, in the last five years, there has been a significant increase in high innovation level start-ups, which made it the third region in the Italian innovation geography, with regard to sustainability issues, circular economy, energy and renewable sources efficiency, and culture.

With respect to the socio-cultural context, the project aims to promote a process of social-centred urban regeneration, in which well-being and social equity are integrated into cultural heritage reuse and valorization strategies [39–41]. Furthermore, this project recognizes the crucial role of energy conversion, the development of renewable energy sources, the implementation of the principles and models of circular economy and bio-economy, the promotion of social inclusion, and the implementation of the 2030 Agenda principles.

About the planned functions, the project intends to develop a pole of technological innovation and new economies, a "Hub & Spoke" structure, connected to a wide network of university laboratories, public and private research centers, local, national and international companies. The project is structured into the following main infrastructures:

– First demonstrator of a Shared pilot facility for the circular bio-economy;
– Experimental Bio-foundry Infrastructure;
– Experimental infrastructure for testing and demonstrating hydrogen technologies;
– Critical Materials Technology Platform - Strategic Raw Materials;
– Bio-Cloud infrastructure.

These infrastructures are integrated by Incubator and Accelerator for Start-ups and innovative SMEs, public and private laboratories, training areas, convention and dissemination areas. Through its functional program, the project intends to oppose the phenomena of migration of qualified individuals, raise the participation rates of young people in tertiary training courses, to integrate an area that is currently marginalized in a process of territorial regeneration, to support the circular transition production chains and improve the quality of life of the community of the San Giovanni a Teduccio district and the city of Naples.

4.2 Selected Indicators for Demand Analysis

Demand analysis was carried on taking into account the multiple functions provided by the project and three macro-categories of actors who drive the demand, identified with: 1. businesses; 2. researchers, young professionals and students; 3. target population and the general public. The information has been obtained from different sources, such as a statistical analysis of historical data, companies-related databases, new patents and innovation, literature and existing studies analysis, comparison with similar projects, surveys and interviews with stakeholders, expert advice, scientometric analysis of publications and citations in the project sectors.

The following table (Table 1) has been articulated by identifying the driving factors (businesses; researchers, young professionals and students; target population and the general public) and the extractable data, which made it possible to estimate specific values relating to the entire infrastructure.

Table 1. Selected indicators for demand analysis

Driving factors	Extractable data	Estimated data
Businesses		
Average growth of the industrial base in the RDI project field in the last years	Annual number of spin-offs/start-ups expected to be generated/supported by the project	19 spin-offs/start-up
Average annual profitability of the industrial base in the RDI project field	Expected number of businesses using the infrastructure to develop new/improved products and processes	969 businesses
Knowledge intensity of businesses in sectors related to the RDI project	Expected annual number of patents registered by project users	50 patents
Researchers, young professionals and students		
Number of scientists operating in the RDI project field and in the geographical area targeted by the infrastructure	Annual number of researchers who will directly use the RDI infrastructure	2.500 researchers, PhD students and professors
	Number of scientific publications expected to be produced by the project users	3 publications/year for each researcher: 7.500 publications/year

(*continued*)

Table 1. (*continued*)

Driving factors	Extractable data	Estimated data
Number of scientists operating in the RDI project field and in the geographical area targeted by the infrastructure	Stable personnel unit	103 units
Current number of students in the RDI project field or related fields in the geographical area targeted by the infrastructure	Average duration of the training programme at the RDI infrastructure	PhD e Master: 2 years Business training: 2 months
Potential for generating income through student fees and private sponsoring	Revenues from student fees	405.000 €/year, calculated on 150 students
Target population and the general public		
Number of people affected by environmental and health risks and virtually targeted by the RDI project	Annual number of people potentially targeted by the project	3227 persons/year[4]

5   Results

Financial sustainability was drawn up according to the indications provided by regulatory and guidelines, both national and international. They are:

Guidelines for the drafting of the technical and economic feasibility project to be used as a basis for the awarding of public works contracts of the National Recovery and Resilience Plan and National Complementary Plan;
Guide to Cost-Benefit Analysis of Investment Projects. Economic appraisal tool for Cohesion Policy 2014–2020 (European Commission);
Legislative Decree No. 50/2016 (Italian Public Contracts Code).

The financial sustainability of the project - with reference to both the restoration and conservative redevelopment of the industrial archaeology building (the ex-Corradini complex) and the construction of the infrastructural connection - has been assessed, in accordance with the aforementioned regulatory references, through the two financial indicators comparing costs and related revenues: Net Present Value (NPV) and Internal Rate of Return (IRR). The reference time horizon is 20 years (long term), generally adopted for interventions in the research and innovation sector.

In order to estimate NPV and IRR, the project costs and revenues have been identified. In particular, three macro-categories of costs have been considered: contract works and charges; sums available to the contracting station; equipment for the innovation ecosystem. These categories include, among others, the following costs: construction costs, executive design costs, waste disposal, road connections, taxes and fees. Furthermore, the maintenance and management costs related to the operational phase of the project have been considered.

The total cost of the work, determined through the economic framework consistent with the aforementioned regulatory references, is 25,000,000 euros, while the cost of the technological equipment of the innovation systems has been estimated at 10,000,000 euros.

Revenues from the identified functions were determined parametrically, starting from existing similar cases.

After determining costs and revenues, and taking into account the planned time schedule related to the implementation and operation of the interventions, a financial analysis was carried out to identify the project performance.

The results obtained show that the NPV is negative, and the IRR is lower than the discount rate considered. Therefore, the revenues generated by the project partially compensate for the costs incurred. In this condition, the Guide to the Cost-Benefit Analysis of Investment Projects highlights the need for co-financing.

In addition to the above-described financial assessment, also an economic analysis, taking into account the social, cultural, territorial and environmental benefits of the project, was performed. To this aim, starting from financial data, according to [2], the following phases were developed: i) fiscal corrections; ii) implementation of shadow prices; iii) evaluation of externalities. As regards the estimation of externalities, both a quantitative and a qualitative analysis were carried out. In particular, the monetisation procedure was performed by identifying indicators representing a proxy for users' WTP/WTA (i.e., Willingness-To-Pay/Willingness-To-Accept), as well as by implementing the benefit transfer method, after an accurate evaluation of the comparability between the selected studies and the analysed case.

The markedly innovative and sustainable features of the research laboratories involved ensure considerable benefits from an environmental perspective and in terms of upgrading the industrial know-how.

The main benefit is certainly given by the amount of CO_2 emissions saved thanks to the adoption of innovative circular bio-economy techniques, bio-tech methods and hydrogen exploitation technologies. In this regard, the savings in terms of avoided emissions have been estimated and appropriately monetised through the adoption of a unit shadow price expressed in €/tonCO_2. It is worth noting that a correction coefficient was also considered thus taking into account the growing importance of environmental sustainability during the analysed time horizon. The high-efficient energy-saving technological solutions, implemented as building elements, further contribute to the aforementioned environmental benefits. Indeed, in addition to a certain amount of avoided emissions, they also generate benefits in terms of demand reduction and energy self-conversion chances. In environmental terms, the only factor that generates a negative externality is related to an increase in mobility flows in the area, given the level of attractiveness that the hub aims to achieve. They were characterised not only in terms of passenger flows, but also as heavy vehicle flows connected to the logistic systems supporting the research activities in the hub (e.g., hydrogen cylinder wagons, articulated trucks for moving equipment, etc.). Specifically, such an effect has been monetised by computing the increase in demand trends, in terms of vehicles-km, and considering the unitary emission factor for each vehicle category and each micropollutant element analysed. Then, the global warming effect was added and, finally, a proper unit shadow price expressed in terms of €/tonCO$_2$-eq has been applied [23]. However, this negative effect can be considered significantly lower than the benefit, in terms of CO$_2$ saved, due to the factors highlighted above. Further, it is worth noting that the modal share concerning private transport modes (cars and motorbikes) will experience a significant decrease thanks to the presence of the designed cycle-pedestrian promenade which strongly enhances the accessibility levels of the area.

Another significant benefit, as already highlighted, is represented by the increase in the know-how level of the companies operating in the bio-economic and bio-tech field which generates an avoided cost in terms of acquisition of the technologies developed in the hub. Indeed, the research activities of the hub will lead to an increase in the so-called readiness level of the technologies involved, with a consequent benefit in terms of the development of new products and processes which, in some cases, could lead to patents or other forms of intellectual property protection. These innovations also have a positive impact on the number and survival rate of start-ups belonging to the reference sectors. All such benefits were monetised in terms of shadow prices and profit values by analysing similar business contexts.

Other benefits, in terms of knowledge spillover and learning-by-doing, concern researchers, young professionals and students involved. Indeed, on the one hand, researchers will benefit in terms of quality and number of publications; on the other hand, students may benefit from supporting actions to experimental and highly innovative thesis activities.

In addition to the above, there are several benefits that, although not directly monetisable, turn out to be crucial for the social profitability of the project and, therefore, were properly considered. Firstly, significant urban regeneration effects will be obtained thanks to functional and territorial synergies between the external spaces of the renovated hub and the surrounding urban environment, as well as in terms of purchase and rental prices of properties (residential or commercial) and lands. Further, benefits in terms of increased levels of employment and improvements in accommodation facilities in the area were properly assessed. Anyway, from a social perspective, the resident population, and especially young people, will gain confidence in their own territory and, potentially, they will not necessarily have to move to meet professional growth and fulfilment chances. This is closely linked to the benefits that the technological pole could provide to the cultural ecosystem of the area, such as synergies between companies and research institutions, networking chances, intellectual osmosis phenomena, etc.

The above-mentioned evaluations, related to indirect and external costs and benefits of the project, allow, starting from the financial analysis, to carry out the economic assessment.

Socio-economic attractiveness, instead, was assessed on the basis of the results of the economic analysis and, in particular, by the verification of a positive Economic Net Present Value (ENPV). As far as the financial coverage plan is concerned, it is necessary the public capital participation, which will allow the construction of infrastructure, and the private capital, which will be involved in the implementation of civil works.

As already explained, according to the methodology suggested by the aforementioned documents, the economic analysis is based on the financial analysis regarding the determination of direct economic costs and benefits and is complemented by determining the indirect and external costs and benefits.

In general terms, it can be observed that the "direct" internal economic costs and benefits are those related to the subject in charge of the realisation and management of the work. "Indirect" costs and benefits are those generated indirectly by the construction and management of the work, to which a market price can be attributed. "External" costs and benefits (or externalities) are those costs and benefits to the community from the construction and management of the work.

Similarly, to the financial analysis, the Economic Net Present Value and the economic Internal Rate of Return have been determined. The analysis of the socio-economic convenience is structurally similar to financial analysis, but cannot be based on market prices and has to take into account non-financial economic costs and benefits. In fact, direct costs and benefits are assessed by multiplying the financial values of projects by coefficients, called "conversion factors", in order to "purify" them of positive transfers to the public administration (basically taxes and social charges on labour) or negative transfers (subsidies and other forms of financial or real benefits). From the results obtained (Positive ENPV, EIRR at 20%), it can be deduced that the planned interventions provide a very wide range of social, cultural, environmental and economic benefits, which in turn lead to an improvement in the quality of life, compensating for the investment costs.

Upon completion of the proposed application, a qualitative risk analysis was performed by identifying possible hostile events linked to the significant involved variables; i.e., costs, revenues and externalities. Specifically, investment costs, operation and maintenance costs, as well as the number of years required for the building process, were considered among the factors which could affect cost items.

Regarding revenues, instead, significant adverse events may be related to income from consultancy services and research agreements, together with entry fees to the laboratory and for the use of research equipment charged to researchers and businesses. Further, student and PhD fees, as well as rental income for using spaces for dissemination and outreach activities to the wider public, need to be analysed. Finally, in the case of externalities, the following factors have been identified as liable to a certain risk degree: i) amount of CO_2 emissions saved, ii) benefits for businesses (enterprises and start-ups), iii) benefits for researchers, young professionals and students, iv) impacts on the real market estate, v) technological readiness of the involved innovative processes.

As instructed by [2], each one of the above-mentioned factors was characterised in terms of probability of occurrence (P) and severity (S), whose product represents the related risk level. Finally, according to the determined risk level, prevention and/or mitigation measures were identified and, consequently, the degree of susceptibility left (i.e., the residual risk) was derived. The outcome shows that the success of the initiative is strongly affected by the level of readiness and reliability of the technologies developed, as well as by their availability when the market requires them or when potential fundings occur. Properly meeting such opportunities in a timely manner, therefore, results being the main critical issue to be monitored and tackled.

6   Discussion and Conclusions

This study aimed to show an application of CBA method in the case of industrial heritage reuse through innovation and research functions.

The advantages of CBA are inherent to the efforts to estimate the social values of a project, the use of multi-criteria methods to manage the conflicts related to the choice of best-fit scenarios, the risk analysis implementation to deal with the uncertainties, the knowledge transfer and the in-depth comprehension of the multi-dimensional impacts of choices in decision-making.

The disadvantages are related to several aspects. First, data downscaling can be very difficult to manage since available data referred to national and regional scopes need hard statistical adjustments to adapt to the scale required for the evaluation.

Furthermore, the huge amount of data to be managed certainly makes the process complex and uncertain. The significant number of variables to be managed makes data collection procedures time-consuming.

The specificity of the data needed requires skills that make the application of this tool unfriendly to those who are not technicians.

Specialists should also search for ways to generate acceptable shadow prices for short-run consequences. Using empirical data from different sources - e.g. participatory workshops, surveys, and in-depth interviews with the most-influential stakeholders - can support the determination of shadow pricing linked to the hot topics in social terms.

The follow-up of this application is oriented to the knowledge transfer of skills and techniques to be implemented not only for the CBA investments projects, but also to expand the methodology to larger areas and neighboring.

Author Contributions.   The authors jointly conceived and developed the approach and decided on the overall objective and structure of the paper. Conceptualisation, M.C. and P.D.T.; methodology, M.B, M.C., P.D.T., F.N., and G.P.; case study analysis, E.M. and S.S.; validation, M.B, M.C., P.D.T., F.N., and G.P.; formal analysis, M.B, F.N., G.P., E.M. and S.S.; investigation, M.B, F.N., G.P., E.M. and S.S.; data curation, M.B, F.N., G.P., E.M. and S.S.; writing - original draft preparation, M.B, F.N., G.P., E.M. and S.S.; writing - review and editing, all the authors; visualization, E.M. and S.S.; supervision, M.C. and P.D.T. All authors have read and agreed to the published version of the manuscript.

References

1. Prest, A.R., Turvey, R.: Cost-benefit analysis: a survey. Econ. J. **75**, 683–735 (1965)
2. European Commission: Guide to Cost-Benefit Analysis of Investment Projects. Economic appraisal tool for Cohesion Policy 2014–2020. Bruxelles, Belgium (2014). https://ec.europa.eu/regional_policy/en/information/publications/guides/2014/guide-to-cost-benefit-analysis-of-investment-projects-for-cohesion-policy-2014-2020. Accessed 18 May 2022
3. Fusco Girard, L., Cerreta, M., De Toro, P.: Integrated assessment for sustainable choices. Scienze regionali **1**(Suppl./2014), 111–141 (2014)
4. Cerreta, M., De Toro, P.: Integrated spatial assessment for a creative decision-making process: a combined methodological approach to strategic environmental assessment. Int. J. Sustain. Dev. **13**(1–2), 17–30 (2010)

5. Mouter, N., Koster, P., Dekker, T.: Contrasting the recommendations of participatory value evaluation and cost-benefit analysis in the context of urban mobility investments. Transp. Res. Part A Policy Pract. **144**, 54–73 (2021)

6. Li, M., Liu, S., Wang, F., Liu, H., Liu, Y., Wang, Q.: Cost-benefit analysis of ecological restoration based on land use scenario simulation and ecosystem service on the Qinghai-Tibet Plateau. Global Ecol. Conserv. **34**(4), e02006 (2022)

7. Mele, R., Poli, G.: The effectiveness of geographical data in multi-criteria evaluation of landscape services †. Data **2**(1), 9 (2017)

8. Cerreta, M., Poli, G.: Landscape services assessment: a hybrid multi-criteria spatial decision support system (MC-SDSS). Sustainability **9**(8), 1311 (2017)

9. Saganeiti, L., Favale, A., Pilogallo, A., Scorza, F., Murgante, B.: Assessing urban fragmentation at regional scale using sprinkling indexes. Sustainability **10**(9), 3274 (2018)

10. Nocca, F., De Toro, P., Voytsekhovska, V.: Circular economy and cultural heritage conservation: a proposal for integrating Level(s) evaluation tool. Aestimum **78**, 105–143 (2021)

11. Marika, G., Beatrice, M., Francesca, A.: Adaptive reuse and sustainability protocols in Italy: relationship with circular economy. Sustainability **13**(14), 8077 (2021)

12. De Medici, S., De Toro, P., Nocca, F.: Cultural heritage and sustainable development: impact assessment of two adaptive reuse projects in Siracusa, Sicily. Sustainability **12**(1), 311 (2020)

13. Balena, P., Sannicandro, V., Torre, C.M.: Spatial analysis of soil consumption and as support to transfer development rights mechanisms. In: Murgante, B., Misra, S., Carlini, M., Torre, C.M., Nguyen, H.-Q., Taniar, D., Apduhan, B.O., Gervasi, O. (eds.) ICCSA 2013. LNCS, vol. 7974, pp. 587–599. Springer, Heidelberg (2013). https://doi.org/10.1007/978-3-642-39649-6_42

14. Torre, C.M., Morano, P., Tajani, F.: Saving soil for sustainable land use. Sustainability **9**(3), 350 (2017)

15. Browne, D., Ryan, L.: Comparative analysis of evaluation techniques for transport policies. Environ. Impact Assess. Rev. **31**(3), 226–233 (2011)

16. Bueno, P.C., Vassallo, J.M., Cheung. K.: Sustainability assessment of transport infrastructure projects: a review of existing tools and methods. Transp. Rev. **35**(5), 622–649 (2015)

17. Marleau Donais, F., Abi-Zeid, I., Waygood, E.O.D., Lavoie, R.: A review of cost–benefit analysis and multicriteria decision analysis from the perspective of sustainable transport in project evaluation. EURO J. Decis. Process. **7**(3–4), 327–358 (2019). https://doi.org/10.1007/s40070-019-00098-1

18. Cascetta, E.: Transportation Systems Analysis: Models and Applications. Springer, New York (2009). https://doi.org/10.1007/978-0-387-75857-2

19. D'Acierno, L., Ciccarelli, R., Montella, B., Gallo, M.: A multimodal multiuser approach for analysing pricing policies in urban contexts. J. Appl. Sci. **11**(4), 599–609 (2011)

20. D'Acierno, L., Gallo, M., Montella, B., Placido, A.: Analysis of the interaction between travel demand and rail capacity constraints. WIT Trans. Built Environ. **128**, 197–207 (2012)

21. D'Acierno, L., Gallo, M., Montella, B., Placido, A.: The definition of a model framework for managing rail systems in the case of breakdowns. In: 16th International IEEE Annual Conference on Intelligent Transportation Systems (IEEE ITSC 2013), pp. 1059–1064. The Hague, The Netherlands (2013)

22. Cartenì, A., Henke, I.: Consenso pubblico ed analisi economico-finanziaria nel progetto di fattibilità: Linee guida ed applicazione al progetto di riqualificazione della linea ferroviaria Formia-Gaeta (in Italian). Lulu editions (2016)

23. European Commission: Handbook on the External Costs of Transport, version 2019 – 1.1. Publications Office, Luxembourg (2019)

24. Bağdatli, M.E.C., Akbiyikli, R., Papageorgiou, E.I.: A fuzzy cognitive map approach applied in cost-benefit analysis for highway projects. Int. J. Fuzzy Syst. **19**, 1512–1527 (2017)

25. Henke, I., Cartenì, A., Di Francesco, L.: A sustainable evaluation processes for investments in the transport sector: a combined multi-criteria and cost–benefit analysis for a new highway in Italy. Sustainability **12**(23), 9854 (2020)
26. Federal Railroad Administration: Benefit-cost analysis guidance for rail projects. Technical report, Washington, DC (2016). https://www.dot.ny.gov/divisions/operating/opdm/pas senger-rail/passenger-rail-repository/FRA%20Benefit-Cost%20Analysis%20Guidance% 20for%20Rail%20Projects.pdf. Accessed 18 May 2022
27. Siciliano, G., Barontini, F., Islam, D.M.Z., Zunder, T.H., Mahler, S., Grossoni, I.: Adapted cost-benefit analysis methodology for innovative railway services. Eur. Transp. Res. Rev. **8**(4), 1–14 (2016). https://doi.org/10.1007/s12544-016-0209-5
28. Jorge, J.D., De Rus, G.: Cost–benefit analysis of investments in airport infrastructure: a practical approach. J. Air Transp. Manag. **10**(5), 311–326 (2004)
29. Midwest Transportation Center: Cost-benefit analysis: Substituting ground transportation for subsidized essential air services. Technical report. Iowa State University, Ames (IA), USA (2015). https://intrans.iastate.edu/app/uploads/2018/03/substituting_GTS_for_EAS_w_cvr. pdf. Accessed 18 May 2022
30. Galletebeitia, A.: Short-sea shipping cost benefit analysis using mathematical modeling. Master thesis, Florida Atlantic University, Boca Raton (FL), USA (2011)
31. Andersson, P., Ivehammar, P.: Cost benefit analysis of dynamic route planning at sea. Transp. Res. Procedia **14**, 193–202 (2016)
32. Commission for Territorial Cohesion Policy and EU Budge: Regeneration of Port Cities and Port Areas. In: 121st Plenary Session, 8 and 9 February 2017, European Committee of the Regions: Bruxelles, Belgium (2017)
33. United Nations Resolution adopted by the General Assembly on 6 July 2017. https://unstats. un.org/sdgs/. Accessed 9 May 2022
34. AIVP Agenda 2030, 10 goals for sustainable port cities. www.aivpagenda2030.com. Accessed 9 May 2022
35. Olivier, D., Slack, B.: Rethinking the port. Environ. Plan. A **38**, 1409–1427 (2006)
36. Asteris, M., Collins, A.: Developing Britain's port infrastructure: markets, policy, and location. Environ. Plan. A **39**, 2271–2286 (2007)
37. Saz-Salazar, S., García-Menéndez, L., Merk, O.: The Port and its Environment: Methodological Approach for Economic Appraisal. OECD Regional Development Working Papers, 2013/24. OECD Publishing (2013)
38. Voogd, H.: Multicriteria Evaluation for Urban and Regional Plannning. Pion, London (1983)
39. Cerreta, M., Daldanise, G., La Rocca, L., Panaro, S.: Triggering active communities for cultural creative cities: the "Hack the City" play ReCH mission in the Salerno Historic Centre (Italy). Sustainability **13**(21), 11877 (2021)
40. Cerreta, M., Giovene di Girasole, E., Poli, G., Regalbuto, S.: Operationalizing the circular city model for Naples' City-Port: a hybrid development strategy. Sustainability **12**(7), 2927 (2020)
41. Cerreta, M., Mazzarella, C., Somma, M.: Opportunities and challenges of a geodesign based platform for waste management in the circular economy perspective. In: Gervasi, O., et al. (eds.) ICCSA 2020. LNCS, vol. 12252, pp. 317–331. Springer, Cham (2020). https://doi.org/ 10.1007/978-3-030-58811-3_23

Unraveling the Role Played by Energy Rating Bands in Shaping Property Prices Using a Multi-criteria Optimization Approach: The Case Study of Padua's Housing Market

Sergio Copiello and Edda Donati[✉]

Department of Architecture and Arts, University IUAV of Venice, Santa Croce 191, 30135 Venezia, Italy
{sergio.copiello,edonati}@iuav.it

Abstract. As the topic of energy efficiency is still in the spotlight after a few decades of studies delving into it, several dozens of recent publications deal with the potential occurrence of a price premium for green buildings, namely, those significantly outperforming the best conventional ones. Almost all of those investigations find that a price premium indeed occurs and is statistically significant - though its magnitude is still debated - using regression analysis. Especially, the traditional hedonic price model is widely used in early studies, while the spatial autoregressive and spatial error models are ever extensively employed in the newest ones. Here we suggest using a different approach by turning to multi-criteria analysis. In particular, we propose combining an application of the analytical hierarchy process and linear optimization to identify the role played by energy rating bands and the energy performance index in shaping property prices. As a result, we expect to estimate the likely magnitude of the price premium for building energy efficiency. We test the approach on the local housing market in Padua, North-eastern Italy. A strong influence is found to be exerted on property prices by the energy performance as expressed by the rating bands. After discussing the empirical findings and the limitations of the analytical procedure, we also outline further developments for our multi-criteria approach.

Keywords: Multi-criteria decision support systems · Green buildings · Building energy efficiency · Energy performance certificates · Energy rating bands · Property prices

1 Introduction and Background Literature

The topic of energy efficiency has been in the spotlight at least for four decades or so, roughly starting from the oil shocks experienced during the seventies [1, 2], and energy efficiency of buildings is one of the most important research branches [3] since they contribute to a large part of energy consumption and greenhouse gas emissions [4, 5]. Two intriguing issues about building energy efficiency are as follows [6]: whether pursuing higher energy performance implies the need to incur higher construction costs -

O. Gervasi et al. (Eds.): ICCSA 2022 Workshops, LNCS 13378, pp. 600–614, 2022.
https://doi.org/10.1007/978-3-031-10562-3_42

hence, the occurrence of a cost premium - is the first; whether higher energy performance is rewarded by the players in the real estate market with a higher willingness to pay - hence, the occurrence of a price premium - is the second.

As far as the first issue is concerned, green buildings are often considered more expensive than traditional ones, roughly in the range between 5% and 20% [7–12], depending on the performance rating [13] or energy saving [14]. The higher upfront cost would represent a market barrier to their spread so as to slow down the desirable renovation of the building stock [15]. As a matter of fact, still recently, the renovation rate is found to remain very low across the EU Member States - about 1% on average, with an upper bound of only 1.2% and a lower bound as small as 0.4% [16] - while the optimal rate should be double, at least [17, 18]. Nevertheless, the literature has not come to compelling and conclusive evidence of a cost premium yet. Several studies point out that the additional cost, if any, could be pretty small [7, 19]. Furthermore, research on LEED and Green Star certified properties finds no difference in the construction costs or a small gap, yet not statistically significant [20, 21].

Concerning the second issue, the financial benefits related to improving building energy performance are thought to be the ground of successful regeneration projects [22, 23], and a much broader corpus of studies deal with the investigation of the occurrence of a price premium [1, 24, 25]. While early analyses date back to the late eighties [26, 27], most of the studies on the topic have been published during the last decade or so. Though a few works cast doubt on the occurrence of a price premium for high-performance properties [28–30], a higher willingness to pay for green building units has been found in dozens of studies published so far. Market players are likely to pay more both for residential [31–35] and commercial [36–41] properties. As regards dwellings, several analyses focus on the additional price related to the Green Mark (GM) certification scheme adopted in Singapore, finding premiums ranging between 1.3–9.1% for GM Certified buildings and 8.0–39.7% for GM Gold or Platinum units [42–46]. Besides, many studies take advantage of the Energy Performance Certificate (EPC) system introduced in the European Union by the Directive 2010/31/EU. The EPC rating scheme is thought to lack the ability to lessen market inefficiencies, such as information asymmetries, and push energy efficiency [47–50]; it is also found to have flaws and drawbacks [51–53], especially misjudgments and assessment distortions are mentioned as major issues [54–56]. Nevertheless, as the scheme implementation goes on, it makes it possible to investigate whether or not the energy rating bands command a price premium when building units are sold or rented. The works focusing on EPCs agree in finding that a positive price premium occurs, though its magnitude is still debated. There is broad consensus towards a premium within 10% when comparing the A and B energy rating bands to the D one [57–62], but sometimes it is also found to go up to 20% [63–65] and even higher in a few studies [24, 66, 67].

The analysis into the occurrence of a price premium for energy efficiency in buildings poses several methodological issues. One of them is as follows. Most of the literature dealing with the topic makes use of regression analysis to find out whether it occurs and whether it is statistically significant or not. Early studies employ the traditional hedonic price model [58, 68, 69], seldom adopting a double-log functional form of the model [70–73], much more often using a semi-log functional form [74–76], also controlling for time and space fixed effects [77–82]. Recently, ever extensive use of spatial data analysis [83, 84] has been seen, especially the spatial autoregressive model [33, 85–89] and the spatial error model [46, 90]. Occasionally, different models have been tested, such as logistic regression [57], quantile regression [62], geographically weighted regression [88], and hierarchical Bayes model [43]. Though there are well-founded reasons why regression models are usually employed in this research strand, the lack of testing for different approaches is somewhat surprising.

Here we propose to turn to multi-criteria analysis to start filling this gap. Multi-criteria methodologies have been adopted in other studies involving the energy efficiency of the real estate market, especially for selection and ranking purposes as far as efficiency measures and policies are concerned [91, 92]. However, to the best of the authors' knowledge, it has still to be fitted into a research issue such as the measurement of the magnitude of the price premium for building energy efficiency, which is generally addressed using different approaches, and this is precisely the original contribution this paper is meant to provide.

The remainder of this paper is organized as follows. Section 2 provides an overview of the multi-criteria optimization approach we propose here. Section 3 introduces the case study focused on Padua's housing market and the data we gathered and processed to test the method. Section 4 presents and discusses the results of the analysis. Section 5 discusses the limitations of the approach we use here and outlines further developments. Section 6 draws the conclusions.

2   Method and Models

2.1   Multivariate Regression Model

Essentially, the studies analyzing the occurrence of a price premium in green buildings using multiple regression models come on the heels of the seminal work authored by Rosen in 1974 [93], where the price P of a property is modeled as a function of k so-called "utility bearing attributes" (p. 34) x:

$$P = f(x_k) \tag{1}$$

The model derived from Rosen's insight into how differentiation of products - according to a set of significant characteristics - affects their value can be written in full as follows:

$$P = \alpha + \beta_k x_k + \varepsilon \tag{2}$$

where α is the constant, β are the coefficients associated with each of the k significant attributes or characteristics x, and ε is the error term. The focus is on energy efficiency as one of the significant attributes; it can be included in the model of Eq. (2) using the energy performance index as an independent variable or instead several dichotomous variables, each of them representing an energy rating band.

Most of the literature cited earlier transforms Eq. (2) into the semi-logarithmic (semi-log or log-linear) functional form by taking the logarithm of the dependent variable, which is common in econometric analysis because one can then easily interpret the β coefficients:

$$\ln P = \alpha + \beta_k x_k + \varepsilon \tag{3}$$

Thanks to Eq. (3), the β_k coefficients can be used to identify the percentage changes in the dependent variable commanded by a unit variation in attributes or characteristics x, according to the following calculation as suggested in the literature [94]:

$$\Delta P = e^{\beta_k} - 1 \tag{4}$$

2.2 Multi-criteria Optimization Model

Here we aim to match the multivariate regression model introduced in the previous subsection, especially the log-linear model of Eq. (3), with the multi-criteria model of the analytic hierarchy process (AHP) outlined by Saaty between the late seventies and the early eighties [95–97]. Let us define the hierarchical value tree as follows:

- the goal is to rank a set of real estate assets according to their market value, with the most valuable properties at the top of the ranking and the less valuable ones at the bottom;
- each criterion corresponds to the kth attribute x_k (with $k = 1, ..., l$), which is thought to play a role in shaping the market value of the analyzed real estate assets;
- the alternatives are i properties; we primarily focus on housing and take into consideration dwellings characterized by different structural, locational, and other attributes [98].

Let us suppose we build a pair-wise comparison matrix of size $n \times n$ to compare the ith and jth properties (with $i, j = 1, ..., n$) using the ratios between their prices, so that:

$$m_{i,j} = P_i/P_j \tag{5}$$

$$m_{i,i} = P_i/P_i = 1 \tag{6}$$

$$m_{j,i} = \frac{1}{P_i/P_j} = P_j/P_i \tag{7}$$

As usual in the AHP, the entries in the main diagonal are all equal to one according to Eq. (6), while the values in the lower triangle are the reciprocal of the elements in the upper triangle according to Eq. (7).

In order to get the rating for each property, the nth root of the product of the values in each row of the pair-wise comparison matrix is calculated and then normalized:

$$w_i = \sqrt[n]{\prod_{j=1}^{n} m_{i,j}} \Big/ \sum_{i=1}^{n} \sqrt[n]{\prod_{j=1}^{n} m_{i,j}} \tag{8}$$

Separately, we also build m pair-wise comparison matrices, each of size $n \times n$, to compare the ith and jth properties using the ratios between the values x_{ki} and x_{kj} for each kth attribute. The priority p_{ki} for each property i on each attribute k is then derived according to the calculation in Eq. (8), while the weighting $\widehat{\beta}_k$ for the attributes - that is to say, the criteria - is derived again according to the calculation in Eq. (8) from the specific pair-wise comparison matrix where they are compared. An estimate of the rating for each property is finally calculated as follows:

$$\widehat{w}_i = \sum\nolimits_{i=1}^{n} p_{ki} \widehat{\beta}_k \tag{9}$$

What is left is to vary and adjust $\widehat{\beta}_k$ so that $\widehat{w}_i$ is overall as close as possible to w_i. We pursue this goal by minimizing a loss function based on the notion of squared distance that resembles the least square method as in regression analysis:

$$\min_{\widehat{\beta}_k} \sum\nolimits_{i=1}^{n} \left(w_i - \widehat{w}_i\right)^2 \tag{10}$$

The weights $\widehat{\beta}_k$ for the attributes - or criteria, if you will - resulting from the optimization in Eq. (10) can be interpreted as a sort of proxy of the β_k coefficients in Eqs. (2)–(3); thus, an estimate - expressed in percentage - of the contribution of each attribute in shaping the property price.

3 Case Study: Padua's Housing Market

We focus on Padua's housing market - previously analyzed in terms of building energy efficiency [24, 60, 99] - to test the multi-criteria optimization approach outlined in the previous section. Below are further details about the data gathering process and sources.

As far as the data gathering process is concerned, we follow the approach adopted in a previous publication [24] and pay exclusive attention to a housing submarket primarily defined by typology, size, and location. We checked several dozens of real estate advertisements to identify a set of properties for sale meeting the following criteria: flats, terraced houses, and two-family houses as far as the typology is concerned; small to average size of the dwelling, namely, up to 130 square meters; location outside the old town, in inner-ring neighborhoods, thus at a distance greater than or equal to 3.4 km from the center. Fifteen properties were selected to be further processed.

Based on the information included in real estate advertisements, we collected the following data for each property (Table 1): listing price (LP); saleable area (SA); unit price (UP), that to say, the ratio between the listing price and saleable area; building typology; number of bedrooms (RM) and bathrooms (BR); presence of a garage or privately-owned parking lots; presence of "out of the ordinary" equipment. Also, in compliance with the national rules adopted following the European Directive on the energy performance of buildings (2010/31/UE), real estate advertisements are supposed to include both the energy rating band of the building unit and its energy performance index (EPI), which are crucial for the analysis we propose. Finally, we kept track of the geolocation of each property, so that to be able to calculate additional variables based on distances: to the city center, the nearest shopping center, and the nearest ramp of the beltway.

Table 1. Descriptive statistics for quantitative variables.

	Unit of measure	Mean	St. dev.	Min.	Max.
LP	Euros	167,667	54,964	98,000	300,000
SA	m^2	98	25	25	129
UP	Euros/m^2	1,822	706	940	3,920
RM	n	3.0	1.1	1.0	5.0
BR	n	1.8	0.4	1.0	2.0
EPI	kWh/m^2 y	147.70	97.56	3.51	292.00

4   Results and Discussion

4.1   Weights of the Attributes with Regard to the Listing Price of the Properties

As far as the listing price is concerned, we get comforting results, in line with expectations (Fig. 1). In particular, on the whole, the location attributes contribute to about one-third (34%) of the listing price, most of all because of the proximity to the city center (15%) and the nearest shopping center (19%). This outcome is in line with the parameter of land leverage [100, 101] - that is to say, the ratio between the land value and the overall market value of a property - which is known to range in the city of Padua from 23% for suburban neighborhoods, to 27% for inner-ring districts, to 41% for downtown areas [102, 103]. Building attributes cover the remaining share of the listing price (nearly 66%), about half of which is related to the energy rating band (36%). As pointed out in a previous study [24], the significant role played by energy efficiency in shaping property prices may be symptomatic of the fact that the energy rating bands themselves absorb other meaningful features, such as the use of state-of-the-art building materials, cladding, insulation, finishes, and systems.

Another interesting result, which is worth mentioning, occurs when replacing the energy rating band with the energy performance index. While the former weighs more than a third (36.0%), the latter is found to represent just a tenth (10.1%) of the listing price. The discrepancy is hardly explainable since there is almost an inverse relationship between the two variables at hand (Fig. 2), unless assuming that market players have a deeper understanding of the energy rating bands than they have of the energy performance index. In other words, aside from the fact that a nonlinear function may best fit the relationship between rating bands and index, the meaning conveyed by the energy rating bands - as well as by all the building features behind them - may be easier to grasp than the information brought by the energy performance index, so that they may be more suitable to represent the willingness to pay for dwellings characterized by high performance.

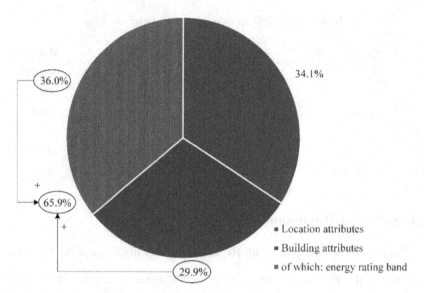

Fig. 1. Contribution of the attributes to the listing price, with a focus on the energy rating band.

4.2 Weights of the Attributes with Regard to the Unit Price of the Properties

Turning to the study of the unit price of the properties (Fig. 3), the most remarkable novelty lies in the fact that - as expected - the saleable area plays a major role. The weight this attribute takes on as a criterion means that it contributes to shaping up to nearly half the price (46.5%).

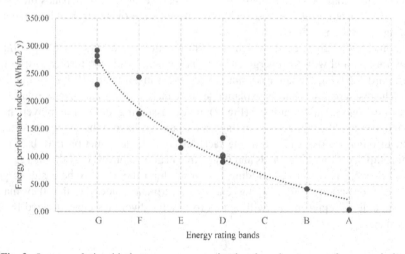

Fig. 2. Inverse relationship between energy rating bands and energy performance index.

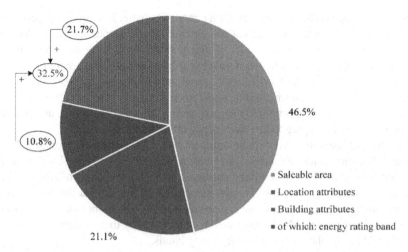

Fig. 3. Contribution of the attributes to the unit price, with a focus on the energy rating band.

Aside from that, location attributes shape about a fifth of the unit price (21.1%), again primarily because of the proximity to the city center (12.9%) or the nearest mall (7.7%), while the building characteristics explain the remaining third of the unit price (32.5%). We get additional evidence that energy rating bands are among the most prominent building features since they shape alone more than twenty percent (21.7%) of the market value. Also, in keeping with the discussion in the previous subsection, we find that the unit price is much less sensitive to the energy performance index (8.2%).

5   Limitations and Further Developments

This study suffers from a few limitations, which we mean to address in further developments of the research strand.

In the first place, we choose to limit the sample size to fifteen residential properties mostly because the case study is used here as a testbed. A brief digression is in order here. It would be helpful to apply traditional regression methods to the same case study to verify the effectiveness of the model we propose. That would enable us to compare the results and judge the computational performance of the different approaches. Unfortunately, the small sample size prevents this test from being performed. However, the small sample size enables us to take under control the issues related to the consistency of pair-wise comparison matrices. In fact, the bigger the size of such matrices, the more likely are violations of transitivity of preferences, which is a fundamental property of the approach we adopt. Nevertheless, since we use ratios as pair-wise comparison judgments, those matrices are perfectly consistent; namely, transitivity of preferences is always met. The latter observation paves the way to the use of far bigger samples, as the size of the matrices is no more an issue. Perhaps, larger sample sizes could help address the gap in weights between the energy rating bands and the energy performance index, which is a crucial issue. For that matter, the use of additional building attributes - hence, criteria - in the model sinks its roots in the same reason.

Secondly, once settled that energy rating bands do matter in shaping property prices and identified their overall weight, it seems essential to dig deeper in order to determine the marginal price associated with each change in rating. In fact, the fundamental issues to address are, for instance, whether A-rated properties are sold at a premium compared to the B-rated ones, whether there is a price premium distinguishing B-rated and C-rated properties, and so forth. The magnitude of the price premium, if any, can be unraveled by the method and model we propose here, performing the analysis on subsamples, which include buildings belonging to just two energy rating bands at a time.

Thirdly and lastly, it is worth mentioning another issue, which is distinguishing and isolating the contribution to property price brought by the energy rating band on its own from that conveyed by related building attributes such as the presence of state-of-the-art materials, finishes, and systems. Though there is no guarantee that the model we use here can provide significant outcomes, it seems worth trying to include those attributes in the model and perform the analysis posing constraints on their weights to check how this affects the role played by the energy rating bands.

6   Conclusions

This study aims to verify the role that energy efficiency - as expressed by the rating bands and the energy performance index - plays in shaping the price of real estate assets. Instead of using traditional multivariate regression models, here we propose a multi-criteria optimization approach, building on the method of analytic hierarchy process. In essence, each statistically significant building characteristic in the regression model is translated into an assessment criterion in the multi-criteria approach. Then, the weights of the criteria - namely, the building characteristics - that match the ranking of the properties according to their prices are found.

Using Padua's housing market as a testbed, we process the data concerning fifteen properties and their main location attributes (e.g., distance to the city center and the nearest shopping center) and building features (e.g., typology, number of rooms and bathrooms). As far as the results are concerned, we find that the overall weights of location and building factors align with the expectations. Furthermore, as expressed by the energy rating bands, energy efficiency is significant among the building attributes, which means it plays a significant role in shaping property prices. Notably, despite the strong relationship that ties them together, the energy performance index is not as important as the energy rating bands are.

We see further developments for the approach we propose here. In particular, the model lends itself to being used on much larger samples, as in our case, the sample size is not an issue for constructing consistent pair-wise comparison matrices. It is also suitable to be used in order to identify the marginal price related to changes in the energy rating band.

References

1. Copiello, S.: Building energy efficiency: a research branch made of paradoxes. Renew. Sustain. Energy Rev. **69**, 1064–1076 (2017). https://doi.org/10.1016/j.rser.2016.09.094

2. Copiello, S.: Economic parameters in the evaluation studies focusing on building energy efficiency: a review of the underlying rationale, data sources, and assumptions. Energy Procedia **157**, 180–192 (2019). https://doi.org/10.1016/j.egypro.2018.11.179

3. Economidou, M., Todeschi, V., Bertoldi, P., D'Agostino, D., Zangheri, P., Castellazzi, L.: Review of 50 years of EU energy efficiency policies for buildings. Energy Build. **225**, 110322 (2020). https://doi.org/10.1016/j.enbuild.2020.110322

4. von Thadden, G.: PPPs and energy efficiency in Europe. Eur. Today **5**, 13–14 (2011)

5. European Commission, Eurostat: Energy, transport and environment indicators: 2011 edition. Publications Office, Luxembourg (2011). https://doi.org/10.2785/17571

6. Copiello, S., Gabrielli, L., Micelli, E.: Building industry and energy efficiency: a review of three major issues at stake. In: Gervasi, O., et al. (eds.) ICCSA 2021. LNCS, vol. 12954, pp. 226–240. Springer, Cham (2021). https://doi.org/10.1007/978-3-030-86979-3_17

7. Dwaikat, L.N., Ali, K.N.: Green buildings cost premium: a review of empirical evidence. Energy Build. **110**, 396–403 (2016). https://doi.org/10.1016/j.enbuild.2015.11.021

8. Brennan, M.C., Cotgrave, A.J.: Sustainable development. Struct. Surv. **32**, 315–330 (2014). https://doi.org/10.1108/SS-02-2014-0010

9. Russ, N.M., Hanid, M., Ye, K.M.: Literature review on green cost premium elements of sustainable building construction. Int. J. Technol. **9**, 1715–1725 (2018). https://doi.org/10.14716/ijtech.v9i8.2762

10. Bartlett, E., Howard, N.: Informing the decision makers on the cost and value of green building. Build. Res. Inf. **28**, 315–324 (2000). https://doi.org/10.1080/096132100418474

11. Uğur, L.O., Leblebici, N.: An examination of the LEED green building certification system in terms of construction costs. Renew. Sustain. Energy Rev. **81**, 1476–1483 (2018). https://doi.org/10.1016/j.rser.2017.05.210

12. Hwang, B.G., Zhu, L., Wang, Y., Cheong, X.: Green building construction projects in Singapore: cost premiums and cost performance. Proj. Manag. J. **48**, 67–79 (2017). https://doi.org/10.1177/875697281704800406

13. Taemthong, W., Chaisaard, N.: An analysis of green building costs using a minimum cost concept. J. Green Build. **14**, 53–78 (2019). https://doi.org/10.3992/1943-4618.14.1.53

14. Zalejska-Jonsson, A., Lind, H., Hintze, S.: Low-energy versus conventional residential buildings: cost and profit. J. Eur. Real Estate Res. **5**, 211–228 (2012). https://doi.org/10.1108/17539261211282064

15. Pohoryles, D.A., Maduta, C., Bournas, D.A., Kouris, L.A.: Energy performance of existing residential buildings in Europe: a novel approach combining energy with seismic retrofitting. Energy Build. **223**, 110024 (2020). https://doi.org/10.1016/j.enbuild.2020.110024

16. European Commission: 2019 assessment of the progress made by Member States towards the national energy efficiency targets for 2020 and towards the implementation of the Energy Efficiency Directive as required by Article 24(3) of the Energy Efficiency Directive 2012/27/EU, Brussels (2020)

17. European Commission: CORDIS Results Pack on Deep Renovation New approaches to transform the renovation market. Publications Office, Luxembourg (2018). https://doi.org/10.2830/370893

18. European Commission: The European Green Deal, Brussels (2019)

19. Chegut, A., Eichholtz, P., Kok, N.: The price of innovation: an analysis of the marginal cost of green buildings. J. Environ. Econ. Manage. **98**, 102248 (2019). https://doi.org/10.1016/j.jeem.2019.07.003

20. Matthiessen, L.F., Morris, P.: Costing Green: A Comprehensive Cost Database and Budgeting Methodology. Davis Langdon, London (2014)

21. Rehm, M., Ade, R.: Construction costs comparison between 'green' and conventional office buildings. Build. Res. Inf. **41**, 198–208 (2013). https://doi.org/10.1080/09613218.2013.769145

22. Copiello, S.: Achieving affordable housing through energy efficiency strategy. Energy Policy **85**, 288–298 (2015). https://doi.org/10.1016/j.enpol.2015.06.017

23. Antonini, E., Longo, D., Gianfrate, V., Copiello, S.: Challenges for public-private partnerships in improving energy efficiency of building sector. Int. J. Hous. Sci. Appl. **40**, 99–109 (2016)

24. Copiello, S., Donati, E.: Is investing in energy efficiency worth it? Evidence for substantial price premiums but limited profitability in the housing sector. Energy Build. **251**, 111371 (2021). https://doi.org/10.1016/j.enbuild.2021.111371

25. Mudgal, S., Lyons, L., Cohen, F., Lyons, R., Fedrigo-Fazio, D.: Energy performance certificates in buildings and their impact on transaction prices and rents in selected EU countries. Final report prepared for European Commission (DG Energy) (2013)

26. Gilmer, R.W.: Energy labels and economic search. Energy Econ. **11**, 213–218 (1989). https://doi.org/10.1016/0140-9883(89)90026-1

27. Dinan, T.M., Miranowski, J.A.: Estimating the implicit price of energy efficiency improvements in the residential housing market: a hedonic approach. J. Urban Econ. **25**, 52–67 (1989). https://doi.org/10.1016/0094-1190(89)90043-0

28. Yoshida, J., Sugiura, A.: The effects of multiple green factors on condominium prices. J. Real Estate Finance Econ. **50**(3), 412–437 (2014). https://doi.org/10.1007/s11146-014-9462-3

29. Feige, A., Mcallister, P., Wallbaum, H.: Rental price and sustainability ratings: which sustainability criteria are really paying back? Constr. Manag. Econ. **31**, 322–334 (2013). https://doi.org/10.1080/01446193.2013.769686

30. Fuerst, F., McAllister, P.: The impact of energy performance certificates on the rental and capital values of commercial property assets. Energy Policy **39**, 6608–6614 (2011). https://doi.org/10.1016/j.enpol.2011.08.005

31. Kahn, M.E., Kok, N.: The capitalization of green labels in the California housing market. Reg. Sci. Urban Econ. **47**, 25–34 (2014). https://doi.org/10.1016/j.regsciurbeco.2013.07.001

32. Koirala, B.S., Bohara, A.K., Berrens, R.P.: Estimating the net implicit price of energy efficient building codes on U.S. households. Energy Policy **73**, 667–675 (2014). https://doi.org/10.1016/j.enpol.2014.06.022

33. Freybote, J., Sun, H., Yang, X.: The impact of LEED neighborhood certification on condo prices. Real Estate Econ. **43**, 586–608 (2015). https://doi.org/10.1111/1540-6229.12078

34. Bruegge, C., Carrión-Flores, C., Pope, J.C.: Does the housing market value energy efficient homes? Evidence from the energy star program. Reg. Sci. Urban Econ. **57**, 63–76 (2016). https://doi.org/10.1016/j.regsciurbeco.2015.12.001

35. Bond, S.A., Devine, A.: Certification matters: is green talk cheap talk? J. Real Estate Finance Econ. **52**(2), 117–140 (2015). https://doi.org/10.1007/s11146-015-9499-y

36. Eichholtz, P., Kok, N., Quigley, J.M.: Doing well by doing good? Green office buildings. Am. Econ. Rev. **100**, 2492–2509 (2010). https://doi.org/10.1257/aer.100.5.2492

37. Wiley, J.A., Benefield, J.D., Johnson, K.H.: Green design and the market for commercial office space. J. Real Estate Financ. Econ. **41**, 228–243 (2010). https://doi.org/10.1007/s11146-008-9142-2

38. Kok, N., Jennen, M.: The impact of energy labels and accessibility on office rents. Energy Policy **46**, 489–497 (2012). https://doi.org/10.1016/j.enpol.2012.04.015

39. Eichholtz, P., Kok, N., Quigley, J.M.: The economics of green building. Rev. Econ. Stat. **95**, 50–63 (2013). https://doi.org/10.1162/REST_a_00291

40. Chegut, A., Eichholtz, P., Kok, N.: Supply, demand and the value of green buildings. Urban Stud. **51**, 22–43 (2014). https://doi.org/10.1177/0042098013484526

41. Devine, A., Kok, N.: Green certification and building performance: implications for tangibles and intangibles. J. Portf. Manag. **41**, 151–163 (2015). https://doi.org/10.3905/jpm.2015.41.6.151

42. Deng, Y., Li, Z., Quigley, J.M.: Economic returns to energy-efficient investments in the housing market: evidence from Singapore. Reg. Sci. Urban Econ. **42**, 506–515 (2012). https://doi.org/10.1016/j.regsciurbeco.2011.04.004

43. Heinzle, S.L., Boey Ying Yip, A., Low Yu Xing, M.: The influence of green building certification schemes on real estate investor behaviour: evidence from Singapore. Urban Stud. **50**, 1970–1987 (2013). https://doi.org/10.1177/0042098013477693

44. Deng, Y., Wu, J.: Economic returns to residential green building investment: the developers' perspective. Reg. Sci. Urban Econ. **47**, 35–44 (2014). https://doi.org/10.1016/j.regsciurbeco.2013.09.015

45. Fesselmeyer, E.: The value of green certification in the Singapore housing market. Econ. Lett. **163**, 36–39 (2018). https://doi.org/10.1016/j.econlet.2017.11.033

46. Dell'Anna, F., Bottero, M.: Green premium in buildings: evidence from the real estate market of Singapore. J. Clean. Prod. **286**, 125327 (2021). https://doi.org/10.1016/j.jclepro.2020.125327

47. Aydin, E., Correa, S.B., Brounen, D.: Energy performance certification and time on the market. J. Environ. Econ. Manage. **98**, 102270 (2019). https://doi.org/10.1016/j.jeem.2019.102270

48. Comerford, D.A., Lange, I., Moro, M.: Proof of concept that requiring energy labels for dwellings can induce retrofitting. Energy Econ. **69**, 204–212 (2018). https://doi.org/10.1016/j.eneco.2017.11.013

49. Fleckinger, P., Glachant, M., Tamokoué Kamga, P.-H.: Energy performance certificates and investments in building energy efficiency: a theoretical analysis. Energy Econ. **84**, 104604 (2019). https://doi.org/10.1016/j.eneco.2019.104604

50. Broberg, T., Egüez, A., Kažukauskas, A.: Effects of energy performance certificates on investment: a quasi-natural experiment approach. Energy Econ. **84**, 104480 (2019). https://doi.org/10.1016/j.eneco.2019.104480

51. Gonzalez-Caceres, A., Lassen, A.K., Nielsen, T.R.: Barriers and challenges of the recommendation list of measures under the EPBD scheme: a critical review. Energy Build. **223**, 110065 (2020). https://doi.org/10.1016/j.enbuild.2020.110065

52. Li, Y., Kubicki, S., Guerriero, A., Rezgui, Y.: Review of building energy performance certification schemes towards future improvement. Renew. Sustain. Energy Rev. **113**, 109244 (2019). https://doi.org/10.1016/j.rser.2019.109244

53. Organ, S.: Minimum energy efficiency – is the energy performance certificate a suitable foundation? Int. J. Build. Pathol. Adapt. **39**, 581–601 (2021). https://doi.org/10.1108/IJBPA-03-2020-0016

54. Ahern, C., Norton, B.: Energy performance certification: misassessment due to assuming default heat losses. Energy Build. **224**, 110229 (2020). https://doi.org/10.1016/j.enbuild.2020.110229

55. Hardy, A., Glew, D.: An analysis of errors in the energy performance certificate database. Energy Policy **129**, 1168–1178 (2019). https://doi.org/10.1016/j.enpol.2019.03.022

56. Tronchin, L., Fabbri, K.: Energy performance certificate of building and confidence interval in assessment: an Italian case study. Energy Policy **48**, 176–184 (2012). https://doi.org/10.1016/j.enpol.2012.05.011

57. Brounen, D., Kok, N.: On the economics of energy labels in the housing market. J. Environ. Econ. Manage. **62**, 166–179 (2011). https://doi.org/10.1016/j.jeem.2010.11.006

58. Hyland, M., Lyons, R.C., Lyons, S.: The value of domestic building energy efficiency — evidence from Ireland. Energy Econ. **40**, 943–952 (2013). https://doi.org/10.1016/j.eneco.2013.07.020

59. Fuerst, F., McAllister, P., Nanda, A., Wyatt, P.: Does energy efficiency matter to homebuyers? An investigation of EPC ratings and transaction prices in England. Energy Econ. **48**, 145–156 (2015). https://doi.org/10.1016/j.eneco.2014.12.012

60. Bonifaci, P., Copiello, S.: Price premium for buildings energy efficiency: empirical findings from a hedonic model. Valori e Valutazioni **14**, 5–15 (2015)

61. Jensen, O.M., Hansen, A.R., Kragh, J.: Market response to the public display of energy performance rating at property sales. Energy Policy **93**, 229–235 (2016). https://doi.org/10.1016/j.enpol.2016.02.029

62. McCord, M., Haran, M., Davis, P., McCord, J.: Energy performance certificates and house prices: a quantile regression approach. J. Eur. Real Estate Res. **13**, 409–434 (2020). https://doi.org/10.1108/JERER-06-2020-0033

63. Fuerst, F., McAllister, P., Nanda, A., Wyatt, P.: Energy performance ratings and house prices in Wales: an empirical study. Energy Policy **92**, 20–33 (2016). https://doi.org/10.1016/j.enpol.2016.01.024

64. Fuerst, F., Oikarinen, E., Harjunen, O.: Green signalling effects in the market for energy-efficient residential buildings. Appl. Energy. **180**, 560–571 (2016). https://doi.org/10.1016/j.apenergy.2016.07.076

65. Evangelista, R., Ramalho, E.A., Andrade e Silva, J.: On the use of hedonic regression models to measure the effect of energy efficiency on residential property transaction prices: evidence for Portugal and selected data issues. Energy Econ. **86**, 104699 (2020). https://doi.org/10.1016/j.eneco.2020.104699

66. Manganelli, B., Morano, P., Tajani, F., Salvo, F.: Affordability assessment of energy-efficient building construction in Italy. Sustainability **11**, 249 (2019). https://doi.org/10.3390/su11010249

67. Davis, P.T., McCord, J.A., McCord, M., Haran, M.: Modelling the effect of energy performance certificate rating on property value in the Belfast housing market. Int. J. Hous. Mark. Anal. **8**, 292–317 (2015). https://doi.org/10.1108/IJHMA-09-2014-0035

68. Bloom, B., Nobe, M., Nobe, M.: Valuing green home designs: a study of ENERGY STAR ® homes. J. Sustain. Real Estate **3**, 109–126 (2011). https://doi.org/10.1080/10835547.2011.12091818

69. Kholodilin, K.A., Michelsen, C.: The market value of energy efficiency in buildings and the mode of tenure. SSRN Electron. J. (2014). https://doi.org/10.2139/ssrn.2472938

70. Högberg, L.: The impact of energy performance on single-family home selling prices in Sweden. J. Eur. Real Estate Res. **6**, 242–261 (2013). https://doi.org/10.1108/JERER-09-2012-0024

71. Stanley, S., Lyons, R.C., Lyons, S.: The price effect of building energy ratings in the Dublin residential market. Energy Effic. **9**(4), 875–885 (2015). https://doi.org/10.1007/s12053-015-9396-5

72. Aydin, E., Brounen, D., Kok, N.: The capitalization of energy efficiency: evidence from the housing market. J. Urban Econ. **117**, 103243 (2020). https://doi.org/10.1016/j.jue.2020.103243

73. Wahlström, M.H.: Doing good but not that well? A dilemma for energy conserving homeowners. Energy Econ. **60**, 197–205 (2016). https://doi.org/10.1016/j.eneco.2016.09.025

74. Mesthrige Jayantha, W., Sze Man, W.: Effect of green labelling on residential property price: a case study in Hong Kong. J. Facil. Manag. **11**, 31–51 (2013). https://doi.org/10.1108/14725961311301457

75. de Ayala, A., Galarraga, I., Spadaro, J.V.: The price of energy efficiency in the Spanish housing market. Energy Policy **94**, 16–24 (2016). https://doi.org/10.1016/j.enpol.2016.03.032

76. Hui, E.C.M., Tse, C., Yu, K.: The effect of BEAM Plus certification on property price in Hong Kong. Int. J. Strateg. Prop. Manag. **21**, 384–400 (2017). https://doi.org/10.3846/1648715X.2017.1409290

77. Cajias, M., Piazolo, D.: Green performs better: energy efficiency and financial return on buildings. J. Corp. Real Estate **15**, 53–72 (2013). https://doi.org/10.1108/JCRE-12-2012-0031
78. Chegut, A., Eichholtz, P., Holtermans, R.: Energy efficiency and economic value in affordable housing. Energy Policy **97**, 39–49 (2016). https://doi.org/10.1016/j.enpol.2016.06.043
79. Kholodilin, K.A., Mense, A., Michelsen, C.: The market value of energy efficiency in buildings and the mode of tenure. Urban Stud. **54**, 3218–3238 (2017). https://doi.org/10.1177/0042098016669464
80. Aroul, R.R., Rodriguez, M.: The increasing value of green for residential real estate. J. Sustain. Real Estate **9**, 112–130 (2017). https://doi.org/10.1080/10835547.2017.12091894
81. Fuerst, F., Warren-Myers, G.: Does voluntary disclosure create a green lemon problem? Energy-efficiency ratings and house prices. Energy Econ. **74**, 1–12 (2018). https://doi.org/10.1016/j.eneco.2018.04.041
82. Pride, D., Little, J., Mueller-Stoffels, M.: The value of residential energy efficiency in interior Alaska: a hedonic pricing analysis. Energy Policy **123**, 450–460 (2018). https://doi.org/10.1016/j.enpol.2018.09.017
83. Copiello, S.: Spatial dependence of housing values in Northeastern Italy. Cities **96**, 102444 (2020). https://doi.org/10.1016/j.cities.2019.102444
84. Copiello, S., Grillenzoni, C.: Is the cold the only reason why we heat our homes? Empirical evidence from spatial series data. Appl. Energy. **193**, 491–506 (2017). https://doi.org/10.1016/j.apenergy.2017.02.013
85. Zhang, L., Liu, H., Wu, J.: The price premium for green-labelled housing: evidence from China. Urban Stud. **54**, 3524–3541 (2017). https://doi.org/10.1177/0042098016668288
86. Taltavull, P., Anghel, I., Ciora, C.: Impact of energy performance on transaction prices. J. Eur. Real Estate Res. **10**, 57–72 (2017). https://doi.org/10.1108/JERER-12-2016-0046
87. Walls, M., Gerarden, T., Palmer, K., Bak, X.F.: Is energy efficiency capitalized into home prices? Evidence from three U.S. cities. J. Environ. Econ. Manage. **82**, 104–124 (2017). https://doi.org/10.1016/j.jeem.2016.11.006
88. McCord, M., Lo, D., Davis, P.T., Hemphill, L., McCord, J., Haran, M.: A spatial analysis of EPCs in the Belfast Metropolitan Area housing market. J. Prop. Res. **37**, 25–61 (2020). https://doi.org/10.1080/09599916.2019.1697345
89. Bisello, A., Antoniucci, V., Marella, G.: Measuring the price premium of energy efficiency: a two-step analysis in the Italian housing market. Energy Build. **208**, 109670 (2020). https://doi.org/10.1016/j.enbuild.2019.109670
90. Dell'Anna, F., Bravi, M., Marmolejo-Duarte, C., Bottero, M.C., Chen, A.: EPC green premium in two different European climate zones: a comparative study between Barcelona and Turin. Sustainability **11**, 5605 (2019). https://doi.org/10.3390/su11205605
91. D'Agostino, D., Parker, D., Melià, P.: Environmental and economic implications of energy efficiency in new residential buildings: a multi-criteria selection approach. Energy Strateg. Rev. **26**, 100412 (2019). https://doi.org/10.1016/j.esr.2019.100412
92. D'Alpaos, C., Chiara Brangolusi, P.: Multicriteria prioritization of policy instruments in buildings energy retrofit. Valori e Valutazioni **21**, 15–24 (2018)
93. Rosen, S.: Hedonic prices and implicit markets: product differentiation in pure competition. J. Polit. Econ. **82**, 34–55 (1974). https://doi.org/10.1086/260169
94. Halvorsen, R., Palmquist, R.: The interpretation of dummy variables in semilogarithmic equations. Am. Econ. Rev. **70**, 474–475 (1980)
95. Saaty, T.L.: A scaling method for priorities in hierarchical structures. J. Math. Psychol. **15**, 234–281 (1977). https://doi.org/10.1016/0022-2496(77)90033-5
96. Saaty, T.L.: Priority setting in completing problems. IEEE Trans. Eng. Manag. **30**, 140–155 (1983)

97. Saaty, T.L.: The analytic hierarchy process: decision making in complex environments. In: Quantitative Assessment in Arms Control. pp. 285–308. Springer, Boston (1984). https://doi.org/10.1007/978-1-4613-2805-6_12

98. Copiello, S., Cecchinato, F., Salih, M.H.: The effect of hybrid attributes on property prices. Real Estate Manag. Valuat. **29**, 36–52 (2021). https://doi.org/10.2478/remav-2021-0028

99. Bonifaci, P., Copiello, S.: Real estate market and building energy performance: data for a mass appraisal approach. Data Br. **5**, 1060–1065 (2015). https://doi.org/10.1016/j.dib.2015.11.027

100. Bostic, R.W., Longhofer, S.D., Redfearn, C.L.: Land leverage: decomposing home price dynamics. Real Estate Econ. **35**, 183–208 (2007). https://doi.org/10.1111/j.1540-6229.2007.00187.x

101. Bourassa, S.C., Hoesli, M., Scognamiglio, D., Zhang, S.: Land leverage and house prices. Reg. Sci. Urban Econ. **41**, 134–144 (2011). https://doi.org/10.1016/j.regsciurbeco.2010.11.002

102. Quotazioni autunno 2016. Consul. Immob., vol. 1011, pp. 2069–2092 (2016)

103. Copiello, S.: An empirical study of land leverage as a function of market value using a spatial autoregressive model. In: Morano, P., Oppio, A., Rosato, P., Sdino, L., Tajani, F. (eds.) Appraisal and Valuation. GET, pp. 29–41. Springer, Cham (2021). https://doi.org/10.1007/978-3-030-49579-4_3

Explicit and Implicit Weighting Schemes in Multi-criteria Decision Support Systems: The Case of the National Innovative Housing Quality Program in Italy

Aurora Ballarini, Sergio Copiello$^{(\boxtimes)}$, and Edda Donati

Department of Architecture and Arts, University IUAV of Venice, Santa Croce 191, 30135 Venezia, Italy
a.ballarini1@stud.iuav.it, {sergio.copiello,edonati}@iuav.it

Abstract. While institutionalized and purely contractual public-private partnerships (PPPs) are supposed to be helpful to carry out urban regeneration interventions, many of such projects - especially in Italy - are developed under the framework of negotiation-based PPPs, also known as negotiating partnerships. The structuring process of negotiation-based PPPs extensively uses various valuation approaches and methods, with a remarkable role played by multi-criteria decision support systems. The literature focuses on using valuation approaches and methods in this field, pointing out the potentialities that can be exploited and the limitations that should be addressed. This paper places itself within this debate and tries to address an inherent issue in multi-criteria decision aid (MCDA) techniques. Here we show that there could be a gap between the explicit weighting system used in MCDA analysis and the implicit weighting system actually employed. The issue is discussed using the National Innovative Housing Quality program, lastly adopted in Italy, as a testbed. As a case study, we consider the program proposal defined by the municipality of Treviso, North-eastern Italy. Implicit weights are identified ex-post according to the allocation of funding to the intervention projects. While confirming a difference in comparison to the importance of the criteria identified a priori by the public body that governs the program, we also argue that the gap narrows if assuming that the project options match the criteria according to a nonlinear relationship. The originality and value of this paper lie in addressing a topic that is underestimated in the reference literature. Instead, its thorough consideration might help structure program proposals based on negotiating partnerships that are more effective.

Keywords: Multi-criteria decision support systems · Multi-criteria decision aid · Weights · Weighting system · PINQuA

1 Introduction and Background Literature

The Green Paper on public-private partnerships and Community law on public contracts and concessions, published in 2004 by the European Commission [1], identifies two

O. Gervasi et al. (Eds.): ICCSA 2022 Workshops, LNCS 13378, pp. 615–628, 2022.
https://doi.org/10.1007/978-3-031-10562-3_43

main partnership models involving public and private stakeholders. On the one hand, the so-called institutionalized public-private partnership (PPP) [2, 3] is characterized by establishing a mixed capital company, the equity of which is jointly owned by a public body and a private entity. On the other hand, the purely contractual PPP features an extension of the scope of public sector procurement, beyond the traditional competitive bidding process. It builds on concession agreements meant to assign the design, funding, execution, or exploitation of facilities and related services - not to mention the associated risks [4–7] - to the private partner, depending on the specific model adopted [8, 9]. Both kinds of PPPs are widely used to deliver infrastructure works and, even more often, supply services for the benefit of a local community [10].

PPPs are put forth as suitable frameworks for urban renewal projects in several European countries, signally through the establishment of urban regeneration companies [11–16] as far as institutionalized PPP is concerned, and - even though sometimes subject to criticism [17–20] - employing concessions based on the private finance initiative model [21–24] as regards purely contractual PPP. Nonetheless, in the Italian context, the potentialities of PPPs are exploited much less in urban development or redevelopment projects than in providing local public services [25, 26].

At the same time, similarly to trends observed elsewhere in Europe [27, 28], urban renewal needs have led to the rise of different kinds of partnership transactions. They are subsumed in the expression negotiation-based PPP (or negotiating partnership), which distinctive features are as follows: private ownership of land to be redeveloped or buildings to be regenerated; cooperation between public bodies and private entities based on negotiation rather than competition; negotiations - as well as the subsequent partnership agreements - are ruled by the national or regional town planning legislation [11, 29].

The structuring of the above-mentioned PPP-based regeneration programs makes extensive use of a variety of valuation approaches and methods [30], such as: discounted cash flow analysis and derived models [31–38], multi-objective linear programming [39–43], Multi-criteria decision aid (MCDA) techniques and multi-criteria decision support systems [44–50], and others [51, 52]. In particular, multi-criteria decision support systems are used to rank and identify the intervention proposals suitable to be included in the urban regeneration programs [53, 54], provided the result of the negotiation process is seen as socially acceptable by the public stakeholders - namely, a share of the benefits goes back to the community on the ground [55] - and financially viable by the private ones [56, 57].

Building on the framework above, this study addresses a crucial issue when using MCDA techniques to rank and select the project proposals to be included in PPP-based regeneration programs. Below we show that there could be a gap between the explicit weighting system used in MCDA analysis - namely, the weights of the selection criteria identified a priori by the public body that governs the process - and the implicit weighting system actually employed - that is, the importance of the same criteria that one can infer ex-post according to the allocation of funding to the projects. We also argue that the gap narrows if assuming that the project options match the criteria according to a nonlinear relationship.

The remainder of this paper is organized as follows. Section 2 provides an overview of the National Innovative Housing Quality program, lastly adopted in Italy and referred to by the acronym PINQuA. Section 3 introduces the case study, which is the Innovative

Housing Quality program adopted by the city of Treviso, North-eastern Italy. Section 4 presents and discusses the results of the simulations we performed on the weighting system used in the case study. Section 5 draws the conclusions.

2 Overview of the National Innovative Housing Quality (PINQuA) Program

2.1 Subject Matter and Objective

The National Innovative Housing Quality (PINQuA is the Italian acronym) program is a public-private partnership tool jointly promoted by the Ministry of sustainable infrastructure and mobility and the Ministry for cultural heritage and activities and tourism.

The program aims to pursue urban regeneration, relieve housing distress and other settlement issues, and improve social inclusion, with a specific focus on the suburbs and urban fringe areas. To this end, five areas of intervention are identified as priorities: redevelopment and increase of the outdated public housing heritage as well as of the social housing stock; economic and social regeneration in areas outside the city historical centers; enhancement of accessibility and safety; re-functionalization of abandoned and degraded public spaces and buildings; experimenting with new models of housing services management.

The program has also been identified as one of the operational tools in the National recovery and resilience plan, which is part of the Next Generation EU program, that is to say, the strategy for the modernization of the country in many respects, such as sustainable development and mobility, environment and climate, improvement of the health chain, reform of both public administration and justice system, and promotion of competition.

2.2 Operating Model

Public bodies such as regions, the so-called metropolitan cities (namely, 14 large core cities and their closely related, surrounding towns, which constitute a kind of sub-provinces), and municipalities with more than 60 thousand inhabitants are allowed to submit a program proposal to the Ministry for public co-financing. The proposal must include a technical report, a general plan, as well as technical and economic feasibility projects for the works included.

Approximately, 2.8 billion Euros in public funding have been allocated to implement the Program. Each proposal can obtain maximum public funding of 15 million Euros (though a narrow set of pilot projects are granted up to 100 million Euros), provided it attracts and leverages additional private resources. Additional constraints are as follows: at least a proposal for each region must be approved and financed; 34% of total public funding is destined to proposals located in the southern regions.

The evaluation of the program proposals submitted to the Ministry makes use of a multi-criteria decision support system. Six classes of criteria - environmental impact, social impact, cultural added value, land use improvement, economic viability, and technological advance - each divided into five or six sub-criteria provide the ranking of the proposals, for the purpose of public funding allocation (see the hierarchical value tree of goals, criteria, and sub-criteria in Fig. 1). Out of 290 applications presented by the entitled public bodies, 159 has been financed. The deadline for completing the works is set for March 31, 2026.

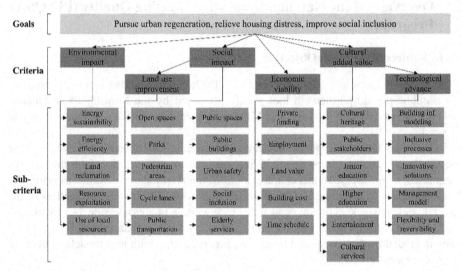

Fig. 1. Hierarchical value tree of goals, criteria, and sub-criteria in the National Innovative Housing Quality (PINQuA) program (source: authors' study and Ministry of Infrastructure).

3 Case Study: The Program Established by the Municipality of Treviso

3.1 Founding Idea of the Program Proposal

The Treviso municipality issued a notice to acquire expressions of interest by other public bodies or private entities in participating in the program proposal. It subsequently approved and submitted to the Ministry the program named after the peripheral district it focuses on: "Treviso San Liberale: the inhabited park."

The program proposal involves projects arranged into four action lines (see Fig. 2). The first action is defined as "City to be re-inhabited", which essentially means the demolition and reconstruction of the outdated public housing stock (see the upper left panel in Fig. 3). Action two - "Acts of cohesion" - provides for the conversion of some properties intended for sale as new facilities serving the public residential area; in addition, it is meant to strengthen the structures and buildings of the public welfare and charity body and the institute for hospitalization and assistance services for the elderly (see the upper right panel in Fig. 3). The third action is hinged upon the notion of a "Network of 15 min on foot", where the network is made by three pedestrian and cycling lanes connecting the parks inside the district and the district itself to the city center (see the lower left panel in Fig. 3). Action four - "Heritage of the'900" - provides for the expansion of

Fig. 2. Layout of the actions included in the program for San Liberale district, North-western of the city center of Treviso (source: authors' study, Municipality of Treviso, Google Earth as the background layer).

existing drainage surfaces and the redesign of the courtyards (see the lower right panel in Fig. 3).

The program proposal is expected to have an overall cost of 55,117,024 Euros. Main financing sources are as follows: 15 million Euros (27.2%) are the public funding provided by the central government; 21.893 million Euros (39.7%) are additional public funding made available by local public bodies, mainly Treviso municipality and the local public housing agency; 1.650 million Euros (3.0%) are expected to be invested by public-private bodies; finally, 16.574 million Euros (30.1%) are the funding attracted though the partnerships with private entities.

3.2 Matching Between Projects and Criteria

The consultancy and design team (see the Acknowledgments section for detail) has built a self-evaluation table where the correspondence between the local projects belonging to the four action lines and the evaluation criteria and sub-criteria defined by the central government to assess program proposals is represented. The self-evaluation table clusters the projects according to three matching levels, which are particularly useful for the purpose of this study since they enable to check for the weighting scheme implicit in the program proposal: none to low is the first matching level, average (or weak) the second, and high (or strong) the third (see Fig. 4).

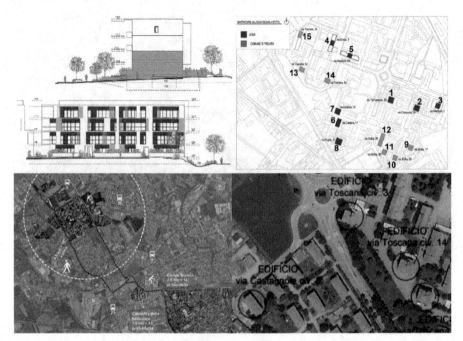

Fig. 3. Actions shaping the program for San Liberale district, North-western of the city center of Treviso (source: Municipality of Treviso).

		Environmental impact					Social impact					Cultural added value						Land use improvement					Economic viability					Technological advance				
		Energy sustainability	Energy efficiency	Land reclamation	Resource exploitation	Use of local resources	Public spaces	Public buildings	Urban safety	Social inclusion	Elderly services	Cultural heritage	Public stakeholders	Junior education	Higher education	Entertainment	Cultural services	Open spaces	Parks	Pedestrian areas	Cycle lanes	Public transportation	Private funding	Employment	Land value	Building cost	Time schedule	Building inf. modeling	Inclusive processes	Innovative solutions	Management model	Flexibility and reversibility
Action line 1	Public housing (32 flats)	◐	●	◐	●	●	●	●	●	○	○	○	○	○	○	○	○	◐	◐	◐	◐	●	◐	○	●	●	●	◐	○	●	○	○
	Public housing (16 flats)	◐	●	◐	●	●	●	●	●	○	○	○	○	○	○	○	○	◐	◐	◐	◐	●	◐	○	○	●	●	◐	○	●	○	○
	Social housing (36 flats)	◐	●	◐	●	●	●	○	●	○	○	○	○	○	○	○	○	◐	◐	◐	◐	●	●	○	○	●	●	○	○	●	○	○
	Social housing (39 flats)	◐	●	◐	●	●	●	○	●	○	○	○	○	○	○	○	○	◐	◐	◐	◐	●	●	○	○	●	●	○	○	●	○	○
	…																															
Action line 2	Nursing home (maint.)	◐	○	○	◐	◐	◐	◐	●	◐	●	○	○	○	○	○	○	○	○	○	○	●	◐	◐	○	●	◐	○	○	◐	○	○
	New sport facilities	◐	◐	○	◐	○	●	●	●	●	◐	○	○	○	○	●	○	●	◐	●	◐	●	●	◐	◐	●	◐	○	●	○	○	○
	New multi-purpose room	◐	◐	○	◐	○	●	●	●	●	◐	○	○	○	○	●	○	●	◐	●	◐	●	●	◐	◐	●	◐	○	●	○	○	○
	…																															

Matching levels: ○ None to low ◐ Average ● High

Fig. 4. An excerpt concerning different matching levels between criteria and projects (source: courtesy of architects Paolo Miotto, Mauro Sarti, and Elena Orsanelli, of the architect firm Archpiùdue).

4 Explicit and Implicit Weighting Schemes: Method, Results, and Discussion

4.1 Objective Function

Let us denote by w_i (with $i = 1, ..., 6$) the (explicit) weight of each ith criterion, as defined by the Ministry in the call for proposals. More to the point, the weights w_i are as follows: 15% for the environmental criterion, 25% for social impact, 10% for cultural added value, 15% for land use improvement, 25% for financial viability, and finally, a weight of 10% is allocated to the criterion of technological advance.

Let us denote by $\widehat{w}_i$ the (implicit) weight of each ith criterion, as it can be inferred from the allocation of public and private funding to the projects included in the program proposal. The estimated (implicit) weights $\widehat{w}_i$ are calculated as follows:

$$\widehat{w}_i = \sum\nolimits_{j=1}^{20} c_{.j} \times m_{i,j} \bigg/ \sum\nolimits_{i=1}^{6} \sum\nolimits_{j=1}^{20} c_{.j} \times m_{i,j} \tag{1}$$

In other words, we first multiply the expected cost $c_{.j}$ of each jth project (with $j = 1, ..., 20$, as twenty local projects are included in the program proposal) by the matching level $m_{i,j}$ between the ith criterion and the jth project, then we sum up all the products, and finally, we normalize them to unity. We let $m_{i,j}$ vary, and the objective function to minimize is a simple sum of squared differences, or squared errors, if you will. That is to say, we use a loss function strictly resembling the method of least squares, which is standard in regression analysis:

$$\min_{m_{i,j}} \sum\nolimits_{i=1}^{6} \left(w_i - \widehat{w}_i \right)^2 \tag{2}$$

4.2 Comparison Between the Explicit and Implicit Weighting Schemes

We first assume that a linear relationship describes the matching levels between projects and criteria. Accordingly, the matching levels $m_{i,j}$ translate into the following values: a none to low matching is given the value 0, 0.5 for an average (or weak) matching, 1 for a high (or strong) matching. That is a basic assumption we make simply to test whether there could be a gap between the explicit and the implicit weights or not. As a matter of fact, we find that the implicit weighting scheme departs from the explicit one (see Fig. 5). As far as the fifth and sixth criteria are concerned, there is no noticeable difference: financial viability is the fifth criterion, 0,25 and 0.22 are its explicit and implicit weights, respectively; technological advance is the sixth criterion, with a relative importance of 0.10 against 0.12.

Nonetheless, the discrepancy is much more apparent when we turn to the other items of the value tree. The implicit weight for the third criterion (cultural added value) is more than half the explicit one, whereas the opposite holds for the first criterion (environmental impact). Similarly, the second criterion (social impact) is underweighted (0.17 versus 0.25), while the fourth (land use improvement) is overweighted instead (0.21 versus 0.15).

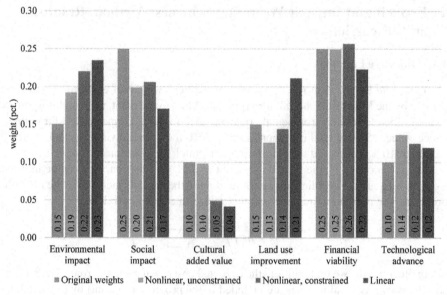

Fig. 5. Comparison between the explicit and implicit weighting schemes, implicit weights inferred according to project costs (source: authors' study).

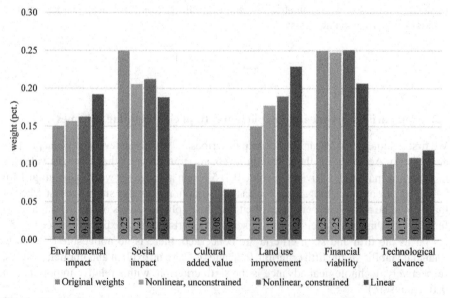

Fig. 6. Comparison between the explicit and implicit weighting schemes, implicit weights inferred regardless of project costs (source: authors' study).

It is worth mentioning that the gap in the weighting schemes might tell much about where local priorities lean towards. In other words, it looks like the program proposal pays a lot of attention to addressing environmental and land use issues while leaving social and cultural aspects somewhat in the background. Nevertheless, the evidence can be biased. On the one hand, upfront costs are likely to be higher for the projects matching the first and the fourth criteria (environmental impact and land use improvement, respectively). On the other hand, the investment costs are likely to turn out lower for the projects meant to meet the second and the third criteria (social impact and cultural added value, respectively). To control for this potential confounder, we also detect the implicit weights regardless of the project costs (see Fig. 6). The latter results show that the gap between the explicit and implicit weighting schemes narrows, suggesting that the program proposal, though focused on environmental and land use matters, does not disregard other areas of intervention such as the social and cultural ones.

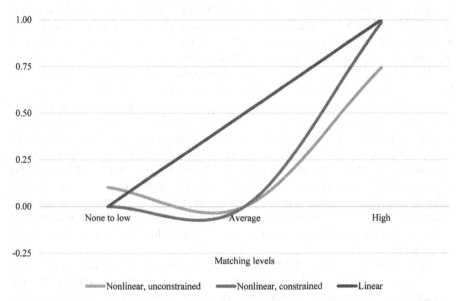

Fig. 7. Comparison between linear and nonlinear functions for the definition of the matching levels between projects and criteria (source: authors' study).

Once we drop the hypothesis that the matching level values are aligned in a straight line, further intriguing results come out of the analysis we perform here. In particular, we test two nonlinear relationships. The first is subject to the constraint that the parameter $m_{i,j}$ must be zero in case of no to low matching between projects and criteria, while the second is unconstrained. The results are not as different as one might expect. To begin with, in comparison to the linear assumption above, both the nonlinear relationships - constrained and unconstrained - provide estimations of the implicit weights $\hat{w}_i$ closer to the original explicit weights w_i (see Figs. 5 and 6 again). That is particularly evident for the third and fifth criteria: cultural added value and financial viability, respectively. There still are differences, but the gap narrows down to a few percentage points, as can be seen for the first and fourth criteria: environmental impact and land use improvement, respectively.

Another interesting outcome is that the distinction between three matching levels does not make sense anymore once we lessen the constraint of a linear relationship. The high (or strong) matching level is essential to shape an implicit weighting scheme close to the explicit one. Instead, the average matching level is as irrelevant as the none-to-low one (see Fig. 7). This empirical evidence may be symptomatic that the program proposal has been designed by focusing on the actions and interventions that fit the evaluation criteria well, and that could be why the implicit weighting scheme - simply driven by the high matching level - gets close to the explicit one.

5 Conclusions

This paper addresses a crucial issue when using MCDA techniques to rank and select the project proposals to be included in PPP-based regeneration programs. Here we consider one of such PPP-based regeneration tools lastly adopted in Italy, namely, the National Innovative Housing Quality program (PINQuA). Moreover, we analyze the program "Treviso San Liberale: the inhabited park" - proposed by the municipality of Treviso, North-eastern Italy, approved by the Ministry of sustainable infrastructure and mobility - and use it as a testbed. The case study is arranged into four action lines, each of which includes several specific projects to be carried out by the public bodies and private entities involved in the partnership. The local projects are meant to meet the evaluation criteria and sub-criteria defined by the central government to evaluate and rank the program proposals, and those criteria are pre-assigned different weights. Accordingly, we aim to assess whether there is a gap between the explicit and implicit weighting systems used in MCDA analysis or not, where the former is the weighting scheme of the selection criteria identified a priori by the public body that governs the process, while the latter is the weighting scheme one can infer ex-post according to the allocation of funding to the projects.

We find that there could indeed be a gap between explicit and implicit weights. In the case study we analyze, the difference is negligible for four out of six criteria, but it is also noticeable for the remaining two criteria. Nevertheless, this holds only when relaxing the constraint that projects match criteria according to a linear relationship, and the implicit weights are detected regardless of the project costs. Otherwise, the gap between explicit and implicit weights becomes more prominent and worthy of attention under linear constraints. Also, inferring the implicit weights according to project costs leads to strong differences between the explicit and the implicit weighting schemes. Essentially, we find that the allocation of public investments and, most of all, total investments is disproportional in comparison to the weights of the criteria.

The kind of analysis we carry out here addresses a much underestimated topic in the field of multi-criteria decision support systems. Especially, as we point out, the fact that an implicit weighting scheme diverges from a set of weights identified a priori is a warning sign that local priorities may differ - in part, at least - from expectations. The thorough consideration of the issue we highlight might help structure program proposals based on more effective negotiating partnerships.

Acknowledgments. The authors are grateful to the architect firm Archpiùdue, particularly architects Paolo Miotto, Mauro Sarti, and Elena Orsanelli, for their valuable support in the data gathering process.

References

1. Commission of the European Communities: Green Paper on public-private partnerships and community law on public contracts and concessions, Brussels (2004)
2. da Cruz, N.F., Marques, R.C., Marra, A., Pozzi, C.: Local mixed companies: the theory and practice in an international perspective. Ann. Public Coop. Econ. **85**, 1–9 (2014). https://doi.org/10.1111/apce.12032
3. da Cruz, N.F., Marques, R.C.: Mixed companies as local utilities. In: Proceedings of the Institution of Civil Engineers-Municipal Engineer, vol. 167, pp. 3–10 (2014). https://doi.org/10.1680/muen.12.00041
4. Marques, R.C., Berg, S.: Public-private partnership contracts: a tale of two cities with different contractual arrangements. Public Adm. **89**, 1585–1603 (2011). https://doi.org/10.1111/j.1467-9299.2011.01944.x
5. Marques, R.C., Berg, S.: Risks, contracts, and private-sector participation in infrastructure. J. Constr. Eng. Manag. **137**, 925–932 (2011). https://doi.org/10.1061/(asce)co.1943-7862.0000347
6. Ke, Y., Wang, S., Chan, A.P.C.: Risk allocation in public-private partnership infrastructure projects: comparative study. J. Infrastruct. Syst. **16**, 343–351 (2010). https://doi.org/10.1061/(ASCE)IS.1943-555X.0000030
7. Bing, L., Akintoye, A., Edwards, P.J., Hardcastle, C.: The allocation of risk in PPP/PFI construction projects in the UK. Int. J. Proj. Manag. **23**, 25–35 (2005). https://doi.org/10.1016/j.ijproman.2004.04.006
8. Yescombe, E.R.: Public-Private Partnerships - Principles of Policy and Finance. Elsevier, Amsterdam (2007). https://doi.org/10.1016/B978-0-7506-8054-7.X5022-9
9. McCarthy, S.C., Tiong, R.L.: Financial and contractual aspects of build-operate-transfer projects. Int. J. Proj. Manag. **9**, 222–227 (1991). https://doi.org/10.1016/0263-7863(91)90030-Y
10. da Cruz, N.F., Marques, R.C.: Mixed companies and local governance: no man can serve two masters. Public Adm. **90**, 737–758 (2012). https://doi.org/10.1111/j.1467-9299.2011.02020.x
11. Stanghellini, S., Copiello, S.: Urban models in Italy: partnership forms, territorial contexts, tools, results. In: Dalla Longa, R. (ed.) Urban Models and Public-Private Partnership. pp. 47–130. Springer, Heidelberg (2011). https://doi.org/10.1007/978-3-540-70508-6_3
12. Kort, M., Klijn, E.-H.: Public-private partnerships in urban regeneration projects: organizational form or managerial capacity? Public Adm. Rev. **71**, 618–626 (2011). https://doi.org/10.1111/j.1540-6210.2011.02393.x
13. Neto, L., Pinto, N., Burns, M.: Evaluating the impacts of urban regeneration companies in Portugal: the case of Porto. Plan. Pract. Res. **29**, 525–542 (2014). https://doi.org/10.1080/02697459.2014.973685
14. Jones, P., Hillier, D., Comfort, D.: Urban regeneration companies and city centres. Manag. Res. News **26**, 54–63 (2003). https://doi.org/10.1108/01409170310783411
15. Evans, B.: The politics of partnership: urban regeneration in New East Manchester. Public Policy Adm. **22**, 201–215 (2007). https://doi.org/10.1177/0952076707075896
16. Henderson, S.R.: Urban regeneration companies and their institutional setting: prevailing instabilities within the West Midlands, England. Local Econ. J. Local Econ. Policy Unit. **29**, 635–656 (2014). https://doi.org/10.1177/0269094214550271

17. Coulson, A.: A plague on all your partnerships: theory and practice in regeneration. Int. J. Public Sect. Manag. **18**, 151–163 (2005). https://doi.org/10.1108/09513550510584973

18. Hodkinson, S.: Housing regeneration and the private finance initiative in England: unstitching the neoliberal urban straitjacket. Antipode **43**, 358–383 (2011). https://doi.org/10.1111/j.1467-8330.2010.00819.x

19. Hodkinson, S.: The private finance initiative in English council housing regeneration: a privatisation too far? Hous. Stud. **26**, 911–932 (2011). https://doi.org/10.1080/02673037.2011.593133

20. Hodkinson, S., Essen, C.: Grounding accumulation by dispossession in everyday life. Int. J. Law Built Environ. **7**, 72–91 (2015). https://doi.org/10.1108/IJLBE-01-2014-0007

21. Adair, A., Berry, J., McGreal, S., Deddis, B., Hirst, S.: The financing of urban regeneration. Land Use Policy **17**, 147–156 (2000). https://doi.org/10.1016/S0264-8377(00)00004-1

22. De Marco, A., Mangano, G., Michelucci, F.V., Zenezini, G.: Using the private finance initiative for energy efficiency projects at the urban scale. Int. J. Energy Sect. Manag. **10**, 99–117 (2016). https://doi.org/10.1108/IJESM-12-2014-0005

23. Copiello, S.: Achieving affordable housing through energy efficiency strategy. Energy Policy **85**, 288–298 (2015). https://doi.org/10.1016/j.enpol.2015.06.017

24. Antonini, E., Longo, D., Gianfrate, V., Copiello, S.: Challenges for public-private partnerships in improving energy efficiency of building sector. Int. J. Hous. Sci. Appl. **40**, 99–109 (2016)

25. Bognetti, G., Robotti, L.: The provision of local public services through mixed enterprises: the Italian case. Ann. Public Coop. Econ. **78**, 415–437 (2007). https://doi.org/10.1111/j.1467-8292.2007.00340.x

26. Kyvelou, S., Karaiskou, E.: Urban development through PPPs in the Euro-Mediterranean region. Manag. Environ. Qual. Int. J. **17**, 599–610 (2006). https://doi.org/10.1108/147778306 10684567

27. Keil, A.: New urban governance processes on the level of neighbourhoods. Eur. Plan. Stud. **14**, 335–364 (2006). https://doi.org/10.1080/09654310500420826

28. McDonald, S., Malys, N., Maliené, V.: Urban regeneration for sustainable communities: a case study. Technol. Econ. Dev. Econ. **15**, 49–59 (2009). https://doi.org/10.3846/1392-8619.2009.15.49-59

29. Copiello, S., Cecchinato, F., Salih, M.H.: The effect of hybrid attributes on property prices. Real Estate Manag. Valuat. **29**, 36–52 (2021). https://doi.org/10.2478/remav-2021-0028

30. Morano, P., Tajani, F., Guarini, M.R., Sica, F.: A systematic review of the existing literature for the evaluation of sustainable urban projects. Sustainability **13**, 4782 (2021). https://doi.org/10.3390/su13094782

31. Copiello, S.: Urban renewal projects: detecting optimal public-to-private benefit ratio through discounted cash flow analysis. In: Sojkova, K., Tywoniak, J., L.A.H.P. (ed.) CESB 2016 - Central Europe Towards Sustainable Building 2016: Innovations for Sustainable Future. pp. 491–498. Grada Publishing, Prague (2016)

32. Copiello, S.: A discounted cash flow variant to detect the optimal amount of additional burdens in public-private partnership transactions. MethodsX **3**, 195–204 (2016). https://doi.org/10.1016/j.mex.2016.03.003

33. Manganelli, B., Pontrandolfi, P.: A model of good practice for urban regeneration as a balance between different requests. In: Proceedings of the 7th International Conference on Computational Methods in Structural Dynamics and Earthquake Engineering (COMPDYN 2015), pp. 5456–5463. Institute of Structural Analysis and Antiseismic Research School of Civil Engineering National Technical University of Athens (NTUA), Greece, Athens (2019). https://doi.org/10.7712/120119.7317.19371

34. Guarini, M.R., Morano, P., Micheli, A., Sica, F.: Public-private negotiation of the increase in land or property value by urban variant: an analytical approach tested on a case of real estate development. Sustainability **13** (2021). https://doi.org/10.3390/su131910958

35. Morano, P., Tajani, F., Di Liddo, F., Amoruso, P.: The public role for the effectiveness of the territorial enhancement initiatives: a case study on the redevelopment of a building in disuse in an Italian small town. Buildings **11**, 1–22 (2021). https://doi.org/10.3390/buildings 11030087

36. Morano, P., Tajani, F., Guarini, M.R., Di Liddo, F.: An evaluation model for the definition of priority lists in PPP redevelopment initiatives. In: Bevilacqua, C., Calabrò, F., Della Spina, L. (eds.) NMP 2020. SIST, vol. 178, pp. 451–461. Springer, Cham (2021). https://doi.org/10. 1007/978-3-030-48279-4_43

37. Tajani, F., Morano, P., Di Liddo, F., Locurcio, M.: An innovative interpretation of the DCFA evaluation criteria in the public-private partnership for the enhancement of the public property assets. In: Calabrò, F., Della Spina, L., Bevilacqua, C. (eds.) ISHT 2018. SIST, vol. 100, pp. 305–313. Springer, Cham (2019). https://doi.org/10.1007/978-3-319-92099-3_36

38. Della Spina, L., Calabrò, F., Rugolo, A.: Social housing: an appraisal model of the economic benefits in Urban regeneration programs. Sustainability **12** (2020). https://doi.org/10.3390/ su12020609

39. Manganelli, B., Tataranna, S., Pontrandolfi, P.: A model to support the decision-making in urban regeneration. Land Use Policy **99**, 104865 (2020). https://doi.org/10.1016/j.landusepol. 2020.104865

40. Morano, P., Tajani, F., del Giudice, V., De Paola, P., Anelli, D.: Urban transformation interventions: a decision support model for a fair rent gap recapture. In: Gervasi, O., et al. (eds.) ICCSA 2021. LNCS, vol. 12954, pp. 253–264. Springer, Cham (2021). https://doi.org/10. 1007/978-3-030-86979-3_19

41. Nesticò, A., Morano, P., Sica, F.: A model to support the public administration decisions for the investments selection on historic buildings. J. Cult. Herit. **33**, 201–207 (2018). https://doi. org/10.1016/j.culher.2018.03.008

42. Nesticò, A., Sica, F.: The sustainability of urban renewal projects: a model for economic multi-criteria analysis. J. Prop. Invest. Finance **35**, 397–409 (2017). https://doi.org/10.1108/ JPIF-01-2017-0003

43. Nesticò, A., Elia, C., Naddeo, V.: Sustainability of urban regeneration projects: novel selection model based on analytic network process and zero-one goal programming. Land Use Policy **99**, 104831 (2020). https://doi.org/10.1016/j.landusepol.2020.104831

44. Morano, P., Locurcio, M., Tajani, F., Guarini, M.R.: Fuzzy logic and coherence control in multi-criteria evaluation of urban redevelopment projects. Int. J. Bus. Intell. Data Min. **10**, 73 (2015). https://doi.org/10.1504/IJBIDM.2015.069041

45. Locurcio, M., Tajani, F., Morano, P., Torre, C.M.: A fuzzy multi-criteria decision model for the regeneration of the urban peripheries. In: Calabrò, F., Della Spina, L., Bevilacqua, C. (eds.) ISHT 2018. SIST, vol. 100, pp. 681–690. Springer, Cham (2019). https://doi.org/10. 1007/978-3-319-92099-3_76

46. Caprioli, C., Bottero, M.: Addressing complex challenges in transformations and planning: a fuzzy spatial multicriteria analysis for identifying suitable locations for urban infrastructures. Land Use Policy **102**, 105147 (2021). https://doi.org/10.1016/j.landusepol.2020.105147

47. Lami, I.M., Bottero, M., Abastante, F.: Multiple criteria decision analysis to assess urban and territorial transformations: insights from practical applications. In: Rezaei, J. (ed.) Strategic Decision Making for Sustainable Management of Industrial Networks. GINS, vol. 8, pp. 93–117. Springer, Cham (2021). https://doi.org/10.1007/978-3-030-55385-2_6

48. Abastante, F., Bottero, M., Greco, S., Lami, I.M.: Dominance-based rough set approach and analytic network process for assessing urban transformation scenarios. Int. J. Multicriteria Decis. Mak. **3**, 212 (2013). https://doi.org/10.1504/IJMCDM.2013.053728

49. Bottero, M., Datola, G.: Addressing social sustainability in urban regeneration processes. An application of the social multi-criteria evaluation. Sustainability **12**, 7579 (2020). https://doi. org/10.3390/su12187579

50. Bottero, M., D'Alpaos, C., Oppio, A.: Multicriteria evaluation of urban regeneration processes: an application of PROMETHEE method in Northern Italy. Adv. Oper. Res. **2018**, 1–12 (2018). https://doi.org/10.1155/2018/9276075

51. La Rosa, D., Privitera, R., Barbarossa, L., La Greca, P.: Assessing spatial benefits of urban regeneration programs in a highly vulnerable urban context: a case study in Catania. Italy. Landsc. Urban Plan. **157**, 180–192 (2017). https://doi.org/10.1016/j.landurbplan.2016.05.031

52. Bottero, M., Bragaglia, F., Caruso, N., Datola, G., Dell'Anna, F.: Experimenting Community Impact Evaluation (CIE) for assessing urban regeneration programmes: the case study of the area 22@ Barcelona. Cities **99**, 102464 (2020). https://doi.org/10.1016/j.cities.2019.102464

53. Guarini, M.R., Battisti, F.: Benchmarking multi-criteria evaluation methodology's application for the definition of benchmarks in a negotiation-type public-private partnership. A case of study: the integrated action programmes of the Lazio Region. Int. J. Bus. Intell. Data Min. **9**, 271 (2014). https://doi.org/10.1504/IJBIDM.2014.068456

54. Guarini, M.R., Chiovitti, A., Battisti, F., Morano, P.: An integrated approach for the assessment of urban transformation proposals in historic and consolidated tissues. In: Gervasi, O., et al. (eds.) ICCSA 2017. LNCS, vol. 10406, pp. 562–574. Springer, Cham (2017). https://doi.org/10.1007/978-3-319-62398-6_40

55. Calabrò, F., Della Spina, L.: The public-private partnerships in buildings regeneration: a model appraisal of the benefits and for land value capture. Adv. Mater. Res. **931–932**, 555–559 (2014). https://doi.org/10.4028/www.scientific.net/AMR.931-932.555

56. Guarini, M.R., Battisti, F.: Evaluation and management of land-development processes based on the public-private partnership. Adv. Mater. Res. **869–870**, 154–161 (2014). https://doi.org/10.4028/www.scientific.net/AMR.869-870.154

57. Battisti, F., Guarini, M.R.: Public interest evaluation in negotiated public-private partnership. Int. J. Multicriteria Decis. Mak. **7**, 54 (2017). https://doi.org/10.1504/IJMCDM.2017.085163

Analysis of the Difference Between Asking Price and Selling Price in the Housing Market

Benedetto Manganelli[1]([✉]) [iD], Francesco Paolo Del Giudice[2] [iD], and Debora Anelli[3] [iD]

[1] School of Engineering, University of Basilicata, via dell'Ateneo Lucano, 85100 Potenza, Italy
benedetto.manganelli@unibas.it
[2] Department of Architecture and Design, "Sapienza" University of Rome, Piazza Borghese 9, 00186 Rome, Italy
francescopaolo.delgiudice@uniroma1.it
[3] Department of Civil, Environmental, Land, Building Engineering and Chemistry, Polytechnic University of Bari, Via Orabona 4, 70125 Bari, Italy
debora.anelli@poliba.it

Abstract. In Italy, the opacity of the real estate market, which often does not reveal the real consistency of selling prices, or, in relation to the phase of the economic cycle, the low number of transactions force appraisers to use asking prices as comparables in the market approach. The international literature recognizes the importance of analyzing the relationship between asking prices and selling prices or time on market for the interpretation of the real estate market. In this work, the analysis of the difference is aimed at interpreting its variance by identifying the variables that have greater weight. For this purpose, a multivariate analysis model is built on a sample of data over a 12 years interval recorded in the housing market of the city of Potenza, Italy.

Keywords: Housing market · Asking price · Selling price · Market approach · Stepwise regression

1 Introduction

The estimate of the market value of an urban property intended for sale or to satisfy other possible purposes, if carried out with a market approach, is expressed in the comparison between the asset to be estimated and similar assets of known price (comparables). However, the search for comparables is made very complex by the segmentation and relative lack of transparency of the market. In Italy, the price declared in the notarial deed of sale, except for cases which are in derogation of DPR 131/86, is considerably lower than the selling price. Based on the norm, in fact, the verification of the congruity of the declared value takes place automatically, multiplying the cadastral income of the property by the coefficient of conversion of the income itself into real estate value. The cadastral value, declared in the deed, is however inferior, to a greater or lesser extent, with respect to the actual market value and therefore to the amount paid for the transfer.

The expedient, in most cases, is used to minimize transfer taxes and at the same time to avoid tax assessment. The trace of the real price of sale remains in the settlement,

O. Gervasi et al. (Eds.): ICCSA 2022 Workshops, LNCS 13378, pp. 629–640, 2022.
https://doi.org/10.1007/978-3-031-10562-3_44

whose accessibility is almost zero. The only exception are sales between individuals who do not act in the exercise of commercial, artistic or professional activities, having as their object property for residential use and related appurtenances that, by art. 1, paragraph 497, Law 23 December 2005 n. 266, go in derogation of the legislation already mentioned. In this case, at the time of the transfer and at the request of the purchaser, the taxable base is constituted by the cadastral value, regardless of what was agreed and written in the deed. The rule forces the purchaser to declare in the deed the selling price. In fact, a false declaration, if discovered during the subsequent assessment phase, entails the payment of taxes on the entire amount, and of an administrative sanction from fifty to one hundred percent of the difference between the tax due and that already applied based on the declared value.

The need to have numerous and reliable comparables, in the face of a market which is not very transparent or not very active, often leads appraiser to replace selling prices with asking prices coming from brokers and intermediary agencies. This method of construction or integration of the comparison sample is now common in estimation practice and in some ways legitimized both by case law (Civil Cass. Ord. Sec. 1 Num. 20307 of 31/07/2018) and by the Revenue Agency (July 27, 2007 "Provisions on the identification of criteria useful for the determination of the value of a property").

These sources recognize the asking prices as useful references for estimating the market value of a property, provided that the properties which are the subject of them present an undoubted character of homogeneity with the property to be estimated.

These are asking prices which, as is well known, generally differ in excess, to a greater or lesser extent, from the selling prices. Therefore, the use in the estimate of the asking prices requires the ascertainment of their probable difference from the selling prices. This operation cannot disregard a preliminary and in-depth knowledge of the phenomenon, which must lead to results that can be logically differentiated according to the trends taking place in the market, the intrinsic and extrinsic characteristics of the goods and the behavior of the operators in its various segments.

The size of the difference between the asking and selling price (spread) can depend on many factors. Some of these can be traced back to the determinants of the economic situation, which can therefore condition the real estate market. These macro-economic factors, by altering the balance between supply and demand for real estate, cause fluctuations in the amount of the difference between the asking price and the selling price, especially in the pre or post-economic transition phase. The microeconomic factors are to be identified among extrinsic characteristics, essentially traceable to the location of the property, and intrinsic ones. Above those listed, there is an element which is largely linked to them but which, more than any other, can condition the spread: this is the strategic behavior of the intermediary.

This work aims at investigating the effects of micro and macro-economic factors on the difference between asking price and selling price. This objective, if on the one hand it can constitute an argumentative matter for a deeper study of the phenomenon, on the other hand it makes explicit the logical and methodological premises essential for an estimate in quantitative terms of both the spread and therefore the market value.

2 Literature Review

As already mentioned, a fundamental determinant of the spread is the strategic behavior of the seller who, in choosing the asking price, evaluates the macro and microeconomic conditions, knowing full well that an incorrect definition of the initial request is costly both in terms of the time required to sell the property and the relative final price [1].

In almost all segments of the real estate market, it is common experience to set a list price that differs from the expected sale price. Time on market (TOM) is also influenced by the skills and motivations of sellers. Setting an initial price too high or too low affects the marketability of the property. The asking price clearly plays a critical role in the transaction, and acts as a benchmark for potential buyers [2]. An initial price that is too high can discourage potential buyers and lengthen the time on market. But it has also been shown that while a higher asking price makes the process by which the offered price tends to the asking price less rapid, it also increases the probability of higher selling prices [3, 4]. Conversely, low asking prices correspond to more potential buyers, thus a quick sale, lower transaction costs but also lower probability of high transaction prices [5–10].

The seller must usually address the two conflicting goals: maximizing the transaction price and minimizing the time on the market [11, 12].

In real estate markets, a trade-off occurs between time on market and selling price, and the "pricing strategy" is the balancing act between the two [13, 14].

The strategy of sale can therefore be conditioned from micro and macroeconomic factors that act on the TOM and on the difference between asking price and selling price. Assuming homogeneous behavior of sellers, the variance of the TOM and therefore of the difference must be linked directly to these factors. Macroeconomic factors include those that can affect the housing cycle [2, 5, 7, 15, 16]. In down markets or normal economic times, the asking price generally exceeds the selling price, however when the real estate market is growing, residences can also command selling prices above list prices [17]. The determinants of the real estate cycle are certainly disposable income [18] and mortgage rates that can affect supply and demand [19–22].

Illiquidity is an inherent characteristic of the real estate market. The relationship between TOM and price dynamics is a manifestation of real estate illiquidity [15, 23].

The literature has also shown that the marketability of a residential property and therefore the TOM and the difference between asking and selling price can depend on the geographic area in which the property is located [24] and its quality characteristics [25]. TOM is significantly shorter for newer homes, particularly those in the middle or high price range, while the size of the home has no significant effect [20].

Based on the indications provided by the literature, the model presented in this paper was constructed to measure how macro and microeconomic variables can affect the difference between the asking and selling price.

To eliminate the effect of the factor that the literature considers to be one of the major determinants on the TOM and therefore on the difference between asking price and selling price, the sample was built on data from a single real estate agency.

3   The Case Study

The data were collected by a single real estate agency in the city of Potenza. The city of Potenza is the capital of the Basilicata Region (Italy), and the most populous municipality in the entire region with 67,122 inhabitants, corresponding to a population density of 382.6 inhabitants/sq. km. The real estate market in the last 15 years has had a dynamic that has uniformly characterized both the residential and non-residential segments. The crisis in the real estate market that began in 2008 and continued until 2013 was followed by a phase of slight upturn, which, however, stopped in 2019. 2020, due to the Covid19 pandemic and its effects on the housing market as well, showed a phase of generalized and consistent decline.

The analysis is carried out on a sample consisting of 202 residential properties located in the city of Potenza bought and sold between 2010 and 2021 (12 years). The diversification of the sampled data reflects the need to capture the effects explained on the spread by micro and macroeconomic factors.

In general, the real estate agencies and, as far as this research is concerned, also the intermediary approached, about the contracts concluded, conserve and manage in appropriate data banks the following information.

a) generalities of the contracting parties;
b) planimetry of the property;
c) destination of the surfaces;
d) location, floor level, orientation, year of construction;
e) state of maintenance;
f) cadastral details (n. rooms or square meters, category, class, income);
g) annotations about the condominium charges, services and comfort details;
h) title of provenance and possible lack of administrative authorizations;
i) information regarding the availability of the property (free or leased);
j) asking price and bid value (and any subsequent downward adjustments);
k) time on market;
l) selling price.

In the case under investigation, the analyses carried out on the data of each property in the sample only covered some of the elements indicated, partly due to the difficulty of accessing all the information listed. The information acquired was translated into the explained variable (difference between the requested price and the offered price) and into other extrinsic and intrinsic (microeconomic) variables of the property, each of which was expressed with an adequate measurement system. It is immediate to note that the spread in the sample has an average of 16.86%, a minimum value of 2.86% and a maximum of 42.86%, with a standard deviation of 8.4%.

The measures of some macroeconomic variables were acquired from national databases. The variables considered and the corresponding methods of measurement are described below.

- Spread [Δ], difference between selling price and asking price, expressed as a percentage [%];

- Selling Price [SP], measured in euro;
- Construction Year [CY], a variable measured in number of retrospective years starting from the year 2021;
- Saleable Floor Area [SFA], is defined as the floor area exclusively allocated to a residential unit including balconies, verandahs, utility platforms and other similar features but excluding common areas;
- Preservation and Maintenance [PM], state of preservation and maintenance, expressed on the following scale: 1 = to be renovated, 3 = habitable, 5 = renovated/new;
- Zones [OMIz], variable that defines the Homogeneous Market Zone in which the property is located. Reference is made to the geographical subdivision of the City of Potenza operated by the Revenue Agency (Observatory of the Real Estate Market OMI). The homogeneous zones are 11, the variable is assigned a numerical scale that in an increasing sense goes from the most central to the most peripheral;
- Trend [T], the variable is obtained as the variation of the average quotation of the market value with respect to the previous year (n-1) expressed as a percentage [%]; the average quotation of the market value is obtained as the arithmetic average of the minimum and maximum quotations of the market value recorded, for the homogeneous zone of reference, by the OMI;
- Disposable Income [DI], gross disposable income of consumer households in real terms, obtained using the deflator of household final consumption expenditure;
- Save [S], share of gross savings in gross disposable income of consumer households;
- Invest [I], Incidence of gross fixed investments on gross disposable income of consumer households.
- Note: DI, S and I are variables surveyed quarterly by the Bank of Italy, the value used is the average annual value expressed in euros.

Table 1 shows the minimum, maximum, mean, and standard deviation for each variable.

Table 1. Variables statistics

Variable	Measuring scale	Min.	Max.	Med.	Standard deviation
Δ	%	2.86	42.86	16.86	8.4
SP	€	35,000.00	320,000.00	135,064.36	58,778.28
CY	Years	11	80	51	12.6
SFA	mq	40.00	250.00	101.06	35.70
PM	Ordinal	1	5	2.73	1.58
T	%	0.1190	0.4878	0.0095	0.0236
OMIz	Ordinal	1	11	3.10	2.57
DI	€	276,064.50	294,018.00	283,771.65	5,925.61
S	€	9.20	17.35	11.03	1.82
I	€	7.22	10.35	8.36	0.94

Table 2 reports the Pearson correlation measure between the variables. The statistical significance of the correlations is congruent with what was deductively to be expected about the causal links between the spread and the explanatory variables.

Table 2. Pearson correlation measure between the variables

Variable	Δ	SP	CY	SFA	PM	T	OMIz	DI	S	I
Δ	1	$-.598^{**}$	0.083	$-.309^{**}$	$-.324^{**}$	$-.166^*$	$-.241^{**}$	-0.063	-0.104	$-.274^{**}$
SP	$-.598^{**}$	1	$-.389^{**}$	$.651^{**}$	$.327^{**}$	-0.027	$.227^{**}$	0.071	-0.124	$.192^{**}$
CY	0.083	$-.389^{**}$	1	$-.262^{**}$	$-.377^{**}$	-0.101	$-.498^{**}$	0.099	$-.224^{**}$	0.073
SFA	$-.309^{**}$	$.651^{**}$	-262^{**}	1	0.028	-0.045	$.180^*$	0.071	-0.031	-0.067
PM	$-.324^{**}$	$.327^{**}$	$-.377^{**}$	0.028	1	$.143^*$	$.164^*$	0.069	0.087	0.058
T	$-.166^*$	-0.027	-0.101	-0.045	$.143^*$	1	0.075	$.366^{**}$	0.085	$.195^{**}$
OMIz	$-.241^{**}$	$.227^{**}$	$-.498^{**}$	$.180^*$	$.164^*$	0.075	1	0.072	0.041	0.022
DI	-0.063	0.071	0.099	0.071	0.069	$.366^{**}$	0.072	1	$-.195^{**}$	$.448^{**}$
S	-0.104	-0.124	$-.224^{**}$	-0.031	0.087	0.085	0.041	$-.195^{**}$	1	-0.133
I	$-.274^{**}$	$.192^{**}$	0.073	-0.067	0.058	$.195^{**}$	0.022	$.448^{**}$	-0.133	1

**. correlation is significant at the 0.01 level

*. correlation is significant at the 0.05 level

Table 2 shows that the main factors correlated with the difference between the asking and selling price, in order of decreasing importance, are: selling price [SP]; state of preservation and maintenance [PM]; saleable floor area [SFA]; investment [I]; homogeneous real estate market zones [OMIz] and trend [T]. Correlations between spread and construction year [CY], disposable income [DI] and saving [S] are not statistically significant.

The variables sale price [SP], state of preservation and maintenance [PM], saleable floor area [SFA], invest [I], zones [OMIz], and trend [T] all show a direct relationship with the spread.

4   The Model

The stepwise linear multiple regression model was chosen to analyze the data. The procedure begins with the assumption that there are no regressors in the model other than the constant. At the next steps, one variable at a time is added to the model. A significance value is set for the variable F, which indicates the ratio of the variance explained by the model to the residual variance below which its contribution is not considered significant. The variable with the highest significance is selected first, i.e., the one with the highest F value for simple linear regression, which therefore has the highest correlation with the explained variable.

The addition of other explanatory variables in the model is investigated through the Extra Sum of Squares Test applied individually for each of the variables still to be considered. At each step all explanatory variables considered previously in the model are tested again through the evaluation of the relative F. A variable already introduced in the model can in fact turn out redundant because of the introduction of new variables.

5 Results

In the first elaboration of the model 8 variables (8 steps) have been inserted. The saleable floor area is excluded. In the Table 3 the results are illustrated. The ANOVA test shows that in the 8 steps the introduced variables improve the interpretation of the phenomenon, the F test is always significant. Table 4 shows the coefficients of the variables and the significance (t student test) for the last two steps.

Table 3. Stepwise regression summary

Step	R	R-squared	Adjusted R-squared	Standard error of the estimate	Durbin-Watson
1	0.598	0.357	0.354	6.744114432	
2	0.625	0.390	0.384	6.584983358	
3	0.652	0.424	0.416	6.413419061	
4	0.691	0.477	0.466	6.130480460	
5	0.724	0.524	0.512	5.863853130	
6	0.742	0.550	0.536	5.712959513	
7	0.748	0.560	0.544	5.664818552	
8	0.759	0.577	0.559	5.570847213	1.959

1. regressors: (constant), SP
2. regressors: (constant), SP, T
3. regressors: (constant), SP T, CY
4. regressors: (constant), SP T, CY, S
5. regressors: (constant), SP T, CY, S, OMIz
6. regressors: (constant), SP T, CY, S, OMIz, PM
7. regressors: (constant), SP T, CY, S, OMIz, PM, I
8. regressors: (constant), SP T, CY, S, OMIz, PM, I, DI

However, in the last step, the introduction of the disposable income, if on the one hand improves R2, on the other produces an unconvincing result. In fact, disposable income has a negative coefficient. This is not consistent with expectations that an increase in income should induce greater demand and therefore a reduction in the spread. This observation should, among other things, be combined with the strong positive correlation that exists between the savings rate and disposable income. The inclusion of both variables distorts the result. In fact, the constant becomes non-significant. Therefore, it was decided to stop the development of the regression at step 7, excluding disposable income from the explanatory variables.

The Durbin-Watson statistic shows that there is no autocorrelation among the model residuals, which is crucial given that these are observations over time (12 years). About 56% of the variability in the response is explained by the variables included. Observing the residual scatter plot, the hypothesis of homoscedasticity is not violated (see Fig. 1). The hypothesis of normal distribution of errors is also verified (see Fig. 2). For all variables the observed p-value is smaller than the theoretical one (<0.05), each explaining a significant proportion of the variance of the spread.

Table 4. Coefficients and significance of the explanatory variables

Step		Non-standardized coefficients		Standardized coefficients	t	Sign.
		B	Standard error	Beta		
1	(Constant)	28.381	1.192		23.817	0.000
	SP	−8.531E−05	0.000	−0.598	−10.542	0.000
2	(Constant)	27.856	1.174		23.719	0.000
	SP	−8.601E−05	0.000	−0.603	−10.880	0.000
	T	−64.584	19.668	−0.182	−3.284	0.001
3	(Constant)	36.144	2.671		13.532	0.000
	SP	−9.732E−05	0.000	−0.682	−11.621	0.000
	T	−72.603	19.297	−0.204	−3.762	0.000
	CY	−0.135	0.039	−0.202	−3.434	0.001
4	(Constant)	52.201	4.428		11.789	0.000
	SP	0.000	0.000	−0.741	−12.856	0.000
	T	−68.563	18.468	−0.193	−3.712	0.000
	CY	−0.185	0.039	−0.279	−4.730	0.000
	S	−1.114	0.251	−0.242	−4.438	0.000
5	(Constant)	59.799	4.575		13.072	0.000
	SP	0.000	0.000	−0.735	−13.325	0.000
	T	−65.792	17.676	−0.185	−3.722	0.000
	CY	−0.269	0.042	−0.404	−6.399	0.000
	S	−1.196	0.241	−0.260	−4.966	0.000
	OMIz	−0.820	0.187	−0.251	−4.396	0.000
6	(Constant)	63.116	4.563		13.832	0.000
	SP	−9.883E−05	0.000	−0.692	−12.543	0.000
	T	−58.037	17.373	−0.163	−3.341	0.001
	CY	−0.303	0.042	−0.456	−7.192	0.000
	S	−1.160	0.235	−0.252	−4.936	0.000

(continued)

Table 4. (*continued*)

Step		Non-standardized coefficients		Standardized coefficients	t	Sign.
		B	Standard error	Beta		
	OMIz	−0.846	0.182	−0.259	−4.647	0.000
	MP	−0.966	0.285	−0.182	−3.390	0.001
7	(Constant)	70.127	5.642		12.430	0.000
	SP	−9.505E−05	0.000	−0.666	−11.851	0.000
	T	−49.692	17.687	−0.140	−2.810	0.005
	CY	−0.288	0.042	−0.434	−6.802	0.000
	S	−1.198	0.234	−0.260	−5.125	0.000
	OMIz	−0.827	0.181	−0.253	−4.579	0.000
	MP	−0.954	0.283	−0.180	−3.375	0.001
	I	−0.940	0.452	−0.106	−2.080	0.039
8	(Constant)	10.801	22.223		0.486	0.628
	SP	−9.528E−05	0.000	−0.667	−12.079	0.000
	T	−67.004	18.493	−0.189	−3.623	0.000
	CY	−0.304	0.042	−0.458	−7.228	0.000
	S	−1.090	0.233	−0.236	−4.676	0.000
	OMIz	−0.885	0.179	−0.271	−4.946	0.000
	MP	−0.998	0.278	−0.188	−3.585	0.000
	I	−1.428	0.478	−0.160	−2.986	0.003
	DI	0.000	0.000	0.157	2.757	0.006

The coefficients of the regression are all negative. The model indicates that when the price of the property, the age of the building, the state of maintenance increase, and when one moves away from the urban center, the difference between the asking price and the selling sale price decreases; as well as the spread decreases with the increase of the saving share, the investment share, and the values of the quotations (real estate cycle in expansion).

The model indicates that as the price of the property, the age of the building, and the state of maintenance increase, and as one moves away from the urban center, the difference between the offered price and the final sales price decreases. As well as also the spread diminishes when the rate of saving and the rate of investment increase and when the values of the quotations grow (the real estate cycle is in expansion).

Looking at the standardized coefficients, the model shows that the greatest weight in explaining the variance of the spread is provided by the selling price, then in descending order: the year of construction, the savings, the location, the state of maintenance, the dynamics of the real estate cycle and the investment rate.

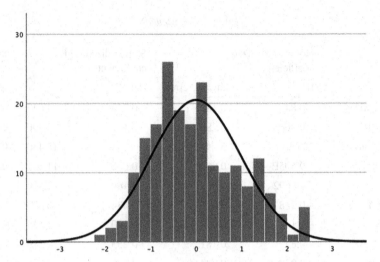

Fig. 1. Regression standardized residual histogram

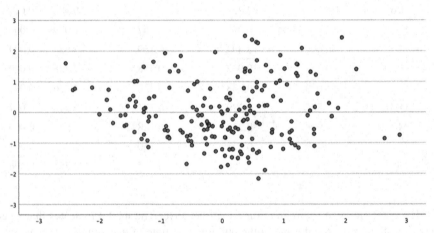

Fig. 2. Scatter plot for linearity and homocedasticity

In relation to the premises we can conclude that the reason the spread decreases if the price and the quality of the finishes (state of maintenance) increase, is to be attributed to the strategic behavior of the brokerage agency. Properties of greater value (in relation to larger size or better quality) are placed in a market segment that has limited demand. The greater difficulty of selling and the need not to lengthen the time on market, with the risk of losing the mandate, pushes the agency to make the seller accept an initial asking price close to the bid price.

This therefore justifies a lower spread. Looking at the microeconomic characteristics, the lower spread for more peripheral properties and those located in older buildings is since in the period covered by the analysis the greatest demand for residential properties was precisely in this segment. In fact, young couples or families with a lower affordability index, understood as the degree of accessibility to the purchase of a home, fed the demand.

Higher demand justifies lower discounts on the asking price. Higher demand also underlies the lower spread resulting from a growing savings and investment rate and an expanding real estate cycle.

6 Conclusions

The Italian real estate market has always been characterized by a lack of transparency in transactions, although regulations have been enacted in recent years that attempt to reduce this opacity. The lack of transparency limits the possibility of constructing statistically significant samples of selling prices that can be used as comparables in estimates. Since there are no reliable sources of selling prices, often in the practice of real estate appraisers, readily available asking prices become the reference for estimation. The use of the latter, however, has the limitation of the probable discount that then occurs in the transaction.

The proposed model tries to provide an explanation of the variability of the spread between selling price and asking price, identifying and selecting the main micro and macroeconomic variables on which it may depend. The future goal of the research is to quantitatively measure the contribution of these variables in order to make the use of asking prices as a proxy for selling prices in real estate estimation through a measure of the likely adjustment.

References

1. Knight, J.R.: Listing price, time on market, and ultimate selling price: causes and effects of listing price changes. Real Estate Econ. **30**(2), 213–237 (2002)
2. Glower, M., Haurin, D.R., Hendershott, P.H.: Selling time and selling price: the influence of seller motivation. Real Estate Econ. **26**(4), 719–740 (1998)
3. Lippman, S., McCall, J.: An operational measure of liquidity. Am. Econ. Rev. **76**(1), 43–55 (1986)
4. Haurin, D.R., Haurin, J.L., Nadauld, T., Sanders, A.: List prices sale prices and marketing time: an application to U.S. housing markets. Real Estate Econ. **38**(4), 659–685 (2010)
5. Anglin, P.M., Rutherford, R., Springer, T.M.: The trade-off between the selling price of residential properties and time-on-the-market: the impact of price setting. J. Real Estate Financ. Econ. **26**(1), 95–111 (2003)
6. Arnold, M.A.: Search, bargaining and optimal asking price. Real Estate Econ. **27**(3), 453–481 (1999)
7. Yavas, A., Yang, S.: The strategic role of listing price in marketing real estate: theory and evidence. AREUEA J. **23**, 347–368 (1995)
8. Deng, Y., Gabriel, S.A., Nishimura, K.G., Zheng, D.D.: Optimal pricing strategy in the case of price dispersion: new evidence from the Tokyo housing market: optimal pricing strategy in the case of price dispersion. Real Estate Econ. **40**, S234–S272 (2012)
9. Horowitz, J.L.: The role of the list price in housing markets: theory and an econometric model. J. Appl. Economet. **7**(2), 115–129 (1992)
10. Knight, J.R., Sirmans, C.F., Turnbull, G.K.: List price signaling and buyer behavior in the housing market. J. Real Estate Financ. Econ. **9**(3), 177–192 (1994)

11. Miller, N.G.: Time on the market and selling price. AREUEA J. **6**, 164–174 (1978)
12. Trippi, R.R.: Estimating the relationship between price and time of sale for investment property. Manage. Sci. **23**(4), 838–842 (1977)
13. Liu, N.: Market buoyancy, information transparency and pricing strategy in the Scottish housing market. Urban Stud. **58**(16), 3388–3406 (2021)
14. Li, W.F.: The impact of pricing on time-on-market in high-rise multiple-unit residential developments. Pac. Rim Prop. Res. J. **10**(3), 305–327 (2004)
15. Cirman, A., Pahor, M., Verbic, M.: Determinants of time on the market in a thin real estate market. Econ. Eng. Decis. **26**(1), 4–11 (2015)
16. McGreal, S., Taltavull, P., de La Paz, V., Kupke, P.R., Kershaw, P.: Measuring the influence of space and time effects on time on the market. Urban Stud. **53**(13), 2867–2884 (2016)
17. Haurin, D., McGreal, S., Alastair, A., Brown, L., Webb, J.R.: List price and sales prices of residential properties during booms. J. Hous. Econ. **22**, 1–10 (2013)
18. Sirmans, G.S., MacDonald, L., Macpherson, D.A.: A meta-analysis of selling price and time-on-the-market. J. Hous. Res. **19**(2), 139–152 (2010)
19. Leung, C.K.Y., Leong, Y.C.F., Chan, I.Y.S.: TOM: why isn't price enough? Int. Real Estate Rev. **5**(1), 91–115 (2002)
20. Kang, H.B., Gardner, M.J.: Selling price and marketing time in the residential real estate market. J. Real Estate Res. **4**(1), 21–35 (1989)
21. Ferreira, E.J., Sirmans, G.S.: Selling price, financing premiums, and days on the market. J. Real Estate Financ. Econ. **2**(3), 209–222 (1989)
22. Manganelli, B., Tajani, F.: Macroeconomic variables and real estate in Italy and in the USA (Variabili macroeconomiche e mercato immobiliare in Italia e negli USA). Scienze Regionali **14**(3), 31–48 (2015)
23. Jud, G.D., Seaks, T.G., Winkler, D.T.: Time on the Market: The impact of residential brokerage. J. Real Estate Res. **12**(3), 447–458 (1996)
24. Curto, R., Fregonara, E., Semeraro, P.: Prezzi di offerta vs prezzi di mercato: un'analisi empirica: asking prices vs market prices: an empirical analysis. Territorio Italia **12**(1), 53–72 (2012)
25. Ong, S.E., Koh, Y.C.: Time on-market and price trade-offs in high-rise housing sub-markets. Urban Stud. **37**(11), 2057–2071 (2000)

Spatial Statistical Model for the Analysis of Poverty in Italy According to Sustainable Development Goals

Paola Perchinunno[1]([⊠]), Antonella Massari[1], Samuela L'Abbate[1], and Lucia Mongelli[2]

[1] Department of Economics, Management and Business Law, University of Bari "Aldo Moro", Bari, Italy
{paola.perchinunno,antonella.massari,samuela.labbate}@uniba.it
[2] ISTAT - Dipartimento per la produzione statistica, Direzione Centrale della Raccolta Dati - Servizio per la Raccolta Dati per le Statistiche Economiche e Ambientali, Piazza A. Moro, 61, 70122 Bari, Italy
mongelli@istat.it

Abstract. The "Sustainable Development Goals" indicate which changes nations and people of the world are committed to achieve, by virtue of a global consensus, obtained through a long, complex, and difficult path of dialogue and international and interdisciplinary collaboration. Ending poverty, in all its manifestations including its most extreme forms, through interconnected strategies, is the theme of Goal 1. Providing people all over the world with the support they need, as through promotion of social protection systems, is, in fact, the very essence of sustainable development. The objective of this work is the statistical analysis of the indicators useful for achieving the "No Poverty" Goal 1 through multidimensional statistical analysis methodologies (Totally fuzzy and relative) to understand which Italian regions need more government intervention.

Keywords: Sustainable development · Spatial statistical model · Fuzzy approach · Poverty

1 Introduction

The United Nations General Assembly on 25 September 2015 adopted the Report on the Sustainable Development Goals (SDGs) with the 2030 Agenda for Sustainable Development in which the global objectives to end poverty, protect the planet and ensure prosperity for all by 2030 are set.

The "Sustainable Development Goals" indicate what changes the nations and peoples of the world are committed to achieving, by virtue of a global consensus, obtained through a long, complex, and difficult path of dialogue and international and interdisciplinary

---

The contribution is the result of joint reflections by the authors, with the following contributions attributed to L. Mongelli (Sect. 1, 4), to A. Massari (Sect. 5), to P. Perchinunno (Sect. 3.1, 3.2) and to S. L'Abbate (Sect. 2and 3.3).

© The Author(s), under exclusive license to Springer Nature Switzerland AG 2022
O. Gervasi et al. (Eds.): ICCSA 2022 Workshops, LNCS 13378, pp. 641–654, 2022.
https://doi.org/10.1007/978-3-031-10562-3_45

collaboration; moreover, numerous references are aimed at the well-being of people and the equitable distribution of the benefits of intra and intergenerational development. Each goal has specific goals to achieve over the next few years. It is a system of indicators of great complexity that sees within it both consolidated indicators available for most countries, and indicators that are not currently produced or that have not yet been exactly defined at the international level. Statistics is called to contribute to progress towards sustainability by guaranteeing its own service, which consists in providing ever better evidence, to accompany every phase of the construction of sustainable development.

Istat, like the other National Statistical Institutes, is called by the United Nations Statistical Commission to play an active role of national coordination in the production of indicators for the measurement of sustainable development and the monitoring of its objectives. The 17 "Sustainable Development Goals (SDGs)" and the related 169 targets with which the three dimensions of sustainable development are declined, have extended the 2030 Agenda from the social pillar alone to the economic and environmental one, to these is added the institutional dimension. Every year Istat publishes the Report on the SDGs, the last update of the fourth Report dates to February 2022. The update summarizes 367 statistical measures (of which 338 are different) for 135 United Nations Inter Agency Expert Group (UN-IAEG) indicators, which constitute the global reference framework with the usual regional analysis, particularly useful for the observation of territorial imbalances. The Report offers a first representation of the impact of the pandemic on SDGs indicators. Below is a summary of the 17 objectives (Fig. 1).

Fig. 1. Sustainable Development Goals

Ending poverty, in all its manifestations including its most extreme forms, through interconnected strategies, is the theme of Goal 1. Providing people all over the world with the support they need, including through the promotion of social protection systems, is, in fact, the very essence of sustainable development.

The scientific research options have therefore been oriented towards a multidimensional approach, expanding the analysis to a wide range of indicators of living conditions and at the same time adopting tools that allow to adequately consider the complexities.

The objective of this work is the statistical analysis of the indicators useful for achieving the Goal 1 *"No Poverty"* goal through multidimensional statistical analysis methodologies (Totally fuzzy and relative) to understand which regions, at national level, need more government intervention.

2 Goal 1: No Poverty

Goal 1 "No Poverty" aims to end all forms of poverty in the world by 2030 [1]. Enabling people to lift them out of poverty require equal rights, access to economic and natural resources, technological resources, property, and basic services. It is also necessary to guarantee all the necessary aid to communities affected by climate-related disasters.

The statistical measures that populate this objective are twenty-one referring to eight UN-IAEG-SDGs indicators (Table 1).

Table 1. List of statistical measures disseminated by Istat, taxonomy with respect to the SDGs indicators relating to Goal 1 No poverty

1.1.1 Percentage of population below the international poverty line, by sex, age, employment status and geographical distribution
- Risk of poverty for those in employment (18 years and over) (Istat, 2019, percentage values)
1.2.1 Percentage of population living below the national poverty line, by sex and age
- Absolute poverty (incidence) (Istat, 2020, percentage values)
1.2.2 Percentage of men, women and children of all ages living in poverty (in all its dimensions) according to national definitions
- Risk of poverty or social exclusion (Istat, 2019, percentage values) - Severe material deprivation (Istat, 2019, percentage values) - Low work intensity (Istat, 2019, percentage values) - Risk of poverty (Istat, 2019, percentage values)
1.3.1 Percentage of population covered by social protection plans/systems by sex, distinguished between children, unemployed, elderly, persons with disabilities, pregnant women, infants, victims of accidents at work, poor and vulnerable
- Waiver of health services (Istat, 2020, percentage values)
1.4.1 Percentage of population/households with access to basic services

(continued)

Table 1. (*continued*)

- Overload of the cost of housing (Istat, 2019, percentage values)
- Families very or quite satisfied with the continuity of the electricity service (Istat, 2020, percentage values)
- People who cannot afford to adequately heat the house (Istat, 2019, percentage values)
- Families who declare difficulties in connecting with public transport in the area in which they reside (Istat, 2020, percentage values)
- Delivery of municipal waste to landfill (Ispra, 2019, percentage values)
- Irregularities in water distribution (Istat, 2020, percentage values)
- Households with fixed and/or mobile broadband connection (Istat, 2020, percentage values)
- People aged 6 and over who use their mobile phones every day, per 100 people with the same characteristics (Istat, 2020, percentage values)

1.5.1 Number of deaths, missing and people affected by disasters per 100,000 people

- Number of deaths and people missing due to landslides (Ispra, 2018, N.)
- Number of deaths and people missing due to floods/floods (Ispra, 2018, N.)
- Number of injured by landslides (Ispra, 2018, N.)
- Number of injuries from floods/floods (Ispra, 2018, N.)

1.a.1 Total Official Development Assistance (ODA) from all donors focusing on poverty reduction as a percentage of the beneficiary country's gross national income

- Official Development Assistance for Education, Health and Social Protection on bilateral ODA allocatable by sector (Ministry of Foreign Affairs and International Cooperation, 2019, percentage values)

1.a.2 Percentage of total government expenditure on essential services (education, health and social protection)

- Share of essential services (health, education and social protection) in general government expenditure (Istat, 2019, percentage values)

3 Methodological Frameworks

3.1 The Set of Indicators

For the purposes of this work, 11 indicators divided into 3 sets have been selected based on the following characterizations:

Set 1 - Indicators of economic poverty
Risk of poverty or social exclusion
Severe material deprivation
Low work intensity
Risk of poverty
Set 2 - Indicators related to the lack of basic services
Waiver of health benefits

(*continued*)

(*continued*)

Families who declare difficulties connecting with public transport in the area in which they reside
Landfilling of municipal waste
Set 3 - Indicators related to the lack of housing services
Overload of the cost of housing
Irregularities in water distribution
Families not satisfied with the continuity of the electricity service
Households without a fixed and/or mobile broadband connection

Let's now analyse in detail the values of the individual indicators detected in 2019 at national level and on the basis of the territorial distribution (Northeast, North West, Centre, South and Islands).

Regarding the first set relating to economic poverty, it should be noted that the indicator linked to the risk of poverty or social exclusion is a multidimensional indicator corresponding to the share of people who have at least one of the following situations:

1) people at risk of income poverty,
2) persons who are seriously materially deprived,
3) people living in families with a very low work intensity.

In 2019, 20.1% of people residing in Italy are at risk of poverty, 7.4% in conditions of severe material deprivation and 10.0% in low-work-intensive families.

The composite indicator built on these three components is equal to 25.6% (about 15 million and 390 thousand people), improving for the third consecutive year (27.3% in 2018, 28.9% in 2017, 30.0% in 2016).

The reduction is mainly attributable to the improvements marked by the indicator of low work intensity and that of severe material deprivation that have been significantly reduced in the last 3 years, while the risk of poverty is substantially stable in the three-year period 2016–2019.

At the level of territorial distribution, there is a clear difference between the Northern area (West and East) which has low levels of risk of poverty and social exclusion (respectively equal to 16.4 and 13.2%) and the South and the Islands with much higher values (respectively 41.6 and 43.6).

The same situation also occurs for the other indicators for which it emerges that the North and Central Italy have values below the average while the South values decidedly above the national average (Table 2).

Table 2. Set 1 indicators of economic poverty. Year 2019 (values%)

Division territorial	Risk of poverty or social exclusion	Severe material deprivation	Low work intensity	Risk of poverty
North West	16.4	4.1	6.0	12.4
Northeast	13.2	2.9	4.4	9.5
Center	21.4	5.5	7.6	15.3
South	41.6	12.7	14.8	33.7
Islands	43.6	15.4	22.8	36.8
Italy	**25.6**	**7.4**	**10.0**	**20.1**

Source: Istat. SDGs Report 2021. Statistical Information for the 2030 Agenda in Italy.

Regarding the lack of basic services, the spread of COVID-19 has had a significant impact on the lack of demand for health services. In 2019, 6.3% of people say they have given up a medical examination despite needing it, among them about half reported a problem related to COVID-19 as the cause. There are no major differences in territorial distribution.

Much higher is the percentage of families who declare difficulties in connecting with public transport in the area in which they reside which at national level is equal to 33.5% with peaks in the South equal to 42.3%. On the other hand, the territorial gap in the delivery of municipal waste to landfills is decidedly high, with values equal to 49.5% in the Islands, compared to a national average of 20.9% (Table 3).

Table 3. Set 2 indicators related to the lack of basic services. Year 2019 (values%)

Division territorial	Waiver of health benefits	Families who declare difficulties connecting with public transport in the area in which they reside	Landfilling of municipal waste
North west	5.4	31.1	10.1
Northeast	4.7	26.4	11.3
Center	6.9	33.1	29.3
South	7.3	42.3	22.4
Islands	8.0	36.3	49.5
Italy	**6.3**	**33.5**	**20.9**

Source: Istat. SDGs Report 2021. Statistical Information for the 2030 Agenda in Italy.

Regarding housing services at the national level, 8.7% of households in 2019 had an overload of the cost of housing, 8.6% irregularities in water distribution and 6.5% discontinuity of the electricity service. The percentage of households without a fixed and/or mobile broadband connection is much higher (Table 4). At the level of territorial distribution, serious problems occur in the Islands about the overload of the cost of housing (14.8%) and the distribution of water (24%). The South also has values higher than the national average, while Central Italy has values above the average only regarding water distribution (9%) and continuity of electricity service (7.6%).

Table 4. Set 3 indicators related to the lack of housing services. Year 2019 (values%)

Division territorial	Overload of the cost of housing	Irregularities in water distribution	Families not satisfied with the continuity of the electricity service	Households without a fixed and/or mobile broadband connection
North west	8.4	3.0	6.5	25.3
Northeast	5.6	3.2	4.4	24.6
Center	6.8	9.0	7.6	22.3
South	10.3	12.9	8.1	29.7
Islands	14.8	24.0	9.9	30.1
Italy	**8.7**	**8.6**	**6.5**	**25.3**

Source: Istat. SDGs Report 2021. Statistical Information for the 2030 Agenda in Italy.

3.2 The Fuzzy Approach

The development of fuzzy theory initially stems from the work of Zadeh [2] and subsequently was conducted by Dubois and Prade [3]. The fuzzy theory develops starting from the assumption that each unit is not univocally associated with only one but simultaneously with all the categories identified based on links of different intensity (degrees of association).

The first measurement based on the fuzzy set theory, named TF (Totally Fuzzy), was suggested by Cerioli and Zani [4]. This logic can be applied to both continuous and ordinal variable cases. However, in the latter case, the maximum and minimum values can be determined by assuming the value of the lowest category as minimum and the highest as maximum.

Cheli and Lemmi [5] have proposed a generalization of this approach, called Totally Fuzzy and Relative (TFR). This method is also called "totally relative" because the value of the membership function is entirely determined by the relative position of the individual in the distribution of the population. The fuzzy TFR approach consists in defining the measurement of an individual's degree of belonging to the totality fuzzy, included in the interval between 0 (with an individual who does not demonstrate a clear belonging) and 1 (with an individual who demonstrates a clear belonging).

If we suppose to observe k indicators for each family, the function of belonging of the i-sima family to the blurred subset, can be defined as follows [4]:

$$f(x_{i.}) = \frac{\sum_{j=1}^{k} g(x_{ij}) \cdot w_j}{\sum_{j=1}^{k} w_j} \quad i = 1, \ldots, n \tag{1}$$

where $w_1, \ldots, w_k$ represent a generic system of weights. The $f(x_{i.})$ is in practice a global poverty index, while $g(x_{ij})$ measures the specific deprivation of the i-sima unit according to the j-th indicator. Following the Totally Blurred and Relative Approach (TFR) recently proposed by Cheli and Lemmi (1995), the function $g(x_{ij})$ is defined in terms of the partition function H (·) of the indicator X_j as follows:

$$g(x_{ij}) = \begin{cases} H(x_{ij}) & \text{if the risk of poverty increases with increasing of } x_j \\ 1 - H(x_{ij}) & \text{if the risk of poverty decreases with increasing of } x_j \end{cases}$$

The indices were chosen in order to identify different levels, related to the aspects of poverty [6, 7]. The indices were grouped into three sets characterized by different situations in the different components considered: economic poverty, lack of basic services, lack of housing-related services. The Total Fuzzy and Relative method was applied on the data of all the Italian regions obtaining a value of the individual w_i weights, which varies according to the level of importance in determining the degree of quality of the situation.

3.3 The Results of the Application of the Totally Fuzzy and Relative Approach

Once the sets of indicators were identified, the minimum, maximum and average values for each indicator of the different sets were detected (Table 5).

Table 5. Results of the application of the TFR method in relation to the distribution function and the weights of the various indices.

Indicators set	Indicators	Minimum	Maximum	Mean	Gmean	Weight w_j
Economic poverty	Risk of poverty or social exclusion	8.1	49.7	23.4	0.4	1.0
	Severe material deprivation	1.7	17.8	6.4	0.3	1.2
	Low work intensity	3.2	25.0	8.8	0.3	1.3
	Risk of poverty	6.1	41.4	17.9	0.4	0.9
Lack of basic services	Waiver of health benefits	3.2	11.7	6.4	0.5	0.7

(*continued*)

Table 5. (*continued*)

Indicators set	Indicators	Minimum	Maximum	Mean	Gmean	Weight w_j
	Families who declare difficulties connecting with public transport in the area in which they reside	13.3	55.9	30.8	0.6	0.5
	Landfilling of municipal waste	1.3	90.0	27.5	0.4	0.8
Lack of housing-related services	Overload of the cost of housing	3.7	16.7	7.5	0.4	1.0
	Irregularities in water distribution	1.4	31.2	8.5	0.4	1.0
	Families not satisfied with the continuity of the electricity service	2.1	11.7	6.2	0.5	0.7
	Households without a fixed and/or mobile broadband connection	18.8	33.3	25.5	0.6	0.6

Source: Our data processing Istat. Rapporto SDGs 2021.

For each set of indicators, the fuzzy value and its values attached to it have been calculated. Of particular interest is the analysis of w_i weights that indicate the relevance of the indicator on the considered set. As already specified, high values of this indicator denote a strongly discriminating condition in the determination of the result. In our case the values of the weights of the indicators relating to low work intensity ($w_i = 1.3$), and those relating to severe material deprivation ($w_i = 1.2$) are particularly discriminating.

In the third set relating to the lack of services related to housing, high values are found for overload of the cost of housing and irregularity in the distribution of water ($w_i = 1.0$). Conversely, the weights of the indicators relating to the difficulty of connecting with public transport in the area in which they reside (equal to $w_i = 0.5$) appear less discriminating.

As a result of the application, we have classified the Italian regions based on fuzzy values obtaining the classification illustrated in Table 6. Recall that high values are significant of regional situations of poverty, conversely low values are significant of situations of absence of difficulties.

Table 6. Composition in absolute values and percentages of regions by fuzzy class membership

Fuzzy value	Number of regions			%		
	Economic poverty	Lack of basic services	Lack of housing-related services	Economic poverty	Lack of basic services	Lack of housing-related services
0,0 ⌐ 0,2	10	3	4	50	15	20
0,2 ⌐ 0,4	2	4	5	10	20	25
0,4 ⌐ 0,6	2	2	4	10	10	20
0,6 ⌐ 0,8	3	6	5	15	30	25
0,8 ⌐ 1,0	3	5	2	15	25	10
Total	**20**	**20**	**20**	**100**	**100**	**100**

Source: Our data processing Istat. Rapporto SDGs 2021.

In particular, the percentage of regions with a high shortage of basic services (between 0.8 and 1) is higher (25%) than that which has a strong shortage of housing services (10%).

The same applies to regions with a high shortage of basic services (between 0.6% and 0.8) with a percentage of regions equal to 30% compared to 15% for regions with a high level of economic poverty.

4 Spatial Distribution of Fuzzy Values

In this paper, the representation of the values attributed to the individual geographic areas, corresponding to the regions, occurs through cartograms, associated with "natural" interval classes, defined within the distribution. By applying the *Total Fuzzy and Relative* method on the data of all Italian regions, as described in the previous section, three synthetic indices were obtained that describe the territorial distribution of the indicators.

Regarding the set of indicators related to *economic poverty*, individuals with severe difficulties live mainly in the southern regions, Sicily, Campania and Puglia and have values between 1.0 and 0.8, followed by Molise, Calabria and Sardinia with values between 0.8 and 0.6. Economic poverty has average indicators between 0.6 and 0.4 in the Regions of Basilicata and Lazio and lows between 0.4 and 0.2 in Abruzzo and Liguria. The individuals who show less difficulty live in the other 10 regions all located in the Centre-North, as shown in Fig. 2.

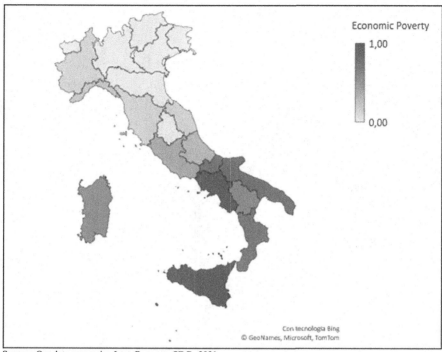

Source: Our data processing Istat. Rapporto SDGs 2021.

Fig. 2. Spatial distribution economic poverty fuzzy value

Regarding the second set of indicators summarized as *a lack of basic services*, the territorial analysis by regions shown in Fig. 3 is particularly interesting. The regions in conditions of great difficulty with values between 1.0 and 0.8 are Calabria, Sicily, Valle d'Aosta, Marche, and Molise. The presence of Valle d'Aosta in this first group lies in the high value of the indicator that measures families who have had to give up health services. It is important to underline, however, that many have said that they have given up for reasons related to COVID-19, fear on the part of the population of contracting infections, closure of many outpatient facilities and suspension of the provision of postponable health services. Values between 0.8 and 0.6 are found in the regions of Umbria, Puglia, Basilicata, Lazio, Sardinia and Abruzzo, followed by the regions of Tuscany and Liguria with values between averages between 0.6 and 0.4. We find with values between 0.4 and 0.2 regions such as Campania, Piedmont, Lombardy and Veneto. We point out in the last group Friuli-Venezia-Giulia, Emilia-Romagna and Trentino-Alto-Adige regions where there are no complaints about shortages of basic services.

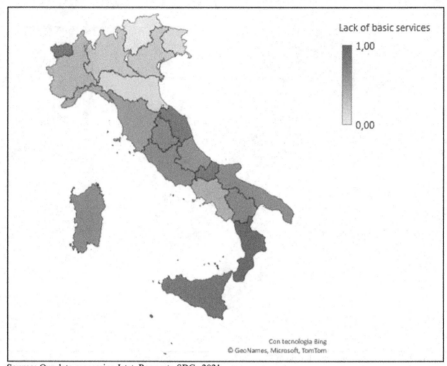

Source: Our data processing Istat. Rapporto SDGs 2021.

Fig. 3. Spatial distribution shortage of basic services fuzzy values

Finally, about the third set of indicators, summarized **in the lack of services related to housing**, we observe in Fig. 4 that the 2 regions with situations considered problematic by families are Sicily and Calabria with values between 1.0 and 0.8. With values between 0.8 and 0.6, there are regions such as Sardinia, Molise, Campania, Puglia and Lazio with always complex situations, but considered less problematic. Followed by 4 other regions, Abruzzo, Basilicata, Umbria and Liguria, with values ranging from 0.6 to 0.4. Lower shortage of housing-related services is highlighted by those who live in the regions of Piedmont, Lombardy, Emilia-Romagna, Marche and Tuscany, with values between 0.4 and 0.2. Almost no shortage is detected by the inhabitants in the regions of Valle d'Aosta, Friuli-Venezia Giulia, Veneto and Trentino-Alto Adige.

The overall analysis of the 3 sets of indicators reveals an overall interesting situation, where territorial differences are evident. The only aspect to highlight is the presence of the Trentino-Alto Adige and Friuli-Venezia Giulia regions in the lowest class, between 0 and 0.2, for all 3 sets of indicators analyzed.

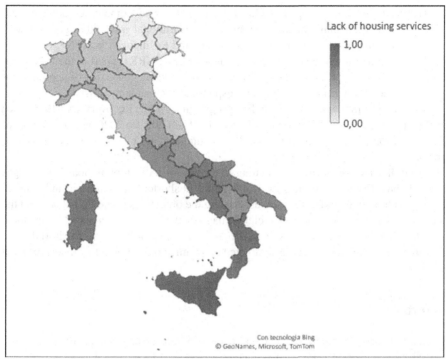

Source: Our data processing Istat. Rapporto SDGs 2021.

Fig. 4. Spatial distribution shortage of services related to housing fuzzy values

5   Conclusions

The profound economic and social transformations that have taken place in recent years place the problem of poverty at the centre of a broad scientific debate. Poverty connotes a discomfort that does not necessarily end in the lack of monetary resources, but that involves a plurality of dimensions of a social and cultural nature such as education, health, housing and is concretized as a lack of equitable access to a plurality of essential goods and services.

Both in the academic literature and in international scientific reports, however, the multidimensional nature of poverty is emphasized, which entails the need to no longer resort to a single indicator but to a group of indicators, useful for better delineating the living conditions of the different subjects.

A first emerging idea from the present work is related to the ability to describe territorial phenomena through an integrated model, which starts from the construction of socio-economic indicators, of a multidimensional nature, and then adopts models capable of identifying areas at risk of discomfort. To enable people to lift them out of poverty, it is necessary to guarantee equal rights, to have access to economic and natural resources, technological resources, property and basic services.

The Sustainable Development Goals refer to different domains of development related to environmental, social, economic and institutional issues that go well with the Missions envisaged by the National Recovery and Resilience Plan (PNRR).

The Italian National Recovery and Resilience Plan (PNRR) is part of the Next Generation EU (NGEU) program, the 750-billion-euro package, consisting of about half of grants, agreed by the European Union in response to the pandemic crisis.

The main component of the NGEU programme is the Recovery and Resilience Facility (RRF), which has a duration of six years, from 2021 to 2026, and a total size of €672.5 billion (€312.5 billion grants, the remaining €360 billion loans at subsidised rates).

From the analysis of the data relating to the sets of indicators of Goal 1 it emerges that Italy has strong territorial gaps, with significant shortcomings in basic and housing services in the South and in the Islands. Among the objectives of the PNRR we find the need to repair the economic and social damage of the pandemic crisis and contribute to addressing the structural weaknesses of the Italian economy and the territorial gaps. This implies the need for increased targeted economic commitment to support regional policies.

References

1. Istat. Rapporto SDGs. Informazioni Statistiche per l'agenda 2030 in Italia. Ed. Anno 2021. Collana: Letture statistiche – Temi (2021)
2. Zadeh, L.A.: Fuzzy sets. Inf. Control **8**(3), 338–353 (1965)
3. Dubois, D., Prade, H.: Fuzzy Sets and Systems. Academic Press, Boston, New York (1980)
4. Cerioli, A., Zani, S.: A fuzzy approach to the measurement of poverty. In: Dagum, C., Zenga, M. (eds.) Income and Wealth Distribution, Inequality and Poverty. Springer, Heidelberg (1990). https://doi.org/10.1007/978-3-642-84250-4_18
5. Cheli, B., Lemmi, A.: A "totally" fuzzy and relative approach to the multidimensional analysis of poverty. Econ. Notes **24**(1), 115–134 (1995)
6. Montrone, S., Perchinunno, P., Rotondo, F., Torre, C.M., Di Giuro, A.: Identification of hot spots of social and housing difficulty in urban areas: scan statistic for housing market and urban planning policies. In: Murgante, B., Borruso, G., Lapucci, A. (eds.) Geocomputation and Urban Planning. Studies in Computational Intelligence, vol. 176, pp. 57–78, Springer, Heidelberg: (2009). https://doi.org/10.1007/978-3-540-89930-3_4
7. Perchinunno, P., Rotondo, F., Torre, C.M.: A multivariate fuzzy analysis for the regeneration of urban poverty areas. In: Gervasi, O., Murgante, B., Laganà, A., Taniar, D., Mun, Y., Gavrilova, M.L. (eds.) ICCSA 2008. LNCS, vol. 5072, pp. 137–152. Springer, Heidelberg (2008). https://doi.org/10.1007/978-3-540-69839-5_11

Real Estate Sales and "Customer Satisfaction": Assessing Transparency of Market Advising

Carmelo Maria Torre[✉], Debora Anelli, Felicia Di Liddo, and Marco Locurcio

Department of Civil, Environmental, Land, Building Engineering and Chemistry,
Polytechnic University of Bari, 70126 Bari, Italy
carmelomaria.torre@poliba.it

Abstract. The paper start describing the change of relationship between the evolution of real estate market, housing services, and economic crisis due to the covid pandemic event.

Looking more in detail housing and hosting services we can discover the evolution of concept of home-service as the reference point for a different number of activity.

In the same time the housing services, considered as a facility for external city users changed their characters and relevance.

The transition of urban costumes from the pre-pandemic era to the post pandemic ones generated new use values as a consequence of new costumes.

The use of the networks for communicating and collaborating from home instead that from offices and clerk's workplaces changed office attitudes, but in the same time reduced the usefulness provided by housing stocks addicted to guest foreign city user.

The paper describes the evolution of the interpretation of role of housing services, and the consequent change of use-values, in a city that is characterized by the presence of many guests workers, that use housing services by renting flats that have been unused in the most recent years, in the pandemic era, that compelled renters to become travellers.

Keywords: Real estate market · Housing services · Flats on rent · City users · Pandemic events

1 Introduction

The aim of this paper is to show some issues due to the pandemic phenomenon [1–4], that pushed city users, of travelers from home to their worksites, to change their costumes. In some way this change affected the real estate values, when a relevant stock of dwelling [5, 6], that have been devoted to guest temporary city users, faced with a decrease of their market values, when the rental activity stopped dramatically [7–9].

This paper describes the phenomena that changed the interpretation of dwelling function for city guests and users, generating as a consequence the decrease of use of flats and every kind of housing services [10].

O. Gervasi et al. (Eds.): ICCSA 2022 Workshops, LNCS 13378, pp. 655–667, 2022.
https://doi.org/10.1007/978-3-031-10562-3_46

The results of some reasoning generated the need of thinking about the increase of the role of mediation agencies inside the urban fringe: The role of mediators became more important, facing with the increase of emptying of housing services offered for rent to city users.

The paper in the first part describe a short reminding of the consequence on real estate de to the pandemic era. In the second part the increasing relevance of real estate agents for mediation between sellers and buyers and/or renters.

In detail, after the introduction, it is shown and explained why the incidence of renters decreased in urban environments, in front of an increase of an empty stock of housing services, and why the regime of real estate values was modified when the role of housing services, changed.

The case of study is a main zone of the metropolitan city of Bari, the capital town of the Apulian Region, guesting students, workers and temporary city users.

2   Main Issues of the Research

In the "Covid Era", Europe and, more in detail, Italy, started to look at a "Green New Deal", reminding, in some sense, the rooseveltian strategies mainly based on the "economic flywheel" theory. The Green new deal in Italy has been implemented according with the directives of the "National Recovery and Resilience Plan" (Piano Nazionale di Ripresa e Resilienza - PNRR as the Italian acronymous).

Thanks to PNRR, 107.7 billion euros (on 330 billion, moreless half of the total) have been allocated to support the construction sector, but about only one for every ten companies (14%) managed the incentives in detail; anyway there are some parts that appeared more attractive than others.

In other few terms, by the support of public fundings, a set of governmental actions were finalized against the potential and existing increase of economic crisis due to pandemic events. As regards companies operating in the Italian context, the 53% of incentives and bonuses related to energy efficiency and safety of private and public built stocks (judicial seats, municipal buildings, care homes etc.) supported the supply chain as better is possible. In the same time a scheme of public financial support was set up, aimed at the preservation of market dynamics in several sectors.

The PNRR allocated 107.7 billion euros for construction, favoring only about more than one on ten companies (14%). The dynamic was known in detail, but there are some parts that arouse more interest than others. In few words, as regards construction and housing sector, the global pandemic invasion affected strongly the dynamics and trends of real estate market.

As regards the real estate main sector -that is to say the housing market- the consequent issues that we underline are:

1) the retain of potential crisis -as a consequence- due to the relevance of construction market for the entire national gross domestic product.
2) the rough increase of several construction works regarding the improvement of quality of the built stock, that affected, and still is affecting, the dynamics of construction market

3) the increase of the offer in the housing market, and as a consequence, the reduction of purchase prices.

More in detail, when looking at dynamics referable to the third point, an interesting and peculiar phenomenon is represented by the increase of "offer on sale" regarding the housing market.

The existence of a wide unused and unretained housing stock was transformed in an opportunity to improve the quality of built environment, pursuing the performance in terms of domestic energy saving systems, and improving the quality of lodgments funded by the government [11, 12].

Thanks to PNRR, the incremental dynamics of construction and refurbishment market and housing sales, shows signs of strong recovery that do not concern only the future but also the present: just for an example, when looking at the results of investigations referring the activity of operators in the field of housing market, the 83% of respondents are already satisfied with their "order book".

The "bonus funds" for supporting the rehabilitation and for improving the energy saving of housing stocks, favored several interventions translated in an economic discounted support according with the needs of construction companies and of house-owners.

Just to give some detail deriving from social investigations about incentives in the construction sector, the "Restructuring Bonus" has been statistically judged positive by 74% of operators in the field of housing market, followed by "Ecobonus" (69% of positive judgement by operators) and "Superbonus" 110% (68% of positive judgement by operators) are the preferred interventions for construction companies.

Due to such set of incentives, the positive revenues of 6 out of 10 companies (in detail the 64%) have encountered the expectation of stakeholders in the last year much more than the previous years (49%).

One of the most interesting consequence is the increase of mediation activities in the field of housing market.

3 A Case of Study in Metropolitan Contexts

An interesting but not sufficient investigated aspect of the description provided in the introduction is the repositioning of housing stock in the real estate market, that favored the work of a great number of mediation agencies. The result of easier funding for young couples or families improved the increase of home-demand.

The paper focuses therefore on such above questions, by looking at and describing the dynamics of real estate market in metropolitan city of Bari, in the most recently and current "eras" [13, 14]. The work of agents in last years is more aiming to facilitate the search and purchase of some typology of home, that is considered a "first lodgment", to be acquired for living in, and/or to be acquired for refurbishing and put again on sale at a higher price.

It is clear that the growth of real estate agencies in terms of roles and number, affected (and still affects) the behavior of sellers, buyers, and intermediate actors inside the urban context.

The mediator roles were and are joining with buyer needs, both searching for appropriate proposal of market-segments inside the housing stock. Among metropolitan urban contexts, a good case of study can be read through the market dynamics of the metropolitan city of Bari, in the middle south of Italy.

The attention towards such dynamics, invites to study and to apply an investigation about the meeting points between owner, mediator and seller and client, and the effectiveness of actions for each player' role.

The relevance of such connection is readable in the number of mediation activities, and in the different diffusion inside the city quarters (see Figs. 1, 2 and 3).

It is easily understandable that the location of agencies is in some way related with the variation of offered real estate, according to several features referable to housing stock in different areas.

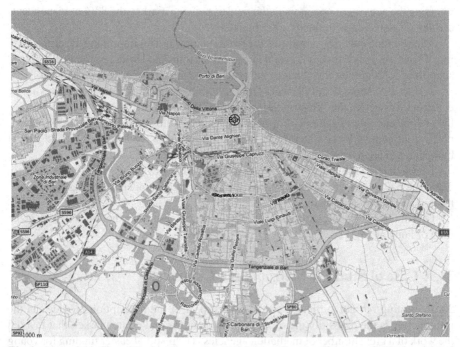

Fig. 1. Map of the City of Bari, at the center of "Metropolitan Area" (source OMI: the national Observatory of Real Estate Market)

The distribution of mediation activities can be considered aa a "geographical indicator" of different dynamics inside various parts of the urbanized context.

It appears obvious that the increase of mediation services nearby some well defined quarters of the city denotes an increase of some peculiar housing market dynamics, favoring the encounter between offers and demands.

Just to give an example, during the years of domestic segregation due to covid, and of work "on line".

Figure 2 put on evidence that the higher presence of housing mediators activity is located in the core of the urban tissue (named "Murat" quarter), and in the closer neighborhoods to the core area of the city (named "Carrassi" quarter), both crossed by main radial axes from south towards north: (the main road axes "Viale della Repubblica" and "Viale Unità d'Italia"), characterized by a number of support services for commercial housing activities.

Obviously, such concentration in well identified location of new activities favored by PNRR, represents an indicator of the seek for interface with potential owners, ready to rent or to put on sale their properties, ready for welcoming commerce and economic services.

In the same time the reduction of presence in the city represented by commuters, like university students ad occasional monthly workers, on one side started stimulating the "home-work on line", produced a shock on real estate market.

The "covid era", in fact, made empty a relevant amount of those apartments that until two years ago were devoted to be rented, and now are devoted to be sold.

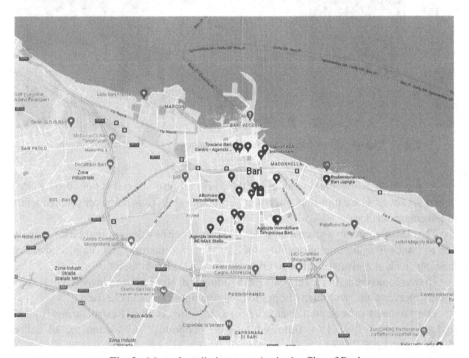

Fig. 2. Map of mediation agencies in the City of Bari

The map of mediation agencies is corresponding perfectly with location of universities, hospital, and with area of concentration of professional offices.

Figure 3, instead, shows the classifications of urban zones according to the attribution of various real estate values, reported by the surveys of the national Observatory of Real Estate Market (OMI).

In some way, the identification of some parts of the urban fringe as the seat of concentrating mediation activity, gives the idea of competitiveness.

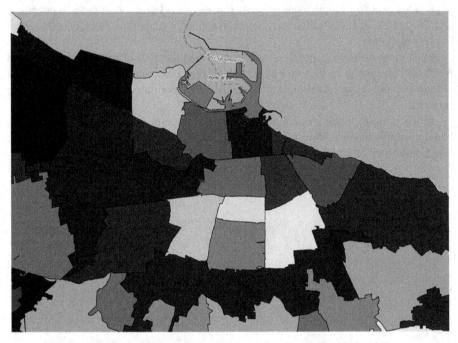

Fig. 3. Classification of areas and real estate values by the Tax National Agency

As regards housing market, therefore, the whole urban context is subdivided by OMI in various zones with a min-max interval of unitarian values (e.g. euros per square meters of pavement surface).

The classification in a given area is defined by a range of values of real estate, usually due to issues like the internal quality, the age (intrinsic factors) and the placement of housing stock at a given distance to urban amenities (external factors).

To overcome them, in recent years the building industry has been involved in several public interventions, with many hopes to continue until now and overpass.

The favoured governmental support for companies is the "Restructuring Bonus" (judged positively by 74%), followed by the "Ecobonus" (69%) and the so called "Superbonus 110%" (68%).

The revenues of 64% of companies (6 out of 10) were coming by these incentives, much more than the previous years (49%).

It could be useful to structure a valuative logical path for assessing some peculiarity if the context:

– on one side we find the dimension of mediatory activities related with housing stocks, and

– on the other side we discover a connection among the characterization of real estate values in various city contexts.

In a phase of cost-effectiveness judgments, there are frequently instances of the construction of some self-judgment, which can be more generally traced to decision theory problems represented by a choice between different alternative solutions, options, rankings, and related with multiple issues that lead to multidimensional assessments.

Multi-criteria analysis/Multiple criteria assessment (MCA) represents a methodology of decision making that supports the collocation of various alternative solutions of a problem, in a ranking/scale of importance derived from a multiple composed classification.

Multi-criteria analysis therefore represents a methodology for choosing between alternative solutions to a problem.

The evaluation is necessarily multidimensional, and therefore is framed in methodologies that take into account different points of view, which can refer to objectives of equity, environmental protection and increase of economic well-being. Some general issues of MCA are listed as follows

– the result of a multicriterial assessment is usually a ranking of the options that appears more eligible.
– the estimation of values that classify the alternatives is necessarily qualitative and/or multidimensional.

The ranking, therefore, is framed within a number of approaches that take into account different viewpoints, which may be related whit objectives of equity, environment and economic welfare enhancement.

The deals for which the use of a multiple dimension approaches can be considered usually:

– location questions, that is to say, related with different hypotheses for the location of a sort of facilities, in the face of multiple possible alternatives provided,
– infrastructure choices: questions related to the choice of routes for primary urbanization works, or for major infrastructure nodes (e.g., ports, airports).
– problems of land use: problems related to the choice of allocation of an intended use for an area, e.g., related to the establishment of proportionate buildings or production zones, defined by a result of a masterplan forecasts, which different locations within a possible urban area.
– questions about reuse of housing stock, related with the restoration of building characterized by historical-architectural relevance (for which different hypotheses of reuse for activities of collective interest are possible).
– ranking based on a multidimensional assessment, referring to some structured features
– problems of redevelopment: e.g. problems related to the choice of works of collective interest to be located in an urbanized area as part of an urban redevelopment plan.
– spatial multidimensional analyses, based on a multicriterial approach that is set up by a classification that consider spatial issues of different locations, generating a geography of contextual/environmental values.

This last option, the classification is the more suitable in our case of study.

The first singular and relevant aspect is related with the diffusion of mediation agencies in some areas classified by the Geographic Information System of Tax national agency (named "Geopoi"), in the following zones: B9, B11, C4, and C2.

Such zones are characterized by relevant asset due to real estate values, that stimulate the selling/buying activities that increase with the support of mediation agencies.

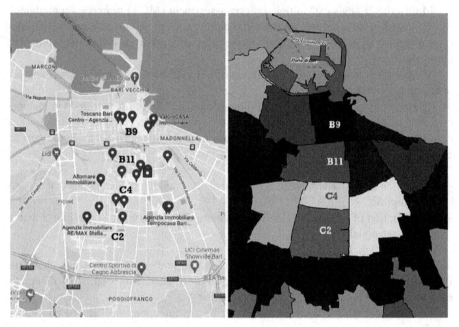

Fig. 4. Overlapping of agencies for real estate mediation and corresponding zones of the Observatory of Real Estate Market

The level of quality of the housing stock is revealed by numbers related with information on real estate value unit (euros per square meters varying since, and the spatial location of the built-up zones are represented by letter (B = demi central and C = demi external).

The aim of this paper is to test a new way of valuing/looking real estate market data, based on an external factor.

The hypothesis is that incentives provided by governmental actions, in order to valorize and to qualify the built housing stock, "pushed" more fast sales and acquisitions in various context of real estate market.

In the case of study, the most interesting phenomena can be described as follows:

– the hypothesis of "external causal factor" is put on evidence by the increase of the number of real estate mediation agencies (see Fig. 5);

- the presence of mediator inside the context is a representative indicator revealing complex dynamics in real estate market, testified by demi-high values of some suburban areas, that can be readable by the "geography" of real estate values for each city context, provided by the Tax National Agency (as shown in Figs. 2, 3 and 4);
- the number of mediation agencies has been moreless constant during past times, but nowadays is characterized by a low increase of activity when looking towards the city center;
- the increase of real estate values is more evident and also related with the parallel increase of nearness to the city center.

In the contest of case of study, the evaluation process can be based on a multidimensional analysis able to connect some relevant parameters with variation of real estate values.

4 The Approach to Evaluation

It is useful to remind the concept of preference modelling for multi-criteria assessments. If our choice is linked to n evaluation criteria, each of them may lead to a different preference. That is, there will be n comparisons between ai and aj, based on n functions of different utility. The Total Utility shall represent a summary of the uk partial utilities, with k = 1....n.

These evaluation models are therefore based on the assumption that it is possible to express a function of total utility U, composition of partial utilities (u1, u2... a), such that:

$$U\ [(uk\ (ai)]\ >\ U\ [(uk\ (aj)]\hat{U}\ ai\ P\ aj$$

the U(ai) function determines the total utility resulting from the choice of the "ai" solution;
the function U(aj) determines the total utility resulting from the choice of the solution aj;
the uk(ai) function determines the partial utility resulting from the choice of solution to ai;
the uk(aj) function determines the partial k-ma utility resulting from the choice of solution aj;

If the general rule of comparison is

$$U\ [ai]\ >\ U\ [aj]\hat{U}\ ai\ P\ aj\ with\ U\ =\ f(u1,\ u2\ldots a),$$

It is clear that the problem is addressed towards the relationship between the general comparison rule and the combination of the partial utilities uk with the total one U.

The different multicriteria methods derive from the construction of different hypotheses with respect to these combinatorial rules.

Therefore, it could be justified that prosecution of study was based on a Multiple Criteria approach.

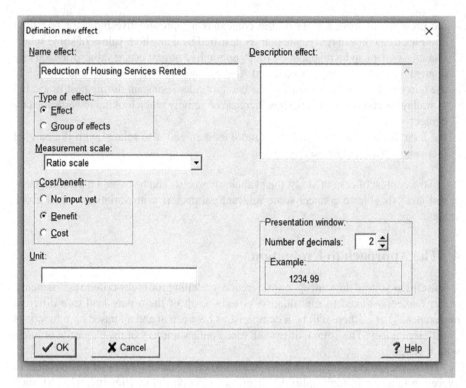

Fig. 5. Example of comparative rule definition for the criterion

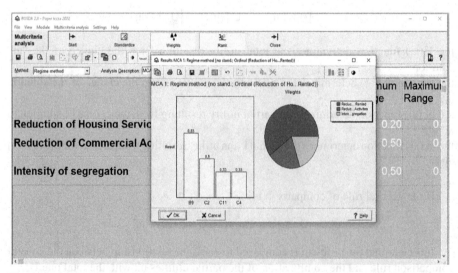

Fig. 6. Rough score obtained with software "definite", representing various reductions of values from the center towards the neighbors

The use of multicriterial assessment was based on "regimes" method, developed with the support of the Software Definite (see Figs. 5 and 6).

The economic consequences are connected with several factors: the reduction of housing stock used on rent (as a consequence of covid-segregation) due to the reduction of the number of city users and guests living inside the quarters: they are corresponding to the perimeters of the areas object of study, during ordinary working period (university student, "from and to" addicted to work, weekly travelers).

- the closure of some economic activities, facilities, commercial services, due to the need to contrasting the pandemic diffusion,
- the limitation of travels and exchange from a public transport service to another
- the impact on all the risky facilities and services that can facilitate a contagion of workers (e.g. medical staff, employers addicted to nursing, clerk in offices with proximity contacts and so on)

According with the above reasoning, therefore, the criteria to assess are:

- reduction of housing services for renters, due to the disappearing of offside students and workers,
- reduction of commerce activities, due to the reduction of expenditure not related with primary goods
- reduction of services activities, related with the increase of home-works

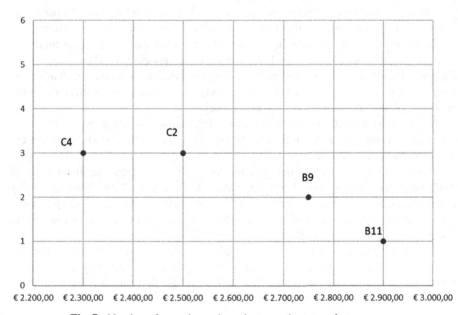

Fig. 7. Number of agencies and maximum real estate values per areas

The combination of such impacts generates an uncertain environment as regards real estate markets and housing services (Fig. 7).

The above diagram represents the co-presence among number of mediation agencies for real estate, and the "peak values" for each classified zone according to the forecast of the national Observatory of Real estate Market.

5 Conclusions: Relationship Among Advising, Mediation and Perception of Post-pandemic Quality of Life

The last step before to reach the final considerations is still related with the presence of a number of mediation agents operating on real estate; more in detail, the number of advisors cannot be forgotten, just because their services increased in the more recent years, but especially in the last ones, after the crisis due to Covid pandemic era.

Such phenomenon is clearly related with:

- the loss of guests (as already reminded, workers and students from outside) that increase the use of unmaterial facilities for communication and exchange of services, information, procedures etc.,
- last but not least, the dramatic loss of a quantity of "fragile human beings", that left empty a number of flats and home-guest services.

Such last considerations testify the development of dynamics connected with a phenomenon of increase of real estate home-services, in an unpredicted scenario.

The paper shows an experimental/pioneer proposal to assess the impact of pandemic events according with the quality of life expressed by some signal of real estate market.

Some habits disappeared during the pandemic period, and in the same time new costumes related with a "life as a recluse" that affected every citizen and city user pushed the dynamics of the market related with housing services. Such phenomena changed the behavior of real state agencies, that can be readable if looking at the variation of the incidence of mediators inside the real estate market.

In some way it could be useful to see in a different way a future "use of the city". This future vision arose to the fore at the beginning of pandemic era, when the concept of home changed its features.

Homes began a protected environment, and in the same travelling from home to workplaces began a risky experience. Several parts of a city had been observed as risky places that pushed city user to change their attitudes. The different use of city generated different use-values. The need of staying in a flat near the work places was re-considered, and became not necessary, evolving in a vision of risky place.

The paper represent just a base for rethinking the urban costumes and to re-interpreting dynamics of real estate market looking at home snot only as a "eternal real estate", but also as a place with flexible kind of use that can shift from "housing stocks" to "temporary guest-places".

References

1. Tajani, F., Morano, P., Di Liddo, F., Guarini, M.R., Ranieri, R.: The effects of Covid-19 pandemic on the housing market: a case study in Rome (Italy). In: Gervasi, O., et al. (eds.) ICCSA 2021. LNCS, vol. 12954, pp. 50–62. Springer, Cham (2021). https://doi.org/10.1007/978-3-030-86979-3_4
2. Grybauskas, A., Pilinkienė, V., Stundžienė, A.: Predictive analytics using Big Data for the real estate market during the COVID-19 pandemic. J. Big Data **8**(1), 1–20 (2021). https://doi.org/10.1186/s40537-021-00476-0
3. Ambrus, A., Field, E., Gonzalez, R.: Loss in the time of cholera: long-run impact of a disease epidemic on the urban landscape. Am. Econ. Rev. **110**(2), 475–525 (2020)
4. Wong, G.: Has SARS infected the property market? Evidence from Hong Kong. J. Urban Econ. **63**(1), 74–95 (2008)
5. Balemi, N., Füss, R., Weigand, A.: COVID-19's impact on real estate markets: review and outlook. Fin. Markets. Portfolio Mgmt. **35**(4), 495–513 (2021). https://doi.org/10.1007/s11408-021-00384-6
6. Francke, M., Matthijs, K.: Housing markets in a pandemic: evidence from historical outbreaks. J. Urban Econ. **123**, 103333 (2020). https://doi.org/10.2139/ssrn.3566909
7. Giudice, V.D., De Paola, P., Giudice, F.P.D.: COVID-19 infects real estate markets: short and mid-run effects on housing prices in Campania Region (Italy). Soc. Sci. **9**(7), 114 (2020)
8. Gupta, A., Mittal, V., Peeters, J., Nieuwerburgh, S.: Flattening the curve: pandemic-induced revaluation of urban real estate (2021). https://doi.org/10.2139/ssrn.3780012
9. Liu, S., Su, Y.: The impact of the COVID-19 pandemic on the demand for density: evidence from the US housing market. SSRN Electron. J. (2021)
10. Huang, J., Palmquist, R.B.: Environmental conditions, reservation prices, and time on the market for housing. J. Real Estate Financ. Econ. **22**(2/3), 203–219 (2001)
11. Attardi, R., Cerreta, M., Sannicandro, V., Torre, C.M.: Non-compensatory composite indicators for the evaluation of urban planning policy: the land-use policy efficiency index (LUPEI). Eur. J. Oper. Res. **264**(2), 491–507 (2018)
12. Oust, A., Hansen, S.N., Pettrem, T.R.: Combining property price predictions from repeat sales and spatially enhanced hedonic regressions. J. Real Estate Financ. Econ. **61**, 183–207 (2020)
13. Rosiers, F.D., Laganà, A., Theriault, M.: Size and proximity effects of primary schools on surrounding house values. J. Property Res. **18**(2), 149–168 (2001)
14. Yinger, J.: A search model of real estate broker behavior. Am. Econ. Rev. **71**, 591–605 (1981)

Author Index

Alvelos, Filipe 55
Alves, Filipe 140
Amaral, Paula 219
Amoura, Yahia 201
Anelli, Debora 629, 655
Azevedo, Beatriz Flamia 201

Ballarini, Aurora 615
Balletto, Ginevra 473, 485
Balucani, Nadia 233, 260
Banegas-Luna, Antonio J. 127
Benvenuti, Martina 97
Blanchi, Alicia 520
Boeres, Maria Claudia Silva 81
Bollini, Letizia 457
Borruso, Giuseppe 473, 485
Botte, Marilisa 585

Cadorel, Lucie 520
Caracciolo, Adriana 260
Carboni, Luca 473
Casado, L. G. 113
Casavecchia, Piergiorgio 260
Cavallotti, Carlo 260
Cerreta, Maria 572, 585
Copiello, Sergio 600, 615
Costa, Lino A. 55, 157
Cruz, N. C. 188

de Andrade, Gustavo Vargas 494
de Aragão, Emília Valença Ferreira 233, 246
de Doncker, Elise 343, 388
de Moraes, Renato Elias Nunes 81
de Oliveira, Francisco Henrique 494
de Souza, Israel Pereira 81
De Toro, Pasquale 585
de Tuesta, J. L. Diaz 140
Del Giudice, Francesco Paolo 629
Di Dato, Chiara 427
Di Felice, Paolino 541
Di Liddo, Felicia 655
Di Valerio, Daniele 541
Dimas, Isabel Dórdio 97
Do, Le Phuc Tam 507

Donati, Edda 600, 615
dos Santos, Francisco 157
Durães, Dalila 43, 68

Emsellem, Karine 520
Espinosa, J. Miguel 219

Facchini, Chiara 457
Falcinelli, Stefano 270
Fernandes, Florbela P. 3, 201
Ferreira, Ângela P. 15
Fusco, Giovanni 520

G.-Tóth, B. 113
Gadylshin, Kirill 295
Gadylshina, Kseniia 295
Garzón, Ester M. 127
Gillet, Valerie J. 127
Goedhart, Joost 29
Gomes, H. T. 140
Guarini, Maria Rosaria 561
Guerrero-García, Pablo 174

Hendrix, Eligius M. T. 29, 113, 174

Kakkar, Harjasnoor 281
Khachkova, Tatyana 328
Kolyukhin, Dmitry 328
Kouya, Tomonori 358, 406

L'Abbate, Samuela 641
Ladu, Mara 473
Lagarias, Apostolos 439
Le, Thi Hanh An 507
Leitão, Paulo 140
Lima, José 3
Lima, Laires A. 3
Lisitsa, Vadim 295, 310, 328
Locurcio, Marco 655

Machado, José 43, 68
Mancini, Luca 233, 246
Manganelli, Benedetto 629
Martínez-Bachs, Berta 281

Marucci, Alessandro   427
Massari, Antonella   641
Matos, Marina A.   55
Mazzoni, Elvis   97
Mendes, Ana   219
Messine, F.   113
Milesi, Alessandra   473
Mongelli, Lucia   641
Morano, Pierluigi   561
Mori, Masao   373
Muccio, Eugenio   585
Mura, Roberto   485

Nocca, Francesca   585
Novikov, Mikhail   310

Ortigosa, E. M.   188
Ortigosa, Pilar M.   127, 188
Otaki, Koki   373

Pacheco, Maria F.   3, 201
Padilha, Victor Luis   494
Paolone, Gaetanino   541
Parriani, Marco   270
Perchinunno, Paola   641
Pereira, Ana I.   3, 140, 201
Pérez-Sánchez, Horacio   127
Pham, Nguyen Hoai   507
Phan, Thi Bich Nguyet   507
Pilotti, Francesco   541
Pirani, Fernando   270
Poli, Giuliano   585
Puertas-Martín, Savíns   127

Ranieri, Rossana   561
Recio, Pedro   260
Redondo, Juana L.   127, 188
Rhodes, James   343
Rimola, Albert   281

Rocha, Ana Maria A. C.   43, 55, 68, 140, 201
Rocha, Humberto   97
Rosi, Marzio   233, 260

Sacco, Sabrina   572, 585
Sciamanna, Matteo   541
Sena, Inês   3
Sica, Francesco   561
Silva, A. M. T.   140
Silva, A. S.   140
Silva, António   43, 68
Silva, Felipe G.   3
Silva, Gonçalo O.   43, 68
Skouteris, Dimitrios   260
Solovyev, Sergey   310
Stratigea, Anastasia   439

Tagliapietra, Davide   473
Tajani, Francesco   561
Tanda, Giulia   485
Thompson, João Vinicius Corrêa   81
Torre, Carmelo Maria   655
Trinari, Marco   233
Trinh, Tu Anh   507

Utsugiri, Taiga   406

Vahldick, Adilson   494
Vanuzzo, Gianmarco   246
Varela, Leonilde   157
Vaz, Clara B.   3, 15
Vecchiocattivi, Franco   270
Vishnevsky, Dmitry   295

Witeck, Gabriela R.   43, 68

Yuasa, F.   388

Zacharakis, Ioannis   439

Printed in the United States
by Baker & Taylor Publisher Services